Methods in Enzymology

Volume 404
GTPases REGULATING MEMBRANE DYNAMICS

METHODS IN ENZYMOLOGY

EDITORS-IN-CHIEF

John N. Abelson   Melvin I. Simon

DIVISION OF BIOLOGY
CALIFORNIA INSTITUTE OF TECHNOLOGY
PASADENA, CALIFORNIA

FOUNDING EDITORS

Sidney P. Colowick and Nathan O. Kaplan

Methods in Enzymology

Volume 404

GTPases Regulating Membrane Dynamics

EDITED BY

William E. Balch

DEPARTMENT OF CELL BIOLOGY
THE SCRIPPS RESEARCH INSTITUTE
LA JOLLA, CALIFORNIA

Channing J. Der

DEPARTMENT OF PHARMACOLOGY
THE UNIVERSITY OF NORTH CAROLINA AT CHAPEL HILL
CHAPEL HILL, NORTH CAROLINA

Alan Hall

CRC ONCOGENE AND SIGNAL TRANSDUCTION GROUP
MRC LABORATORY FOR MOLECULAR CELL BIOLOGY
UNIVERSITY COLLEGE LONDON
LONDON, UNITED KINGDOM

AMSTERDAM • BOSTON • HEIDELBERG • LONDON
NEW YORK • OXFORD • PARIS • SAN DIEGO
SAN FRANCISCO • SINGAPORE • SYDNEY • TOKYO
Academic Press is an imprint of Elsevier

ELSEVIER

Elsevier Academic Press
525 B Street, Suite 1900, San Diego, California 92101-4495, USA
84 Theobald's Road, London WC1X 8RR, UK

This book is printed on acid-free paper. ∞

For all information on all Elsevier Academic Press publications
visit our Web site at www.books.elsevier.com

ISBN-13: 978-0-12-182809-7
ISBN-10: 0-12-182809-3

PRINTED IN THE UNITED STATES OF AMERICA
05   06   07   08   09   9   8   7   6   5   4   3   2   1

Working together to grow
libraries in developing countries

www.elsevier.com | www.bookaid.org | www.sabre.org

ELSEVIER   BOOK AID
             International   Sabre Foundation

Table of Contents

Contributors to Volume 404

Article numbers are in parentheses following the name of contributors.
Affiliations listed are current.

YOSHIKATSU AIKAWA (37), *Department of Pharmaceutical Technology, Faculty of Pharmaceutical Sciences at Kagawa, Tokushima-bunri University, Saniki-ku, Kagawa, Japan*

BRUNO ANTONNY (10), *Institut de Pharmacologie, Moleculaire et Cellulaire, CNRS, Valbonne, France*

MEIR ARIDOR (11), *Department of Cell Biology and Physiology, University of Pittsburgh School of Medicine, Pittsburgh, Pennsylvania*

BRIAN J. BACSKAI (50), *Alzheimer's Disease Research Laboratory, Department of Medicine, Harvard Medical School, Massachusetts General Hospital, Boston, Massachusetts*

WILLIAM E. BALCH (6, 7), *The Scripps Research Institute, Departments of Cell and Molecular Biology, and the Institute for childhood and Neglected Disease, La Jolla, California*

RENE BARTZ (42), *Department of Cell Biology, University of Texas, Southwestern Medical Center, Dallas, Texas*

MATTHEW BEARD (26), *Department of Physiology and Cellular Biophysics, Columbia University, New York*

CORINNE BENZING (42), *Institute of Reconstructive Neurobiology, University of Bonn, Life & Brain Center, Bonn, Germany*

GALINA V. BEZNOUSSENKO (5), *Department of Cell Biology and Oncology, Consorzio Mario Negri Sud, Santa Maria Imbaro (Chieti), Italy*

JÖELLE BIGAY (10), *Institut de Pharmacologie, Moleculaire et Cellulaire, CNRS, Valbonne, France*

ANDREE BLAUKAT (36), *Merck KGaA, Oncology Research Darmstadt, Darmstadt, Germany*

JUAN S. BONIFACINO (28), *Laboratory of Cellular Oncology, Center for Cancer Research, National Cancer Institute, National Institutes of Health, Bethesda, Maryland*

J. BRADFORD BOWZARD (40), *Department of Biochemistry, Emory University School of Medicine, Atlanta, Georgia*

WILLIAM J. BROWN (12), *Department of Molecular Biology and Genetics, Cornell University, Ithaca, New York*

SVEN R. CARLSSON (48), *Department of Medical Biochemistry and Biophysics, Umeå University, Umeå, Sweden*

GIANLUCA CESTRA (47), *1 Department of Cell Biology, Yale University School of Medicine, New Haven, Connecticut*

CIHAN CETIN (2), *Cell Biology and Cell Biophysics Program, European Molecular Biology Laboratory, Heidelberg, Germany*

J. PAUL CHAPPLE (41), *Institute of Ophthalmology, UCL, London, United Kingdom*

MANGUS MUTAH CHE (15), *Laboratory of Cellular Oncology, Center for Cancer Research, National Cancer Institute, National Institutes of Health, Bethesda, Maryland*

MICHAEL E. CHEETHAM (41), *Institute of Ophthalmology, University College London, London, United Kingdom*

JI-LONG CHEN (30), *Department of Physiology and Biophysics, Carver College of Medicine, University of Iowa, Iowa City, Iowa*

DANIELA CORDA (27), *Department of Cell Biology and Oncology, Consorzio Mario Negri Sud, Santa Maria Imbaro (Chieti), Italy*

DGANIT DANINO (55), *Department of Biotechnology and Food Engineering, Technion, Haifa, Israel*

PIETRO DE CAMILLI (47), *1 Department of Cell Biology, Yale University School of Medicine, New Haven, Connecticut*

IVAN DE CURTIS (25), *Unit of Cell Adhesion, San Raaele Scientific Institute, Department of Molecular Biology and Functional Genomics, Via Olgettina, Milano, Italy*

CRISLYN D'SOUZA-SCHOREY (14, 33), *Department of Biological Sciences, University of Notre Dame, Notre Dame, Indiana*

HOLGER ERFLE (1), *Cell Biology and Cell Biophysics Program, European Molecular Biology Laboratory, Heidelberg, Germany*

R. JANE EVANS (41), *Institute of Ophthalmology, University College London, London, United Kingdom*

JOHN H. EXTON (31), *Department of Molecular Physiology and Biophysics, Vanderbilt University School of Medicine, Nashville, Tennessee*

KOHJI FUKUNAGA (22), *Department of Pharmacology, Tohoku University, Graduate School of Pharmaceutical Sciences, Sendai, Japan*

EUGENE FUTAI (8), *Department of Molecular and Cell Biology, University of California, Berkeley, California*

FRANK GERTLER (47), *Department of Biology, Massachusetts Institute of Technology, Cambridge, Massachusetts*

STEPHEN J. GOULD (51), *Department of Biological Chemistry, The Johns Hopkins University School of Medicine, Baltimore, Maryland*

CELENE GRAYSON (41), *Institute of Ophthalmology, University College London, London, United Kingdom*

LORENA GRIPARIC (54), *Department of Biological Chemistry, David Geffen School of Medicine at University of California at Los Angeles, Los Angeles, California*

CEMAL GURKAN (6), *The Scripps Research Institute, Department of Cell Biology, La Jolla, California*

VI LUAN HA (16), *Laboratory of Cellular Oncology, Center for Cancer Research, National Cancer Institute, National Institutes of Health, Bethesda, Maryland*

OTTO HALLER (55), *Department of Virology, University of Freiburg, Freiburg, Germany*

ALISON J. HARDCASTLE (41), *Institute of Ophthalmology, University College London, London, United Kingdom*

ARI HASHIMOTO (21), *Department of Molecular Biology, Osaka Bioscience Institute, Osaka, Japan*

SHIGERU HASHIMOTO (21), *Department of Molecular Biology, Osaka Bioscience Institute, Osaka, Japan*

VOLKER HAUCKE (34), *Department of Biochemistry II, Center for Biochemistry and Molecular Cell Biology, University of Göttingen, Göttingen, Germany*

IRA M. HERMAN (33), *Department of Physiology, Tufts University School of Medicine, Boston, Massachusetts*

DELIA J. HERNÁNDEZ-DEVIEZ (23), *Department of Cell Biology and Anatomy, College of Medicine, University of Arizona, Tucson, Arizona*

CHRISTIAN HERRMANN (45), *Physikalische Chemie 1, Ruhr-Universität Bochum, Bochum, Germany*

MATTHEW K. HIGGINS (52), *Medical Research Council, Laboratory of Molecular Biology, Cambridge, United Kingdom*

JENNY E. HINSHAW (55), *Laboratory of Cell Biochemistry and Biology, NIDDK, NIH Bethesda, Maryland*

WANJIN HONG (38, 39), *Membrane Biology Laboratory, Institute of Molecular and Cell Biology, Singapore*

HOLLY HOOVER (14), *Department of Biological Sciences, University of Notre Dame, Notre Dame, Indiana*

ELENA INGERMAN (53), *Section of Molecular and Cellular Biology, Center of Genetics and Development, University of California, Davis, California*

LUDGER JOHANNES (39), *Laboratoire Trafic et Signalisation, Paris, Cedex 05, France*

HEATHER D. JONES (17), *Pulmonary-Critical Care Medicine Branch, National Heart, Lung, and Blood Institute, National Institutes of Health, Bethesda, Maryland*

RICHARD A. KAHN (40), *Department of Biochemistry, Emory University School of Medicine, Atlanta, Georgia*

AKIFUMI KAMATA (22), *Department of Pharmacology, Tohoku University, Graduate School of Pharmaceutical Sciences, Sendai, Japan*

BJOERN KINDLER (3), *Cell Biology and Cell Biophysics Program, European Molecular Biology Laboratory, Heidelberg, Germany*

GEORG KOCHS (55), *Department of Virology, University of Freiburg, Freiburg, Germany*

HISATAKE KONDO (22), *Department of Cell Biology, Tohoku University Graduate School of Medicine, Sendai, Japan*

MICHAEL KRAUSS (34), *Department of Biochemistry II, Center for Biochemistry and Molecular Cell Biology, University of Göttingen, Göttingen, Germany*

ANAMARIJA KRULJAC-LETUNIC (36), *European Molecular Biology Laboratory, Cell Biology and Biophysics Unit, Heidelburg, Germany*

SIMONE KUNZELMANN (45), *Physikalische Chemie 1, Ruhr-Universitat Bochum, Bochum, Germany*

ADAM KWIATKOWSKI (47), *Department of Biology, Massachusetts Institute of Technology, Cambridge, Massachusetts*

PAUL LAPOINTE (7), *The Scripps Research Institute, Department of Cell Biology, La Jolla, California*

MARILYN LEONARD (43), *The Scripps Research Institute, Department of Cell Biology, La Jolla, California*

URBAN LIEBEL (3), *Cell Biology and Cell Biophysics Program, European Molecular Biology Laboratory, Heidelberg, Germany*

LEI LU (38, 39), *Membrane Biology Laboratory, Institute of Molecular and Cell Biology, Singapore*

ALBERTO LUINI (5, 27), *Department of Cell Biology and Oncology, Consorzio Mario Negri Sud, Santa Maria Imbaro (Chieti), Italy*

RICHARD LUNDMARK (48), *Department of Medical Biochemistry and Biophysics, Umeå University, Umeå, Sweden*

FRÉDÉRIC LUTON (29), *Institut de Pharmacologie, Moléculaire et Cellulaire, CNRS, Valbonne Sophia-Antipolis, France*

JÖRG MALSAM (13), *Max Planck Institute for Molecular Cell Biology and Genetics, Dresden, Germany*

THOMAS F. J. MARTIN (37), *Department of Biochemistry, University of Wisconsin, Madison, Wisconsin*

HARVEY T. MCMAHON (52), *Medical Research Council, Laboratory of Molecular Biology, Cambridge, United Kingdom*

ALEXANDER A. MIRONOV (5), *Department of Cell Biology and Oncology, Consorzio Mario Negri Sud, Santa Maria Imbaro (Chieti), Italy*

ISHIDO MIWAKO (44), *The Scripps Research Institute, Department of Cell Biology, La Jolla, California*

JOEL MOSS (17, 18, 19), *Pulmonary-Critical Care Medicine Branch, National Heart, Lung, and Blood Institute, National Institutes of Health, Bethesda, Maryland*

VANDHANA MURALIDHARAN-CHARI (14), *Department of Biological Sciences, University of Notre Dame, Notre Dame, Indiana*

AKIHIKO NAKANO (9), *Molecular Membrane Biology Laboratory, RIKEN Discovery Research Institute, Saitama, Japan*

KAZUHISA NAKAYAMA (20, 32), *Department of Physiological Chemistry, Graduate School of Pharmaceutical Sciences, Kyoto University, Sakyo-ku, Kyoto, Japan*

SHERRI L. NEWMYER (50), *G. W. Hooper Foundation, The University of California, San Francisco, California*

ZHONGZHEN NIE (15), *Laboratory of Cellular Oncology, Center for Cancer Research, National Cancer Institute, National Institutes of Health, Bethesda, Maryland*

JODI NUNNARI (53), *Section of Molecular and Cellular Biology, Center of Genetics and Development, University of California, Davis, California*

YASUHITO ONODERA (21), *Department of Molecular Biology, Osaka Bioscience Institute, Osaka, Japan*

GUSTAVO PACHECO-RODRIGUEZ (18), *Pulmonary-Critical Care Medicine Branch, National Heart, Lung, and Blood Institute, National Institutes of Health, Bethesda, Maryland*

SIMONA PARIS (25), *San Raaele Scientific Institute, Department of Molecular Biology and Functional Genomics, Via Olgettina, Milano, Italy*

RAINER PEPPERKOK (1, 2, 3), *Cell Biology and Cell Biophysics Program, European Molecular Biology Laboratory, Heidelberg, Germany*

TREVOR R. PETTITT (35), *CR United Kingdom Institute for Cancer Studies, Birmingham University, Birmingham, United Kingdom*

ROMAN S. POLISHCHUK (5), *Department of Cell Biology and Oncology, Consorzio Mario Negri Sud, Santa Maria Imbaro (Chieti), Italy*

DALE J. POWNER (35), *CR United Kingdom Institute for Cancer Studies, Birmingham University, Birmingham, United Kingdom*

GERRIT J. K. PRAEFCKE (45), *Zentrum für Molekulare Medizin Köln (CMMC), Institut für Genetik, Köln, Germany*

ANNIE QUAN (49), *Cell Signalling Unit, Children's Medical Research Institute, Sydney, Australia*

RAJESH RAMACHANDRAN (43), *The Scripps Research Institute, Department of Cell Biology, La Jolla, California*

PAUL A. RANDAZZO (15, 16, 28), *Laboratory of Cellular Oncology, Center for Cancer Research, National Cancer Institute, National Institutes of Health, Bethesda, Maryland*

MIKE REICHELT (55), *Department of Virology, University of Freiburg, Freiburg, Germany*

JENS RIETDORF (2), *Cell Biology and Cell Biophysics Program, European Molecular Biology Laboratory, Heidelberg, Germany*

KATHLEEN N. RILEY (33), *Department of Physiology, Tufts University School of Medicine, Boston, Massachusetts*

PHILLIP J. ROBINSON (49), *Cell Signalling Unit, Children's Medical Research Institute, Sydney, Australia*

GUILLERMO ROMERO (11), *Department of Cell Biology and Physiology, University of Pittsburgh School of Medicine, Pittsburgh, Pennsylvania*

HISATAKA SABE (21), *Department of Molecular Biology, Osaka Bioscience Institute, Osaka, Japan*

HIROYUKI SAKAGAMI (22), *Department of Cell Biology, Tohoku University Graduate School of Medicine, Sendai, Japan*

MARCO SALAZAR (47), *1 Department of Cell Biology, Yale University School of Medicine, New Haven, Connecticut*

KEN SATO (9), *Molecular Membrane Biology Laboratory, RIKEN Discovery Research Institute, Saitama, Japan*

AYANO SATOH (13, 26), *Department of Cell Biology, Yale University School of Medicine, New Haven, Connecticut*

RANDY SCHEKMAN (8), *Department of Molecular and Cell Biology, Howard Hughes Medical Institute, University of California, Berkeley, California*

SANDRA L. SCHMID (43, 44), *The Scripps Research Institute, Department of Cell Biology, La Jolla, California*

JOHN A. SCHMIDT (12), *Department of Molecular Biology and Genetics, Cornell University, Ithaca, New York*

MICHAEL SCHRADER (51), *Department of Cell Biology and Cell Pathology, University of Marburg, Marburg, Germany*

SANJA SEVER (50), *Department of Medicine, Harvard Medical School, Renal Unit, Massachusetts General Hospital, Boston, Massachusetts*

J. DANIEL SHARER (40), *Department of Genetics, University of Alabama at Birmingham, Birmingham, Alabama*

HYE-WON SHIN (20), *Department of Physiological Chemistry, Graduate School of Pharmaceutical Sciences, Kyoto University, Sakyo-ku, Kyoto, Japan*

OK-HO SHIN (31), *Center for Basic Neuroscience, University of Texas Southwestern Medical Center, Dallas, Texas*

CHISA SHINOTSUKA (20), *Department of Internal Medicine, Cell Biology and Pathology, Yale University School of Medicine, New Haven, Connecticut*

KUNTALA SHOME (11), *Department of Cell Biology and Physiology, University of Pittsburgh School of Medicine, Pittsburgh, Pennsylvania*

JEREMY C. SIMPSON (2), *Cell Biology and Cell Biophysics Program, European Molecular Biology Laboratory, Heidelberg, Germany*

JESSE SKOCH (50), *Alzheimer's Disease Research Laboratory, Department of Medicine, Harvard Medical School, Massachusetts General Hospital, Boston, Massachusetts*

BYEONG DOO SONG (43), *The Scripps Research Institute, Department of Cell Biology, La Jolla, California*

STEFANIA SPANÒ (27), *Department of Cell Biology and Oncology, Consorzio Mario Negri Sud, Santa Maria Imbaro (Chieti), Italy*

MARK STAMNES (30), *Department of Physiology and Biophysics, Carver College of Medicine, University of Iowa, Iowa City, Iowa*

STACEY STAUFFER (16), *Laboratory of Cellular Oncology, Center for Cancer Research, National Cancer Institute, National Institutes of Health, Bethesda, Maryland*

BRIAN STORRIE (4), *Department of Physiology and Biophysics, University of Arkansas for Medical Sciences, Little Rock, Arkansas*

SARAH TAGUE (14), *Department of Biological Sciences, University of Notre Dame, Notre Dame, Indiana*

GUIHUA TAI (38, 39), *Membrane Biology Laboratory, Institute of Molecular and Cell Biology, Singapore*

HIROYUKI TAKATSU (32), *Research Center for Allergy and Immunology, RIKEN, Yokohama, Kanagawa, Japan*

KOHJI TAKEI (46), *Department of Neuroscience, Okayama University Graduate School of Medicine and Dentistry, Okayamashi, Okayama, Japan*

STEFAN TERJUNG (2), *Cell Biology and Cell Biophysics Program, European Molecular Biology Laboratory, Heidelberg, Germany*

GERAINT M. H. THOMAS (16), *Laboratory of Cellular Oncology, Center for Cancer Research, National Cancer Institute, National Institutes of Health, Bethesda, Maryland*

OLIVER ULLRICH (42), *Hochschule für Angewandte, Wissenschaften Hamburg, Lohbrügger Kirchstr, Hamburg, Germany*

CARMEN VALENTE (27), *Department of Cell Biology and Oncology, Consorzio Mario Negri Sud, Santa Maria Imbaro (Chieti), Italy*

ALEXANDER M. VAN DER BLIEK (54), *Department of Biological Chemistry, David Geffen School of Medicine at University of California at Los Angeles, Los Angeles, California*

MARTHA VAUGHAN (17, 18, 19), *Pulmonary-Critical Care Medicine Branch, National Heart, Lung, and Blood Institute, National Institutes of Health, Bethesda, Maryland*

KANAMARLAPUDI VENKATESWARLU (24), *Department of Pharmacology, School of Medical Sciences, The University of Bristol, Bristol, United Kingdom*

ALESSANDRO VICHI (19), *Pulmonary-Critical Care Medicine Branch, National Heart, Lung, and Blood Institute, National Institutes of Health, Bethesda, Maryland*

MICHAEL J. O. WAKELAM (35), *CR United Kingdom Institute for Cancer Studies, Birmingham University, Birmingham, United Kingdom*

GRAHAM WARREN (13, 26), *Department of Cell Biology, Yale University School of Medicine, New Haven, Connecticut*

ALICE WELCH (33), *NIH-NIAID office of Technology Development, Bethesda, Maryland*

JEAN M. WILSON (23), *Department of Cell Biology and Anatomy, College of Medicine, University of Arizona, Tucson, Arizona*

WEIDONG XU (30), *Department of Physiology and Biophysics, Carver College of Medicine, University of Iowa, Iowa City, Iowa*

ATSUKO YAMADA (21), *Department of Molecular Biology, Osaka Bioscience Institute, Osaka, Japan*

HYE-YOUNG YOON (28), *Laboratory of Cellular Oncology, Center for Cancer Research, National Cancer Institute, National Institutes of Health, Bethesda, Maryland*

YUMI YOSHIDA (46), *Department of Neuroscience, Okayama University Graduate School of Medicine and Dentistry, Okayamashi, Okayama, Japan*

TIMO ZIMMERMAN (2), *Cell Biology and Cell Biophysics Program, European Molecular Biology Laboratory, Heidelberg, Germany*

Preface

The Ras superfamily (>150 human members) encompasses Ras GTPases involved in cell proliferation; Rho GTPases involved in regulating the cytoskeleton; Rab GTPases involved in membrane targeting/fusion; and a group of GTPases including Sar1, Arf, Arl, and dynamin involved in vesicle budding/fission. These GTPases act as molecular switches and their activities are controlled by a large number of regulatory molecules that affect either GTP loading (guanine nucleotide exchange factors or GEFs) or GTP hydrolysis (GTPase activating proteins or GAPs). In their active state, they interact with a continually increasing, functionally complex array of downstream effectors.

In this new series of *Methods in Enzymology*, we have strived to bring together the latest thinking, approaches, and techniques in this area. Two volumes (403 and 404) focus on membrane regulating GTPases; the first is dedicated to those involved in budding and fission (Sar1, Arf, Arl, and dynamin), and the second focuses on those that control targeting and fusion (Rabs). Volumes 406 and 407 focus on the Rho and Ras families, respectively. It is important to emphasize that even though each of these volumes deals with a different GTPase family, they contain a wealth of common methodologies. As such, the techniques and approaches pioneered with respect to one class of GTPase are likely to be equally applicable to other classes. Furthermore, the functional distinctions that have been classically associated with the distinct branches of the superfamily are beginning to blur. There is now considerable evidence for biological and biochemical interplay and crosstalk among seemingly divergent family members. The compilation of a database of regulators and effectors of the whole superfamily by Bernards (Vol. 407) reflects some of these complex interrelationships. In addition to fostering cross-talk among investigators who study different GTPases, these volumes will also aid the entry of new investigators into the field.

In Volume 404, which is focused on GTPases involved in membrane dynamics, we highlight a broad group of GTPases, including both the Sar1 involved in vesicle budding from the endoplasmic reticulum and the Arf and Arf-like (Arl) GTPases controlling COPI vesicle formation and other activities in exocytic and endocytic pathways. Although they are not members of the Ras superfamily, because of their biochemical and functional relationship, we also include the dynamin GTPases involved in regulating membrane fission in exocytic and potentially endocytic pathways, as well as homologs that control fission of mitochondria. Each plays a critical role in defining the exocytic and endocytic trafficking pathways and distribution of membranes during mitosis. Like Rab GTPases, a large number of gene families defining regulators (GEFs and GAPs) and effectors are now evident. However, knowledge of their many roles remains a challenge. Their link to the activity of Rab GTPases is evident in the membrane database (see Vol. 404). However, from a systems biology perspective, their activities appear to be far more varied in scope. Introductory chapters in this volume focus on general methodologies for annotating the genome for GTPase function and exploring function, including interfering RNA screening and imaging platforms. General methods for use of microinjection to explore GTPase function by, and methodologies to integrate video fluorescence with, three-dimensional electron microscopy, should be applied across the entire group.

Many additional chapters in this volume by experts in the field highlight the methodologies and proteins involved in the analysis of individual GTPase function in vesicle budding and fission. The knowledge base presented in these articles should provide an important foundation for further rapid growth of our understanding of budding and fission GTPase activities and their role in controlling eukaryotic membrane architecture.

We are extremely grateful to the many investigators who have generously contributed their time and expertise to bring this wealth of technical expertise into this and other volumes comprising the Ras superfamily series.

WILLIAM E. BALCH
CHANNING J. DER
ALAN HALL

METHODS IN ENZYMOLOGY

VOLUME 72. Lipids (Part D)
Edited by JOHN M. LOWENSTEIN

VOLUME 73. Immunochemical Techniques (Part B)
Edited by JOHN J. LANGONE AND HELEN VAN VUNAKIS

VOLUME 74. Immunochemical Techniques (Part C)
Edited by JOHN J. LANGONE AND HELEN VAN VUNAKIS

VOLUME 75. Cumulative Subject Index Volumes XXXI, XXXII, XXXIV–LX
Edited by EDWARD A. DENNIS AND MARTHA G. DENNIS

VOLUME 76. Hemoglobins
Edited by ERALDO ANTONINI, LUIGI ROSSI-BERNARDI, AND EMILIA CHIANCONE

VOLUME 77. Detoxication and Drug Metabolism
Edited by WILLIAM B. JAKOBY

VOLUME 78. Interferons (Part A)
Edited by SIDNEY PESTKA

VOLUME 79. Interferons (Part B)
Edited by SIDNEY PESTKA

VOLUME 80. Proteolytic Enzymes (Part C)
Edited by LASZLO LORAND

VOLUME 81. Biomembranes (Part H: Visual Pigments and Purple Membranes, I)
Edited by LESTER PACKER

VOLUME 82. Structural and Contractile Proteins (Part A: Extracellular Matrix)
Edited by LEON W. CUNNINGHAM AND DIXIE W. FREDERIKSEN

VOLUME 83. Complex Carbohydrates (Part D)
Edited by VICTOR GINSBURG

VOLUME 84. Immunochemical Techniques (Part D: Selected Immunoassays)
Edited by JOHN J. LANGONE AND HELEN VAN VUNAKIS

VOLUME 85. Structural and Contractile Proteins (Part B: The Contractile Apparatus and the Cytoskeleton)
Edited by DIXIE W. FREDERIKSEN AND LEON W. CUNNINGHAM

VOLUME 86. Prostaglandins and Arachidonate Metabolites
Edited by WILLIAM E. M. LANDS AND WILLIAM L. SMITH

VOLUME 87. Enzyme Kinetics and Mechanism (Part C: Intermediates, Stereo-chemistry, and Rate Studies)
Edited by DANIEL L. PURICH

VOLUME 88. Biomembranes (Part I: Visual Pigments and Purple Membranes, II)
Edited by LESTER PACKER

VOLUME 89. Carbohydrate Metabolism (Part D)
Edited by WILLIS A. WOOD

VOLUME 210. Numerical Computer Methods
Edited by LUDWIG BRAND AND MICHAEL L. JOHNSON

VOLUME 211. DNA Structures (Part A: Synthesis and Physical Analysis of DNA)
Edited by DAVID M. J. LILLEY AND JAMES E. DAHLBERG

VOLUME 212. DNA Structures (Part B: Chemical and Electrophoretic Analysis of DNA)
Edited by DAVID M. J. LILLEY AND JAMES E. DAHLBERG

VOLUME 213. Carotenoids (Part A: Chemistry, Separation, Quantitation, and Antioxidation)
Edited by LESTER PACKER

VOLUME 214. Carotenoids (Part B: Metabolism, Genetics, and Biosynthesis)
Edited by LESTER PACKER

VOLUME 215. Platelets: Receptors, Adhesion, Secretion (Part B)
Edited by JACEK J. HAWIGER

VOLUME 216. Recombinant DNA (Part G)
Edited by RAY WU

VOLUME 217. Recombinant DNA (Part H)
Edited by RAY WU

VOLUME 218. Recombinant DNA (Part I)
Edited by RAY WU

VOLUME 219. Reconstitution of Intracellular Transport
Edited by JAMES E. ROTHMAN

VOLUME 220. Membrane Fusion Techniques (Part A)
Edited by NEJAT DÜZGÜNEŞ

VOLUME 221. Membrane Fusion Techniques (Part B)
Edited by NEJAT DÜZGÜNEŞ

VOLUME 222. Proteolytic Enzymes in Coagulation, Fibrinolysis, and Complement Activation (Part A: Mammalian Blood Coagulation Factors and Inhibitors)
Edited by LASZLO LORAND AND KENNETH G. MANN

VOLUME 223. Proteolytic Enzymes in Coagulation, Fibrinolysis, and Complement Activation (Part B: Complement Activation, Fibrinolysis, and Nonmammalian Blood Coagulation Factors)
Edited by LASZLO LORAND AND KENNETH G. MANN

VOLUME 224. Molecular Evolution: Producing the Biochemical Data
Edited by ELIZABETH ANNE ZIMMER, THOMAS J. WHITE, REBECCA L. CANN, AND ALLAN C. WILSON

VOLUME 225. Guide to Techniques in Mouse Development
Edited by PAUL M. WASSARMAN AND MELVIN L. DEPAMPHILIS

VOLUME 335. Flavonoids and Other Polyphenols
Edited by LESTER PACKER

VOLUME 336. Microbial Growth in Biofilms (Part A: Developmental and Molecular Biological Aspects)
Edited by RON J. DOYLE

VOLUME 337. Microbial Growth in Biofilms (Part B: Special Environments and Physicochemical Aspects)
Edited by RON J. DOYLE

VOLUME 338. Nuclear Magnetic Resonance of Biological Macromolecules (Part A)
Edited by THOMAS L. JAMES, VOLKER DÖTSCH, AND ULI SCHMITZ

VOLUME 339. Nuclear Magnetic Resonance of Biological Macromolecules (Part B)
Edited by THOMAS L. JAMES, VOLKER DÖTSCH, AND ULI SCHMITZ

VOLUME 340. Drug–Nucleic Acid Interactions
Edited by JONATHAN B. CHAIRES AND MICHAEL J. WARING

VOLUME 341. Ribonucleases (Part A)
Edited by ALLEN W. NICHOLSON

VOLUME 342. Ribonucleases (Part B)
Edited by ALLEN W. NICHOLSON

VOLUME 343. G Protein Pathways (Part A: Receptors)
Edited by RAVI IYENGAR AND JOHN D. HILDEBRANDT

VOLUME 344. G Protein Pathways (Part B: G Proteins and Their Regulators)
Edited by RAVI IYENGAR AND JOHN D. HILDEBRANDT

VOLUME 345. G Protein Pathways (Part C: Effector Mechanisms)
Edited by RAVI IYENGAR AND JOHN D. HILDEBRANDT

VOLUME 346. Gene Therapy Methods
Edited by M. IAN PHILLIPS

VOLUME 347. Protein Sensors and Reactive Oxygen Species (Part A: Selenoproteins and Thioredoxin)
Edited by HELMUT SIES AND LESTER PACKER

VOLUME 348. Protein Sensors and Reactive Oxygen Species (Part B: Thiol Enzymes and Proteins)
Edited by HELMUT SIES AND LESTER PACKER

VOLUME 349. Superoxide Dismutase
Edited by LESTER PACKER

VOLUME 350. Guide to Yeast Genetics and Molecular and Cell Biology (Part B)
Edited by CHRISTINE GUTHRIE AND GERALD R. FINK

VOLUME 351. Guide to Yeast Genetics and Molecular and Cell Biology (Part C)
Edited by CHRISTINE GUTHRIE AND GERALD R. FINK

[1] Arrays of Transfected Mammalian Cells for High Content Screening Microscopy

By HOLGER ERFLE and RAINER PEPPERKOK

Abstract

In this chapter we describe protocols for reverse transfection to generate mammalian cell arrays for systematic gene knock-downs by RNAi or knock-ins by ectopic cDNA expression. The method is suitable for high content screening microscopy at a high spatial and temporal resolution allowing even time-lapse analysis of hundreds of samples in parallel.

Introduction

The information of complete genome sequences and the identification and systematic cloning of human cDNAs provide the challenging opportunity to analyze the complexity of biological processes on a large scale. For this purpose high-throughput techniques such as protein analysis by mass spectrometry (Smith *et al.*, 2001) or expression and transcription profiling by protein or DNA microarrays (DeRisi *et al.*, 1996; MacBeath *et al.*, 2000; Schena *et al.*, 1995) have been developed and successfully applied to diverse biological questions. An elegant high-throughput method allowing parallel analysis of gene function in intact living cells has been recently introduced by Ziauddin and Sabatini (Ziauddin *et al.*, 2001). In this method expression plasmids encoding for example, GFP-tagged proteins, are printed together with the appropriate transfection reagents in a gelatin matrix at defined locations on glass slides. Tissue culture cells are subsequently plated on these slides resulting in clusters of living cells expressing the respective cDNAs at each location. The approach is called "reverse transfection," as in comparison to conventional transfection the order of addition of DNA and cells is reversed.

The method, originally introduced for ectopically expressing genes, has recently been adapted to transfections of siRNAs to knock down target genes of interest for functional analysis (Elbashir *et al.*, 2001; Erfle *et al.*, 2004; Kumar *et al.*, 2003; Mousses *et al.*, 2003). Several other extensions of the original transfected cell array protocol have been introduced such as the arraying of smaller grids of siRNAs or cDNAs into the wells of 96 well plates (Mishina *et al.*, 2004), enabling "ultra-high"-throughput applications like drug screens in living cells. The method has also been extended

METHODS IN ENZYMOLOGY, VOL. 404
0076-6879/05 $35.00
DOI: 10.1016/S0076-6879(05)04001-2

to perform transfections of human primary (Yoshikawa *et al.*, 2004) and non-adherent cells (Kato *et al.*, 2004).

Recent advances in automated fluorescence scanning microscopy and image processing (see e.g., Liebel *et al.*, 2003; Starkuviene *et al.*, 2004) allow now rapid analysis of transfected cell arrays in large scale screening applications. In the following we describe the method of reverse transfection on cell arrays as we use it in our laboratory to examine gene function by RNAi or overexpression of plasmid DNAs with high content screening microscopy.

Methodology

The method comprises five individual steps (see Fig. 1), including the preparation of the transfection solutions, followed by their spotting onto a cell substrate (e.g., Lab-Tek culture dishes, Nalge Nunc International, Rochester, NY), plating of the cells onto the arrays of spotted transfection solutions, preparation of the transfected cells for functional analysis, and finally the analysis of transfected cells by high content screening microscopy.

Although we describe the method for adherent tissue culture cells on noncoated chambered cover-glass tissue culture dishes, Lab-Tek chambered cover-glass, the very same protocols also work well with different cell substrates such as MatTek (MatTek, Ashland, MA) culture dishes or glass slides. The protocol has been equally successful for transfections of synthetic siRNAs or plasmid cDNAs.

As transfection reagent we use Effectene (Qiagen, Hilden, Germany) or Lipofectamine 2000 (Invitrogen, La Jolla, CA), which give optimal transfection efficiencies for both siRNAs and plasmid DNAs in MCF7, HeLa, COS7L, or HEK 293 cells.

Preparation of Transfection Solutions

The siRNA (plasmid cDNA) gelatin transfection solutions are prepared in 384-well plates (Nalge-Nunc).

1. Add 1 μl of the respective siRNA stock solution (20 μM in RNA dilution buffer as supplied by the manufacturer) to each well. For plasmid transfections 1 μl of plasmid DNA at a stock concentration of 500 ng/μl is added.
2. Add 7.5 μl EC buffer (EC Buffer is part of the Effectene Transfection kit, Qiagen) containing 0.2 M sucrose and mix thoroughly by pipetting three times up and down.
3. Incubate the mixture for 10 min at room temperature.

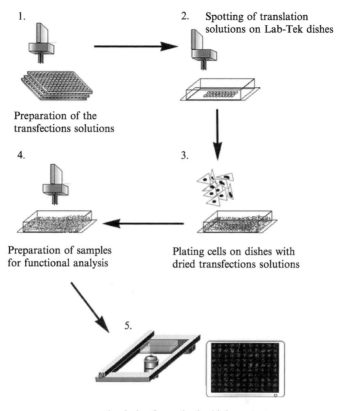

1. Preparation of the transfections solutions

2. Spotting of translation solutions on Lab-Tek dishes

3. Plating cells on dishes with dried transfections solutions

4. Preparation of samples for functional analysis

5. Analysis of samples by high content screening microscopy

FIG. 1. The five steps to produce arrays of transfected mammalian cells for high content screening microscopy. 1. Preparation of the transfection solutions on an automated liquid handler. 2. Spotting of the transfection solutions with a spotting Robot, for example, ChipWriter Compact on a cell substrate, for example, Lab-Tek tissue culture dishes. 3. Plating of the cells on dishes with dried transfection solutions. 4. Preparation of samples for functional analysis, for example, immunostaining. 5. Analysis of samples by high content screening microscopy.

4. Add 4.5 μl of the Effectene transfection reagent (Qiagen).
5. Incubate for 10 min at room temperature.
6. Add 7.25 μl of 0.08% gelatin (G-9391, Sigma-Aldrich, St. Louis, MO) containing 3.5×10^{-4}% fibronectin (Sigma-Aldrich). The final solution is now ready for the spotting process.

Comments

It is important to mix the transfection components just prior to the spotting, to achieve a high reproducibility in transfection efficiency. The optimal incubation times and amounts of transfection reagent are determined empirically for different transfection reagents. However, for optimizing the transfection mix for transfection reagents different from Effectene, the protocol described previously is a good starting point for optimization by simply replacing Effectene with equal amounts of the alternative transfection reagent. The EC buffer from the Effectene kit can be replaced by water without significant loss of transfection efficiencies, when tranfection reagents different from Effectene are used. The presence of sucrose in the EC buffer reduces the loss in transfection efficiencies when the dried arrays are stored prior to their use (see below). Sucrose also facilitates considerably the transfer of the siRNA (cDNA)-gelatin transfection solution to the substrate during the spotting procedure.

The presence of fibronectin in the gelatin solution increases cell adherence to the spot region and reduces the migration of transfected cells away from it.

In order to retrieve the spot regions and to highlight successfully transfected cells for siRNA transfections a Cy3-labeled DNA oligonucleotide is used as a transfection marker. In this case 0.5 μl of a 40 μM marker solution is included in step 1 of the protocol above resulting in a total oligonucleotide volume of 1.5 μl (1 μl siRNA plus 0.5 μl Cy3 labeled oligonucleotide).

With the protocol described previously it is possible to cotransfect plasmid cDNA and siRNA. In this case 1 μl siRNA plus 1 μl plasmid cDNA are added in step 1, resulting in a total oligo-nucleotide volume of 2 μl.

Spotting the Transfection Solution on Lab-Tek Dishes

For the spotting of the transfection solutions onto chambered glass coverslips, we use a ChipWriter Compact robot (Bio-Rad Laboratories, Hercules, CA) equipped with either SNS10 (TeleChem International, Sunnyvale, CA) or PTS 600 (Point Technologies, Boulder, CO) solid pins. These pins show a high reliability of the spotting process, in particular in the spot geometry achieved. Also, they deliver a sufficiently high volume (approximately 4 nl) to give rise to spots of approximately 400 μm in size, which each can be overlaid with approximately 100 to 200 Hela or MCF7 cells. Transfecting fewer cells per spot suffers in our hands from poor statistics of the results achieved in functional assays. The spot-to-spot distance is adjusted to 1125 μm and allows all 384 samples of a 384 well plate to be delivered onto one Lab-Tek chambered coverglass (NalgeNunc). The spotted solutions on the Lab-Tek chamber are then dried at room temperature for at least 12 hours after printing before cells are plated onto them. The solutions of one 384-well plate (see preceding) are sufficient to spot at least 50 identical Lab-Tek chambers, which can be stored for later use in a dry environment for several months without significant loss in transfection efficiency. After the tranfection solutions have dried they become visible on the LabTek chambers, which we routinely use to monitor the successful spotting.

Comments

It is crucial that the spotting robot used has to be able to pass the walls of the Lab-Tek chamber. The spotting procedure for 384 samples on 50 Lab-Tek chambers in parallel using 8 solid pins typically lasts 6 hours. In order to avoid evaporation of the small sample volumes in the 384 well plate during spotting, it is cooled with an in-house built water-cooled plate. In order to deposit more sample volume, which for some cell types improves transfection efficiencies, we spot repeatedly to the same position.

Plating of Cells on Lab-Tek Dishes with Dried Transfection Solution

The density of the cells plated on the spotted Lab-Tek chambers is always a compromise between the improved statistics that can be achieved with high cell densities and the quality of microscopic analyses that one expects.

Typically, we plate 1.25×10^5 actively growing HeLa, MCF7, COS7L, or HEK 293 cells in 2.5 ml culture medium (DMEM containing 10% heat-inactivated fetal calf serum, 2 mM glutamine, 100 U/ml penicillin, and 100 μg/ml streptomycin) on the dried spots of one Lab-Tek culture dish. This results in about 100 to 200 cells residing on one spot.

Typical incubation times (at 37° and 5% CO_2) for the successful expression of plasmid cDNAs vary between 12 and 24 h. The incubation time for RNAi experiments varies between 20 to 50 h and strongly depends on the stability of the proteins targeted by the siRNAs spotted. For long-term experiments lasting several days, for example, RNAi experiments targeting stable proteins, the cell density needs to be lowered as with the cell density typically used (see preceding), the cells may stop growing due to contact inhibition, which makes experiments addressing cell cycle or signal transduction related questions difficult to interpret.

Transfection efficiencies of plasmid cDNAs depend strongly, as with standard transfection protocols, on the protein expressed. However, tranfection efficiencies of up to 90% can be achieved in some cases for plasmid cDNAs in HEK293 cells. Those for the transfection of siRNAs show less variation and are close to 100% as determined by the presence of rhodamine-labeled siRNA oligo-nucleotides in cells residing on a spot.

Preparation of Samples for Functional Analysis

For functional analysis involving high content screening microscopy, we frequently use immunofluorescence to monitor molecule specific morphological and biochemical parameters. It always has the advantage that features of different genes can be analyzed in parallel (Pepperkok *et al.*, 2000). The immunostaining procedure in Lab-Tek chambered glass coverslips is very effective and cheap as it can be performed with the same antibody for hundreds of target genes in parallel.

Specifically for Lab-Teks, we apply 250 μl of the corresponding antibody by carefully distributing the fluid over the spotted area. We incubate for 10 min with the lid closed followed by 2 washes with 2 ml PBS (30 min each). We routinely include reliable stains highlighting cell nuclei (e.g., Hoechst or Dapi) which facilitates automated focusing and image acquisition (Liebel *et al.*, 2003). The stained samples are stored at 4° either embedded in Mowiol or in PBS solution containing azide after a brief poststaining fixation of the samples with paraformaldehyde for 2 min.

Analysis of the Samples by High Content Screening Microscopy

In principle, images of the cells on the spots can be acquired with any commercially available inverted microscope. We use a ScanR (Olympus Biosystems, Planegg, Germany; Liebel *et al.*, 2003) scanning microscope, with automated focus, allowing time-lapse data acquisition. This microscope is equipped with a 10 × /0.4 air and 40 × /0.95 air PlanApo objective (Olympus, Melville, NY). The 10× objective allows the imaging of all cells of one spot in one image with a reasonable resolution. Both objectives used are air objectives, as this facilitates the sample scanning process in a reliable manner, which is often difficult when oil immersion objectives are used. A key point of the whole imaging process is to find the first spot of the array. For this purpose we use the cotransfected Cy3 DNA oligonucleotide. In addition we mark the first spot manually with a thin and water-resistant black marker pen on the opposite side of the coverglass before cells are seeded.

Additional Comments

As the accuracy of the spot-to-spot distance is extremely important, precise positioning of the spots is needed, demanding high-resolution spotting robots. Starting cell plating densities resulting in 100 to 200 cells per spot is a good compromise between the demands on sample statistics for functional assays and cells remaining in the logarithmic growth phase.

Optimizing Conditions and Troubleshooting

To optimize transfection efficiencies for different cell lines, a labeled siRNA resulting in a known phenotype or a GFP tagged cDNA plasmid with known subcellular localization can be used. The ratio of siRNA (cDNA) to transfection reagent for different transfection reagents needs to be adjusted individually. A good starting point for optimization is the ratio recommended by the supplier of the transfection reagent. Thorough mixing of transfection reagent and siRNA (cDNA) is crucial for successful transfection. A possible source for variations in transfection efficiency is batch-to-batch variations in the quality of the transfection reagent.

Summary and Perspectives

We present in this article a protocol to perform parallel analysis of hundreds of different genes on a single Lab-Tek chamber. We focused on a transfection protocol, as it has been applied successfully in our laboratory in combination with high content screening microscopy analysis. It may serve as a basis for any subsequent adaptations of the method to more complicated analysis methods or cell systems.

References

DeRisi, J., Penland, L., Brown, P. O., Bittner, M. L., Meltzer, P. S., Ray, M., Chen, Y., Su, Y. A., and Trent, J. M. (1996). Use of a cDNA microarray to analyse gene expression patterns in human cancer. *Nat. Genet.* **14,** 457–460.

Elbashir, S. M., Harborth, J., Lendeckel, W., Yalcin, A., Weber, K., and Tuschl, T. (2001). Duplexes of 21-nucleotide RNAs mediate RNA interference in cultured mammalian cells. *Nature* **411,** 494–498.

Erfle, H., Simpson, J. C., Bastiaens, P. I., and Pepperkok, R. (2004). siRNA cell arrays for high-content screening microscopy. *Biotechniques* **37,** 454–462.

Kato, K., Umezawa, K., Miyake, M., Miyake, J., and Nagamune, T. (2004). Transfection microarray of nonadherent cells on an oleyl poly(ethylene glycol) ether-modified glass slide. *Biotechniques* **37,** 444–452.

Kumar, R., Conklin, D. S., and Mittal, V. (2003). High-throughput selection of effective RNAi probes for gene silencing. *Genome Res.* **13,** 2333–2340.

Liebel, U., Starkuviene, V., Erfle, H., Simpson, J. C., Poustka, A., Wiemann, S., and Pepperkok, R. (2003). A microscope-based screening platform for large scale functional analysis in intact cells. *FEBS Lett.* **554,** 394–398.

MacBeath, G., and Schreiber, S. L. (2000). Printing proteins as microarrays for high-throughput function determination. *Science* **289,** 1760–1763.

Mishina, Y. M., Wilson, C. J., Bruett, L., Smith, J. J., Stoop-Myer, C., Jong, S., Amaral, L. P., Pedersen, R., Lyman, S. K., Myer, V. E., Kreidler, B. L., and Thompson, C. M. (2004). Multiplex GPCR assay in reverse transfection cell microarray. *J.Biomol. Screening* **9,** 196–207.

Mousses, S., Caplen, N. J., Cornelison, R., Weaver, D., Basik, M., Hautaniemi, S., Elkahloun, A. G., Lotufo, R. A., Choudary, A., Dougherty, E. R., Suh, E., and Kallioniemi, O. (2003). RNAi microarray analysis in cultured mammalian cells. *Genome Res.* **13,** 2341–2347.

Pepperkok, R., Girod, A., Simpson, J. C., and Rietdorf, J. (2000). Imunofluorescence microscopy. *In* "Monoclonal Antibodies: A practical approach" (P. Shepard and C. Dean, eds.), pp. 355–370. Oxford University Press, New York.

Schena, M., Shalon, D., Davis, R. W., and Brown, P. O. (1995). Quantitative monitoring of gene expression patterns with a complementary DNA microarray. *Science* **270,** 467–470.

Smith, R. D., Pasa-Tolic, L., Lipton, M. S., Jensen, P. K., Anderson, G. A., Shen, Y., Conrads, T. P., Udseth, H. R., Harkewicz, R., Belov, M. E., Masselon, C., and Veenstra, T. D. (2001). Rapid quantitative measurements of proteomes by Fourier transform ion cyclotron resonance mass spectrometry. *Electrophoresis* **22,** 1652–1668.

Starkuviene, V., Liebel, U., Simpson, J. C., Erfle, H., Poustka, A., Wiemann, S., and Pepperkok, R. (2004). High-content screening microscopy identifies novel proteins with a putative role in secretory membrane traffic. *Genome Res.* **14,** 1948–1956.

Yoshikawa, T., Uchimura, E., Kishi, M., Funeriu, D. P., Miyake, M., and Miyake, J. (2004). Transfection microarray of human mesenchymal stem cells and on-chip siRNA gene knockdown. *J. Control Rel.* **96,** 227–232.

Ziauddin, J., and Sabatini, D. M. (2001). Microarrays of cells expressing defined cDNAs. *Nature* **411,** 107–110.

[2] Imaging Platforms for Measurement of Membrane Trafficking

By RAINER PEPPERKOK, JEREMY C. SIMPSON, JENS RIETDORF, CIHAN CETIN, URBAN LIEBEL, STEFAN TERJUNG, and TIMO ZIMMERMANN

Abstract

In this chapter we describe automated imaging methods used to measure the transport of an established membrane transport marker from the endoplasmic reticulum to the plasma membrane. The method is fast and significantly robust to be applied in systematic studies on a large scale such as genome-wide screening projects. We further describe the use of software macros and plugins in Image J that allow the quantification of the

METHODS IN ENZYMOLOGY, VOL. 404
0076-6879/05 $35.00
DOI: 10.1016/S0076-6879(05)04002-4

kinetics of membrane transport intermediates in fluorescence microscopy time-lapse sequences.

Introduction

Membrane traffic enables cells to distribute proteins, lipids, and carbohydrates between membrane compartments and is thus vital for many cellular processes such as cell growth, homeostasis, and differentiation. Genetic and biochemical approaches have been very efficiently applied in the past to identify and characterize individual molecular components involved in the regulation of membrane traffic in the secretory and endocytic pathways. Central to many of these studies has been the reconstitution of the particular transport step of interest *in vitro* using purified components. Although this has led to an enormous body of information on how membrane traffic is organized at the molecular level, such simplified *in vitro* systems are lacking important regulatory elements relating to the spatial organization that occurs in living cells. More recently, systematic approaches, such as organelle proteomics or yeast two hybrid screening, have attempted to identify structural and regulatory components of membrane traffic with the goal of reaching a more complete description of its molecular regulation (see for example Bell *et al.*, 2001; Calero *et al.*, 2002; Monier *et al.*, 2002). However, despite their great potential, these techniques have limitations, not least of which is their lack of demonstrating a functional involvement of the molecules identified in the particular trafficking step under investigation.

Functional microscope-based assays in intact living cells with the potential for large-scale analyses have been recently developed and applied to problems in membrane trafficking (Ghosh *et al.*, 2000; Liebel *et al.*, 2003; Pelkmans *et al.*, 2005; Starkuviene *et al.*, 2004). These techniques provide single cell or even subcellular resolution. They promise, in combination with genome-wide RNA interference (RNAi, Elbashir *et al.*, 2001) or overexpression strategies (Starkuviene *et al.*, 2004), to help to reveal comprehensively the regulatory networks underlying membrane traffic in intact cells. Here we describe such assays used to quantitatively monitor secretory transport to the plasma membrane and the subcellular kinetics of transport carriers in time-lapse series.

Methodology

The problem of performing functional microscope-based assays to address membrane trafficking on a large set of proteins in cells is presently still a challenge. It requires automation and coordination of various steps such as sample preparation, image acquisition, the handling and analysis of large sets of image data, and the integration of the results with existing knowledge provided by bioinformatic databases. A number of automated microscope-based image acquisition systems are already available on the market. Limitations of such commercially available systems are often that they have been designed and optimized for specific applications, which restricts their adaptation to new assays. Systems with ultra-high throughput capacities are mostly lacking single cell or subcellular resolution and thus provide only specialized information. Therefore, it is important when setting up an imaging platform to choose equipment that is flexible enough to be compatible with several protocols for sample preparation, data handling, and analyses. The protocols we describe in this chapter are based on the use of cell arrays for high-content screening microscopy (see Chapter 1 of this issue). However, they will also work for other sample preparation methods with modifications.

Measurement of Transport from the Endoplasmic Reticulum to the Plasma Membrane

As a transport marker we use the temperature-sensitive membrane protein ts-O45-G from the vesicular stomatitis virus (Zilberstein *et al.*, 1980). This transmembrane protein has the feature of accumulating in the ER at 39.5°, but moves vectorially through the secretory pathway to the plasma membrane (PM) at the permissive temperature of 32°, where an antibody recognizing an external epitope can be used to detect it. This has the principal advantage that transport in individual cells is highly synchronized.

For the expression of this marker in cells we use a recombinant adenovirus encoding ts-O45-G tagged with either CFP or YFP (Keller *et al.*, 2001). For automated image acquisition we use a Scan^R high content screening microscope (Olympus Biosystems, Munich, Germany, see also Liebel *et al.*, 2003 for a detailed description of the features of this microscope). An outline of the rationale of the transport assay is shown in Fig. 1.

Expression and Transport of ts-O45-G to the Plasma Membrane

1. Plate 1.25×10^5 actively growing HeLa cells in 2.5 ml culture medium (DMEM containing 10% heat-inactivated fetal calf serum, 2 mM glutamine, 100 U/ml penicillin, and 100 μg/ml streptomycin) on one Lab-Tek culture dish containing spotted siRNAs or plasmid DNAs as described in Chapter 1 of this issue.
2. Incubate the cells for 24 h (plasmid DNAs) or 36 h (siRNAs) at 37°.
3. Overlay the cells with recombinant adenovirus encoding the secretory marker protein ts-O45-G tagged with CFP or YFP and incubate for further 1 h at 37°.

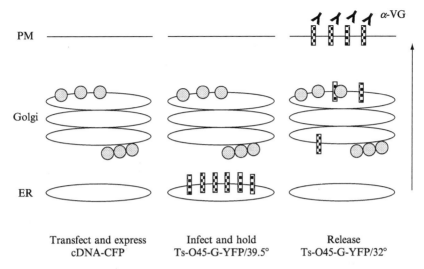

Transfect and express
cDNA-CFP

Infect and hold
Ts-O45-G-YFP/39.5°

Release
Ts-O45-G-YFP/32°

FIG. 1. Rationale of the ts-O45-G transport assay. The assay consists of three steps. (I) cells are transfected with either siRNAs or, as shown in the figure, with plasmid DNAs tagged with a GFP version complementary to the tag of the ts-O45-G. (II) 24 h later cells are infected with adenovirus encoding GFP-tagged ts-O45-G and incubated at 39.5° to accumulate ts-O45-G in the ER. After a further 16 h of incubation at 39.5°, (III) the temperature is shifted to 32° to induce transport of ts-O45-G to the plasma membrane, where it is detected after fixation by immunostaining using a monoclonal antibody recognizing an external ts-O45-G epitope. In this way the relative amount of total ts-O45-G (e.g., YFP-signal) transported to the plasma membrane (α-VG staining using, e.g., a Cy3 conjugated secondary antibody) can be determined by the ratio of Cy3/YFP signals. Transfected cells can be distinguished from nontransfected cells by their CFP signal.

4. Wash cells three times for 2 min with culture medium to remove unbound virus.

5. Transfer cells to 39.5° and incubate at this temperature for 16 h in order to accumulate YFP- or CFP-tagged ts-O45-G in the endoplasmic reticulum.

6. Transfer cells to 32° in the presence of 100 μg/ml cycloheximide (Calbiochem, San Diego, USA) to release ts-O45-G from the endoplasmic reticulum.

7. Fix cells at various time-points after transfer to 32° with 3% paraformaldehyde. Saturation of the arrival of ts-O45-G at the plasma membrane typically occurs 1 h after the temperature shift to 32°.

8. Detect the arrival of ts-O45-G at the plasma membrane by immunofluorescence using a monoclonal antibody recognizing an extracellular epitope of ts-O45-G (obtained as a gift from Kai

Simons, Dresden, Germany; see Pepperkok *et al.*, 2000 for detailed protocols for immunostaining).

9. Stain cell nuclei with Hoechst 33342 stain (final concentration 0.1 μg/ml) for 5 min at room temperature.

Comments

The stained samples are stored at 4° either embedded in Mowiol or in PBS solution containing 0.01% azide after a brief post-staining fixation of the samples with 3% paraformaldehyde for 2 min. The initial cell density after plating of the cells on LabTek dishes is critical for the success of the experiments and needs to be adjusted when different cell types are used. If cell densities are too high, the efficiency of adenovirus transfection decreases considerably and the number of cells expressing the transport marker is low. In cell cultures that are too sparse, the number of cells transfected per spot (see Chapter 1) with either siRNA or plasmid DNAs are too low for significant statistical analyses of the results.

Measurement of ts-O45-G Transport to the Plasma Membrane

For automated image acquisition we use a Scan^R system from Olympus Biosystems (Munich, Germany, see also Liebel *et al.*, 2003). This microscope is equipped with standard filter sets for discriminating between DAPI, CFP, GFP, YFP, and Cy3 in sequential imaging mode. Typically we use a 10 × /0.4 air PlanApo objective (Olympus, Melville, NY) to image one entire spot on the LabTek array. For data analysis we use our own software packages (see Liebel *et al.*, 2003) or the analysis software supplied with the Scan^R system by Olympus Biosystems. Both packages give identical results. The individual steps of data analysis described below can also be executed by using the freeware image processing software Image J (available at: http://rsb.info.nih.gov/ij/).

1. Images of the samples prepared as described in 1.1 are acquired automatically using the Scan^R system. An example of one multicolor data set acquired is shown in Fig. 2(A–C).
2. All images are background corrected by using a user defined fixed gray value that is subtracted from the images.
3. Subsequently images are low-pass filtered by a 3 × 3 median filter to reduce image noise.
4. Cell nuclei are then identified in each image data set by simple thresholding of the image acquired with the DAPI filter set (D in

Fig. 2). The respective thresholding value is set by the operator at the beginning of an analysis session.

5. Cell nuclei touching each other or the image borders are disregarded from further analysis.

6. The image containing the thresholded nuclei is then used to generate a binary image mask with the nuclei having associated gray value "1" and background is set to "0" (Fig. 2D).

7. The areas of the cell nuclei are then dilated to also extend into the cytoplasm of each cell (Fig. 2E). This image is then used to generate a second mask.

8. This mask is then multiplied with the image containing information of the cell surface staining (Fig. 2B) resulting in image Fig. 2H.

9. A cytoplasmic mask is generated by subtracting image Fig. 2D from image Fig. 2E, resulting in a mask shown in Fig. 2F.

10. This cytoplasmic mask is then multiplied with the image containing information on the total ts-O45-G expressed (Fig. 2C) resulting in the image shown in Fig. 2I.

11. The average intensities of each object in images Fig. 2H and Fig. 2I are then determined and the values corresponding to the same cell are used to determine the ratio R defined as the intensity at the plasma membrane (Fig. 2H) divided by the intensity representing the amount of YFP- or CFP-ts-O45-G expressed in the cell (Fig. 2I). This ratio is proportional to the relative amount of ts-O45-G transported to the plasma membrane.

Comments

It is important to test at the beginning of the image acquisition that images are not saturated as this distorts the quantification of fluorescence. Therefore, to adjust the systems parameters appropriately, the operator should take a few test images from different areas on the sample and check the images for saturation. Parameters should then be adjusted such that even the brightest signal can still be acquired at nonsaturating conditions.

Figure 2G shows the results of an example analysis of siRNA transfected cells. Should however GFP-tagged plasmids be expressed instead of siRNAs, then tranfected cells can be distinguished from nontransfected cells by multiplying the mask shown in Fig. 2E with the image containing the signal for the transfected cells. The average fluorescence intensity of the objects of the resulting image is then determined and transfected cells are discriminated from nontransfected ones by a threshold defined by the operator.

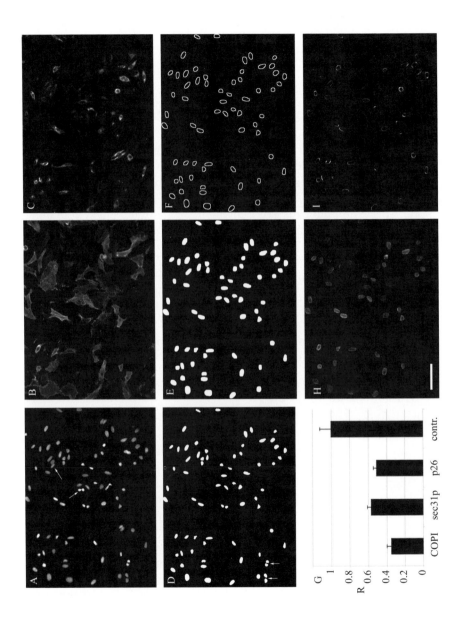

Analyzing the Kinetics of Transport Carriers Using Kymographs

Kymograph "wave drawer," an instrument to monitor signal changes over time, was first described in 1845. Since then a number of devices and programs to plot events over time have inherited this name. Here we describe a variant we developed to plot dynamic events along an arbitrary line region selection in time-lapse sequences to analyze the intracellular movements of fluorescently labeled structures.

In a time-lapse recording, fluorescently labeled transport intermediates often appear as small dots or tubules of variable size and intensity (see Fig. 3A). Tracks of moving transport carriers can be obtained from these time series in a two-step protocol. First subsequent images are subtracted from each other, to eliminate in the resulting images those structures, which are not moving. Second, images in the resulting sequence are projected onto each other highlighting the trajectories the moving structures in the time-lapse series have taken (e.g., Fig 3B). Monitoring the intensities along each of these trajectories for every time-point in the image sequence results in time-space plots (Fig. 3I). In these, the x-axis represents the intensity profile of the selected track and the y-axis relates to time points of the analyzed series. In these time-space plots a moving fluorescent structure is represented by a slope (see arrows in Fig. 3I) from which the velocity of the moving structure can be determined.

The time-space plotting is a very sensitive, accurate, and fast interactive method to determine velocities of moving carriers from time-lapse recordings. In the following we describe the use of a collection of macros and plugins (available at: www.embl.de/eamnet/downloads) we have developed for the freeware multiplatform image analysis software ImageJ (http://rsb.info.nih.gov/ij/). These macros aim at the automation of the required tasks to determine transport carrier kinetics by kymographs. In addition they offer a variety of functions to improve the image quality in time-lapse sequences.

FIG. 2. Image acquisition and processing steps to determine the transport of ts-O45-G to the plasma membrane. HeLa cells were transfected with siRNAs on LabTek arrays as they are described in Chapter 1 of this issue. The ts-O45-G transport assay was carried out as described in protocol 1.1 as described earlier in this chapter. Images were acquired sequentially using a 10X objective on a Scan^R system using filters to detect specifically DAPI stained nuclei (A), Cy3 stained ts-O45-G at the plasma membrane (B), and CFP-tagged ts-O45-G (C). Images D-I were generated as described in protocol 1.2. earlier in this chapter. "R" in (G) is the ratio of ts-O45-G at the plasma membrane (measured in H) to ts-O45-G expressed in cells (measured in I). Results for siRNAs targeting the COPI component β-COP, the COPII component Sec31p, and a p24 related membrane protein p26 are shown. The values are the average of two independent experiments (Bar = 50 μm).

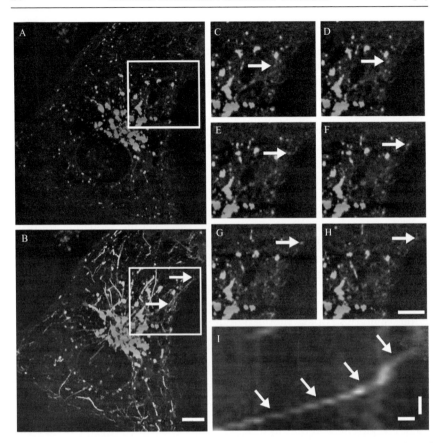

Fig. 3. Generation of a kymograph or time-space-plot to measure the dynamics of transport carrier molecules. (A) First frame of a time-lapse recording of a cell expressing YFP-tagged ts-O45-G at 20 min after temperature shift from 39.5° to 32° (the entire QuickTime movie of the experiment shown is available at: http://www.embl.de/eamnet/downloads/vsv-tsO45G.mov). (B) Maximum intensity projection of 30 frames covering 6 sec of the time-lapse sequence. The arrow indicates the track of a carrier analyzed. (C-H) Zoom in of images in the region surrounded by the box in A and B. Images are 1 sec apart from each other. (I) Time space plot of the track indicated by arrows in B. The arrows in (I) point to the slope describing the kinetics of the moving particle. Scalebars in B and H:5 μm, Scalebar in I: horizontal 1 μm, vertical 1 sec.

Installation of the Macros and Plugins in Image J

1. Install the Image J program (available at: http://rsb.info.nih.gov/ij/).
2. Copy the plugins "WalkingAverage_.class, StackDifference_.class, MultipleOverlay_.class, MultipleKymograph_.class" (all available

at: www.embl.de/eamnet/downloads) to the Image J plugin folder. Alternatively the collection of macros tsp040421.txt may be loaded into Image J.

Analyzing the Kinetics of Transport Carriers

1. Load the time-lapse image sequence into Image J.
2. If necessary crop the image sequence using the "crop" tool of Image J (edit menu).
3. Reduce image noise in all images of the sequence by using the "walking average" plugin.
4. Generate tracks of moving objects in the image sequence by running the macro "track" or the plugin "StackDifference."
5. Select individual tracks with the "segmented line" tool of Image J or use the plugin "MultipleOverlay" to select and store more than one track.
6. Execute the macro "kymograph" or the plugin "Multiple Kymograph" to generate a time-space plot (e.g., Fig. 3I).
7. Mark in the time-space plot the trace of a moving particle by using the "segmented line" tool of Image J.
8. Determine the speed of the moving particle at each time-point by using the macro "read velocities from tsp."

Comments

The speed of image acquisition is a critical parameter for the success of the time-space plot approach and needs to occur at a frequency high enough to result in a continuous trajectory in the projection image (refer to Step 4 in the protocol described earlier). The macro "read velocities from tsp" calculates a number of parameters such as, the entire length of the trajectory taken by the structure under view, average, and instantaneous speed. The speed is returned in units of pixels/frame, which needs to be converted to, for instance, μm/sec using image acquisition calibration parameters (e.g., physical distance represented by one pixel and the time two subsequent images in the sequence are apart from each other).

Summary and Perspectives

We described two methods to quantify membrane transport in single cells using automated fluorescence microscopy and image analysis. The assay to measure the plasma membrane transport of ts-O45-G is robust and has already been used in systematic analyses to identify new proteins involved in the regulation of the secretory pathway (e.g., Starkuviene *et al.*, 2004). Applying this approach to genome-wide siRNA screens in intact cells, for example, may help to reveal the machinery and interaction networks underlying the regulation of membrane traffic in the secretory pathway.

References

Bell, A. W., Ward, M. A., Blackstock, W. P., Freeman, H. N., Choudhary, J. S., Lewis, A. P., Chotai, D., Fazel, A., Gushue, J. N., Paiement, J., Palcy, S., Chevet, E., Lafreniere-Roula, M., Solari, R., Thomas, D. Y., Rowley, A., and Bergeron, J. J. (2001). Proteomics characterization of abundant Golgi membrane proteins. *J. Biol. Chem.* **276**, 5152–5165.

Calero, M., Winand, N. J., and Collins, R. N. (2002). Identification of the novel proteins Yip4p and Yip5p as Rab GTPase interacting factors. *FEBS Lett.* **515**, 89–98.

Elbashir, S. M., Harborth, J., Lendeckel, W., Yalcin, A., Weber, K., and Tuschl, T. (2001). Duplexes of 21-nucleotide RNAs mediate RNA interference in cultured mammalian cells. *Nature* **411**, 494–498.

Keller, P., Toomre, D., Diaz, E., White, J., and Simons, K. (2001). Multicolour imaging of post-Golgi sorting and trafficking in live cells. *Nat. Cell. Biol.* **3**, 140–149.

Liebel, U., Starkuviene, V., Erfle, H., Simpson, J. C., Poustka, A., Wiemann, S., and Pepperkok, R. (2003). A microscope-based screening platform for large-scale functional protein analysis in intact cells. *FEBS Lett.* **554**, 394–398.

Monier, S., Jollivet, F., Janoueix-Lerosey, I., Johannes, L., and Goud, B. (2002). Characterization of novel Rab6-interacting proteins involved in endosome-to-TGN transport. *Traffic* **3**, 289–297.

Pelkmans, L., Fava, E., Grabner, H., Hannus, M., Habermann, B., Krausz, E., and Zerial, M. (2005). Genome-wide analysis of human kinases in clathrin- and caveolae/raft-mediated endocytosis. *Nature* **436**, 78–86.

Pepperkok, R., Girod, A., Simpson, J. C., and Rietdorf, J. (2000). Imunofluorescence microscopy. *In* "Monoclonal Antibodies: A Practical Approach," pp. 355–370. Oxford University Press, New York.

Starkuviene, V., Liebel, U., Simpson, J. C., Erfle, H., Poustka, A., Wiemann, S., and Pepperkok, R. (2004). High-content screening microscopy identifies novel proteins with a putative role in secretory membrane traffic. *Genome Res.* **14**, 1948–1956.

Zilberstein, A., Snider, M. D., Porter, M., and Lodish, H. F. (1980). Mutants of vesicular stomatitis virus blocked at different stages in maturation of the viral glycoprotein. *Cell* **21**, 417–427.

[3]   Bioinformatic "Harvester": A Search Engine for
Genome-Wide Human, Mouse, and Rat Protein Resources

By URBAN LIEBEL, BJOERN KINDLER, and RAINER PEPPERKOK

Abstract

Harvester is a meta search engine for gene and protein information. It searches 16 major databases and prediction servers and combines the results on pregenerated HTML pages. In this way Harvester can provide comprehensive gene–protein information from different servers in a convenient and fast manner. The Harvester search engine works similar to Google, offering genome-wide ranked results at very high speed. Here we describe how to use this bioinformatic tool along with selected examples.

Introduction

The continuously growing amount of gene and protein associated information spread over numerous databases worldwide makes it difficult to find and evaluate relevant information of the genes or proteins of interest. Some databases provide small amounts of manually generated but high quality data, others offer genome wide annotations that have been generated automatically. To obtain the most comprehensive knowledge on the genes–proteins under view, it is essential to combine and compare the information of these different sources (Liebel *et al.*, 2004). However, querying several databases manually is labor intensive, especially if lists of genes–proteins are under investigation, and often uncertainty remains if all relevant information has been collected. In addition different sources often provide different information on the same genes–proteins, and it becomes necessary to filter the information accordingly. To overcome these problems we developed Harvester, a meta search engine that cross-links and caches 16 major bioinformatics resources. Many of the databases combine information from several subservers. Caching and precomputing the "harvested" information locally allows access to it conveniently and quickly. A ranking system based on simple rules sorts cached and precomputed results similar to Google's page rank on a genome-wide scale. The following material describes some examples for effective use of Harvester and the interpretation of the results obtained with it. Harvester currently works exclusively with human and mouse proteins.

METHODS IN ENZYMOLOGY, VOL. 404   0076-6879/05 $35.00
          DOI: 10.1016/S0076-6879(05)04003-6

How Does the Harvester Search Engine Work?

Harvester collects bioinformatic information from 16 resources. The current collection offers information for 64,000 human and 47,000 mouse proteins. All information regarding one protein is saved on a single static HTML page. The pages are hierarchically linked from the servers' main page.

Harvester collects two types of information: pure text-based information and information rich in graphical elements. Text-based information is retrieved and indexed from the following public databases and prediction servers: UNIPROT, SOURCE, SMART, SOSUI, PSORT, RZPD, Homologene, gfp-cDNA, IPI, CDART, STRING. For optimal search engine indexing, redundant text information is removed by server-specific converter modules.

Graphical rich information is cross-linked via "iframes." Iframes are "transparent windows" within an HTML page and allow the display of information from one server into an HTML page of another server (here the Harvester pages) in real time. Several of those iframes are combined on every individual Harvester page. The information displayed via these iframes is therefore always up to date. Currently Harvester has implemented six servers via "iframes" (NCBI BLAST, Genome-Browser, Ensembl, RZPD, CDART, STRING).

Data from computing-intensive prediction servers is precomputed, collected, and indexed (e.g., PSORT II [Nakai and Horton, 1999] or SMART [Letunic et al., 2004]). Harvester provides these data instantaneously and thus increases the speed of searches drastically compared to searching the respective databases manually one after the other. Pages older than 21 days are continuously updated. The quality of the collected information automatically increases with the quality of the crawled data servers, which themselves collect their information from several sub sources but quality-check them prior to publication (Apweiler et al., 2004; Diehn et al., 2003).

Because different servers use distinct prediction algorithms, they often return contradictory results on the same request. Since all acquired data on one protein are collected on a single page by Harvester, it has been possible to address this problem by introducing a ranking system in Harvester, where predictions that are identical on two or more servers are scored higher than predictions returned by only one server. In this way the levels of confidence for predictions presented by Harvester become transparent. At the top of every search results page, a link "description of the search index" guides the user to a detailed explanation of how the Harvester ranking works.

Examples of How to Query the Harvester Search Engine

To query the Harvester data, a simple three-step protocol is sufficient to return all protein- or gene-specific data in a few seconds.

1. Point your browser to the Harvester server at http://harvester.embl.de.
2. Find the main search interface (human or mouse) underneath the umbrella-like graphical representation of all supported databases and type your search term (e.g., "sar1a").
   Shortly after hitting the SEARCH button, a list of all Harvester pages found along with the search score is presented.
3. Click on one of the results and the corresponding Harvester page will be displayed.

Examples of queries and their corresponding search phrases are summarized in Table I.

As Harvester comprises various protein and gene nomenclatures of several major bioinformatic resources searching with gene aliases like "golgin-160" or "golga3," it is possible to obtain information of the same protein from the different databases using the distinct aliases. As Harvester also extracts information extracted from the GO consortium (Gene Ontology project; see http://www.geneontology.org/ for details), which has the goal of producing a controlled vocabulary for gene–protein function, searches for molecular functions or biological processes such as "intra-Golgi transport" or "protein kinase binding" are possible. SOURCE and other databases provide information on the chromosomal localization of genes, and with Harvester, it is possible to find all genes on a certain cytoband, for example, "17p13," if the corresponding cytoband has been linked to a particular disease, phenotype, or event. A Harvester-search for "17p13" automatically retrieves information

TABLE I
EXAMPLES OF HARVESTER QUERIES

Query	Search phrase
gene name and aliases	"golgin-160" "golga3"
molecular function	"intra golgi transport"
chromosomal location	"17p13"
protein domains	"sh2" "sar"
localization	"golgi""plasma membrane"
disease-related information	"brain malformation"
disease-related information	"occurs in hemizygous males"
paper title	"Cloning and expression of a human CDC42..."
author	"Barfold"
automatic–manual annotation:	"highly expressed in heart"

from subcytoband "17p13.2" and "17p13.1." A more specific search ("17p13.2") reduces the number of results accordingly.

Disease-related information is mostly extracted from the OMIM database. (Wheeler *et al.*, 2004). OMIM is a catalogue of human genes and genetic disorders offering many disease-related data such as disease occurrences (e.g., search term "occurs in hemizygous males"), phenotype (e.g., search term "brain malformation") or the disease name itself (e.g. "Alzheimer").

Harvester also collects protein domain analyses from the SMART protein server (Letunic *et al.*, 2004). Information from the UNIPROT (Apweiler *et al.*, 2004) and SOURCE (Diehn *et al.*, 2003) databases confirm or complete the SMART protein domain information.

The PSORT server (Nakai and Horton, 1999) contributes a prediction algorithm for the subcellular localizations for all known proteins, which is complemented by experimental data on the subcellular localization of GFP-tagged human cDNAs (Simpson and Pepperkok, 2003).

Database Identifier Cross-links

Many databases use their own protein or nucleotide identification system. Therefore it is difficult to map the information obtained on the database-specific identifier from one database to another. Every Harvester page therefore contains database identifiers from various sources as shown in Table II. Among the most frequently used IDs are:

a. Uniprot IDs (example "Q9NPD3"), which is a central repository of protein sequence and function created by joining the information contained in Swiss-Prot, TrEMBL (translated EMBL), and PIR (Protein Information Resource at Georgetown University).

b. Unigene Cluster IDs (example "Hs.449360"). Unigene is an experimental system for automatically partitioning GenBank/EMBL/DDBJ sequences into a nonredundant set of gene-oriented clusters. Each Unigene cluster contains sequences that represent a

TABLE II
DATABASE CROSS-LINKS USED BY HARVESTER

Database identifier	Example
Uniprot identifier	"Q9NPD3"
OMIM	"16478"
UniGene Cluster	"Hs.449360"
ensEMBL genes	"ENSG00000157933"
RefSeq identifier	"NM_022650" (NM_ = NCBI mRNA)

unique gene, as well as related information such as the tissue types in which the gene has been expressed and map location.

c. Ensembl gene IDs (example "ENSG00000157933"). Ensembl produces and maintains automatic annotation on metazoan genomes.

d. RefSeq IDs (example "NM_022650" NM for NCBI mRNA). The Reference Sequence collection aims to provide a comprehensive, nonredundant set of sequences, including genomic DNA, transcript (RNA), and protein products, for major research organisms.

e. Interpro IDs (example "IPR00906"). InterPro is a database of protein families, domains, and functional sites in which identifiable features found in known proteins can be applied to unknown protein sequences.

Advanced Searches

The number of search terms is not limited in Harvester and even a list of 30 and more search terms for one query session is possible. By default the search terms are connected with the OR operator, but this can be switched to AND or combined with a NOT operator. If one or several of the search terms do not exist on a Harvester page, the ranking score for this page is not zero as Harvester scores by default every search term separately on every page and returns the sum of all scores for every page.

An example "long list search" excluding all hypothetical predicted proteins could be "KDEL transmembrane signal peptide golgi endoplasmic reticulum sec0013 se23 sec0030 transport" NOT "hypothetical."

An example "long list search" for small GTPases with homologues in mouse could be"ras rho raf growth differentiation movement lipid vesicle transport G-protein mus musculus."

All Harvester pages are indexed from Google and other search engines as well. A "search term" in combination with the words "bioinformatic harvester" retrieves the corresponding Harvester pages. However, it should be noted that Google is not well suited for long lists of search terms as it always uses the AND operator to connect search terms.

A Closer Look at the Search Results

The results returned for a Harvester search with the query "GTPase transport Golgi" NOT "rab" with the search terms linked via the AND operator are shown in Table III.

The first column ("Score") displays the sum of all word scores. Proteins are sorted according to this score. For determination of the page score Harvester counts the number of occurrences of each of the search terms individually, divides it by the total number of occurrences on all Harvester pages, and multiplies this number with 100,000 (for better readability). The

TABLE III
Part of the Result Table Returned by Harvester for the Search "GTPase and Transport and Golgi" but not "Rab"

Score	Title	Unigene	c	s	0	1	2
611.850	Human protein: Q8N6T3 – ADP-ribosylation factor GTPase activating protein 1 (ADP-ribosylation factor 1 GTPase activating protein) (ARF1 GAP) (ARF1-directed GTPase-activating protein) (GAP protein)	Hs.25584	3	44	192.43(12)	384.86(24)	34.56(8)
494.037	Human protein: Q08379 – Golgi autoantigen, golgin subfamily A member 2 (Golgi matrix protein GM130) (Gm130 autoantigen) (Golgin-95)	Hs.155827	3	33	465.04(29)	16.04(1)	12.96(3)
355.276	Human protein: Q9HD26 – PIST (Golgi associated PDZ and coiled-coil motif containing)	Hs.191539	3	28	304.68(19)	16.04(1)	34.56(8)
317.641	Human protein: Q92834 – X-linked retinitis pigmentosa GTPase regulator	Hs.61438	3	22	80.18(5)	224.50(14)	12.96(3)
270.776	Human protein: Q9NP61 – ADP-ribosylation factor GTPase-activating protein 3 (ARF GAP 3)	Hs.13014	3	22	48.11(3)	192.43(12)	30.24(7)
254.740	Human protein: Q9NR31 – GTP-binding protein SAR1a (COPII-associated small GTPase)	Hs.499960	3	21	96.22(6)	128.29(8)	30.24(7)
214.029	Human protein: Q9Y6B6 – GTP-binding protein SAR1b (GTBPB)	Hs.279582	3	17	128.29(8)	64.14(4)	21.60(5)
212.787	Human protein: O43150 – Development and differentiation-enhancing factor 2 (Pyk2 C-terminus associated protein) (PAP) (Paxillin-associated protein with ARFGAP activity 3) (PAG3)	Hs.467662	3	14	96.22(6)	112.25(7)	4.32(1)
196.751	Human protein: P36406 – GTP-binding protein ARD-1 (ADP-ribosylation factor domain protein 1) (Tripartite motif protein 23) (RING finger protein 46)	Hs.792	3	13	64.14(4)	128.29(8)	4.32(1)
196.751	Human protein: Q13795 – ARF-related protein 1 (ARP)	Hs.389277	3	13	80.18(5)	112.25(7)	4.32(1)
157.282	Human protein: P18085 – ADP-ribosylation factor 4	Hs.148330	3	12	64.14(4)	80.18(5)	12.96(3)

page score is then the sum of all individual word scores (for more details, see http://harvester.embl.de). If the Harvester search option "single word score display" is chosen, additional single word score for each search term is listed on the right side of the column (e.g., column 0 for "GTPase," 1 for "transport," and 2 for "Golgi" in Table III).

In the column "Title" (Table III) a short feature summary of the protein corresponding to the particular Harvester page is shown next to the UNIPROT identifier.

Currently, there are over 64,000 human proteins in the UNIPROT database comprising manually and automatically translated proteins and fragments from many different projects. The "Unigene" column in Table III maps these proteins and fragments to a certain gene. Unigene is an experimental system for automatically partitioning GenBank/EMBL/DDBJ sequences into a nonredundant set of gene-oriented clusters and each Unigene cluster contains sequences that represent a unique gene.

Summary and Perspectives

Harvester unifies and cross-links bioinformatic information from 16 resources and thereby offers a fast search engine that sorts results based on a Google-like ranking system. As the amount of new information resources, such as the biological data obtained from genome-wide large scale projects, increases steadily, it will become even more critical in the future to link and integrate the information of several subservers into a few central information servers. Such meta search engines and cross-linking sites will not only collect all the information for fast access but can also help to compare and evaluate information and most importantly provide filters to remove redundant information and extract the high-quality information.

Acknowledgments

We thank Lars Juhl Jensen for the idea of using "iframes" to look at several databases in parallel and in real time. We also appreciate all those users of Harvester who have returned suggestions and criticism to us, which all have been useful to improve Harvester. We also thank the EMBL Heidelberg Computer and Networking group (CNG) for providing indexing and web server capacity.

References

Apweiler, R., Bairoch, A., Wu, C. H., Barker, W. C., Boeckmann, B., Ferro, S., Gasteiger, E., Huang, H., Lopez, R., Magrane, M., Martin, M. J., Natale, D. A., O'Donovan, C., Redaschi, N., and Yeh, L. S. (2004). UniProt: The Universal Protein knowledgebase. *Nucleic Acids Res.* **32,** D115–119.
Diehn, M., Sherlock, G., Binkley, G., Jin, H., Matese, J. C., Hernandez-Boussard, T., Rees, C. A., Cherry, J. M., Botstein, D., Brown, P. O., and Alizadeh, A. A. (2003). SOURCE: A

unified genomic resource of functional annotations, ontologies, and gene expression data. *Nucleic Acids Res.* **31,** 219–223.

Letunic, I., Copley, R. R., Schmidt, S., Ciccarelli, F. D., Doerks, T., Schultz, J., Ponting, C. P., and Bork, P. (2004). SMART 4.0: Towards genomic data integration. *Nucleic Acids Res.* **32,** D142–144.

Liebel, U., Kindler, B., and Pepperkok, R. (2004). Harvester: A fast meta search engine of human protein resources. *Bioinformatics* **20,** 1962–1963.

Nakai, K., and Horton, P. (1999). PSORT: A program for detecting sorting signals in proteins and predicting their subcellular localization. *Trends Biochem. Sci.* **24,** 34–6.

Simpson, J. C., and Pepperkok, R. (2003). Localizing the proteome. *Genome Biol.* **4,** 240.

Wheeler, D. L., Church, D. M., Edgar, R., Federhen, S., Helmberg, W., Madden, T. L., Pontius, J. U., Schuler, G. D., Schriml, L. M., Sequeira, E., Suzek, T. O., Tatusova, T. A., and Wagner, L. (2004). Database resources of the National Center for Biotechnology Information: Update. *Nucleic Acids Res.* **32,** D35–40.

[4] Microinjection as a Tool to Explore Small GTPase Function

By BRIAN STORRIE

Abstract

Microinjection overcomes the plasma membrane barrier to the introduction of charged or large nonlipid soluble molecules into cells by the direct insertion of a hollow capillary micropipette into the cell. With the application of pressure, aqueous solution is then directly transferred into either the cytosol or the nucleus. I give specific examples of the application of this approach to the functional study of small GTPases of the Sar1, ARF, and rab family in membrane trafficking between the Golgi apparatus and endoplasmic reticulum (ER). The principles illustrated by these examples should be generally applicable to other small GTPases. Detailed protocols for capillary microinjection using semiautomated equipment are given.

Introduction

Microinjection is a tool to overcome the plasma membrane permeability barrier to the introduction of charged molecules, polypeptides, or DNA plasmids into cells. In essence, microinjection treats the cell as the test tube and uses a microinjection capillary needle as the pipette to add small volumes of solution to the cytoplasm or nucleus. The approach has major advantages: (i) the technique is highly synchronous with a few hundred cells being injected over 10–20 minutes, (ii) intracellular environment and cell morphology is preserved, (iii) the reaction vessel is small, that is, the size of a cell, and correspondingly reagent dilution is confined to the

METHODS IN ENZYMOLOGY, VOL. 404
0076-6879/05 $35.00
DOI: 10.1016/S0076-6879(05)04004-8

volume of the cell, and (iv) reagent consumption is limited. No more than 0.5 to 1.0 μl of solution is required to load the microcapillary. Injection volumes are on the order of 1 pl, $^{-12}$ liters or ~5–10% of cell volume. Both flat substratum-attached and rounded-to-free floating cells can be injected. The technique is popular with an early March 2005 PubMed search against "microinjection" yielding 15,762 hits and perhaps underappreciated in the small GTPase field with the corresponding search against "microinjection small GTPase" giving 134 hits.

Microinjection is a light microscope based technique. The microinjection capillary and its positioning relative to cells are tracked by phase contrast. Moreover, phase contrast is used to identify cell nucleus versus cytoplasm. Microinjection requires in addition to an inverted phase contrast microscope at least two committed pieces of equipment (Fig. 1). These are a micromanipulator to position the injection capillary and a pressure regulator to increase pressure to the capillary selectively as it

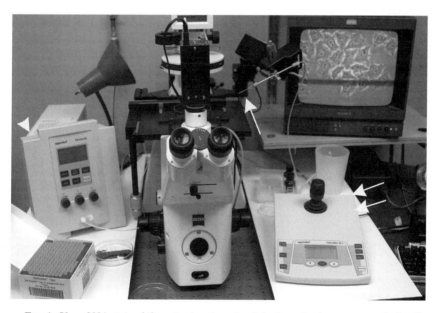

FIG. 1. Year 2004 state-of-the-art setup for microinjection of substratum attached cells. Left to right: Box of Geloader® pipette tips for backloading capillary micropipettes, glass Petri dish bottom with modeling clay bridge for storing newly fabricated capillary micropipettes, Eppendorf FemtoJet pressure regulator (arrowhead), research grade inverted microscope, Eppendorf XYZ motor drive attached to right side of microscope stage (arrow points to capillary micropipette in holder), Eppendorf InjectMan® joystick controller for micromanipulation/injection (double arrows), and black and white video monitor showing substratum attached HeLa cells.

enters a cell. The increased pressure is what expels fluid from the capillary within cells. The micromanipulator is mounted onto the stage of the inverted microscope. Ideally the micromanipulator and pressure regulator work together under microprocessor control. In fact, the microprocessor and the possibility of programmed injection is what places microinjection as a practical technique in the hands of many investigators. In programmed injection, the microprocessor controls the actual dipping of the capillary into the cell and coordinates rapid changes in capillary pressure with capillary position. Hence, the skill level required is reduced and the success rate of injections greatly increased. Microinjection was introduced as a technique in the 1940s (e.g., de Fonbrune, 1949). Microprocessor control was introduced in the 1980s through instrument development efforts at the European Molecular Biology Laboratory (Ansorge, 1982; Pepperkok *et al.*, 1988) and commercialized by Carl Zeiss and Eppendorf. The microprocessor controlled Eppendorf injection system used by my laboratory may be mounted on any of a number of different inverted microscopes. Microcapillaries suitable for injection into either the cytoplasm or nucleus may be purchased from Eppendorf. Microcapillaries may also be pulled to individual requirements using commercially available equipment such as the Flaming-Brown puller from Sutter Instruments.

The end purpose of microinjection is to produce phenotype. I consider in detail here the situation of substratum attached interphase cells and experiments designed to elucidate small GTPase function in secretion. The principles developed are applicable to small GTPases in other metabolic pathways. I will first illustrate approaches taking examples primarily from the work of my laboratory, second use the strengths and weaknesses of these examples to propose a set of good practice standards for microinjection experiments, and finally present a set of detailed protocols for injection experiments. In general, microinjection experiments involve the introduction of either increased amounts of a normal or mutant protein or a potential inhibitor into cells. For the consideration of alternate approaches such as vaccinia virus for rapid, synchronous, high-level transient transfections or siRNA for slow, but effective knockdowns of wild-type protein expression, the reader is directed to other articles within the Methods in Enzymology series.

Example Uses

GTPases act as conformation-dependent molecular switches. In the GTP-bound state, the switch is on. In the GDP-bound state, the switch is off. When switched on, small GTPase act to recruit effectors. GTPγS as a non-hydrolyzable analogue of GTP is an inhibitor of all small GTPases. Activity of individual GTPases may be selectively altered by amino acid substitutions. For example, the activity state of small GTPases may be altered by amino acid substitutions in their primary sequence that affect GTP hydrolysis, guanine nucleotide exchange, nucleotide binding, and potentially specific effector interactions. In addition to altering the amount or activity state of small GTPases, alterations in the amount or activity state of effectors is also an attractive avenue for probing function. Dominant negative phenotypes may be produced by amino acid substitutions that either activate the GTPase switch, for example, mutations that slow GTP hydrolysis, so-called GTP-restricted mutations, or those that inactivate the GTPase switch, for example, mutations that stabilize the association of GDP with the protein, so-called GDP-restricted mutations. Increased amounts of GTP-restricted mutations act as dominant negative inhibitors because they slow the reutilization of effectors. GDP-restricted mutant proteins act not by competing for effectors, but by competing with wild-type protein for binding to guanine nucleotide exchange factors (GEFs). In fact, GDP-restricted small GTPases in general bind more tightly to GEFs than wild type proteins and hence form "dead-end" complexes with the GEF (Feig, 1999). Of course, how specific the effects of a GDP-restricted rab6a, for example, is dependent on how specific the GEF is.

With this background in mind, I now consider a number of specific examples relating to membrane trafficking between the ER and Golgi apparatus. These examples concern Sar1a, a small GTPase required for COPII coat protein recruitment to ER membranes; ARF1, a small GTPase required for COPI coat protein and clathrin recruitment to Golgi apparatus membranes; rab6a/a', small GTPases involved in membrane trafficking at the trans Golgi apparatus/trans Golgi network (TGN); and rab33b, a small GTPase of the medial Golgi apparatus. In all cases, phenotype is assayed on a single cell basis by assessing effects on protein distribution. Lastly antibody microinjection to inhibit cell intoxication by Pseudomonas exotoxin or Shiga-like toxin will be considered as an example of outcome assessment by incorporation of radiolabeled amino acid.

1. GTPγS as a general inhibitor of small GTPase dependent processes – We used microinjected GTPγS to probe the dependence of Golgi apparatus scattering in response to microtubule depolymerizaton to small GTPases (Yang and Storrie, 1998). Vero cells were microinjected with GTPγS from a 500 μM stock solution. As the microinjection volume is about 5–10% of that of the cell, the intracellular concentrations were between 25 and 50 μM. These concentrations had no effect on Golgi scattering in a 4.5 h endpoint assay. Higher concentrations led to cell damage. As a positive control, we found that these concentrations of GTPγS were sufficient to disperse the juxtanuclear, Golgi apparatus

associated concentration of the COPI coat protein complex. Protein distributions were assessed by immunofluorescence.

Do negative results in such experiments truly indicate that GTPases play no role in a process? The answer may well be no in any given situation. As Pepperkok et al. (1998) showed, GTPγS microinjection fails to block Sar1a function and hence to affect ER exit of vesicular stomatitis virus G protein. Contrary to the expectations of our GTPγS microinjection experiments, we later showed that overexpression of GDP-rab6a delays but does not block microtubule–depolymerization-induced Golgi scattering (Jiang and Storrie, 2005), that is, a small GTPase does have some role in Golgi scattering. The bottom line is that such experiments require careful analysis of protein distribution over time, that is, kinetic analysis of phenotype, and the acknowledgement that some GTPases may be comparatively insensitive to GTPγS.

2. Overexpression of wild-type protein – The levels of small GTPases and effectors must be exquisitely balanced to produce normal resident protein distributions within the secretory pathway. Nevertheless, the finding that overexpression of wild type rab6a (Martinez et al., 1997, transfection) or rab33b (Valsdottir et al., 2001, microinjection) induced the redistribution of Golgi glycosyltransferases to the ER was remarkable. In such experiments, the advantage of microinjection versus transfection was the high degree of synchrony that comes from the narrow time window over which the plasmids are injected into cells. Plasmid injection into the nucleus is often used because it is much simpler to purify a plasmid than to purify from an animal cell source a protein expressed in small amounts or to purify the same protein in an active, properly modified protein from bacteria. Small GTPases are post-translationally modified by fatty acid addition.

3. Overexpression of mutant protein – Considering that either GTP- or GDP-restricted rab6/rab33b, Sar1, or Arf1 all can produce a dominant negative phenotype when introduced into cells, does it really matter which allele is used? Whether these experiments are done as plasmid expression or purified protein microinjection, the answer is very much "Yes." GTP-restricted rab6a or rab33b both induce the redistribution of Golgi resident proteins to the ER while respective GDP-restricted isoform induces, if anything, a more compact, juxtanuclear Golgi apparatus (Girod et al., 1999; Jiang and Storrie, 2005; Martinez et al., 1997; Valsdottir, 2001; Young et al., 2005). Correspondingly, expression of GTP-restricted Arf1 results in the stabilization of COPI association with Golgi membranes while expression of GDP-restricted Arf1 results in the failure to recruit new rounds of COPI coat proteins to Golgi membranes and hence produces decoated Golgi membranes that are absorbed into the ER in a brefeldin A like phenotype (Dascher and Balch, 1994).

In the case of Sar1p, the phenotypic differences between alternate mutations are subtle with respect to Golgi apparatus organization and have led to controversy and confusion in the literature. When microin-jected into HeLa cells in the same plasmid background, both isoforms are effective in inhibiting the transport of VSV-G protein from the ER to the Golgi apparatus (Stroud et al., 2003). At moderate inhibitory protein levels, GTP-restricted Sar1 stabilizes the association of COPII coat proteins with ER membranes and ER exit sites. In some cell types, ER exit sites are juxtanuclearly clustered as is the Golgi apparatus. At high GTP-restricted Sar1 levels, this association is destabilized (Stroud et al., 2003). While on the other hand, GDP-restricted Sar1 prevents COPII coat protein recruit-ment to ER membranes at any inhibitory concentration. The net pheno-type is in either case a block in protein transport from the ER to the Golgi apparatus. However, in the case of the GTP-restricted mutant, the juxta-nuclear localization of Golgi matrix proteins appears more stable than that of recycling Golgi cisternal enzymes (Miles et al., 2001; Seemann et al., 2000; Ward et al., 2001). In the case of the GDP-restricted mutant, no such difference is apparent (Stroud et al., 2003; Ward et al., 2001). All Golgi proteins disperse in a manner consistent with protein cycling between the Golgi apparatus and ER.

In sum, the experimenter must know the small GTPase and the ins and outs of how phenotype might be affected.

4. Direct microinjection of mutant protein – In many ways, this is conceptually the most straightforward experiment. The purified protein is microinjected into the cytoplasm, often in the presence of cycloheximide to inhibit new proteins synthesis, and phenotype is assessed with no need to wait for plasmid-encoded protein expression. As one example, I cite our use of microinjected GTP-Sar1H79N protein to establish that accumula-tion of Golgi resident proteins in the ER in response to an ER exit block was due to protein recycling, not accumulation of newly synthesized pro-teins (Girod et al., 1999; Miles et al., 2001; Storrie et al., 1998). In these experiments, HeLa cells were microinjected with purified histidine (his)-tagged Sar1H79N protein at stock concentrations of ∼0.75 mg/ml or final end cellular concentration of ∼35–75 μg/ml in the presence of either cycloheximide or emetine as strong inhibitors of protein synthesis. (The his-tag is included in the sequence of the engineered protein because it simplifies the isolation of the recombinant protein from E. coli by Ni-affinity chromatography.) The kinetics of Golgi protein redistribution to the ER was then scored over a few hours. The comparative strengths and weaknesses of the approach are apparent from the outcome of the experi-ments. The onset of the ER exit block is quick and there is no need for protein synthesis. However, the actual level of Sar1 mutant protein is likely low in comparison to what can be achieved by plasmid expression. This point is important in the case of the GTP-restricted protein because it is only at high concentrations that the GTP-restricted protein destabilizes

COPII association with ER membranes (Stroud *et al.*, 2003). Furthermore, it should be noted that in the case of Sar1 the GTP-restricted protein has been the microinjected protein of choice not because it is the right mutant protein but because it is easier to isolate than the GDP-restricted mutant in stable, high activity form. As a final buyer beware note, the reader is reminded that small GTPase are in general lipid modified and that these modifications typically are not produced when isolating the recombinant, his-tagged protein from *E. coli*. The experimenter hopes that the injected cell will do what *E. coli* did not.

In sum, direct mutant protein microinjection is an attractive choice for which the major hurdle is isolation of the mutant protein as an active, stable protein.

5. Direct microinjection of inhibitory antibodies or protein fragments – In the secretory pathway, small GTPases act to recruit effector proteins to membranes. One class of effectors is coat proteins such as COPI and COPII. COPI coat proteins are one example of ARF1 effectors. Micro-injected antibodies to ß-COPI, one of seven different subunits of COPI, provide a specific tool to probe the role of COPI in a given process, parti-cularly as ARF1 has been implicated in many processes. As with other purified proteins, the microinjection is into the cytoplasm. Antibodies directed against a ß-COPI peptide containing the sequence EAGE react with an exposed region of COPI *in vivo* and inhibit protein transport from the ER to the plasma membrane (Pepperkok *et al.*, 1993). Importantly they do this by stabilizing the association of COPI coat proteins with Golgi apparatus membranes. Monovalent Fab fragments of these divalent IgG antibodies are effective, indicating that inhibition is due directly to anti-body binding to a specific site on ß-COPI rather than to divalent antibody-induced crosslinking of COPI components. In our experiments, we used affinity purified EAGE Fab fragments at a stock concentration of ~1.4 mg/ml and found that these had no effect on recycling of resident Golgi glycosyltransferases to the ER (Girod *et al.*, 1999). Importantly, microin-jected EAGE Fab fragments strongly inhibited the recycling of KDEL receptor from the cis Golgi apparatus to the ER. The negative result on resident glycosyltransferases recycling to the ER is only meaningful be-cause of the successful positive control. We concluded from the Fab frag-ment experiments together with supporting data with GTP-restricted ARF1 expression that resident Golgi protein recycling to the ER is a COPI independent process.

6. Co- and sequential microinjection experiments – Coinjection or sequential microinjection experiments provide an excellent approach to probe for the role of multiple gene products in a process. Here I give one example. Over expresssion of rab6a and rab6a', two rab6 isoforms differing

in only three amino acids, both cause redistribution of Golgi resident proteins to the ER (Jiang and Storrie, 2005; Young *et al.*, 2005). If the two rab6 isoforms are redundant, then coexpression of the GDP-restricted form of one should inhibit the redistribution induced by the GTP-restricted form of the other isoform (Jiang and Storrie, 2005). This is exactly what is observed indicating that indeed the two are redundant in the pathway. A simple biochemical explanation for the cross-competition between rab6a and rab6a' is that the two use the same GEF or other machinery in their cycling between GTP- and GDP-bound states. Coinjection–coexpression experiments are technically fairly simple. Sequential microinjection experiments are more technically demanding. If each round of injection has a 40% success rate, that is, successful injection per capillary dip into cell, then the overall successful rate of two sequential injections is the product of each injection or 16%. In practice, I note that the success rate for cytoplasmic injections is higher than that of nuclear injections.

7. Single cell versus biochemical assays for phenotype – The outcome of most microinjection experiments is assayed on a single cell basis. For example, Cascade blue dextran might be used as a coinjection marker to identify microinjected cells and antibody staining might be used to determine the distribution of the protein of interest. Alternatively, an antibody might be used to identify the microinjected or expressed protein. Antibody identification of the injected cells has the major advantage of allowing for assessment of protein–expression levels through staining intensity.

With skill, biochemical assays are possible. For example, Girod *et al.* (1999) assayed for the ability of anti-EAGE or anti-KDEL receptor antibodies to inhibit cell intoxication by Pseudomonas exotoxin or Shiga-like toxin by assessing the scintillation counting of [^{35}S] methionine incorporation by the injected cells. These experiments used cells cultured on cover-glass fragments and the coverslips were processed for trichloroacetic acid precipitation *in situ*. Such experiments require mock injection controls for injection-damaged cells act like dead cells with respect to amino acid incorporation.

Some Practical Comments and Suggested Good Practice Standards for Microinjection

Let me be straightforward: microinjection, even with present state-of-the-art semi-automatic injection equipment is a skill that does require a time investment. Based on my experience of having taught microinjection to ~15 people in a research laboratory setting over the last several years, almost anyone can do acceptable microinjection for research purposes with a week of learning. However, true skill, in being able to handle a range of samples, coping with pipette clogging, and being fast, takes more experience than that. These words are not meant to be discouraging. Rather they are meant to be realistic. To date, I have met only one person who did not have the patience or persistence to learn microinjection. The skill once learned is like riding a bicycle, something where proficiency is rapidly regained.

Microinjection can produce remarkably nice quantitative concentration and time responses (see, Jiang and Storrie, 2005; Stroud *et al.*, 2003; Young *et al.*, 2005 for example data). However, doing so requires treating the microinjection capillary (needle) with respect and being quick. In the real world even after centrifugation, samples are prone to aggregation within the narrow confines of the capillary, typical tip diameter 0.5 to 1.0 μm. Getting the capillary once sample loaded quickly into the cell culture media and down to the level of the cells is important. Injection is done with continuous holding pressure to the capillary (Pc). Pc is set to prevent backflow into the capillary and hence high enough to give a slow continuous flow to the sample. One does not want drying at the tip by being slow to get the capillary into the media. Often there is a bubble at the tip after loading. This bubble should be cleared quickly after bringing the capillary down to cell level. Sufficient Pc, typically 50 hPa, gives a slow positive flow important to limit clogging. If the capillary tends to clog, reload sample and start with a new capillary or recentrifuge the sample and reload. If clogging is a persistent problem, increase capillary diameter by using less heat when pulling the capillary pipette. Make a better pipette for the sample. Do not arbitrarily enlarge the capillary, thus breaking the tip. A broken tip is irregular in shape and hence does not seal well as it enters the cell and is irreproducible as one does a concentration or time series.

Injections are done under phase contrast microscopy. Good cytoplasmic injection produces a gentle wave of movement in the cytosol as the injection occurs. Good nuclear injection produces a slight wave of movement in the nucleus. The cell does not bleb during injection. The cell looks normal by phase contrast after the injection. Cell health is good both during, immediately after, and several hours later by phase contrast microscopy. Decrease injection pressures or capillary diameter if these things are not true.

Ideally, in all microinjection experiments, whether plasmid expression or direct protein injection, the injected cells can be identified by antibody staining against the expressed or injected polypeptide. This allows for quantification of expression levels or polypeptide delivery.

Ideally, the concentration of injected polypeptide or plasmid needed to produce phenotype is known from an actual titration experiment, and actual kinetics for the onset of phenotype are determined in preliminary

experiments. Ideally, experimental design includes both positive and negative controls.

Finally, as cited in these examples, experience suggests that the experimenter must understand the consequences of using the GDP- versus GTP-
restricted isoform as dominant negative mutant.

A Suggested Equipment Setup and Materials List

1. A limited personnel access room for injections. Injections are done
   into cells in open media tissue culture containers. Contamination is
   least if people are not continuously coming in and out of the room.
   There is no need for sterile room conditions or restriction of the
   room to a single purpose.
2. A stable research-grade inverted microscope such as the Zeiss
   Axiovert 200. An intermediate lens optovar is useful for fine-tuning
   microsope magnification to injection situation. The microscope
   may need to be mounted on a vibration table or not. Frequently
   this is required in modern forced draft air handling buildings to
   eliminate building transmitted vibration to the injection capillary.
3. Phase contrast optics: $10\times$ and $32\times$ long-working distance, phase 1
   objectives.
4. Capillary pipette puller such as the Sutter Model 97 Flaming/
   Brown micropipette puller.
5. Thin wall borosilicate glass capillaries for fabrication of micropipettes. We use a thin wall borosilicate glass capillary with
   filament from Sutter Instruments (catalog number, BF120-94-10,
   1.20 mm outside diameter, 0.94 mm inside diameter, 10 cm overall
   length). We use Geloader™ tips from Eppendorf (catalog number
   22 35 165-6, various distributors) to backload solutions into the
   fabricated micropipettes.
6. Integrated microprocessor controlled micromanipulator and pressure regulator system. We use the items as a system (InjectMan®
   NI 2 and FemtoJet®) from Eppendorf AG (Hamburg, Germany)
   that operates together under microprocessor control. This is
   the optimal situation. This system is designed for the injection of
   substratum-attached cells.
7. Video rate CCD camera with high quality monitor. It is much more
   pleasant to view the injection process on a monitor than through the
   microscope oculars. Video rate, 25 or 30 frames per second, is
   important to give a "real-time" sense of micropipette position. We
   have used at times a C-mounted Cohu 9010 black and white CCD
   camera paired with a Panasonic monitor or a Dage-MTI black and

white CCD camera paired to a Sony monitor. The Dage-MTI camera is the flexible camera but an exercise in overkill for the purpose.

8. CO_2 independent culture media for maintaining the cells during the microinjection process. One source is In Vitrogen. Other suppliers are possible. Phenol red free media is best. Remember to prewarm the media to at least room temperature before use. Animal cells round in the cold.

9. Substratum attached cells. My experience is entirely with either substratum attached HeLa or Vero cells. The more spread the cell the better. Depending on purpose, the cells may be grown either on tissue plastic, glass coverglass onto which cells are seeded in a tissue culture dish, or cover glass bottomed tissue culture dishes (MatTek). Cells should ideally be 50–70% confluent for short-term, 6–10 h post injection experiments. Microinjection can be done with suspended cells. Successful suspended cell injection requires a two micropipette-system in which one pipette is used to hold the cell and another to inject the cell.

10. It is convenient to have a CO_2 incubator close to the injection setup for culture incubation.

11. Purified plasmid DNA, polypeptides, or membrane impermeant solutions. We routinely use Qiagen Maxi- or Mini-prep prepared plasmids. For microinjection, we prepare plasmid inserts in pCMUIV, a plasmid designed for high-level expression in human cells (Nilsson *et al.*, 1989). We directly resuspend purified DNA in molecular biology grade water. Cells do not react well to EDTA and Tris. Obviously, micoinjection into live cells mandates the avoidance of azide or other preservative in protein solutions.

Detailed Protocols

1. Pulling a capillary pipette—Prepulled and mounted capillary pipettes for either cytoplasmic or nuclear injection may be purchased commercially from Eppendorf. Many experimenters pull their own pipettes. A popular pipette puller is the Model 97, Flaming/Brown micropipette puller from Sutter Instrument Company, Novato, CA (Fig. 2A). Steps—The capillary (arrow, Fig. 2B) is inserted through a box heating filament (arrowhead, Fig. 2B) of 2.5×2.5 mm box size and 2.5 mm width in the puller, centered, and clamped into place by tightening the thumb wheel of each of the capillary carriers. The capillary is then placed under tension and heat applied to the box filament. A starting heat value is found by the ramp test. In the ramp test, the box filament is gradually heated to a value at which the heated glass just begins to flow. Tension is applied to the glass

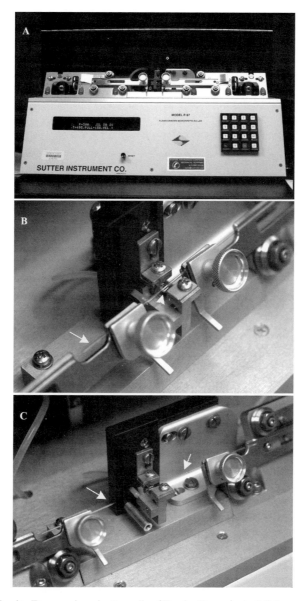

FIG. 2. Flaming/Brown micropipette puller (Flaming/Brown). A. full frame view with lid open to show top interior of puller. B. Capillary carriers with inserted capillary (arrow). Arrowhead points to box filament for heating glass. C. Two newly fabricated capillary micropipettes (arrows) after heating the glass to the point of rupture under tension. The safety cover over the box filament has been removed to give better visual access in the photographs.

during the ramp test. We then take the value from the ramp test, reduce it by 10–20 and try this first for production micropipettes. Production pipettes have a narrower opening at higher heating and a larger opening at lower heating. What works best is determined by effectiveness in injections into cells. The taper length of the microinjection pipette should be in the range of 5–7 mm. As shown in Fig. 2C, two essentially identical micropipettes are produced each heating cycle.

Newly fabricated pipettes are best used within days of being pulled. Other Sutter specific settings—Pull, 150; Velocity, 100; Time, 135; Pressure, 300 for thin-walled glass

2. Backloading a capillary micropipette—We use Geloader™ plastic pipette tips to introduce precentrifuged solutions for microinjection into capillary micropipettes. The solution is centrifuged for 15–20 min in the cold at full speed in a refrigerated microcentrifuge. We typically prepare final solutions in an end volume of 12 μl. The final solution consists of diluted plasmid (common end stock concentrations of 10 to 100 ng/μl for pCMUIV inserts) or protein (common end stock concentrations of ~1 mg/ml), any co-injection marker (e.g., Cascade blue dextran [Molecular Probes, Eugene, OR] at an end stock concentration of 3 mg/ml or so), and molecular biology grade water to volume. An example of backloading is shown in Fig. 3A. A left-handed person holds the capillary micropipette and the arrow points to the thin extended tip of the Geloader™. We load 1.5 to 2.5 μl of solution into the capillary micropipette. This is sufficient for the injection of many thousands of cells. Air pressure applied to the capillary micropipette will force solution down to the end of the capillary. Obviously, the fabricated microcapillary has too wide a tip when its resistance is not sufficient to retain within the micropipette solution at a pressure of 5000 hPa, a typical pressure for clearing clogged particulates from the capillary micropipette.

3. Bringing the capillary micropipette down to cell level—The overall Eppendorf InJectMan manipulator system consists of two parts: a microscope stage-mounted XYZ motorized assembly and a joystick control box (see Fig. 1 for overview picture and Fig. 3B for closeup of stage-mounted assembly). The loaded capillary micropipette is mounted into the holder on the motorized microscope unit, the micropipette pressurized to the continuous or holding pressure (Pc), and the capillary brought down into the cell culture media as fast as possible. Using phase contrast illumination, a bright phase halo forms where the micropipette makes contact with the culture media (Fig. 3C). As shown in Fig. 3C for a right mounted XYZ assembly, the capillary micropipette tip will be to the left of the phase bright spot (arrow pointing towards tip, blurry capillary image, Fig. 3C). The phase spot is useful for first getting the capillary micropipette into the

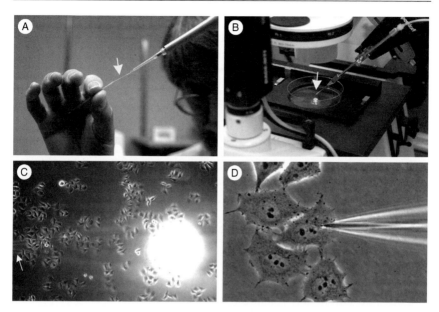

FIG. 3. Steps in capillary micropipette manipulation. A. Backloading of capillary micropipette with solution (arrow points to Geloader® tip inserted into back opening of micropipette). B. Capillary micropipette in place over open cell culture media (arrow points to capillary micropipette). C. Low power appearance of capillary micropipette in culture media (arrow points towards slightly out of focus capillary tip and phase bright spot is where capillary enters the culture media). 10× objective, phase contrast. D. Appearance of capillary micropipette as it enters the cytoplasm of a HeLa cell. Cytoplasmic microinjections are near the nucleus as the cell is thickest in this area and hence the target depth is greatest.

media. After that, all attention should be on the actual capillary tip, not the spot. Under a 10× objective, bring the capillary tip down to almost in focus with the cells, shift to a 32× objective for final fine adjustments of the tip relative to cells. It is important to shift phase contrast objectives while the capillary tip is out of focus above the cells. If the tip is too near cell level, it may be broken from vibration as the 32× objective is clicked into place.

Most people break several capillary micropipettes in the course of learning the coarse micromanipulation to place the capillary near cell level. Experienced personnel do all coarse manipulation steps with the 32× objective rather than first with a 10× objective.

4. Injecting substratum attached cells – The Eppendorf system uses a microprocessor to dip the capillary micropipette into the cell, increase air pressure within the pipette to injection pressure (Pi), and then withdraw the capillary from the cell. The machine like the autostart system on a Formula 1 race car is better than a person. The operator sets or maintains

several parameters. First the operator must set the Z-limit. Using the fine movement setting on the manipulator, the micropipette tip is either brought down to focus at the same height as the cell or actually inserted into the nucleus or cytoplasm (Fig. 3D) and then Z-limit keypad button pressed. The capillary micropipette is then brought up to a "cruising" height above the cells. The tip will be slightly out of focus with respect to the cells. It will be sufficiently high above the cells to avoid collisions as it is moved from cell to cell. The operator sets the injection time. We routinely use 0.3 sec. When the operator presses the button on top of the joy stick, the capillary micropipette tip is automatically moved down into the cells to the Z-limit. Injector pressure automatically increases to Pi, stays there for the injection time, and then the pipette is removed from the cell. All of this is done at a 45-degree pipette entry angle into and out of the cell. Ideally all is set now for repeated nuclear or cytoplasmic injections. Small wave-like movement of cytoplasm or nucleoplasm will be seen during the injection process.

Unfortunately, the surface of the culture dish or glass coverglass is not perfectly flat. This means that as the micropipette is moved from cell to cell the Z-limit position must be adjusted up or down by keypad. Moreover, cell height differs with phase of the cell cycle. Furthermore, in practice, capillary micropipettes clog (see following).

An experienced person can inject a few hundred cells in 10 to 20 min— from starting to load the capillary to bringing the capillary back up out of the culture media after having completed injecting cells. We typically start injecting at a pressure of 300 hPa. We then adjust Pi downwards depending on whether damage is being done to the cells. My experience is that it is better to kill a few cells initially and be sure that transfer is occurring than to be too careful. Being too careful often results in nondetectable transfer as the capillary is dipped into the cell. For a good flowing solution, a Pi of ~100 hPa produces successful transfer with about 50% of the downward capillary dips. Some solutions require a higher Pi than 300 hPa to produce good flow into the cell.

Clearing clogged capillary micropipettes. In the real world, capillary micropipettes clog during microinjection. The 5–7 mm glass taper of the micropipette tends to catalyze particulate formation in the injection solution. The faster the individual injections are done the less the problem. Antibodies at high concentration tend to be a problem. Sar1 mutants tend to be a problem. For a well-behaved solution, a single capillary micro-pipette is sufficient to inject a few hundred cells with no clogs. More commonly there may be a bubble from loading the capillary that needs to be cleared or an actual particulate clog. Particulate clogs appear as white spots in the taper of the tip. The first approach to either is increasing the pressure to cleaning pressure, 5000 hPa or more, briefly. This almost always suffices for bubbles and often for clogs. When doing this, the micropipette should be over a cell-free region and sufficiently above the substratum to allow for slight pipette movement under the high pressure. If high pressure does not suffice, high pressure combined with tapping on the pipette holder often will suffice. Again, the pipette needs to be somewhat high and over a cell-free region to allow for vibration. There is nothing gained in having the pipette collide with cells. Impaled cells can often be dislodged by the same approach. Recentrifuging the injection solution can help. Clogging is more of a problem at high plasmid or protein concentrations. Increasing capillary pipette diameter can help.

In the real world, I have sometimes used 3–4 capillary micropipettes to successfully inject cells for a single time point or single concentration.

Assaying for phenotype in injected cells. The microinjected cells are identified independently of phenotype by a fluorescent coinjection marker or antibody to expressed or injected protein. Phenotype assay depends on experimental design. Examples of different assays and experiment situations are given in earlier sections of this article. Protein expression levels may be assessed by antibody staining intensities using current technology CDD cameras and software (e.g., Jiang and Storrie, 2005; Young et al., 2005).

Acknowledgments

I express my sincere appreciation to Rainer Pepperkok (European Molecular Biology Laboratory, EMBL, Heidelberg) for teaching me microinjection on the now-discontinued Zeiss Automated Microinjection System in the early 1990s. I also express appreciation to Rainer Saffrich (EMBL) for help in setting up my first microinjection system in the United States. Work in the author's laboratory has been supported by grants from the National Science Foundation and the National Institutes of Health.

References

Ansorge, W. (1982). Improved system for capillary microinjection into living cells. *Exp. Cell Res.* **140**, 31–37.

Dascher, C., and Balch, W. E. (1994). Dominant inhibitory mutants of ARF1 block endoplasmic reticulum to Golgi transport and trigger disassembly of the Golgi apparatus. *J. Biol. Chem.* **269**, 1437–1448.

de Fonbrune, P. (1949). "Technique de micromanipulation." Monographies de l'Institut Pasteur. Masson et Cie, Paris.

Feig, L. A. (1999). Tools of the trade: Use of dominant-inhibitory mutants of ras-family GTPases. *Nat. Cell Biol.* **1**, E25–E27.

Girod, A., Storrie, B., Simpson, J. C., Johannes, L., Goud, B., Roberts, L. M., Lord, J. M., Nilsson, T., and Pepperkok, R. (1999). Evidence for a COP-I-independent transport route from the Golgi complex to the endoplasmic reticulum. *Nat. Cell Biol.* **1,** 423–430.

Jiang, S., and Storrie, B. (2005). Cisternal rab proteins regulate Golgi apparatus redistribution in response to hypotonic stress. *Mol. Biol. Cell* **16,** 2586–2596.

Martinez, O., Antony, C., Pehau-Arnaudet, G., Berger, E. G., Salamero, J., and Goud, B (1997). GTP-bound forms of rab6 induce the redistribution of Golgi proteins into the endoplasmic reticulum. *Proc. Natl. Acad. Sci. USA* **94,** 1828–1833.

Miles, S., McManus, H., Forsten, K. E., and Storrie, B. (2001). Evidence that the entire Golgi apparatus cycles in interphase HeLa cells: Sensitivity of Golgi matrix proteins to an ER exit block. *J. Cell Biol.* **155,** 543–555.

Nilsson, T., Jackson, M., and Pederson, P. A. (1989). Short cytoplasmic sequences serve as retention signals for transmembrane proteins in the endoplasmic reticulum. *Cell* **58,** 707–718.

Pepperkok, R., Schneider, C., Philipson, L., and Ansorge, W. (1988). Single cell assay with an automated capillary microinjection system. *Exp. Cell Res.* **178,** 369–376.

Pepperkok, R., Scheel, J., Horstmann, H., Hauri, H. P., Griffiths, G., and Kreis, T. E. (1993). ß-COPI is essential for biosynthetic membrane transport from the endoplasmic reticulum to the Golgi complex *in vivo. Cell* **16,** 71–82.

Pepperkok, R., Lowe, M., Burke, B., and Kreis, T. E. (1998). Three distinct steps in transport of vesicular stomatitis virus glycoprotein from the ER to the cell surface *in vivo* with differential sensitivities to GTP_S. *J. Cell Sci.* **111,** 1877–1888.

Seemann, J., Jokitalo, E., Pypaert, M., and Warren, G. (2000). Matrix proteins can generate the higher order architecture of the Golgi apparatus. *Nature* **407,** 1022–1026.

Storrie, B., White, J., Röttger, S., Stelzer, E., Suganuma, T., and Nilsson, T. (1998). Recycling of Golgi-resident glycosyltransferases through the ER reveals a novel pathway and provides an explanation for nocodazole-induced Golgi scattering. *J. Cell Biol.* **143,** 1505–1521.

Stroud, W. J., Jiang, S., Jacks, G., and Storrie, B. (2003). Persistence of Golgi matrix distribution exhibits the same dependence on Sar1p activity as a Golgi glycosyltransferase. *Traffic* **4,** 631–641.

Valsdottir, R., Hashimoto, H., Ashman, K., Koda, T., Storrie, B., and Nilsson, T. (2001). Identification of rabaptin-5, rabex-5, and GM130 as putative effectors of rab33b, a regulator of retrograde traffic between the Golgi apparatus and ER. *FEBS Lett.* **508,** 201–209.

Ward, T. H., Polishchuk, R. S., Caplan, S., Hirschberg, K., and Lippincott-Schwartz, J. (2001). Maintenance of Golgi structure and function depends on the integrity of ER export. *J. Cell Biol.* **155,** 557–570.

Yang, W., and Storrie, B. (1998). Scattered Golgi elements during microtubule disruption are initially enriched in trans-Golgi proteins. *Mol. Biol. Cell* **9,** 191–207.

Young, J., Stauber, T., Vernos, I., Pepperkok, R., and Nilsson, T. (2005). Regulation of microtubule-dependent recycling at the TGN by rab6a and rab6a'. *Mol. Biol. Cell* **16,** 162–177.

[5] Visualizing Intracellular Events *In Vivo* by Combined Video Fluorescence and 3-D Electron Microscopy

By ALEXANDER A. MIRONOV, GALINA V. BEZNOUSSENKO, ALBERTO LUINI, and ROMAN S. POLISHCHUK

Abstract

The combination of the capability of *in vivo* fluorescence video microscopy with the power of resolution of electron microscopy (EM) has been described. This approach is based on such an association of two techniques. An individual intracellular structure can be monitored *in vivo*, typically through the use of markers fused with green fluorescent protein (GFP), and a "snapshot" of its three-dimensional (3-D) ultrastructure and especially tomographic reconstruction can then be taken at any chosen time during its life cycle. The pitfalls and potential of this approach are discussed.

Introduction

Many cellular functions are very fast and some events within these functions could be extremely rare. Rapid translocations or shape changes of specific intracellular organelles include intracellular traffic, cytokinesis, and cell migration, among others. To understand how such functions, that is, a budding transport carrier, an elongation of microtubule or a developing mitotic spindle, fusion and subsequent detachment of two membrane compartments, are organized and executed *in vivo*, it is necessary to apply the degree of spatial resolution afforded by EM. However, routine EM deals with an average image of the structure of interest. Moreover, EM usually examines only sections, which do not allow 3-D view. Thus, one should initially observe the dynamic structures in real time in living cells and then examine the same structures using EM combined with tomography.

The most suitable methodology to achieve this is conceptually simple, yet powerful; we refer to this as correlative video-light EM (CVLEM), by which observations of the *in vivo* dynamics and ultrastructure of intracellular objects can indeed be combined to achieve this result. We also illustrate here the kinds of questions that the CVLEM approach was designed to address, as well as the particular know-how that is important for the successful application of this technique.

Correlative light-EM (CLEM) was developed several years ago, and it has been used in cases in which the analysis of immunofluorescently

METHODS IN ENZYMOLOGY, VOL. 404
Copyright 2005, Elsevier Inc. All rights reserved.

labeled structures need a better-than-light-microscopy resolution (Powell *et al.*, 1998; Svitkina and Borisy, 1998; Tokuyasu and Maher, 1987). CLEM can also be combined with microinjection (Kweon *et al.*, 2004). Despite its potential, CLEM has not been used very often in the past, probably because the ability to correlate two static images, one fluorescent and one under EM, is of interest only in a limited number of situations, and in particular for the examination of the cell cytoskeleton (Svitkina *et al.*, 2003). CLEM has also been used to characterize the structures that are formed by cells to facilitate the degradation of the extracellular matrix (Baldassarre *et al.*, 2003), and to confirm the direct fusion of an endosome and a lysosome (Bright *et al.*, 2005).

However, the most important gains from CLEM come, we believe, from its combination with the kind of dynamic observations obtainable from GFP video microscopy in living cells (i.e., from its use in CVLEM). The CVLEM approach is potentially valuable in any area of cell biology where the elucidation of the 3-D ultrastructures of individual dynamic cellular objects at times of choice can be informative (Mironov *et al.*, 2000; Polishchuk *et al.*, 2000). For example, the growth of a subset of microtubules can be visualized *in vivo* (Perez *et al.*, 1999). By CVLEM, it is now possible to study at EM resolution the environment and the interactions of the microtubule tips with other cytoskeletal elements, or intracellular organelles, at various stages of the tip growth. In addition to cytoskeletal dynamics, other fields where the application of CVLEM can be easily imagined are those of cell division and cell-cell interactions (although the specific questions to be addressed are best left to the specialists). CVLEM has been successfully applied to the characterization of the ultrastructure of membrane carriers transporting secretory proteins to (Marra *et al.*, 2001; Mironov *et al.*, 2003), through (Mironov *et al.*, 2001), and from (Polishchuk *et al.*, 2000, 2004) the Golgi complex, to the analysis of endocytic structures (Caplan *et al.*, 2002), and to the morphology of the Golgi complex in mitosis (Altan-Bonnet *et al.*, 2003).

One limitation, and at the same time, attraction, of CVLEM, is its complexity. The use of this technique is demanding, and to master its various steps requires a whole array of skills. However, microscopy is developing quickly both in the field of living-cell imaging and in the field of EM, and new powerful technologies are rapidly becoming available in more user-friendly versions. This should make the use of CVLEM more appealing to a number of cell biologists. Through the use of fluorescent proteins of different colors, several marker proteins can now be observed simultaneously (Ellenberg *et al.*, 1999). This will allow the analysis of the interactions between different organelles and organelle subdomains (Ellenberg *et al.*, 1999; Pollok and Heim, 1999). Combining these and other methods with the quickly developing EM tomography (Ladinsky *et al.*, 1994) will increase the subtlety and range of the questions that can be answered by the CVLEM approach. While we await the microscopy of the future, which it is hoped will be endowed with the "magical" power to show cellular structures with EM resolution *in vivo* in real time, CVLEM offers a useful chance to look deeper inside living cells.

As illustrated in Figs. 1 and 2, the CVLEM procedure includes several stages: (1) observation of the structures labeled with green fluorescent protein (GFP) in living cells (Figs. 1A, 2A); (2) immuno-labeling and embedding for EM (Fig. 1B); (3) identification of the cell on the resin block and cutting of serial sections (Fig. 1C); (4) EM analysis and structure identification (Figs. 1D, 2B–E); and (5) digital 3-D reconstruction of the structure of interest (Figs. 1E, 2F). During the first step, the cells are transfected with the cDNA encoding the GFP fusion protein of choice and the fluorescence of the associated structures in living cells is followed (Lippincott-Schwartz and Smith, 1997). In this way, it is possible to gain information about the dynamic properties of these structures (i.e., motility speed and direction, changes in size and shape, etc.). At the end of this stage, the cells are killed by the addition of fixative, capturing the fluorescent object at the moment of interest. As GFP is not visible under EM, immuno-staining allows for the identification of the GFP-labeled structure at the EM level. The immuno-gold and immuno-peroxidase protocols to perform this staining for EM are described below. Usually, the immuno-gold protocol (Burry *et al.*, 1992) is suitable for the labeling of the large majority of antigens, while immuno-peroxidase only allows the labeling of antigens that reside within small membrane-enclosed compartments, because the electron-dense product of the peroxidase reaction tends to diffuse from the actual location of the antibody binding (Brown and Farquhar, 1989; Deerinck *et al.*, 1994). Once stained, the cells must be prepared for EM by traditional epoxy embedding, and the cell and structure of interest must be identified in sections under EM. The finding of individual subcellular structures in single thin sections can be complex, and sometimes even impossible, simply because most of the cellular organelles are bigger than the thickness of the section and lie along a plane that is different from that of the section. So an analysis of serial sections from the whole cell is required for the identification of the structure(s) observed previously *in vivo*. An example of this identification is given in Fig. 2A–E. Finally, the EM analysis of serial sections can be supported by high-voltage EM tomography and/or digital 3-D reconstruction (Fig. 2F).

A   GFP-based time-lapse confocal microscopy
    followed by chemical fixation.

B   Immunoperoxidase labeling and
    embedding in resin.

C   Cutting of serial sections.

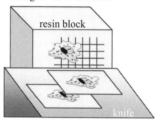

D   Identification of the structure
    in sections for EM.

E   Digital 3D reconstruction.

Fig. 1. The main steps in the CVLEM procedure. A. The structure of interest (circled; in this case a transport carrier) is monitored *in vivo* using an appropriate marker tagged with GFP and time-lapse confocal microscopy. The cell is then fixed at a time chosen by the

Method Steps

Observation and Fixation of Living Cells

  Materials

    Cells of interest, DNA and transfection reagents;
    MatTek petri dishes with CELLocate cover slip (MatTek Corporation, Ashland, MA);
    HEPES buffer (0.2 M). Dissolve 4.77 g HEPES in 100 ml distilled water and add 1 N HCl to provide a pH of ~7.2–7.4;
    Fixative (0.1% glutaraldehyde–8% paraformaldehyde). Dissolve 8 g paraformaldehyde powder in 50 ml HEPES buffer, stirring and heating the solution to 60°. Add drops of 1 N NaOH to clarify the solution. Add 1.25 ml 8% glutaraldehyde and 50 ml HEPES buffer;
    4% paraformaldehyde. Dissolve 4 g paraformaldehyde powder in 100 ml HEPES buffer, stirring and heating the solution to 60°. Add drops of 1 N NaOH to clarify the solution.

The cells are plated for CVLEM on MatTek petri dishes that have CELLocate cover slips attached to their base. The CELLocate cover slips have etched grids with coordinates that allow the cells of interest to be found easily through all of the steps of the procedure. Transfect the cells with the cDNA of the GFP fusion protein of choice using any method available in your laboratory. As illustrated in Fig. 3, after a transfected cell has been chosen and located on the CELLocate grid, its position is drawn on the map of the CELLocate grid (available from MatTek; see the example of the map in Fig. 3A). This living cell can then be observed for the GFP-labeled structures using confocal or light microscopy, which allows the grabbing of a time-lapse series of images by a computer. At the moment of interest, the fixative is added to the cell culture medium while still grabbing images (fixative: medium volume ratio of 1:1). The fixation usually induces the fast fading of the GFP fluorescence and blocks the motion of the labeled structures in the cells. Particular attention must also

experimenter (e.g., during budding, translocation, or fusion; in this case during translocation). B. The cell is labeled by the immuno-HRP technique with an antibody against the GFP-tagged protein marker. The patterns of peroxidase labeling and of GFP fluorescence must coincide. C. The cell in A and B is identified in the resin block by using the system of spatial coordinates (also described in Polishchuk *et al.*, 2000), and serial sections are cut. D. The carrier previously monitored *in vivo* (circled) is identified in each section by using specific cellular structures as spatial landmarks (see example in Fig. 2). E. The images in the serial sections are used for the 3-D computer-aided reconstruction. (See color insert.)

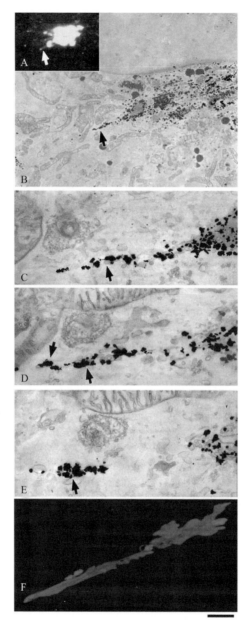

FIG. 2. Ultrastructure of a Golgi-to-plasma-membrane carrier (GPC) formation site. A. A VSVG-GFP-transfected Cos7 cell was fixed at the moment of formation of a GPC (arrow), labeled with an antibody against VSVG using the immuno-gold protocol, and processed for

be paid to the temperature of the fixative, as the addition of a colder fixative to the cell medium will induce a shift in the focal plane during the time-lapse observations. The grabbing of the time-lapse images is then stopped, and the cells are kept in the fixative for 10 min. Wash once with 4% paraformaldehyde and leave the cells in 4% paraformaldehyde for 30 min.

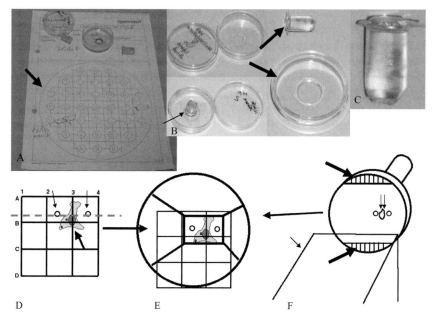

FIG. 3. Orientation of the sample. A. The sample map (arrow). B. The position of the block before (thin arrow) and after (thick arrows) the detachment from the cover slip. C. The sample before its trimming; the external view of the block. D. Scheme of the grid and the position of the cell (green, thick arrow) and cavities (circles, thin arrows). The cavities should form a horizontal line (red broken line). E. Position of the pyramid. F. The position of the sample after the orientation and trimming of the upper and lower edges (thick arrows) of the sample. The cell of interest is shown by the double arrow. The small arrow indicates the position of the glass knife related to the pyramid.

CVEM. B–E. The same GPC formation site (arrows and arrowhead) at low EM magnification (B) and under higher magnification in serial sections (C–E). F. 3D reconstruction from the serial sections. Bar: 10.5 μm (A), 2.1 μm (B), 490 nm (C–E). (See color insert.)

Immuno-labeling for Electron Microscopy

The immuno-labeling of fixed cells for EM during the CLEM procedure can be performed using both immuno-gold and immuno-peroxidase protocols (see Introduction). However, preliminary experiments should be done to determine whether the antibodies selected for labeling of the GFP-fusion protein work with the immuno-EM protocol. Many antibodies that give perfect results for immuno-fluorescence do not work for immuno-EM staining. This happens because the glutaraldehyde used in most EM fixatives tends to cross-link the amino groups of the antigen epitopes, and therefore to decrease the antigenicity of the target protein. However, the decreasing or removing of the glutaraldehyde in a fixative can result in poor preservation of the ultrastructure of the intracellular organelles. So, if there are problems with immuno-EM labeling, it is possible to optimize the concentration of glutaraldehyde in the fixative or to use a periodate-lysine-paraformaldehyde fixative (Brown and Farquhar, 1989). After this, it is important to select the immuno-peroxidase or immuno-gold protocol to label the structure of interest. We would advise the use of only the immuno-gold protocol to label epitopes of GFP-fusion proteins located in the cytosol (see Strategy), while for other epitopes, horseradish peroxidase (HRP) labeling is also suitable.

Materials

Blocking solution. Dissolve 0.50 g BSA, 0.10 g saponin, 0.27 g NH_4Cl in 100 ml of PBS;

Fab fragments of the secondary antibodies conjugated with HRP (Rockland, Gilbertsville, PA);

NANOGOLD conjugated Fab fragments of the secondary antibodies (Nanoprobes Inc., Yaphank, NY);

TRIS-HCl buffer (0.1 M). Dissolve 1.21 g TRIZMA base in 100 ml distilled water and add 1 N HCl to provide a pH of ~7.2–7.4;

Diaminobenzidine (DAB) solution. Dissolve 0.01 g DAB in 20 ml TRIS-HCl buffer. Add 13.3 μl 30% H_2O_2 solution just before use;

Gold-enhance mixture. Use a gold-enhance kit from Nanoprobes. Using equal amounts of the four components (Solutions A, B, C, and D), prepare about 200 μl reagent per petri dish (a convenient method is to use an equal number of drops from each bottle). First mix Solution A (enhancer; green cap) and Solution B (activator; yellow cap). Wait for 5 min, and then add Solution C (initiator; purple cap), and finally Solution D (buffer; white cap). Mix well.

Immuno-labeling for EM with NANOGOLD. Wash the fixed cells for 3 × 5 min with PBS. Incubate the cells with the blocking solution for 30 min, and then with the primary antibodies diluted in blocking solution, overnight. Wash the cells for 6 × 2 min with PBS. Dilute the NANO-GOLD-conjugated Fab fragments of the secondary antibodies ~50 times in the blocking solution and add this to the cells; incubate for 2 h. Wash the cells again for 6 × 2 min with PBS. Fix the cells with 1% glutaraldehyde in 0.2 M HEPES buffer for 5 min. Wash the cells for 3 × 5 min with PBS, and then for 3 × 5 min in distilled water. Incubate the cells with the gold-enhancement mixture for 6–10 min. The cells will become violet-grey in color if the gold enhancement is successful. Finally, wash cells for 3 × 5 min with distilled water, and proceed with the embedding.

Immuno-labeling for EM with HRP. Wash the fixed cells for 3 × 2 min with PBS. Incubate the cells with the blocking solution for 30 min, and then with the primary antibodies diluted in blocking solution, overnight. Wash the cells for 6 × 2 min with PBS. Incubate the cells with the HRP-conjugated Fab fragments of the secondary antibody for 2 h. Wash the cells again for 6 × 2 min with PBS. Fix the cells with 1% glutaraldehyde in 0.2 M HEPES buffer for 5 min. Wash the cells 3 × 5 min with PBS, and then incubate them with the DAB solution. A successful peroxidase reaction results in a slightly brown staining of the cells. Finally, wash the cells for 3 × 2 min with PBS.

Embedding

Materials

Cacodylate buffer *(0.2 M)*. Dissolve 2.12 g sodium cacodylate in 100 ml distilled water add 1 N HCl to provide a pH of ~7.2–7.4;
OsO_4 (Electron Microscopy Sciences, Fort Washington, PA);
Potassium ferrocyanide;
EPON. Put 20.0 g EPON, 13.0 g Dodecenyl succinic anhydride (DDSA) and 11.5 g Methyl nadic anhydride (MNA) into the same test-tube. Heat the tube in the oven for 2–3 min at 60° and then vortex it well. Add 0.9 g tri-Dimethylaminomethyl phenol (DMP-30, all from Electron Microscopy Sciences, Fort Washington, PA) and immediately vortex the tube again. It is possible to freeze the EPON in aliquots and to store it for a long time at –20° before use.

Wash the cells 5×/6× with distilled water. Be careful, because the residual phosphate may precipitate when it is mixed with OsO_4. Incubate the cells in the 1:1 mixture of 2% OsO_4 and 3% potassium ferrocyanide in 0.2 M cacodylate buffer for 1 h. Wash the cells once with distilled water, and then incubate them for 10 min each with the following ethanol solutions: 50% (once), 70% (once), 90% (once), 100% (3×), for the dehydration of the specimens. Keep the cells in a mixture of EPON and 100% ethanol (1:1; v/v) for 1–2 h. Keep the cells in EPON for 1–2 h at room temperature, and then leave them in an oven at 60° overnight.

Locating of the Cell on the Resin Block

Materials

Hydrofluoric acid (HF)
EPON
Stereomicroscope.

After 12 h for the polymerization of the EPON, place a small droplet of a fresh resin on the site where the examined cell is located, and insert a resin cylinder (prepared before by polymerization of the resin in a cylindrical mold) with a flat lower surface; leave the samples for an additional 18 h in the oven at 60°. Carefully pick up the resin from the petri dish and glass; this is easy to do by gentle bending of the resin cylinder to and fro. The resin block and the empty MatTek petri dish after block detachment are shown in Fig. 3A–C. If the cover glass with a coordinated grid cannot be detached from the cells included into the resin, the latter should be placed into commercially available HF (do not use glassware for this) for 30–60 min. Control the completeness of the glass dissolution under a stereomicroscope. Wash the samples in water after the complete removal of the glass. Leave the samples in 0.1 M HEPES buffer (pH 7.3) for 60 min to neutralize the HF. Wash the samples in water and allow them to dry. The final resin block is shown in Fig. 3C.

Cutting

Materials and Equipment

Glass and diamond knives;
Ultratome.

Find the cell of interest among the cells within the sample according to the coordinated grid, and put the resin block into the holder of an ultratome. Using a steel needle and rotating the sample in the holder, make two small cavities in such a way that they (thin arrows in Fig. 3D) form a horizontal line (broken red line in Fig. 3D) with the cell appearing in the center of the sample (see Figs. 3E and F). Introduce the holder into the ultratome in such a way that the segment arc of the ultratome is in the vertical position and the two cavities form a horizontal line. By rotating the glass knife stage, align the bottom edge of the pyramid parallel to the knife-edge. Using the segment arc, orient the plane of the sample vertically. Bring the sample as close as possible towards the glass knife. Adjust the gap (which is visible as a bright band if all three of the lamps of an ultratome are switched on) between the knife-edge and the surface of the sample. The gap has to be identical in width between the most upper and lower edges of the sample during the up and down movement of the resin block. This ascertains that every point of the sample surface containing the cell of interest is at the same distance from the knife edge. Slowly moving the sample up and down, continue its approach until the knife begins to cut one of the edges of the sample. The sectioning begins from either the upper or the lower part of the sample; the middle part of the sample where the cell of interest is situated will be unaffected because the length of the radius passing through the cell is shorter than the radii passing through the upper and the lower edges of the sample. If the sectioning is to begin from the upper part of the sample, tilt the segment arc to approach the lower edge towards the knife. If the sectioning is to begin from the lower edge of the sample, tilt the segment arc and approach the upper edge of the sample towards the knife. A vertically oriented sample should produce equal sections from both the upper and lower edges of the samples (Fig. 3F, arrows). Note down precisely all of parameters relating to the position of the sample in the ultratome, that is, the degree of rotation of the sample in the holder, the degree of tilting of the segment arc, and the degree of rotation of the knife in its stage. Take the sample and trim it to provide a narrow horizontal pyramid of about 0.2×1.5 mm in size with the cell of interest at its center (Fig. 3E). Do not take the sample from the holder and do not rotate the sample inside the holder. The pyramid should be as narrow as possible (no wider than 200 μm), and the cell of interest should be at the center of the pyramid. An experienced person can trim a pyramid directly with a razor blade. Introduce the sample back into the ultratome, and lock it in exactly the same position as before (preserving all of the parameters of sample positioning; this is very important). Replace the glass knife with the diamond one, and position the latter towards the pyramid. If the sample is not parallel to the knife, adjust the angle of the diamond knife by rotating the knife stage to make its edge parallel to the plane of the pyramid. Do not change any other parameters of the sample position. Approach the sample towards the edge of the knife until the gap is extremely narrow. Using a 200-nm approaching step, begin the sectioning. Take serial 200-nm sections according to the instructions with the ultratome. It is enough to take only 10 sections to pass 2 μm from the height of the cell. Remember the position of the organelle of interest according to the Z-stacking, and select those thick sections that should correspond to this position. For instance, if the organelle of interest is situated at 500 nm from the bottom of the cell, it is enough to collect only the first four 200-nm serial sections. If the position is at 1 μm in height, it will be necessary to collect from the fourth to the eighth serial sections.

Picking up the Sections

Materials and Equipment

Pick-up loop (Agar, Cambridge, England);
Slot grids covered with carbon-formvar supporting film (Electron Microscopy Sciences, Fort Washington, PA).

For the picking-up of the sections, stop the motor, and divide the band of the sections into pieces of a suitable size for collection with the pick-up loop, using two eyelashes. Touch the surface of the water with the band of sections with the pick-up loop in such a way that the band is completely inside the inner circle of the loop, without touching it. Raise the loop with the droplet of water with the sections on it, and place the loop inside the tripod near the microscope. The loop should be visible under the stereomicroscope of the ultratome. Take the slot grid coated with the formvar (or preferably butvar)/carbon supporting film and gently touch sections on the water (do not touch the loop) with the carbon-coated surface of the grid. Very slowly, move the slot grid away from the loop. If the movement is slow enough, the water is eliminated from the surface of the supporting film, and only a very small droplet of water remains on the grid, which will not represent an obstacle for the placement of the grid directly into the grid container.

Electron Microscopy Analysis

Place the slot grid under the electron microscope, and using the traces of the coordinated grid filled with the resin on the first few sections, or the central position of the cell of interest within the pyramid, identify the cell on the sections. Take consecutive photographs (or grab the images with a computer using a video camera) of the serial sections until the organelle of interest (just observed under the LSCM) is no longer seen. If EM tomography is to be used, take a tilting series of the organelle of interest and produce the electron tomogram according to the instruction for the IMOD software (available at the following web site: http://bio3d. colorado.edu/imod/). Using the software for the 3-D reconstruction, align the images and then make a 3-D model according to the instructions with the software.

Critical Parameters and Troubleshooting

Taken together, all of the steps for CVLEM represent quite a long procedure (see following), and they require significant effort of the experimenter. Therefore, it would be particularly disappointing to lose such *tour de force* experiments because of small problems with specimen handling. To apply CVLEM successfully, several important parameters should always be taken into account by the experimenter.

First, it is extremely important to be able to find the cell of interest during all of the steps of the CVLEM procedure. So, only cells located on the grid of the MatTek petri dish should be selected for time-lapse observations. The position of the cell of interest on the grid must be noted; otherwise it would be difficult to find it again. The low magnification images showing the field surrounding the cell of interest can help greatly to be able to trim the resin block around the right cells, and to find them later under the electron microscope. In this case, neighboring cells can be used as landmarks to identify the cell of interest. For this reason, the cells for experimentation should be plated with a lower confluence (50–60%) than usual. During the analysis of the serial sections under the electron microscope, it is useful to have the fluorescent and phase-contrast images of the target cell because particular structures (microvilli, pseudopodia, inclusions, etc.) can help greatly to find both the cell and the structure of interest.

Secondly, during the cutting of the specimen, the thickness of the serial sections needs to be selected. This should be about 80 nm for routine work, 50 nm (or less) for very precise 3-D reconstruction, or 250 nm for EM tomography.

Acknowledgments

We thank Dr. C. P. Berrie for critical reading of the manuscript, Dr. E. V. Polishchuk for providing CVLEM images, and Telethon Italia, the Italian Association for Cancer Research, and the National Research Council for financial support.

References

Altan-Bonnet, N., Phair, R. D., Polishchuk, R. S., Weigert, R., and Lippincott-Schwartz, J. (2003). A role for Arf1 in mitotic Golgi disassembly, chromosome segregation, and cytokinesis. *Proc. Natl. Acad. Sci. USA* **100,** 13314–13319.

Baldassarre, M., Pompeo, A., Beznoussenko, G., Castaldi, C., Cortellino, S., McNiven, M. A., Luini, A., and Buccione, R. (2003). Dynamin participates in focal extracellular matrix degradation by invasive cells. *Mol. Biol. Cell* **14,** 1074–1084.

Bright, N. A., Gratian, M. J., and Luzio, J. P. (2005). Endocytic delivery to lysosomes mediated by concurrent fusion and kissing events in living cells. *Curr. Biol.* **15,** 360–365.

Brown, W. J., and Farquhar, M. G. (1989). Immunoperoxidase methods for the localization of antigens in cultured cells and tissue sections by electron microscopy. *Methods Cell Biol.* **31,** 553–569.

Burry, R. W., Vandre, D. D., and Hayes, D. M. (1992). Silver enhancement of gold antibody probes in pre-embedding electron microscopic immunocytochemistry. *J. Histochem. Cytochem.* **40,** 1849–1856.

Caplan, S., Naslavsky, N., Hartnell, L. M., Lodge, R., Polishchuk, R. S., Donaldson, J. G., and Bonifacino, J. S. (2002). A tubular EHD1-containing compartment involved in the recycling of major histocompatibility complex class I molecules to the plasma membrane. *EMBO J.* **21,** 2557–2567.

Deerinck, T. J., Martone, M. E., Lev-Ram, V., Green, D. P., Tsien, R. Y., Spector, D. L., Huang, S., and Ellisman, M. H. (1994). Fluorescence photooxidation with eosin: A method for high resolution immunolocalization and *in situ* hybridization detection for light and electron microscopy. *J. Cell Biol.* **126,** 901–910.

Ellenberg, J., Lippincott-Schwartz, J., and Presley, J. F. (1999). Dual-colour imaging with GFP variants. *Trends Cell Biol.* **9,** 52–56.

Kweon, H.-S., Beznoussenko, G. V., Micaroni, M., Polishchuk, R. S., Trucco, A., Martella, O., Di Giandomenico, D., Marra, P., Fusella, A., Di Pentima, A., Berger, E. G., Geerts, W. J., Koster, A. J., Burger, K. N., Luini, A., and Mironov, A. A. (2004). Golgi enzymes are enriched in perforated zones of Golgi cisternae but are depleted in COPI vesicles. *Mol. Biol. Cell* **15,** 4710–4724.

Ladinsky, M. S., Kremer, J. R., Furcinitti, P. S., McIntosh, J. R., and Howell, K. E. (1994). HVEM tomography of the trans-Golgi network: Structural insights and identification of a lace-like vesicle coat. *J. Cell Biol.* **127,** 29–38.

Lippincott-Schwartz, J., and Smith, C. L. (1997). Insights into secretory and endocytic membrane traffic using green fluorescent protein chimeras. *Curr. Opin. Neurobiol.* **7,** 631–639.

Marra, P., Maffucci, T., Daniele, T., Tullio, G. D., Ikehara, Y., Chan, E. K., Luini, A., Beznoussenko, G., Mironov, A., and De Matteis, M. A. (2001). The GM130 and GRASP65 Golgi proteins cycle through and define a subdomain of the intermediate compartment. *Nat. Cell Biol.* **3,** 1101–1113.

Mironov, A. A., Polishchuk, R. S., and Luini, A. (2000). Visualising membrane traffic *in vivo* by combined video fluorescence and 3-D-electron microscopy. *Trends Cell Biol.* **10,** 349–353.

Mironov, A. A., Beznoussenko, G. V., Nicoziani, P., Martella, O., Trucco, A., Kweon, H. S., Di Giandomenico, D., Polishchuk, R. S., Fusella, A., Lupetti, P., Berger, E. G., Geerts,

W. J., Koster, A. J., Burger, K. N., and Luini, A. (2001). Small cargo proteins and large aggregates can traverse the Golgi by a common mechanism without leaving the lumen of cisternae. *J. Cell Biol.* **155,** 1225–1238.

Mironov, A. A., Mironov, A. A., Jr., Beznoussenko, G. V., Trucco, A., Lupetti, P., Smith, J. D., Geerts, W. J., Koster, A. J., Burger, K. N., Martone, M. E., Deerinck, T. J., Ellisman, M. H., and Luini, A. (2003). ER-to-Golgi carriers arise through direct en bloc protrusion and multistage maturation of specialized ER exit domains. *Dev. Cell* **5,** 583–594.

Perez, F., Diamantopoulos, G. S., Stalder, R., and Kreis, T. E. (1999). CLIP-170 highlights growing microtubule ends *in vivo. Cell* **96,** 517–527.

Polishchuk, R. S., Polishchuk, E. V., Marra, P., Buccione, R., Alberti, S., Luini, A., and Mironov, A. A. (2000). GFP-based correlative light-electron microscopy reveals the saccular-tubular ultrastructure of carriers in transit from the Golgi apparatus to the plasma membrane. *J. Cell Biol.* **148,** 45–58.

Polishchuk, R., Di Pentima, A., and Lippincott-Schwartz, J. (2004). Delivery of raft-associated, GPI-anchored proteins to the apical surface of polarized MDCK cells by a transcytotic pathway. *Nat. Cell Biol.* **6,** 297–307.

Pollok, B. A., and Heim, R. (1999). Using GFP in FRET-based applications. *Trends Cell Biol.* **9,** 57–60.

Powell, R. D., Halsey, C. M., and Hainfeld, J. F. (1998). Combined fluorescent and gold immunoprobes: Reagents and methods for correlative light and electron microscopy. *Microsc. Res. Tech.* **42,** 2–12.

Svitkina, T. M., and Borisy, G. G. (1998). Correlative light and electron microscopy of the cytoskeleton of cultured cells. *Methods Enzymol.* **298,** 570–592.

Svitkina, T. M., Bulanova, E. A., Chaga, O. Y., Vignjevic, D. M., Kojima, S., Vasiliev, J. M., and Borisy, G. G. (2003). Mechanism of filopodia initiation by reorganization of a dendritic network. *J. Cell Biol.* **160,** 409–421.

Tokuyasu, K. T., and Maher, P. A. (1987). Immunocytochemical studies of cardiac myofibrillogenesis in early chick embryos. II. Generation of alpha-actinin dots within titin spots at the time of the first myofibril formation. *J. Cell Biol.* **105,** 2795–2801.

[6]   Recombinant Production in Baculovirus-Infected Insect Cells and Purification of the Mammalian Sec13/Sec31 Complex

By CEMAL GURKAN and WILLIAM E. BALCH

Abstract

Membrane traffic along the eukaryotic secretory pathway starts with the selective packing of biosynthetic cargo into nascent vesicles that are forming on the endoplasmic reticulum (ER). This process is mediated by the coat protein complex II (COPII) machinery, which at the minimum, comprises the Sar1 GTPase and the cytosolic protein complexes Sec23/Sec24 (Sec23/24) and Sec13/Sec31 (Sec13/31). While the components of the basic COPII machinery are highly conserved from yeast to human, it is now clearly evident that the overall process is under tighter spatial and temporal regulation in higher eukaryotes. Here we describe recombinant production in baculovirus-infected insect cells and subsequent purification to homogeneity of the mammalian Sec13/31 complex for biochemical and biophysical characterization.

Introduction

Anterograde membrane traffic along the secretory pathway starts with the selective packaging of biosynthetic cargo into vesicles budding from the ER (Lee *et al.*, 2004). Biogenesis of these nascent vesicles is in turn mediated by the evolutionarily conserved COPII (coat protein complex II) machinery, assembly of which is initiated and regulated by the small Sar1 GTPase (Barlowe *et al.*, 1993; Kuge *et al.*, 1994). Normally present in an inactive, cytosolic form when bound to GDP, Sar1 is activated upon loading with GTP and recruited to the ER membranes. Activation of Sar1 is facilitated by Sec12, an ER-resident guanine nucleotide exchange factor (GEF) (Barlowe and Schekman, 1993; Weissman *et al.*, 2001). Membrane recruitment of Sar1 is followed by that of the Sec23/Sec24 (Sec23/24) complex, comprising the Sar1-specific GTPase activating protein (GAP), Sec23 (Yoshihisa *et al.*, 1993), and Sec24 that recognizes cargo destined for ER export (Nishimura and Balch, 1997). Finally, recruitment of the Sec13/Sec31 (Sec13/31) complex leads to the COPII coat polymerization and subsequent fission of the cargo-bearing vesicles (Lee *et al.*, 2004).

METHODS IN ENZYMOLOGY, VOL. 404
0076-6879/05 $35.00
DOI: 10.1016/S0076-6879(05)04006-1

Like many other Sec proteins, Sec13 was first isolated as a temperature-sensitive yeast secretory (sec) mutant (Novick *et al.*, 1980), and further genetic studies implicated a role in ER vesicle formation (Kaiser and Schekman, 1990). Initial biochemical isolation of the Sec13 protein from the yeast cytosol revealed that it largely exists as a complex with another protein that was subsequently cloned and characterized as Sec31 (Pryer *et al.*, 1993; Salama *et al.*, 1997). *In vitro* biochemical recapitulation of the ER cargo export confirmed that Sar1, Sec23/Sec24, and Sec13/31 are the minimal cytosolic components required for COPII function, at least in yeast (Matsuoka *et al.*, 1998). Orthologs of these yeast genes have also been shown to play an equally important role in the mammalian COPII function (Aridor *et al.*, 1998), albeit recent results suggest that additional, possibly hitherto unknown factor(s) may also be required (Aridor and Balch, 2000; Kapetanovich *et al.*, 2005; Kim *et al.*, 2004; LaPointe *et al.*, 2004).

Both Sec13 and Sec31 define limited gene families. Like in yeast, there are two mammalian Sec13 isoforms, Sec13L1/Sec13R (~35.5 kDa) with a transcription variant that differs by three extra N-terminal residues (MGK) (Shaywitz *et al.*, 1995; Swaroop *et al.*, 1994; Tang *et al.*, 1997) and a larger Sec13-like/Seh1 (39.6 kDa) (Siniossoglou *et al.*, 1996). While both Sec13 isoforms have been shown to play key structural roles in the nuclear pore complexes (NPCs) (Siniossoglou *et al.*, 1996, 2000), Sec13-like/Seh1 does not seem to have a direct role in the yeast or mammalian COPII function. Furthermore, Sec13 has been recently shown to interact with Rabphilin-11/Rab11BP, which is already known to bind the GTP-bound form of Rab11, a small GTPase associated with the endocytic pathways (Mammoto *et al.*, 1999, 2000). Consequently, it is now evident that Sec13 function is not limited to that in COPII machinery as initially thought. Like Sec13, there are also two mammalian Sec31 isoforms, Sec31L1 (~133 kDa) and Sec31L2 (~129 kDa) (Shugrue *et al.*, 1999; Tang *et al.*, 2000), each with at least a single transcription variant that is missing two in-frame coding exons and the C-terminal 830 residues, respectively. To investigate the mRNA expression profiles of the mammalian Sec13 and Sec31 isoforms across numerous tissues and also to obtain further annotation details, we encourage the readers to use the SymAtlas web-application (http://symatlas.gnf.org) as recently described in a separate volume of the *Methods in Enzymology* series (Gurkan *et al.*, 2005), (for quick access, use the following GenBank® accession numbers for the human isoforms: Sec13L1/Sec13R, NM_030673; Sec13-like/Seh1, NM_031216; Sec31L1, NM_014933; and Sec31L2, NM_015490).

Cumulative data from biochemical and electron microscopy studies now suggest that the Sec13/31 complex exists as an heterotetramer of two

Sec13 and two Sec31 monomers that are presumably arranged in an elongated structure with two fold symmetry (Sec31-Sec13-Sec13-Sec31) (Lederkremer *et al.*, 2001). However, despite recent reports (Antonny *et al.*, 2003; Lederkremer *et al.*, 2001; Matsuoka *et al.*, 2001), the exact role played by the Sec13/31 complex in COPII function still remains poorly defined. Here we describe recombinant production and purification of the mammalian Sec13/31 complex from baculovirus-infected insect cells, to allow its further biochemical and biophysical characterization.

Materials

Unless otherwise specified, all chemicals used were of the highest purity and obtained from either Sigma (St. Louis, MO) or Fisher Scientific (Pittsburgh, PA). Mouse monoclonal anti-pentahis antibody (Cat. #: 34660) was purchased from Qiagen (Valencia, CA). Rabbit polyclonal anti-Sec31 antibody was raised against an *E. coli*-produced GST-fusion peptide corresponding to the amino acids 865 to 897 of human Sec31L1 (NP_055748)   (HTQVPPYPQPQPYQPAQPYPFGTGGSAMYRPQQ) (Weissman, 2000). ImmunoPure, peroxidase conjugated goat anti-mouse (Cat. #: 31430) and goat anti-rabbit (Cat. #: 31460) polyclonal antibodies were from Pierce Biotechnology (Rockford, IL).

Recombinant Expression

The mammalian Sec13 and Sec31 clones used in this study corresponded to the human Sec13-like 1 (Sec13L1/Sec13R transcript variant 2, GenBank® Accession #: NM_183352) and Sec31-like 1 (Sec31L1 transcript variant 1, Genbank® Accession #: NM_014933) genes, respectively. Recombinant production of the human Sec13R/Sec31L1 (Sec13R/31L1) complex was achieved using the Bac-To-Bac® Baculovirus Expression System (Cat. #: 10359–016) from Invitrogen (Carlsbad, CA) following the manufacturer's instructions. The human Sec13R and Sec31L1 genes were cloned into the pFastBac™ DUAL expression vector, and recombinant production was carried out in BTI-TN-5B1–4 (Tn5 or High Five™) insect cells (Cat. #: B85502) also obtained from Invitrogen (Weissman, 2000). This strategy allowed construction of recombinant baculoviruses containing two expression cassettes that corresponded to the human Sec13R and Sec31L1 genes. Coexpression of Sec13R and Sec31L1 in baculovirus-infected insect cells leads to the formation of an apparently native-like, recombinant Sec13R/31L1 (rSec13R/31L1) complex in cytosol. Finally, introduction of an N-terminal hexa-histidine tag on Sec13R allows initial purification of rSec13R/31L1 by immobilized metal affinity chromatography (IMAC) as described in the next section.

Purification

All chromatography steps were carried out using an ÄKTA™ FPLC™ system with a Frac-950 fraction collector, and using propriety chromatography columns provided by Amersham Biosciences (Piscataway, NJ). Prior to various stages of purification, samples were concentrated or conditioned using Amicon® Centrifugal Filter Units (Cat. #: UFC803024 and UFC901024) from Millipore (Billerica, MA) and Slide-A-Lyzer Dialysis Cassettes (Cat. #: 66415 and 66425) from Pierce Biotechnology, respectively. All buffers were freshly supplemented with Complete® EDTA-free protease inhibitor cocktail tablets (Cat. #: 1873580) from Roche (Indianapolis, IN) and either DTT or β-mercaptoethanol at the given concentrations.

Chromatography Columns and Buffers

Ni-IDA immobilized metal affinity chromatography (IMAC):
HiTrap Chelating HP (1 ml) Column (Cat. #: 17–0408–01), freshly
   charged with Ni^{2+}
Buffer A: 20 mM NaH_2PO_4, pH7.5; 1 M NaCl; 1 mM MgOAc; 5 mM
   β-mercaptoethanol
Buffer B: 20 mM NaH_2PO_4, pH7.5; 0.5 M imidazole; 1 M NaCl; 1 mM
   MgOAc; 5 mM β-mercaptoethanol
Mono Q anion exchange chromatography:
Mono Q HR 5/5 (1 ml) Column (Cat. #: 17–0546–011)
Buffer A: 20 mM Tris-HCl, pH 7.5; 1 mM MgOAc; 10 mM DTT
Buffer B: 20 mM Tris-HCl, pH 7.5; 1 M NaCl; 1 mM MgOAc; 10 mM
   DTT

Protocol

All chromatography, centrifugation, and sample conditioning steps are carried out at 4° (or in the cold room) to minimize sample proteolysis. In a typical purification strategy, a frozen pellet from 0.5 l of rSec13R/ 31L1-expressing insect cell culture is thawed from −80° by the addition of 25 ml of ice-cold *Ni-IDA Buffer A* and with occasional gentle vortexing to ensure a homogenous suspension. Since high NaCl concentration of this buffer precipitates insect cell DNA during lysis, bovine pancreas DNase I (Roche, Cat. # 104159) can be added at 100 μg/ml to minimize sample viscosity that would otherwise interfere with the subsequent

IMAC step. Lysis is achieved by nitrogen cavitation using a nitrogen bomb (Fike Metal Products, Blue Springs, MO) at 500 psi for 45 min, and cell debris is removed by ultracentrifugation using a Beckman Ti60 rotor (40,000 rpm, at 4°, 90 min). The cleared cell lysate is then loaded and fractionated over a Ni^{2+}-charged HiTrap Chelating HP column at a flow-rate of 0.2 ml/min and 0.5 ml fractions are collected. Typically, the Ni-IDA column is pre-equilibrated with 10 column volumes (*cv*) of the *Ni-IDA Buffer A* prior to the sample loading, and unbound proteins are washed with 15 *cv* of the *Ni-IDA Buffer A*. Fractionation is achieved using the *Ni-IDA Buffer B* through subsequent step gradients of 20 m*M* and 50 m*M* imidazole over 20 *cv* each, and then a continuous imidazole gradient from 50 m*M* to 300 m*M* over 10 *cv*. Silver-stained SDS/PAGE and the corresponding immunoblot analyses reveal that the rSec13R/31L1 complex largely elutes at 50 m*M* imidazole.

The Sec13R/31L1-rich fractions from the IMAC step are pooled and conditioned by 10- to 15-fold dilution with the *Mono Q Buffer A*, and fractionated over a Mono Q HR 5/5 column at a flow-rate of 1 ml/min and 0.25 ml fractions are collected. Typically, the Mono Q column is pre-equilibrated with 5 *cv* of the *Mono Q Buffer A* prior to the sample loading, and unbound proteins are washed with 15 *cv* of the *Mono Q Buffer A*. Fractionation is achieved using the *Mono Q Buffer B* through a continuous NaCl gradient from 50 m*M* to 450 m*M* over 20 *cv*. Silver-stained SDS/PAGE and the corresponding immunoblot analyses reveal two major peaks corresponding to the recombinant Sec13R-alone (rSec13R) and the rSec13R/31L1 complex that elute at ~200 m*M* and ~280 m*M* NaCl, respectively. Fractions rich in rSec13R-alone and rSec13R/31L1 are separately pooled and concentrated 5- to 10-fold if the need be (see *Further Highlights*). Size exclusion chromatography of the rSec13R/31L1-rich Mono Q pool using a Superoseb HR 10/30 column (Cat.#: 17-053701) and the *Mono Q Buffer A* with 280m*M* NaCl can further resolve rSec13R/31L1 and Free rSec13R, which fractionate as $>670 \times 10^3 Mr$ and $>44 \times 10^3 Mr$ species, respectively, as judged by previous calibration of this column using Bio-Rad® (Hercules, CA) gel filtration standards (Cat.#: 151-1901). Alternatively, rSec13R-alone and rSec13R/31L1-rich Mono Q pods are dialyzed twice against the *Sec13/31 Buffer* (25 m*M* Hepes, pH 7.5; 700 m*M* KOAc; 1 m*M* MgOAc; 1 m*M* DTT). Finally, all samples are quantified using the Pierce Coomassie® (Bradford) Protein Assay Kit (Cat.#: 23200), aliquoted, and snap-frozen in liquid nitrogen prior to storage at −80°. Figure 1 depicts silver-stained SDS-PAGE and the corresponding immunoblot analyses of typical rSec13R-alone and rSec13R/31L1 complex samples purified by the strategy described in this chapter. Final yields of rSec13R-alone and rSec13R/31L1 samples are often ~0.8 mg and ~1.2 mg per liter of insect cell culture, respectively.

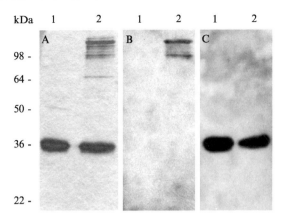

FIG. 1. Silver-stained SDS/PAGE (A) and the corresponding immunoblot (B and C) analyses of the recombinant Sec13R and the Sec13R/31L1 complex purified from baculovirus-infected insect cells. Lanes 1 and 2: The final rSec13R and rSec13R/31L1 samples, respectively. The immunoblots B and C were probed with anti-Sec31 rabbit polyclonal antibodies (raised against a GST-fusion peptide corresponding to the amino acids 865 to 897 of human Sec31L1) and anti-pentaHis mouse monoclonal antibody (recognizes the N-terminal hexa-histidine tag present on recombinant Sec13R), respectively, and finally visualized using enhanced chemiluminescence (ECL) substrate. The molecular masses of the protein standards used are given to the left of Panel A.

Further Highlights

While the overall purification strategy for the recombinant mammalian Sec13/31 complex is relatively straightforward, there are a few caveats that should be considered:

1. Despite the inclusion of protease inhibitors in all of our buffers and working at 4° at all times, we find that human Sec31L1 is particularly prone to proteolysis, as it has also been reported for yeast and mammalian Sec31 from both recombinant and native sources (Shimoni and Schekman, 2002; Shugrue *et al.*, 1999).

2. Despite the addition of DNAse I prior to lysis, we find that cleared cell lysates can still be viscous, therefore we recommend flow rates lower than 0.2 ml/min during the initial stages of IMAC (e.g., sample loading) to minimize high back pressures. Alternatively, HisTrap FF crude (1 mL) columns can be used to minimalize the back pressure due to high sample viscosity. In this case, we also recommend changing the second step gradient from 50 mM to 100 mM imidazole during fractionation, and using a flow rate of 0.5 mL/min throughout the IMAC protocol.

3. When rSec13R/31L1 samples at concentrations higher than 0.5 mg/ml are required, Amicon® Centrifugal Filter Units can be used, preferably

prior to the dialysis step against the *Sec13/31 Buffer* to minimize sample losses by aggregation.

4. Use of highly stringent conditions, such as the inclusion of 1 *M* NaCl and 10 m*M* DTT during the IMAC and Mono Q chromatography steps, respectively, ensures minimal copurification of contaminating proteins. These stringent conditions do not dissociate the rSec13R/31L1 complex and also have no apparent adverse effects on the overall integrity of rSec13R/31L1, as judged by dynamic light scattering, gel filtration/size exclusion chromatography and electron microscopy analyses (Gurkan *et al.*, 2005). Indeed, routine mass spectroscopy (LC MS/MS) analyses of purified rSec13R/31L1 confirm the identity of Sec13R and Sec31L1 with up to 76% and 48% sequence coverage, respectively, as well as validating the overall homogeneity of our samples.

5. As discussed earlier, while minimal COPII function in yeast can be reconstituted *in vitro* using Sar1, and the Sec23/24 and Sec13/31 complexes (Matsuoka *et al.*, 1998), additional, yet hitherto uncharacterized cytosolic factor(s) may be needed for the mammalian COPII function (Aridor and Balch, 2000; Kim *et al.*, 2004). Consequently, in the absence of these apparently requisite additional factor(s), it has been particularly difficult to assess the biological activity of the purified recombinant Sec13R/Sec31L1 using traditional *in vitro* biochemical assays. While not shown here, we have recently developed alternative analysis methods to directly assess Sec13/31 function *in vitro* using dynamic light scattering, gel filtration–size exclusion chromatography, and electron microscopy (Gurkan *et al.*, 2005).

Acknowledgments

These studies are supported by grants from the National Institutes of Health (DK-51870, HL-067201, GM-42336) to W.E.B, and a Cystic Fibrosis Foundation Postdoctoral Research Fellowship to C. G. We thank Greg Cantin and John Venable for LC MS/MS analysis, and Jacques T. Weissman, Kimberly Straley, and Sarah J. Lloyd for setting up or helping with recombinant expression of the human Sec13R/31L1 complex in baculovirus-infected insect cells. This is TSRI Manuscript No.: 17373-CB.

References

Antonny, B., Gounon, P., Schekman, R., and Orci, L. (2003). Self-assembly of minimal COPII cages. *EMBO Rep* **4**, 419–424.

Aridor, M., and Balch, W. E. (2000). Kinase signaling initiates coat complex II (COPII) recruitment and export from the mammalian endoplasmic reticulum. *J. Biol. Chem.* **275**, 35673–35676.

Aridor, M., Weissman, J., Bannykh, S., Nuoffer, C., and Balch, W. E. (1998). Cargo selection by the COPII budding machinery during export from the ER. *J. Cell Biol.* **141**, 61–70.

Barlowe, C., d'Enfert, C., and Schekman, R. (1993). Purification and characterization of SAR1p, a small GTP-binding protein required for transport vesicle formation from the endoplasmic reticulum. *J. Biol. Chem.* **268,** 873–879.

Barlowe, C., and Schekman, R. (1993). SEC12 encodes a guanine-nucleotide-exchange factor essential for transport vesicle budding from the ER. *Nature* **365,** 347–349.

Gurkan, C., Lapp, H., Hogenesch, J. B., and Bakh, W. E. (2005). Exploring trafficking GTPase function by mRNA expression profiling: Use of SymAtlas web-application and the membrome datasets. *Methods Enzymol.* **403,** 1–10.

Kaiser, C. A., and Schekman, R. (1990). Distinct sets of SEC genes govern transport vesicle formation and fusion early in the secretory pathway. *Cell* **61,** 723–733.

Kapetanovich, L., Baughman, C., and Lee, T. H. (2005). Nm23H2 facilitates coat protein complex II assembly and endoplasmic reticulum export in mammalian cells. *Mol. Biol. Cell* **16,** 835–848.

Kim, J., Hamamoto, S., Ravazzola, M., Orci, L., and Schekman, R. (2005). Uncoupled packaging of amyloid precursor protein and presenilin 1 into coat protein complex II vesicles. *J. Biol. Chem.* **280,** 7758–7768.

Kuge, O., Dascher, C., Orci, L., Rowe, T., Amherdt, M., Plutner, H., Ravazzola, M., Tanigawa, G., Rothman, J. E., and Balch, W. E. (1994). Sar1 promotes vesicle budding from the endoplasmic reticulum but not Golgi compartments. *J. Cell Biol.* **125,** 51–65.

Lederkremer, G. Z., Cheng, Y., Petre, B. M., Vogan, E., Springer, S., Schekman, R., Walz, T., and Kirchhausen, T. (2001). Structure of the Sec23p/24p and Sec13p/31p complexes of COPII. *Proc. Natl. Acad. Sci. USA* **98,** 10704–10709.

Lee, M. C., Miller, E. A., Goldberg, J., Orci, L., and Schekman, R. (2004). Bi-directional protein transport between the ER and Golgi. *Annu. Rev. Cell Dev. Biol.* **20,** 87–123.

Mammoto, A., Ohtsuka, T., Hotta, I., Sasaki, T., and Takai, Y. (1999). Rab11BP/Rabphilin-11, a downstream target of rab11 small G protein implicated in vesicle recycling. *J. Biol. Chem.* **274,** 25517–25524.

Mammoto, A., Sasaki, T., Kim, Y., and Takai, Y. (2000). Physical and functional interaction of rabphilin-11 with mammalian Sec13 protein. Implication in vesicle trafficking. *J. Biol. Chem.* **275,** 13167–13170.

Matsuoka, K., Orci, L., Amherdt, M., Bednarek, S. Y., Hamamoto, S., Schekman, R., and Yeung, T. (1998). COPII-coated vesicle formation reconstituted with purified coat proteins and chemically defined liposomes. *Cell* **93,** 263–275.

Matsuoka, K., Schekman, R., Orci, L., and Heuser, J. E. (2001). Surface structure of the COPII-coated vesicle. *Proc. Natl. Acad. Sci. USA* **98,** 13705–13709.

Nishimura, N., and Balch, W. E. (1997). A di-acidic signal required for selective export from the endoplasmic reticulum. *Science* **277,** 556–558.

Novick, P., Field, C., and Schekman, R. (1980). Identification of 23 complementation groups required for post-translational events in the yeast secretory pathway. *Cell* **21,** 205–215.

Pryer, N. K., Salama, N. R., Schekman, R., and Kaiser, C. A. (1993). Cytosolic Sec13p complex is required for vesicle formation from the endoplasmic reticulum *in vitro*. *J. Cell Biol.* **120,** 865–875.

Salama, N. R., Chuang, J. S., and Schekman, R. W. (1997). Sec31 encodes an essential component of the COPII coat required for transport vesicle budding from the endoplasmic reticulum. *Mol. Biol. Cell* **8,** 205–217.

Shaywitz, D. A., Orci, L., Ravazzola, M., Swaroop, A., and Kaiser, C. A. (1995). Human SEC13Rp functions in yeast and is located on transport vesicles budding from the endoplasmic reticulum. *J. Cell Biol.* **128,** 769–777.

Shimoni, Y., and Schekman, R. (2002). Vesicle budding from endoplasmic reticulum. *Methods Enzymol.* **351,** 258–278.

Shugrue, C. A., Kolen, E. R., Peters, H., Czernik, A., Kaiser, C., Matovcik, L., Hubbard, A. L., and Gorelick, F. (1999). Identification of the putative mammalian orthologue of Sec31P, a component of the COPII coat. *J. Cell Sci.* **112**(Pt 24), 4547–4556.

Siniossoglou, S., Lutzmann, M., Santos-Rosa, H., Leonard, K., Mueller, S., Aebi, U., and Hurt, E. (2000). Structure and assembly of the Nup84p complex. *J. Cell Biol.* **149**, 41–54.

Siniossoglou, S., Wimmer, C., Rieger, M., Doye, V., Tekotte, H., Weise, C., Emig, S., Segref, A., and Hurt, E. C. (1996). A novel complex of nucleoporins, which includes Sec13p and a Sec13p homolog, is essential for normal nuclear pores. *Cell* **84**, 265–275.

Swaroop, A., Yang-Feng, T. L., Liu, W., Gieser, L., Barrow, L. L., Chen, K. C., Agarwal, N., Meisler, M. H., and Smith, D. I. (1994). Molecular characterization of a novel human gene, SEC13R, related to the yeast secretory pathway gene SEC13, and mapping to a conserved linkage group on human chromosome 3p24-p25 and mouse chromosome 6. *Hum. Mol. Genet.* **3**, 1281–1286.

Tang, B. L., Peter, F., Krijnse-Locker, J., Low, S. H., Griffiths, G., and Hong, W. (1997). The mammalian homolog of yeast Sec13p is enriched in the intermediate compartment and is essential for protein transport from the endoplasmic reticulum to the Golgi apparatus. *Mol. Cell. Biol.* **17**, 256–266.

Tang, B. L., Zhang, T., Low, D. Y., Wong, E. T., Horstmann, H., and Hong, W. (2000). Mammalian homologues of yeast sec31p. An ubiquitously expressed form is localized to endoplasmic reticulum (ER) exit sites and is essential for ER-Golgi transport. *J. Biol. Chem.* **275**, 13597–13604.

Weissman, J. T. (2000). Regulation of mammalian cargo ER export. Ph.D. Thesis, Department of Cell Biology. The Scripps Research Institute, La Jolla, CA.

Weissman, J. T., Plutner, H., and Balch, W. E. (2001). The mammalian guanine nucleotide exchange factor mSec12 is essential for activation of the Sar1 GTPase directing endoplasmic reticulum export. *Traffic* **2**, 465–475.

Yoshihisa, T., Barlowe, C., and Schekman, R. (1993). Requirement for a GTPase-activating protein in vesicle budding from the endoplasmic reticulum. *Science* **259**, 1466–1468.

[7] Purification and Properties of Mammalian Sec23/24 from Insect Cells

By PAUL LAPOINTE and WILLIAM E. BALCH

Abstract

The Sec23/24 complex is a large heterodimeric protein involved in COPII vesicle biogenesis. The individual mammalian protein subunits are too large for expression in bacterial systems. This article details the use of the Bac-to-Bac baculovirus coexpression system in insect cells for both the human Sec23A and Sec24C. This strategy results in high yields of pure, functional protein and can be adapted for the purification of other Sec23/24 isoforms for their biochemical and biological characterization.

METHODS IN ENZYMOLOGY, VOL. 404   0076-6879/05 $35.00
Copyright 2005, Elsevier Inc. All rights reserved.   DOI: 10.1016/S0076-6879(05)04007-3

Introduction

Background

Vesicular traffic from the ER is regulated by the small GTPase, Sar1, and involves the concentration and packaging of cargo molecules into COPII transport vesicles (Aridor *et al.*, 2001; Kuge *et al.*, 1994). The cytosolic, GDP-bound form of Sar1 is activated by its guanine nucleotide exchange factor (GEF), Sec12, resulting in the association of the GTP-bound Sar1 with the ER membrane (Weissman *et al.*, 2001). This activated Sar1 then recruits two protein heterocomplexes that comprise the COPII coat. The first complex recruited to the membrane is a heterodimer consisting of Sec23 (80 kDa) and Sec24 (120 kDa). Sec23 interacts directly with Sar1 and acts as its GTPase activating protein (GAP), while Sec24 has been shown to interact directly with ER cargo possessing ER exit motifs (Nishimura and Balch, 1997; Yoshihisa *et al.*, 1993). After recruitment of Sec23/24 to the ER membrane, two more proteins involved in COPII budding are recruited, Sec13 and Sec31 (Matsuoka *et al.*, 1998, 2001). The subsequent recruitment of Sec13/31 to the Sar1–Sec23/24 prebudding complex is thought to polymerize the coat and drive vesicle budding. Uncoating of COPII vesicles occurs as a result of GTP hydrolysis by Sar1, which is reported to be accelerated tenfold with the recruitment of Sec13/31 (Antonny *et al.*, 2001).

The ability to express and purify the Sec23/24 heterocomplex is critical to understanding its role in cargo selection and COPII vesicle biogenesis, and to characterize its biochemical properties. Expression in bacterial systems is difficult due to the size of the individual subunits and their susceptibility to proteolysis. As such, insect cell expression systems are ideally suited for expressing large complexes such as the mammalian Sec23/24 complex. This article describes a general approach for the purification of the complete Sec23/24 complex from insect cells.

Genomics—Practical Considerations

There are multiple isoforms of Sec23 and Sec24 in most organisms from yeast to humans. In mammalian cells there are two Sec23 isoforms and four Sec24 isoforms (Table I). Little is known of how these different isoforms are naturally combined in living cells, or how these heterodimers may behave functionally or during purification. In this article, the expression and purification of the human Sec23A and Sec24C is described. If other Sec23/24 pairs are desired, the same strategy can be employed but the solubility, expression, and heterodimer formation/stability must be examined.

TABLE I
HUMAN SEC23 AND SEC24 VARIANTS

Species	Sec23 variants	Accession	Sec24 variants	Accession
Homo sapiens	Sec23A	Q15436	Sec24A	O95486
	Sec23B	Q15437	Sec24B	O95487
			Sec24C	P53992
			Sec24D	O94855

Cloning and Virus Amplification

The details of gene cloning, bacmid construction, and virus production are not the main focus of this review and will not be discussed here. Detailed information can be obtained from the Invitrogen website for the BactoBac Baculovirus expression system (Cat.#: 10359-016). Briefly, the coding sequences for the human Sec23A and Sec24C genes (accession numbers Q15436 and P53992, respectively) were amplified by polymerase chain reaction (PCR) and cloned into the pFastBac™ DUAL vector (Gibco) and P1 virus stocks were prepared according to the manufacturer's protocol. Plaque-purified P1 virus were amplified to make P2 and P3 stocks or virus in Sf9 cells (Cat.#: 11496-015) and used for expression in BTI-TN-5B1-4 (Tn5) cells (Cat.#: B85502). The viral titer of each amplified stock will vary and can be titered using standard techniques. It is recommended that a large stock of P3 or P4 virus is prepared in Sf9 cells so that several large-scale expressions can be carried out and the resultant pellets stored at $-80°$ for later purifications. Virus not used for infection and expression in Tn5 cells will lose potency over time if stored at $4°$ or $-80°$ for long periods. The infection–expression conditions for a virus that has been stored for several months may have to be re-established.

Expression

General Considerations for Expression in Tn5 Cells

The amount of virus (multiplicity of infection or MOI) necessary to achieve the optimum expression of the Sec23/24 complex will vary between different amplifications of virus in Sf9 cells. The amount of virus necessary to achieve optimal expression as well as the time allowed for the infection to progress should be established empirically. Typically, Tn5 cells should be infected at a cell density of 2×10^6 cells/mL. Small cultures of Tn5 cells should be infected with dilutions of the virus stock and grown for at least three days. A time-course examining expression as well as cell death should be conducted to determine the optimum growth/infection time. The expression samples will likely show a band corresponding to the full-length protein, as well as bands representing breakdown products of Sec24C, which is susceptible to proteolysis, particularly during the cell lysis phase. The optimum time of infection can be identified when the band representing the full-length Sec24C no longer significantly increases in intensity but the breakdown products do (Fig. 1). Ideally, the time at which Sec24C expression is maximal should occur before the majority of the cells lyse.

This purification strategy involves the purification of the Sec23/24 complex through an N-terminal hexa-Histidine tag on Sec23A. Proteolysis of Sec23A does not appear to be as much of a problem as it is for Sec24C. When the optimum expression conditions have been defined, the expression can be scaled up to the desired level. The cells in the large-scale expression cultures should be harvested by centrifugation, washed with PBS, and centrifuged again in 50 mL conical tubes for storage at $-80°$ until purification.

Purification Strategy

Overview

The strategy described here involves immobilized metal affinity chromatography followed by ion exchange chromatography with Mono Q resin. The Sec23A/24C complex is very sensitive to pH and ionic environment during

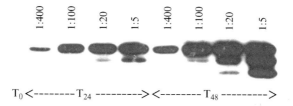

FIG. 1. Expression of the Sec23/24 complex in insect cells. Cell lysates from Tn5 cells grown for 0, 24 or 48 hrs after infection with a P3 stock of virus at the indicated dilutions run on SDS-PAGE and analyzed by western blot with an anti-hSec24C antibody. At T_0 (before infection) there is no expression of Sec24C. Maximum expression of Sec24 occurs at a P3 dilution of 1:100 as the band corresponding to the full length Sec24C (top band) does not increase in intensity with increasing virus (1:20 or 1:5 dilution). In this case, breakdown products of Sec24C appear as the infection time reaches 48 hrs with P3 dilutions of 1:20 or 1:5.

the purification. Careful attention should be paid to the buffers used as they will greatly influence the overall yield and activity of the final product.

Protocol

Lysis of the Harvested Cells. Pellets from 500 mL expression cultures should be resuspended in 50 mL of immobilized metal affinity chromatography Buffer A (IMAC-A, Table II) supplemented with an EDTA-free protease inhibitor tablet (Complete EDTA-free, Cat.#: 1873580) from Roche (Indianapolis, IN). This is a convenient volume of IMAC-A to use for an expression culture of that size because the viscosity of the resultant lysate is a factor during the nitrogen bomb lysis technique as well as subsequent sample loading by FPLC. When the sample is completely resuspended, lysis should be carried out with a nitrogen cavitator (Fike Metal Products, Blue Springs, MO) at 500 psi for 45 min. The resultant lysate should be ultracentrifuged at 45,000 rpms for 30 min in a Beckman Ti60 rotor. The cleared lysate will have a cloudy phase near the pellet and near the top of the supernatant. Avoiding the cloudy–lipid phase has reduced yields significantly in our hands, which suggests that the Sec23/24 complex associates with the lipid after lysis. It is recommended to collect as much of the supernatant as possible (including the cloudy lipid phase) without disturbing the pellet. The lipid is easily removed during the subsequent His-tag purification. This supernatant will be used in the initial His-tag purification by FPLC.

TABLE II
BUFFERS FOR IMAC CHROMATOGRAPHY

Buffer	Chemical	Chemical formula/MW	g/L - mL/L
IMAC A	20 mM Sodium Phosphate pH 7.6	NaH$_2$PO4/119.96	2.40 g
	1 M Sodium Chloride	NaCl/58.44	58.44 g
	1 mM Magnesium Acetate	C$_4$H$_6$O$_4$Mg·4H$_2$O/214.5	0.2145 g
	5 mM β-Mercaptoethanol	C$_2$H$_6$OS/78.13 (12.8 M)	0.391 mL
IMAC B	20 mM Sodium Phosphate pH 7.6	NaH$_2$PO$_4$/119.96	2.40 g
	1 M Sodium Chloride	NaCl/58.44	58.44 g
	1 mM Magnesium Acetate	C$_4$H$_6$O$_4$Mg·4H$_2$O/214.5	0.2145 g
	1 M Imidazole	C$_3$N$_2$H$_4$/68.08	68.08 g
	5 mM β-Mercaptoethanol	C$_2$H$_6$OS/78.13 (12.8 M)	0.391 mL

Immobilized Metal Affinity Chromatography. The buffers used in the initial His-tag purification are detailed in Table II. The 50 mL lysate is loaded onto a 1 mL Ni^{2+}-charged HiTrap Chelating HP column (Amersham Pharmacia, Cat.# 17-0408-01) using the ÄKTA™ FPLC™ system with Frac-950 fraction collector (Amersham Pharmacia). The purification scheme is detailed in Table III. The His-tagged Sec23/24 complex should elute at an Imidizole concentration around 150 mM with a typical chromatogram and resultant western blot of fractions shown in Fig. 2.

Mono Q Anion Exchange Chromatography. The desired fractions from the HiTrap column should be pooled and diluted in Mono Q Buffer A (Table IV) to a final volume of 50 mL. This will dilute the salt concentration

TABLE III
SCHEMATIC OF THE IMAC PURIFICATION PROTOCOL

Sample Injection	50 mL injection; 0.2 mL/min flow rate
Step Gradient 1	2% IMAC-B for 15 column volumes
Step Gradient 2	5% IMAC-B for 15 column volumes
Linear Gradient 3	5% to 30% IMAC-B over 20 column volumes
Fraction Collection	1 mL fractions

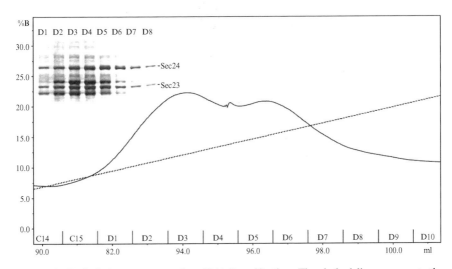

FIG. 2. Typical chromatogram of an IMAC purification. The dashed line represents the gradient used in the purification. The solid line is the absorbance (A280) of the eluted protein with the fractions containing the Sec23/24 complex shown on the inset silver stained gel. The relative percent of IMAC Buffer B is shown with the dashed line.

TABLE IV
Buffers for Mono Q Chromatography

Buffer	Chemical	Chemical formula/MW	g/L
Mono QA	20 mM Tris HCl pH 7.5	$NH_2C(CH_2OH)_3 \cdot HCl/157.6$	3.152 g
	1 mM Dithiothreitol	$C_4H_{10}O_2S_2/154.25$	0.154 g
Mono QB	20 mM Tris HCl pH 7.5	$NH_2C(CH_2OH)_3 \cdot HCl/157.6$	3.152 g
	1 M Sodium Chloride	NaCl/58.44	58.44 g
	1 mM Dithiothreitol	$C_4H_{10}O_2S_2/154.25$	0.154 g

TABLE V
Schematic of the MonoQ Purification Protocol

Sample injection	50 mL injection; 1 mL/min flow rate
Linear Gradient 1	0% to 100% Mono QB over 20 column volumes
Fraction collection	0.25 mL fractions

of the sample and allow binding to the Mono Q resin. (The Sec23/24 complex will begin to elute from the Mono Q resin at NaCl concentrations around 150 mM.) The diluted 50 mL sample is loaded onto a Mono Q HR 5/5 1mL column (Cat.#: 17-0546-011) using the ÄKTA™ FPLC™ system and purified with the program detailed in Table V. A typical result from the Mono Q purification is shown in Fig. 3. Modifications can be made to the overall scheme to increase the concentration of the protein in the elution fractions or to purify Sec23/24 complexes consisting of different isoforms. However, the linear gradient shown in Fig. 2 should be appropriate for most purifications.

Storage and Handling of the Final Product. The Mono Q fractions containing the Sec23/24 complex should be pooled, quantified by protein assay, and supplemented with 5% glycerol before aliquoting and snap-freezing in liquid nitrogen. The glycerol will help to maintain function and solubility in future assays.

Behavior of the Sec23/24 Complex

Analysis of the Final Product

The Sec23 subunit of the Sec23/24 complex is the GAP for the GTPase, Sar1 (Aridor *et al.*, 2001; Kuge *et al.*, 1994). This property can be used to assess the function of the Sec23/24 complex in an *in vitro* GTPase assay.

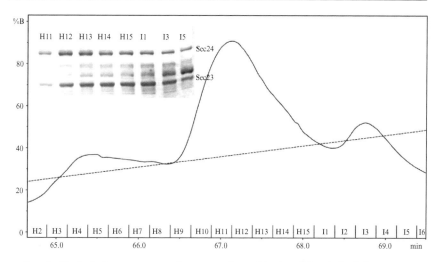

FIG. 3. Typical chromatogram of a mono Q purification. The dashed line represents the gradient used in the purification. The solid line represents the absorbance (A280) of the eluted protein with the fractions containing the Sec23/24 complex shown on the inset silver stained gel.

The conversion of GTP labeled with ^{32}P at the alpha position to GDP can be monitored using thin layer chromatography (TLC) (Aridor *et al.*, 1998). Sar1 has very low intrinsic GTPase activity, which is activated through its interaction with Sec23, providing an indication of the functionality of the purified Sec23/24 complex.

Sec23/24 has also been shown to interact with exit codes in the cytoplasmic tails of ER cargo molecules in a Sar1-dependent fashion. GST-pulldown assays can be performed to show that the purified Sec23/24 heterodimer is competent to interact with GST-fusions to the cytoplasmic tail of the yeast protein, Bet1p, or that of the viral glycoprotein, VSV-G (Aridor *et al.*, 2001; Miller *et al.*, 2002, 2003).

These simple tests provide a reliable means to monitor the quality of the mammalian Sec23/24 complex from insect cells.

References

Antonny, B., Madden, D., Hamamoto, S., Orci, L., and Schekman, R. (2001). Dynamics of the COPII coat with GTP and stable analogues. *Nat. Cell Biol.* **3**, 531–537.
Aridor, M., Fish, K. N., Bannykh, S., Weissman, J., Roberts, T. H., Lippincott-Schwartz, J., and Balch, W. E. (2001). The Sar1 GTPase coordinates biosynthetic cargo selection with endoplasmic reticulum export site assembly. *J. Cell Biol.* **152**, 213–229.

Aridor, M., Weissman, J., Bannykh, S., Nuoffer, C., and Balch, W. E. (1998). Cargo selection by the COPII budding machinery during export from the ER. *J. Cell Biol.* **141,** 61–70.

Kuge, O., Dascher, C., Orci, L., Rowe, T., Amherdt, M., Plutner, H., Ravazzola, M., Tanigawa, G., Rothman, J. E., and Balch, W. E. (1994). Sar1 promotes vesicle budding from the endoplasmic reticulum but not Golgi compartments. *J. Cell Biol.* **125,** 51–65.

Matsuoka, K., Orci, L., Amherdt, M., Bednarek, S. Y., Hamamoto, S., Schekman, R., and Yeung, T. (1998). COPII-coated vesicle formation reconstituted with purified coat proteins and chemically defined liposomes. *Cell* **93,** 263–275.

Matsuoka, K., Schekman, R., Orci, L., and Heuser, J. E. (2001). Surface structure of the COPII-coated vesicle. *Proc. Natl. Acad. Sci. USA* **98,** 13705–13709.

Miller, E., Antonny, B., Hamamoto, S., and Schekman, R. (2002). Cargo selection into COPII vesicles is driven by the Sec24p subunit. *EMBO J.* **21,** 6105–6113.

Miller, E. A., Beilharz, T. H., Malkus, P. N., Lee, M. C., Hamamoto, S., Orci, L., and Schekman, R. (2003). Multiple cargo binding sites on the COPII subunit Sec24p ensure capture of diverse membrane proteins into transport vesicles. *Cell* **114,** 497–509.

Nishimura, N., and Balch, W. E. (1997). A di-acidic signal required for selective export from the endoplasmic reticulum. *Science* **277,** 556–558.

Weissman, J. T., Plutner, H., and Balch, W. E. (2001). The mammalian guanine nucleotide exchange factor mSec12 is essential for activation of the Sar1 GTPase directing endoplasmic reticulum export. *Traffic* **2,** 465–475.

Yoshihisa, T., Barlowe, C., and Schekman, R. (1993). Requirement for a GTPase-activating protein in vesicle budding from the endoplasmic reticulum. *Science* **259,** 1466–1468.

[8] Purification and Functional Properties of Yeast Sec12 GEF

By Eugene Futai and Randy Schekman

Abstract

In order to reconstitute the generation of COPII vesicles from synthetic liposomes, the minimum requirements are the coat components, Sar1p GTPase, Sec23/24p, Sec13/31p, and a nonhydrolyzable GTP analog such as GMP-PNP. However, in the presence of GTP, nucleotide hydrolysis by Sar1p renders the coat insufficiently stable to sustain vesicle budding. Sar1p GTPase activity was activated by the Sec23/24p GTPase-activating protein (GAP), and further accelerated 10-fold by Sec13/31p. In order to study GTP-dependent budding, we introduced the Sar1p guanine nucleotide exchange factor (GEF), Sec12p. We evaluated Sar1p activation by Sec12p and the dynamics of coat assembly and disassembly in the presence of both Sec12p and Sec23/24p. The cytoplasmic domain of Sec12p activated Sar1p with a turnover 10-fold higher than the GAP activity of Sec23p in the presence of Sec13/31p. As a result, the entire COPII coat remains stable in

0076-6879/05 $35.00
DOI: 10.1016/S0076-6879(05)04008-5

the presence of GTP. Here, we describe methods to purify Sec12p, real-time fluorescence assays to evaluate COPII coat formation, and the relevant kinetic analyses.

Introduction

A combination of genetic and biochemical approaches using *Saccharomyces cerevisiae* revealed the essential genes for COPII vesicle formation on the endoplasmic reticulum (ER): the COPII coat (Sar1p small GTPase, Sec23/24p heterodimer, and Sec13/31p heterotetramer), Sec16p, and Sec12p. Sec12p is involved in vesicle formation as an ER-resident transmembrane protein (Bonifacino and Glick, 2004). Sec12p is a 70kD glycoprotein with a 40kD N-terminal cytosolic domain, a single transmembrane domain, and a 30kD glycosylated luminal domain (Nakano and Muramatsu, 1989). The cytosolic domain catalyzes guanine-nucleotide exchange (GEF) on the Sar1p GTPase in a process that recruits Sar1 to the ER, where COPII-coated vesicles are produced (Barlowe and Schekman, 1993). The COPII coat is formed with GTP-bound Sar1p, Sec23/24p, and Sec13/31p heteromeric complexes. Sec23p acts as a GTPase-activating protein (GAP) specific for Sar1p (Yoshihisa *et al.*, 1993). The Sec23/24p complex is also responsible for the selection and concentration of cargo proteins for packaging, providing binding sites for multiple cargo proteins (Miller *et al.*, 2003 and Mossessova *et al.*, 2003). Sec13/31p is thought to bridge Sec23/24p and Sar1p-GTP to create a coat that envelops the membrane, shaping it into a bud (Matsuoka *et al.*, 2001). In order to reproduce aspects of COPII vesicle formation, we have developed assays with synthetic membranes (liposomes). COPII vesicles are generated from liposomes with the minimum coat components Sar1p, Sec23/24p, Sec13/31p, and a nonhydrolyzable GTP analog such as GMP-PNP (Matsuoka *et al.*, 1998). However, with GTP and the full complement of coat subunits, Sar1 nucleotide hydrolysis, activated by Sec23GAP and stimulated 10-fold by Sec13/31p, rendered the coat very unstable (Antonny *et al.*, 2001). More recently, we reconstituted COPII coat formation on liposomes supplemented with Sec12p (Futai *et al.*, 2004). The catalytically potent Sec12GEF stabilizes the coat complex formed on the membrane surface (Futai *et al.*, 2004). In this chapter, we describe the purification of Sec12p and enzymatic analyses of COPII coat formation using real time fluorescence and light scattering assays.

Protein Purification

Sar1p, Sec23/24(His$_6$)p, and Sec13/31(His$_6$)p were purified as described (Barlowe *et al.*, 1994 and Salama *et al.*, 1997). Sar1p was purified in the presence of 5 μM GDP to avoid protein aggregation. The coding sequences of Sec12ΔC(1–354)p and Sec12TM(1–373)p were amplified by PCR from yeast chromosomal DNA (S388C strain) using the following primers: primer 1 (GAAGATCTATGAAGTTCGTGACAGCTAGTTATAACG) and primer 3 (GAAGATCTTCAGTGGTGGTGGTGGTGGTGGTGGTGA TAGGAGAACTGTAAAATGAAAGAAAGCAG), and primer 2 (GAA GATCTTCATTTAGAGATTTTTTGTTTCATTGAGGT GTAGTTGGCG) and primer 3 combinations of primers, respectively. The PCR products were digested with Bgl II and ligated into pGEX2T BamHI site (Pharmacia, Piscataway, NJ), resulting in the plasmids pGEXSec12ΔC and pGEXSec12TMHis$_8$. Point mutations in Sec12ΔCp were introduced using Quickchange (Stratagene, La Jolla, CA). The Sec12p expression plasmids were transformed into BL21-CodonPlus(DE3)-RIL strain (Stratagene). An overnight culture in 2× YT broth (1.6% tryptone, 1% yeast extract, 0.5% NaCl) containing 50 μg/ml carbenicillin and 50 μg/ml chloramphenicol was used to start a 1 L culture in 2× YT culture containing 50 μg/ml carbenicillin. Cells grown at 37° to an optical density (OD$_{600}$) of 0.8 were induced with 1 mM IPTG for 4 h. For Sec12ΔCp purification, the cells (∼3g) were collected by sedimentation, treated with 0.5 mg/ml lysozyme for 20 min on ice in 30 ml buffer A (25 mM Tris pH 8.0, 400 mM KCl, 10% glycerol, 5 mM β-mercaptoethanol, 1 mM phenylmethylsulphonyl fluoride (PMSF), and lysed by sonication. Lysates were clarified by centrifugation at 20,000g for 10 min twice and incubated with prewashed glutathione agarose (2 ml) for 1 h at 4°.

Sec12ΔCp bound to beads was washed with 25 ml buffer B (25 mM Tris pH 8.0, 400 mM KCl, 10% glycerol, 2 mM β-mercaptoethanol) three times and 25 ml TCB (50 mM Tris pH 8.0, 250 mM KOAc, 5 mM CaCl$_2$), and eluted after thrombin cleavage (2 unit/ml thrombin) (Boehringer, Mannheim, Germany) in 3 ml TCB at room temperature for 2 h. Sec12ΔCp retained on the beads was eluted with the additional TCB (2 ml). The eluate in TCB (5 ml, combined) was applied to the NAP-25 column (Amersham, Piscataway, NJ) (2.5 ml each to two columns) equilibrated with buffer B (25 ml), and eluted with 3.5 ml buffer B. For Sec12TMp purification, buffer A containing 4% Triton X-100 and buffer B containing 1% β-octyl glucoside (OG) were used instead of buffer A and B. GSTSec12TMp was eluted with 5 ml buffer B (1%OG) containing 50 mM glutathione and exchanged buffers to TCB (1% OG). The GST tag was cleaved by thrombin in solution and removed by passing the treated material over a 2 ml glutathione agarose column. On-resin cleavage caused protein aggregation. BET1, BOS1, and UFE1 were cloned as above in vectors pGEX-2T (for GST-Bos1p, GST-Ufe1p) (Pharmacia, Piscataway, NJ) or pETGEX (for GST-Bet1p) (Sharrocks, 1994), produced as N-terminal GST fusion proteins, and purified as described for Sec12TM.

Liposome Preparation

Liposomes were prepared as described with modifications (Antonny *et al.*, 2001). Briefly, a lipid film was formed by evaporating 500 µl of a 2 mM phospholipid mixture (in chloroform) in a glass tube by rotary evaporation at 37°, and hydrated in 1 ml HK buffer (20 mM Hepes-K, pH 6.8, 160 mM KOAc) at room temperature for more than 1 h to generate multilamellar liposomes. The lipid suspension (1 mM phospholipids) was extruded through a 400 nm polycarbonate filter (Osmonics Inc., Minnetonka, MI) 19 times. The optimal lipid composition (major-minor mix) for COPII budding from synthetic liposomes was established by Matsuoka *et al.* (1998). Major-minor mix liposomes contained: (in mol%) phosphatidylcholine, 50:phosphatidylethanol amine 21; phosphatidylserine, 8; phosphatidic acid, 5; phosphatidylinositol, 9; phosphatidylinositol-4-phosphate, 2.2; phosphatidylinositol-4,5-bisphosphate, 0.8; cytidine-diphosphate-dioleoylglycerol, 2. PC/PE mix contained phosphatidylcholine, 53; phosphatidylethanolamine 46. Major-minor liposomes were prepared with 1,2-dioleoyl-derivatives of phospholipids. Liposomes also contained 10% (w/w) cholesterol and 1% (mol/mol) Texas red-phosphatidylethanolamine for determination of lipid concentrations.

Proteoliposomes containing Sec12TMp, Bet1p, Bos1p, and GSTUfe1p were prepared by dialysis (Parlati *et al.*, 2000). A lipid film was hydrated in 600 µl buffer B containing 1% OG and protein (1 mM phospholipids and 0.33 µM protein). The lipid suspension was incubated for 15 min at room temperature. Octylglucoside was removed by extensive dialysis against 2 L reconstitution buffer (25 mM Hepes-K, pH 7.4, 400 mM KCl, 10% glycerol, 1 mM DTT) at 4° for 30 h with four buffer changes (500 ml buffer for each), using 10,000 MWCO dialysis cassettes (Pierce, Rockford, IL). The dialysate (500 µl) was mixed with 330 µl of 2.5 M sucrose in HK buffer (final 1 M sucrose) and overlaid with 660 µl 0.75 M sucrose in HK buffer and 200 µl HK buffer. The resulting gradients were centrifuged in a Beckman SW55 rotor (13 × 51 mm polycarbonate tube) at 55,000 rpm for 4 h at 4°. Proteoliposomes were collected from the top fraction and extruded through a 400 nm polycarbonate filter. Fluorescence was used to quantify lipid concentration.

Assay of Functional Sec12p

Light scattering and Sar1p tryptophan fluorescence measurements were performed as previously described (Antonny *et al.*, 2001; Robbe and Antonny, 2003). We used Fluoromax-3 or Fluorolog-3 (Jobin-Yvon, 90° format) fluoremeters with thermostatted cuvette folders set at 27°. Tryptophan fluorescence was recorded at 340 nm (10 nm bandwidth). Because large amounts of nucleotide were used, $\lambda = 297.5$ nm (1.5 nm bandwidth) was used to excite the sample. For light scattering measurements, the excitation and emission were set at 355 nm (1 nm bandwidth). Data were sampled at 1/s with 0.3 s time response. For better temporal resolution, all experiments were performed in a cylindrical cuvette (sample volume 550 μl) in which reaction mixtures were mixed by a magnetic stir bar. HKM buffer (20 mM Hepes-K, pH 6.8, 160 mM KOAc, 1 mM MgCl$_2$) filtered and degassed with a Steriflip device (Millipore, Billerica, MA) was used for the assays. Solutions of COPII proteins were centrifuged briefly (50,000 rpm, 5 min, TLA100 rotor) before use to remove aggregates. Coat components, Sec12ΔC, and GTP were added to major-minor liposome mixture as in Figs. 1 and 2 (Futai *et al.*, 2004). The recording was not interrupted by the injections and the mixing time was <2 s, allowing kinetic measurement in the range of few seconds.

Sec12 activity was also assessed by a GST pull-down assay to detect interaction between Sec12ΔCp and Sar1p. GSTSec12ΔC (3 μM) was incubated with Sar1p (3 μM) and guanine nucleotide (300 μM), or Sar1p, EDTA (2 mM) and alkaline phosphatase (calf intestine, Promega, Madison, WI) for 90 min at 4$°$ in 100 μl HKM buffer containing 0.2% OG. Reaction samples were added to 10 μl prewashed glutathione agarose beads, and incubation was continued for 30 min. Beads were washed with 500 μl HKM buffer (0.2% OG) three times. Sar1p recovered with GSTSec12ΔC was analyzed by SDS-PAGE. EDTA and alkaline phosphatase render Sar1p nucleotide-free and yield the highest recovery.

Kinetic Measurements

Activation of Sar1p upon GDP/GTP exchange was analyzed as described elsewhere (Beraud-Dufour *et al.*, 1998). The apparent rate constant of Sar1p activation (k_{act}) was determined by fitting the Sar1p fluorescence change with a single exponential and plotted as a function of [Sec12ΔCp] (Fig. 1A, B). k_{act} increased linearly with the concentration of Sec12ΔCp according to the equation:

$$k_{act} = \mathbf{a} + \mathbf{b}[Sec12\Delta Cp] \tag{1}$$

where $\mathbf{a}$ is the rate constant of spontaneous nucleotide exchange, which was measured in the absence of Sec12ΔCp, and $\mathbf{b}$ is the specific exchange activity of Sec12ΔCp, that is, the rate constant of catalyzed GDP/GTP exchange normalized to the concentration of Sec12ΔCp ($k_{exchange}$/[Sec12ΔCp]) (Fig. 2C). For inactivation of Sar1p upon GAP-catalyzed GTP hydrolysis, the apparent rate constant for Sar1p inactivation was determined by the value τ_{inact} ($k_{inact} = 1/\tau_{inact}$) graphically, and plotted as a function of [Sec23/24p] (Figs. 1C, D). Assuming k_{inact} is 0 in the absence of

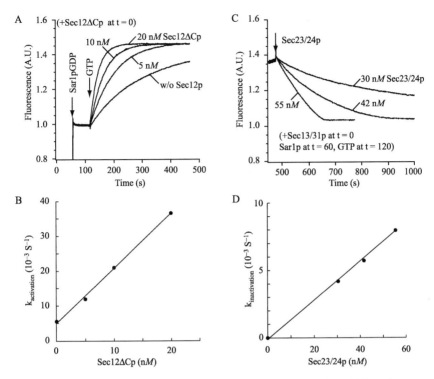

FIG. 1. Tryptophan fluorescence assay for Sec12GEF catalyzed GDP-GTP exchange on Sar1p (A and B) and Sec23GAP catalyzed GTP hydrolysis (C and D). (A) Sar1p-GDP (2 μM) was added to the initial mixture; major-minor liposomes (300 μg ml^{-1}) in HKM buffer with various concentrations of Sec12ΔCp (0–20 nM), and activated by the addition of GTP (30 μM). (B) The rate constants for Sar1p activation were plotted against Sec12ΔCp concentration to determine the nucleotide exchange activity of Sec12p. (C) Sar1p was activated as in (A) with no Sec12ΔCp and 90 nM Sec13/31p in the initial mixture, which stimulates GAP activity at saturation. The concentration of Sec23/24p was varied (30–55 nM) to activate GTP hydrolysis. (D) The rate constant for the Sec23/24p GAP activity with saturating Sec13/31p was plotted against Sec23/24p concentration to determine catalytic activity.

Sec23/24p, k_{inact} increases linearly with the concentration of Sec23/24p according to the equation:

$$k_{inact} = c[Sec23/24p] \tag{2}$$

where c is the specific catalytic activity of Sec23/24p, that is, the rate constant of catalyzed GTP hydrolysis normalized to the concentration of Sec23/24p saturated with Sec13/31p ($k_{catalyze}$/[Sec23/24p]) (Fig. 2C). These parameters ($k_{exchange}$, $k_{catalyze}$) correspond to the apparent second order rate constant k_{cat}/K_m of a Michaelis–Menten mechanism. The rate of exchange for Sec12TM(1–373)p on proteoliposomes was also determined.

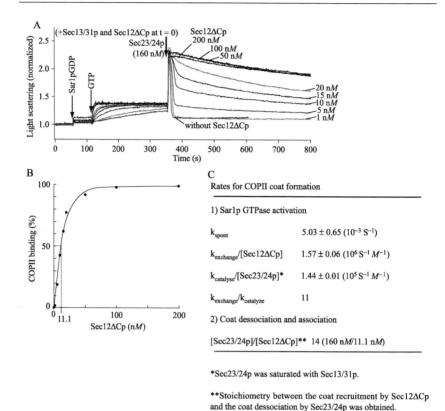

FIG. 2. Light scattering assay of the assembly of COPII coat with Sec12ΔCp and GTP. (A) The light scattering of a suspension of major-minor liposomes (100 μg ml⁻¹) in HKM buffer was continuously monitored upon the addition of 950 nM Sar1p GDP, 100 μM GTP, and 160 nM Sec23/24p at specific time points as indicated by the arrows. The initial mixture contained different concentrations of Sec12ΔCp (0–200 nM) and 260 nM Sec13/31p. (B) In (A), Sec12ΔCp stabilized GTP-driven COPII coat assembly on liposomes after Sec23/24p addition. The intensity of the signal was measured at 400 s and plotted against Sec12ΔCp concentration as a percentage of maximal COPII binding. Half maximal binding was obtained at 11.1 nM Sec12ΔCp. (C) Rates for Sar1p activation/inactivation and coat formation/ deformation determined from Figs. 1 and 2 are listed.

Compared to the activity of Sec12ΔCp, Sec12TMp proteoliposomes displayed 62% of full activity, suggesting that most of the cytoplasmic domain was accessible on the liposome surface. The activity of Sec12p was also analyzed using a light scattering assay. Addition of Sec12ΔCp to liposomes incubated with Sar1p and GTP resulted in a rapid rise in the light scattering signal, consistent with accelerated Sar1p nucleotide exchange and recruitment to the membrane. Furthermore, a stable increase in signal on addition of Sec23/24p indicated that the full coat was maintained by continuous generation of Sar1GTP by Sec12ΔCp (Fig. 2A). Titration of the Sec12ΔCp concentration established a stoichiometry of 14:1 Sec23/24p to Sec12ΔCp as sufficient to maintain coat stability. This ratio approximates the 10-fold difference between $k_{exchange}$ and $k_{catalyze}$ of Sec12 and Sec23/24.

Real-time assays were used to characterize Sec12p mutants. Four residues conserved from yeast to human were tested (P13A, N40A, H308A, and T313A). N40 appears to be a critical residue, as the N40A mutation has ~2% of wildtype activity. N40 resides in a predicted loop region of the Sec12p β-propeller domain. Further analysis of the nucleotide exchange mechanism awaits a determination of Sec12 structure.

Electron Microscopic Analysis

The morphology of liposomes with coat proteins and Sec12ΔCp was analyzed. Liposome mixtures were processed for thin-section EM as described (Matsuoka et al., 1998). Coating and budding on liposomes were observed for the Sec12-stabilized COPII coat.

COPII Vesicle Budding Assay

COPII vesicle budding from liposome membranes may be assessed by thin section electron microscopy and by sucrose density gradient centrifugation (Matsuoka et al., 1998). COPII vesicles generated with GMPPNP are separated from partly coated liposomes or uncoated liposomes on sucrose gradients (2.2 M–0.2 M sucrose). Synthetic COPII vesicles formed from proteoliposomes are enriched for cargo proteins (Bet1p or Bos1p), but Sec12TMp and the ER resident protein GSTUfe1p are not concentrated into COPII vesicles, suggesting proper cargo selection occurred. Unfortunately, synthetic COPII vesicles formed in incubations of liposomes with COPII proteins, GTP, Sar1p, Sec12TMp, and with or without cargo proteins fail to stably retain coat material during the course of sucrose gradient centrifugation. For this purpose, thin section electron microscopy remains the only technique suitable for quantitative and qualitative analysis of synthetic vesicle budding reactions.

Acknowledgements

We thank Crystal Chan and Robert Lesch for COPII proteins and Matthew Welsh and David G. Drubin for sharing their equipment. We thank Chris Fromme for improving the manuscript and Bruno Antonny for advice on kinetic analysis of real-time assays. We thank members of the Schekman lab for discussion and encouragement. This work was supported by the HHMI (R.S.) and postdoctoral research fellowships from JSPS and ACS (E.F.).

References

Antonny, B., Madden, D., Hamamoto, S., Orci, L., and Schekman, R. (2001). Dynamics of the COPII coat with GTP and stable analogues. *Nat. Cell Biol.* **3**, 531–537.

Barlowe, C., and Schekman, R. (1993). *SEC12* encodes a guanine nucleotide exchange factor essential for transport vesicle budding from ER. *Nature* **365**, 347–349.

Barlowe, C., Orci, L., Yeung, T., Hosobuchi, M., Hamamoto, S., Salama, N., Rexach, M. F., Ravazzola, M., Amherdt, M., and Schekman, R. (1994). COPII: A membrane coat formed by Sec proteins that drive vesicle budding from the endoplasmic reticulum. *Cell* **77**, 895–907.

Beraud-Dufour, S., Robineau, S., Chardin, P., Paris, S., Chabre, M., Cherfils, J., and Antonny, B. (1998). A glutamic finger in the guanine nucleotide exchange factor ARNO displaces Mg^{2+} and the beta-phosphate to destabilize GDP on ARF1. *EMBO J.* **17**, 3651–3659.

Bonifacino, J. S., and Glick, B. S. (2004). The mechanism of vesicle budding and fusion. *Cell* **116**, 153–166.

Futai, E., Hamamoto, S., Orci, L., and Schekman, R. (2004). GTP/GDP exchange by Sec12 enables COPII vesicle bud formation on synthetic liposomes. *EMBO J.* **23**, 4146–4155.

Matsuoka, K., Orci, L., Amherdt, M., Bednarek, S. Y., Hamamoto, S., Schekman, R., and Yeung, T. (1998). COPII-coated vesicle formation reconstituted with purified coat proteins and chemically defined liposomes. *Cell* **93**, 263–275.

Matsuoka, K., Schekman, R., Orci, L., and Heuser, J. E. (2001). Surface structure of the COPII-coated vesicle. *Proc. Natl. Acad. Sci. USA* **98**, 13705–13709.

Miller, E. A., Beilharz, T. H., Malkus, P. N., Lee, M. C., Hamamoto, S., Orci, L., and Schekman, R. (2003). Multiple cargo binding sites on the COPII subunit Sec24p ensure capture of diverse membrane proteins into transport vesicles. *Cell* **114**, 497–509.

Mossessova, E., Bickford, L. C., and Goldberg, J. (2003). SNARE selectivity of COPII coat. *Cell* **114**, 483–495.

Nakano, A., and Muramatsu, M. (1989). A novel GTP-binding protein, Sar1p, is involved in transport from the endoplasmic reticulum to the Golgi apparatus. *J. Cell Biol.* **109**, 2677–2691.

Parlati, F., McNew, J. A., Fukuda, R., Miller, R., Sollner, T. H., and Rothman, J. E. (2000). Topological restriction of SNARE dependent membrane fusion. *Nature* **407**, 194–198.

Robbe, K., and Antonny, B. (2003). Liposomes in the study of GDP/GTP cycle of Arf and related small G proteins. *Methods Enzymol.* **372**, 151–166.

Salama, N. R., Chuang, J. S., and Schekman, R. W. (1997). Sec31 encodes an essential component of the COPII coat required for transport vesicle budding from the endoplasmic reticulum. *Mol. Biol. Cell.* **8**, 205–217.

Sharrocks, A. D. (1994). A T7 expression vector for providing N- and C-terminal fusion proteins with glutathione S-transferase. *Gene* **138**, 105–108.

Yoshihisa, T., Barlowe, C., and Schekman, R. (1993). Requirement for a GTPase-activating protein in vesicle budding from the endoplasmic reticulum. *Science* **259**, 1466–1468.

[9] Reconstitution of Cargo-Dependent COPII Coat Assembly on Proteoliposomes

By KEN SATO and AKIHIKO NAKANO

Abstract

The coat protein complex II (COPII) coat is responsible for direct capture of membrane cargo proteins and for the physical deformation of the endoplasmic reticulum (ER) membrane that drives the transport vesicle formation. The use of an *in vitro* reconstitution system comprising purified components is desirable for studies aimed at elucidating the role(s) of individual proteins in a specific biochemical reaction. To investigate the assembly–disassembly of COPII coats in a completely reconstituted reaction, we have developed a synthetic budding reaction involving purified coat proteins and cargo-reconstituted proteoliposomes. We describe here a fluorescence resonance energy transfer (FRET)-based method for monitoring the kinetics of COPII coat complex assembly and disassembly in cargo-reconstituted proteoliposomes. This assay allows comparison of the time course of the coat disassembly from the cargo as monitored by FRET signal with the time course of accompanying Sar1p GTP hydrolysis by tryptophan fluorescence.

Introduction

Intracellular protein and lipid transport between the organelles of the secretory pathway is mediated by transport vesicles, which bud from a donor organelle and fuse with an appropriate acceptor membrane of a different compartment (Bonifacino and Glick, 2004). The formation of transport vesicles and sorting of cargo into the emerging vesicles are mediated by coat proteins associated with the cytoplasmic face of the organelles (Bonifacino and Lippincott-Schwartz, 2003; Kirchhausen, 2000). A common feature of vesicular carriers is that most of them employ small GTPases to regulate coat assembly at the donor membranes. The COPII generates vesicles that mediate protein transport from the ER toward the Golgi complex (Aridor and Balch, 1996; Schekman and Orci, 1996). The COPII coat consists of the small GTPase Sar1p (Nakano and Muramatsu, 1989) and heterodimeric protein complexes Sec23/24p and Sec13/31p (Barlowe *et al.*, 1994) that sequentially bind to the ER membrane.

METHODS IN ENZYMOLOGY, VOL. 404
0076-6879/05 $35.00
DOI: 10.1016/S0076-6879(05)04009-7

Assembly/disassembly of the COPII coat on the ER membrane is coupled to the Sar1p GTPase cycle. Assembly of the COPII coat on the ER membrane is initiated by the exchange of GDP for GTP on Sar1p, catalyzed by the transmembrane guanine nucleotide exchange factor (GEF) Sec12p (Barlowe and Schekman, 1993). Sec12p is strictly regulated to localize to the ER, and thereby COPII assembly is restricted to the ER (Nishikawa and Nakano, 1993; Sato et al., 1996). Activated Sar1p-GTP then recruits the Sec23/24p complex by binding to Sec23p (Bi et al., 2002), and the cytoplasmically exposed signal of the transmembrane cargo is captured by direct contact with Sec24p (Miller et al., 2003; Mossessova et al., 2003) to form a so-called "prebudding complex" (Kuehn et al., 1998). Formation of the prebudding complex is a cargo recognition step prior to coat polymerization, followed by clustering of the prebudding complexes by the Sec13/31p complex to form COPII vesicles (Antonny and Schekman, 2001). Conversely, COPII coat disassembly requires the catalysis of GTP hydrolysis in Sar1p by a specific GTPase-activating protein (GAP), Sec23/24p, which reverses the assembly process (Antonny et al., 2001).

Membrane targeting is catalyzed by specific pairing of v-SNARE on the vesicle membrane with target membrane t-SNARE (Jahn et al., 2003). In *Saccharomyces cerevisiae*, v-SNARE Bet1p and the cognate t-SNARE consisting of Sed5p, Bos1p, and Sec22p are required for transport from the ER to the Golgi complex and these proteins are selectively packaged into COPII vesicles (Cao and Barlowe, 2000; Hardwick and Pelham, 1992; Newman et al., 1990; Parlati et al., 2000). We chose these SNAREs as model cargos for this study, and FRET donor–acceptor pairs CFP and YFP were fused to the N-termini of SNARE and Sec24p, respectively. FRET between liposome-reconstituted CFP-SNARE and YFP-Sec24/23p in the presence of activated Sar1p-GTP was monitored in real time, allowing observation of a single round of prebudding complex formation and dissociation. The presence of GDP/GTP exchange factor Sec12p in proteoliposomes allows the observation of multiple rounds of Sec23/24p-mediated Sar1p GTP hydrolysis.

Materials

Chloroform solutions of dioleoylphosphatidylcholine (DOPC), dioleoyl-phosphatidylethanolamine (DOPE), dioleoylphosphatidylserine (DOPS), dioleoylphosphatidic acid (DOPA), and phosphatidylinositol (PI) from soybean are purchased from Avanti Polar Lipids (Alabaster, AL). Phosphatidylinositol 4-phosphate (PI4P), phosphatidylinositol 4,5-biphosphate (PIP$_2$), and cytidine 5′-diphosphate-diacylglycerol (CDP-DAG) are obtained from Sigma (St. Louis, MO). Cholesterol is purchased from Nacalai

Tesque (Kyoto, Japan). PI4P, PIP$_2$, and CDP-DAG are dissolved in water-saturated chloroform; cholesterol is dissolved in chloroform. Lipids solutions are stored at $-20°$ under nitrogen in glass tubes with Teflon-lined screw caps. N-octyl-β-D-glucopyranoside (OG) is obtained from Dojindo Laboratories (Kumamoto, Japan). Streptactin-Sepharose and desthiobiotin are obtained from IBA (Goettingen, Germany). All other chemicals are of the highest quality available commercially. The N-terminal His$_6$-tagged version of YFP-Sec24p is co-overexpressed with Sec23p in *S. cerevisiae* KUY211 (*MATα∆pep4::ADE2 ura3 lys2 ade2 trp1 his3 leu2*) and purified as a complex by Ni-NTA affinity chromatography followed by DEAE chromatography as described previously for wild-type Sec23/24p (Barlowe *et al.*, 1994). Sar1p is prepared as previously reported (Saito *et al.*, 1998).

Fluorescence Resonance Energy Transfer Assay Design

FRET is a phenomenon whereby a donor fluorophore transfers energy to an acceptor fluorophore. Efficient energy transfer depends on the quantum yield of donor fluorophore, orientation of the acceptor adsorption and donor emission dipoles, and physical distance between fluorescence donor and acceptor. Energy transfer requires that the emission spectrum of the donor fluorophore has significant overlap with the excitation spectrum of the acceptor fluorophore. The GFP variants cyan fluorescent protein (CFP) and yellow fluorescent protein (YFP) have been optimized for FRET pair (Miyawaki and Tsien, 2000). We have previously shown that CFP/YFP tagging at the N-termini of SNARE and Sec24p preserves the ability of the CFP-SNARE to be packaged into COPII-coated vesicles with YFP-Sec24/23p as verified by *in vitro* COPII budding assay (Sato and Nakano, 2005). To study the dynamics of key players involved in COPII coat assembly and disassembly, including Sar1p, Sec23/24p and transmembrane cargo, SNARE, purified CFP-SNARE is reconstituted into liposomes with a lipid composition similar to that of yeast ER membranes. In theory, when CFP-SNAREs and YFP-Sec24/23ps are in the prebudding complex that responds to activated Sar1p-GTP, excitation of CFP should lead to fluorescence emission from YFP through resonance energy transfer, as outlined in Fig. 1.

Construction of Expression Vectors

The *E. coli* expression vector pASK-IBA5 (IBA) with a tetA promoter–operator is used to express CFP-Bet1p and CFP-Sec22p with Strep-tagII (WSHPQFEK). Expression of the gene of interest occurs upon addition of a low concentration of anhydrotetracycline to the culture medium. The open reading frame of Bet1p is generated by PCR from *S. cerevisiae* genomic DNA and inserted into the *Bam*HI site of pASK-IBA5. The

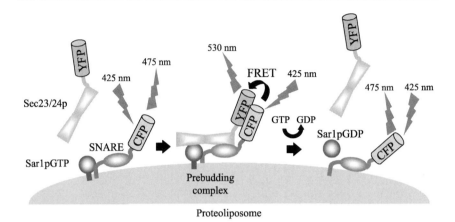

FIG. 1. Schematic of FRET assay for monitoring the kinetics of COPII coat complex assembly and disassembly on SNARE reconstituted proteoliposomes. Sar1p-GTP leads to association of YFP-Sec24/23p with CFP-SNARE on proteoliposomes, with an accompanying increase in FRET between CFP and YFP. Subsequent GTP hydrolysis on Sar1p reverses the assembly process.

coding sequence for CFP with an N-terminal Strep-tagII is amplified by PCR from pECFP (Clonetech, Palo Alto, CA) and inserted into the *Eco*RI-*Bam*HI sites of the above plasmid to generate an expression vector for the production of Strepx2-CFP-Bet1p (pKSE165). The coding sequence of Sec22p with an N-terminal Strep-tagII is generated by PCR from genomic DNA and inserted into the *Eco*RI-*Bam*HI sites of pASK-IBA5. The open reading frame of CFP is amplified by PCR from pECFP and inserted into the *Eco*RI site of the above plasmid to yield an expression vector for production of Strep-CFP-Strep-Sec22p (pKSE168). The Sec12p without the luminal domain (amino acids 1–373) is amplified by PCR from genomic DNA and cloned into the *Eco*RI-*Xho*I sites of pGEX-6P-1 (Amersham Biosciences, Tokyo, Japan) to generate an expression vector for the GST-Sec12Δlum (pKSE176). An expression vector for the production of the cytoplasmic domain of N-terminal His_6-tagged Bos1p (Bos1ΔC, amino acids 1–222) and Sed5p (Sed5ΔC, amino acids 1–316) with an N-terminal His_6 tag is generated from a PCR product coding for the corresponding region and inserted into the *Bam*HI-*Sal*I sites of pQE-30 (Qiagen) and the *Nde*I-*Xho*I sites of pET-21a (Novagen, Tokyo, Japan), respectively. *SEC24* and its flanking regions are amplified by PCR from genomic DNA and inserted into the *Not*I-*Sma*I sites of multicopy vector pYO326 (Qadota *et al.*, 1992). A *Sph*I site is created just before the start codon in the above plasmid and the PCR fragment generated from pEYFP (Clonetech) encoding YFP with an N-terminal His_6 tag is inserted into this *Sph*I site to yield the yeast expression construct containing the N-terminal His$_6$-tagged version of YFP-Sec24p (pKSY193).

Protein Expression and Purification

Purification of CFP-Bet1p and CFP-Sec22p

CFP-Bet1p and CFP-Sec22p are expressed in *E. coli* Rosetta(DE3) (Novagen) cells containing the respective plasmids. Eight 3-liter cultures are incubated in LB medium containing 50 μg/ml ampicillin and 10 μg/ml chloramphenicol at 30° to an OD$_{600}$ of ~0.6 with vigorous shaking. Cells are induced with a final concentration of 200 μg/liter (anhydrotetracycline), and incubation is continued for an additional 4~5 hr. All subsequent steps are performed at 4°. Cells are harvested by centrifugation at 5000 rpm for 5 min, washed once with buffer A (20 mM HEPES-KOH, pH 8.0, 160 mM potassium acetate), pelleted, and resuspended in 120~150 ml of buffer A containing 5 mM EDTA and a protease inhibitor cocktail (Roche Complete, EDTA-free, Tokyo, Japan). The cell suspension is evenly dividedinto four 50-ml tubes and sonicated in ice water at a setting of 5 for twenty 20-sec pulses with a 1-min rest in between each pulse (Astrason Model XL2020, Farmingdale, NY). The lysate is centrifuged at 5000g for 10 min and the resulting supernatant is centrifuged at 42,000 rpm for 2 hr (Hitachi Ultracentrifuge with a P45AT rotor, Tokyo, Japan). The resultant pellet, representing the total membrane fraction of the cell lysate, is frozen in liquid N$_2$ and stored at −80°. Frozen membranes are resuspended in 50 ml cold buffer A containing 1 mM EDTA and a protease inhibitor cocktail, and then OG is added to a final concentration of 2.5% (w/v). Avidin (2 mg) is added to the sample to block any biotin-containing proteins and the sample is incubated at 4° for 1 hr. After removal of insoluble material by centrifugation at 11,000g for 10 min, the supernatant is combined with 2 ml of Streptactin-Sepharose and rotated for 1 hr at 4°. The loaded resin is transferred to a column and washed with buffer A containing 1.25% (w/v) OG. We routinely find that the 70-kDa band (probably *E. coli* hsp70 chaperone DnaK) copurified with the CFP-Bet1p. This protein can be removed by incubating the loaded resin with 2 mM ATP and 10 mM MgCl$_2$ in wash buffer at 30° for 15 min. Strep-tagged protein is eluted with 1~2 ml of buffer B (20 mM HEPES-KOH, pH6.8, 160 mM KOAc) containing 1.25% (w/v) OG and 5 mM desthiobiotin.

Purification of Sec12Δlum

Sec12Δlum is purified from *E. coli* JM109 cells carrying the plasmid pKSE176. Eight 3-liter cultures are incubated in LB medium containing 50 μg/ml ampicillin at 30° with vigorous shaking to an OD$_{600}$ of ~0.6.

Expression is induced by addition of IPTG to a final concentration of 1 mM, and incubation is continued for 4~5 hr. Cells are harvested by centrifugation at 5000 rpm for 5 min, and the total membrane fraction is solubilized as described above. The solubilized membrane is incubated with 2 ml of glutathione-Sepharose (Amersham Biosciences, Tokyo, Japan) at 4° for 1 hr. The beads are washed with buffer A containing 1.25% (w/v) OG, and then GST-Sec12Δlum is eluted with 2.5 ml of the same buffer containing 10 mM glutathione. The resulting GST-Sec12Δlum is incubated with PreScission protease (30 U/ml) (Amersham Bioscience) at 4° for 3.5 hr and then desalted using a PD-10 column (Amersham Bioscience) equilibrated with buffer B containing 1.25% (w/v) OG. The sample (3.5 ml) is incubated with 1 ml of glutathione-Sepharose at 4° for 1 hr, and purified Sec12Δlum, cleaved from the GST moiety is collected from the flow-through of the glutathione-Sepharose column.

Purification of Bos1ΔC and Sed5ΔC

Bos1ΔC and Sed5ΔC are purified from *E. coli* JM109 and BL21(DE3) cells, respectively. Cells carrying the expression plasmid are cultured in four 3-liter LB medium containing 50 μg/ml ampicillin at 30° with vigorous shaking to an OD_{600} of ~0.6. IPTG then is added to the culture at a final concentration of 1 mM and the cells are induced for 4~5 hr. Cells are harvested by centrifugation at 5000 rpm for 5 min, and washed once with buffer A (20 mM HEPES-KOH, pH 8.0, 160 mM potassium acetate). The cell pellet is resuspended in 60~75 ml of buffer C (20 mM HEPES-KOH, pH8.0, 0.5 M potassium acetate) containing 10 mM imidazole, 0.2% (w/v) Triton X-100, and a protease inhibitor cocktail, and is stored for later use at −80°. Cells are disrupted by sonication as above, and the lysate is cleared by centrifugation at 11,000g for 10 min. The supernatant is incubated with 2 ml of Ni-NTA Sepharose (Qiagen, Tokyo, Japan) at 4° for 1 hr. The beads are washed with buffer C containing 40 mM imidazole and 1.25% (w/v) OG, and then His$_6$-tagged protein is eluted with 2.5 ml of buffer B containing 400 mM imidazole and 1.25% (w/v) OG, followed by desalting with a PD-10 column equilibrated with buffer B containing 1.25% (w/v) OG.

Comment

Frequently, some of the SNAREs and Sec12Δlum precipitate out of solution during the elution step, appearing as cloudy suspensions. If this occurs, the eluate can be centrifuged at 15,000g to remove the insoluble material.

Preparation of Proteoliposomes

A total of 0.24 mg cholesterol and 0.99 mg of lipids in chloroform in the ratio of 51 mol% DOPC, 23 mol% DOPE, 8 mol% PI, 8 mol% DOPS, 5 mol% DOPA, 2.2 mol% PI4P, 0.8 mol% PIP$_2$, and 2 mol% CDP-DAG are dried down in a 13 ×100-mm glass tube using a gentle stream of nitrogen at room temperature to yield a lipid film. The lipids are dissolved in buffer B containing 1.25% (w/v) OG by bath sonication, and then 0.1~0.3 mg of CFP-Bet1p or CFP-Sec22p is added (total 1 ml). Equimolar amounts of Sec12Δlum are included in some cases. To reconstitute the t-SNARE complex (CFP-Sec22p/Bos1ΔC/Sed5ΔC), Bos1ΔC and Sed5ΔC are added in 6-fold molar excess over the CFP-Sec22p at this point. After 20–30 min incubation at room temperature, the solution is dialyzed against 2 liters of buffer B containing 2 g of Bio-Beads SM-2 (Bio-Rad, Tokyo, Japan) at room temperature for 3 hr and then overnight at 4°. Proteoliposomes are recovered by flotation in a Nycodenz (Sigma, St. Louis, MO) step gradient. A volume of 1 ml dialysate is mixed with the same amount of 80% (w/v) Nycodenz in buffer B and placed in a 13 × 52-mm centrifuge tube (Hitachi). The sample is overlaid with 2 ml of 30% (w/v) Nycodenz in buffer B followed by 1 ml of buffer B, and then centrifuged in an RPS65T rotor (Hitachi) at 60,000 rpm for 1 hr at 4°. The proteoliposomes are collected from the 0/30% Nycodenz interface (~1 ml) and diluted to 4 ml with buffer B. Proteoliposomes are recovered and concentrated by centrifugation at 70,000 rpm for 20 min (Hitachi Ultracentrifuge with a P100AT2 rotor), then resuspended in buffer B and stored at –80°. Eighty to 95% of the CFP-SNAREs and Sec12Δlum are reconstituted with their cytoplasmic domains on the outside of the proteoliposomes, as determined by the accessibility to digestion by proteases. Sar1p requires phospholipids to exchange GDP for GTP, which is easily detected by the amplitude of the tryptophan fluorescence change of Sar1p (Antonny et al., 2001). Lipid concentration of the resulting proteoliposomes is estimated by comparing the amplitude of the GMP-PNP-induced fluorescence change of Sar1p in the presence of known concentrations of liposomes to its value with proteoliposomes. Size distribution profiles of the resulting proteoliposomes were measured by dynamic light scattering using LB-500 (Horiba) or transmission electron microscopy as described previously (Matsuoka and Schekman, 2000; Sato and Nakano, 2004). The mean diameter of proteoliposomes prepared by this method is about 200 nm. In certain preparations, the resulting proteoliposomes possess a mean diameter less than 100 nm. Ca^{2+}-induced fusion is performed to increase the diameter of these proteoliposomes (Miller and Racker, 1976). Fusion is induced by adding (final concentration) 20 mM CaCl$_2$ at room temperature and allowing the solution to stand for 30 min. The fused vesicles are collected by flotation then followed by ultracentrifugation as above to recover proteoliposomes. These Ca^{2+}-fused proteoliposomes have a diameter of 200 to 300 nm.

Comment

The purified CFP-Bet1p contains degradation fragments. Most of these impurities are without transmembrane domains and do not incorporate into the liposomes; they can be reproducibly removed during reconstitution. The CFP-Bet1p protein after membrane reconstitution shows a single band in SDS-PAGE (Fig. 2).

Real-time Measurement of Prebudding Complex Formation and Dissociation Monitored by FRET

Here, we describe an example of real-time assay based on the measurement of FRET signal to monitor the interaction between YFP-Sec24/23p and liposome-reconstituted CFP-SNARE. We have optimized the donor excitation wavelength (425 nm) such that direct excitation of YFP is minimized, because the wavelength typically used to excite CFP (433 nm) results in significant direct excitation of YFP. Fluorescence (excitation wavelength 425 nm, excitation slit width 2.5 nm; emission wavelength 530 nm, emission slit width 20 nm) is measured in a small quartz cuvette (5 mm path length) with a Hitachi fluorescence spectrophotometer (F-2500) equipped with a thermostatically controlled cell holder and magnetic stirrer bar (3 × 3-mm). The reaction temperature is 25°. Injection of

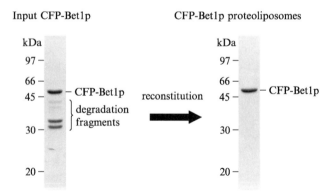

FIG. 2. Reconstitution of proteoliposomes from purified CFP-Bet1p. Purified CFP-Bet1p (left) and reconstituted proteoliposomes were analyzed by SDS-PAGE and stained with Coomassie Blue.

YFP-Sec24/23p is performed with a Hamilton syringe through a guide in the cover of the fluorescence spectrophotometer.

The cuvette initially contained a suspension of proteoliposomes in buffer B containing 1 mM MgCl$_2$ (sample volume 200 μl). Final SNARE concentrations ranged from 60 to 75 nM at a lipid concentration of 40 μg/ml. Sar1p (830 nM) and guanine nucleotide (0.1 mM) are added to the mixture followed by at least a 5 min incubation to allow nucleotide exchange. YFP-Sec24/23p (160 nM) is added from a concentrated stock solution and the FRET signal between CFP-SNARE and YFP-Sec24/23p at 530 nm is continuously monitored. The sampling rate is one measurement per second.

Figure 3A shows the time course of YFP-Sec24/23p-CFP-Bet1p complex assembly and disassembly as measured by FRET in the presence of GDP, GMP-PNP, or GTP. The large increase induced by the addition of YFP-Sec24/23p with Sar1p-GDP is due to direct excitation of YFP and nonspecific interaction between CFP-Bet1p and YFP-Sec24/23p. The addition of YFP-Sec24/23p preloaded with Sar1p-GMP-PNP induces an increase in the FRET signal compared with that of basal FRET with Sar1p-GDP. The observed FRET decrease in the presence of Sar1p-GTP monitors YFP-Sec24/23p release from CFP-Bet1p, which then declines to reach the level of Sar1p-GDP.

If the results of FRET experiments indicate that COPII components and SNARE can associate, it is important to show that the FRET signal is the result of a specific interaction. This can be demonstrated by showing that FRET does not occur between YFP-Sec24/23p and CFP-tagged unrelated membrane protein. We have performed FRET experiments using the ER resident protein Ufe1p. Proteoliposomes reconstituted with CFP-tagged form of MBP-Ufe1p (Sato and Nakano, 2005) fails to produce an increase in FRET in the presence of YFP-Sec24/23p and Sar1p-GTP (or GMP-PNP) (Fig. 3B).

Real-time Measurement of Sar1p GTP Hydrolysis Monitored by Tryptophan Fluorescence

To monitor the GTPase activity of Sar1p during prebudding complex dissociation, we take advantage of the large tryptophan fluorescence change that distinguishes Sar1p-GTP from Sar1p-GDP (Antonny et al., 2001). Sar1p has a tryptophan residue in a switch region (Trp 84), which acts as an intrinsic fluorescence probe of the conformation of Sar1p. The Sar1p-GDP shows a higher intrinsic fluorescence than the Sar1p-GTP, thereby GDP to GTP exchange and GTP hydrolysis can be monitored by tryptophan fluorescence. Tryptophan fluorescence is measured by the same

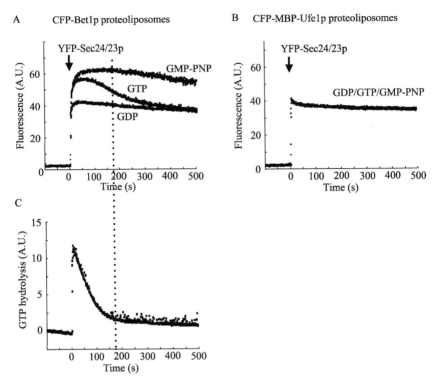

FIG. 3. Assembly/disassembly kinetics of COPII coat complex as detected by FRET and tryptophan fluorescence assay of GAP-catalyzed GTP hydrolysis on Sar1p with proteoliposomes. The reaction initially contained proteoliposomes (40 μg lipids/ml) reconstituted with CFP-Bet1p (60 nM) (A) or CFP-MBP-Ufe1p (65 nM) (B) and Sar1p (830 nM) loaded with 0.1 mM of GDP, GTP or GMP-PNP. After 10-min incubation, YFP-Sec24/23p (160 nM) was added and the FRET signal at 530 nm was continuously monitored at 25°. (C) Tryptophan fluorescence assay of Sar1p GTP hydrolysis with CFP-Bet1p reconstituted proteoliposomes. Tryptophan fluorescence was measured by the same procedure as (A), except that fluorescence was recorded at 340 nm upon excitation at 298 nm.

procedure as described above, except that the fluorescence is recorded at 340 nm upon excitation at 298 nm.

Figure 3C shows the time course of Sar1p GTP hydrolysis under the same condition as above FRET measurement. The fluorescence increase after the addition of YFP-Sec24/23p is due to tryptophan fluorescence from YFP-Sec24/23p. On YFP-Sec24/23p addition, a decrease in the tryptophan fluorescence of Sar1p is observed that reflects Sec23/24p-GAP catalyzed GTP hydrolysis on Sar1p. Notably, the time course of Sar1p GTP hydrolysis is faster than that observed for the FRET decline, suggesting that the complex of Sec23/24p and Bet1p is stable on membranes for a short period after hydrolysis of Sar1p-GTP.

Concluding Remarks

We have described protocols for the kinetic measurement of the interaction between COPII coat subunit and liposome-reconstituted cargo using FRET. This assay is effective in comparing the behavior of both Sar1p GTP hydrolysis and coat assembly–disassembly. There remains the potential for this assay to be modified and utilized *in vitro* to detect COPII coat assembly and its modulation by cargo. Our FRET-based assay is likely to provide a facile means of dissecting cargo-capture and coat disassembly events during the transport vesicle formation, thereby providing novel tools to examine other coat systems on membranes.

Acknowledgments

This work was supported by PRESTO, Japan Science and Technology Agency, by Grants-in-Aid from the Ministry of Education, Science, Sports and Culture of Japan, and by a fund from the Bioarchitect Research Project of RIKEN.

References

Antonny, B., Madden, D., Hamamoto, S., Orci, L., and Schekman, R. (2001). Dynamics of the COPII coat with GTP and stable analogues. *Nat. Cell. Biol.* **3,** 531–537.

Antonny, B., and Schekman, R. (2001). ER export: Public transportation by the COPII coach. *Curr. Opin. Cell. Biol.* **13,** 438–443.

Aridor, M., and Balch, W. E. (1996). Principles of selective transport: Coat complexes hold the key. *Trends Cell. Biol.* **6,** 315–320.

Barlowe, C., Orci, L., Yeung, T., Hosobuchi, M., Hamamoto, S., Salama, N., Rexach, M. F., Ravazzola, M., Amherdt, M., and Schekman, R. (1994). COPII: A membrane coat formed by Sec proteins that drive vesicle budding from the endoplasmic reticulum. *Cell* **77,** 895–907.

Barlowe, C., and Schekman, R. (1993). *SEC12* encodes a guanine-nucleotide-exchange factor essential for transport vesicle budding from the ER. *Nature* **365,** 347–349.

Bi, X., Corpina, R. A., and Goldberg, J. (2002). Structure of the Sec23/24-Sar1 pre-budding complex of the COPII vesicle coat. *Nature* **419,** 271–277.

Bonifacino, J. S., and Glick, B. S. (2004). The mechanisms of vesicle budding and fusion. *Cell* **116,** 153–166.

Bonifacino, J. S., and Lippincott-Schwartz, J. (2003). Coat proteins: Shaping membrane transport. *Nat. Rev. Mol. Cell. Biol.* **4,** 409–414.

Cao, X., and Barlowe, C. (2000). Asymmetric requirements for a Rab GTPase and SNARE proteins in fusion of COPII vesicles with acceptor membranes. *J. Cell. Biol.* **149,** 55–66.

Hardwick, K. G., and Pelham, H. R. (1992). SED5 encodes a 39-kD integral membrane protein required for vesicular transport between the ER and the Golgi complex. *J. Cell. Biol.* **119,** 513–521.

Jahn, R., Lang, T., and Sudhof, T. C. (2003). Membrane fusion. *Cell* **112,** 519–533.

Kirchhausen, T. (2000). Three ways to make a vesicle. *Nat. Rev. Mol. Cell. Biol.* **1,** 187–198.

Kuehn, M. J., Herrmann, J. M., and Schekman, R. (1998). COPII-cargo interactions direct protein sorting into ER-derived transport vesicles. *Nature* **391,** 187–190.

Matsuoka, K., and Schekman, R. (2000). The use of liposomes to study COPII- and COPI-coated vesicle formation and membrane protein sorting. *Methods* **20,** 417–428.

Miller, C., and Racker, E. (1976). Fusion of phospholipid vesicles reconstituted with cytochrome c oxidase and mitochondrial hydrophobic protein. *J. Membr. Biol.* **26,** 319–333.

Miller, E. A., Beilharz, T. H., Malkus, P. N., Lee, M. C., Hamamoto, S., Orci, L., and Schekman, R. (2003). Multiple cargo binding sites on the COPII subunit Sec24p ensure capture of diverse membrane proteins into transport vesicles. *Cell* **114,** 497–509.

Miyawaki, A., and Tsien, R. Y. (2000). Monitoring protein conformations and interactions by fluorescence resonance energy transfer between mutants of green fluorescent protein. *Methods Enzymol.* **327,** 472–500.

Mossessova, E., Bickford, L. C., and Goldberg, J. (2003). SNARE selectivity of the COPII coat. *Cell* **114,** 483–495.

Nakano, A., and Muramatsu, M. (1989). A novel GTP-binding protein, Sar1p, is involved in transport from the endoplasmic reticulum to the Golgi apparatus. *J. Cell. Biol.* **109,** 2677–2691.

Newman, A. P., Shim, J., and Ferro-Novick, S. (1990). BET1, BOS1, and SEC22 are members of a group of interacting yeast genes required for transport from the endoplasmic reticulum to the Golgi complex. *Mol. Cell. Biol.* **10,** 3405–3414.

Nishikawa, S., and Nakano, A. (1993). Identification of a gene required for membrane protein retention in the early secretory pathway. *Proc. Natl. Acad. Sci. USA* **90,** 8179–8183.

Parlati, F., McNew, J. A., Fukuda, R., Miller, R., Sollner, T. H., and Rothman, J. E. (2000). Topological restriction of SNARE-dependent membrane fusion. *Nature* **407,** 194–198.

Qadota, H., Ishii, I., Fujiyama, A., Ohya, Y., and Anraku, Y. (1992). *RHO* gene products, putative small GTP-binding proteins, are important for activation of the *CAL1/CDC43* gene product, a protein geranylgeranyltransferase in *Saccharomyces cerevisiae*. *Yeast* **8,** 735–741.

Saito, Y., Kimura, K., Oka, T., and Nakano, A. (1998). Activities of mutant Sar1 proteins in guanine nucleotide binding, GTP hydrolysis, and cell-free transport from the endoplasmic reticulum to the Golgi apparatus. *J. Biochem. (Tokyo)* **124,** 816–823.

Sato, K., and Nakano, A. (2004). Reconstitution of coat protein complex II (COPII) vesicle formation from cargo-reconstituted proteoliposomes reveals the potential role of GTP hydrolysis by Sar1p in protein sorting. *J. Biol. Chem.* **279,** 1330–1335.

Sato, K., and Nakano, A. (2005). Dissection of COPII subunit-cargo assembly and disassembly kinetics during Sar1p-GTP hydrolysis. *Nat. Struct. Mol. Biol.* **12,** 167–174.

Sato, M., Sato, K., and Nakano, A. (1996). Endoplasmic reticulum localization of Sec12p is achieved by two mechanisms: Rer1p-dependent retrieval that requires the transmembrane domain and Rer1p-independent retention that involves the cytoplasmic domain. *J. Cell. Biol.* **134,** 279–293.

Schekman, R., and Orci, L. (1996). Coat proteins and vesicle budding. *Science* **271,** 1526–1533.

[10]   Real-Time Assays for the Assembly–Disassembly Cycle of COP Coats on Liposomes of Defined Size

By Joëlle Bigay and Bruno Antonny

Abstract

The assembly–disassembly cycle of COPI and COPII coats is controlled by the GTPase cycle of the small G proteins Arf1 and Sar. We describe here two spectroscopic assays that enable real-time studies of some elementary steps of coat assembly and disassembly on artificial liposomes of defined composition and curvature. A flotation assay to assess the effect of membrane curvature on protein adsorption to liposomes is also presented.

Introduction

COPI and COPII vesicles mediate anterograde or retrograde traffic between the endoplasmic reticulum and the Golgi apparatus (Lee *et al.*, 2004). Generation of COP vesicles is a multi-event process that starts with the recruitment of small G-proteins and large coat complexes on the Golgi or the endoplasmic reticulum (ER) membrane. At the membrane surface, coat proteins collect transmembrane proteins and polymerize into a curved lattice. The lattice shapes the underlying membrane into a bud, which by membrane fission leads to the formation of an individual transport vesicle. After vesicle formation, the coat depolymerizes and the COP components are recycled in the cytosol for another round.

Despite the overall complexity of coat assembly and vesicle formation, major advances have been made in the understanding of COP machineries. One breakthrough was the reconstitution of COPI and COPII assembly using purified components and artificial liposomes of defined composition (Bremser *et al.*, 1999; Matsuoka *et al.*, 1998; Spang *et al.*, 1998). In this chapter we describe two spectroscopic assays that complement the biochemical reconstitution and that enable the study of some dynamics aspects of protein coats, notably their assembly–disassembly cycle under the control of the small G-proteins Arf and Sar. In addition a biochemical flotation assay is detailed that permits fair determination of protein binding to liposomes of increasing curvature.

METHODS IN ENZYMOLOGY, VOL. 404   0076-6879/05 $35.00
Copyright 2005, Elsevier Inc. All rights reserved.   DOI: 10.1016/S0076-6879(05)04010-3

Proteins and Nucleotides

The COPI coat is made by the small G-protein Arf1 and the heptameric complex coatomer. The COPII coat consists of the small G-protein Sar1 and two large complexes, Sec23/24 and Sec13/31 (Lee *et al.*, 2004). Detailed protocols for the purification of these proteins can be found elsewhere (Nickel and Wieland, 2001; Shimoni and Schekman, 2002). The spectroscopic methods presented here required relatively pure proteins (>80%) devoid of aggregates. It is a good idea to remove aggregated material from stock protein solutions by ultracentrifugation (100,000g for 15 min) before use. The small G-protein Arf1 and Sar1 should have their N-terminus intact and properly modified. Indeed, Arf and Sar bind in a GTP-dependent manner to lipid membranes through the switch of their N-terminus (a myristoylated amphipathic helix in the case of Arf1 and a longer amphipathic helix in the case of Sar1). This precludes the use of N-tagged constructs or GST fusions. It is important to know the ionic composition of the protein stock solutions in order to evaluate their contributions in the final ionic strength of the sample. It is preferable to use concentrated (>10X) stock solutions so as to minimize dilution effects when the proteins are sequentially added to the liposome suspension during real-time measurements. Last, detergents should be avoided in protein purification to keep the liposomes intact.

Some commercial preparations of GTP contain a significant amount of GDP, which competes with GTP during the nucleotide-loading step. Therefore, GTP and analogues should be of the highest purity available. We use lithium salt solutions of nucleotides from Roche (www.roche-diagnostics. com).

Buffers

Experiments on Arf1 and on the COPI coat are performed in 50 mM Hepes-KOH, pH 7.2, 120 mM Kacetate, 1 mM MgCl$_2$ and 1 mM DTT (buffer A). Experiments on Sar1 and on the COPII coat are performed in 20 mM Hepes-KOH, pH 7.0, 160 mM Kacetate, 1 mM MgCl$_2$, 1 mM DTT (buffer B). These solutions are classical isotonic buffers but some points should be underlined.

Mg^{++} interacts with the nucleotide bound to small G-proteins and thereby determines the rate of spontaneous nucleotide dissociation. By controlling the concentration of free Mg^{++} one can either promote or stop the replacement of the bound nucleotide by the nucleotide added in excess. Thus, once GTP is added to Arf1-GDP mixed with liposomes in buffer A, which contains 1 mM MgCl$_2$, Arf1 remains essentially bound to GDP for minutes. When 2 mM EDTA is added (from a 100 X solution), the concentration of free Mg^{++} drops approximately at 1 μM, and GDP is fully replaced by GTP within 10 minutes. After GDP to GTP exchange, [Mg^{++}] free is raised back at 1 mM by adding 2 mM $MgCl_2$.

Buffer A contains a high concentration of Hepes to minimize pH changes that result from the chelation of Mg^{++} by EDTA. One carboxyl group of EDTA is indeed protonated at neutral pH, resulting in the release of one proton when one Mg^{++} ion is chelated. Sar shows a substantial rate of nucleotide exchange at millimolar Mg^{++}. We generally do not chelate Mg^{++} to promote nucleotide exchange on Sar and thus buffer B contains less Hepes than buffer A.

The nature and the concentration of the monovalent salt used in the assay buffer are important parameters. We use Kacetate as we noticed less liposome recruitment of coatomer in experiments conducted with KCl or NaCl (see following). For the COPII coat, it is important to keep the ionic strength of the sample within 180–240 mM (taking into account the contribution of the COPII components). The Sec23/24 complex aggregates below 160 mM salt.

Liposomes Preparation

Unilamellar liposomes are prepared by the extrusion method (MacDonald *et al.*,1991; Mayer *et al.*, 1986; Mui *et al.*, 2003). In this method a suspension of multilamellar liposomes is forced to pass several times through a polycarbonate filter of defined pore size. The mechanical stress results in membrane fragmentation into smaller and more unilamellar liposomes. The size of the liposomes is well controlled and can be varied from 150 to 30 nm (radius) depending on the pore size of the filter. Thanks to the development of commercial hand extruders (from Avanti polar lipids or from Avestin) this method is adapted to the preparation of small volumes of liposomes. Note that recent volumes of *Methods in Enzymology* (Volumes 367, 372, 373, and 387) are dedicated to liposome preparations and their uses.

Most natural and synthetic lipids are purchased from Avanti Polar Lipids (Alabaster, AL; http://www.avantilipids.com) as chloroform solutions. The fluorescent lipid NBD-PE (1-Oleoyl-2-[6-[(7-nitro-2–1,3-benzoxadiazol-4-yl)amino]hexanoyl]-sn-Glycero-3-Phosphoethanolamine) is from molecular probes (Eugene, OR; http://probes.invitrogen.com/).

Phophoinositides are from Echelon (Salt Lake City, UT; http://www.echelon-inc.com/) or from Matreya (State College, PA; http://www.matreya.com). Stock solutions of pure lipids (0.5–20 mM) are stored and aliquoted in 2 ml glass vials under argon at –20°.

Experiments on the COPI coat are performed with liposomes containing in mol %: egg phosphatidylcholine (PC), 50; egg or liver phosphatidylethanolamine (PE), 19; brain phosphatidylserine (PS), 5; liver phosphatidylinositol (PI), 10; cholesterol, 16 and NBD-PE, 0.2. This mixture will be referred here as to "Golgi mix" liposomes. In some experiments, lipids of natural source are replaced by synthetic lipids such as dioleoyl (C18:1–C18:1) or palmitoyl-oleoyl (C16:1–C18:1) lipids. The composition of the liposomes used in experiments with the COPII coat is (mol %) PC, 50; PE, 21; PS, 8; phosphatidic acid, 5; PI, 9; PI(4)P, 2.2; PI (4,5)P_2, 0.8; cytidine-diphosphate-diacylglycerol, 2; NBD-PE, 2. These liposomes also contain ergosterol at 20% w/w. Compositions given here should be considered as reasonable starting mixtures, which can be modified according to the molecular mechanism studied. It is important to note that COPII assembly on liposomes is favored by the presence of anionic lipids, whereas COPI assembly is relatively insensitive to the charge of the membrane surface (Matsuoka et al., 1998; Spang et al., 1998). Also of note is the fact that acyl chain unsaturation has strong impact on some steps of COP assembly or disassembly. Thus, the recruitment of the small G protein Sar1 is strongly favored by lipids with mono-unsaturated acyl chains and an even more dramatic effect is observed on ArfGAP1, a GTPase activating protein for Arf1 (Bigay et al., 2003; Matsuoka et al., 1998).

To prepare liposomes of defined composition, lipids are mixed at the desired molar ratio in a pear-shaped glass flask using glass syringes (Hamilton, www.hamiltoncompany.com). The total amount of lipids ranges from 1 to 5 μmoles and the volume of solvent, mostly chloroform, is 1 ml. For mixtures containing phosphoinosides (up to 5%), a cloudy suspension is observed reflecting the poor solubility of these lipids in chloroform. Adding 50% vol methanol leads to a clear solution. The solvent is removed in a rotary evaporator at 20° to 25° at 500 rpm. A lipid film rapidly forms on the glass surface. After 30 min, the evaporator is filled with argon and the glass flask is removed and placed in a vacuum chamber to eliminate traces of solvent. The lipid film is resuspended at 1.25 to 5 mM in the appropriate liposome buffer (see following) to give a suspension of multilamellar liposomes. The suspension is submitted to 5 cycles of freezing and thawing using liquid nitrogen and a water bath (40°). The suspension can be stored at −20° or immediately extruded.

The buffer used for the preparation of liposomes should be isoosmotic with the assay buffers in which proteins and liposomes are finally mixed. It is preferable to omit divalent salt (MgCl$_2$) in the liposome buffer to prevent long-term fusion. Therefore the liposomes used for experiments with COPI proteins are prepared in 50 mM Hepes-KOH, pH 7.2, 120 mM Kacetate, whereas liposomes used for experiments with COPII proteins are prepared in 20 mM Hepes-KOH, pH 7.0, 160 mM Kacetate.

For extrusion we use the mini-extruder from Avanti Polar Lipids in which a 19 mm polycarbonate filter (Corning [www.corning.com] or Millipore [www.millipore.com]) is sandwiched between two Teflon parts. The liposome suspension is extruded 19 times through the filter using 1 ml or 0.25 ml Hamilton syringes. The extrusion is first performed on a 0.4 μm (pore size) filter. If liposomes with a smaller diameter are required, the extrusion is repeated with filters of decreasing pore size: 0.2, 0.1, 0.05, and 0.03 μm. With this complete sequence of extrusions, one obtains liposomes of well-defined size distribution. The 0.25 ml Hamilton syringes should be used in the last extrusion steps because of the high manual pressure required.

The size distribution of the liposomes is determined by dynamic light scattering (DLS) with a Dynapro apparatus (http://www.wyatt.com). DLS is a hydrodynamic method by which one determines the rate of diffusion of particles through the solvent. The hydrodynamic radius is defined as the radius of a theoretical hard sphere that diffuses with the same speed as the particle under examination. The measurement is performed at 25° and requires about 2 μl of the extruded liposome suspension diluted in 18 μl of liposome buffer (final lipid concentration in the range of 0.1 mM). Ten autocorrelation functions are sequentially measured, from which the size distribution of the liposome is determined using the Dynamics v5 software from Dynapro. A complete measurement takes a few minutes. Figure 1A shows typical size distributions of extruded liposomes as determined by DLS. Figure 1B shows how the actual hydrodynamic radius of the liposomes varies with the pore size of the polycarbonate filter.

After extrusion, the liposomes are stored at room temperature in the dark under argon and can be used within a few days. Small liposomes

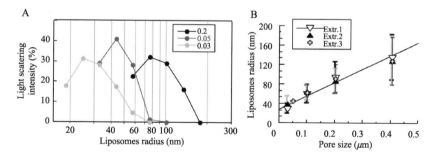

FIG. 1. Determination of the size of extruded liposomes by dynamic light scattering. (A) Size distribution of Golgi-mix liposomes obtained by sequential extrusion through polycarbonate filters of 0.2, 0.05, and 0.03 μm pore size diameter. (B) Mean hydrodynamic radius and polydispersity (error bars) of extruded liposomes as a function of the pore size of the polycarbonate filters. Three experiments of sequential extrusions are shown.

should be used within two days, due to slow lipid flip-flop, which tends to compensate the high membrane curvature.

In some experiments, the liposomes are supplemented with lipopeptides that imitate the cytosolic tail of the transmembrane protein p23. Lipopeptide synthesis has been described elsewhere (Bremser *et al.*, 1999; Nickel and Wieland, 2001). The lipopeptide (1 m*M* in DMSO) is incorporated by a 100-fold dilution in the extruded liposomes suspension (1 m*M*), giving a surface concentration of 2 mol % (considering the outer leaflet of the liposomes).

Tryptophan Fluorescence Measurements of the GDP/GTP Cycle of the Small G Protein Arf and Sar

Small G-proteins contain two regions named switch I and switch II, which undergo a binary conformational change upon nucleotide exchange and GTP hydrolysis (Vetter and Wittinghofer, 2001). The switch II region of Arf and Sar contains a conserved tryptophan residue, which acts as an intrinsic fluorescent probe of the protein conformation. The intrinsic fluorescence of Arf1 and Sar1 increases (by +100% and +200%, respectively) when GDP is replaced by GTP. Tryptophan fluorescence is thus a convenient way to follow the activation–inactivation cycle of these small G-proteins in real-time (Antonny *et al.*, 1997, 2001; Bigay *et al.*, 2003; Futai *et al.*, 2004).

Tryptophan fluorescence measurements are performed in a standard fluorimeter (90° format; e.g., Shimadzu RF5301PC). For accurate kinetics measurements, it is strongly recommended to use a cuvette holder equipped with a magnetic stirrer and connected to a thermostat; 10×10 or 4×10 mm quartz cells adapted to small magnetic bars can be purchased from Hellma (catalogue numbers 109.000F and 109.004F, www.hellma-worldwide.com). For most experiments, we used custom-made cylindrical quartz cuvettes (internal diameter 8 mm) inserted in a 3-window metal holder. This minimizes the sample volume (600 μl) and allows efficient and continuous mixing of the sample with a small magnetic bar (2×7 mm, Hellma). Injections from stock solutions are done with Hamilton syringes (10 to 50 μl) through a guide in the cover of the fluorimeter. The guide is positioned such that the tip of the needle touches the meniscus of the sample without crossing the light beam. Thus the recording is not interrupted by injections and kinetics in the range of a few seconds can be accurately monitored. Alternatively experiments can be performed in small cuvettes (50–100 μl; Hellma, catalogue number 105251QS). In this case, injections and mixing are performed manually with Gilson pipettes. This interrupts the recording for about 10 seconds.

Experiments on mammalian Arf1 and on yeast Sar1 are performed at 37° and 27°, respectively. Tryptophan fluorescence is recorded at 340 nm with a large (10 to 30 nm) bandwidth to enhance the signal to noise ratio. To bypass nucleotide absorption it is recommended to excite the sample at $\lambda > 290$ nm (e.g., 297.5 nm; bandwidth 1 to 5 nm). The fluorescence cuvette initially contains a suspension of liposomes in buffer A or B. The buffer must be freshly filtered and degassed to minimize light-scattering artifacts due to bubbles and dust particles. Arf1-GDP (or Sar-GDP), guanine nucleotides, EDTA, $MgCl_2$, and enzymes that catalyze guanine nucleotide exchange or GTP hydrolysis (GEF and GAP) are added sequentially from concentrated stock solutions using Hamilton syringes. Activation (GDP to GTP exchange) is seen as a fluorescence increase and inactivation (GTP hydrolysis) as a fluorescence decrease. Figure 2 shows a complete cycle of Arf1 activation/inactivation as measured by tryptophan fluorescence. Here Arf1 activation is promoted by the addition of GTP and by lowering temporally the concentration of free Mg^{++} with EDTA. After the activation step, the concentration of free Mg^{++} is restored at the millimolar range by adding an excess of $MgCl_2$. Last, ArfGAP1 (50 nM) is added to promote Arf1 inactivation by GTP hydrolysis. A similar tryptophan fluorescence-based assay can be used to study the catalysis of GDP to GTP exchange by GEFs. In this case, the GEF is simply added after the addition of GTP and Mg^{++} is not chelated by EDTA (Beraud-Dufour *et al.*, 1998; Futai *et al.*, 2004).

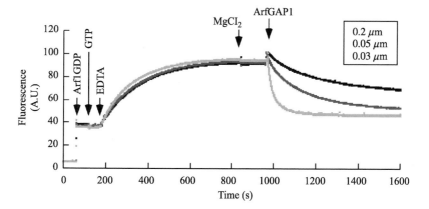

FIG. 2. Tryptophan fluorescence assay of the activation–deactivation cycle of Arf1. Myristoylated Arf1-GDP (0.5 μM) is added to extruded Golgi mix liposomes (0.2 mM lipids) of decreasing size. Arf1 activation is induced by addition of 0.1 mM GTP and by lowering temporally the concentration of free Mg^{++} with EDTA. Arf1 inactivation is initiated by addition of a truncated form of ArfGAP1 (a.a. 1–257, 50 nM).

The fluorescence experiments can be performed over a large range of protein and lipid concentration. Arf and Sar are generally used in the range of 0.2 to 1 μM, but the GDP/GTP fluorescence changes are so large that an acceptable signal–noise ratio should be obtained with much less protein. The liposome concentration is in the range of 100 μM to 1 mM lipid. At lower lipid concentration, the membrane surface can limit the amount of Arf and Sar that translocates upon GDP/GTP exchange. Note that in the absence of liposome, no activation of Arf and Sar is observed since activation and membrane translocation are interdependent events due to the hydrophobicity of the GTP-bound form.

The fluorescence assay is useful to dissect physical and chemical factors of the lipid membrane that influence the Arf1 activation–inactivation cycle. Some GEFs or GAPs harbor PH domains and introducing phosphoinositides at few % in the liposome formulation can have a dramatic effect on the rate of Arf activation or inactivation. As shown in Fig. 2, the tryptophan fluorescence assay can also be used to study the effect of membrane curvature (Bigay *et al.*, 2003). In these experiments, Arf1 has been activated by artificial nucleotide exchange (low Mg^{++} concentration) on small or large liposomes obtained by extrusion.

Inactivation is then initiated by the addition of ArfGAP1. The rate of Arf1 inactivation is much faster on small liposomes than on large liposomes, reflecting the avidity of ArfGAP1 for highly curved membranes.

Light Scattering Measurements of the Assembly–Disassembly Cycle of the COPI and COPII Coat on Liposomes

The scattering of light by a suspension of liposomes depends on several parameters including the size of the liposomes, their shape, and the refractive index of the lipid membrane. When a protein coat assembles at the liposome surface, the refractive index of the membrane increases, which results in an increase in light scattering. This can be measured in a standard fluorimeter by setting the excitation and emission monochromators to the same wavelength. It is important to note that light-scattering by objects having a size comparable to that of the wavelength of light (which is the case here considering the submicrometer size of the liposomes) is a complex process that can result in non-linear signals. However, it turns out that the assembly of COPI and COPII coats on liposomes results in simple and large light-scattering changes, which at first approximation are roughly proportional to the protein mass that is gained or lost by the liposome membrane (Antonny *et al.*, 2001; Bigay *et al.*, 2003; Futai *et al.*, 2004).

Light-scattering experiments are performed with the same instrument and with the same facilities (cuvettes, injection, stirring, and controlled temperature) as for tryptophan fluorescence experiments. The excitation and emission monochromators are set at 350 nm with small (e.g., 1–3 nm) bandwidths. The gain, the sampling rate, and the time response of the fluorimeter should be adjusted to optimize the linearity, the sensitivity, and the time resolution of the measurement. Because any dust or bubble will result in large light scattering artifacts, special care must be taken. The solutions should be freshly degassed and filtered; the magnetic bar in the cuvette should rotate in a very regular manner; the light beam should be quite far from the magnetic bar and the meniscus; the Hamilton syringes should be purged to prevent air bubbles during injections. Alternatively, the experiments can be performed in a small cuvette with no stirring.

Figure 3A shows light-scattering recordings of the assembly–disassembly cycle of the COPI coat on liposomes. The order of additions has been chosen to illustrate the three elementary signals that can be recorded with this technique: Arf1 binding, COP binding, and COP disassembly upon GTP hydrolysis. The small quartz cuvette (sample volume 100 μl) contained extruded (0.4 μm pore size) "Golgi mix" liposomes (0.1 mM) supplemented with 2 mol% p23 lipopeptide. Arf1-GDP (0.5 μM, grey trace or 1 μM, black trace) is added, followed by the addition of GTP (100 μM) and EDTA (2 mM). Upon EDTA addition, a slow light-scattering increase is observed. The amplitude of the signal is proportional to the amount of

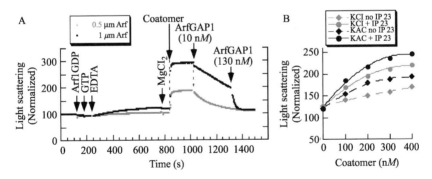

FIG. 3. Light scattering assay of COPI assembly and disassembly on liposomes. An Arf1-GDP (0.5 μM grey trace, 1 μM black trace) was added to extruded Golgi mix liposomes (0.1 mM lipids; pore size: 0.4 mm; lp p23: 2 mol %). Arf1 activation was induced by the addition of 2 mM EDTA in the presence of 0.1 mM GTP. The small increase in light scattering reflects the recruitment of Arf1 upon GDP/GTP exchange. Coatomer (0.15 μM grey trace, 0.3 μM black trace) is then added, followed by the addition of 10 nM or 130 nM ArfGAP1. (B) Titration of the coatomer binding signal with liposomes supplemented with (circles) or without (diamonds) p23 lipopeptide and preloaded with 1 μM Arf and GTP as in panel A. Experiments were performed in a buffer consisting of 50 mM Hepes (pH 7.2), 1 mM MgCl$_2$, 1 mM DTT and with either 120 mM KCl (grey curves) or 120 mM Kacetate (black curves).

Arf1 used. This signal reflects the binding of Arf1 to the liposome surface, which is concomitant to GDP/GTP exchange reaction. (The kinetics of the light-scattering signal is identical to the kinetics of Arf1 activation as measured by tryptophan fluorescence.) After Arf1 activation, addition of purified coatomer (150 or 300 nM) induces a large light-scattering increase, which reflects the instantaneous recruitment of the complex on the liposome surface. This signal depends on the amount of bound Arf-GTP and the amount of added coatomer and thus reflects their stoichiometric assembly at the membrane surface. Figure 3B compares four titrations of the coatomer signal. In all cases, the liposomes were preloaded with 1 μM Arf1 and GTP as in panel A. The experiments were performed in the classical Kacetate buffer (black curves) or in a KCl buffer (grey curves). In addition, the Golgi mix liposomes were either supplemented with the p23 lipopeptide (circles) or not (diamonds). p23 lipopeptide clearly facilitates co-atomer recruitment, but substantial coatomer binding is observed in the absence of the p23 lipopeptide, notably in the Kacetate buffer. We have no explanation for the inhibitory effect of chloride (half-effect in the range of 50 mM). These experiments suggest, however, that the p23–coatomer interaction helps but is not essential for coatomer recruitment, in contrast to the interaction with Arf1-GTP.

The last event that can be recorded by light-scattering is coat disassembly. This is shown in Fig. 3A where, after Arf1 activation and coatomer recruitment, ArfGAP1 is added and induces reversal of the binding signal. This signal can be used to study the effect of many parameters on COPI disassembly. This includes membrane curvature and could be extended to the study of cargo export motifs, which are imitated by synthetic lipopeptides (such as the p23 lipopeptide).

It is important to note that the fluorescence and light scattering assays are non-invasive techniques and thus can be coupled to other techniques such as biochemical fractionation or electron microscopy. The sample size and concentration of reagents are in the same range as those used in other techniques, thus facilitating comparison.

A Flotation Assay for the Binding of ArfGAP1 to Liposomes of Various Diameters

McLaughlin and coauthors have developed a straightforward method to assess the binding of proteins or peptides to liposomes (Buser and McLaughlin, 1998). They use sucrose-loaded liposomes obtained by extrusion through 0.4 μm pore size filter. When such liposomes are mixed with proteins in an aqueous buffer, they can be readily pelleted by ultracentrifugation. By comparing the amount of protein in the pellet and in the supernatant, one can determine the avidity of proteins for liposomes of defined composition. However, this technique has two drawbacks: (1) it is not adapted to large protein complexes, which tend to sedimentate by their own; and (2) it is not adapted to very small liposomes, which show slow sedimentation velocity. Here we describe a flotation assay that is adapted to the use of liposomes of various sizes and that permits studying the effect of membrane curvature on protein binding. This assay lies on sedimentation, but the buoyant force now exceeds the centrifugal force since the liposomes do not contain sucrose and are mixed with proteins in a dense sucrose solution. This protocol is similar to the one described by Matsuoka *et al.* (Matsuoka and Schekman, 2000) but has been improved by the use of a fluorescence imaging system to detect the liposomes.

Proteins (0.5 to 1 mM) are incubated at room temperature with liposomes (0.5 to 1 mM) in buffer A (final volume 150\ ml) in a polycarbonate

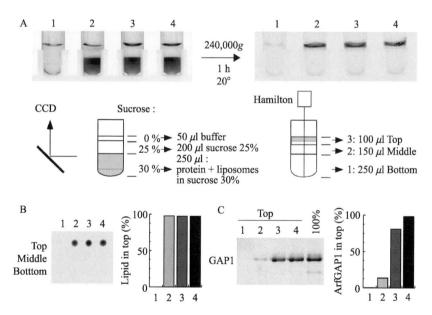

Fig. 4. Flotation assay. (A) ArfGAP1 (0.7 mM) was incubated with fluorescent-labeled liposomes (0.7 mM) extruded through polycarbonate filters of decreasing pore size (tube 2: 0.2 mm, tube 3: 0.05 mm, tube 4: 0.03 mm) or without liposomes (tube 1). The sucrose gradient was generated as described. The fluorescent lipids were visualized before and after centrifugation (1h, 20° at 240,000g in a TLS 55 rotor) with a fluorescent imaging system. Fractions were carefully collected from bottom to top with a Hamilton syringe. (B) Lipid fluorescence in each fraction was quantified from 5 ml spots. (C) Lipid-bound proteins recovered in the top fraction were analyzed by SDS-PAGE and SYPRO-orange staining.

tube adapted to a TLS 55 Beckman rotor (http://beckmancoulter.com). The liposomes contain 0.2% NBD-PE and have been obtained by sequential extrusion as described previously. The protein–lipid suspension is adjusted to 30% w/vol sucrose by adding and mixing 100 ml of 2.2 M sucrose in buffer A and overlaid with two cushions: 200 ml of 0.75 M sucrose and 50 μl of buffer A with no sucrose. A first photograph of the centrifuge tubes is taken using a fluorescence imaging system (see following). The tubes are then centrifuged at 240,000g (55,000 rpm) for one hour in a swing rotor (TLS 55, Beckman). After centrifugation a second photograph of the centrifuge tubes is taken, and three fractions are collected from bottom to top using a Hamilton syringe. The fraction volume is: bottom, 250 ml; middle, 150 ml; top, 100 ml. Five microliters of each fraction are deposited on a black plastic plate for the determination of lipid recovery by NBD fluorescence and 30 μl is used for SDS-PAGE analysis and protein recovery analysis.

The fluorescence imaging system (FUJIFILM LAS-3000, http://home. fujifilm.com) consists of a black box with blue diodes for sample illumination and a vertically positioned CCD camera positioned after a filter-wheel to monitor the emitted fluorescence light. This system can be used to visualize and quantify protein gels stained with SYPRO Orange as well as NBD-labeled liposomes. The images of the centrifuge tubes are taken using a small mirror inclined at 45°, which directs to the camera the side image of the tubes positioned vertically. Both SYPRO Orange and NBD fluorescence are measured using the 515 DI nm filter of the imaging system.

Figure 4 shows a complete analysis of the binding of the protein ArfGAP1 to Golgi mix liposomes of decreasing radius. The position of NBD fluorescence before and after centrifugation shows that whatever their size, the liposomes move completely from the bottom of the tube to the top after centrifugation. As a result, the efficiency of liposome recovery is independent of liposome size and the SDS-PAGE analysis gives a fair estimate of the dramatic effect of membrane curvature on the avidity of ArfGAP1 for lipid membranes.

References

Antonny, B., Beraud-Dufour, S., Chardin, P., and Chabre, M. (1997). N-terminal hydrophobic residues of the G-protein ADP-ribosylation factor-1 insert into membrane phospholipids upon GDP to GTP exchange. *Biochemistry* **36,** 4675–4684.

Antonny, B., Madden, D., Hamamoto, S., Orci, L., and Schekman, R. (2001). Dynamics of the COPII coat with GTP and stable analogues. *Nat. Cell. Biol.* **3,** 531–537.

Beraud-Dufour, S., Robineau, S., Chardin, P., Paris, S., Chabre, M., Cherfils, J., and Antonny, B. (1998). A glutamic finger in the guanine nucleotide exchange factor ARNO-displaces Mg2+ and the beta-phosphate to destabilize GDP on ARF1. *EMBO J.* **17,** 3651–3659.

Bigay, J., Gounon, P., Robineau, S., and Antonny, B. (2003). Lipid packing sensed by ArfGAP1 couples COPI coat disassembly to membrane bilayer curvature. *Nature* **426**, 563–566.

Bremser, M., Nickel, W., Schweikert, M., Ravazzola, M., Amherdt, M., Hughes, C. A., Sollner, T. H., Rothman, J. E., and Wieland, F. T. (1999). Coupling of coat assembly and vesicle budding to packaging of putative cargo receptors. *Cell* **96**, 495–506.

Buser, C. A., and McLaughlin, S. (1998). Ultracentrifugation technique for measuring the binding of peptides and proteins to sucrose-loaded phospholipid vesicles. *Methods Mol. Biol.* **84**, 267–281.

Futai, E., Hamamoto, S., Orci, L., and Schekman, R. (2004). GTP/GDP exchange by Sec12p enables COPII vesicle bud formation on synthetic liposomes. *EMBO J.* **23**, 4146–4155.

Lee, M. C., Miller, E. A., Goldberg, J., Orci, L., and Schekman, R. (2004). Bi-directional protein transport between the ER and Golgi. *Annu. Rev. Cell. Dev. Biol.* **20**, 87–123.

MacDonald, R. C., Mac Donald, R. I., Menco, B. P., Takeshita, K., Subbarao, N. K., and Hu, L. R. (1991). Small-volume extrusion apparatus for preparation of large, unilamellar vesicles. *Biochim. Biophys. Acta* **1061**, 297–303.

Matsuoka, K., Orci, L., Amherdt, M., Bednarek, S. Y., Hamamoto, S., Schekman, R., and Yeung, T. (1998). COPII-coated vesicle formation reconstituted with purified coat proteins and chemically defined liposomes. *Cell* **93**, 263–275.

Matsuoka, K., and Schekman, R. (2000). The use of liposomes to study COPII- and COPI-coated vesicle formation and membrane protein sorting. *Methods* **20**, 417–428.

Mayer, L. D., Hope, M. J., and Cullis, P. R. (1986). Vesicles of variable sizes produced by a rapid extrusion procedure. *Biochim. Biophys. Acta* **858**, 161–168.

Mui, B., Chow, L., and Hope, M. J. (2003). Extrusion technique to generate liposomes of defined size. *Methods Enzymol.* **367**, 3–14.

Nickel, W., and Wieland, F. T. (2001). Receptor-dependent formation of COPI-coated vesicles from chemically defined donor liposomes. *Methods Enzymol.* **329**, 388–404.

Shimoni, Y., and Schekman, R. (2002). Vesicle budding from endoplasmic reticulum. *Methods Enzymol.* **351**, 258–278.

Spang, A., Matsuoka, K., Hamamoto, S., Schekman, R., and Orci, L. (1998). Coatomer, Arf1p, and nucleotide are required to bud coat protein complex I-coated vesicles from large synthetic liposomes. *Proc. Natl. Acad. Sci. USA* **95**, 11199–11204.

Vetter, I. R., and Wittinghofer, A. (2001). The guanine nucleotide-binding switch in three dimensions. *Science* **294**, 1299–1304.

[11] Assay and Measurement of Phospholipase D Activation by Sar1

By Meir Aridor, Kuntala Shome, and Guillermo Romero

Introduction

Dynamic physical changes in the lipid membranes of secretory pathway compartments promote vesicular traffic. The membranous tubular/vesicular carriers that bud from donor compartments mediate the selective packaging and shuttling of cargo and machinery proteins within the secretory pathway while preserving compartment integrity. Carrier formation is driven by the activity of cytosolic coat protein complexes. In the secretory pathway, coat complexes are recruited to the membrane by the activity of cytosolic small GTPases of the ARF/Sar1 family (Kirchhausen, 2000). These GTPases are translocated to the membrane upon activation. The membrane-bound GTPases interact with coat components to direct their recruitment on target membranes. ARF1 mediates the recruitment of COPI and AP-1 coats to cis-Golgi and trans-Golgi network (TGN) respectively, whereas Sar1 proteins recruit the COPII coat to endoplasmic reticulum (ER) exit sites. Re-modeling of the lipid layer at the sites of budding is required to promote the formation of vesicular tubular carriers (Wenk *et al.*, 2004). The formation of phosphatidic acid (PA), phosphatydimositol 4-phosphate (PtdIns4P), phosphatydimositol 4,5 bis phosphate (PtdIns(4,5) P2), and diacylglycerol (DAG) is required to regulate protein export from the TGN (Baron and Malhotra, 2002; Bruns *et al.*, 2002; Chen *et al.*, 1997; Godi *et al.*, 2004; Wang *et al.*, 2003). PtdIns4P and PtdIns(4,5)P_2 regulate the ARF1 dependent assembly of the clathrin adaptor complex AP-1 (Crottet *et al.*, 2002; Wang *et al.*, 2003; Zhu *et al.*, 1999). DAG regulates protein kinase D controlled activities that mediate vesicular/tubular fission at the TGN (Baron *et al.*, 2003; Liljedahl *et al.*, 2001). Fapp1 proteins, which are targeted to the TGN by PtdIns4P are implicated in the regulation of protein transport from the TGN to the plasma membrane (Godi *et al.*, 2004). Accumulating evidence suggests that lipid remodeling regulates ER export. Putative lipid modifying proteins such as DAG kinase delta or p125, a Sec23 binding protein that is homologous to PA specific phospholipase A1 proteins, are involved in the control of exit site assembly and ER export (Nagaya *et al.*, 2002; Shimoi *et al.*, 2005). Pharmacological studies implicated DAG in regulation of ER export (Fabbri *et al.*, 1994). The formation of PA and PtdIns4P is required to

METHODS IN ENZYMOLOGY, VOL. 404
Copyright 2005, Elsevier Inc. All rights reserved.

0076-6879/05 $35.00
DOI: 10.1016/S0076-6879(05)04011-5

support COPII mediated ER export (Blumental-Perry *et al.*, 2005; Pathre *et al.*, 2003). Lipid remodeling supports two related functions during vesicle formation: In one, lipid remodeling promotes the physical transformation of the lipid layer to promote or stabilize curved vesicular tubular structures and support vesicle fission. The unique biophysical properties of the formed lipids can lead to changes in bi-layer packing and lipid geometry that support membrane fission. In the other, regulatory and targeting signals are transiently generated on localized surfaces of the membrane to direct recruitment and activation of the vesicle budding machinery. The regulatory function of lipid remodeling is characterized for the cargo selection activity of the COPII coat. The COPII Sec23/Sec24 subunits, which are recruited to the membranes by activation of Sar1, expose specific binding sites to the membrane surface that can interact with defined ER export motifs presented by cargo proteins and their receptors (Aridor *et al.*, 2001; Miller *et al.*, 2003; Mossessova *et al.*, 2003). This binding mediates the selection and incorporation of cargo into COPII vesicles. The affinities of COPII-cargo interactions are relatively low and do not support coat-cargo binding on synthetic liposomes or ER membranes. However, these low affinity interactions can be transformed into high avidity multivalent protein-lipid and protein-protein interactions on the membranes to enable ER export. The avidity buildup is mediated by the provision of low affinity and perhaps low specificity lipid binding sites that support coat-membrane binding to promote cargo recognition. The acidic phospholipids PA, PtdIns4P, and PtdIns(4,5)P_2 support Sar1 dependent COPII assembly and cargo binding on synthetic liposomes. Likewise, PA formation on ER membranes is required to support COPII recruitment and ER export (Matsuoka *et al.*, 1998a,b; Pathre *et al.*, 2003).

The fluid nature of lipid membranes requires that lipid signals will be generated dynamically to provide local support for the budding machinery. Moreover, for lipid re-modeling to be effective, elevated concentration of lipid signals should be generated at the sites of vesicle formation. The ability of small GTPases of the ARF/Sar1 family to regulate lipid remodeling activities provides the required spatial and temporal regulation, and this regulation is essential to support the dynamic and localized nature of coat assembly and disassembly (De Matteis and Godi, 2004; Exton, 1999, 2002; Rizzo and Romero, 2002). At the ER, Sar1 activates phospholipase D (PLD) activity leading to PA formation (Pathre *et al.*, 2003). PLD enzymes catalyze the hydrolysis of phosphatidylcholine (PC) to generate choline and PA. PA promotes membrane curvature and indeed supports the Sar1 induced membrane tubulation observed at ER exit sites. PA formation also provides acidic binding sites that support COPII assembly and participates as a regulatory component of the lipid re-modeling cascade that promotes vesicle formation (Pathre *et al.*, 2003).

PA is a minor component of the ER membrane that accounts for less than 1% of total ER membrane lipids (Allan, 1996). Formed PA is rapidly consumed by the activity of phosphatidate phosphohydrolase (PAP). In order to measure the formation of PA, the dynamics of PA formation and consumption has to be controlled. This is achieved by exploiting a unique transphosphatidylation reaction that is catalyzed by PLD enzymes. In this reaction, the aliphatic chain of a primary alcohol is transferred to the phosphatidyl moiety of the phosphatidic acid product. In the presence of low concentrations of primary alcohols, PLD enzymes generate phosphatidylalcohols, which are not recognized by PAP and are not efficiently consumed (Morris *et al.*, 1997). Therefore the measurement of transphosphatidylation activity of PLD provides a convenient assay that avoids the otherwise highly dynamic nature of the lipid remodeling cascade induced by Sar1 to support COPII mediated ER export.

Methods

Microsome Membranes Preparation

Microsomes are prepared from normal rat kidney (NRK) cells essentially as previously detailed (Rowe *et al.*, 1996) (with one exception, NRK cells utilized are not infected with tsO45 VSV). Briefly, NRK cells are grown on 150 mm tissue culture dishes in DMEM containing 5% fetal bovine serum and supplemented with antibiotics at 37°. Confluent cells are transferred to ice and washed twice with ice-cold Ca^{2+}/Mg^{2+}-free phosphate buffered saline (PBS, 12 ml). Homogenization buffer (0.375 M sorbitol 20 mM Hepes, pH: 7.4, 5 ml) is added and rubber policeman is utilized to scrape the cells from the dish. The procedure is repeated to ensure efficient collection of the cells. Scraped cells are transferred to 50 ml tubes and collected by centrifugation (720g for 3 min at 4°). Cell pellets are resuspended in homogenization buffer supplemented with commercially available protease inhibitor cocktail (PIC, sigma P 8340). Usually, 4 plates are collected together and resuspended in 0.9 ml homogenization buffer supplemented with PIC. Cells are homogenized by one complete pass in a 1 ml ball-bearing homogenizer. The homogenate is centrifuged for 5 min at 720g and postnuclear supernatant (PNS) is carefully removed and diluted with half a volume of salt buffer (0.21 M KOAc, 3 mM Mg(OAc)$_2$ 20 mM Hepes, pH 7.4). PNS membranes are collected by centrifugation (3 min at 12,200g in a refrigerated Sorval microfuge using slow acceleration deceleration cycles). The membranes are resuspended in transport buffer (0.25 M

sorbitol, 70 mM KOAc, 1 mM Mg(OAc)$_2$, 20 mM Hepes, pH 7.4 supplemented with PIC) and further washed by repeating the centrifugation step. Membranes are resuspended in PIC containing transport buffer at a final concentration of 3–4 mg/ml and aliquots are frozen in liquid nitrogen and stored at −80°. Note: for the study of dynamic lipid remodeling activities such as PtdIns kinase activities, we found that further fractionation of the membranes on sucrose gradients to enrich for ER exit site containing membranes was required (Blumental-Perry et al., 2005). However, for PLD assays no further purification of the membranes was utilized.

Purification of Sar1 Proteins

His-tagged hamster Sar1a wild type, Sar1a (H79G) mutant, and Sar1a (T39N) mutant expression was directed from pET11dHis vectors (Rowe and Balch, 1995). Sar1a (H79G) is deficient in GTP hydrolysis and thus termed Sar1-GTP, whereas Sar1a (T39N) is deficient in GTP binding and is termed Sar1-GDP. The proteins are expressed in E. coli strain BL21 (DE3) essentially as previously described with the following modification (Rowe and Balch, 1995). Protein expression is achieved without induction; transformed bacteria are inoculated into liquid cultures and grown overnight at 28° with gentle shaking. After overnight growth the cultures are chilled and the cells are harvested for protein purification (Rowe and Balch, 1995). We find that the yield of soluble monomer protein under these conditions is improved. His-tagged Sar1 proteins are purified using affinity binding to Ni^{2+}-nitrilotriacetic acid-agarose beads (Ni-NTA-agarose, Qiagen, Chatsworth, CA), followed by fractionation on Sephacryl S-100 (Pharmacia) gel filtration column as previously described (Rowe and Balch, 1995). Two peaks of Sar1 are resolved on gel filtration, the first peak represents high molecular weight Sar1 (<140 kDa) and the second represents the monomer protein, which is eluted at <17 kDa. The monomer Sar1 proteins are collected, dialyzed into buffer containing 25 mM Hepes pH 7.2, 125 mM KOAc, and 1 mM Mg(OAc)$_2$, and further concentrated to generate a stock of 0.5–1.5 mg/ml. Protein aliquots are snap frozen in liquid nitrogen and stored at −80°.

Assaying the Activation of PLD by Sar1

PLD cleaves phosphatidylcholine (PC) to generate choline and phosphatidic acid (PA) via a transphosphatidylation reaction. However, because PLD is a phosphotransferase, short primary alcohols, such as ethanol, 1-propanol, and 1-butanol, may efficiently substitute for water in the reaction, which then leads to the production of phosphatidylalcohol.

Most cell- and test tube-based assays for PLD activity take advantage of this property of the enzyme (reviewed in Exton, 1999, 2002; Rizzo and Romero, 2002).

Substrate Preparation

Substrate liposomes can be prepared by several different procedures. The simplest method is to dissolve each individual lipid in chloroform at a concentration of 1–10 mg/ml and mix the lipid solutions at the desired mol fractions. The chloroform is removed by evaporation using nitrogen gas and a short spin on a SpeedVac centrifuge. Liposomes are prepared by hydration of the dried lipid films using buffer containing 25 mM Hepes, pH 7.4, 100 mM KCl, 3 mM NaCl, 1 mM EGTA, and 1 mM DTT (H/K buffer). Since results may vary depending on the relative concentrations of the various lipids, the lipid composition of the liposomes should be kept constant. For most assays, we used a mixture of phosphatidylethanolamine (PE), PtdIns (4,5)P2, and PC (molar ratios: 16:1.4:1) containing L-α-dipalmitoyl-12-[2-palmitoyl-9,10-^{3}H]-PC (5 μCi/sample). Liposomes are used at a final concentration of 10 μM PC in the assay (see below). The second substrate for the PLD reaction, ethanol (1.5%) can also be included in the liposome preparation (Brown et al., 1993; Iyer and Kusner, 1999). Substrate vesicles are sonicated for 10 min at 25° in a bath sonicator just before their use.

Activation of Sar1 and ARF

These small GTPases can be preloaded with a GTP analog prior to their use. This is done by incubating recombinant proteins (Sar1 or ARF, which we used as a control) with cell membranes in the presence of GTPγS and ATP. The membranes are both a source of PLD and contain the guanine nucleotide exchange factors required for the activation of the small GTPases. Note: PLD assays that measure Sar1 activity can also be conducted in one-stage incubation, allowing the Sar1 dedicated GDP-GTP exchange factor Sec12 to activate the added Sar1 on ER membranes while measuring PLD activity. Sar1-GTP, a constitutively active mutant that cannot efficiently hydrolyze GTP is utilized in the assay, thus GTP can be utilized instead of GTPγS under these conditions. Sar1-GDP, a dominant negative mutant, can serve as a control in these reactions.

PLD Assay

We routinely measure PLD activity using ethanol as a substrate. A 2-stage assay is recommended to measure the effects of small GTPases on PLD activity. In first stage of the assay, recombinant GTPases (Sar1 or

ARF; 8 μg) are preincubated with microsomal membranes (75 μg), ATP (1 mM), and GTPγS (100 μM) for 30 min at 37° in 100 μl of H/K buffer supplemented with 5 mM MgCl$_2$. At the end of the first incubation, the membranes are collected by centrifugation using a table-top microfuge and resuspended in 75 μl of MgCl$_2$ containing H/K buffer. The second stage of the reaction measures the PLD activity of the sample. The reaction is started by addition of 25 μl of freshly sonicated liposomes (10 μM final concentration of PC) and ethanol (1.5%) and the incubation is continued for 60 min at 37°. GTPγS (100 μM) can be added to all samples at the second stage as additional control for the specific effects of Sar1 and ARF proteins (Iyer and Kusner, 1999). The reaction is stopped by the addition of 0.5 ml of chloroform:methanol (2:1) on ice. Lipids are extracted with vigorous shaking followed by centrifugation of the samples. The chloroform (lower) phase is removed using a glass pipette, transferred to a clean centrifuge tube, dried, and dissolved in 20 μl of chloroform. The lipids are then spotted onto Silica Gel 60 thin layer chromatography plates, dried, and developed using ethyl acetate:trimethylpentane:acetic acid (9:5:2 vol/vol). In this solvent, the product phosphatidylethanol (PEtOH), runs near the leading edge of the chromatogram, whereas the substrate, PC, remains at the origin. The position of the product is determined using true standards (Avanti Polar Lipids), which are visualized with iodine. The position of the radiolabeled spots is determined by autoradiography or using a phosphorimaging system. A scintillation fluid, such as En[3]Hance, should be used for rapid development of the films. PLD activity is quantified by scraping the spots corresponding to PEtOH and PC followed by liquid scintillation counting. Data are expressed as the percentage of total PC converted to PEtOH in each sample.

References

Allan, D. (1996). Mapping the lipid distribution in the membranes of BHK cells (mini-review). *Mol. Membr. Biol.* **13,** 81–84.

Aridor, M., Fish, K. N., Bannykh, S., Weissman, J., Roberts, T. H., Lippincott-Schwartz, J., and Balch, W. E. (2001). The Sar1 GTPase coordinates biosynthetic cargo selection with endoplasmic reticulum export site assembly. *J. Cell Biol.* **152,** 213–229.

Baron, C. L., and Malhotra, V. (2002). Role of diacylglycerol in PKD recruitment to the TGN and protein transport to the plasma membrane. *Science* **295,** 325–328.

Blumental-Perry, A., Honey, C. J., Weixel, K. M., Watkins, S. C., Weis, O. A., and Aridor, M. (2005). Phosphatidylinositol 4-phosphate formation at ER exit sites regulates ER export. *Submitted.*

Brown, H. A., Gutowski, S., Moomaw, C. R., Slaughter, C., and Sternweis, P. C. (1993). ADP-ribosylation factor, a small GTP-dependent regulatory protein, stimulates phospholipase D activity. *Cell* **75,** 1137–1144.

Bruns, J. R., Ellis, M. A., Jeromin, A., and Weisz, O. A. (2002). Multiple roles for phosphatidylinositol 4-kinase in biosynthetic transport in polarized Madin-Darby canine kidney cells. *J. Biol. Chem.* **277,** 2012–2018.

Chen, Y. G., Siddhanta, A., Austin, C. D., Hammond, S. M., Sung, T. C., Frohman, M. A., Morris, A. J., and Shields, D. (1997). Phospholipase D stimulates release of nascent secretory vesicles from the trans-Golgi network. *J. Cell. Biol.* **138,** 495–504.

Crottet, P., Meyer, D. M., Rohrer, J., and Spiess, M. (2002). ARF1.GTP, tyrosine-based signals, and phosphatidylinositol 4,5-bisphosphate constitute a minimal machinery to recruit the AP-1 clathrin adaptor to membranes. *Mol. Biol. Cell.* **13,** 3672–3682.

De Matteis, M. A., and Godi, A. (2004). PI-loting membrane traffic. *Nat. Cell. Biol.* **6,** 487–492.

Exton, J. H. (1999). Regulation of phospholipase D. *Biochim. Biophys. Acta.* **1439,** 121–133.

Exton, J. H. (2002). Regulation of phospholipase D. *FEBS Lett.* **531,** 58–61.

Fabbri, M., Bannykh, S., and Balch, W. E. (1994). Export of protein from the endoplasmic reticulum is regulated by a diacylglycerol/phorbol ester binding protein. *J. Biol. Chem.* **269,** 26848–26857.

Godi, A., Di Campli, A., Konstantakopoulos, A., Di Tullio, G., Alessi, D. R., Kular, G. S., Daniele, T., Marra, P., Lucocq, J. M., and De Matteis, M. A. (2004). FAPPs control Golgi-to-cell-surface membrane traffic by binding to ARF and PtdIns(4)P. *Nat. Cell. Biol.* **6,** 393–404.

Iyer, S. S., and Kusner, D. J. (1999). Association of phospholipase D activity with the detergent-insoluble cytoskeleton of U937 promonocytic leukocytes. *J. Biol. Chem.* **274,** 2350–2359.

Kirchhausen, T. (2000). Three ways to make a vesicle. *Nat. Rev. Mol. Cell. Biol.* **1,** 187–198.

Liljedahl, M., Maeda, Y., Colanzi, A., Ayala, I., Van Lint, J., and Malhotra, V. (2001). Protein kinase D regulates the fission of cell surface destined transport carriers from the trans-Golgi network. *Cell* **104,** 409–420.

Matsuoka, K., Morimitsu, Y., Uchida, K., and Schekman, R. (1998a). Coat assembly directs v-SNARE concentration into synthetic COPII vesicles. *Mol. Cell.* **2,** 703–708.

Matsuoka, K., Orci, L., Amherdt, M., Bednarek, S. Y., Hamamoto, S., Schekman, R., and Yeung, T. (1998b). COPII-coated vesicle formation reconstituted with purified coat proteins and chemically defined liposomes. *Cell* **93,** 263–275.

Miller, E. A., Beilharz, T. H., Malkus, P. N., Lee, M. C., Hamamoto, S., Orci, L., and Schekman, R. (2003). Multiple cargo binding sites on the COPII subunit Sec24p ensure capture of diverse membrane proteins into transport vesicles. *Cell* **114,** 497–509.

Morris, A. J., Frohman, M. A., and Engebrecht, J. (1997). Measurement of phospholipase D activity. *Anal. Biochem.* **252,** 1–9.

Mossessova, E., Bickford, L. C., and Goldberg, J. (2003). SNARE selectivity of the COPII coat. *Cell* **114,** 483–495.

Nagaya, H., Wada, I., Jia, Y. J., and Kanoh, H. (2002). Diacylglycerol kinase delta suppresses ER-to-Golgi traffic via its SAM and PH domains. *Mol. Biol. Cell.* **13,** 302–316.

Pathre, P., Shome, K., Blumental-Perry, A., Bielli, A., Haney, C. J., Alber, S., Watkins, S. C., Romero, G., and Aridor, M. (2003). Activation of phospholipase D by the small GTPase Sar1p is required to support COPII assembly and ER export. *EMBO J.* **22,** 4059–4069.

Rizzo, M., and Romero, G. (2002). Pharmacological importance of phospholipase D and phosphatidic acid in the regulation of the mitogen-activated protein kinase cascade. *Pharmacol. Ther.* **94,** 35–50.

Rowe, T., and Balch, W. E. (1995). Expression and purification of mammalian Sar1. *Methods Enzymol.* **257,** 49–53.

Rowe, T., Aridor, M., McCaffery, J. M., Plutner, H., Nuoffer, C., and Balch, W. E. (1996). COPII vesicles derived from mammalian endoplasmic reticulum microsomes recruit COPI. *J. Cell. Biol.* **135**, 895–911.

Shimoi, W., Ezawa, I., Nakamoto, K., Uesaki, S., Gabreski, G., Aridor, M., Yamamoto, A., Nagahama, M., Tagaya, M., and Tani, K. (2005). p125 is localized in endoplasmic reticulum exit sites and involved in their organization. *J. Biol. Chem.* **18**, 10141–10148.

Wang, Y. J., Wang, J., Sun, H. Q., Martinez, M., Sun, Y. X., Macia, E., Kirchhausen, T., Albanesi, J. P., Roth, M. G., and Yin, H. L. (2003). Phosphatidylinositol 4 phosphate regulates targeting of clathrin adaptor AP-1 complexes to the Golgi. *Cell* **114**, 299–310.

Wenk, M. R., and De Camilli, P. (2004). Protein-lipid interactions and phosphoinositide metabolism in membrane traffic: Insights from vesicle recycling in nerve terminals. *Proc. Natl. Acad. Sci. USA* **101**, 8262–8269.

Zhu, Y., Drake, M. T., and Kornfeld, S. (1999). ADP-ribosylation factor 1 dependent clathrin-coat assembly on synthetic liposomes. *Proc. Natl. Acad. Sci. USA* **96**, 5013–5018.

[12] Use of Acyltransferase Inhibitors to Block Vesicular Traffic Between the ER and Golgi Complex

By WILLIAM J. BROWN and JOHN A. SCHMIDT

Abstract

This article describes the use of acyltransferase inhibitors as probes for studying the potential role of lysophospholipid acyltransferases (LPAT) in intracellular membrane trafficking in the secretory and endocytic pathways. The small molecule inhibitors that are described here were originally found as acyl-CoA:cholesterol acyltransferase (ACAT) inhibitors. One of these, CI-976 (2,2-methyl-N-(2,4,6,-trimethoxyphenyl)dodecanamide), was also found to be a potent LPAT inhibitor. CI-976 is a small, hydrophobic, membrane-permeant compound and both *in vivo* and *in vitro* studies have shown that it, but not other ACAT inhibitors, has a profound effect on multiple membrane trafficking pathways in eukaryotic cells including: (1) inhibition of COPII vesicle budding from the endoplasmic reticulum (ER), (2) inhibition of transferrin and transferrin receptor export from the endocytic recycling compartment, and (3) stimulation of tubule-mediated retrograde trafficking of Golgi membranes to the ER. Here we describe the use of CI-976 and other ACAT inhibitors for studies with both cultured mammalian cells and *in vitro* reconstitution assays, with a particular emphasis on COPII vesicle budding from the ER. All of these studies strongly suggest that CI-976-sensitive LPATs play a role in coated vesicle fission, and therefore, CI-976 is a valuable addition to the arsenal of small molecule inhibitors that can be used to study secretory and endocytic membrane trafficking pathways.

METHODS IN ENZYMOLOGY, VOL. 404
         DOI: 10.1016/S0076-6879(05)04012-7

Introduction

Lysophospholipid acyltransferases (LPAT) are a group of enzymes that transfer fatty acids from acyl-CoA donors to different species of lysophospholipid acceptors. The LPAT family contains both polytopic transmembrane proteins and soluble cytoplasmic enzymes. The human genome encodes at least six integral membrane LPATs, two of which have been characterized as polytopic lysophosphatidic acid-specific LPATs, abbreviated LPAATs (E.C. 2.3.1.51), that catalyze the formation of phosphatidic acid (PA) for *de novo* phospholipid biosynthesis in the endoplasmic reticulum (ER) (Leung, 2001; Lewin *et al.*, 1999). The other four integral membrane LPATs remain uncharacterized. Soluble cytoplasmic LPAATs, such as CtBP3/BARS and the endophilins, are known to be involved in membrane fission in late secretory and endocytic trafficking events, although perhaps independent of their LPAAT activities (Carcedo *et al.*, 2004; Schmidt *et al.*, 1999; Weigert *et al.*, 1999).

It has been suggested that LPATs might also contribute to membrane trafficking events by altering membrane curvature. First, LPAATs could be involved in the generation of negative-curve inducing PA, which might aid in the fission reaction that occurs at the neck of budding coated vesicles (Corda *et al.*, 2002; Fuller and Rand, 2001). Second, LPATs that utilize lysophosphatidylcholine (LPC) as an acceptor might negatively regulate the accumulation of positive-curve inducing LPC, whose production by phospholipase A_2 enzymes appears to be involved in the formation of membrane tubules (Brown *et al.*, 2003; de Figueiredo *et al.*, 1998).

Evidence for a role of LPATs in both coated vesicle formation and regulation of membrane tubule formation recently came from small molecule inhibitor studies. Currently, there are no known LPAT-specific inhibitors; however, a very large number of antagonists have been developed against the acyl-CoA:cholesterol acyltransferase, E.C. 2.3.1.1.26, (ACAT) enzymes, which catalyze the formation of cholesterol esters and are thus of therapeutic interest (Patankar and Jurs, 2000). We reasoned that because LPATs and ACATs catalyze very similar reactions, but with different acceptors, it is possible that some ACAT inhibitors might also be active against LPATs. Indeed, using isolated Golgi complexes as a source of a membrane-associated LPAT, we identified one such compound, CI-976, that is also a potent LPAT inhibitor (Chambers and Brown, 2004; Drecktrah *et al.*, 2003) (Fig. 1). CI-976 is a membrane-permeant small molecule, so we have been able to characterize its effects on a variety of membrane trafficking events in living cells. For example, CI-976 was shown to induce the tubule-mediated retrograde transport of Golgi membranes to

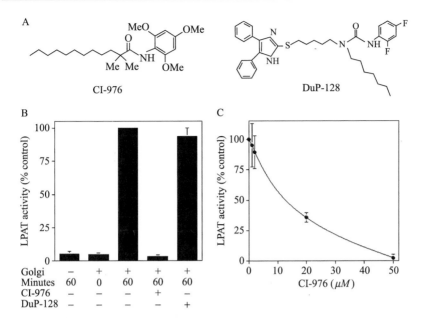

Fig. 1. (A) Chemical structures of CI-976 and DuP-128. (B) The effect of CI-976 (50 μM) and DuP-128 (5 μM) on Golgi membrane-associated LPAT activity. LPAT activity is represented as the percentage of the control (Golgi membranes alone for 1 h at 37°). (C) Dose response of CI-976 inhibition of LPAT activity. The conversion of LPC to PC was monitored by incubating isolated rat liver Golgi membranes, arachidonyl-CoA, inhibitors, and [^{14}C]LPC for 1 h at 37°. Each data point represents the mean plus 1 S.D. of quadruplicate samples. Figure was taken from Drecktrah et al. (2003) and reprinted here with permission from *Molecular Biology of the Cell.*

the ER, likely by preventing LPAT-mediated reacylation of positive-curve inducing LPC (Chambers and Brown, 2004; Drecktrah et al., 2003).

More recently, however, CI-976 has also been used to implicate LPAATs in coated vesicle formation. For example, following redistribution to the ER, Golgi membrane proteins remain in the ER as long as CI-976 is kept in the media, suggesting that CI-976 may also inhibit export via COPII vesicles. Indeed, we have recently found that CI-976 potently inhibits a very late step, possibly the fission step, in the formation of COPII vesicles at ER exit sites (ERESs) (Brown et al., submitted). In addition, CI-976 also inhibits the recycling of transferrin and transferrin receptors from endocytic recycling compartment to the cell surface (Chambers et al., 2005).

Thus, CI-976 affects multiple membrane trafficking pathways, suggesting a role for LPATs at several distinct steps in both membrane vesicle and tubule formation. This article describes the general properties of CI-976, and several other acyltransferase inhibitors, utilizing COPII-mediated ER to Golgi trafficking *in vivo* and *in vitro* as models.

Properties of CI-976

Chemical and Biophysical Properties

CI-976, 2,2-methyl-N-(2,4,6,-trimethoxyphenyl)dodecanamide, ($C_{23}H_{39}$ NO_4, molecular weight 393.57) (Fig. 1) is a white powder that can be stored at room temperature or 4° in a light tight container for at least 1 year. CI-976 is insoluble in water but soluble in ethanol to 50 mM or DMSO to 100 mM. CI-976 is hydrophobic and is highly partitioned into membranes (Homan and Hamelehle, 2001). Stock solutions, generally 25–50 mM, should be stored tightly sealed at –0° or below and used within one month.

Biological Properties

CI-976 was synthesized as a fatty acid anilide derivative designed to mimic fatty acyl-CoA, the fatty acid donor for ACAT enzymes, and it has been most extensively studied in this regard as a competitive ACAT inhibitor (Field *et al.*, 1991; Roth *et al.*, 1992). Various animal studies have shown that CI-976 lowers plasma low density lipoprotein (LDL)-cholesterol and raises high density lipoprotein-cholesterol by inhibiting both liver and intestinal ACAT activities; CI-976 also lowers liver cholesterol esters (CE) and decreases CE secretion (Carr *et al.*, 1995; Krause *et al.*, 1993). The metabolic fate of CI-976 has been studied in both whole animals and isolated hepatocytes, and it is oxidized to numerous metabolites, likely by cytochrome P_{450} pathways (Sinz *et al.*, 1997). The biological activities of these metabolites are unknown.

As discussed above, we identified CI-976 in a screen of ACAT inhibitors that might also be active against LPATs, a result that was already suggested by earlier studies (Krause *et al.*, 1993). Characterization of this CI-976-sensitive Golgi LPAT showed that it greatly prefers LPC or lysophosphatidylethanolamine compared to lysophosphatidic acid, has little specificity for acyl chain length, and can transfer fatty acids to either exogenous or endogenous lysophospholipid acceptors (Chambers and Brown, 2004). To date, CI-976 is the only ACAT inhibitor we have found that also inhibits the Golgi-associated LPAT. Others that we have tried include DuP-128, PKF 058 035, and TS-962.

Sources of CI-976 and Other Acyltransferase Inhibitors

We have successfully used CI-976 that was originally obtained through Material Transfer Agreements with GlaxoSmithKline Pharmaceuticals

(Essex UK) and more recently with Pfizer Global Research & Development (Groton, CT). CI-976 is also commercially available from Tocris Cookson Inc (Ellisville, MO) and Tocris Cookson Ltd (Bristol, UK) (Cat. No. 2227), and material from this source is active in our hands. However, we have not used material from this source. Other ACAT inhibitors were: DuP-128 (GlaxoSmithKline); PFK 058 035 (Novartis, Summit, NJ), and TS-962 (Taisho Pharmaceutical Co., LTD, Tokyo, Japan).

Suggestions for Using Acyltransferase Inhibitors on Cultured Cells

We used CI-976 to examine the potential role of LPATs in a number of *in vivo* intracellular trafficking steps including Golgi-to-ER retrograde, endocytic recycling, and more recently, COPII-mediated vesicular export from ERES. During the course of these studies, we discovered several important factors when using CI-976 and below we will describe how we have used this compound for studies on cultured cells.

Adding CI-976 to Cultured Cells

a. For most *in vivo* studies (immunofluorescence, metabolic labeling, etc), cultured cells are grown on glass coverslips in 35 or 60 mm Petri dishes to ~50–70% confluence.

b. The potency of CI-976 is greatly reduced by fetal bovine serum (FBS) (or serum substitutes such as Nu-Serum™), so cells should be thoroughly rinsed with MEM or other media devoid of FBS.

c. CI-976 works best when added directly to media that is bathing the cells. Cells are covered with a small amount of media, for example, 1 ml for a 35 mm dish), CI-976 is added from a 1000-fold stock solution, and the media is quickly swirled to dilute the drug. For the *in vivo* trafficking steps we have studied, CI-976 exhibits an IC_{50} in the low micromolar range (Chambers *et al.*, 2005; Drecktrah *et al.*, 2003).

d. CI-976 has a very high membrane partitioning coefficient, therefore its potency is significantly affected by the amount of membrane present (Homan and Hamelehle, 2001). Therefore, care should be taken to ensure that cell cultures are at the same density for each experiment.

e. CI-976 is relatively fast acting and a 10–15 min pre-incubation is generally sufficient to achieve maximal inhibition.

f. Cells remain viable and drug effects are reversible with concentrations of CI-976 up to 50 μM for at least 1 h. However, prolonged treatment (>2 h) lowers cell viability.

g. CI-976 has been shown to induce Golgi membrane tubulation and to block COPII budding in a variety of cells including rat clone 9 hepatocytes,

NRK, and HeLa cells (Drecktrah *et al.*, 2003), although some slight cell-type differences were noted, for example, NRK cell Golgi complexes are slower to tubulate than Clone 9 cells.

Reversibility of CI-976

Reversibility from CI-976 inhibition was variable depending on the trafficking step. CI-976 washout was most effectively achieved by rapidly rinsing cells multiple times in media containing FBS. For COPII vesicle budding from the ER, transport to the Golgi complex was restored within 15 min following removal of the drug. However, inhibition of transferrin and transferrin receptor export from the endocytic recycling compartment was very slowly reversible in Clone 9 rat hepatocytes (little within 24 h), but appeared slightly more rapid in HeLa cells (Chambers *et al.*, 2005).

Critical Controls with ACAT Inhibitors

Because CI-976 is a both an LPAT and ACAT inhibitor, it is crucial to determine if any observed drug effect is due to one or the other enzyme. The simplest way to determine this is by using other ACAT inhibitors that do not have any activity against LPATs. We have examined several other ACAT inhibitors including DuP-128 (Fig. 1), TS-962, and PFK-058–035, and find that none produce CI-976 effects (Chambers and Brown, 2004; Chambers *et al.*, 2005). Therefore, at least one of these other ACAT inhibitors must be used in any initial experiments to determine if the observed effects might be on an ACAT enzyme, instead of an LPAT.

Specific Use of CI-976 for Studying COPII Vesicle Budding

In Vivo *Studies of COPII Budding*

Recent studies have shown that CI-976 produces a potent but reversible block in the export of vesicular stomatitis virus (VSV) G membrane glycoprotein in COPII coated vesicles from the ER (Brown *et al.*, submitted). These studies utilized the well-established thermo-reversible mutant (ts045) of VSV-G to stage the synchronous export of VSV-G from the ER (Bannykh *et al.*, 1998; Fabbri *et al.*, 1994). For these studies, cells were infected with ts045 VSV and incubated at the restrictive temperature for ~3.5 h to accumulate VSV-G in the ER (Fig. 2A, B). In solvent-treated control cells shifted to the permissive temperature for 15 min, VSV-G rapidly exited the ER and was transported to the Golgi complex (Fig. 2C, D). In contrast, when CI-976 was included following shift to the permissive

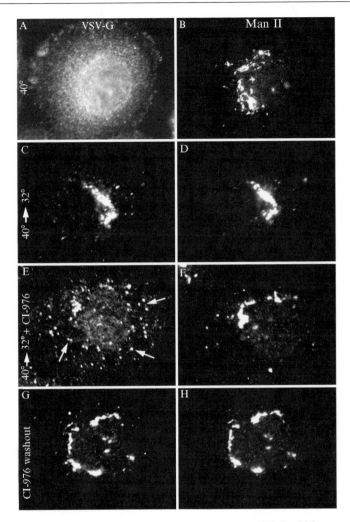

FIG. 2. Treatment of cells with CI-976 inhibits ER export of VSV-G, which accumulates in COPII buds at ERES. Cells were infected with ts045 VSV at 40° to accumulate VSV-G in the ER, subjected to various chase protocols, and then stained by double-immunofluorescence for VSV-G (left panels) and the Golgi marker α-mannosidase II (Man II) (right panes). (A, B) No chase; (C, D) Cells shifted to 32° for 15 min; (E, F) Cells shifted to 32° in the presence of CI-976; (G, H) Cells shifted to 32° in the presence of CI-976 for 15 min and then washed free of the drug for an additional 15 min. Arrows in E indicate accumulation of VSV-G in foci corresponding to ERESs.

temperature, VSV-G did not reach the Golgi complex and instead moved from a diffuse ER staining pattern to a discrete number of foci (Fig. E, F). Other studies revealed these VSV-G-enriched foci to be ERESs, as shown by staining with antibodies against the COPII components sec13 and sec31 (data not shown). Accumulation at these ERESs was reversible because following washout of the drug VSV-G was rapidly (within 15 min) delivered to the Golgi complex (Fig. 2G, H). These studies indicate that CI-976 does not prohibit the assembly of COPII coats at ERESs, but inhibits a very late step in COPII budding, probably at the level of vesicle fission.

In Vitro *Studies of COPII Budding*

Consistent with its ability to inhibit COPII budding *in vivo*, CI-976 has been found to inhibit the budding of COPII vesicles in an *in vitro* reconstitution assay. The procedures for conducting this assay are detailed elsewhere (Rowe *et al.*, 1996), so we will describe our modifications for use with CI-976.

a. Prepare total microsomes from NRK cells infected with ts045-VSV under conditions that accumulate VSV-G in the ER (as above). Prepare rat liver cytosol (RLC) (Davidson and Balch, 1993), which supplies COPII coat proteins, store at $-80°$, and clarify immediately prior to use by centrifugation at $100,000g$ for 10 min. Prepare the Budding Cocktail consisting of 25 mM Hepes, pH 7.2, 40 mM KOAc, 0.18 M sorbitol, 2.5 mM MgOAc, 1.8 mM Ca(OAc)$_2$, 5 mM EDTA, and an ATP regenerating system (Davidson and Balch, 1993).

b. Aliquots of Budding Cocktail (30 μl minus the volume needed for CI-976) and microsomes (15 μl) are added together and mixed on ice. CI-976 is then added from a concentrated stock solution (in DMSO), so that the final concentration after adding cytosol (in the next step) is $\leq$50 μM. Gently mix contents. A similar volume of DMSO should be added to other tubes for the solvent control. RLC (15 μl) is then added with gentle mixing.

c. To initiate COPII budding, transfer tubes to a 32° water bath and incubate for $\leq$15 min.

d. After incubation in the budding reaction, place tubes on ice and centrifuge at 14,000 RPM for 2 min in an Eppendorf 5417C refrigerated centrifuge at 4°. Collect the supernatant (50 μl) containing released COPII vesicles and sediment them to a pellet by centrifugation at $100,000g$ for 20 min at 4°. Carefully remove the supernatant and dissolve the vesicle-containing pellet in SDS sample buffer.

e. The amount of VSV-G released with the vesicles can then be determined by Western blotting with polyclonal anti-VSV-G as described (Rowe *et al.*, 1996).

Several Aspects of this Assay Deserve Additional Discussion

a. A critical aspect of this assay is the order in which components are added (Step 2b above) because we found that CI-976 was able to inhibit COPII vesicle budding only if it was added to the budding cocktail and microsome mixture before adding RLC. If CI-976 was added to a mixture of budding cocktail plus RLC before adding microsomes, then no inhibition was observed. These results indicate that the CI-976 target is associated with the microsomal membranes and that RLC "inactivates" the drug, perhaps by simply being nonspecifically bound to cytosolic proteins.

b. We found it necessary to titrate both the microsome and RLC preparations each time a new batch of either was made to ensure that CI-976 was added to samples that were still within the linear range of the assay. The exact concentrations of either will depend on purity of microsomes and quality of the RLC preps. The amount of microsomes is particularly important because, as stated above, CI-976 has a high membrane partitioning coefficient, so the presence of additional membrane will raise the effective concentration.

Summary

CI-976 has proven to be a valuable reagent for demonstrating the potential role of LPATs in regulating various membrane trafficking events. The evidence to date shows that CI-976 inhibits integral membrane LPATs; however, it may also be active against soluble cytoplasmic enzymes too. It is hoped that future studies will be able to identify the specific LPATs that are inhibited by CI-976. CI-976 should also be useful for elucidating the mechanisms by which LPATs contribute to coated vesicle budding, for example, perhaps by facilitating membrane fission.

Acknowledgments

We would also like to thank Dr. Brian Jackson GlaxoSmithKline Pharmaceuticals for supplying the original batch of CI-976 and DuP-128 for our studies. This work was supported by NIH grant DK 51596 (to W.J.B.).

References

Bannykh, S. I., Nishimura, N., and Balch, W. E. (1998). Getting into the Golgi. *Trends Cell. Biol.* **8,** 21–25.
Brown, W. J., Chambers, K., and Doody, A. (2003). Phospholipase A_2 (PLA_2) enzymes in membrane trafficking: Mediators of membrane shape and function. *Traffic* **4,** 214–221.

Carcedo, C. H., Bonazzi, M., Spano, S., Turacchio, G., Colanzi, A., Luini, A., and Corda, D. (2004). Mitotic Golgi partitioning is driven by the membrane-fissioning protein CtBP3/ BARS. *Science* **305**, 93–96.

Carr, T. P., Hamilton, R. L., Jr., and Rudel, L. L. (1995). ACAT inhibitors decrease secretion of cholesteryl esters and apolipoprotein B by perfused livers of African green monkeys. *J. Lipid Res.* **36**, 25–36.

Chambers, K., and Brown, W. J. (2004). Characterization of a novel CI-976-sensitive lysophospholipid acyltransferase that is associated with the Golgi complex. *Biochem. Biophys. Res. Commun.* **313**, 681–686.

Chambers, K., Judson, B., and Brown, W. J. (2005). A unique lysophospholipid acyltransferase (LPAT) antagonist, CI-976, affects secretory and endocytic membrane trafficking pathways. *J. Cell Sci.* **118**, 3061–3071.

Corda, D., Hidalgo Carcedo, C., Bonazzi, M., Luini, A., and Spano, S. (2002). Molecular aspects of membrane fission in the secretory pathway. *Cell. Mol. Life Sci.* **59**, 1819–1832.

Davidson, H. W., and Balch, W. E. (1993). Differential inhibition of multiple vesicular transport steps between the endoplasmic reticulum and trans Golgi network. *J. Biol. Chem.* **268**, 4216–4226.

de Figueiredo, P., Drecktrah, D., Katzenellenbogen, J. A., Strang, M., and Brown, W. J. (1998). Evidence that phospholipase A$_2$ activity is required for Golgi complex and trans Golgi network membrane tubulation. *Proc. Natl. Acad. Sci. USA* **95**, 8642–8647.

Drecktrah, D., Chambers, K., Racoosin, E. L., Cluett, E. B., Gucwa, A., Jackson, B., and Brown, W. J. (2003). Inhibition of a Golgi complex lysophospholipid acyltransferase induces membrane tubule formation and retrograde trafficking. *Mol. Biol. Cell.* **14**, 3459–3469.

Fabbri, M., Bannykh, S., and Balch, W. E. (1994). Export of protein from the endoplasmic reticulum is regulated by a diacylglycerol/phorbol ester binding protein. *J. Biol. Chem.* **269**, 26848–26857.

Field, F. J., Albright, E., and Mathur, S. (1991). Inhibition of acylcoenzyme A: Cholesterol acyltransferase activity by PD128042: Effect on cholesterol metabolism and secretion in CaCo-2 cells. *Lipids* **26**, 1–8.

Fuller, N., and Rand, R. P. (2001). The influence of lysolipids on the spontaneous curvature and bending elasticity of phospholipid membranes. *Biophys. J.* **81**, 243–254.

Homan, R., and Hamelehle, K. L. (2001). Influence of membrane partitioning on inhibitors of membrane-bound enzymes. *J. Pharm. Sci.* **90**, 1859–1867.

Krause, B. R., Anderson, M., Bisgaier, C. L., Bocan, T., Bousley, R., De Hart, P., Essenburg, A., Hamelehle, K., Homan, R., Kieft, K., McNally, W., Stanfield, R., and Newton, R. S. (1993). *In vivo* evidence that the lipid-regulating activity of the ACAT inhibitor CI-976 in rats is due to inhibition of both intestinal and liver ACAT. *J. Lipid. Res.* **34**, 279–294.

Leung, D. W. (2001). The structure and functions of human lysophosphatidic acid acyltransferases. *Front. Biosci.* **6**, D944–D953.

Lewin, T. M., Wang, P., and Coleman, R. A. (1999). Analysis of amino acid motifs diagnostic for the sn-glycerol-3-phosphate acyltransferase reaction. *Biochemistry* **38**, 5764–5771.

Patankar, S. J., and Jurs, P. C. (2000). Prediction of IC50 values for ACAT inhibitors from molecular structure. *J. Chem. Inf. Comput. Sci.* **40**, 706–723.

Roth, B. D., Blankley, C. J., Hoefle, M. L., Holmes, A., Roark, W. H., Trivedi, B. K., Essenburg, A. D., Kieft, K. A., Krause, B. R., and Stanfield, R. L. (1992). Inhibitors of acyl-CoA: Cholesterol acyltransferase. 1. Identification and structure-activity relationships of a novel series of fatty acid anilide hypocholesterolemic agents. *J. Med. Chem.* **35**, 1609–1617.

Rowe, T., Aridor, M., McCaffery, J. M., Plutner, H., Nuoffer, C., and Balch, W. E. (1996). COPII vesicles derived from mammalian endoplasmic reticulum microsomes recruit COPI. *J. Cell. Biol.* **135**, 895–911.

Schmidt, A., Wolde, M., Thiele, C., Fest, W., Kratzin, H., Podtelejnikov, A. V., Witke, W., Huttner, W. B., and Soling, H. D. (1999). Endophilin I mediates synaptic vesicle formation by transfer of arachidonate to lysophosphatidic acid. *Nature* **401,** 133–141.

Sinz, M. W., Black, A. E., Bjorge, S. M., Holmes, A., Trivedi, B. K., and Woolf, T. F. (1997). *In vitro* and *in vivo* disposition of 2,2-dimethyl-N-(2,4,6-trimethoxyphenyl)dodecanamide (CI-976). Identification of a novel five-carbon cleavage metabolite in rats. *Drug Metab. Dispos.* **25,** 123–130.

Weigert, R., Silletta, M. G., Spano, S., Turacchio, G., Cericola, C., Colanzi, A., Senatore, S., Mancini, R., Polishchuk, E. V., Salmona, M., Facchiano, F., Burger, K. N., Mironov, A., Luini, A., and Corda, D. (1999). CtBP/BARS induces fission of Golgi membranes by acylating lysophosphatidic acid. *Nature* **402,** 429–433.

[13] Tethering Assays for COPI Vesicles Mediated by Golgins

By Ayano Satoh, Jörg Malsam, and Graham Warren

Abstract

A method is described that allows the attachment of COPI vesicles and Golgi membranes to glass slides that can then be analyzed using electron microscopy (EM) and immuno-EM methods. Subpopulations of COPI vesicles can be bound selectively using recombinant golgins. Alternatively, COPI vesicles can be attached to prebound Golgi membranes. Marking these vesicles selectively with biotin allows their site of attachment to be identified.

Introduction

The flow of material within the Golgi apparatus is mediated by COPI vesicles. Newly synthesized cargo proteins and lipids undergo serial modifications to the bound oligosaccharides as they move through this organelle. Large cargo is thought to move through cisternae that are remodeled around it, COPI vesicles removing one set of Golgi enzymes and bringing in the next set needed to carry out the modifications. Smaller cargo is thought to move in COPI vesicles, from one cisterna to the next, undergoing appropriate modifications in each. Simultaneous transport of different cargo therefore requires the coordinated action of different types of COPI vesicle, to ensure the correct intersection of cargo and enzymes in both time and space, so that each undergoes the correct sequence of post-translational modifications (Palade, 1975; Pelham and Rothman, 2000).

Determining the flow patterns for COPI vesicles has been hampered by the lack of available methods to subfractionate them biochemically, so that

METHODS IN ENZYMOLOGY, VOL. 404
0076-6879/05 $35.00
DOI: 10.1016/S0076-6879(05)04013-9

the origin and destination of each type can be determined using EM methods. One recent approach has been to exploit the specificity of membrane tethers to isolate subpopulations (Malsam *et al.*, 2005). These tethers are thought to help determine the destination of COPI vesicles, providing the initial attachment for the vesicle on the recipient cisternal membrane. There are many types of tethers, and we have focused on the golgin tethers, coiled-coil proteins that were originally identified as auto-antigens in patients with Sjögren's syndrome (Fritzler *et al.*, 1993). There are at least a dozen golgins, arrayed across the Golgi, so they could help provide the initial attachment at all levels of the stack (Barr and Short, 2003; Gillingham and Munro, 2003). Golgins interact with each other as well as directly (via transmembrane domains) or indirectly with membranes, as part of the tethering process (Shorter and Warren, 2002). Many also interact with small GTPases of the Ypt/Rab or Arl families (Barr and Short, 2003; Gillingham and Munro, 2003; Jackson, 2004).

A number of golgins have been characterized as well as several tethers, made up of different golgins. The latest is the CASP-golgin-84 tether that we recently showed is involved in the retrograde transport of Golgi enzymes within the Golgi stack (Malsam *et al.*, 2005). Both are membrane-anchored proteins (Bascom *et al.*, 1999; Gillingham *et al.*, 2002), golgin-84 in the COPI vesicle interacting with CASP on cisternal membranes (Malsam *et al.*, 2005). As part of the characterization process, we devised methods to reconstitute this tether attached to glass beads. This allowed us to study the composition of this subpopulation. We have also devised methods to attach COPI vesicles to glass slides coated with the cognate tether (in this case CASP) or Golgi membranes, so that the targeting site(s) of these vesicles could be ascertained. These are described in this chapter.

In Vitro COPI Vesicle Tethering Assay Using Purified Golgins

Materials

Purified rat liver Golgi: 300 μg (Hui *et al.*, 1997)

Rat liver cytosol (35% ammonium sulfate cut): 1 mg (Rabouille *et al.*, 1995)

Purified golgins: >1 mg/ml, ~10 μl (Satoh *et al.*, 2005)

Purified early endosomal antigen 1 (EEA1): >1 mg/ml, ~10 μl (Simonsen *et al.*, 1998)

Blocking reagents: BSA and soybean trypsin inhibitor (STI, Sigma, St. Louis, MO)

Assay buffer: 25 mM HEPES (pH 7.0), 2.5 mM magnesium acetate, 100 mM KCl, 1 mM DTT

GTP: 40 mM stock, neutralized with KOH
ATP-regenerating system (Rabouille *et al.*, 1995)
Glass slides: Esco Superfrost, precleaned (Erie Scientific Company, Portsmouth, NH)
Pap-pen (Zymed, San Francisco, CA)
All membranes and reagents should be prepared in, or buffer-exchanged to, assay buffer.

Methods

Preparation of COPI Vesicle-containing Supernatant

Rat liver Golgi membranes (300 μg) are incubated with rat liver cytosol (1 mg), 1 mM GTP, and an ATP-regenerating system in a final volume of 120 μl assay buffer for 10 min at 37°. The mixture is centrifuged at 14,000g at 4° for 10 min to remove larger membranes. The supernatants are used for the following tethering assay. They contain sufficient vesicles without the need to release more from the membranes using KCl (Malhotra *et al.*, 1989).

In Vitro Tethering of COPI Vesicles to Golgin-Coated Glass Slides

Microscope glass slides should be precleaned with 2 M NaOH for 30 min at room temperature, washed extensively with distilled water, and dried using an air duster. Circles should be drawn on the slides using a paraffin pen (Pap-pen, Zymed). Ten μl of recombinant proteins (1 mg/ml) are then allowed to attach for 30 min at 4° in a wet chamber. After washing with assay buffer, the slides are blocked with 1% BSA and 1% STI in assay buffer containing 9% (w/w) sucrose. Ten μl of the COPI vesicle-containing supernatant (above) is then placed in the circles and incubated in a wet chamber for 30 min at 4°. After washing, the samples are processed for EM (Seemann *et al.*, 2000). Briefly, the samples are embedded in Epon, which is then removed by immersion of the slide into liquid nitrogen for a couple of seconds followed by rapid warming using submersion into lukewarm water. Repeating this procedure several times will eventually remove all glass residues from the Epon surface. Blowing air onto the Epon surface using an air duster helps to get rid of all glass remnants and residual water. The Epon block is then cut parallel to the surface to generate 60 nm thick sections. For competition experiments, recombinant proteins (1 mg/ml final concentration) should be added prior to incubations on the coated slides. Results are quantitated as the number of vesicles/μm^2 ± S.D. Five areas on each of a total of three grids per condition should be counted.

An example is shown in Fig. 1. The input (Fig. 1F) contained a mixture of membranes including a minor fraction of uncoated COPI-sized

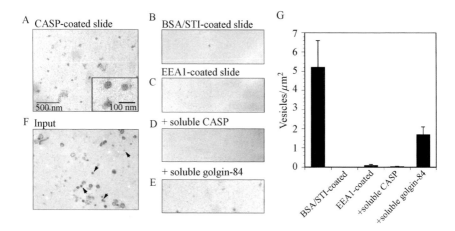

Fig. 1. Tethering of COPI vesicles to CASP-coated slides. (A–E) The vesicle-containing supernatant was incubated for 30 min at 4° on glass slides precoated with (A, D, E) recombinant CASP, (B) 1% BSA and STI, or (C) recombinant EEA1 in the absence (A, B, C) or presence of (D) soluble CASP or (E) soluble golgin-84. Glass slides were embedded in Epon resin and processed for conventional EM. (F) Input membranes. (G) Quantitation of the results in (A–E) presented as the mean of the number of the bound vesicles per $\mu m^2 \pm$ S. D.(n=5). (Reprinted with permission from *Science*.)

vesicles (~50 nm diameter, arrows). These were selectively bound to the CASP-coated slides (Fig. 1A) but not to those coated with BSA/STI or an irrelevant coiled-coil protein, EEA1 (Fig. 1B, C). The binding was specifically inhibited when carried out in the presence of soluble CASP or golgin-84 (Fig. 1D, E).

Note that uncoating the COPI vesicles is required for binding to golgin tethers (or Golgi membranes, see below) since binding was not observed when an Arf1Q71L mutant was used for preparation of COPI vesicles (Malsam *et al.*, 2005).

This method is not limited to CASP. Other tethers, such as GM130, have also been used to retrieve COPI vesicles from the vesicle-containing supernatant. GM130 binds to vesicles via p115 and giantin in the vesicle. In this case, p115-depleted cytosol should be used for the preparation of the COPI vesicle-containing supernatant since the cytoplasmic pool of p115 may mask the tether (giantin, in this case) on the vesicles (Malsam *et al.*, 2005).

Tethering of COPI Vesicles to Golgi Membranes Bound to Glass Slides

To approximate more the *in vivo* tethering of COPI vesicles to Golgi membranes, we have devised a method to attach Golgi membranes to glass slides and then incubate them with purified COPI vesicles. To distinguish the added COP vesicles from the attached Golgi membranes they need to be marked. Serendipitously, this proved possible using a biotinylation reagent that, remarkably, labeled, almost exclusively, golgin-84 in COPI vesicles.

(A) Purification of Biotinylated COPI Vesicles

Materials

Purified rat liver Golgi: 1 mg (Hui *et al.*, 1997)
Sulfo-NHS-LC-biotin (sulfo-succinimidyl 6-(biotinamido) hexanoate; Pierce Chemical Co., Rockford, IL)
Purified rabbit liver coatomer: 200 μg (Pavel *et al.*, 1998)
Recombinant myristoylated Arf1: 200 μg (Franco *et al.*, 1995)
Assay buffer: 25 mM HEPES (pH 7.0), 2.5 mM magnesium acetate, 100 mM KCl, 1 mM DTT
GTP: 40 mM stock, neutralized with KOH
ATP-regenerating system (Rabouille *et al.*, 1995)
Beckman Vti 65.1 rotor, 6.3 ml tubes (#345830) and spacers (#349289)
All membranes and reagents should be prepared in, or buffer-exchanged to, assay buffer.

Methods

Biotinylation of Rat Liver Golgi Membranes

Golgi membranes (1 mg) are incubated with 0.5 μg/μl sulfo-NHS-LC-biotin for 1 h on ice in 200 μl assay buffer containing 500 mM KCl. The excess sulfo-NHS-LC-biotin is then quenched with NH$_4$Cl (final concentration 50 mM) for another hour on ice. The membranes are recovered by centrifugation at 13,000g at 4° for 10 min onto a sucrose cushion (10 μl of 23% (w/w) sucrose in assay buffer).

Purification of Labeled COPI Vesicles

The biotinylated Golgi membranes ($\sim$1 mg) are incubated with purified coatomer (200 μg), myristoylated Arf (200 μg), 1 mM GTP, and an ATP regenerating system in assay buffer containing 9% (w/w) sucrose in a final volume of 500 μl. After incubation for 10 min at 37°, 500 μl of 400 mM KCl in cold assay buffer is added to stop the reaction and release the vesicles generated. Large Golgi remnants are removed at 14,000g for 10 min at 4°. The supernatant is layered on a step gradient comprising 1.5 ml 30%, 1.5 ml 35%, 0.25 ml 37.5%, 0.25 ml 50%, and 2 ml 55% (w/w) sucrose in assay buffer containing 250 mM KCl. Membranes are centrifuged to equilibrium at 400,000g at 4° for 2.5 h in a Vti65.1 vertical rotor. After puncturing the bottom of the tube, 0.5 ml fractions are collected as water is pumped (at 0.5 ml/min) on to the top of the gradient. COPI coated vesicles typically peak at 40–43% (w/w) sucrose (fractions 6–8) (Malhotra *et al.*, 1989). Coated cisternal remnants peak at about 35% (w/w) sucrose, and uncoated remnants peak at 30% (w/w) sucrose. The purity of the COPI vesicles can be confirmed by SDS-PAGE (an example is shown in Fig. 2A) and by electron microscopy by directly applying droplets of fractions on grids followed by

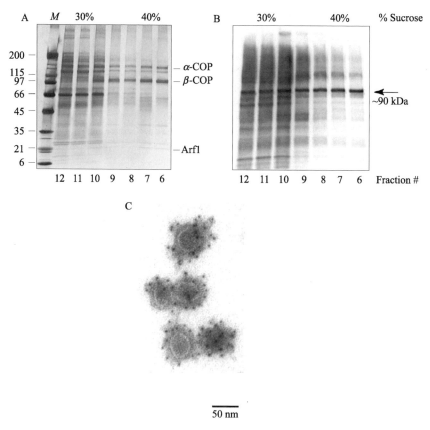

FIG. 2. Purification of biotinylated COPI vesicles. Biotinylated Golgi membranes were incubated with coatomer and Arf for 10 min at 37° to generate COPI vesicles. Membranes were centrifuged to equilibrium on a sucrose gradient and gradient fractions were analyzed by SDS-PAGE and (A) silver staining (M = Molecular weight markers) or (B) immunoblotting using anti-biotin antibodies. (C) Droplets from the gradient corresponding to 40% sucrose were processed for pre-embedding labeling using anti-biotin antibodies. Note that in the total population ~65% vesicles were labeled with anti-biotin.

negative staining, or immuno-labeled using antibodies to biotin followed by 10 nm Protein A-gold, and negatively-stained (an example is shown in Fig. 2C). The biotinylated protein on COPI vesicles (Fig. 2B, ~90 kDa) was identified as golgin-84 by MALDI-mass spectrometry.

Instead of biotinylation, fluorescently labeled COPI vesicles can be purified using this method. However, reagents such as NHS-Alexa-488 (Molecular Probes, A-10235) label many proteins in the COPI vesicles in addition to golgin-84.

(B) *In Vitro* Tethering of COPI Vesicles to Golgi Membranes

Materials

> Purified rat liver Golgi: 1 mg/ml, ~10 μl (Hui *et al.*, 1997)
> Rat liver cytosol (35% ammonium sulfate cut): 200 μg (Rabouille *et al.*, 1995)
> Purified biotinylated COPI vesicles (above): ~500 μl
> Blocking reagents: BSA and soybean trypsin inhibitor (STI, Sigma)
> Assay buffer: 25 mM HEPES (pH 7.0), 2.5 mM magnesium acetate, 100 mM KCl, 1 mM DTT
> GTP, ATP: 40 mM stock, neutralized with KOH
> Glass slides: Esco Superfrost, precleaned (Erie Scientific Company)
> Pap-pen (Zymed)
> Beckman TLS55 rotor
> All membranes and reagents should be prepared in, or buffer-exchanged to, assay buffer.

Methods

Microscope glass slides should be precleaned with 2 M NaOH for 30 min at room temperature, washed extensively with distilled water, and dried using an air duster. Incubation wells should be drawn on cleaned microscope glass slides using a Pap-pen. Ten μl of Golgi membranes are allowed to attach for 10 min at 4° and blocked with 1% BSA and 1% STI in 10 μl of assay buffer containing 9% (w/w) sucrose. Biotinylated COPI vesicles (~500 μl as collected from the gradient) are diluted with 2 volumes of assay buffer and transferred to a tube into which a small step gradient is introduced by sequentially underlaying 5 μl each of 38/40/45/50% (w/w) sucrose. After centrifugation at 100,000g in a TLS55 rotor for 1 h at 4°, around 10 μl of the visible band is sampled.

For uncoating, the concentrated COPI vesicles are incubated with cytosol (200 μg) in 120 μl assay buffer containing 1 mM GTP and 1 mM ATP for 10 min at 37°. BSA (1 mg/ml) is also added to this reaction to

prevent aggregation. Large membranes are then removed by brief centrifugation at 14,000*g*. After incubation with the vesicles in a wet chamber
at 4° for 30 min, the slides are washed and fixed with 8% (w/w) buffered
PFA, containing 10% (w/w) sucrose, and processed for immuno-EM
(Satoh *et al.*, 2003).

Figure 3 shows examples of the types of results that have been obtained.
Biotinylated COPI vesicles (5 nm gold) bind to cisternal membranes containing CASP (Fig. 3A, 10 nm gold) and p24 (Fig. 3B, C, 10 nm gold), the

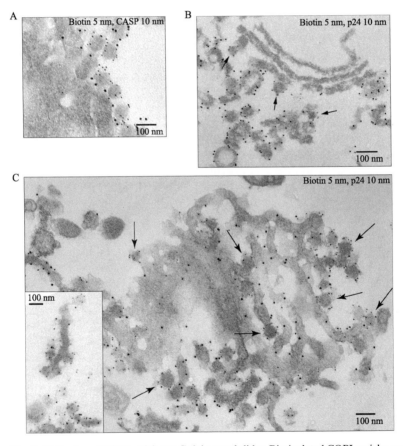

Fig. 3. Tethering of COPI vesicles to Golgi-coated slides. Biotinylated COPI vesicles were
purified and incubated with cytosol for 10 min at 37° to allow uncoating. (A–C) Vesicles were
incubated for 30 min at 4° on glass slides precoated with purified Golgi membranes (Inset in
C). Additional incubation for 30 min at 25° was carried out to allow membrane fusion. After
washing, slides were fixed and double-labeled for biotin (5 nm gold) and either CASP (A) or
p24 (B, C, and inset) (both 10 nm gold), and processed for EM. Arrows in (B and C) indicate
biotinylated COPI vesicles, binding to p24-containing membranes.

latter identifying them as cis-Golgi membranes (Rojo *et al.*, 1997). Warming the slides to 25° for 30 min led to fusion of the vesicles with the membrane (Fig. 3C, inset) showing that biotinylation had not inactivated them.

Acknowledgments

We thank all members of the Warren, Mellman, and Toomre laboratories for helpful comments and discussions, and Marino Zerial for generous provision of purified EEA1. This work was supported by the NIH and the Ludwig Institute for Cancer Research. A.S. was supported by the American Heart Association.

References

Barr, F. A., and Short, B. (2003). Golgins in the structure and dynamics of the Golgi apparatus. *Curr. Opin. Cell Biol.* **15,** 405–413.

Bascom, R. A., Srinivasan, S., and Nussbaum, R. L. (1999). Identification and characterization of golgin-84, a novel Golgi integral membrane protein with a cytoplasmic coiled-coil domain. *J. Biol. Chem.* **274,** 2953–2962.

Franco, M., Paris, S., and Chabre, M. (1995). The small G-protein ARF1GDP binds to the Gt beta gamma subunit of transducin, but not to Gt alpha GDP-Gt beta gamma. *FEBS Lett.* **362,** 286–290.

Fritzler, M. J., Hamel, J. C., Ochs, R. L., and Chan, E. K. (1993). Molecular characterization of two human autoantigens: Unique cDNAs encoding 95- and 160-kD proteins of a putative family in the Golgi complex. *J. Exp. Med.* **178,** 49–62.

Gillingham, A. K., and Munro, S. (2003). Long coiled-coil proteins and membrane traffic. *Biochim. Biophys. Acta* **1641,** 71–85.

Gillingham, A. K., Pfeifer, A. C., and Munro, S. (2002). CASP, the alternatively spliced product of the gene encoding the CCAAT-displacement protein transcription factor, is a Golgi membrane protein related to giantin. *Mol. Biol. Cell.* **13,** 3761–3774.

Hui, N., Nakamura, N., Slusarewicz, P., and Warren, G. (1998). Purification of rat liver Golgi stacks. *In* "Cell Biology: A Laboratory Handbook" (J. Celis, ed.). Academic Press.

Jackson, C. L. (2004). N-terminal acetylation targets GTPases to membranes. *Nat. Cell. Biol.* **6,** 379–380.

Malhotra, V., Serafini, T., Orci, L., Shepherd, J. C., and Rothman, J. E. (1989). Purification of a novel class of coated vesicles mediating biosynthetic protein transport through the Golgi stack. *Cell* **58,** 329–336.

Malsam, J., Satoh, A., Pelletier, L., and Warren, G. (2005). Golgin tethers define subpopulations of COPI vesicles. *Science* **307,** 1095–1098.

Palade, G. (1975). Intracellular aspects of the process of protein synthesis. *Science* **189,** 347–358.

Pavel, J., Harter, C., and Wieland, F. T. (1998). Reversible dissociation of coatomer: Functional characterization of a beta/delta-coat protein subcomplex. *Proc. Natl. Acad. Sci. USA* **95,** 2140–2145.

Pelham, H. R., and Rothman, J. E. (2000). The debate about transport in the Golgi—two sides of the same coin? *Cell* **102,** 713–719.

Rabouille, C., Misteli, T., Watson, R., and Warren, G. (1995). Reassembly of Golgi stacks from mitotic Golgi fragments in a cell-free system. *J. Cell. Biol.* **129,** 605–618.

Rojo, M., Pepperkok, R., Emery, G., Kellner, R., Stang, E., Parton, R. G., and Gruenberg, J. (1997). Involvement of the transmembrane protein p23 in biosynthetic protein transport. *J. Cell. Biol.* **139**, 1119–1135.

Satoh, A., Beard, M. B., and Warren, G. (2005). Preparation and characterization of recombinant golgin tethers. *Methods in Enzymology* **404**, 279–296.

Satoh, A., Wang, Y., Malsam, J., Beard, M. B., and Warren, G. (2003). Golgin-84 is a rab1 binding partner involved in Golgi structure. *Traffic* **4**, 153–161.

Seemann, J., Jokitalo, E., Pypaert, M., and Warren, G. (2000). Matrix proteins can generate the higher order architecture of the Golgi apparatus. *Nature* **407**, 1022–1026.

Shorter, J., and Warren, G. (2002). Golgi architecture and inheritance. *Annu. Rev. Cell. Dev. Biol.* **18**, 379–420.

Simonsen, A., Lippe, R., Christoforidis, S., Gaullier, J. M., Brech, A., Callaghan, J., Toh, B. H., Murphy, C., Zerial, M., and Stenmark, H. (1998). EEA1 links PI(3)K function to Rab5 regulation of endosome fusion. *Nature* **394**, 494–498.

[14] Investigating the Role of ADP-Ribosylation Factor 6 in Tumor Cell Invasion and Extracellular Signal-Regulated Kinase Activation

By HOLLY HOOVER, VANDHANA MURALIDHARAN-CHARI, SARAH TAGUE, and CRISLYN D'SOUZA-SCHOREY

Abstract

Tumor cell invasion is a coordinated process involving the formation of invadopodia and the localized degradation of the extracellular matrix (ECM). The process of cell invasion is regulated by cell-signaling proteins such as Ras-related GTPases and members of the mitogen-activated protein kinase (MAPK) family. Our studies have focused on the role of the ADP-ribosylation factor 6 (ARF6) GTPase in the process of tumor cell invasion. Using activated and dominant negative mutants of ARF6 in a tumor cell culture model, our laboratory has demonstrated that the GTPase cycle of ARF6 regulates invadopodia formation and matrix degradation. Furthermore, ARF6-mediated cell invasion was found to be dependent on the activation of the extracellular signal-regulated kinase (ERK). These findings demonstrate a critical role for ARF6 in ERK activation and tumor cell invasion.

To investigate the role of ARF6 in tumor cell invasion and ERK activation, a number of methods were employed. These procedures include transfection of LOX cells, *in vitro* matrix-degradation assays, immunofluorescence microscopy, and biochemical assays. These approaches can be applied effectively to measure the degree of invasiveness fostered by

METHODS IN ENZYMOLOGY, VOL. 404
0076-6879/05 $35.00
DOI: 10.1016/S0076-6879(05)04014-0

ARF6 and/or other GTPases and to examine the subcellular distribution of the molecular players that are trafficked or recruited to sites of cell invasion.

Introduction

The determining characteristic of an invasive tumor cell is its ability to reorganize the surrounding matrix with the help of proteolytic or matrix degrading enzymes and migrate into the surrounding tissues. As cells invade the extracellular matrix (ECM), they form actin-rich membrane protrusions at the adherent cell surface called "invadopodia." Invadopodia were first described in localized matrix degradation by chicken embryo fibroblasts transformed by Rous sarcoma virus (Chen et al., 1984). Invadopodia recruit various enzymes such as metalloproteinases and serine-proteases to degrade matrix proteins and form the invasive front of the cell (Chen, 1996). Studies have shown that the invasive potential of a tumor cell is directly correlated with the ECM degrading activity of its invadopodia (Chen et al., 1994; Coopman et al., 1998).

The process of cell invasion is regulated by cell-signaling proteins including Ras-related GTPases and members of the mitogen-activated protein kinase (MAPK) family (Hernandez-Alcoceba et al., 2000; Klemke et al., 1997). ADP-ribosylation factor 6 (ARF6), a member of the Ras superfamily of small GTPases, cycles between an active GTP-bound form and an inactive GDP-bound form. Research from several laboratories has conferred roles for ARF6 in endosomal trafficking, regulated exocytosis, actin rearrangements at the cell surface (Chavrier and Goud, 1999; Donaldson, 2003; Sabe, 2003), epithelial to mesenchymal transitions, and cell motility (Palacios and D'Souza-Schorey, 2003; Palacios et al., 2002; Santy and Casanova, 2001). These processes can impinge upon the acquisition of an invasive phenotype. Studies with breast cancer and melanoma cell lines have demonstrated a critical role for ARF6 in tumor cell invasion (Hashimoto et al., 2004; Tague et al., 2004). Using the invasive human melanoma cell line, LOX, our laboratory has shown that ARF6 localizes to invadopodia and that invasion is regulated by the GTPase cycle of ARF6. Expression of ARF6(Q67L), an activated ARF6 mutant, enhanced LOX cell invasion, while expression of ARF6(T27N), a dominant negative mutant of ARF6, abolished the invasive capacity of LOX cells, as measured by in vitro gelatin degradation assays. Furthermore, ARF6-induced invasion is dependent on ERK activity, since inhibiting MAPK/ERK kinase (MEK) by use of PD98059 or a dominant negative MEK mutant, inhibited the invasion mediated by ARF6(Q67L). In contrast, expression of ARF6(T27N) had no effect on invasion induced by a constitutively active MEK mutant.

To examine the invasive capabilities of cells *in vitro*, a number of methods to qualitatively determine and quantitatively measure tumor cell invasion in a LOX cell culture model were employed. This chapter describes the invasion assays and biochemical procedures that our laboratory has utilized to examine the functional properties of ARF6 in ERK activation and tumor cell invasion.

LOX Melanoma Cell Line as Model System

LOX is an amelanotic human melanoma cell line. The cells are highly invasive and form prominent invadopodia that are capable of degrading ECM (Monsky *et al.*, 1994; Nakahara *et al.*, 1998), making LOX a good *in vitro* model system to study tumor cell invasion. Also, in several *in vivo* models, LOX was shown to be highly invasive, as the cells had metastasized to lungs and other secondary sites (Kjonniksen *et al.*, 1989; Yang *et al.*, 1999).

Maintenance of LOX Cells

LOX cells are cultured in RPMI 1640 medium (Gibco, Gaithersburg, MD #31800–022) supplemented with 10% FBS, 2 mM L-glutamine and 100 μg/ml of penicillin and streptomycin (complete RPMI) in a 37°, 5% CO_2 humidified incubator. Cells should be maintained in log growth phase and split before cultures are confluent. LOX cells are typically passaged at a 1:6 to 1:8 dilution twice a week with 0.05% trypsin/EDTA (Gibco, #15400–054) in PBS.

Transfection of LOX Cells

Transfection by Metafectene Reagent. Materials: LOX cells in log phase, 6-well tissue culture treated plates, 1X Trypsin/EDTA, serum free (SF) RPMI, complete RPMI, Metafectene reagent (Biontex Laboratories, Munich, Germany #T020), 1 mg ARF6(Q67L)-HA, and ARF6(T27N)-HA plasmids (Boshans *et al.*, 2000) and Hemacytometer.
Procedure:

1. Harvest LOX cells by incubation with 1X trypsin/EDTA and determine cell count.
2. Seed approximately 175,000 LOX cells in each well of the tissue culture plate with 2 ml complete RPMI and incubate overnight to allow cells to adhere.
3. Aspirate complete RPMI from wells and add 2 ml SF RPMI.
4. Add 1 μg of DNA to 100 μl of SF RPMI and 6 μl of Metafectene reagent to another 100 μl of SF RPMI. Combine diluted DNA and diluted

reagent and incubate at room temperature (RT) for 15 min to allow DNA-Metafectene complexes to form.
5. Add DNA-Metafectene complexes dropwise to the cells and incubate for 4 hr. Incubation with the Metafectene-DNA complexes for longer periods can increase toxicity.
6. Aspirate transfection mix and add 2 ml of complete RPMI and incubate for 24 hr.

Transfection by Electroporation. Materials: 15 mg of plasmid DNA, 0.4 cm Electroporation cuvettes (BioRad, Hercules, CA, 165–2088), Electroporator (BioRad Gene Pulser II) and Electroporation media (EM) (20% FBS/SF RPMI), 6-well tissue culture plates, SF RPMI, and complete RPMI. Procedure:

1. Trypsinize and wash cells in SF RPMI by centrifugation at 250 g for 3 min.
2. Resuspend cells in SF RPMI and determine concentration of cells.
3. Spin down 1.5×10^6 cells and resuspend in 400 μl EM. Add 15 μg DNA and mix by pipetting. Transfer cells to cuvette and electroporate for 15 s at 230 V and 950 μF.
4. Transfer cells to a well with 2 ml complete RPMI and allow cells to recover for 48 hr.

Comments: LOX cells can be successfully transfected using either transfection method. However, the Metafectene protocol is recommended for the following reasons: (1) transfection efficiency ranges from 30 to 40% with Metafectene while it is ≤10% with electroporation; (2) transfection with Metafectene requires less DNA compared to electroporation; (3) fewer cells are required with Metafectene due to low level of toxicity; and (4) recovery time is 24 hr with Metafectene and 48 hr with electroporation.

In Vitro Invasion Assays/Gelatin Degradation Assays

To investigate the role of ARF6 in invadopodia formation and ECM degradation, previously published protocols were adapted. Briefly, localized matrix degradation by LOX is observed by seeding cells on a gelatin matrix labeled with green or red fluorescence. Regions of gelatin degradation appear as dark spots on a fluorescent background.

CFDSE and TRSE Labeled Gelatin Coated Coverslips

[This protocol is a modified version of a previously published procedure (Chen *et al.*, 1994).]

<u>Materials</u>: Gelatin (Sigma, St. Louis, MO, G-2500), 12-well glass coverslips (Fisher, Pittsburg, PA, 12-545-100 18CIR.-1), 12-well tissue culture plates, 1% paraformaldehyde, sterile water, SF RPMI, 5-6-carboxyl-fluorescein diacetate succinimidyl ester (CFDSE) (Molecular Probes, Eugene, OR, C-1157), Texas-red-X, succinimidyl ester (TRSE) (Molecular Probes, T-20175), DMSO, PBS, and 1 M NaHCO$_3$ (pH 8.3).

<u>Procedure:</u>

Preparation of Gelatin-Coated Coverslips

1. Prepare 2% gelatin in PBS. To aid dissolution, warm gelatin/PBS slurry at 37° and vortex until all gelatin crystals have dissolved. Repeat if necessary,
2. Flame individual circular 12-well glass coverslips to sterilize and place in 12-well plate. Allow coverslips to cool completely, as this will aid in even spreading of gelatin.
3. Coat the entire surface of each coverslip with 200 μl of gelatin by surface tension. Immediately aspirate off excess gelatin by tilting the plate and leaving just a small bead at the bottom edge. Tilt the plate back so the bead of gelatin spreads evenly over the entire coverslip.
4. Place coated coverslips at 4° to air dry overnight. It is important to dry gelatin slowly at 4° or it will dry unevenly. Do not over dry or crystallization may occur.
5. Rehydrate the gelatin for 30 min with 1 ml sterile water at 4°.
6. Fix for 30 min in 1% paraformaldehyde at RT.
7. Wash the coated coverslips 3 times with PBS, 5 min/wash. To ensure sterility, the coverslips can be exposed to UV light during the washes.

Fluorescent Labeling of Gelatin with CFDSE

1. Prepare a 10 mM stock of CFDSE in DMSO.
2. From the stock prepare 0.5 μM of CFDSE in PBS and add 1 ml to each gelatin-coated coverslip. Incubate at RT for 10 min in the dark.
3. Wash 3 times in PBS, 5 min/wash.
4. Quench coverslips in SF RPMI for 1 hr in the incubator before seeding cells.

Fluorescent Labeling of Gelatin with TRSE

1. Prepare 5 mg/ml stock of TRSE in DMSO.
2. Add 0.6 μl of stock TRSE to 12 ml of buffer (0.1 M NaHCO$_3$/PBS) and immediately add 1 ml to each gelatin-coated coverslip. Incubate 15 min at RT in the dark.
3. Wash 3 times in PBS, 5 min/wash.

4. Quench coverslips in SF RPMI for 1 hr in the incubator before seeding cells.

FITC-Conjugated Gelatin Coverslips

[This protocol is adapted from a previously published protocol (Bowden *et al.*, 2001).]

Preparation of FITC-Conjugated Gelatin. Materials: Low-salt conjugation buffer (50 mM Na$_2$B$_4$O$_7$, pH 9.3), High-salt conjugation buffer (50 mM Na$_2$B$_4$O$_7$, 40 mM NaCl, pH 9.3), 6 L PBS, Dialysis tubing (MWCO 6000–8000), Fluorescein 5isothiocyanate (FITC) (Molecular Probes, F-1906), 300 bloom gelatin type A from porcine skin (Sigma, G-2500) and Sucrose.
Procedure:

1. Mix 10 mg FITC with 333 ml low-salt conjugation buffer.
2. Dissolve 0.5 g gelatin in 25 ml high-salt conjugation buffer at 37°.
3. Wash prepared dialysis tubing inside and out with sterile distilled water and then with high-salt conjugation buffer.
4. Dialyze the gelatin solution against FITC/low-salt conjugation buffer for at least 90 min in complete darkness at 37° with constant stirring.
5. Replace the FITC/low-salt conjugation buffer with PBS at 37° and continue dialysis for 3 days with PBS changes twice per day.
6. Carefully remove FITC-gelatin from the dialysis tubing.
7. Add sucrose to 2% (w/v) final concentration.
8. Divide into 0.5 ml aliquots and store at 4° in the dark.

Comments: FITC is hazardous. Proper handling and disposal should be taken. FITC-conjugated gelatin is stable for 4 months.

Preparation of FITC-Gelatin Coated Coverslips. Materials: FITC-conjugated gelatin, 12-well glass coverslips, 12-well tissue culture plates, glutaraldehyde, PBS, sodium borohydride (Sigma, S-9125), and SF RPMI.

1. De-solidify FITC-gelatin for 15 min in the dark at 37°.
2. Flame 12-well coverslips and place in 12-well plate. Spread approximately 200 μl of FITC-gelatin to coat the entire surface of the coverslip and aspirate off excess gelatin.
3. Place coverslips at 4° for 5–10 min to harden gelatin.
4. Carefully overlay at least 500 μl of ice-cold 0.5% glutaraldehyde/PBS onto each coverslip and incubate for 15 min at 4°.
5. Wash coverslips 3 times with PBS, 5 min/wash.
6. Incubate coverslips with 1 ml freshly made sodium borohydride (5 mg/ml PBS) for 3 min to quench auto-fluorescence. Since

hydrogen bubbles are produced during this incubation step, ensure that the coverslips stay submerged.

7. Wash coverslips 3 times with PBS, 5 min/wash.
8. Equilibrate coverslips with SF RPMI for 30 min at 37° in the incubator before seeding cells.

Seeding of LOX Cells onto Gelatin-Coated Coverslips

Seed 10^5 LOX cells per labeled gelatin-coated coverslip in a 12-well plate, 1–2 days after transfection. Generally, cells from each transfected well of a 6-well tissue culture plate are trypsinized and re-seeded at a 1:6 to 1:8 dilution onto each coverslip in 1 ml complete RPMI. LOX cells are cultured for 16–24 hrs on gelatin coverslips for degradation/invasion assays. Experiments are repeated three times with two separate coverslips per experimental condition.

Comments: To quantitate the degree of invasion, the FITC-conjugated gelatin coverslips are recommended. On these coverslips, 25–30% of control LOX cells exhibit degradation spots, while on CFDSE/TRSE labeled coverslips the degree of invasion is lower and more variable, ranging from 9 to 13%. The reason for this discrepancy is unclear, but possibilities include the differences in fixation procedures and/or gelatin labeling. While the CFDSE/TRSE labeling method allows for a green or red matrix, FITC-gelatin is limited to observing cells under green fluorescence.

Immunostaining for HA-Tagged ARF6 Mutants

Materials: Murine monoclonal HA antibody (Covance, Princeton, NJ, MMS-101P), rhodamine-phalloidin (Molecular Probes, R-415) or fluorescein-phalloidin (Molecular Probes, F-432), and goat-anti-mouse Cy 5 antibody (Amersham, PA45002).

Procedure:

1. Fix cells with 2% paraformaldehyde followed by blocking and permeabilization.
2. Add anti-HA antibody (1:200) to coverslips by surface tension and incubate for 2 hr.
3. Wash and incubate cells with Cy 5 secondary antibody (1:400) and rhodamine-phalloidin (1:100) or fluorescein-phalloidin for 1 hr to stain for actin. Fluorophores should be distinct from gelatin fluorescence.
4. Wash cells, mount coverslips onto slides, and seal edges with clear fingernail polish.

Comments: LOX cells exhibit a low level of red auto-fluorescence.

Image Capture and Confocal Microscopy

Immunofluorescent imaging is performed on a Bio-Rad MRC 1024 scanning confocal 3-channel system that houses a krypton-argon laser with excitation filters for 488, 568, and 647 nm. Images are captured using the Bio-Rad Lasersharp 2000 software and processed in Adobe Photoshop. Confocal images are taken along the x/y axis to visualize the tips of invadopodia at the sites of degradation (Fig. 1A–C). Invadopodia projecting into the gelatin matrix is observed along the x/z axis. For this, consecutive x/y planes are taken along the entire length of invadopodia projecting from the cell into the matrix and visualized as a stacked x/z image (Fig. 1D and E). To quantitate invasion, cells are counted on a single cell basis, and those exhibiting degradation spots are represented as a percentage of invasion.

Observations: Degradation spots are typically seen as a cluster of small dark spots against the fluorescent matrix (Fig. 1A). They are usually found underneath or at the edges of an invasive cell. Occasionally, more aggressive degradation patches can be seen as a "trail" through the gelatin (Fig. 2A). These degradation trails appear to have formed as cells first

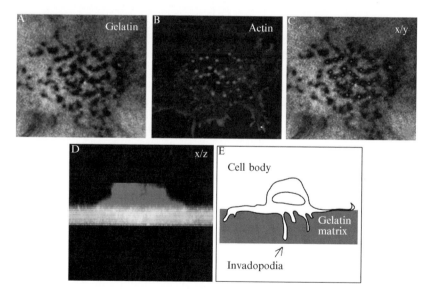

FIG. 1. LOX cells were seeded on CFDSE-labeled gelatin for 24 hr, fixed and stained for actin with rhodamine-phalloidin. Images are representative single confocal sections of an invading cell. (A) Single confocal plane of gelatin with degradation spots and actin (B) along the x/y axis at tips of invadopodia. (C) Merged x/y images of A and B. (D) Stacked side projection of same cell along x/z axis showing invadopodia extending into the gelatin matrix. (E) Schematic representation of D. (See color insert.)

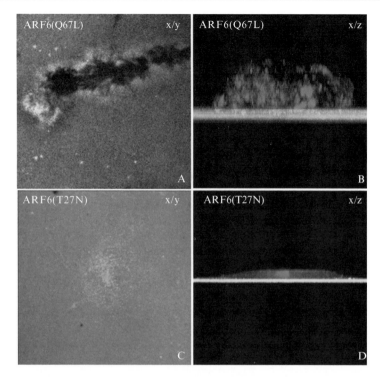

Fig. 2. LOX cells were transfected with HA-tagged ARF6 mutants, seeded on CFDSE-labeled gelatin and immunofluorescently labeled red. Images are shown along the x/y or x/z axis. (A) Representative confocal image of an ARF6(Q67L)-transfected LOX cell exhibiting an aggressive degradation trail. (B) Image of an ARF6(Q67L)-transfected LOX cell with a rounded phenotype and invadopodia extending into the gelatin matrix. (C and D) Images of an ARF6(T27N)-expressing cell that shows a spread phenotype and does not form invadopodia or degrade gelatin. (See color insert.)

invaded down into the gelatin and then continued to degrade through the matrix while moving horizontally.

ARF6(Q67L)-HA is diffuse throughout the cell with increased localization at the cell surface and invadopodia. ARF6(Q67L)-transfected cells exhibit a more rounded phenotype with an enhancement of invadopodia formation (Fig. 2B) and large degradation spots under them. Approximately 3–5% of these cells create "degradation trails," and thus, qualitatively demonstrate the aggressive nature of ARF6(Q67L)-transfected cells (Fig. 2A). In contrast, LOX cells expressing ARF6(T27N) have a broad, flattened appearance exhibiting little or no gelatin degradation (Fig. 2C and D). ARF6(T27N)-HA localizes predominantly to the perinuclear cytoplasm.

Comments: Not all degradation spots are associated with invado-podia, as the cells are dynamic and constantly form and reform invadopo-dia. Hence, actin counterstaining is particularly useful to label individual cells to confirm the association of degradation spots with their respective cells (Fig. 1A–C). If the gelatin coating is too thick, cells will sink into the matrix, leaving outlines of the cells as large dark areas and should not be mistaken for degradation.

In addition to the gelatin substrates described here, dye-quenched (DQ) gelatin can also be used to visualize matrix degradation (Artym et al., 2002). This gelatin is so heavily labeled with fluorescein molecules that the fluorescence is quenched. Cleavage of DQ gelatin by proteolytic activity produces fluorescent gelatin-FITC peptides, and thus, the fluorescence localizes to sites of degradation. This is another technique to monitor invasion; however, the degradation assays outlined in this section may facilitate trafficking–localization studies in invadopodia and quantitative analysis of invasion due to the absence of fluorescence at sites of degradation.

Detection of Endogenous ARF6 Activation and ARF6 Localization in Invasive LOX Cells

Measurement of Endogenous ARF6-GTP Levels

As previously mentioned, the ARF6 GTPase cycle regulates the invasive ability of LOX cells. To investigate whether ARF6 might also mediate cell invasiveness induced by a physiologically relevant stimulus, LOX cells were treated with 40 ng/ml hepatocyte growth factor (HGF). HGF induced invasion in LOX cells by 2.5 fold over control. This increase was not seen in the presence of ARF6-GDP, indicating that ARF6 is required for events downstream of growth factor signaling that lead to melanoma cell invasion (Tague et al., 2004). Endogenous ARF6-GTP levels increased with HGF treatment, as measured by a novel in vitro GST pull-down assay developed in our laboratory (Schweitzer and D'Souza-Schorey, 2002). This assay was developed based on the specific binding of metallothionein 2 (MT-2) to ARF6-GTP. GST-MT-2 fusion protein conjugated glutathi-one-agarose beads are incubated with cell lysates, washed, and the bound proteins are subjected to Western blotting procedures with antibodies against ARF6.

Comments: ARF6-GTP is sensitive to freeze-thaw cycles. It is best to use lysates as soon as they are collected; however, they can be snap-frozen in liquid N_2 and stored at $-70°$.

Localization of Endogenous ARF6

The DS1 antibody can be used to detect endogenous ARF6 by standard immunofluorescence microscopy techniques. This antibody was generated from a 12-aa peptide close to the amino-terminal end of ARF6 as previously described (Schweitzer and D'Souza-Schorey, 2002). In addition to the tubular endosomal compartments, endogenous ARF6 localizes to the tips of invadopodia along with actin, paxillin, and phosphotyrosine (Tague *et al.*, 2004), which were previously identified at the invadopodia (Bowden *et al.*, 1999; Nakahara *et al.*, 1998).

ARF6 Regulates ERK Activation in LOX Cells

Measurement and Localization of Endogenous ERK

Activation of the MEK/ERK signaling pathway has been associated with increased cell motility, matrix metalloproteinase production, and invasion (Klemke *et al.*, 1997; Satyamoorthy *et al.*, 2003; Smalley, 2003). Activation of ERK is concomitant with its phosphorylation. To analyze the distribution of activated ERK in LOX cells transfected with ARF6(Q67L) and ARF6(T27N), the mouse monoclonal anti-phospho-p44/42 MAPK (Thr-202/Tyr-204) antibody, E10 (Cell Signaling, Beverly, MA, #9106), was utilized according to manufacturer's instructions. Phosphorylated ERK partially colocalized with activated ARF6 at the invadopodia (Tague *et al.*, 2004). Furthermore, phospho-ERK levels were augmented in ARF6 (Q67L)-expressing cells and decreased in ARF6(T27N)-expressing cells compared to non-transfected cells. Additionally, results from Western blotting procedures using the anti-phospho-p44/42 antibody revealed a similar increase in phospho-ERK levels from cultures transfected with ARF6(Q67L), but the inhibitory effect of ARF6(T27N) on ERK activation was not observed due to masking of endogenous phospho-ERK from non-transfected cells. However, ARF6(T27N) inhibited HGF-induced ERK activation. These mutants did not affect total ERK levels as detected by anti-p44/42 antibody (Cell Signaling, #9102). To confirm transgene expression, membranes were also probed with the mouse anti-HA antibody.

Measurement of Endogenous ERK Kinase Activity

To complement the phospho-ERK results above, assessment of endogenous ERK kinase activity was carried out using a nonradioactive p44/42 MAP kinase assay kit (Cell Signaling, #9800). Briefly, the p44/42 MAP kinase is immunoprecipitated from lysates and incubated with the Elk-1 transcription factor as the substrate. Phosphorylated Elk-1 is detected using a phospho-specific Elk-1 antibody by immunoblotting procedures. In LOX cultures transfected with ARF6(Q67L), a significant increase in the levels of Elk-1 phosphorylation is observed. In contrast, ARF6(T27N)-expressing LOX cells show little ERK phosphorylation, and likewise, Elk-1 phosphorylation (unpublished results).

Use of Inhibitors to Block MEK/ERK Signaling

Studies involving the inhibition or activation of ERK have been useful in shedding light on the mechanism by which ARF6 regulates tumor cell invasion. To this end, PD98059 and MEK mutants are used to regulate ERK activation. In our studies, treatment of LOX cells with 18 μM PD98059 significantly reduced the levels of ERK activation and invasion mediated by HGF and ARF6(Q67L). Also, co-expression of dominant negative (S218A) MEK completely abolished the invasiveness of ARF6 (Q67L)-transfected cells. In contrast, ARF6(T27N) had no effect on the invasive capacity of cells expressing constitutively activated (S218/222D) MEK (Tague et al., 2004). (MEK mutants were kindly provided by Andrew Catling and Mike Weber [University of Virginia, Charlottesville].) These findings document a link between ARF6-mediated signaling and ERK activation.

Concluding Remarks

The methods described in this chapter have been used successfully to investigate the role of ARF6 in ERK activation and invasion using a tumor cell culture model. These approaches can be applied effectively to examine the subcellular distribution of ARF6 and other GTPases and their effectors to further elucidate the mechanisms that regulate invadopodia formation and localized matrix degradation. Moreover, extended studies conducted with these protocols would foster a greater understanding of the molecular players involved in tumor cell invasion, such as proteases and integrin receptors, which are trafficked or recruited to sites of cell invasion and are essential for matrix degradation.

Acknowledgments

This work was supported as a subproject of a Program Project Grant to the Notre Dame-Walther Cancer Center from the Department of Defense, U.S. Army Medical Research and Materiel Command.

References

Artym, V. V., Kindzelskii, A. L., Chen, W. T., and Petty, H. R. (2002). Molecular proximity of seprase and the urokinase-type plasminogen activator receptor on malignant melanoma cell membranes: Dependence on beta1 integrins and the cytoskeleton. *Carcinogenesis* **23**, 1593–1601.

Boshans, R. L., Szanto, S., van Aelst, L., and D'Souza-Schorey, C. (2000). ADP-ribosylation factor 6 regulates actin cytoskeleton remodeling in coordination with Rac1 and RhoA. *Mol. Cell. Biol.* **20**, 3685–3694.

Bowden, E. T., Barth, M., Thomas, D., Glazer, R. I., and Mueller, S. C. (1999). An invasion-related complex of cortactin, paxillin and PKCmu associates with invadopodia at sites of extracellular matrix degradation. *Oncogene* **18**, 4440–4449.

Bowden, E. T., Coopman, P. J., and Mueller, S. C. (2001). Invadopodia: Unique methods for measurement of extracellular matrix degradation *in vitro*. *Methods Cell. Biol.* **63**, 613–627.

Chavrier, P., and Goud, B. (1999). The role of ARF and Rab GTPases in membrane transport. *Curr. Opin. Cell. Biol.* **11**, 466–475.

Chen, W. T. (1996). Proteases associated with invadopodia, and their role in degradation of extracellular matrix. *Enzyme. Protein.* **49**, 59–71.

Chen, W. T., Olden, K., Bernard, B. A., and Chu, F. F. (1984). Expression of transformation-associated protease(s) that degrade fibronectin at cell contact sites. *J. Cell. Biol.* **98**, 1546–1555.

Chen, W. T., Lee, C. C., Goldstein, L., Bernier, S., Liu, C. H., Lin, C. Y., Yeh, Y., Monsky, W. L., Kelly, T., Dai, M., *et al.* (1994). Membrane proteases as potential diagnostic and therapeutic targets for breast malignancy. *Breast Cancer Res. Treat.* **31**, 217–226.

Chen, W. T., Yeh, Y., and Nakahara, H. (1994). An *in vitro* cell invasion: Determination of cell surface proteolytic activity that degrades extracellular matrix. *J. Tissue. Cult. Methods* **16**, 177–181.

Coopman, P. J., Do, M. T., Thompson, E. W., and Mueller, S. C. (1998). Phagocytosis of cross-linked gelatin matrix by human breast carcinoma cells correlates with their invasive capacity. *Clin. Cancer Res.* **4**, 507–515.

Donaldson, J. G. (2003). Multiple roles for Arf6: sorting, structuring, and signaling at the plasma membrane. *J. Biol. Chem.* **278**, 41573–41576.

Hashimoto, S., Onodera, Y., Hashimoto, A., Tanaka, M., Hamaguchi, M., Yamada, A., and Sabe, H. (2004). Requirement for Arf6 in breast cancer invasive activities. *Proc. Natl. Acad. Sci. USA* **101**, 6647–6652.

Hernandez-Alcoceba, R., del Peso, L., and Lacal, J. C. (2000). The Ras family of GTPases in cancer cell invasion. *Cell. Mol. Life. Sci.* **57**, 65–76.

Kjonniksen, I., Storeng, R., Pihl, A., McLemore, T. L., and Fodstad, O. (1989). A human tumor lung metastasis model in athymic nude rats. *Cancer. Res.* **49**, 5148–5152.

Klemke, R. L., Cai, S., Giannini, A. L., Gallagher, P. J., de Lanerolle, P., and Cheresh, D. A. (1997). Regulation of cell motility by mitogen-activated protein kinase. *J. Cell. Biol.* **137**, 481–492.

Monsky, W. L., Lin, C. Y., Aoyama, A., Kelly, T., Akiyama, S. K., Mueller, S. C., and Chen, W. T. (1994). A potential marker protease of invasiveness, seprase, is localized on invadopodia of human malignant melanoma cells. *Cancer Res.* **54**, 5702–5710.

Nakahara, H., Mueller, S. C., Nomizu, M., Yamada, Y., Yeh, Y., and Chen, W. T. (1998). Activation of beta1 integrin signaling stimulates tyrosine phosphorylation of p190Rho-GAP and membrane-protrusive activities at invadopodia. *J. Biol. Chem.* **273**, 9–12.

Palacios, F., and D'Souza-Schorey, C. (2003). Modulation of Rac1 and ARF6 activation during epithelial cell scattering. *J. Biol. Chem.* **278**, 17395–17400.

Palacios, F., Schweitzer, J. K., Boshans, R. L., and D'Souza-Schorey, C. (2002). ARF6-GTP recruits Nm23-H1 to facilitate dynamin-mediated endocytosis during adherens junctions disassembly. *Nat. Cell. Biol.* **4,** 929–936.

Sabe, H. (2003). Requirement for Arf6 in cell adhesion, migration, and cancer cell invasion. *J. Biochem. (Tokyo)* **134,** 485–489.

Santy, L. C., and Casanova, J. E. (2001). Activation of ARF6 by ARNO stimulates epithelial cell migration through downstream activation of both Rac1 and phospholipase D. *J. Cell. Biol.* **154,** 599–610.

Satyamoorthy, K., Li, G., Gerrero, M. R., Brose, M. S., Volpe, P., Weber, B. L., Van Belle, P., Elder, D. E., and Herlyn, M. (2003). Constitutive mitogen-activated protein kinase activation in melanoma is mediated by both BRAF mutations and autocrine growth factor stimulation. *Cancer Res.* **63,** 756–759.

Schweitzer, J. K., and D'Souza-Schorey, C. (2002). Localization and activation of the ARF6 GTPase during cleavage furrow ingression and cytokinesis. *J. Biol. Chem.* **277,** 27210–27216.

Smalley, K. S. (2003). A pivotal role for ERK in the oncogenic behaviour of malignant melanoma? *Int. J. Cancer* **104,** 527–532.

Tague, S. E., Muralidharan, V., and D'Souza-Schorey, C. (2004). ADP-ribosylation factor 6 regulates tumor cell invasion through the activation of the MEK/ERK signaling pathway. *Proc. Natl. Acad. Sci. USA* **101,** 9671–9676.

Yang, M., Jiang, P., An, Z., Baranov, E., Li, L., Hasegawa, S., Al-Tuwaijri, M., Chishima, T., Shimada, H., Moossa, A. R., and Hoffman, R. M. (1999). Genetically fluorescent melanoma bone and organ metastasis models. *Clin. Cancer Res.* **5,** 3549–3559.

[15] Assays and Properties of the Arf GAPs AGAP1, ASAP1, and Arf GAP1

By MAGNUS MUTAH CHE, ZHONGZHEN NIE, and PAUL A. RANDAZZO

Abstract

ADP-ribosylation factors (Arfs) are Ras-like GTP-binding proteins that regulate membrane traffic and actin remodeling. Arf function requires GTP hydrolysis but Arf lacks GTPase activity; consequently, Arf function is dependent on Arf GTPase-activating proteins (GAPs). The Arf GAPs are a structurally diverse group of at least 16 proteins. Several Arf GAPs use a single Arf isoform. However, due to structural differences, the conditions supporting productive interactions between Arf and different Arf GAPs vary. Here, we describe preparation and basic properties of three Arf GAPs. We use these proteins to illustrate assays for Arf GAP activity. Conditions that optimize activity for each GAP are discussed. These methods can be used for the further characterization of Arf–Arf GAP interaction that is necessary for understanding the function of Arf in cellular physiology.

METHODS IN ENZYMOLOGY, VOL. 404   0076-6879/05 $35.00
                                                   DOI: 10.1016/S0076-6879(05)04015-2

Introduction

ADP-ribosylation factors (Arfs) are members of a family of Ras-like GTP-binding proteins. Arf proteins are ubiquitously expressed in eukaryotic cells and highly conserved (reviewed in Randazzo et al., 2000; Moss and Vaughan, 1998; Logsdon and Kahn, 2003). The six mammalian Arf proteins are grouped into three classes based on sequence homology: class I (Arf1, 2, and 3), class II (Arf4 and 5), and class III (Arf6). The Arfs were first identified as cofactors for cholera toxin catalyzed ADP-ribosylation of the heterotrimeric G protein Gs. Subsequent work, focusing on Arf1 and Arf6, has established Arfs as regulators of membrane traffic and actin.

As for other G-proteins, the function of Arf1 requires that it bind and hydrolyze GTP. Because Arf1 has low intrinsic nucleotide exchange and GTP hydrolysis rates, it is dependent on accessory proteins to accelerate these reactions. The proteins that accelerate nucleotide exchange are called the guanine nucleotide exchange factors (Donaldson and Jackson, 2000; Jackson and Casanova, 2000). Those accessory proteins that accelerate the hydrolysis of GTP that is bound to Arf are called Arf-directed GTPase activating proteins (Arf GAPs) (Cassel, 2003; Randazzo and Hirsch, 2004; Schmalzigaug and Premont, 2003; Turner et al., 2001).

The Arf GAPs are a diverse group of proteins. Arf GAP1 was the first Arf GAP to be cloned (Cukierman et al., 1995; Makler et al., 1995). A zinc binding motif was found to comprise the catalytic Arf GAP domain. Subsequently, 24 genes have been found to encode proteins with the Arf GAP domain (Randazzo and Hirsch, 2004; Schmalzigaug and Premont, 2003). Of those, three groups, comprising 16 genes, encode proteins with documented GAP activity (Fig. 1). Arf GAP1/3 have the Arf GAP domain at the N terminus of the protein and a targeting domain in the C-terminus (Cukierman et al., 1995; Liu et al., 2001a,b; Makler et al., 1995). Git1/2 also has the Arf GAP domain at the N-terminus but diverges from Arf GAP1/3 in the C-terminal portion of the protein (Bagrodia et al., 1999; Premont et al., 1998; Turner et al., 1999). The AZAPs have a catalytic core of a PH, Arf GAP, and ANK repeat domains (Andreev et al., 1999; Brown et al., 1998; Jackson et al., 2000; Krugmann et al., 2002; Miura et al., 2002; Nie et al., 2003, 2002; Stacey et al., 2004). The AZAPs can be further grouped into four classes, the ASAPs, AGAPs, ARAPs, and ACAPs, based on protein domains outside the catalytic core. Both ASAPs and ACAPs have a BAR domain in the N-terminus (Peter et al., 2004; Nie and Randazzo, unpublished), but the ASAPs also have a proline-rich domain and an SH3 domain in the C-terminus (Brown et al., 1998; Jackson et al., 2000). The ARAPs have five PH domains, a Rho GAP domain, and Ras-association domain. The AGAPs have a split PH domain and a

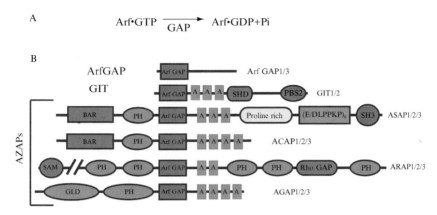

FIG. 1. Proteins with Arf GAP activity: (A) Reaction catalyzed by Arf GAPs. (B) Schematic of proteins with Arf GAP activity. Key: A, ankyrin repeat, BAR, Bin Amphiphysin Rsv161,167 domain; GLD, GTP-binding protein-like domain; PBS2, paxillin binding site; PH, pleckstrin homology domain; SAM, sterile α-motif domain; SH3, Src-homology 3 domain; SHD, spa homology domain. Accession #s: rat Arf GAP1 = U35776; human Arf GAP3 = NM_014570; rat Git1 = AF085693; mouse Git2 = NM_019834; mouse ASAP1 = AF075461, AF075462; human ASAP2 = NM_003887; human ASAP3 = NM_017707; human ACAP1 = D30758, NM_014716; human ACAP2 = AJ238248; human ACAP3 = KIAA1716; human ARAP1 = AB018325, AY049732; human ARAP2 = AY049733; human ARAP3 = NM_022481; human AGAP1 = NM_014770; human AGAP2 = NM_014914; human AGAP3 = AF359283. (See color insert.)

GTP-binding protein-like domain. Some of the domains on these structurally complex proteins are involved in regulating enzymatic activity.

Lipids have been found to be critical regulators of Arf GAP activity. Both the substrate Arf1•GTP and the catalytic Arf GAPs associate with lipids. We have identified point mutants of Arf1 that do not require the presence of lipids to bind GTP (Yoon et al., 2004); however, even with soluble Arf1 mutants, each Arf GAP requires lipid for optimal activity (Bigay et al., 2003; Brown et al., 1998; Jackson et al., 2000; Kam et al., 2000; Nie et al., 2002; Randazzo and Kahn, 1994). Furthermore, the lipid requirements are unique to particular Arf GAPs, with proteins being sensitive to both composition and form of the lipids.

The purpose of this chapter is to describe methods for determining GAP activity, the reaction presented in Fig. 1A. We describe the properties and preparation of three Arf GAPs and the substrate Arf1•GTP. In addition, because the reaction is affected by both the composition and physical form of lipids, methods for preparing lipids for the assays are described. Details of the assays are then given, using data obtained with ASAP1 to illustrate typical results.

Properties of Arf GAP1, ASAP1 and AGAP1 Relevant to Determination
of GAP Activity

Arf GAP1 (Cukierman *et al.*, 1995) has the catalytic domain at the
N-terminus and a C-terminal targeting domain (see schematic in Fig. 1B).
A recombinant protein comprised of the N-terminal half, residues 1–257,
is active as a GAP. When first purified, the protein was reported to be
phosphatidylinositol 4,5-bisphosphate (PIP2) dependent (Makler *et al.*,
1995). Although this dependence is observed when using nonmyristoylated
Arf1 as a substrate[1], Arf GAP1 is not affected by PIP2 when myrArf1 is
used as a substrate. Nevertheless, ArfGAP1 does associate with lipids
and activity is greatest in the presence of vesicles approximating the
size of membrane trafficking transport vesicles (Bigay *et al.*, 2003). For
optimal activity, assays use large unilamellar vesicles (LUVs) prepared by
extrusion as described below.

ASAP1 and AGAP1 are members of the AZAP family of Arf GAPs
(Randazzo and Hirsch, 2004) with a catalytic core of PH, Arf GAP, and
Ank repeat domains. Proteins lacking the PH domain are devoid of activi-
ty. Proteins lacking the Ank repeats are insoluble. Therefore, the minimum
structure for a recombinant protein to have activity is comprised of the PH,
Arf GAP, and ANK domains of the AZAP family member. The AZAPs
have additional domains that define individual classes (see Fig. 1B for
schematic of ASAP1 and AGAP1) and that may affect GAP activity.
When examining these proteins, the influence of other domains on the
Arf GAP domain should be considered.

ASAP1 and AGAP1 are dependent on acid phospholipid binding to
their PH domains for activity (Che *et al.*, 2005; Kam *et al.*, 2001). Phospholipids can be presented either as mixed micelles with Triton X-100 or
liposomes. Both total acid phospholipid content and the presence of specific phospholipids are important. For [325–724]ASAP1, mixed micelles of
0.1% Triton X-100 with 180–360 μM phosphatidylserine (or phosphatidic
acid) and 45–90 μM phosphatidylinositol 4,5-bisphosphate (PIP2) or LUVs
containing 15% PS and 1–5% PIP2, at a total lipid concentration of 500–
1000 μM efficiently support activity. The same conditions can be used for
full-length AGAP1 and [347–804]AGAP1.

---

[1] In eukaryotic cells, Arf1 is cotranslationally modified with myristic acid at the N-terminal
glycine. Bacteria lack the enzyme involved in myristoylation. Consequently, recombinant
Arf1 expressed and purified from bacteria is not typically modified (note: in Chapter 10 of
this volume, "Preparation of Myristoylated Arf1 and Arf6," we describe the preparation of
myristoylated Arf1 in bacteria). Arf1 without myristate is recruited to hydrophobic surfaces
by acid phospholipids and specific association with PIP2 (Randazzo, 1997a; Seidel *et al.*,
2004).

Methods

Preparation of Large Unilamellar Vesicles (LUVs) by Extrusion

To prepare LUVs by extrusion, lipids dissolved in chloroform/methanol are mixed in a 12×75 mm siliconized glass tube in a quantity sufficient to make a 5–10 fold concentrate of the vesicles to be added to an assay. For instance, for a reaction that will contain 500 μM total phospholipids, add 2.50 μmol total lipid to make 500 μl of a 5 mM phospholipid suspension. Typically, the mixture contains 40% phosphatidylcholine, 25% phosphatidylethanolamine, 15% phosphatidylserine, 10% phosphatidylinositol, and 10% cholesterol. For ASAP1 and AGAP1, the amount of PI is reduced to 9–5% and replaced with 1–5% PIP2. Chloroform is evaporated under a stream of nitrogen for 30–60 min. Then, residual chloroform is removed in vacuo for one hour. We use a lyophilizer and achieve a pressure of <100 μm of mercury. Add buffer to the dried lipids. Tris-buffered (TBS) or phosphate-buffered saline (PBS) are suitable. Allow the lipids to hydrate at room temperature for 10 min, then mix to make a suspension, removing the dry lipids from the wall of the glass tube. The suspension is frozen and thawed five times using an ethanol/dry ice bath and a water bath. Then, using a lipid extruder (Lu and Nelsestuen, 1996; Macdonald *et al.*, 1991), which can be purchased from Avanti Polar Lipids (Alabama) or made at a machine shop, pass the lipid suspension through a Whatman Nuclepore Track Etched membrane 11 times. For Arf GAP1, we recommend using membranes with 0.1 μm pores. [325–724]ASAP1 and AGAP1 are less sensitive to membrane curvature and LUVs prepared using pores from 0.05 to 1 μm give similar results.

Loading Arf1 with GTP

Although native Arf1 is myristoylated, Arf GAPs use nonmodified Arf as well as myristoylated Arf. The use of a myristoylated or nonmodified Arf depends on which assay will be used and whether the substrate will be used for saturation kinetics. Nonmodified Arfs are simple to prepare following protocols described in this series (Randazzo *et al.*, 1992) and are used in our laboratory for routine assays with ^{32}P labeled GTP. However, nonmyristoylated Arf1 is difficult to load to high concentrations with GTP. Therefore, when examining saturation kinetics or using the assay that depends on tryptophan fluorescence of Arf (third assay described in this chapter), we use either myristoylated Arf1 or an Arf1 point mutant (leucine 8 changed to lysine). By following the protocol in Randazzo (1997b) and in Chapter 10 of this volume ("Preparation of Myristoylated Arf1 and Arf6"), we have been able to routinely prepare

Arf1 that is 100% myristoylated and will bind more than 0.7 mol GTP/ mol Arf. We have also found that [L8K]Arf1 loads efficiently with GTP and is efficiently used by all Arf GAPs that we have examined (Yoon et al., 2004).

Two buffer systems for nucleotide exchange on Arf1 are used. Buffer A (25 mM HEPES, pH 7.4, 100 mM NaCl, 3.5 mM MgCl2, 1 mM EDTA, 1 mM ATP, 25 mM KCl, 1.25 U/ml pyruvate kinase, and 3 mM phosphoenolpyruvate) has a regeneration system to convert GDP, produced by any contaminating nucleotidases in the protein preparations, to GTP. Buffer B (25 mM HEPES, pH 7.4, 100 mM NaCl, 0.5 mM MgCl2, 1 mM EDTA, 1 mM ATP, and 1 mM DTT) contains high concentrations of ATP to prevent the nonspecific hydrolysis of GTP by nucleotidases and low free Mg^{2+} concentrations to maximize the rate and extent of GTP loading.

To load nonmyristoylated Arf1 with radiolabeled GTP, 1 μM of Arf1 is incubated in either buffer A or buffer B containing 0.1–1 μM [α^{32}P]GTP or [γ^{32}P]GTP (SA = 25,000–100,000 cpm/pmol) and 0.1% Triton X-100 for 30–60 min at 30°. The reaction is then chilled to 4° until use. In either buffer system, a small fraction of the Arf1 exchanges GDP for GTP. Triton X-100 supports exchange of nucleotide on nonacylated Arf1 more efficiently than either phospholipid–cholate micelles or liposomes. Little or no exchange occurs without a lipid or detergent.

Either myristoylated Arf1 or [L8K]Arf1 can be used to achieve more efficient exchange and higher concentrations of Arf1•GTP when necessary for saturation kinetics. When loading with a radiolabeled GTP, myristoylated Arf1, or [L8K]Arf1 at concentrations up to 200 μM is incubated in buffer B with up to 400 μM [α^{32}P]GTP or [γ^{32}P]GTP (SA = 10,000–50,000 cpm/pmol) (GTP at a 2 to 5 fold excess of Arf1) and either 3 mM dimyristoylphosphatidylcholine/0.1% (w/v) sodium cholate, pH 7.4, (described in this series, Randazzo et al., 1992) or 500 μM phospholipids in the form of LUVs for 30–60 min at 30°. MgCl2 is then added to adjust the free Mg^{2+} concentration to 1 mM and the reaction mixture is chilled to 4° until use. When loading with GTP for the assay based on tryptophan fluorescence, 0.1 to 1 μM myristoylated Arf1 or [L8K]Arf1 is incubated in buffer B with 100-fold excess of GTP over Arf1 and 100–500 μM phospholipids in LUVs of a composition compatible with the GAP being assayed for 30 to 60 min at 30° as described in more detail below. Both myristoylated Arf1 and [L8K]Arf1 have higher affinities for GTP than GDP and, therefore, large excesses of GTP over Arf1•GDP are not required to load preferentially with GTP. When using [L8K] Arf1, lipids are not necessary in the loading reaction. Binding stoichiometry is determined as described in this series (Randazzo et al., 1992).

The most critical point in the loading reaction is to minimize the activity of nonspecific nucleotidases that may copurify with Arf. If nucleotidases interfere with loading even with a GTP regeneration system and high concentrations of ATP present, Arf1 can be further purified by adsorption to hydroxylapatite and elution with a gradient of 10–150 mM NaPi, pH 7.0. Nonmyristoylated Arf•GTP is not stable at 4° and should be used on the day it is prepared. Myristoylated Arf•GTP is more stable and we have stored it at 4° for up to a week prior to use.

Preparation of Arf GAPs

ASAP1. The expression of [325–724]ASAP1 in bacteria and purification of the expressed protein are described in Randazzo *et al.* (2001).

AGAP1. Full-length or truncated AGAP1 (accession # AB029022), fused to either GST or a His10 tag, expressed in bacteria is soluble and active. The approach for expression and purification is very similar to that used for ASAP1. *Nde*I sites and *Xho*I sites are used to ligate a reading frame encoding amino acids 1–804 or 347–804 into pET19 (Novagen, Madison, WI) for a His10 tagged protein. *Eco*RI and *Not*I restriction sites were used to subclone the cDNA into pGEX4T-1 to generate a protein fused to GST (Amersham Bioscience, Piscataway, NJ). BL21(DE3) strain of *E. coli* is used to express the His10 tagged protein. BL21 is used for the GST fusion protein. Transformed bacteria are streaked on Luria Broth (LB) agar plates containing 50 μg/ml ampicillin. A single colony is used to inoculate 100 ml LB containing 100 μg/ml ampicillin. The culture is grown to an OD600 of 0.6 and refrigerated overnight. The next morning, the bacteria from the culture are collected by centrifugation and used to inoculate 1–2l of LB containing 100 μg of ampicillin. When the OD600 is 0.6, ZnCl2 is added to a concentration of 1 μM and protein expression is induced by the addition of 1 mM isopropyl β-D-thiogalactylpyranoside (IPTG). The incubation is continued for 3–4 hours and the bacteria are collected by centrifugation at 1,500–2,500$\times g$ for 20 min and stored at –70°.

The bacterial pellet from 250–500 ml of culture is resuspended in 10 ml of 20 mM Tris, pH 8.0, 25 mM NaCl, 1 mM β-mercaptoethanol, 10% glycerol, and one tablet of Complete Protease Inhibitor Cocktail (Roche, Indianapolis, IN) is added. The bacteria are lysed using a French press operated at 12,000 psi and the lysate is cleared by centrifugation at 100,000$\times g$ for 60 min at 4°. The protein in the supernatant is purified using Glutathione Sepharose 4B (Amersham Bioscience, Piscataway, NJ) for GST fusion proteins or a metal chelating matrix such as Talon from Clontech (Mountainview, CA) or Ni-NTA from Qiagen (Valencia, CA) for His10 tagged protein. Because of the high cysteine content of the protein, buffers should contain a reducing reagent. β-mercaptoethanol works well with both purification schemes. The proteins are typically 90–95% pure after these simple affinity purification procedures. The proteins are snap frozen in aliquots and stored at –70°.

Arf GAP1. The expression and purification of His6[1–257]Arf GAP1 is described in Huber *et al.* (2001). We use a modified protocol. We express the protein in bacteria using the plasmid described by Huber *et al.* (2001). The expression protocol is the same as that used for AGAP1. The His6[1–257]ArfGAP1 accumulates in inclusion bodies. The purification strategy is to collect the inclusion bodies, use a chaotrope to extract the protein from the inclusion bodies, and then adsorb the protein to metal chelating column. The protein is refolded while adsorbed on the column and then eluted. The bacterial pellet from a 250–500 ml culture is resuspended in 10 ml of 20 mM Tris-HCl, pH 8, and cells are lysed either using a French press operated at 12,000 psi or by sonication (4 × 10 sec) on ice. The lysate is clarified by centrifugation at 10,000×g for 10 min at 4°. The pellet, containing inclusion bodies, is washed twice by resuspending in 8 ml of 2 M urea, 20 mM Tris-HCl, pH 8.0, 0.5 M NaCl, 2% Triton X-100, and sonicating (4 × 10 sec) on ice. The inclusion bodies are collected by centrifugation at 10,000×g × 10 min after each wash. The protein is solubilized from the inclusion bodies by incubating for 45 min at room temperature in 12 ml of 20 mM Tris-HCL, pH 8, 0.5 M NaCl, 5 mM imidazole, 6 M guanidine hydrochloride, 1 mM DTT. The sample centrifuged for 15 min at 10,000×g and passed through a 0.22 μm filter. Protein is adsorbed to a 1 ml HisTrap™ chelating column equilibrated with 20 mM Tris-HCl, pH 8, 0.5 M NaCl, 5 mM imidazole, 6 M guanidine hydrochloride, 1 mM DTT. The column is washed with 10 ml of 20 mM Tris-HCl, pH 8, 0.5 M NaCl, 5 mM imidazole, 6 M guanidine hydrochloride, 1 mM DTT, and then with 10 ml with 20 mM Tris-HCL, pH 8, 0.5 M NaCl, 20 mM imidazole, 6 M urea, 1 mM DTT. A linear decreasing gradient of urea (6 to 0 M) in 20 mM Tris-HCL, pH 8, 0.5 M NaCl, 20 mM imidazole, 1 mM DTT over 35 ml is used to refold ArfGAP1 on the column. ArfGAP1 is then eluted from the HisTrap™ column with a linear imidazole gradient (20 mM–500 mM) in 20 mM Tris-HCL, pH 8, 0.5 M NaCl, 1mM DTT over 20 ml. The protein is stored in 30% glycerol at –20° or, if for greater than 2 weeks, at –70°. The storage buffer should contain glycerol since one freeze-thaw cycle in a buffer lacking glycerol results in a ≥50% loss of activity.

Handling purified proteins. In assays for activity, the Arf GAPs, in particular ASAP1 and AGAP1, are used at nanomolar or subnanomolar concentrations. Because purified proteins are not stable at these low concentrations, the proteins should be diluted into a buffer with a carrier protein such as 100 μg/ml BSA.

Assays

Conversion of $[\alpha^{32}P]GTP$ to $[\alpha^{32}P]GDP$. In the first step of this assay, the substrate Arf1•GTP is prepared by exchanging GDP for $[\alpha^{32}P]GTP$ on myristoylated Arf1 or nonmyristoylated Arf1. The GAP reaction is initiated by the addition of Arf1•$[\alpha^{32}P]GTP$ to a reaction mixture containing GAP in 25 mM HEPES, pH 7.4, 100 mM NaCl, 2 mM MgCl2, 1 mM GTP, 1 mM DTT (henceforth called "GAP cocktail") and, for ASAP1 and AGAP1, either mixed micelles containing 90 μM PIP2 and either 360 μM PA (or PS) in 0.1% Triton X-100, or LUVs. For ArfGAP1, LUVs extruded through filters with 0.1 μm are included. The high concentration of GTP in the GAP cocktail dilutes the radiolabeled GTP carried over from the loading reaction or released from Arf. At the much lower specific activity, any GTP that might be hydrolyzed while free in solution and bind either to Arf or another contaminating GTP-binding protein makes little or no contribution to the signal. Mg^{2+} is required for activity and should be in excess of the nucleotide; however, free Mg^{2+} concentrations of greater than 1 mM inhibit activity of ASAP1. Typically, the GAP assay cocktail is prepared as a 1.43-fold concentrate. Seven parts of this buffer (17.5 μL for a 25 μL reaction) is added to 1 part (2.5 μl) of a 10-fold lipid concentrate and 1 part of a 10-fold concentrate of GAP. The reaction is initiated by the addition of 1 part Arf1•GTP, that is, for a 25 μl reaction, add 2.5 μl of the exchange reaction. The temperature is immediately shifted to 30°. Incubations are continued for 1 to 15 min, and stopped by dilution into stop-buffer (20 mM Tris HCl, pH 8.0, 100 mM NaCl, 10 mM MgCl2 and 1 mM DTT) at 4°. For a 10–100 μl reaction volume, dilute with 2 ml of the stop-buffer.

To measure conversion of GTP to GDP, nucleotide bound to Arf is trapped on nitrocellulose filters (BA85, pore size 0.45 μM, size 25 mm from Schleicher and Schuell, Keene, NH) as described in this series (Randazzo *et al.*, 1992). The nucleotides are released from the protein by extraction with 1 ml 1 M formic acid, conveniently accomplished by placing the nitrocellulose filter to which the protein is adsorbed into a 20 ml glass scintillation vial containing the formic acid solution. Fifty μl of the extract, 10–12.5 μl at a time, are applied to PEI (polyethylenimino)-cellulose thin layer chromatography plates (Fischer Scientific, Hampton, NH; or Macherey-Nagel, Easton, PA) and GTP and GDP are separated by developing the chromatogram in 1 M formic acid, 1 M lithium chloride (see Fig. 2A for example chromatogram of the titration data presented in Fig. 2C). The relative amounts of GTP and GDP from the incubations are quantified either with a phosphorimager or by scintillation spectrometry. When examining saturation kinetics, the reaction products are examined at several time points to determine the initial rate of GTP hydrolysis. We suggest that progress curves be

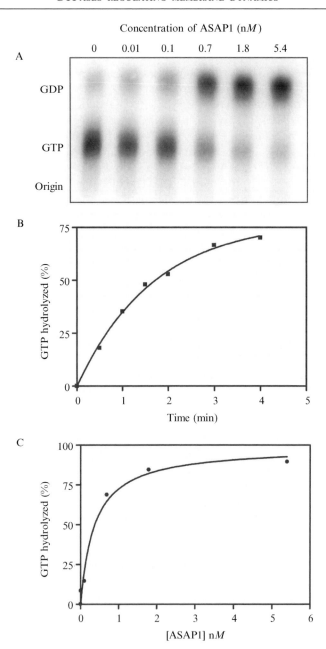

examined to validate the assay even if it is not being used for saturation kinetic analysis. GTP hydrolysis should fit a first order rate equation when Arf1•GTP is present at less than 100 nM. An example time course using ASAP1 as an enzyme is shown in Figure 2B. Once first order rates are established, single time points can be used for routine assay. In this case, the linear range of the assay can be extended by keeping concentrations of Arf •GTP lower than the Km, which is approximately 5 μM (Che *et al.*, 2005), and using a simplified form of the integrated rate equation ($\frac{v=\ln s_0/s}{t}$, Randazzo *et al.*, 2001) to calculate relative enzymatic activity. For comparing the activity of different Arf GAPs or one Arf GAP under different reaction conditions, GAP can be titrated into the reaction, illustrated in Fig. 2C, to determine the amount that induces hydrolysis of 50% of the GTP on Arf1 in a fixed period of time such as 3 min.

Loss of [γ^{32}P]GTP bound to Arf1. This assay is less labor intensive than the assay using [α^{32}P]GTP and, therefore, suitable for performing multiple time courses required for determining initial rates for saturation kinetic analyses (Che *et al.*, 2005). This method will also work for fixed time point analysis. In this assay, 1–200 μM myrArf1 or [L8K]Arf1 is loaded with [γ^{32}P]GTP, with a total GTP concentration in 4-fold excess of Arf1, in buffer B. When using myrArf1, 500 μM of total phospholipids in the form of an LUV is used. Arf1•GTP is added to a reaction mixture containing GAP and 500 μM of the appropriate phospholipids in GAP cocktail. The amount of GAP should be sufficient to hydrolyze about 50% of the GTP on Arf1 in 2–5 min. Samples are taken from 0 to 10 min and diluted into ice-cold stop buffer. The samples are filtered on nitrocellulose as described

---

FIG. 2. Using conversion of [α^{32}P]GTP to [α^{32}P]GDP to follow GAP activity. (A) Separation of GTP and GDP by thin layer chromatography. [L8K] Arf1 was loaded with [α^{32}P]GTP and added to a reaction mixture containing 20 mM HEPES, pH 7.4, 100 mM NaCl, 1 mM GTP, 2 mM MgCl2, 360 μM PA, 90 μM PIP2, 0.1% Triton X-100 and the indicated concentration of [325–724]ASAP1. The reaction was stopped and [L8K]Arf1•[α^{32}P] GTP was trapped on nitrocellulose as described in the text. Nucleotide was extracted from the filters with 2 M formic acid and separated on PEI-cellulose thin layer chromatography plates, cut to 10 cm x 20 cm, developed by ascending chromatography in 1 M formic acid, 1 M lithium chloride. (B) Time course for reaction. Arf1•[α^{32}P]GTP was a substrate for a reaction run under the same conditions as in A but containing 0.5 nM [325–724]ASAP1 and reactions were stopped after the indicated time. The data were quantified by determining the radioactivity associated with GDP and GTP using a phosphorimager and calculating the percentage of total radiolabel in guanine nucleotide associated with GTP, that is, 100*CPM in GTP/(CPM in GTP + CPM in GDP). The percentage of GTP hydrolyzed was plotted against time and the data fit to a first order rate equation (%*hydrolyzed* $=$ max *hydrolyzed* $\cdot$ ($1 - e^{-kt}$)). (C) Titration of ASAP1 into reaction. The percentage of GTP bound to Arf that was hydrolyzed to GDP was determined from the data in part (A) of this figure and plotted against the concentration of [325–724]ASAP1.

above. Radiolabel on the filters is quantified by scintillation spectrometry. GAP activity results in a loss of protein-associated ^{32}P (see example data in Fig. 3A). From a plot of ^{32}P associated with the membranes against time, initial rates can be determined by the same methods described for

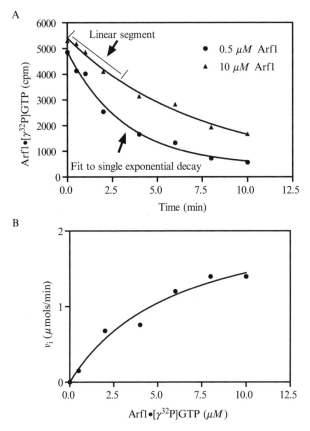

FIG. 3. Use of [γ^{32}P]GTP to follow GAP reaction. (A) Decrease in protein associated ^{32}P induced by [325–724]ASAP1. Two progress curves at different [L8K]Arf1•GTP concentrations are shown from an experiment in which [L8K]Arf1•GTP was loaded with [γ^{32}P]GTP as described in the text. The labeled [L8K]Arf1, at final concentrations ranging from 0.5 μM to 10 μM, was added to a reaction mixture containing 360 μM PA, 90 μM PIP2, 0.1% (w/v) Triton X-100, and 1 nM [325–724]ASAP1. Samples of the reaction were taken at the indicated times following addition of [L8K]Arf1•GTP and protein bound radiolabel was determined. Binding stoichiometries were determined by measuring the amount of bound nucleotide of a known specific activity. The binding stoichiometry was used to calculate the amount of Arf1•GTP present. (B) Replot of initial rate data. The initial rates of loss of protein-associated ^{32}P, taken to be vi, were determined from the plots in (A). The determined vis were plotted against Arf1•GTP concentration. The data were fit to the Michaelis-Menten equation.

determining the rate of dissociation of GTPγS in Chapter 10, "*In vitro* Assays of Arf1 Interaction with GGA," fitting data either to a single exponential decay ($cpm = cpm \cdot e^{-kt}$, where k is the initial fractional rate; the initial rate in amount of Arf1•GTP hydrolyzed per unit time is k· [Arf1•GTP]) or to a line over the initial linear part of the curve[2] (Fig. 3A). The concentration of [γ^{32}P]GTP•Arf1 is determined by measuring the binding stoichiometry (Randazzo *et al.*, 1992). A plot of initial rate vs. [Arf1•GTP] (Fig. 3B), or any desired data transform, is used to determine Km and Vmax. The data can be fit to the Michaelis-Menten rate equation ($v = \frac{V_{\max} \cdot Arf1 \bullet GTP}{K_m + Arf1 \bullet GTP}$) using a nonlinear least squares algorithm implemented with commercially available software.

This assay has the drawback from the first assay described earlier in this chapter in that a nucleotide exchange factor could also accelerate loss of label from the protein and its confounding effect could go undetected. In the assay described previously, the associated radiolabel should be constant, since Arf binds both [α^{32}P]GTP and the conversion product [α^{32}P] GDP. However, this confounding factor is not a problem when using highly purified Arf GAPs. Nonspecific loss of Arf1•GTP, determined from a control reaction lacking GAP, is usually negligible over the time course of this reaction ($\leq$10 min).

Detection of change in tryptophan fluorescence of Arf1 on GTP to GDP conversion. Arf1•GTP has about twice the tryptophan fluorescence of Arf1•GDP. The difference in fluorescence can be used to follow both GTP binding and hydrolysis (Antonny *et al.*, 1997; Paris *et al.*, 1997). In this assay, myrArf1 or [L8K]Arf1, at a concentration of 0.1 to 1 μM, is incubated in buffer B containing a 100-fold excess of GTP (i.e., 10–100 μM) and 100–500 μM phospholipids prepared as LUVs of a composition suitable for the GAP being assayed. Fluorescence is excited at 297 nm and emission at 340 nm is monitored using a spectrofluorimeter with a cuvette thermostatted to 30°. Fluorescence increases with GTP binding to Arf1. When the fluorescence has reached a plateau (typically 30–45 min, see Fig. 4), Mg^{2+} is added to achieve a final concentration of 2 mM and the GAP is added to a concentration that will achieve 50% GTP hydrolysis in 1–6 min. Data collection is continued for 15 min, taking readings every 10–30 sec. GAP activity is detected as a decrease in tryptophan fluorescence (Fig. 4).

This assay has the advantage of real-time data collection. The assay has a disadvantage in that the Arf being used must efficiently bind GTP.

---

[2] Although the data theoretically should fit an equation with two or more exponential terms as the substrate concentration approaches Km, it is impractical to try to fit such an equation to a limited number of data points.

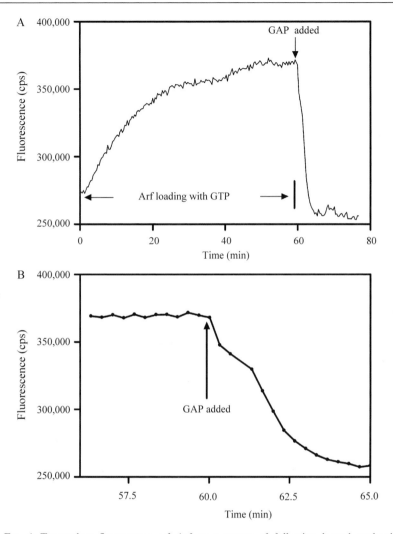

FIG. 4. Tryptophan fluorescence of Arf as a means of following bound nucleotide. (A) Monitoring binding and hydrolysis of GTP by Arf1. In a reaction mixture monitored by excitation with 297 nm light and measuring emission at 340 nm, 1 μM [L8K]Arf1 was incubated with 100 μM GTP in PBS with 0.5 mM MgCl2, 1 mM EDTA, 500 μM LUVs containing 15% PS, and 0.5% PIP2 and 1 mM dithiothreitol for 1 h at 30°. MgCl2 was then added to adjust the total Mg^{2+} concentration to 2 mM. [325–725]ASAP1 was added to achieve a final concentration of 0.5 nM at the time indicated and the incubation was continued for an additional 15 min. (B) Diminished fluorescence due to hydrolysis of GTP. The scale of part (A) is expanded to show the individual data points for the decrease in fluorescence following addition of [325–724]ASAP1.

Introduction of mutations in switch 1 or 2 of Arf1 could reduce GTP binding. Furthermore, the mutations could affect efficiency of myristoylation. Another disadvantage of this method is the limited Arf1 protein concentration range that can be used. Using a Spex Fluoromax-3 with a 100 μl cell and the minimum photobleach option in the fixed wavelength protocol, we saturate the fluorescence signal with Arf concentrations greater than 2–3 μM and, therefore, cannot achieve saturating concentrations of Arf1•GTP (the Km's for the Arf GAPs we have examined are greater than 1 μM). This problem can be solved by using cells with shorter pathlengths and changing the slit widths when the option is available.

Additional Remarks

One consideration not discussed above is the importance of the quality and stability of the GAP and Arf1•GTP for the quantitative characterization of GAP activity. We have described assays with conditions and time scales ensuring stability of the proteins being examined. The purifications described in this chapter, in Chapter 10 (Preparation of Myristoylated Arf1 and Arf6), in Randazzo *et al.* (2000) and in Randazzo *et al.*, (1992) yield proteins relatively free of interfering contaminants. However, because of variability between proteins, the stability and purity of the protein preparations should be assessed when examining other Arf GAPs.

References

Andreev, J., Simon, J. P., Sabatini, D. D., Kam, J., Plowman, G., Randazzo, P. A., and Schlessinger, J. (1999). Identification of a new Pyk2 target protein with Arf-GAP activity. *Mol. Biol. Cell* **19,** 2338–2350.

Antonny, B., Huber, I., Paris, S., Chabre, M., and Cassel, D. (1997). Activation of ADP-ribosylation factor 1 GTPase-activating protein by phosphatidylcholine-derived diacylglycerols. *J. Biol. Chem.* **272,** 30848–30851.

Bagrodia, S., Bailey, D., Lenard, Z., Hart, M., Guan, J. L., Premont, R. T., Taylor, S. J., and Cerione, R. A. (1999). A tyrosine-phosphorylated protein that binds to an important regulatory region on the cool family of p21-activated kinase-binding proteins. *J. Biol. Chem.* **274,** 22393–22400.

Bigay, J., Gounon, P., Robineau, S., and Antonny, B. (2003). Lipid packing sensed by ArfGAP1 couples COPI coat disassembly to membrane bilayer curvature. *Nature* **426,** 563–566.

Brown, M. T., Andrade, J., Radhakrishna, H., Donaldson, J. G., Cooper, J. A., and Randazzo, P. A. (1998). ASAP1, a phospholipid-dependent Arf GTPase-activating protein that associates with and is phosphorylated by Src. *Mol. Cell. Biol.* **18,** 7038–7051.

Cassel, D. (2003). Arf GTPase-activating protein 1. *In* "Arf Family GTPases" (R. A. Kahn, ed.), pp. 137–158. Kluwer Academic Publishers, Dordrecht.

Che, M. M., Boja, E. S., Yoon, H.-Y., Gruschus, J., Jaffe, H., Stauffer, S., Schuck, P., Fales, H. M., and Randazzo, P. A. (2005). Regulation of ASAP1 by phospholipids is dependent on the interface between the PH and Arf GAP domains. *Cell. Signal.* **17,** 1276–1288.

Cukierman, E., Huber, I., Rotman, M., and Cassel, D. (1995). The Arf1 Gtpase-activating protein—zinc-finger motif and Golgi-complex localization. *Science* **270,** 1999–2002.

Donaldson, J. G., and Jackson, C. L. (2000). Regulators and effectors of the ARF GTPases. *Curr. Opin. Cell Biol.* **12,** 475–482.

Huber, I., Rotman, M., Pick, E., Makler, V., Rothem, L., Cukierman, E., and Cassel, D. (2001). Expression, purification, and properties of ADP-ribosylation factor (ARF) GTPase activating protein-1. *Meth. Enzymol.* **329,** 307–316.

Jackson, C. L., and Casanova, J. E. (2000). Turning on ARF: The Sec7 family of guanine-nucleotide-exchange factors. *Trends Cell Biol.* **10,** 60–67.

Jackson, T. R., Brown, F. D., Nie, Z. Z., Miura, K., Foroni, L., Sun, J. L., Hsu, V. W., Donaldson, J. G., and Randazzo, P. A. (2000). ACAPs are Arf6 GTPase-activating proteins that function in the cell periphery. *J. Cell Biol.* **151,** 627–638.

Kam, J. L., Miura, K., Jackson, T. R., Gruschus, J., Roller, P., Stauffer, S., Clark, J., Aneja, R., and Randazzo, P. A. (2000). Phosphoinositide-dependent activation of the ADP-ribosylation factor GTPase-activating protein ASAP1—Evidence for the pleckstrin homology domain functioning as an allosteric site. *J. Biol. Chem.* **275,** 9653–9663.

Krugmann, S., Anderson, K. E., Ridley, S. H., Risso, N., McGregor, A., Coadwell, J., Davidson, K., Eguinoa, A., Ellson, C. D., Lipp, P., Manifava, M., Ktistakis, N., Painter, G., Thuring, J. W., Cooper, M. A., Lim, Z. Y., Holmes, A. B., Dove, S. K., Michell, R. H., Grewal, A., Nazarian, A., Erdjument-Bromage, H., Tempst, P., Stephens, L. R., and Hawkins, P. T. (2002). Identification of ARAP3, a novel PI3K effector regulating both Arf and Rho GTPases, by selective capture on phosphoinositide affinity matrices. *Mol. Cell* **9,** 95–108.

Liu, X. Q., Zhang, C. G., Xing, G. C., Chen, Q. T., and He, F. C. (2001a). Cloning, expression and biochemical activity of ARFGAP3, a regulator of intracellular transport. *Biochem. Biophys.* **28,** 507–513.

Liu, X. Q., Zhang, C. G., Xing, G. C., Chen, Q. T., and He, H. C. (2001b). Functional characterization of novel human ARFGAP3. *FEBS Lett.* **490,** 79–83.

Logsdon, J. M., and Kahn, R. A. (2003). The Arf family tree. *In* "Arf family GTPases" (R. A. Kahn, ed.), pp. 1–21. Kluwer Academic Publishers, Dordrecht.

Lu, Y. F., and Nelsestuen, G. L. (1996). Dynamic features of prothrombin interaction with phospholipid vesicles of different size and composition: Implications for protein-membrane contact. *Biochemistry* **35,** 8193–8200.

Macdonald, R. C., Macdonald, R. I., Menco, B. P. M., Takeshita, K., Subbarao, N. K., and Hu, L. R. (1991). Small-volume extrusion apparatus for preparation of large, unilamellar vesicles. *Biochim. Biophys. Acta* **1061,** 297–303.

Makler, V., Cukierman, E., Rotman, M., Admon, A., and Cassel, D. (1995). ADP-ribosylation factor-directed GTPase-activating protein—purification and partial characterization. *J. Biol. Chem.* **270,** 5232–5237.

Miura, K., Jacques, K. M., Stauffer, S., Kubosaki, A., Zhu, K. J., Hirsch, D. S., Resau, J., Zheng, Y., and Randazzo, P. A. (2002). ARAP1: A point of convergence for Arf and Rho signaling. *Mol. Cell* **9,** 109–119.

Moss, J., and Vaughan, M. (1998). Molecules in the ARF orbit. *J. Biol. Chem.* **273,** 21431–21434.

Nie, Z., Boehm, M., Boja, E., Vass, W., Bonifacino, J., Fales, H., and Randazzo, P. A. (2003). Specific regulation of the adaptor protein complex AP-3 by the Arf GAP AGAP1. *Dev. Cell* **5,** 513–521.

Nie, Z. Z., Stanley, K. T., Stauffer, S., Jacques, K. M., Hirsch, D. S., Takei, J., and Randazzo, P. A. (2002). AGAP1, an endosome-associated, phosphoinositide-dependent ADP-ribosylation factor GTPase-activating protein that affects actin cytoskeleton. *J. Biol. Chem.* **277,** 48965–48975.

Paris, S., Beraud-Dufour, S., Robineau, S., Bigay, J., Antonny, B., Chabre, M., and Chardin, P. (1997). Role of protein-phospholipid interactions in the activation of ARF1 by the guanine nucleotide exchange factor Arno. *J. Biol. Chem.* **272,** 22221–22226.

Peter, B. J., Kent, H. M., Mills, I. G., Vallis, Y., Butler, P. J. G., Evans, P. R., and McMahon, H. T. (2004). BAR domains as sensors of membrane curvature: The amphiphysin BAR structure. *Science* **303,** 495–499.

Premont, R. T., Claing, A., Vitale, N., Freeman, J. L. R., Pitcher, J. A., Patton, W. A., Moss, J., Vaughan, M., and Lefkowitz, R. J. (1998). Beta(2)-adrenergic receptor regulation by GIT1, a G protein-coupled receptor kinase-associated ADP ribosylation factor GTPase-activating protein. *Proc. Natl. Acad. Sci. USA* **95,** 14082–14087.

Randazzo, P. A., Nie, Z., Miura, K., and Hsu, V. (2000). Molecular aspects of the cellular activities of ADP-ribosylation factors. *Sci STKE* **59** *RE1.*

Randazzo, P. A. (1997a). Functional interaction of ADP-ribosylation factor 1 with phosphatidylinositol 4,5-bisphosphate. *J. Biol. Chem.* **272,** 7688–7692.

Randazzo, P. A. (1997b). Resolution of two ADP-ribosylation factor 1 GTPase-activating proteins from rat liver. *Biochem. J.* **324,** 413–419.

Randazzo, P. A., and Hirsch, D. S. (2004). Arf GAPs: Multifunctional proteins that regulate membrane traffic and actin remodelling. *Cell. Signal.* **16,** 401–413.

Randazzo, P. A., and Kahn, R. A. (1994). GTP hydrolysis by ADP-ribosylation factor is dependent on both an ADP-ribosylation factor GTPase-activating protein and acid phospholipids. *J. Biol. Chem.* **269,** 10758–10763.

Randazzo, P. A., Miura, K., and Jackson, T. R. (2001). Assay and purification of phosphoinositide-dependent ADP-ribosylation factor (ARF) GTPase activating proteins. Regulators and effectors of small GTPases. *Meth. Enzymol.* **329,** 343–354.

Randazzo, P. A., Weiss, O., and Kahn, R. A. (1992). Preparation of recombinant ADP-ribosylation factor. *Meth. Enzymol.* **219,** 362–369.

Schmalzigaug, R., and Premont, R. T. (2003). GIT proteins: Arf GAPs and signaling scaffolds. *In* "Arf family GTPases" (R. A. Kahn, ed.), pp. 159–184. Kluwer Academic Publishers, Dordrecht.

Seidel, R. D., Amor, J. C., Kahn, R. A., and Prestegard, J. H. (2004). Structural perturbations in human ADP ribosylation factor-1 accompanying the binding of phosphatidylinositides. *Biochemistry* **43,** 15393–15403.

Stacey, T. T. I., Nie, Z. Z., Stewart, A., Najdovska, M., Hall, N. E., He, H., Randazzo, P. A., and Lock, P. (2004). ARAP3 is transiently tyrosine phosphorylated in cells attaching to fibronectin and inhibits cell spreading in a RhoGAP-dependent manner. *J. Cell Sci.* **117,** 6071–6084.

Turner, C. E., Brown, M. C., Perrotta, J. A., Riedy, M. C., Nikolopoulos, S. N., McDonald, A. R., Bagrodia, S., Thomas, S., and Leventhal, P. S. (1999). Paxillin LD4 motif binds PAK and PIX through a novel 95-kD ankyrin repeat, ARF-GAP protein: A role in cytoskeletal remodeling. *J. Cell Biol.* **145,** 851–863.

Turner, C. E., West, K. A., and Brown, M. C. (2001). Paxillin-ARF GAP signaling and the cytoskeleton. *Curr. Opin. Cell Biol.* **13,** 593–599.

Yoon, H. Y., Jacques, K., Nealon, B., Stauffer, S., Premont, R. T., and Randazzo, P. A. (2004). Differences between AGAP1, ASAP1 and Arf GAP1 in substrate recognition: Interaction with the N-terminus of Arf1. *Cell. Signal.* **16,** 1033–1044.

[16]   Preparation of Myristoylated Arf1 and Arf6

By Vi Luan Ha, Geraint M. H. Thomas, Stacey Stauffer, and Paul A. Randazzo

Abstract

Arf proteins are members of the Arf family of small Ras-like GTP binding proteins. Six Arfs, grouped into three classes, have been identified in mammalian cells and three members have been identified in yeasts. Arf1 and Arf6, more extensively studied than other Arfs, have been found to affect membrane traffic and actin remodeling. A structural feature that distinguishes Arfs from other Ras superfamily members is an N-terminal α-helix, extending from the basic G-protein fold, which is cotranslationally myristoylated. Both the helix and the myristate affect biochemical properties of Arfs, including nucleotide exchange, membrane association, and interaction with some effector proteins. Preparation of myristoylated Arf for *in vitro* studies of Arf function requires consideration of both the reaction yielding myristoylated protein and the properties of the modified Arfs. Here, we describe methods that yield homogeneous preparations of myristoylated Arf1 and Arf6.

Introduction

Arf proteins were first purified and cloned from mammalian liver on the basis of their activity as a cofactor for cholera toxin catalyzed ADP-ribosylation of heterotrimeric G proteins (Lee *et al.*, 1994; Logsdon and Kahn, 2003; Moss and Vaughan, 1998; Randazzo *et al.*, 2000). Subsequently, a family of genes encoding Arf and Arf-like (Arl) proteins, within the Ras-superfamily of GTP-binding proteins, was identified (Logsdon and Kahn, 2003). The Arf family is highly conserved and found only in the eukaryotes. Within the family, Arf proteins are defined as those that have activity as a cofactor for cholera toxin. The Arf proteins are divided into three groups on the basis of sequence similarity (Hosaka *et al.*, 1996; Moss and Vaughan, 1995, 1998; Tsuchiya *et al.*, 1991). Mammalian Arf1, Arf2, and Arf3 are class I. Arf4 and Arf5 are class II. Arf6 is class III. In all cells examined, multiple Arf isoforms are present.

Though identified as cofactors for cholera toxin, the physiologic role of Arfs appears to be unrelated to Gs and ADP-ribosylation. The best de-scribed function for Arfs is as a regulator of membrane traffic (Bonifacino

METHODS IN ENZYMOLOGY, VOL. 404   0076-6879/05 $35.00
DOI: 10.1016/S0076-6879(05)04016-4

and Glick, 2004; Bonifacino and Lippincott-Schwartz, 2003; Randazzo *et al.*, 2000). Arfs have also been implicated in the regulation of actin (Nie *et al.*, 2003; Norman *et al.*, 1998; Randazzo *et al.*, 2000; Radhakrishna *et al.*, 1996). Several biochemical functions have been identified. Arfs, when bound to GTP, bind to vesicle coat proteins and coat protein adaptors including AP-1 (Stamnes and Rothman, 1993; Traub *et al.*, 1993), AP-3 (Ooi *et al.*, 1998), AP-4 (Boehm *et al.*, 2001), GGA (Boman *et al.*, 2000; Dell'Angelica *et al.*, 2000; Hirst *et al.*, 2000), and coatomer (Donaldson *et al.*, 1992; Palmer *et al.*, 1993), mediating recruitment of these proteins to membranes in an early step in the formation of membrane trafficking intermediates (Bonifacino and Glick, 2004; Bonifacino and Lippincott-Schwartz, 2003; Moss and Vaughan, 1998; Nie *et al.*, 2003; Rothman, 2002; Randazzo *et al.*, 2000). Arf•GTP also binds and activates enzymes of phospholipid metabolism including PI(4)P 5-kinase (Honda *et al.*, 1999; Jones *et al.*, 2000) and phospholipase D (Brown *et al.*, 1993; Cockcroft *et al.*, 1994). PI 4-kinase is thought to be affected by Arf1 (Godi *et al.*, 1999). In addition, Arf•GTP binds several proteins whose functions have not yet been as clearly defined, for example, arfaptin and arfophilin (Kanoh *et al.*, 1997; Shin *et al.*, 1999; reviewed in Nie *et al.*, 2003).

Arf proteins have a unique structural feature among Ras family proteins, an N-myristoylated α-helical extension from the basic G-protein fold (Antonny *et al.*, 1997; Balch *et al.*, 1992; Kahn *et al.*, 1992; Losonczi and Prestegard, 1998; Losonczi *et al.*, 2000; Pasqualato *et al.*, 2002; Randazzo *et al.*, 1995). Myristate is a 14-carbon saturated fatty acid that is catalytically transferred from myristoyl CoA to the amino group of the N-terminal glycine in a reaction catalyzed by the enzyme N-terminal myristoyl transferase (Duronio *et al.*, 1992). The reaction occurs cotranslationally and the bond is stable so that all Arf in cells is myristoylated. Several Arf functions are dependent on myristoylation including membrane binding, vesicle coat protein recruitment (Donaldson *et al.*, 1992; Kahn *et al.*, 1992), nucleotide exchange, interaction with guanine nucleotide exchange factors (Franco *et al.*, 1996; Paris *et al.*, 1997), and interaction with PLD (Brown *et al.*, 1993, 1995). Therefore, recapitulation of the physiological activities of Arf *in vitro* is dependent on using appropriately modified protein.

Several types of Arf preparations with different levels of myristoylation have been described in the literature. Arf purified from mammalian tissue is 100% myristoylated. However, it is difficult to obtain a single Arf isoform and the purification from mammalian tissues is labor intensive. Expressing Arf as a recombinant protein in bacteria does not yield myristoylated protein unless N-myristoyl transferase is coexpressed (since bacteria do not have the enzyme). However, myristoylation has not been efficient in this system, with only a small fraction (5–15%) of the Arf being modified. Expression of Arf in Sf9 cells, though the cells do have NMT, also does not yield efficiently modified Arf (Randazzo, unpublished data). In this chapter, we describe the preparation of myristoylated Arf1 and Arf6 using bacterial expression. Homogeneity of the proteins depends on the separation of myristoylated from unmodified protein.

Methods

Preparation of Wild Type Myristoylated Arf1

Bacterial Expression of Myristoylated Arf1. Recombinant Arf1 is myristoylated by coexpression of Arf1 with N-myristoyltransferase (NMT) in *E. coli* since bacteria lack transferase activity. NMT catalyses the *in vivo* transfer of a myristate group to a glycine residue at position 2 of the Arf protein. This system has been previously described. Briefly, cDNA containing the open reading frame for Arf1 is ligated into a pET3 plasmid with the codon for the initiating methionine being part of the NdeI restriction site that is located at the optimal distance from the ribosomal binding site to ensure efficient translation. This plasmid has a T7 promoter directing transcription. BL21(DE3) bacteria contain the T7 structural gene under control of the lacUV5 promoter, allowing the induction of T7 on addition of isopropyl-β-D-thiogalactopyranoside (IPTG). BL21(DE3) are cotransformed with the Arf1 cDNA in pET3, which is under ampicillin selection, and pBB131, a plasmid containing the open reading frame for yeast NMT1 (Duronio *et al.*, 1992) and a marker for kanamycin resistance. Transformed bacteria can be selected for both plasmids with 100 μg/ml ampicillin and 50 μg/ml kanamycin. An optional step that nevertheless is useful for optimizing expression is to select and grow single colonies to an OD$_{600}$ of 0.6 in 10 ml of LB with 100 μg/ml ampicillin and 50 μg/ml kanamycin at 37°.

The cultures are divided into three parts. Incubation at 37° is continued for two samples. IPTG is added to 1 mM to one sample. [^{3}H]myristic acid can also be added at this point. After an overnight incubation at 25°, the cells are collected by centrifugation and expression level, and efficiency of myristate incorporation is compared. Besides using tritium incorporation, the extent of myristoylation can be assessed by migration on SDS-PAGE. The myristoylated protein migrates slightly faster than Arf1 that is not modified (Franco *et al.*, 1995; Randazzo, 1997). However, the difference is small and if proteins are not well resolved on the gel, relative myristoylation rates cannot be determined this way. Tritium incorporation, on the other hand, will give an unambiguous assessment of relative myristoylation levels. Glycerol stocks are made of those colonies that give both optimal soluble protein expression and myristoylation.

Bacteria from glycerol stocks are streaked on LB agar plates containing 100 μg/ml ampicillin and 50 μg/ml kanamycin and incubated overnight at 37°. The next day, 100 ml of Luria broth containing 100 μg/ml ampicillin and 50 μg/ml kanamycin is inoculated with a single colony of the transformed bacteria. The culture is grown at 37° to an OD_{600} of $\approx$0.6 and refrigerated overnight. The bacteria are collected by centrifugation at 1500 $\times g$ for 20 min and used to inoculate a 1- to 2-liter culture with the same antibiotic selection. The bacteria are grown at 37° to an $OD_{600} \approx$ 0.6. Sodium myristate is added to the culture to achieve a concentration of 50 μM and the incubation is continued for an additional 20 min. IPTG is added to a final concentration of 1 mM. The cultures are shifted to 25° and the incubation is continued for an additional 12–16 h. Cells are harvested by centrifugation at $1500-2500 \times g$ for 20 min at 4° and stored at -80°. Prior to induction and after induction, 1-ml samples of the cultures are taken to confirm expression of Arf. The cells in these samples are collected by centrifugation, lysed in 20–100 μl of Laemli's sample buffer and 10–25 μl of the lysate are fractionated on a polyacrylamide gel and stained with Coomassie Blue (Fig. 1A, inset).

An alternative approach has been developed to increase the fraction of Arf1 that is myristoylated in bacteria. Plasmids are used that drive expression of Arf1 and NMT from different promoters and inducers. In this case, the open reading for Arf1 is subcloned into the NcoI and HindIII sites of pMon5840, a plasmid that encodes ampicillin resistance and which has a recA promoter. BL21-pLys bacteria (Novagen) are cotransformed with pMon5480-Arf1 and pBB131 and transformants are selected with 100 μg/ml ampicillin, 25 μg/ml kanamycin. Cells are grown to an OD_{600} of 0.6 and NMT expression is induced by the addition of 1 mM IPTG. Myristic acid is added. One hour later, 0.1 mM nalidixic acid is added to induce expression of Arf1 and the incubation is continued for an additional three hours. Protein expression and efficiency of myristoylation for isolates can be assessed as described above.

Purification of Myristoylated Arf1 from Bacteria. The bacterial pellet is thawed at room temperature and resuspended in 10–20 ml 20 mM Tris, pH 8.0, 100 mM NaCl, 1 mM MgCl2, 1 mM DTT/ in a 2-liter culture. We have lysed cells in two ways. One method uses a French press operated at 12,000 psi. If a French press is not available, we have also used a probe sonicator. In this case, lysozyme is added to a final concentration of 1 mg/ml and the cell suspension is incubated at RT for 30 min. The cells are lysed by 3–10 sec bursts. In either case, the lysate is clarified by centrifugation at 100,000 $\times g$ for 60 min at 4°. The supernatant is used for further purification.

A

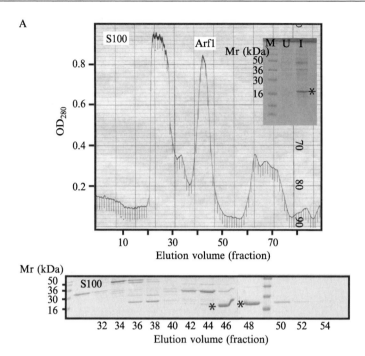

B

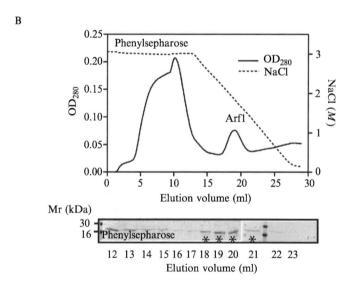

The supernatant is loaded onto a 5 ml HiTrap™ Q column (Amersham Biosciences) that has been equilibrated in 20 mM Tris, pH 8.0, 100 mM NaCl, 1 mM MgCl2, 1 mM DTT, 10% v/v glycerol. (Buffer A). The column eluate is collected in 2 ml fractions. The column is further developed with 20 ml of Buffer A. Arf1 adheres poorly or not at all. Fractions containing protein, determined from small samples ($\approx$10 μl) or by OD$_{280}$, are pooled. The pooled fractions should be about 10 ml and are loaded onto a 330 ml Sephacryl™ S-100 26/60 column (Amersham Biosciences) which has been equilibrated in buffer A. If run at room temperature, the column is developed at 1.2 ml/min, which is linear flow rate of 0.22 cm/min on a column of these dimensions. If run at 4°, the column is developed at 0.3 ml/min. A representative elution profile with relative elution position of Arf1 is shown in Fig. 1A. Fractions of 3.6 ml are collected and those containing Arf1 are identified by SDS-PAGE, detecting proteins with Coomassie Blue. The pooled fractions are concentrated by ultrafiltration to 2–5 ml and equal volume of 20 mM Tris, pH 8.0, 4 M NaCl, 1 mM MgCl2, 1 mM DTT (Buffer B) are added (e.g., add 2 ml of Buffer B to 2 ml of pool). This is repeated once to bring the NaCl concentration to 3 M. The protein is then loaded at a rate of 1 ml/min onto a 1 ml Phenyl Sepharose HP™ column (Amersham Biosciences, Piscataway, NJ) that had been equilibrated in 20 mM Tris, pH 8.0, 3 M NaCl, 1 mM MgCl2, 1 mM DTT (Buffer C). The column is developed at room temperature in a gradient of 3 M to 100 mM NaCl, going from buffer C to 20 mM Tris, pH 8.0, 100 mM NaCl, 1 mM MgCl2, 1 mM DTT in 15 ml. 1-ml fractions are collected. Nonmyristoylated Arf does not adhere to the column whereas myristoylated Arf1 elutes at 2–1.5 M NaCl (Fig. 1B). Nonmyristoylated Arf is not detected in the second peak. Fractions containing Arf can be confirmed using SDS-PAGE, pooled, and both concentrated and exchanged into 20 mM Tris, pH 8.0, 100 mM NaCl, 1 mM MgCl2, 1 mM DTT, 10% v/v glycerol by ultrafiltration. In this case, the protein is concentrated to 500 μl–1 ml, diluted 5–10 fold in the desired buffer, and concentrated again.

FIG. 1. Purification of myristoylated Arf1. (A) Elution profile of Arf1 proteins on Sephacryl™ S-100 26/60 column. The peak containing Arf1 is labeled. Inset shows the expression of myristoylated Arf1 in *E. coli* with NMT. Lane U, uninduced; lane I, induced. *, band containing Arf1. Proteins from the indicated elution fractions were separated by SDS-PAGE in the gel shown below the chromatogram. * marks the band containing Arf1. (B) Resolution of myristoylated ARF1 on Phenyl Sepharose HP™ column. The protein is loaded in the presence of 3 M NaCl and eluted with a descending sodium chloride gradient. In the SDS-PAGE fractionation of the proteins from the elution, shown below the chromatogram, the band containing Arf1 is marked with *.

This cycle is repeated, resulting in dilution of the NaCl present in the column fractions. For example, if 10-fold dilutions are used, two cycles result in a net 100 fold dilution of the original 1.5 M NaCl. Aliquots are snap frozen and stored at $-80°$.

Preparation of Myristoylated Arf6

Bacterial Expression of Myristoylated Arf6. The strategy for expression of myristoylated Arf6 is nearly identical to that for Arf1. DNA encoding the open reading from of Arf6 is subcloned into pET3 as described for Arf1. BL21(DE3) bacteria are transformed with the Arf6 and yeast NMT expression vectors, selecting with ampicillin and kanamycin. Glycerol stocks are prepared from single colonies of bacteria with optimized expression as determined for Arf1. Arf6 expression differs from Arf1 in a single respect. Protein expression is induced for 3 h at 37° with 1 mM IPTG.

Bacterial Lysis and Protein Purification. The bacterial pellet is thawed at room temperature and resuspended in 10 ml of 20 mM Tris, pH 8.0, 100 mM NaCl, 1 mM MgCl2, 1 mM DTT with protease inhibitors. We often use a protease inhibitor cocktail called Complete® from Roche (Indianapolis, IN). The cell suspension is lysed in a French press with a pressure of 12,000 psi. The lysate is centrifuged at $100,000 \times g$ at 4° for 60 min. The pellet contains the myristoylated Arf6 and is resuspended in 20 mM Tris, pH 8.0, 100 mM NaCl, 1 mM MgCl2, 1 mM DTT, 10% (v/v) glycerol and centrifuged $100,000 \times g$ at 4° for 60 min. The pellet from the second centrifugation is resuspended in 20 mM Tris, pH 8.0, 25 mM NaCl, 1 mM MgCl2, 1 mM DTT, 1% (w/v) Triton X-100, 10% (v/v) glycerol. Triton X-100 in this buffer extracts Arf6 from the particulate fraction. The suspension is passed through a French press and centrifuged at $100,000 \times g$ for 60 min at 4°. GDP is added to a final concentration of 10 μM in the supernatant and the proteins precipitated with 35% ammonium sulfate at 4°. The precipitated proteins are collected by centrifugation ($20,000 \times g$ at 4° for 30 min), dissolved in 10 ml of 20 mM Tris, pH 8.0, 25 mM NaCl, 1 mM MgCl2, 1 mM DTT, 1% (w/v) Triton X-100, 10% (v/v) glycerol containing 10 μM GDP. The sample is dialyzed 2 times against 1 L of 20 mM Tris, pH 8.0, 25 mM NaCl, 1 mM MgCl2, 1 mM DTT, 0.1% Triton X-100, 10% v/v glycerol with GDP. The sample is then applied to a 5 ml HiTrap® Q column (Amersham Biosciences, Piscataway, NJ) and the material that does not adhere to the column is collected. The protein is greater than 95% Arf6. SDS-PAGE analysis of the preparation at each step of the purification is shown in Fig. 2. It is critical to snap freeze and store the Arf6 immediately. The protein denatures, as assessed by nucleotide binding, with a t1/2 of approximately 1 day when stored at 4°.

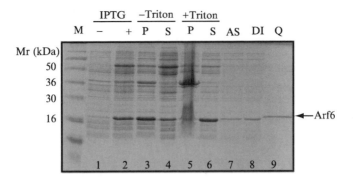

FIG. 2. Purification of myristoylated Arf6. Proteins were resolved on SDS 10–20% polyacrylamide gels and stained with Coomassie blue. Lane 1, uninduced; lane 2, induced; lanes 3 and 4, pellet and supernatant in the absence of Triton; lanes 5 and 6, pellet and supernatant in the presence of Triton; lane 7, precipitated proteins in ammonium sulfate; lane 8, dialyzed proteins prior to loading on HiTrap® Q column; lane 9, purified myristoylated Arf6 from HiTrap™ Q column.

Conclusions

The myristoylated Arf preparations described in this chapter have been used in a number of *in vitro* assays including GAP, GEF, and effector binding assays. The proteins are useful when efficient loading with GTP is needed. More critically, some Arf activities, such as recruitment of vesicle coat protein, are highly dependent on myristoylation. In these cases, use of a partially myristoylated Arf, particularly with preparation-to-preparation differences in the extent of myristoylation, can confound results and their clear interpretation. The methods described in this chapter require only one additional column beyond that required for purification of nonmyristoylated or partially myristoylated Arf and provide homogeneously myristoylated Arf preparations.

References

Antonny, B., Beraud-Dufour, S., Chardin, P., and Chabre, M. (1997). N-terminal hydrophobic residues of the G-protein ADP-ribosylation factor-1 insert into membrane phospholipids upon GDP to GTP exchange. *Biochemistry* **36,** 4675–4684.

Balch, W. E., Kahn, R. A., and Schwaninger, R. (1992). ADP-ribosylation factor is required for vesicular trafficking between the endoplasmic reticulum and the cis-Golgi compartment. *J. Biol. Chem.* **267,** 13053–13061.

Boehm, M., Aguilar, R. C., and Bonifacino, J. S. (2001). Functional and physical interactions of the adaptor protein complex AP-4 with ADP-ribosylation factors (ARFs). *EMBO J.* **20**, 6265–6276.

Boman, A. L., Zhang, C. J., Zhu, X. J., and Kahn, R. A. (2000). A family of ADP-ribosylation factor effectors that can alter membrane transport through the trans-Golgi. *Mol. Biol. Cell* **11**, 1241–1255.

Bonifacino, J. S., and Glick, B. S. (2004). The mechanisms of vesicle budding and fusion. *Cell* **116**, 153–166.

Bonifacino, J. S., and Lippincott-Schwartz, J. (2003). Coat proteins: Shaping membrane transport. *Nat. Rev. Mol. Cell Biol.* **4**, 409–414.

Brown, H. A., Gutowski, S., Kahn, R. A., and Sternweis, P. C. (1995). Partial purification and characterization of Arf-sensitive phospholipase D from porcine brain. *J. Biol. Chem.* **270**, 14935–14943.

Brown, H. A., Gutowski, S., Moomaw, C. R., Slaughter, C., and Sternweis, P. C. (1993). ADPribosylation factor, a small GTP-dependent regulatory protein, stimulates phospholipase D activity. *Cell* **75**, 1137–1144.

Cockcroft, S., Thomas, G. M., Fensome, A., Geny, B., Cunningham, E., Gout, I., Hiles, I., Totty, N. F., Truong, O., and Hsuan, J. J. (1994). Phospholipase D: A downstream effector of ARF in granulocytes. *Science* **263**, 523–526.

Dell'Angelica, E. C., Puertollano, R., Mullins, C., Aguilar, R. C., Vargas, J. D., Hartnell, L. M., and Bonifacino, J. S. (2000). GGAs: A family of ADP ribosylation factor-binding proteins related to adaptors and associated with the Golgi complex. *J. Cell Biol.* **149**, 81–94.

Donaldson, J. G., Cassel, D., Kahn, R. A., and Klausner, R. D. (1992). ADP-ribosylation factor, a small GTP-binding protein, is required for binding of the coatomer protein beta-COP to Golgi membranes. *Proc. Natl. Acad. Sci. USA* **89**, 6408–6412.

Duronio, R. J., Knoll, L. J., and Gordon, J. I. (1992). Isolation of a *Saccharomyces cerevisiae* long chain fatty acyl:CoA synthetase gene (FAA1) and assessment of its role in protein N-myristoylation. *J. Cell Biol.* **117**, 515–529.

Franco, M., Chardin, P., Chabre, M., and Paris, S. (1995). Myristoylation of ADPribosylation factor 1 facilitates nucleotide exchange at physiological Mg2+ levels. *J. Biol. Chem.* **270**, 1337–1341.

Franco, M., Chardin, P., Chabre, M., and Paris, S. (1996). Myristoylation-facilitated binding of the G protein ARF1(GDP) to membrane phospholipids is required for its activation by a soluble nucleotide exchange factor. *J. Biol. Chem.* **271**, 1573–1578.

Godi, A., Pertile, P., Meyers, R., Marra, P., Di Tullio, G., Iurisci, C., Luini, A., Corda, D., and De Matteis, M. A. (1999). ARF mediates recruitment of PtdIns-4-OH kinase-beta and stimulates synthesis of PtdIns(4,5)P-2 on the Golgi complex. *Nat. Cell Biol.* **1**, 280–287.

Hirst, J., Lui, W. W. Y., Bright, N. A., Totty, N., Seaman, M. N. J., and Robinson, M. S. (2000). A family of proteins with gamma-adaptin and VHS domains that facilitate trafficking between the trans-Golgi network and the vacuole/lysosome. *J. Cell Biol.* **149**, 67–79.

Honda, A., Nogami, M., Yokozeki, T., Yamazaki, M., Nakamura, H., Watanabe, H., Kawamoto, K., Nakayama, K., Morris, A. J., Frohman, M. A., and Kanaho, Y. (1999). Phosphatidylinositol 4-phosphate 5-kinase alpha is a downstream effector of the small G protein ARF6 in membrane ruffle formation. *Cell* **99**, 521–532.

Hosaka, M., Toda, K., Takatsu, H., Torii, S., Murakami, K., and Nakayama, K. (1996). Structure and intracellular localization of mouse ADP-ribosylation factors type 1 to type 6 (Arf1-Arf6). *J. Biochem.* **120**, 813–819.

Jones, D. H., Morris, J. B., Morgen, C. P., Kondo, H., Irvine, R. F., and Cockcroft, S. (2000). Type I phosphatidylinositol 4-phosphate 5-kinase directly interacts with ADPribosylation factor 1 and is responsible for phosphatidylinositol 4,5-bisphosphate synthesis in the Golgi compartment. *J. Biol. Chem.* **275,** 13962–13966.

Kahn, R. A., Randazzo, P., Serafini, T., Weiss, O., Rulka, C., Clark, J., Amherdt, M., Roller, P., Orci, L., and Rothman, J. E. (1992). The amino terminus of ADP-ribosylation factor (ARF) is a critical determinant of ARF activities and is a potent and specific inhibitor of protein transport. *J. Biol. Chem.* **267,** 13039–13046.

Kanoh, H., Williger, B. T., and Exton, J. H. (1997). Arfaptin 1, a putative cytosolic target protein of ADP-ribosylation factor, is recruited to Golgi membranes. *J. Biol. Chem.* **272,** 5421–5429.

Lee, F.J, Stevens, L.A, Kao, Y. L., Moss, J., and Vaughan, M. (1994). Characterization of a glucose-repressible ADP-ribosylation factor 3 (ARF3) from *Saccharomyces cerevisiae.* *J. Biol. Chem.* **269,** 20931–20937.

Logsdon, J. M., and Kahn, R. A. (2003). The Arf family tree. *In* "Arf family GTPases" (R. A. Kahn, ed.), pp. 1–21. Kluwer, Netherlands.

Losonczi, J. A., and Prestegard, J. H. (1998). Nuclear magnetic resonance characterization of the myristoylated, N-terminal fragment of ADP-ribosylation factor 1 in a magnetically oriented membrane array. *Biochemistry* **37,** 706–716.

Losonczi, J. A., Tian, F., and Prestegard, J. H. (2000). Nuclear magnetic resonance studies of the N-terminal fragment of adenosine diphosphate ribosylation factor 1 in micelles and bicelles: Influence of N-myristoylation. *Biochemistry* **39,** 3804–3816.

Moss, J., and Vaughan, M. (1998). Molecules in the ARF orbit. *J. Biol. Chem.* **273,** 21431–21434.

Moss, J., and Vaughan, M. (1995). Structure and function of ARF proteins: Activators of cholera toxin and critical components of intracellular vesicular transport processes. *J. Biol. Chem.* **270,** 12327–12330.

Nie, Z. Z., Hirsch, D. S., and Randazzo, P. A. (2003). Arf and its many interactors. *Curr. Opin. Cell Biol.* **15,** 396–404.

Norman, J. C., Jones, D., Barry, S. T., Holt, M. R., Cockcroft, S., and Critchley, D. R. (1998). ARF1 mediates paxillin recruitment to focal adhesions and potentiates Rho-stimulated stress fiber formation in intact and permeabilized Swiss 3T3 fibroblasts. *J. Cell Biol.* **143,** 1981–1995.

Ooi, C. E., Dell'Angelica, E. C., and Bonifacino, J. S. (1998). ADP-ribosylation factor 1 (ARF1) regulates recruitment of the AP-3 adaptor complex to membranes. *J. Cell Biol.* **142,** 391–402.

Palmer, D. J., Helms, J. B., Beckers, C. J. M., Orci, L., and Rothman, J. E. (1993). Binding of coatomer to Golgi membranes requires ADP-ribosylation factor. *J. Cell Biol.* **268,** 12083–12089.

Paris, S., Beraud-Dufour, S., Robineau, S., Bigay, J., Antonny, B., Chabre, M., and Chardin, P. (1997). Role of protein-phospholipid interactions in the activation of ARF1 by the guanine nucleotide exchange factor ARNO. *J. Cell Biol.* **272,** 22221–22226.

Pasqualato, S., Renault, L., and Cherfils, J. (2002). Arf, Arl, Arp and Sar proteins: A family of GTP-binding proteins with a structural device for 'front-back' communication. *EMBO Rep.* **3,** 1035–1041.

Radhakrishna, H., Klausner, R. D., and Donaldson, J. G. (1996). Aluminum fluoride stimulates surface protrusions in cells overexpressing the ARF6 GTPase. *J. Cell Biol.* **134,** 935–947.

Randazzo, P. A., Nie, Z., Miura, K., and Hsu, V. (2000). Molecular aspects of the cellular activities of ADP-ribosylation factors. *Sci. STKE.* **59,** RE1.

Randazzo, P. A. (1997). Resolution of two ADP-ribosylation factor 1 GTPase-activating proteins from rat liver. *Biochem. J.* **324,** 413–419.

Randazzo, P. A., Terui, T., Sturch, S., Fales, H. M., Ferrige, A. G., and Kahn, R. A. (1995). The myristoylated amino terminus of ADP-ribosylation factor 1 is a phospholipid- and GTP-sensitive switch. *J. Biol. Chem.* **270,** 14809–14815.

Rothman, J. E. (2002). The machinery and principles of vesicle transport in the cell. *Nat. Med.* **8,** 1059–1062.

Shin, O. H., Ross, A. H., Mihai, I., and Exton, J. H. (1999). Identification of arfophilin, a target protein for GTP-bound class II ADP-ribosylation factors. *J. Biol. Chem.* **274,** 36609–36615.

Stamnes, M. A., and Rothman, J. E. (1993). The binding of AP-1 clathrin adaptor particles to Golgi membranes requires ADP-ribosylation factor, a small GTP-binding protein. *Cell* **73,** 999–1005.

Traub, L. M., Ostrom, J. A., and Kornfeld, S. (1993). Biochemical dissection of AP-1 recruitment onto Golgi membranes. *J. Cell Biol.* **123,** 561–573.

Tsuchiya, M., Price, S. R., Tsai, S. C., Moss, J., and Vaughan, M. (1991). Molecular identification of ADP-ribosylation factor mRNAs and their expression in mammalian cells. *J. Biol. Chem.* **266,** 2772–2777.

[17] BIG1 and BIG2, Brefeldin A-Inhibited Guanine Nucleotide-Exchange Factors for ADP-Ribosylation Factors

By HEATHER D. JONES, JOEL MOSS, and MARTHA VAUGHAN

Abstract

BIG1 and BIG2 are large (~200 kDa) guanine nucleotide-exchange proteins for ADP-ribosylation factors, or ARFs, that were isolated based on sensitivity of their guanine nucleotide-exchange activity to inhibition by brefeldin A. The intracellular distributions of BIG1 and BIG2 differ from those of other ARF guanine nucleotide-exchange proteins. In addition to its presence in Golgi membranes, BIG2 is seen in peripheral vesicular structures that most likely represent recycling endosomes, and BIG1 is found in nuclei of serum-starved HepG2 cells. Several binding partners for BIG1 and BIG2 that were identified via yeast two-hybrid screens include FKBP13 and myosin IXb for BIG1 and, for BIG2, the regulatory RIα subunit of protein kinase A, Exo70, and the GABA receptor β subunit. Autosomal recessive periventricular heterotopia with microcephaly, a disorder of human embryonic development due to defective vesicular trafficking, has been attributed to mutations in BIG2. Methods for purification of BIG1 and BIG2 from HepG2 cells are presented here, along with a summary of information regarding their structure and function.

METHODS IN ENZYMOLOGY, VOL. 404   0076-6879/05 $35.00
DOI: 10.1016/S0076-6879(05)04017-6

Background

ADP-ribosylation factors, or ARFs, are 20-kDa guanine nucleotide-binding proteins that are critical components of vesicular trafficking systems in eukaryotic cells. ARF activity, like that of all regulatory GTP-ases, is determined by the guanine nucleotide that is bound. In general, ARF-GDP is inactive and largely cytosolic, whereas active ARF-GTP is membrane-bound. Mammalian ARF molecules are grouped into three classes; class I ARFs 1–3 are implicated in vesicular trafficking among the endoplasmic reticulum, Golgi, and plasma membranes, including endosomal organelles. Relatively little is known about class II ARFs 4 and 5 (Moss and Vaughan, 1995). Class III ARF 6 differs the most from other members of the ARF family in both structure and function and is critical in control of plasma membrane ruffling and rearrangements of the actin cytoskeleton in the cell periphery (Donaldson, 2003). ARF activity is regulated by proteins that determine its guanine nucleotide-bound state. The GTPase activity of ARF is undetectable in the absence of a GAP, or GTPase-activating protein, which is required for ARF inactivation. Activation of ARF by GTP binding must be preceded by the release of bound GDP, which requires interaction with a guanine nucleotide-exchange protein, or GEP, all of which are still incompletely characterized (Moss and Vaughan, 1995).

ARF GEPs are a group of otherwise quite different kinds of proteins that have in common the ability to accelerate the replacement of ARF-bound GDP by GTP, reflecting the presence in each of a similar catalytic site in the region of the molecule termed "Sec7 domain." The Sec7 protein had been identified in a *Saccharomyces cerevisiae* genetic screen for mutations in secretory pathways (Novick *et al.*, 1980). This structural element of the molecule was then recognized in additional proteins and later shown to be responsible for ARF activation (Chardin *et al.*, 1996). Mammalian ARF GEPs have been identified through a number of approaches, including genome database mining for ARNO (Chardin *et al.*, 1996), genetic screening to identify a brefeldin A-resistance factor for GBF1 (Claude *et al.*, 1999), and classical purification of a brefeldin A-inhibited GEP from homogenized brain for BIG1 and BIG2 (Morinaga *et al.*, 1996, 1997). GEPs from a variety of species were grouped phylogenetically in six families (Jackson, 2003). The GBF and BIG families, which are the largest molecules, are distinguished most obviously by their difference in sensitivity to inhibition by brefeldin A, a fungal fatty acid metabolite that was used as an inhibitor of virus production (Tamura *et al.*, 1968) and protein secretion (Misumi *et al.*, 1986) by mammalian cells before ARFs were known. This chapter summarizes information on the structure and function of BIG1 and BIG2 along with methods for purification of the two proteins.

BFA-inhibition of BIG1 and BIG2 GEP Activity

BIG1 and BIG2 were isolated as components of a 670-kDa complex from bovine brain cytosol in six steps using assays of BFA-inhibited GEP activity with a partially purified ARF preparation as substrate (Morinaga *et al.*, 1996). Proteins of 200 kDa (BIG1) and 190 kDa (BIG2) separated from the complex of proteins by SDS-PAGE exhibited similar GEP activities. Among tryptic peptides from the 200-kDa protein, amino acid sequence with similarity to those of known Sec7 domains was recognized. Recombinant BIG1 was much more active with ARF1 and ARF3 than with ARF5, and no activity was found with ARF6 (Morinaga *et al.*, 1996). Deduced amino acid sequences of the cloned cDNAs revealed several regions of identity, greater than 90% in the Sec7 domain, with 74% identity overall (Morinaga *et al.*, 1997; Togawa *et al.*, 1999). The human BIG1 gene is located on chromosome 8q13 (Mansour *et al.*, 1999), and BIG2 on chromosome 20 (Togawa *et al.*, 1999).

Kinetics of BFA inhibition of GEP activity of a recombinant BIG1 fragment (amino acids 560–890) defined it as a noncompetitive inhibitor (Mansour *et al.*, 1999). Different pairs of amino acids that were critical for BFA sensitivity were identified in Sec7 domains from yeast Sec7 (Sata *et al.*, 1999) and human ARNO (Peyroche *et al.*, 1999). Peyroche *et al.* (1999) also demonstrated the stoichiometric, stable ARF-GDP-Sec7 domain-BFA complex that results in noncompetitive inhibition and described the effects of specific Sec7-domain amino acids on its formation. Molecular characteristics of BFA inhibition were defined by comparison of [^{3}H]-BFA binding and release by Sec7 in complex with ARF1-GDP, defining two pairs of residues in the Sec7 domain (Y190-S191 and D198-M208) and a single residue in ARF1 (H80) required for stable BFA binding (Robineau *et al.*, 2000).

Cellular Localization of BIG1 and BIG2

To a very large extent, subcellular localization of proteins has relied on immunofluorescence microscopy. Mansour *et al.* (1999), using overexpression of HA-tagged BIG1 protein fragments in 293-HEK and BHK21 cells, mapped a Golgi-localization signal to the N-terminal one-third of the protein. HA-BIG1 colocalized with the Golgi marker mannosidase II, whereas HA-BIG1(560–890), comprising only the Sec7 domain plus 37 residues at its N-terminus, localized in the nucleus. In cultured HepG2 cells, endogenous BIG1 and BIG2 were seen in the perinuclear region and in a punctate distribution throughout the cytoplasm (Yamaji *et al.*, 2000). Endogenous BIG1 appeared colocalized with overexpressed myc-tagged

BIG2 in HeLa and HepG2 cells. Most (70–75%) of endogenous BIG1 and BIG2 was immunoprecipitated from HepG2 cytosol by antibodies against the other protein (Yamaji et al., 2000), which seemed consistent with their copurification as a 670-kDa complex from bovine brain cytosol (Morinaga et al., 1997).

Zhao et al. (2002) showed by immunofluorescence microscopy that endogenous BIG1 and GBF1, the ~160 kDa, BFA-resistant mammalian GEP isolated by Claude et al. (1999) because of its ability to confer BFA resistance to CHO cells, localized in different parts of the Golgi system. There was little overlap between GBF1 and an overexpressed N-terminal fragment of BIG2, which was associated with perinuclear structures distinct from the cis-Golgi. Endogenous BIG1 was not seen with COPI, a component of the coatomer complex, but partially overlapped with clathrin in the perinuclear region. Zhao et al. (2002) suggested that the primary function of BIG1 and BIG2 may be to activate ARFs for recruitment of vesicle components such as adaptins and GGAs in the trans-Golgi network (TGN).

BIG2 immunoreactivity was described in the TGN, as well as in scattered concentrations in the cell periphery (Shinotsuka et al., 2002; Yamaji et al., 2000). Shin et al. (2004) reported that overexpressed BIG2 overlapped with staining for AP1 and transferrin receptors (TfnR), which are markers for early and recycling endosomes. BIG2 and internalized transferrin (Tfn) were colocalized in tubular structures consistent with the TfnR recycling compartment. In cells overexpressing the dominant negative mutant Rab11 (S25N), overexpression of BIG2 (E738K), a mutant without GEP activity, caused tubulation of TfnR-positive structures that lacked markers of early endosomes or lysosomes, that is, BIG2 (E738K) was associated with structures related to recycling that were not early endosomes, late endosomes, or Golgi complex. The altered morphology of the recycling endosome system, which was ascribed to interference with the activation ARF1 and ARF3, did not affect the release of Tfn from the cells. It appears that the recycling endosome structure is maintained by BIG2 activation of class I ARFs (Shin et al., 2004).

The presence of BIG1 in nuclei was recognized only relatively recently. After incubation of HepG2 cells overnight without serum, immunoreactive endogenous BIG1 was seen in essentially all nuclei, whereas that was rare in cells growing in serum (Padilla et al., 2004). BIG1 colocalized with nucleoporin p62 at the nuclear membrane, and with nucleolin in a punctate pattern superimposed on overall nuclear matrix staining. Subcellular and nuclear fractionation and co-immunoprecipitation experiments confirmed the shift of BIG1 from membrane to nuclear fractions after serum starvation. BIG1 was associated with nucleolin and nucleoporin in purified nucleoli (Padilla *et al.*, 2004). The authors suggested a signaling role for BIG1 in the nucleus, and specifically in nucleoli, that is related to the cell cycle and may or may not involve its ARF GEP function (Padilla *et al.*, 2004).

Functional Interactions of BIG1

Because BIG1 and BIG2 were purified and defined by their BFA-inhibited ARF GEP activity, their interactions with individual ARFs and with BFA were among the first to be characterized (Fig. 1). Most other interactions have been recognized via clues from intracellular localization and conditions that influence it, or in two-hybrid and co-immunoprecipitation experiments. In a yeast two-hybrid screen, both full-length BIG1 and its N-terminus (amino acids 1–331), interacted with the FK506-binding protein FKBP13, which was confirmed by co-immunoprecipitation of endogenous BIG1 and FKBP13 from Jurkat cells (Padilla *et al.*, 2003). Subsequent studies demonstrated effects of FK506 on the subcellular distributions of BIG1 and FKBP13, including their translocation to the nucleus, where they were apparently not colocalized (Padilla *et al.*, 2004).

In a yeast two-hybrid screen with the tail of the myosin IXb molecule as "bait," Saeki *et al.* (2005) identified five clones encoding C-terminal

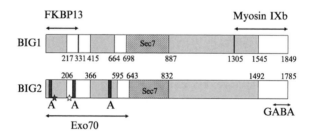

FIG. 1. Schematic representation of human BIG1 and BIG2 molecules with regions of identified function. BIG1(1–331) interacted with FKBP13 (Padilla *et al.*, 2003) and BIG2(1–643) with Exo70 (Xu *et al.*, 2005) in yeast two-hybrid screens. "A" denotes the AKAP (A kinase-anchoring protein) sequences in BIG2, sites of interaction with specific regulatory subunits of protein kinase A (Li *et al.*, 2003). Filled star denotes point and frameshift mutations found in Pedigree 2 of ARPHM at amino acids 81, 83, and 84; open star denotes point mutation found in Pedigree 1 of ARPHM at residue 209 (Sheen *et al.*, 2005). Regions of 80–90% identity in BIG1 and BIG2 are shaded; Sec7 domains are 90% identical (Togawa *et al.*, 1999). A Golgi localization signal is located N-terminal to the Sec7 domain in both BIG1 and BIG2 (Mansour *et al.*, 1999; Zhao *et al.*, 2002), but has not been specifically identified. Human BIG1(1305–1849) interacted with myosin IXb (Saeki *et al.*, 2005) and rat BIG2(1679–1791) with the GABA$_A$ receptor β subunit (GABA) in yeast two-hybrid experiments. The rat BIG2 sequence has significant identity to the human BIG2 C-terminus (Charych *et al.*, 2004).

fragments of BIG1. Both the RhoGAP domain and the zinc-finger motif of myosin IXb were necessary for BIG1 binding. BIG1 inhibited RhoA binding to myosin IXb and RhoGAP activity of the myosin IXb tail in a concentration-dependent manner. The IC50 for inhibition of activity was similar to the dissociation constant for BIG1-myosin IXb binding. Myosin IXb did not alter BIG1 GEP activity or accelerate the hydrolysis of ARF-bound GTP. The authors suggested that BIG1 may be an inhibitor of myosin IXb RhoGAP activity at Golgi membranes, perhaps to allow Rho participation in actin polymerization, and that myosin IXb might function in translocation of BIG1 from the Golgi toward the minus ends of actin filaments (Saeki et al., 2005).

Recent observations are rapidly increasing recognition of the dynamic and diverse character of BIG1 functions in cells, and raising new questions about how they are related and regulated.

Functional Interactions of BIG2

Recognition of novel binding partners for BIG2, as well as for BIG1, is growing. Three AKAP (A kinase-anchoring protein) sequences that bind specific regulatory subunits of cAMP-dependent protein kinase (protein kinase A or PKA) were identified in the N-terminal region of the BIG2 molecule by yeast two-hybrid experiments (Li et al., 2004). Interactions were also demonstrated by co-immunoprecipitation of in vitro-synthesized RIα (retrieved in the initial two-hybrid screen) and BIG2 (amino acids 1–832). Similarly, endogenous BIG2, RIα, and RIIα, as well as BIG1, were co-immunoprecipitated from HepG2 cell cytosol. BIG1 does contain an AKAP sequence identical to one of those in BIG2, but direct association of BIG1 with PKA subunits was not shown. Incubation of HepG2 cells with 8-Br-cAMP or forskolin increased amounts of BIG1 and BIG2 recovered in membrane fractions, consistent with a potential function of the AKAP sequences in BIG2, and perhaps BIG1 (Li et al., 2004).

A clone, which appears to represent a splice variant of Exo70 (a component of the exocyst complex), was retrieved in the same yeast two-hybrid screen that yielded RIα. The protein products of the Exo70 clone and the N-terminus of BIG2 were co-immunoprecipitated after in vitro synthesis (Xu et al., 2005). By immunofluorescence microscopy, endogenous BIG2 and Exo70 were colocalized along microtubules stained for α-tubulin and at the Microtubule Organizing Center (MTOC), which was identified by the presence of immunoreactive γ-tubulin. Microtubules purified by taxol polymerization contained BIG2 (as well as BIG1), Exo70, and Sec6 (another exocyst component). The exocyst complex had been implicated in coordinating signal transduction and cytoskeletal processes by targeting vesicles to plasma membranes for exocytosis (Vega and Hsu, 2001), and the association of BIG2 with subunits of protein kinase A, as well as with the exocyst complex, is consistent with this model. (Xu *et al.*, 2005)

Charych *et al.* (2004) retrieved a clone encoding the C-terminal 110 amino acids of BIG2 from a rat brain cDNA library in a yeast two-hybrid screen using the $\beta3$ GABA$_A$ receptor (GABA$_A$ R) subunit intracellular loop (IL) as a "bait." An 18-amino acid conserved region in the intracellular loops of GABA$_A$ receptor subunits $\beta1$, $\beta2$, and $\beta3$ contains the BIG2-binding site. Truncation of the BIG2 fragment at either end abrogated binding. Endogenous BIG2 and GABA$_A$ receptor were coprecipitated from a detergent-solubilized synaptosome/microsome fraction of rat hippocampus with antibodies against either protein, and interaction of BIG2 with the assembled receptor was demonstrated by its immunoprecipitation using antibodies against the α subunit. In HEK293 cells, the overexpressed HA-tagged C-terminal fragment of BIG2 was colocalized with GABA$_A$ receptor in the perinuclear region. The authors suggested that a dominant-negative effect of the BIG2 C terminus prevented interaction of the GABA$_A$ receptor with endogenous BIG2. In contrast, overexpression of full-length BIG2 decreased the amount of perinuclear GABA$_A$ receptor, consistent with a role for BIG2 in receptor transport out of the endoplasmic reticulum. In rat brain cells, BIG2 was found in TGN cisternae, as well as associated with axonal microtubules and postsynaptic microtubules by electron microscopy with immunogold labeling. Charych *et al.* (2004) concluded that "the molecular interaction between BIG2 and the ILs [intracellular loops] of GABA$_A$ R [receptor] β subunits... might play a role in the transport of newly assembled GABA$_A$ Rs [receptors] by clathrin/AP-1 coated vesicles from the TGN to endosomes and perhaps to the synaptic plasma membrane."

Sheen *et al.* (2004) described BIG2 mutations in two families with autosomal recessive periventricular heterotopia with microcephaly (ARPHM), which is characterized by malformation of the cerebral cortex and severe developmental delay resulting, in part, from the failure of neurons to migrate properly within the embryonic brain. This is reminiscent of the defective embryonic morphology associated with mutation of EMB30, an ARF GEP gene in *Arabadopsis thaliana* (Shevell *et al.*, 1994). Sheen *et al.* (2004) postulated that vesicular trafficking is essential for neuronal proliferation and migration, probably for delivery to the cell surface of proteins required for normal embryonal growth and differentiation. The demonstrated interaction of BIG2 and Exo70 in association with exocytic vesicles and microtubules (Xu *et al.*, 2005) may be part of this critical mechanism. Vega and Hsu (2001), who studied differentiation of

PC-12 (neuro-endocrine) cells, suggested a function for the exocyst complex in signaling and cytoskeletal regulatory pathways that determine sites of exocytosis.

BIG2, like BIG1, is increasingly understood as a multifunctional protein, serving at numerous times and places in cell signaling, regulatory, and mechanical processes that remain to be completely characterized.

Methods

Purification of BIG1 and BIG2 from bovine brain cytosol in a ~670 kDa complex was described in *Methods in Enzymology* by Pacheco-Rodriguez *et al.* (2002). The most convenient assay of activity is still GTPγS binding to a partially purified preparation of ARF1 and ARF3, described in the same reference and in this volume by Pacheco-Rodriguez *et al.* The procedure for purification of BIG1 and BIG2 proteins from HepG2 cells used most recently is outlined here.

Scrape confluent HepG2 cells from 210 10-cm plates in 0.1 ml/plate of TENDS buffer (100 mM Tris, pH 8.0, 1 mM EDTA, 1 mM NaN$_3$, 1 mM dithiothreitol, 0.25 M sucrose) containing 0.5 mM 4-(2-aminoethyl) benzenesulfonyl fluoride (AEBSF), 1 mM benzamidine, and aprotinin, leupeptin, and soybean and lima bean trypsin inhibitors, each 1 μg/ml. Homogenize cells by hand in a glass tissue grinder and centrifuge homogenate (37,000 x g) for 1.5 h. Apply 8 ml samples of supernatant (about 10 mg protein/ml) to columns (2 x 85 cm, 267 ml total volume) of Sephacryl CL-6B equilibrated and elute with buffer A (TENDS plus 2 mM MgCl$_2$, and 0.5 mM AEBSF) containing 0.1 M NaCl. Collect approximately 3 ml fractions (6.4 min). Identify fractions (usually 45–57) containing BIG1 and BIG2 by SDS-PAGE and Western blotting, and pool (~33 mg protein, ~120 ml), concentrate (Amicon XM100) to 30 ml, centrifuge (12,000×g, 1 h), and dilute supernatant with 30 ml of buffer A. Apply 60 ml pool (containing 50 mM NaCl) to column (1.5 x 11.3 cm, 20 ml total volume) of DEAE-Sephacryl that had been equilibrated with 60 ml of 0.5 M Tris buffer, pH 8.0, and then with 100 ml of buffer A containing 50 mM NaCl. During elution with the same buffer, collect fractions 1–10 (~6 ml/10 min) and 11–20 (~5 ml/5 min). Elute with a linear gradient of 50 to 300 mM NaCl in buffer A (100 ml of each) while collecting fractions 21–80 (about 5 ml/5 minutes). Usually, fractions 43–48 contain predominantly BIG2 (~3.6 mg protein in 30 ml) and fractions 49–60 contain predominantly BIG1 (~8.1 mg protein in 60 ml).

Dialyze separate BIG1 and BIG2 pools overnight against 2 L of buffer A before chromatography on heparin-agarose (1-cm diameter column, total volume 4 ml for BIG1 and 2 ml for BIG2) equilibrated with 10 volumes of buffer A. Application of BIG1 pool (~8 mg protein) is followed by buffer A while collecting fractions 1–13 (4.5 ml/10 min), then with seven column volumes of buffer A containing 50 mM NaCl (fractions 14–20, 4 ml/9 min) and finally, (fractions 21–80, 2.5 ml/6 min) a linear gradient of 50 to 400 mM NaCl in buffer A (20 volumes of each). BIG1 is eluted before BIG2 (fractions 44–46 and 48–56, respectively). The BIG2 pool is fractionated similarly on the 2 ml heparin-agarose column with proportionately smaller volumes of eluant and fractions. Separation of BIG1 and BIG2 is better on the larger column.

Analyze fractions by SDS-PAGE and Western blotting to identify those best for the intended use.

References

Chardin, P., Paris, S., Antonny, B., Robineau, S., Beraud-Dufour, S., Jackson, C. L., and Chabre, M. (1996). A human exchange factor for ARF contains Sec7- and pleckstrin-homology domains. *Nature* **384,** 481–484.

Charych, E., Yu, W., Miralles, C., Serwanski, D., Li, X., Rubio, M., and De Blas, A. (2004). The brefeldin A-inhibited GDP/GTP exchange factor 2, a protein involved in vesicular trafficking, interacts with the beta subunits of the GABA receptors. *J. Neurochem.* **90,** 173–189.

Claude, A., Zhao, B. P., Kuziemsky, C. E., Dahan, S., Berger, S. J., Yan, J. P., Arnold, A. D., Sullivan, E. M., and Melancon, P. (1999). GBF1: A novel Golgi-associated BFA-resistant guanine nucleotide exchange factor that displays specificity for ADP-ribosylation factor 5. *J. Cell Biol.* **146,** 71–84.

Donaldson, J. G. (2003). Multiple roles for Arf6: Sorting, structuring, and signaling at the plasma membrane. *J. Biol. Chem.* **278,** 41573–41576.

Jackson, C. (2003). The Sec7 family of ARF guanine nucleotide exchange factors. *In* "ARF Family GTPases" (R. A. Kahn, ed.), Vol. 1, pp. 71–99. Kluwer Academic Publishers, Netherlands.

Li, H., Adamik, R., Pacheco-Rodriguez, G., Moss, J., and Vaughan, M. (2004). Protein kinase A-anchoring (AKAP) domains in brefeldin A-inhibited guanine nucleotide-exchange protein 2 (BIG2). *Proc. Natl. Acad. Sci. USA* **100,** 1627–1632.

Mansour, S., Skaug, J., Xin-Hua, Z., Giordano, J., Scherer, S., and Melancon, P. (1999). p200 ARF-GEP1: A Golgi-localized guanine nucleotide exchange protein whose Sec7 domain is targeted by the drug brefeldin A. *Proc. Natl. Acad. Sci. USA* **96,** 7968–7973.

Misumi, Y., Misumi, Y., Miki, K., Takatsuki, A., Tamura, G., and Ikehara, Y. (1986). Novel blockade by brefeldin A of intracellular transport of secretory proteins in cultured rat hepatocytes. *J. Biol. Chem.* **261,** 11398–11403.

Morinaga, N., Tsai, S., Moss, J., and Vaughan, M. (1996). Isolation of a brefeldin A-inhibited guanine nucleotide-exchange protein for ADP ribosylation factor (ARF) 1 and ARF3 that contains a Sec7-like domain. *Proc. Natl. Acad. Sci. USA* **93,** 12856–12860.

Morinaga, N., Moss, J., and Vaughan, M. (1997). Cloning and expression of a cDNA encoding a bovine brain brefeldin A-sensitive guanine nucleotide-exchange protein for ADP-ribosylation factor. *Proc. Natl. Acad. Sci. USA* **94**, 12926–12931.

Moss, J., and Vaughan, M. (1995). Structure and function of ARF proteins: Activators of cholera toxin and critical components of intracellular vesicular transport processes. *J. Biol. Chem.* **270**, 12327–12330.

Novick, P., Field, C., and Schekman, R. (1980). Identification of 23 complementation groups required for post-translational events in the yeast secretory pathway. *Cell* **21**, 205–215.

Pacheco-Rodriguez, G., Moss, J., and Vaughan, M. (2002). BIG1 and BIG2: Brefeldin A-inhibited guanine nucleotide-exchange proteins for ADP-ribosylation factors. *Methods Enzymol.* **345**, 397–404.

Padilla, P., Pacheco-Rodriguez, G., Moss, J., and Vaughan, M. (2004). Nuclear localization and molecular partners of BIG1, a brefeldin A-inhibited guanine nucleotide-exchange protein for ADP-ribosylation factors. *Proc. Natl. Acad. Sci. USA* **101**, 2752–2757.

Padilla, P., Chang, M., Pacheco-Rodriguez, G., Adamik, R., Moss, J., and Vaughan, M. (2003). Interaction of FK506-binding protein 13 with brefeldin A-inhibited guanine nucleotide-exchange protein 1 (BIG1): Effects of FK506. *Proc. Natl. Acad. Sci. USA* **100**, 2322–2327.

Peyroche, A., Antonny, B., Robineau, S., Acker, J., Cherfils, J., and Jackson, C. (1999). Brefeldin A acts to stabilize an abortive ARF-GDP-Sec7 domain protein complex: Involvement of specific residues in the Sec7 domain. *Molecular Cell* **3**, 275–285.

Robineau, S., Chabre, M., and Antonny, B. (2000). Binding site of brefeldin A at the interface between the small G protein ADP-ribosylation factor 1 (ARF1) and the nucleotide-exchange factor Sec7 domain. *Proc. Natl. Acad. Sci. USA* **97**, 9913–9918.

Saeki, N., Tokuo, H., and Ikebe, M. (2005). BIG1 is a binding partner of myosin IXb and regulates its Rho-GTPase activating protein activity. *J. Biol. Chem.* **280**, 10128–10134.

Sata, M., Moss, J., and Vaughan, M. (1999). Structural basis for the inhibitory effect of brefeldin A on guanine nucleotide-exchange proteins for ADP-ribosylation factors. *Proc. Natl. Acad. Sci. USA* **96**, 2752–2757.

Sheen, V. L., Ganesh, V. S., Topcu, M., Sebire, G., Bodell, A., Hill, R. S., Grant, P. E., Shugart, Y. Y., Imitola, J., Khoury, S. J., Geurrini, R., and Walsh, C. A. (2005). Mutations in *ARFGEF2* implicate vesicle trafficking in neural progenitor proliferation and migration in the human cerebral cortex. *Nature Genetics* **36**, 69–76.

Shevell, D. E., Leu, W. M., Gillmor, C. S., Xia, G., Feldmann, K. A., and Chua, N. H. (1994). EMB30 is essential for normal cell division, cell expansion, and cell adhesion in Arabidopsis and encodes a protein that has similarity to Sec7. *Cell* **77**, 1051–1062.

Shin, H., Morinaga, N., Masatoshi, N., and Nakayama, K. (2004). BIG2, a guanine nucleotide exchange factor for ADP-ribosylation factors: Its localization to recycling endosomes and implication in the endosome integrity. *Mol. Biol. Cell* **15**, 5283–5294.

Shinotsuka, C., Yoshida, Y., Kawamoto, K., Takatsu, H., and Nakayama, K. (2002). Overexpression of an ADP-ribosylation factor-guanine nucleotide exchange factor, BIG2, uncouples brefeldin A-induced adaptor protein-1 coat dissociation and membrane tubulation. *J. Biol. Chem.* **277**, 9468–9473.

Tamura, G., Ando, K., Suzuki, S., Takatsuki, A., and Arima, K. (1968). Antiviral activity of brefeldin A and verrucarin A. *J. Antibiot. (Tokyo)* **21**, 160–161.

Togawa, A., Morinaga, N., Ogasawara, M., Moss, J., and Vaughan, M. (1999). Purification and cloning of a brefeldin A-inhibited guanine nucleotide-exchange protein for ADP-ribosylation factors. *J. Biol. Chem.* **274**, 12308–12315.

Vega, I. E., and Hsu, S. (2001). The exocyst complex associates with microtubules to mediate vesicle targeting and neurite outgrowth. *J. Neurosci.* **21,** 3839–3848.

Xu, K. F., Shen, X., Li, H., Pacheco-Rodriguez, G., Moss, J., and Vaughan, M. (2005). Interaction of BIG2, a brefeldin A-inhibited guanine nucleotide-exchange protein, with exocyst protein Exo70. *Proc. Natl. Acad. Sci. USA* **102,** 2784–2789.

Yamaji, R., Adamik, R., Takeda, K., Togawa, A., Pacheco-Rodriguez, G., Ferrans, V., Moss, J., and Vaughan, M. (2000). Identification and localization of two brefeldin A-inhibited guanine nucleotide-exchange proteins for ADP-ribosylation factors in a macromolecular complex. *Proc. Natl. Acad. Sci. USA* **97,** 2567–2572.

Zhao, X., Lasell, T., and Melancon, P. (2002). Localization of large ADP-ribosylation factor-guanine nucleotide exchange factors to different Golgi compartments: Evidence for distinct functions in protein traffic. *Mol. Biol. Cell.* **13,** 119–133.

[18] Cytohesin-1: Structure, Function, and ARF Activation

By GUSTAVO PACHECO-RODRIGUEZ,
JOEL MOSS, and MARTHA VAUGHAN

Abstract

Mammalian cytohesins are a family of very similar guanine nucleotide-exchange proteins (GEPs) that activate ADP-ribosylation factors (ARFs). Cytohesins are multifunctional molecules comprising a Sec7 domain that is responsible for the GEP activity, a PH domain that binds specific phosphatidylinositol phosphates, and a coiled-coil domain responsible for homodimerization and interaction with other proteins. Cytohesin proteins are ubiquitous and have been implicated in several functions including cell spreading and adhesion, chemotaxis, protein trafficking, and cytoskeletal rearrangements, only some of which appear to depend on their ability to activate ARFs. Unlike the GEP activity of BIG1 and BIG2, the acceleration by cytohesins of guanine nucleotide exchange to generate active ARF-GTP is not inhibited by the fungal metabolite brefeldin, A (BFA). This chapter is concerned for the most part with cytohesin-1 and the assay of its GEP activity.

Structure and Function of Cytohesin-1

Organization of Cytohesin Genes

Human cytohesins are encoded by genes PSCD1 (cytohesin-1/B2–1), PSCD2 (cytohesin-2/ARNO[ARF nucleotide-binding-site opener]), PSCD3 (cytohesin-3/ARNO3/GRP1[general receptor for phosphoinositides 1]), and

METHODS IN ENZYMOLOGY, VOL. 404
0076-6879/05 $35.00
DOI: 10.1016/S0076-6879(05)04018-8

PSCD4 (cytohesin-4) located, respectively on chromosomes 17q25 (Dixon *et al.*, 1993), 19q13.3 (Buchet-Poyau *et al.*, 2002), 7p22.1 (Kim, 1998) and 22q12.3-q13.1 (Ogasawara *et al.*, 2000). Overall, the four human cytohesin mRNAs are ~80% identical.

Cytohesin-1 is encoded in 14 exons (which appears to be true also for cytohesin-2 and cytohesin-3). Exons 1–3 encode the N-terminal coil-coiled domain, exons 4–9 the Sec7 domain, and exons 10–14 the PH domain plus C-terminal sequence (Ogasawara *et al.*, 2000). Exon 10 "GAG" encodes a single glycine. It is lacking in cytohesin-4, which has only 13 exons. Intron 9 of the cytohesin-4 gene does contain the appropriate "GAG" potential exon sequence, but it was not found in any cytohesin-4 mRNA. Alternatively spliced forms of cytohesin-1, -2, and -3 mRNA, which result in two forms of each of the proteins, are present in different proportions in different tissues (Ogasawara *et al.*, 2000). Inclusion of the glycine encoded by exon 10 produces a tri-glycine sequence affecting dramatically the affinity (specificity) of the PH domain for binding phosphatidylinositol 4,5-*bis*phosphate or phosphatidylinositol 3,4,5-*tris*phosphate with major effects on biological function (Klarlund *et al.*, 2000).

Domains of Cytohesin-1 Protein

The cytohesin-1 molecule (diagram in Fig. 1 see for review [Moss and Vaughan, 2002]) is 77% identical in amino acid sequence to cytohesin-4 and ~90% identical to the other two cytohesins. All four human cytohesins contain at the N-terminus a coiled-coil domain of a ~60 amino acids, a centrally located Sec7 domain of ~200 residues, and a PH domain of ~100 residues. The greatest variability among cytohesins is found in the C-terminal ~30 amino acids.

Coiled-coil domain. The coiled-coil domain at the N-terminus of cytohesin-1 is involved in protein-protein interactions that are important for the regulation of GEP activity as well as for membrane and cytoskeleton associations. Coiled-coil structures consist of two to five amphipathic alpha-helices wrapped around each other in a left-handed helix to form a supercoil. Coiled-coil structures are, in general, left-handed with a typical periodicity of seven amino acids. The less common right-handed coiled-coils have a periodicity of 11 residues. The interfaces between coils involve mainly nonpolar residues, allowing for oligomerization, which produces strong interactions (Mason and Arndt, 2004). The N-terminal coiled-coil domain of cytohesin-1 (and cytohesin-2, but not cytohesin-3) interacted with the coiled-coil domain of CASP. CASP has a structure similar to that of GRASP (GRP1-binding partner) (Mansour *et al.*, 2002). GRASP was identified as a retinoic acid-induced protein that interacted with GRP1 and

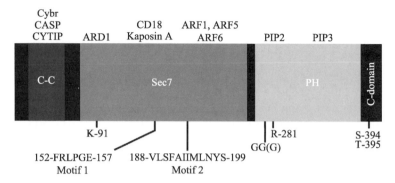

Fig. 1. Structure and function in the human cytohesin-1 molecule. Above the "molecule" are ligands demonstrated to interact with the indicated region of cytohesin-1. The coiled-coil (C-C) domain at the N-terminus interacts with Cybr, CASP, and CYTIP proteins. The central Sec7 domain (Sec7), via motifs 1 and 2, has functional interactions with ARFs and ARD1 that result in their activation. The Sec7 domain also interacts with CD18 (integrin-β2) and kaposin A. The PH domain binds phosphatidylinositol 4,5-*bis*phosphate (PIP2) and phosphatidylino-sitol 3,4,5-*tris*phosphate (PIP3). Below the "molecule" are regions implicated in cytohesin-1 functions. Motif 1 contains the E-157 required for GEP activity. K-91 is critical for the specificity of association with the ARF domain of ARD1. In the pleckstrin homology (PH) domain, a di- or triglycine motif binds specific phosphatidylinositol phosphates that influence cytohesin function. R-281 corresponds to the arginine in Bruton's tyrosine kinase that is required for phospholipid binding. A highly basic segment (C-domain) at the C-terminus, contains phosphorylation sites S-394 and T-395.

colocalized at the plasma membrane with ARFs (Nevrivy *et al.*, 2000). Interactions of cytohesin-1 and -2 with CASP appear to participate in their recruitment to membranes in COS-1 cells incubated with EGF (Mansour *et al.*, 2002). The coiled-coil N-terminus of cytohesin-1 interacted with Cybr, the production of which was increased in NK2.3 cells incubated with IL-2 and IL-12 (Tang *et al.*, 2002). Cytohesin-1 also bound the coiled-coil domain of CYTIP (Boehm *et al.*, 2003). All of the identified cytohe-sin-binding proteins use a coiled-coil structure for interaction with the similar structure in the cytohesin molecule. The cytohesin-binding pro-teins contain, in addition to the coiled-coil structure, a PDZ domain, the relationship of which to cytohesin function has not been defined.

Sec7 Domain. The SEC7 protein was identified in a genetic screen of *Saccharomyces cerevisiae* mutants for defects in protein secretion (Achstetter *et al.*, 1988). Structurally similar domains that were later recognized in other proteins involved in vesicular trafficking were referred as Sec7 domains and subsequently shown to be responsible for ARF activation (Morinaga *et al.*, 1996; Peyroche *et al.*, 1996) as well as its sensitivity to brefeldin A inhibition (Peyroche *et al.*, 1996; Sata *et al.*, 1998). Sec 7 domain crystal structures were determined for the yeast brefeldin A-sensitive protein Gea2 (Mossessova *et al.*, 1998) and for human cytohesin-2 (Cherfils *et al.*, 1998). An NMR structure of the cytohesin-1 Sec7 domain was reported soon thereafter (Betz *et al.*, 1998). All of the Sec7 domain structures have two parts, each with five alpha-helices. The GEP active site is in the C-terminal portion where amino acids critical for catalysis are present in sequences termed motif 1 and 2 that are highly conserved among cytohesins. The Sec7 domain interacts with switch I and II regions in the ARF molecule to accelerate the release of bound GDP. The Sec7 domain of cytohesin-1 interacted also with the intracellular portion of $\beta 2$ integrin CD18 (Kolanus *et al.*, 1996).

Recognition of Sec7 domain sequences in peptides from p200 and p190, now known, respectively, as brefeldin A-inhibited guanine nucleotide-exchange proteins 1 and 2 or BIG1 and BIG2 (Morinaga *et al.*, 1997; Togawa *et al.*, 1999), led to the demonstration that cytohesin-1 is an ARF GEP (Meacci *et al.*, 1997). On comparison of *in vitro* guanine nucleotide-exchange activity with recombinant ARFs of different classes, cytohesin-1 appeared to be more effective with ARF1 than ARF5 or ARF6 (Pacheco-Rodriguez *et al.*, 1998). Cytohesin-4 accelerated guanine nucleotide exchange similarly on ARF1 and ARF5, but did not affect ARF6 (Ogasawara *et al.*, 2000). As shown in Fig. 2, GEP activity of full-length recombinant cytohesin-1 was clearly less that of its Sec7 domain (Pacheco-Rodriguez *et al.*, 1998). The Sec7 domain was, in addition, essentially equally active with ARF1, ARF5, and ARF6 leading to the conclusion that regions outside of the Sec7 domain can have important effects on GEP activity and ARF specificity of the intact protein.

Amino acids critical for GEP activity of the cytohesin-1 Sec7 domain are "FRLPGE" present in motif 1 and "VLSFAIIMLNYS" motif 2. The glutamic acid in motif 1 is required for GEP activity (Béraud-Dufour *et al.*, 1998). The ARF-activating GEPs can be grouped by their sensitivity to inhibition by the fungal metabolite brefeldin A (BFA), as well as by their multidomain structures (Jackson and Casanova, 2000). All of the cytohesins lack the Sec7 domain amino acids that confer sensitivity to BFA (Peyroche *et al.*, 1999; Sata *et al.*, 1999). Chimeric Sec7 domains in which those amino acids were systematically exchanged proved critical for establishing the structural basis of BFA sensitivity (Robineau *et al.*, 2000). Later experiments of Renault *et al.* (2002) provided valuable new insight into the molecular mechanism of Sec7 domain function in guanine nucleotide exchange and the effect of BFA on it.

In addition to activating ARF proteins, cytohesin-1, but not cytohesin-2, activated the ADP-ribosylation factor-domain protein 1 ARD1

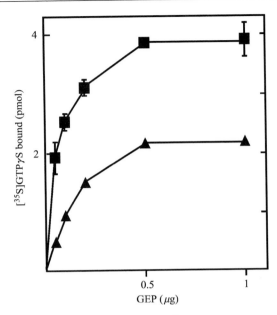

FIG. 2. Effect of cytohesin-1 and C-1 Sec7 on [^{35}S] GTPγS binding by hARF1. Both cytohesin −1 (▲) and its Sec7 domain (■) accelerate the binding of GTPγS in a concentration dependent manner (reprinted with permission from Pacheco-Rodriguez *et al.*, 1998).

(Pacheco-Rodriguez *et al.*, 1998; Vitale *et al.*, 2000). Interaction with the ARF domain of ARD1 required lysine 91 in the cytohesin-1 Sec7 domain (Vitale *et al.*, 2000), which is not within the sequences (motifs 1 and 2) that are critical for GEP activity. The 64-kDa ARD-1 was discovered and cloned because of its C-terminal ~18-kDa ARF domain (Mishima *et al.*, 1993). Its GTPase-activating domain was later recognized (Vitale *et al.*, 1996). The molecular structure of ARD1 identified it as a member of the TRIM (Tripartite motif) or RBCC (RING, B-Box, coiled-coil) protein family and Vichi *et al.*, (2005) demonstrated E3 ubiquitin ligase activity in the N-terminal part of the molecule.

PH Domain. The pleckstrin homology (PH) domain was required for membrane association of cytohesin-1 *via* phosphoinositide binding (Nagel *et al.*, 1998). L-α-phosphatidyl-L-serine (PS) is included in the routine GEP assays described here for cytohesins as well other GEPs (Pacheco-Rodriguez *et al.*, 1998). With or without PS, the Sec7 domain of cytohesin-1 accelerated GTPγS binding to nonmyristoylated ARF, which lacks the propensity to associate with membranes that is characteristic of native ARF. Because both ARF1 and cytohesin-1 appear to require membrane binding for activity in cells, it was suggested that phosphoinositide specificity of the PH domain could define intracellular sites of cytohesin-1 action. Overexpression of the cytohesin-1 PH domain blocked the adhesion of stimulated Jurkat cells to ICAM, leading Kolanus et al. (1996) to suggest that the PH domain of cytohesin-1 could also participate in protein-protein interactions as had been shown for other PH domains. Phosphoinositides appear to influence the ARF specificities of cytohesin-1 and GRP1 (Klarlund et al., 1998; Knorr et al., 2000), which could result from different affinities of the proteins for individual phospholipids.

C Terminus. The cytohesins differ most in structure near the C-terminus. Cytohesin-1 was phosphorylated by protein kinase C at serine 394 and threonine 395, which influenced its interaction with the cytoskeleton, but had no effect on binding to phospholipid membranes (Dierks et al., 2001). In contrast, phosphorylation of a residue at the C-terminus of cytohesin-2 resulted in lack of association with membranes and inefficient ARF activation (Santy et al., 1999). It seems likely that these, and perhaps other phosphorylations, have a role(s) in cytohesin function.

Intracellular Distribution of Cytohesin-1

When overexpressed in COS-7 cells, myc-cytohesin-1was found associated with cytoskeletal structures and in the nucleus (Vitale et al., 2000). The N-terminus of cytohesin-1 appeared to be responsible for its presence in Golgi structures (Lee et al., 2000). Overexpressed cytohesin-1 was associated with the plasma membrane in Jurkat cells (Kolanus et al., 1996). This seemed to be influenced by stimuli like EGF and NGF, as well as integrin interactions with extracellular molecules such as ICAM (Kolanus et al., 1996; Venkateswarlu et al., 1999).

Assay of Cytohesin-1

To quantify GEP activity of cytohesins in vitro, most assays measure ARF binding of guanine nucleotides using either radiolabeled nucleotides (e.g., Pacheco-Rodriguez et al., 1998) or changes of the tryptophan fluorescence of ARF that occur when bound GDP is replaced by GTP (e.g., Béraud-Dufour et al., 1998). GEP action on ARFs in cells has been evaluated by "pull–down" assays to recover activated ARF bound to GGA proteins (Santy and Casanova, 2001).

Several procedures have been used to assay GEP-catalyzed acceleration of GTP binding to ARF and/or the resulting ARF activation. Quantification of the increase in rate of GTP binding to ARF due to the addition of a GEP is the most direct measure of GEP activity and is relatively easily done with radiolabeled GTP and pure proteins. It cannot be used when impure protein preparations contain other GTP-binding proteins. During the early steps of purification of a GEP from cells, for example, the amount of ARF-GTP formed can be determined by measuring its enhancement of cholera toxin ADP-ribosyltransferase activity (Pacheco-Rodriguez et al., 2001).

After partial purification, GEP activity can be quantified directly, either by the rate of GTP binding to ARF as described here, or by the rate of release of ARF-bound nucleotide, a prerequisite for GTP binding (Pacheco-Rodriguez et al., 1998). The measurement of release rate requires, first, the binding by ARF of radiolabeled GDP or GTP by incubation with 4 μM nucleotide in the absence of Mg^{2+} (which facilitates nucleotide exchange), followed by the addition of 3 mM $MgCl_2$, which favors tight nucleotide binding. Unlabeled GDP or GTP (1 mM) is added (to minimize rebinding of released radiolabeled nucleotide) and incubation is continued without and with a GEP preparation to determine its effect on the rate of nucleotide release. Release is calculated as the difference between amounts of nucleotide bound at the beginning and end of the incubation period.

The basic procedure for assay of the effect of a GEP on rate of GTPγS binding to ARF described here has been used with minor variations for more than ten years. Conditions are not designed to maximize the activity of any specific ARF/GEP combination (which should be done for its optimal application in any particular study). For example, Mg^{2+} concentration is an important variable, as are the effects of phospholipids and detergents, which differ greatly with different ARFs and GEPs. Despite marked improvement in the production of recombinant, myristoylated ARF1 and ARF6 by Randazzo and Fales (2002), we do not consistently obtain proteins that are 100% myristoylated. To assay GEP activity, we routinely use recombinant nonmyristoylated ARFs of class I (ARF1), class II (ARF5), and class III (ARF6), or a mixture of partially purified native mammalian ARF1 and ARF3 (Pacheco-Rodriguez et al., 1998, 1999). It is important to remember that utilization of any of these ARFs in vitro does not necessarily reflect the specificity of GEP action in cells, which is influenced by multiple other molecular interactions and functions of these proteins resulting in or from different cellular localizations.

To characterize stable interactions, recombinant Sec7 domain and ARF proteins lacking 13 or 17 amino acids at the N-terminus have been used (e.g., Morinaga et al., 1999; Vitale et al., 2000). In the absence of nucleotides, low concentrations of free magnesium (1–2 μM) favored stable association between the Sec7 domain and ARF. It was more difficult to form stable complexes with full-length proteins, which likely means that the presence of additional molecules (and/or different conditions) contributes to the formation and stability of the native protein complexes in cells. This again emphasizes the necessity for caution in the extrapolation from *in vitro* findings to biological mechanisms. For example, in most of these studies, the ratio of cytohesin to ARF concentrations is not that typically used in enzyme assays and cannot reflect the intracellular conditions.

One of the confounding technical problems, especially with recombinant proteins, is the presence of improperly folded molecules with only a small, usually unknown and difficult to quantify, fraction of active protein. In an attempt to minimize this problem in practice, we induce at least 10 different colonies containing the cytohesin-1 construct and pool the cells for protein purification, which improves the assay reproducibility. In addition, assays with native ARF1/3 are routinely included with those employing recombinant preparations (Fig. 3) (Pacheco-Rodriguez *et al.*, 1999, 2002). Some of the Sec7 domain proteins and especially certain recombinant fragments are unstable under assay conditions (Morinaga *et al.*, 1999; Sata *et al.*, 1999). For any quantitative evaluation of GEP

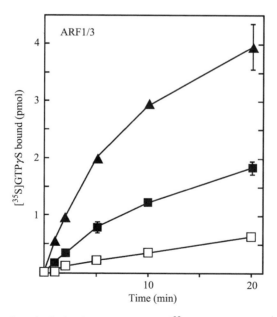

FIG. 3. Effect of cytohesin-1 or its Sec7 domain on [^{35}S]GTPγS binding by native ARF1/3. A partially purified preparation of ARF1 and ARF3 was incubated for the indicated time without (□) or with recombinant cytohesin-1 (■) or its Sec7 domain (▲) (reprinted with permission from Pacheco-Rodriguez *et al.*, 1999).

activity assay conditions must be adjusted to ensure measurement of initial rates. Often, this is most easily accomplished by decreasing the incubation temperature to improve protein stability.

Cytohesins and Disease

Cytohesin-1 was implicated in the action of kaposin A, a protein from herpesvirus 8 (Kliche *et al.*, 2001), which enhanced cytohesin-1-catalyzed acceleration of GTP binding by myristoylated ARF1. Mempel *et al.* (2003) reported the presence of mutant cytohesin-1 (E157K) in a strain of vaccinia virus. More recently Corveleyn *et al.* (2005) described a chromosomal translocation resulting in the production of a Sec7 domain protein that contained a PH-like domain. Levels of magnesium in endothelial cells influenced the expression of cytohesin-1 with associated changes in cell adhesion (Maier *et al.*, 2004) that could contribute to pathogenesis of atherosclerosis, inflammation, and thrombosis.

No genomic abnormalities in the cytohesins have been reported. A polymorphism in the PSCD2 gene that alters the predicted protein sequence (K244R) was identified toward the C-terminus of the Sec7 domain (Buchet-Poyau *et al.*, 2002), but its physiological relevance is unclear.

References

Achstetter, T., Franzusoff, A., Field, C., and Schekman, R. (1988). SEC7 encodes an unusual, high molecular weight protein required for membrane traffic from the yeast Golgi apparatus. *J. Biol. Chem.* **263,** 11711–11717.

Béraud-Dufour, S., Robineau, S., Chardin, P., Paris, S., Chabre, M., Cherfils, J., and Antonny, B. (1998). A glutamic finger in the guanine nucleotide exchange factor ARNO displaces Mg^{2+} and the beta-phosphate to destabilize GDP on ARF1. *EMBO J.* **17,** 3651–3659.

Betz, S. F., Schnuchel, A., Wang, H., Olejniczak, E. T., Meadows, R. P., Lipsky, B. P., Harris, E. A., Staunton, D. E., and Fesik, S. W. (1998). Solution structure of the cytohesin-1 (B2-1) Sec7 domain and its interaction with the GTPase ADP ribosylation factor 1. *Proc. Natl. Acad. Sci. USA* **95,** 7909–7914.

Boehm, T., Hofer, S., Winklehner, P., Kellersch, B., Geiger, C., Trockenbacher, A., Neyer, S., Fiegl, H., Ebner, S., Ivarsson, L., Schneider, R., Kremmer, E., Heufler, C., and Kolanus, W. (2003). Attenuation of cell adhesion in lymphocytes is regulated by CYTIP, a protein which mediates signal complex sequestration. *EMBO J.* **22,** 1014–1024.

Buchet-Poyau, K., Mehenni, H., Radhakrishna, U., and Antonarakis, S. E. (2002). Search for the second Peutz-Jeghers syndrome locus: Exclusion of the STK13, PRKCG, KLK10, and PSCD2 genes on chromosome 19 and the STK11IP gene on chromosome 2. *Cytogenet. Genome Res.* **97,** 171–178.

Cherfils, J., Menetrey, J., Mathieu, M., Le Bras, G., Robineau, S., Béraud-Dufour, S., Antonny, B., and Chardin, P. (1998). Structure of the Sec7 domain of the Arf exchange factor ARNO. *Nature* **392**, 101–105.

Corveleyn, A., Wlodarska, I., Mecucci, C., and Marynen, P. (2005). The der(12)t(12;16) breakpoint in an acute leukaemia case targets a Sec7 domain containing protein. *Int. J. Oncol.* **26**, 1111–1120.

Dierks, H., Kolanus, J., and Kolanus, W. (2001). Actin cytoskeletal association of cytohesin-1 is regulated by specific phosphorylation of its carboxyl-terminal polybasic domain. *J. Biol. Chem.* **276**, 37472–37481.

Dixon, B., Mansour, M., and Pohajdak, B. (1993). Assignment of human B2-1 gene (D17S811E) to chromosome 17qter by PCR analysis of somatic cell hybrids and fluorescence in situ hybridization. *Cytogenet. Cell Genet.* **63**, 42–44.

Jackson, C. L., and Casanova, J. E. (2000). Turning on ARF: The Sec7 family of guanine-nucleotide-exchange factors. *Trends Cell Biol.* **10**, 60–67.

Kim, H. S. (1998). Assignment of the human ARNO3 gene (PSCD3) to chromosome 7p21 by radiation hybrid mapping. *Ann. Hum. Genet.* **62**, 551–553.

Klarlund, J. K., Tsiaras, W., Holik, J. J., Chawla, A., and Czech, M. P. (2000). Distinct polyphosphoinositide binding selectivities for pleckstrin homology domains of GRP1-like proteins based on diglycine versus triglycine motifs. *J. Biol. Chem.* **275**, 32816–32821.

Klarlund, J. K., Rameh, L. E., Cantley, L. C., Buxton, J. M., Holik, J. J., Sakelis, C., Patki, V., Corvera, S., and Czech, M. P. (1998). Regulation of GRP1-catalyzed ADP ribosylation factor guanine nucleotide exchange by phosphatidylinositol 3,4,5-trisphosphate. *J. Biol. Chem.* **273**, 1859–1862.

Kliche, S., Nagel, W., Kremmer, E., Atzler, C., Ege, A., Knorr, T., Koszinowski, U., Kolanus, W., and Haas, J. (2001). Signaling by human herpesvirus 8 kaposin A through direct membrane recruitment of cytohesin-1. *Mol. Cell.* **7**, 833–843.

Knorr, T., Nagel, W., and Kolanus, W. (2000). Phosphoinositides determine specificity of the guanine-nucleotide exchange activity of cytohesin-1 for ADP-ribosylation factors derived from a mammalian expression system. *Eur. J. Biochem.* **267**, 3784–3791.

Kolanus, W., Nage, W., Schiller, B., Zeitlmann, L., Godar, S., Stockinger, H., and Seed, B. (1996). Alpha L beta 2 integrin/LFA-1 binding to ICAM-1 induced by cytohesin-1, a cytoplasmic regulatory molecule. *Cell* **86**, 233–242.

Lee, S. Y., and Pohajdak, B. (2000). N-terminal targeting of guanine nucleotide exchange factors (GEF) for ADP ribosylation factors (ARF) to the Golgi. *J. Cell Sci.* **113**, 1883–1889.

Mansour, M., Lee, S. Y., and Pohajdak, B. (2002). The N-terminal coiled coil domain of the cytohesin/ARNO family of guanine nucleotide exchange factors interacts with the scaffolding protein CASP. *J. Biol. Chem.* **277**, 32302–32309.

Mason, J. M., and Arndt, K. M. (2004). Coiled coil domains: Stability, specificity, and biological implications. *Chembiochem.* **5**, 170–176.

Meacci, E., Tsai, S. C., Adamik, R., Moss, J., and Vaughan, M. (1997). Cytohesin-1, a cytosolic guanine nucleotide-exchange protein for ADP-ribosylation factor. *Proc. Natl. Acad. Sci. USA* **94**, 1745–1748.

Mempel, M., Isa, G., Klugbauer, N., Meyer, H., Wildi, G., Ring, J., Hofmann, F., and Hofmann, H. (2003). Laboratory acquired infection with recombinant vaccinia virus containing an immunomodulating construct. *J. Invest. Dermatol.* **120**, 356–358.

Mishima, K., Tsuchiya, M., Nightingale, M. S., Moss, J., and Vaughan, M. (1993). ARD 1, a 64-kDa guanine nucleotide-binding protein with a carboxyl-terminal ADP-ribosylation factor domain. *J. Biol. Chem.* **268**, 8801–8807.

Morinaga, N., Adamik, R., Moss, J., and Vaughan, M. (1999). Brefeldin A inhibited activity of the sec7 domain of p200, a mammalian guanine nucleotide-exchange protein for ADP-ribosylation factors. *J. Biol. Chem.* **274,** 17417–17423.

Morinaga, N., Moss, J., and Vaughan, M. (1997). Cloning and expression of a cDNA encoding a bovine brain brefeldin A-sensitive guanine nucleotide-exchange protein for ADP-ribosylation factor. *Proc. Natl. Acad. Sci. USA* **94,** 12926–12931.

Morinaga, N., Tsai, S. C., Moss, J., and Vaughan, M. (1996). Isolation of a brefeldin A-inhibited guanine nucleotide-exchange protein for ADP ribosylation factor (ARF) 1 and ARF3 that contains a Sec7-like domain. *Proc. Natl. Acad. Sci. USA* **93,** 12856–12860.

Moss, J., and Vaughan, M. (2002). Cytohesin-1 in 2001. *Arch. Biochem. Biophys.* **397,** 156–161.

Mossessova, E., Gulbis, J. M., and Goldberg, J. (1998). Structure of the guanine nucleotide exchange factor Sec7 domain of human arno and analysis of the interaction with ARF GTPase. *Cell* **92,** 415–423.

Nagel, W., Schilcher, P, Zeitlmann, L., and Kolanus, W. (1998). The PH domain and the polybasic c domain of cytohesin-1 cooperate specifically in plasma membrane association and cellular function. *Mol. Biol. Cell* **9,** 1981–1994.

Nevrivy, D. J., Peterson, V. J., Avram, D., Ishmael, J. E., Hansen, S. G., Dowell, P., Hruby, D. E., Dawson, M. I., and Leid, M. (2000). Interaction of GRASP, a protein encoded by a novel retinoic acid-induced gene, with members of the cytohesin family of guanine nucleotide exchange factors. *J. Biol. Chem.* **275,** 16827–16836.

Ogasawara, M., Kim, S. C., Adamik, R., Togawa, A., Ferrans, V. J., Takeda, K., Kirby, M., Moss, J., and Vaughan, M. (2000). Similarities in function and gene structure of cytohesin-4 and cytohesin-1, guanine nucleotide-exchange proteins for ADP-ribosylation factors. *J. Biol. Chem.* **275,** 3221–3230.

Pacheco-Rodriguez, G., Moss, J., and Vaughan, M. (2002). ARF-directed guanine-nucleotide-exchange (GEP) proteins. *Methods Mol. Biol.* **189,** 181–189.

Pacheco-Rodriguez, G., Moss, J., and Vaughan, M. (2001). Isolation, cloning, and characterization of brefeldin A-inhibited guanine nucleotide-exchange protein for ADP-ribosylation factor. *Methods Enzymol.* **329,** 3000–3006.

Pacheco-Rodriguez, G., Patton, W. A., Adamik, R., Yoo, H. S., Lee, F. J., Zhang, G. F., Moss, J., and Vaughan, M. (1999). Structural elements of ADP-ribosylation factor 1 required for functional interaction with cytohesin-1. *J. Biol. Chem.* **274,** 12438–12444.

Pacheco-Rodriguez, G., Meacci, E., Vitale, N., Moss, J., and Vaughan, M. (1998). Guanine nucleotide exchange on ADP-ribosylation factors catalyzed by cytohesin-1 and its Sec7 domain. *J. Biol. Chem.* **273,** 26543–26548.

Peyroche, A., Antonny, B., Robineau, S., Acker, J., Cherfils, J., and Jackson, C. L. (1999). Brefeldin A acts to stabilize an abortive ARF-GDP-Sec7 domain protein complex: Involvement of specific residues of the Sec7 domain. *Mol. Cell.* **3,** 275–285.

Randazzo, P. A., and Fales, H. M. (2002). Preparation of myristoylated Arf1 and Arf6 proteins. *Methods Mol. Biol.* **189,** 169–179.

Renault, L., Christova, P., Guibert, B., Pasqualato, S., and Cherfils, J. (2002). Mechanism of domain closure of Sec7 domains and role in BFA sensitivity. *Biochemistry* **41,** 3605–3612.

Robineau, S., Chabre, M., and Antonny, B. (2000). Binding site of brefeldin A at the interface between the small G protein ADP-ribosylation factor 1 (ARF1) and the nucleotide-exchange factor Sec7 domain. *Proc. Natl. Acad. Sci. USA* **97,** 9913–9918.

Santy, L. C., Frank, S. R., Hatfield, J. C., and Casanova, J. E. (1999). Regulation of ARNO nucleotide exchange by a PH domain electrostatic switch. *Curr. Biol.* **9,** 1173–1176.

Santy, L. C., and Casanova, J. E. (2001). Activation of ARF6 by ARNO stimulates epithelial cell migration through downstream activation of both Rac1 and phospholipase D. *J. Cell Biol.* **154**, 599–610.

Sata, M., Moss, J., and Vaughan, M. (1999). Structural basis for the inhibitory effect of brefeldin A on guanine nucleotide-exchange proteins for ADP-ribosylation factors. *Proc. Natl. Acad. Sci. USA* **96**, 2752–2757.

Sata, M., Donaldson, J. G., Moss, J., and Vaughan, M. (1998). Brefeldin A-inhibited guanine nucleotide-exchange activity of Sec7 domain from yeast Sec7 with yeast and mammalian ADP ribosylation factors. *Proc. Natl. Acad. Sci. USA* **95**, 4204–4208.

Tang, P., Cheng, T. P., Agnello, D., Wu, C. Y., Hissong, B. D., Watford, W. T., Ahn, H. J., Galon, J., Moss, J., Vaughan, M., O'Shea, J. J., and Gadina, M. (2002). Cybr, a cytokine-inducible protein that binds cytohesin-1 and regulates its activity. *Proc. Natl. Acad. Sci. USA* **99**, 2625–2629.

Togawa, A., Morinaga, N., Ogasawara, M., Moss, J., and Vaughan, M. (1999). Purification and cloning of a brefeldin A-inhibited guanine nucleotide-exchange protein for ADP-ribosylation factors. *J. Biol. Chem.* **274**, 12308–12315.

Venkateswarlu, K., Gunn-Moore, F., Tavare, J. M., and Cullen, P. J. (1999). EGF- and NGF-stimulated translocation of cytohesin-1 to the plasma membrane of PC12 cells requires PI 3-kinase activation and a functional cytohesin-1 PH domain. *J. Cell Sci.* **112**, 1957–1965.

Vichi, A., Payne, D. M., Pacheco-Rodriguez, G., Moss, J., and Vaughan, M. (2005). E3 ubiquitin ligase activity of the trifunctional ARD1 (ADP-ribosylation factor domain protein 1). *Proc. Natl. Acad. Sci. USA* **102**, 1945–1950.

Vitale, N., Pacheco-Rodriguez, G., Ferrans, V. J., Riemenschneider, W., Moss, J., and Vaughan, M. (2000). Specific functional interaction of human cytohesin-1 and ADP-ribosylation factor domain protein (ARD1). *J. Biol. Chem.* **275**, 21331–21339.

Vitale, N., Moss, J., and Vaughan, M. (1996). ARD1, a 64-kDa bifunctional protein containing an 18-kDa GTP-binding ADP-ribosylation factor domain and a 46-kDa GTPase-activating domain. *Proc. Natl. Acad. Sci. USA* **93**, 1941–1944.

[19] ADP-Ribosylation Factor Domain Protein 1 (ARD1), a Multifunctional Protein with Ubiquitin E3 Ligase, GAP, and ARF Domains

By ALESSANDRO VICHI, JOEL MOSS, and MARTHA VAUGHAN

Abstract

ADP-ribosylation factor domain protein 1 (ARD1) is a multifunctional protein that belongs to the family of 20-kDa ARF proteins. The ARD1 gene encodes a 64-kDa protein with a structure comprising an 18-kDa ADP-ribosylation factor (ARF) domain at the C-terminus (amino acids 403–574), and a 46-kDa N-terminal domain (amino acids 1–402) that

METHODS IN ENZYMOLOGY, VOL. 404   0076-6879/05 $35.00
                                                 DOI: 10.1016/S0076-6879(05)04019-X

contains, from the translation start site, a RING finger domain, two predicted B-Boxes, and a coiled-coil protein interaction motif, which places it among the TRIM (tripartite motif) or RBCC (RING, B-Box, coiled-coil) protein families. Recombinant ARD1 (amino acids 1–574) or its RING finger domain (amino acids 1–110) produced polyubiquitylated proteins when incubated *in vitro* with a mammalian E1, an E2 enzyme (UbcH6 or UbcH5a, -5b, or -5c), ATP, and ubiquitin. Via its C-terminal ARF domain, recombinant ARD1 binds guanine nucleotides, through which it can enhance, in a GTP-dependent manner, cholera toxin ADP-ribosyltransferase activity. Unlike ARFs, ARD1, but not its ARF domain, exhibits significant GTPase activity. Hydrolysis of GTP bound to the C-terminal ARF domain was stimulated by addition of the 46-kDa N-terminal domain (amino acids 1–402) via its GTPase activating protein (GAP) activity. The rate of GDP dissociation from the C-terminal ARF domain in ARD1, is slowed by the adjacent 15 amino acids, which act as a GDP-dissociation inhibitor (GDI) domain. Cytohesin-1, known already as a guanine nucleotide-exchange factor (GEF) ARF activator, also specifically activated recombinant human ARD1, via activation of the ARF domain. Overexpressed ARD1 fusion proteins were associated with structures resembling lysosomes and Golgi membranes, as well as the nucleus, in different types of cells, and sequences potentially responsible for the intracellular localizations were identified.

Introduction

ARF was first purified from rabbit liver membranes based on its ability to activate cholera toxin (CT)-catalyzed ADP-ribosylation of Gαs (Kahn and Gilman, 1984), and later demonstrated to be a GTP-binding protein (Kahn and Gilman, 1986). Mammalian ARFs are now known as a family of six 20-kDa GTPases that are grouped in three classes based on amino acid sequence similarity, gene structure, and phylogenetic relationships (Moss and Vaughan, 1998, 1995). ARF action, which depends on whether GTP or GDP is bound, is involved in diverse aspects of cell function including vesicular trafficking, cytoskeletal remodeling, and phospholipids metabolism. Activation of the inactive ARF–GDP requires interaction with a guanine nucleotide-exchange protein that accelerates release of GDP and subsequent GTP binding. Inactivation, that is, hydrolysis of bound GTP, requires a GTPase-activating protein or GAP. ARD1, or ARF-domain protein 1, was cloned during a search for new ARF proteins (Mishima *et al.*, 1993). It has an 18-kDa ARF domain at the C-terminus, but is a 64-kDa molecule with multiple other functional domains.

Structure and Function of the ARD1 Molecule (Fig. 1)

ARF and ARF Regulatory Domains of ARD1

ARD1 is a 64-kDa protein containing an 18-kDa C-terminal ARF domain. The 46-kDa non-ARF molecule has an overall amino acid sequence resembling those referred to as tripartite motif (TRIM) (Reymond *et al.*, 2001) or RBCC (Saurin *et al.*, 1996) proteins, which contain RING (Freemont *et al.*, 1991), B-Box (Reddy *et al.*, 1992), and coiled-coil (Lupas, 1996), structures. With few exceptions, this architecture characterizes all of the TRIM/RBCC proteins (Reymond *et al.*, 2001). The coding region of the ARD1 cDNA comprises 1722 nucleotides, 1200 of which encode the non-ARF domain (Mishima *et al.*, 1993). Figure 1 is a diagram of the structure and the putative functional domains of the translated ARD1 cDNA sequence. The ARF-domain cDNA sequence of human ARD1 is 60–66% identical to those of the six mammalian ARFs, and the amino acid sequence is 55–60% identical (Mishima *et al.*, 1993). As recombinant fusion proteins, ARD1 and its ARF domain, like the ARFs, enhanced CTA-catalyzed ADP-ribosylation of agmatine in a GTP-dependent fashion (Mishima *et al.*, 1993; Vitale *et al.*, 1997b), and, their abilities to bind GTPγS were influenced by MgCl$_2$ and phospholipids or detergents (Vitale *et al.*, 1997a).

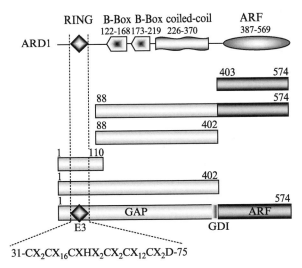

FIG. 1. Predicted protein structure and functional domains of ARD1, a member of the TRIM/RBCC family.

The GDI region is a sequence of 15 amino acids (388–402), which is adjacent to the ARF domain and acts as a GDP-dissociation inhibitor to stabilize ARD1 in a GDP-bound form. Similar to the N-terminus of ARF1, the region contains a core of hydrophobic amino acids (Phe 391, Val 397, Ile 399) that are responsible for the GDI function (Vitale *et al.*, 1997a).

The GTPase-activating (GAP) domain is responsible for a major functional difference between ARD1 and the ARFs, which do not detectably hydrolyze bound GTP without an added GAP (Vitale *et al.*, 1996). Using recombinant proteins synthesized in *E. coli*, it was demonstrated that the GAP activity of ARD1(1–402) is specific for the ARD1 ARF domain; it did not accelerate GTP hydrolysis by several ARFs or other ARF preparations (Ding *et al.*, 1996). The ARD1 fragment comprising only amino acids 101–200 (Vitale *et al.*, 1998b) exhibited GAP activity (Vitale *et al.*, 1996). As shown in Fig. 1, the 100-amino acid GAP sequence corresponds to a large part of the B-Box region (122–219), whereas amino acids 190–333, ~70% of the coiled-coil domain (residues 226–370), were required for stable association of the separately synthesized ARF and non-ARF domains (Vitale *et al.*, 1998b).

The RING Domain and E3 Ubiquitin Ligase Activity of ARD1

The RING (Really Interesting New Gene) domain is a cross-braced structure in which two zinc ions can be chelated by cysteine and histidine residues. The consensus sequence in ARD1 (Fig. 1) is [$CX_2CX_{16}CXHX_2CX_2CX_{12}CX_2D$] (Freemont *et al.*, 1991). Proteins containing a RING finger domain are known to catalyze, either alone (Joazeiro and Weissman, 2000; Joazeiro *et al.*, 1999; Lorick *et al.*, 1999) or in association with other proteins (Petroski and Deshaies, 2005), ubiquitylation of numerous proteins, many of which are unidentified. Ubiquitylation of proteins involves a series of three reactions. First, the activating enzyme E1 catalyzes ATP-dependent automodification of a cysteine with a thiol ester bond to the C-terminus of ubiquitin. The ubiquitin is then ligated through a thiol ester bond to an E2 conjugating enzyme. In the third reaction, which involves an E3 ligase, the activated ubiquitin is ligated to a lysine in E3 itself, in another ubiquitin molecule, or in a third protein substrate. Three major families of E3 ligases, HECT domain, RING domain, and UFD2 domain E3 ligases, are distinguished by the mechanisms of catalysis and the structures of their active sites. Only the HECT family of ligases catalyze direct transfer of the ubiquitin moiety from the thiol ester intermediate to a third protein (Pickart and Eddins, 2004).

E3 ligase activity of recombinant ARD1 and of two fragments that include the RING finger domain was demonstrated *in vitro* with purified recombinant E1 and E2 proteins plus ubiquitin, and ATP. Multiple ubiquitylated products were formed with four different E2s: UbcH5a, UbcH5b, UbcH5c, or UbcH6, but not with UbcH1, UbcH2, UbcH3/CDC34, UbcH7, or UbcH10 (Vichi et al., 2005). As reported for other RING E3 ligases (Joazeiro et al., 1999; Lorick et al., 1999), replacement of a cysteine or the histidine in the RING structure with alanine, abolished ARD1 E3 ligase activity (Vichi et al., 2005). Among proteins that reacted with the monoclonal antibody against ubiquitin, none reacted with antibodies against ARD1 or against the GST-tag on the recombinant ARD1 fusion protein used in E3 assays. Proteins corresponding to mono-, di-, and tri-ubiquitylated E2 did react with monoclonal antibody against UbcH6, the E2 used in those assays. The observed decrease in GST-ARD1 with time in assays was consistent with its ubiquitylation to yield the concomitant increasing amounts of >200-kDa ubiquitin-conjugated proteins (Vichi et al., 2005).

Cellular Localization of ARD1

ARD1 mRNA (4.2 and 3.7 kb) was seen on Northern blots of RNA from several rat tissues (Mishima et al., 1993). The original ARD1 clone was from a cAMP-differentiated HL-60 cell cDNA library but no hybridization was detected when poly(A)$^+$ RNA from undifferentiated HL-60 lymphoblasts, was probed with a cDNA complementary to bases 1270–1299 of the ARD1 clone (Mishima et al., 1993).

Endogenous ARD1 was identified in Golgi and lysosomal membranes from human liver (Vitale et al., 1998a). Recombinant human ARD1 tagged with GFP (green fluorescent protein), when overexpressed in NIH 3T3 (murine) or COS7 (simian) cells was seen in a punctate cytoplasmic distribution and in perinuclear concentrations, resembling Golgi and lysosomes (Vitale et al., 1998a). GFP-ARD1 overexpressed in HeLa and U2OS (human tumor) cells was distributed in cytoplasmic and nuclear structures (Reymond et al., 2001). Two tyrosine-based motifs in the ARF domain were responsible for ARD1 localization in the Golgi, whereas ^{369}KXXXQ^{373} in the GAP domain was required for its localization in lysosomes (Vitale et al., 2000a).

Isolation of Recombinant and Endogenous ARD1

Synthesis and Purification of Recombinant ARD1 and Its Fragments

The procedure is outlined in Fig. 2. To prepare glutathione-S-transferase (GST)-tagged full-length ARD1(1–574) (accession number L04510), the GAP domain ARD1(1–402), and the ARF domain ARD1(403–574) were

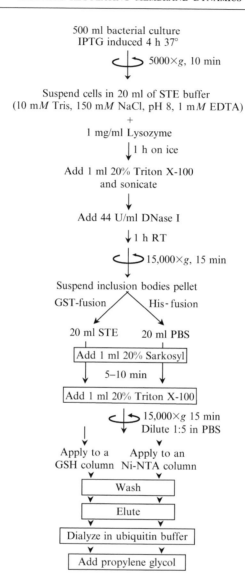

Fig. 2. Procedure for synthesis and purification of GST-tagged or 6 × His-tagged proteins.

ligated to bacterial expression vector pGEX-5G/LIC (Haun and Moss, 1992; Mishima *et al.*, 1993). The same vector was used for deletion mutants derived from the original ARD1 clone. The 6xHis-ARD1 DNA was ligated to a bacterial expression vector p-ET14b (Vitale *et al.*, 2000b). Amino acid replacements in the RING finger were introduced by site-directed mutagenesis (Stratagene kit). Competent XL1 blue cells were transformed with plasmids, sequences of which were checked by DNA sequencing after recovery. Single colonies were transferred to 5 ml of Luria-Bertani broth with ampicillin, 0.1 mg/ml, and incubated overnight at 37° with shaking, before addition to 500 ml of the same medium, which was incubated at 37° with shaking. When $OD_{600} = 600$, incubation was continued for 4 h with 0.2 mM isopropyl β-D-thiogalactoside and cells were collected by centrifugation ($5000 \times g$ 10 min) for storage at $-20°$ until protein purification.

For isolation of recombinant proteins, frozen cells were dispersed in 20 ml of STE buffer (10 mM Tris-HCl, pH 8, 150 mM NaCl, 1 mM EDTA) containing lysozyme, 1 mg/ml, and incubated for 1 h on ice. After addition of 1% Triton X-100, cells were sonified and incubated (1 h, room temperature, shaking) with DNase I, 44 units/ml. The protein was purified from inclusion bodies essentially as described previously (Frangioni and Neel, 1993; Vichi et al., 2005). Briefly, inclusion bodies containing GST-fusion protein, collected by centrifugation ($15,000 \times g$, 15 min), were dispersed in 20 ml of STE. After addition of 1 ml of 20% Sarkosyl, intermittent mixing (vortex mixer, 5 s every 30 s) for 5–10 min, and addition of 1 ml of 20% Triton X-100 followed by centrifugation ($15,000 \times g$, 10 min, 4°) to remove insoluble material, the supernatant was incubated (2 h, 4°) with 0.25 ml of GSH-Sepharose. Beads were collected in a column and washed three times with 10 ml of STE buffer before elution of bound GST-fusion proteins with three 0.5-ml portions of 10 mM GSH in 50 mM Tris-HCl, pH 8. Eluted proteins were concentrated using Microcon centrifugal filter devices (10,000 or 100,000 molecular weight cut-off, Millipore, Bedford, MA).

For purification of 6×His-tagged protein, the inclusion bodies containing the recombinant protein were dispersed in 20 ml of phosphate-buffered saline (PBS), followed by addition of 1 ml of 20% Sarkosyl in H_2O, intermittent mixing for 5–10 min, addition of 1 ml of 20% Triton X-100 in H_2O, and centrifugation ($15,000 \times g$, 10 min, 4°). The supernatant was diluted with five volumes of PBS and applied to a Ni-NTA affinity column (0.25 ml bed volume) previously equilibrated in 10 bed volumes of PBS. After three sequential washes with PBS containing 10, 25, and 50 mM imidazole pH 8, respectively, bound proteins were eluted in 0.5 ml of 0.5M imidazole, pH 8.0, and dialyzed overnight against 2 L of ubiquitylation buffer containing 20 mM Tris-HCl, pH 7.4, 2 mM $MgCl_2$, 0.1 mM dithiothreitol (DTT). Protein concentration was quantified (Bradford, 1976) with bovine serum albumin as standard and purity (Fig. 3) was assessed by silver staining ($0.5\mu g$ of protein) after SDS-PAGE. Protein, 0.1–1 mg/ml in the elution (GST-ARD1) or dialysis (6×His-ARD1) buffer was added with 35% propylene glycol (final concentration) and stored in small portions at $-20°$.

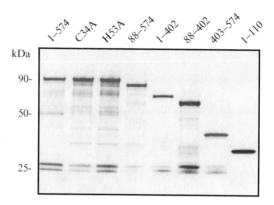

FIG. 3. SDS-PAGE of recombinant GST-ARD1 proteins. Samples (0.5 μg) of purified recombinant GST-proteins were subjected to SDS-PAGE in 4–20% Tris-glycine gels and detected by silver staining. The doublet at 26 kDa represents GST probably, generated via proteolysis of bacterially synthesized GST-fusion proteins.

ARD1 purification from inclusion bodies using the denaturation/renaturation procedure described by Vitale *et al.* (2001) can also be used.

Immunodetection of Endogenous ARD1

Rat liver membranes presumed to be from Golgi and lysosomal structure were isolated by immunoadsorption on Dynabeads M-500 (Dynal) conjugated to, respectively, anti-AP1 and anti-LAMP1 antibodies. After separation of membrane proteins by SDS-PAGE, endogenous ARD1 was detected by immunoblotting with specific polyclonal anti-ARD1 antibodies (Vitale *et al.*, 1998a).

Assays of ARD1 Function

Assays of GTP Binding and GAP Activities

In general, GTP binding to the ARF domain of ARD1 can be demonstrated or quantified in two ways.

GTP binding by proteins after SDS-PAGE separation can be qualitatively evaluated by an "over-lay assay." Recombinant ARD1 (22 pmol, 1.4 μg) was subjected to SDS-polyacrylamide gel electrophoresis and transferred to nitrocellulose membrane, which was incubated (3 h, 25°) in 50 mM TRIS-HCl, pH 7.5/ 150 mM NaCl/ 2 mM DTT/ 2.5 mM EDTA, containing protease inhibitors, bovine serum albumin, 0.3 mg/ml, and cardiolipin, 1 mg/ml. The membrane was transferred to the same buffer plus 10 mM MgCl$_2$ and [$\acute{\alpha}$-^{32}P]GTP (3000 Ci/mmol), 2 μCi/ml, for 2 h. After washing ten times for 5 min in buffer without [$\acute{\alpha}$-^{32}P]GTP and MgCl$_2$, the membrane was air-dried and exposed to X-ray film overnight at $-80°$ (Vitale et al., 2001).

To quantify GTP binding, bound nucleotides were first removed from recombinant ARD1 (30 pmol), by incubation for 30 minutes at 30° in stripping buffer (20 mM TRIS-HCl, pH 8.0, 10 mM DTT, 2.5 mM EDTA, BSA, 0.3 mg/ml, and cardiolipin, 1 mg/ml). Then, 10 mM MgCl$_2$ and 3 μM [^{35}S]GTPγS (ca. 10^6 dpm) were added to initiate binding (150 μl final volume) and incubation at 30° was continued for the time selected to assess rate or extent of binding. Reactions were terminated by transferring the mixture to a nitrocellulose filter in a manifold (Millipore, Bedford, MA) for rapid filtration. After washing five times with 1 ml of ice-cold buffer (25 mM TRIS-HCl, pH 8.0, 1 mM DTT, 1 mM EDTA, 5 mM MgCl$_2$), filters were dried and dissolved in scintillation fluid for radioassay (Tsai et al., 1994). To determine a time course of binding, it is convenient to multiply the total volume of assay mixture and remove replicate samples from duplicate mixtures at several times during incubation (Vitale et al., 2001)

To assess rates of dissociation of previously bound radiolabeled nucleotides, ARD1 (300 pmol in 450 μl) was first stripped of bound nucleotides for 30 min at 30°. Incubation was then continued in the same medium plus 10 mM MgCl$_2$ and 3 μM [^{35}S]GDPβS (2 X 10^7 dpm) or [^{3}H]GDP (1.5 x 10^7 dpm) for 40 min at 30° to allow nucleotide binding (total volume 600 μl). A 100-μl sample was transferred to a nitrocellulose filter which was washed, dried, and dissolved in liquid scintillation fluid for radioassay (zero time bound nucleotide). The remaining 500 μl were diluted with 500 μl of stripping buffer containing 2 mM GDPβS or GDP and incubated at 30° with sampling at appropriate times for quantification of remaining bound nucleotide (Vitale et al., 2001).

To measure the GTPase-activation of the ARF domain by the non-ARF domain of ARD1, samples of the ARD1 ARF domain (typically 50 pmol) were incubated in stripping buffer for 30 min and then, for 40 min in the same medium plus 0.5 μM [α^{32}PGTP] (3000 Ci/mmol) and 10 mM MgCl$_2$ (120 μl final volume). After addition of the non-ARF domain of ARD1 (50 pmol in 40 μl), the mixture (160 μl) was incubated at 25° for 30 min before proteins were collected on nitrocellulose filters in a manifold, washed, and air-dried. Nucleotides were eluted from filters in 2 M formic acid (250 μl) and samples (2–4 μl) analyzed by thin layer chromatography on polyethyleneimine-cellulose plates followed by autoradiography. Radioactivity in the remaining eluate was quantified to determine total protein-bound nucleotide (Vitale et al., 2001) .

Assay of Cholera Toxin-catalyzed ADP-ribosylagmatine Formation

ARF-GTP is required to activate the cholera toxin ADP-ribosyltransferase. GTPase activity of the ARD1 non-ARF domain was evaluated by its effect on the ARF domain enhancement of cholera toxin catalytic activity. Toxin ADP-ribosyltransferase activity was quantified by measurement of ADP-ribosylagmatine formed from agmatine and NAD. Assays contained 50 mM potassium phosphate, pH 7.5, 6 mM MgCl$_2$, 20 mM DTT, ovalbumin, 0.3 mg/ml, 0.2 mM [*adenine*-^{14}C]NAD (0.05 μCi), 20 mM agmatine, 100 μM GTPγS or GTP, and 0.18 μM cholera toxin. To activate ARD1 or its ARF domain, the protein (ca. 200 pmol) was first stripped of bound nucleotides for 30 min at 30° and incubated for 20 min with 100 mM GTPγS or GTP and 10 mM MgCl$_2$ (final concentrations in 60 μl total volume). Components for the ADP-ribosyltransferase assay were added in 10 μl to yield final concentrations of 50 mM potassium phosphate, pH 7.5, 6 mM MgCl$_2$, 20 mM DTT, ovalbumin, 0.3 mg/ml, 0.2 mM [*adenine*-^{14}C]NAD (0.05 μCi), 20 mM agmatine, 100 μM GTPγS or GTP, and 0.18 μM cholera toxin. After incubation for 1 h at 30°, samples were transferred to columns of AG1-X2 and [^{14}C] NAD-ribosylagmatine was collected for radio-assay (Vitale *et al.*, 2001).

Assay of E3 Ubiquitin Ligase Activity

ARD1 ubiquitylation activity (Vichi *et al.*, 2005) was demonstrated in assays containing 0.8 pmol of purified rabbit ubiquitin-activating enzyme (116-kDa, E1), 19 pmol of recombinant human E2 UbcH6 (or UbcH5a, 5b, or 5c), 0.6 nmol of purified bovine or recombinant human ubiquitin, 4 mM ATP, and 1.5–7.5 pmol of ARD1 or its RING domain, in 30 μl of ubiquitylation buffer. After incubation (e.g., 60 min, 30°), reaction was terminated by adding 10 μl of 4× Laemmli buffer (Laemmli, 1970). Proteins were separated by SDS-PAGE in 4–20% Tris-glycine gel, and transferred to Hybond-C extra nitrocellulose membranes (Amersham Biosciences, Piscataway, NJ). After blocking, the membrane was incubated with anti-ubiquitin monoclonal antibody (clone P4D1, Santa Cruz, Santa Cruz, CA), followed by HRP-conjugated secondary antibodies and ECL detection reagent.

References

Bradford, M. M. (1976). A rapid and sensitive method for the quantitation of microgram quantities of protein utilizing the principle of protein-dye binding. *Anal. Biochem.* **72,** 248–254.

Ding, M., Vitale, N., Tsai, S. C., Adamik, R., Moss, J., and Vaughan, M. (1996). Characterization of a GTPase-activating protein that stimulates GTP hydrolysis by both ADP-ribosylation factor (ARF) and ARF-like proteins. *J. Biol. Chem.* **271,** 24005–24009.

Frangioni, J. V., and Neel, B. G. (1993). Solubilization and purification of enzymatically active glutathione S-transferase (pGEX) fusion proteins. *Anal. Biochem.* **210,** 179–187.

Freemont, P. S., Hanson, I. M., and Trowsdale, J. (1991). A novel cysteine-rich sequence motif. *Cell* **64,** 483–484.

Haun, R. S., and Moss, J. (1992). Ligation-independent cloning of glutathione S-transferase fusion genes for expression in *Escherichia coli. Gene* **112,** 37–43.

Joazeiro, C. A., and Weissman, A. M. (2000). RING finger proteins: Mediators of ubiquitin ligase activity. *Cell* **102,** 549–552.

Joazeiro, C. A., Wing, S. S., Huang, H., Leverson, J. D., Hunter, T., and Liu, Y. C. (1999). The tyrosine kinase negative regulator c-Cbl as a RING-type, E2-dependent ubiquitin-protein ligase. *Science* **286,** 309–312.

Kahn, R. A., and Gilman, A. G. (1986). The protein cofactor necessary for ADP-ribosylation of Gs by cholera toxin is itself a GTP binding protein. *J. Biol. Chem.* **261,** 7906–7911.

Kahn, R. A., and Gilman, A. G. (1984). Purification of a protein cofactor required for ADP-ribosylation of the stimulatory regulatory component of adenylate cyclase by cholera toxin. *J. Biol. Chem.* **259,** 6228–6234.

Laemmli, U. K. (1970). Cleavage of structural proteins during the assembly of the head of bacteriophage T4. *Nature* **227,** 680–685.

Lorick, K. L., Jensen, J. P., Fang, S., Ong, A. M., Hatakeyama, S., and Weissman, A. M. (1999). RING fingers mediate ubiquitin-conjugating enzyme (E2)-dependent ubiquitination. *PNAS* **96,** 11364–11369.

Lupas, A. (1996). Coiled coils: New structures and new functions. *Trends Biochem. Sci.* **21,** 375–382.

Mishima, K., Tsuchiya, M., Nightingale, M. S., Moss, J., and Vaughan, M. (1993). ARD 1, a 64-kDa guanine nucleotide-binding protein with a carboxyl-terminal ADP-ribosylation factor domain. *J. Biol. Chem.* **268,** 8801–8807.

Moss, J., and Vaughan, M. (1998). Molecules in the ARF orbit. *J. Biol. Chem.* **273,** 21431–21434.

Moss, J., and Vaughan, M. (1995). Structure and function of ARF proteins: Activators of cholera toxin and critical components of intracellular vesicular transport processes. *J. Biol. Chem.* **270,** 12327–12330.

Petroski, M. D., and Deshaies, R. J. (2005). Function and regulation of cullin-ring ubiquitin ligases. *Nat. Rev. Mol. Cell. Biol.* **6,** 9–20.

Pickart, C. M., and Eddins, M. J. (2004). Ubiquitin: Structures, functions, mechanisms. *Biochim et Biophys Acta (BBA) - Molecular Cell Research* **1695,** 55–72.

Reddy, B. A., Etkin, L. D., and Freemont, P. S. (1992). A novel zinc finger coiled-coil domain in a family of nuclear proteins. *Trends Biochem. Sci.* **17,** 344–345.

Reymond, A., Meroni, G., Fantozzi, A., Merla, G., Cairo, S., Luzi, L., Riganelli, D., Zanaria, E., Messali, S., Cainarca, S., Guffanti, A., Minucci, S., Pelicci, P. G., and Ballabio, A. (2001). The tripartite motif family identifies cell compartments. *EMBO J.* **20,** 2140–2151.

Saurin, A. J., Borden, K. L., Boddy, M. N., and Freemont, P. S. (1996). Does this have a familiar RING? *Trends Biochem. Sci.* **21,** 208–214.

Tsai, S., Adamik, R., Moss, J., and Vaughan, M. (1994). Identification of a brefeldin A-insensitive guanine nucleotide-exchange protein for ADP-ribosylation factor in bovine brain. *PNAS* **91,** 3063–3066.

Vichi, A., Payne, D. M., Pacheco-Rodriguez, G., Moss, J., and Vaughan, M. (2005). E3 ubiquitin ligase activity of the trifunctional ARD1 (ADP-ribosylation factor domain protein 1). *PNAS* **102,** 1945–1950.

Vitale, N., Moss, J., and Vaughan, M. (2001). Purification and properties of ARD1, an ADP-ribosylation factor (ARF)-related protein with GTPase-activating domain. *Methods Enzymol.* **329,** 324–334.

Vitale, N., Ferrans, V. J., Moss, J., and Vaughan, M. (2000a). Identification of lysosomal and Golgi localization signals in GAP and ARF domains of ARF domain protein 1. *Mol. Cell Biol.* **20,** 7342–7352.

Vitale, N., Pacheco-Rodriguez, G., Ferrans, V. J., Riemenschneider, W., Moss, J., and Vaughan, M. (2000b). Specific functional interaction of human cytohesin-1 and ADP-ribosylation factor domain protein (ARD1). *J. Biol. Chem.* **275,** 21331–21339.

Vitale, N., Horiba, K., Ferrans, V. J., Moss, J., and Vaughan, M. (1998a). Localization of ADP-ribosylation factor domain protein 1 (ARD1) in lysosomes and Golgi apparatus. *PNAS* **95,** 8613–8618.

Vitale, N., Moss, J., and Vaughan, M. (1998b). Molecular characterization of the GTPase-activating domain of ADP-ribosylation factor domain protein 1 (ARD1). *J. Biol. Chem.* **273,** 2553–2560.

Vitale, N., Moss, J., and Vaughan, M. (1997a). Characterization of a GDP dissociation inhibitory region of ADP-ribosylation factor domain protein ARD1. *J. Biol. Chem.* **272,** 25077–25082.

Vitale, N., Moss, J., and Vaughan, M. (1997b). Interaction of the GTP-binding and GTPase-activating domains of ARD1 involves the effector region of the ADP-ribosylation factor domain. *J. Biol. Chem.* **272,** 3897–3904.

Vitale, N., Moss, J., and Vaughan, M. (1996). ARD1, a 64-kDa bifunctional protein containing an 18-kDa GTP-binding ADP-ribosylation factor domain and a 46-kDa GTPase-activating domain. *PNAS* **93,** 1941–1944.

[20] Expression of BIG2 and Analysis of Its Function in Mammalian Cells

By Hye-Won Shin, Chisa Shinotsuka, and Kazuhisa Nakayama

Abstract

BIG2 is a member of brefeldin A-inhibited guanine nucleotide exchange factors for ADP-ribosylation factors (ARFs). Although BIG2 is associated mainly with the *trans*-Golgi network, we have recently revealed that some population of BIG2 is associated with the recycling endosome. Moreover, we have found that expression of a catalytically inactive BIG2 mutant, E738K, selectively induces membrane tubules from this compartment. We have also demonstrated that the exchange activity of BIG2 is specific for class I ARFs (ARF1 and ARF3) *in vivo* and inactivation of either ARF enhances the BIG2(E738K)-induced membrane tubulation. Therefore, we have proposed that BIG2 is implicated in the structural integrity of the recycling endosome through activating class I ARFs. This article describes methods used for examining the function of BIG2

METHODS IN ENZYMOLOGY, VOL. 404
0076-6879/05 $35.00
DOI: 10.1016/S0076-6879(05)04020-6

including basic protocols, which are not conventional because of cytotoxicity of BIG2 in *Escherichia coli* cells and its low expression efficiency in mammalian cells.

Introduction

The ADP-ribosylation factor (ARF) family of small GTPases functions to regulate membrane traffic and dynamics in eukaryotic cells. Guanine nucleotide exchange factors (GEFs) for ARFs, which have been well characterized so far, can be grouped into four different subfamilies: the Gea/GBF/GNOM, Sec7/BIG, ARNO/cytohesin/GRP, and EFA6 subfamilies (Shin and Nakayama, 2004). The Sec7/BIG subfamily includes yeast Sec7p and mammalian BIG1 and BIG2. They all share a catalytic region of approximately 200 amino acids termed the Sec7 domain (Peyroche *et al.*, 1996), which is the molecular target of brefeldin A (BFA) (Jackson and Casanova, 2000; Shin and Nakayama, 2004).

We previously showed that BIG2 is responsible for recruitment of the AP-1 clathrin adaptor complex, but not that of the COPI complex, onto Golgi membranes (Shinotsuka *et al.*, 2002b). We also found that expression of a catalytically inactive mutant of BIG2, E738K, induces membrane tubules similar to those induced by BFA treatment of cells. More importantly, we demonstrated that BIG2 localizes to the recycling endosome (RE) as well as to the *trans*-Golgi network (TGN), and the catalytically inactive BIG2(E738K) mutant selectively induces tubulation of the RE, but not that of the early endosome (EE), late endosome (LE), or the Golgi complex (Shin *et al.*, 2004).

There are six ARFs (ARF1–ARF6) in mammals, which are grouped into three classes: class I, ARF1–ARF3; class II, ARF4 and ARF5; and class III, ARF6 (Welsh *et al.*, 1994). Purified BIG1 from bovine brain and recombinant BIG2 have been shown to exhibit *in vitro* GEF activities toward ARF1 and ARF3 (Morinaga *et al.*, 1996) and toward ARF1, ARF5, and ARF6, respectively (Togawa *et al.*, 1999). However, it has not been examined whether BIG2 has the same substrate specificity *in vivo*. To address this issue, we examined activation of ARFs *in vivo* by BIG2 using a pull down assay with the GAT (GGA and Tom1) domain of GGA1, which specifically interacts with a GTP-bound form of ARFs (Dell'Angelica *et al.*, 2000; Santy and Casanova, 2001; Shinotsuka *et al.*, 2002b; Takatsu *et al.*, 2002). In this article, we describe the methods for preparation of expression vectors for BIG2 and its E738K mutant and their expression in mammalian cells, and those for the pull down assay for the guanine-nucleotide exchange activity of BIG2 using the GGA1-GAT domain. Despite the toxicity of the BIG2 plasmid for *E. coli* and its very low expression efficiencies in mammalian cells, we have established methods for reproducible preparation of the BIG2 plasmid and managed to obtain its consistent and reproducible expression levels sufficient to characterize the cellular function of BIG2.

BIG2 Plasmid Preparation

Plasmid Construction

For expression of BIG2 in mammalian cells, we made use of the pcDNA4/HisMax vector (Invitrogen, Carlsbad, CA), which is designed for overproduction of recombinant proteins in mammalian cells. The vector contains a QBI SP163 translational enhancer for increased levels of recombinant protein expression. In the case of BIG2 expression, this vector gives better expression efficiency than other examined vectors, such as pcDNA3, do.

To construct an expression vector for epitope-tagged wild type BIG2 or its E738K mutant (Shinotsuka *et al.*, 2002a,b), a double-stranded oligonucleotide for an epitope sequence for hemagglutinin (HA) was first introduced between the *Hind*III and *Bam*HI sites of pcDNA4/HisMax. The entire coding sequence for BIG2 was then ligated downstream of the epitope sequence of the resulting pcDNA4-HAN vector.

Large-Scale Plasmid Preparation

After transformation of the BIG2 plasmid into *E. coli*, not only the transformants grow very slowly (usually, it takes more than 20 h for the transformants to form visible colonies on ampicillin-containing plates), but also their viability is drastically decreased during storage even at 4°. Therefore, we routinely use fresh transformants for plasmid preparation. The plasmid yield decreases with the progress of the day after transformation.

1. An *E. coli* strain DH5α is transformed with the BIG2 plasmid by a standard heat-shock protocol.
2. After heat-shock at 42° for 45 sec, the transformants are immediately chilled on ice for 2 min and, after the addition of LB medium, incubated for 30–60 min at 37° with gentle shaking for recovery.
3. The transformants (approximately 10^4 transformants per plate) are plated onto an LB plate (8.5 × 13.5 cm) supplemented with 50 μg/ml ampicillin and incubated at 37° for 20 h. (*Note*: the size of colonies is usually small.)
4. On the next day, all colonies are harvested by scraping them from the plate.

5. The BIG2 plasmid is purified using a QIAGEN plasmid purification kit. (*Note*: from colonies of five plates, the plasmid yield is usually 100–200 μg.)

Transient Transfection into Mammalian Cells

1. For transfection into mammalian cells in a well of a 6-well plate, 3 μl of FuGENE6 reagent (Roche Diagnostics, Indianapolis, IN) is suspended in 100 μl of serum free medium (FuGENE6 suspension).
2. The FuGENE6 suspension is mixed with 1 μg of the BIG2 plasmid DNA and left at room temperature for 15 min.
3. The DNA/FuGENE6 mixture is then added to cells in a total volume of 2 ml complete (serum-containing) medium and incubated for 20 h at 37°.

Note 1: When using a 24-well plate for transfection into cells grown on a coverslip, 20 μl of the FuGENE6 suspension is mixed with 200 ng of plasmid DNA and left at room temperature for 15 min. The DNA/Fu-GENE6 mixture is then added to cells in a total volume of 0.5 ml complete (serum-containing) medium and incubated for 20 h at 37° (less than 12 h in the case of cotransfection of the Arf(QL) plasmid with that of BIG2) before immunofluorescence analysis.

Note 2. When coexpressing either Rab or the Arf(TN) mutant with BIG2, 50 ng of Rab or Arf(TN) plasmid and 200 ng of BIG2 plasmid are included in the transfection mixture.

Note 3. When coexpressing Arf(WT) or the Arf(QL) mutant with BIG2, 20 ng of the Arf plasmid and 200 ng of the BIG2 plasmid are included in the transfection mixture.

Note 4. When coexpressing Arf(WT) with BIG2 in cells on a 10-cm dish, 100 ng of the Arf(WT) plasmid and 1 μg of the BIG2 plasmid are mixed with the FuGENE6 suspension (15 μl of FuGENE6 and 500 μl of serum free medium). The DNA/FuGENE6 mixture is added to cells in a total volume of 10 ml complete medium and incubated for 20 h before further analysis.

Effects of BIG2 Mutant on the Endosomal Integrity

We have previously shown that BIG2 is localized not only to the TGN but also to the punctate structures throughout the cytoplasm, where BIG2 is extensively colocalized with the AP-1 clathrin adaptor complex and the transferrin (Tfn) receptor (a marker for the EE and RE) but not with EEA1 (a marker for the EE) or Lamp-1 (a marker for the LE and lysosome) (Shin *et al.*, 2004). Moreover, the membrane tubules, which are induced by expression of BIG2(E738K), contain recycling endosomal Rab GTPases, Rab4, and Rab11, but do not contain early endosomal Rab5 or late endosomal Rab7 (Shin *et al.*, 2004). In line with these observations, endocytosed Tfn, which is recycled back to the plasma membrane via the RE (Dunn *et al.*, 1989; Hopkins, 1983a), but not EGF, which is transported to the lysosome for degradation via the EE and LE, is incorporated into membrane tubules induced by BIG2(E738K) (Fig. 1).

EGF or Tfn Internalization

1. HeLa cells expressing HA-tagged BIG2 are grown on glass coverslips and cultured for 3 h in serum-free medium supplemented with 0.2% BSA to deplete endogenous EGF or Tfn.
2. The cells are incubated with Alexa488-conjugated EGF or Tfn (Molecular Probes, Eugene, OR) for appropriate periods. The cells are then fixed with 3% paraformaldehyde and stained with monoclonal rat anti-HA antibody (3F10, Roche Diagnostics, West Grove, PA) followed by Cy3-conjugated anti-rat secondary antibodies (Jackson ImmunoResearch Laboratory).

Tfn Recycling

1. HeLa cells expressing HA-BIG2 are serum starved for 3 h and incubated with Alexa488-conjugated Tfn for 60 min at 37°.
2. The cells are then extensively washed with PBS to remove unbound Tfn and further incubated at 37° in normal medium for appropriate periods. Immunofluorescence staining was performed as described above.

As shown in Fig. 1, the BIG2(E738K) expression did not affect the internalization of Tfn or EGF. After 2.5 min uptake, Tfn was found mostly in the EE and no morphological difference of the Tfn-containing compartments was detected between the cells with and without the BIG2(E738K) expression (A, A'). After 30 min internalization, when Tfn reaches the Rab4/Rab11-positive RE (Sonnichsen *et al.*, 2000), the majority of the Tfn signals were found on the BIG2(E738K)-positive tubules (B, B'), indicating that Tfn is accessible to the tubular compartment. In contrast, EGF was not accessible to the tubular structures at any time point after internalization (C, C').

The Tfn recycling was largely unaffected in the BIG2(E738K)-expressing cells (Fig. 1, D, D'). The labeled Tfn accumulated in the BIG2 (E738K)-positive tubular structures was externalized out to the medium after 60 min chase as in nontransfected cells (marked by an asterisk).

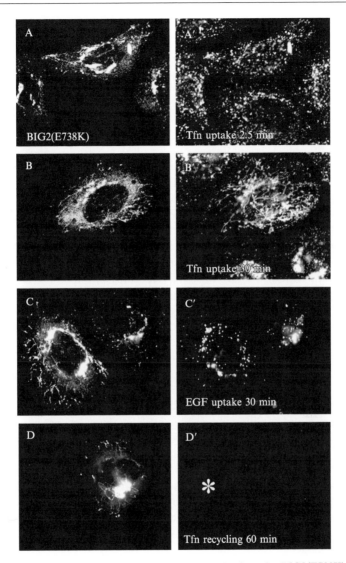

FIG. 1. Incorporation of internalized Tfn but not EGF into the BIG2(E738K)-induced tubular structures. (A–C) HeLa cells expressing HA-BIG2(E738K) were serum-starved for 3 h and incubated with Alexa488-conjugated Tfn for 2.5 or 30 min (A, A', B, B') or EGF at 37° for 30 min (C, C'). The cells were then fixed and stained with anti-HA antibody. (D) HeLa cells expressing HA-tagged BIG2(E738K) were serum-starved for 3 h and incubated with Alexa488-conjugated Tfn at 37° for 60 min. The cells were washed extensively and incubated with medium that did not contain labeled Tfn at 37° for 60 min. The cells were then fixed and stained with anti-HA antibody. Asterisk indicates non-transfected cell. Reprinted from *Molecular Biology of the Cell* (Shin *et al.*, 2004) with the permission of The American Society for Cell Biology.

These observations indicate that the morphological change of the RE induced by BIG2(E738K) does not affect the Tfn recycling through this compartment.

GEF Activity *In Vivo*

The buffers used in the assay are as follows:

GST-fusion protein binding buffer: PBS, pH 7.4, 5 mM β-mercaptoethanol, 5 μg/ml DNase I, 5 μg/ml RNase A, and a protease inhibitor mixture (Complete™–EDTA free, Roche Diagnostics)

GST-fusion protein washing buffer: PBS, pH 7.4, 5 mM β-mercaptoethanol, Complete™–EDTA free

Pull-down binding buffer: 50 mM Tris-Cl, pH 7.5, 100 mM NaCl, 2 mM MgCl$_2$, 0.5% sodium deoxycholate, 1% Triton X-100 10% glycerol, Complete™–EDTA free

Pull-down washing buffer: 50 mM Tris-Cl, pH 7.5, 100 mM NaCl, 2 mM MgCl$_2$, 1% NP40, 10% glycerol, Complete™–EDTA free

1. The GAT domain of GGA1 fused to the C-terminus of GST (Takatsu *et al.*, 2002) is expressed in *E. coli* BL21(DE3) cells and purified using glutathione–Sepharose 4B beads (Amersham Biosciences, Tokyo, Japan) as described in the manufacturer's instructions. The composition of GST-fusion protein binding buffer and washing buffer are described above.

2. HeLa cells grown on a 10-cm dish are transfected with an expression vector for C-terminally HA-tagged ARF (Hosaka *et al.*, 1996) in combination with that for HA-tagged BIG2(WT) or an empty vector using a FuGENE6 reagent as described above.

3. The cells are then lysed in 0.65 ml of pull-down binding buffer for 20 min on ice with constant mixing and centrifuged at maximum speed (15,000g) in a microcentrifuge for 20 min at 4°.

4. The supernatant (cell lysate containing ~500 μg protein) is then precleared with glutathione-Sepharose 4B beads for 30 min at 4° with gentle shaking.

5. Then the mixture is centrifuged at maximum speed in a microcentrifuge for 2 min at 4° to remove the beads.

6. The supernatant is incubated with the GST-GGA1-GAT domain (~40 μg) prebound to glutathione-Sepharose 4B beads for 1 h at 4° with gentle shaking and centrifuged at 2,000×g for 5 min at 4°.

7. The pellet is washed three times with pull-down washing buffer.

8. The bound materials are eluted by boiling the pellet in an SDS-PAGE sample buffer and resolved on a 15% SDS-polyacrylamide gel for

detection of ARF or 7% SDS-polyacrylamide gel for detection of BIG2, and subjected to Western blot analysis. Results are quantified using the Image Gauge software program (Fuji Photo Film, Tokyo, Japan).

As shown in Fig. 2, approximately 3- and 4-fold larger amounts of GTP-bound ARF1 and ARF3, respectively, were estimated to be present in the BIG2-overexpressing cells than those in the control cells. In contrast, only 1.5-fold larger amounts of ARF5 were bound to GGA1-GAT and the ARF6 amount bound to GGA1-GAT did not change by the BIG2 over-expression (Fig. 2). These results suggest that BIG2 specifically activates class I ARFs (ARF1 and ARF3) *in vivo*.

Interestingly, coexpression of dominant negative ARF mutants, ARF1 (T31N) and ARF3(T31N), exaggerated the BIG2(E738K)-induced tubulation (Fig. 3). In contrast, coexpression of ARF6(T27N) did not significantly promote the BIG2(E738K)-induced tubule formation. These observations suggest that the BIG2(E738K)-induced tubulation is ascribed to the inactivation of ARF1 and ARF3, and that expression of BIG2(E738K) is not sufficient to inactivate all class I ARF present in the cells, probably because of the presence of endogenous BIG2 and/or another GEF such as BIG1. Taken together, class I ARFs seem to be implicated in the structural integrity of the RE.

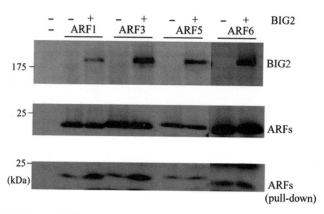

FIG. 2. BIG2 has a GEF activity toward ARF1 and ARF3 *in vivo*. Lysate from HeLa cells expression ARF alone or in combination with BIG2 were subjected to pull-down assay with GST-GGA1-GAT (bottom panel). The top and middle panels show the expression levels of HA-BIG2 and ARF-HA, respectively, by subjecting the cell lysates directly to Western blot analysis using anti-HA antibody. Reprinted from *Molecular Biology of the Cell* (Shin *et al.*, 2004) with the permission of The American Society for Cell Biology.

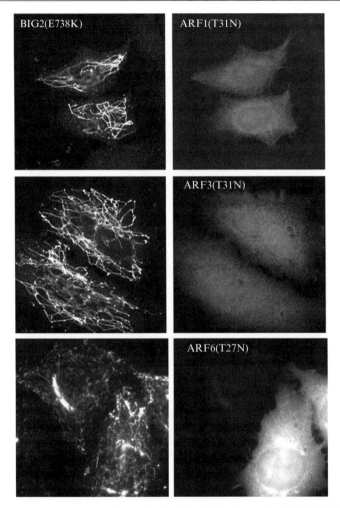

FIG. 3. Dominant negative mutants of ARF1 and ARF3 enhance the BIG2(E738K)-induced tubulation. HeLa cells coexpressing untagged BIG2(E738K) and either HA-tagged ARF1 (T31N), ARF3(T31N), or ARF6(T27N) were fixed and double-stained with polyclonal rabbit antiBIG2 and monoclonal rat anti-HA antibodies. Reprinted from *Molecular Biology of the Cell* (Shin *et al.*, 2004) with the permission of The American Society for Cell Biology.

Acknowledgments

Hye-Won Shin is a research fellow of the 21st Century COE Program. This work was supported in part by grants from the Japan Society for Promotion of Science, from the Ministry of Education, Culture, Sports, Science and Technology of Japan, from the Protein 3000 Project, from the Naito Foundation, from the Takeda Science Foundation, and from the Uehara Memorial Foundation, and by the Sasakawa Scientific Research Grant from the Japan Science Society.

References

Dell'Angelica, E. C., Puertollano, R., Mullins, C., Aguilar, R. C., Vargas, J. D., Hartnell, L. M., and Bonifacino, J. S. (2000). GGAs: A family of ADP ribosylation factor-binding proteins related to adaptors and associated with the Golgi complex. *J. Cell Biol.* **149,** 81–94.

Dunn, K. W., McGraw, T. E., and Maxfield, F. R. (1989). Iterative fractionation of recycling receptors from lysosomally destined ligands in an early sorting endosome. *J. Cell Biol.* **109,** 3303–3314.

Hopkins, C. R. (1983a). Intracellular routing of transferrin and transferrin receptors in epidermoid carcinoma A431 cells. *Cell* **35,** 321–330.

Hosaka, M., Toda, K., Takatsu, H., Torii, S., Murakami, K., and Nakayama, K. (1996). Structure and intracellular localization of mouse ADP-ribosylation factors type 1 to type 6 (ARF1-ARF6). *J. Biochem. (Tokyo)* **120,** 813–819.

Jackson, C. L., and Casanova, J. E. (2000). Turning on ARF: The Sec7 family of guanine-nucleotide-exchange factors. *Trends Cell Biol.* **10,** 60–67.

Morinaga, N., Tsai, S. C., Moss, J., and Vaughan, M. (1996). Isolation of a brefeldin A-inhibited guanine nucleotide-exchange protein for ADP ribosylation factor (ARF) 1 and ARF3 that contains a Sec7-like domain. *Proc. Natl. Acad. Sci. USA* **93,** 12856–12860.

Peyroche, A., Paris, S., and Jackson, C. L. (1996). Nucleotide exchange on ARF mediated by yeast Gea1 protein. *Nature* **384,** 479–481.

Santy, L. C., and Casanova, J. E. (2001). Activation of ARF6 by ARNO stimulates epithelial cell migration through downstream activation of both Rac1 and phospholipase D. *J. Cell Biol.* **154,** 599–610.

Shin, H.-W., Morinaga, N., Noda, M., and Nakayama, K. (2004). BIG2, a guanine nucleotide exchange factor for ADP-ribosylation factors: Its localization to recycling endosomes and implication in the endosome integrity. *Mol. Biol. Cell* **15,** 5283–5294.

Shin, H.-W., and Nakayama, K. (2004). Guanine nucleotide-exchange factors for Arf GTPases: Their diverse functions in membrane traffic. *J. Biochem. (Tokyo)* **136,** 761–767.

Shinotsuka, C., Waguri, S., Wakasugi, M., Uchiyama, Y., and Nakayama, K. (2002a). Dominant-negative mutant of BIG2, an ARF-guanine nucleotide exchange factor, specifically affects membrane trafficking from the trans-Golgi network through inhibiting membrane association of AP-1 and GGA coat proteins. *Biochem. Biophys. Res. Commun.* **294,** 254–260.

Shinotsuka, C., Yoshida, Y., Kawamoto, K., Takatsu, H., and Nakayama, K. (2002b). Overexpression of an ADP-ribosylation factor-guanine nucleotide exchange factor, BIG2, uncouples brefeldin A-induced adaptor protein-1 coat dissociation and membrane tubulation. *J. Biol. Chem.* **277,** 9468–9473.

Sonnichsen, B., De Renzis, S., Nielsen, E., Rietdorf, J., and Zerial, M. (2000). Distinct membrane domains on endosomes in the recycling pathway visualized by multicolor imaging of Rab4, Rab5, and Rab11. *J. Cell Biol.* **149,** 901–914.

Takatsu, H., Yoshino, K., Toda, K., and Nakayama, K. (2002). GGA proteins associate with Golgi membranes through interaction between their GGAH domains and ADP-ribosylation factors. *Biochem. J.* **365,** 369–378.

Togawa, A., Morinaga, N., Ogasawara, M., Moss, J., and Vaughan, M. (1999). Purification and cloning of a brefeldin A-inhibited guanine nucleotide-exchange protein for ADP-ribosylation factors. *J. Biol. Chem.* **274,** 12308–12315.

Welsh, C. F., Moss, J., and Vaughan, M. (1994). ADP-ribosylation factors: A family of approximately 20-kDa guanine nucleotide-binding proteins that activate cholera toxin. *Mol. Cell. Biochem.* **138,** 157–166.

[21] Assays and Properties of the ArfGAPs, AMAP1 and AMAP2, in Arf6 Function

By SHIGERU HASHIMOTO, ARI HASHIMOTO, ATSUKO YAMADA, YASUHITO ONODERA, and HISATAKA SABE

Abstract

The GTPase-activating protein (GAP) domain for Arfs primarily consists of a zinc-finger structure, which is not present in known GAPs for the other Ras-superfamily GTPases. More than 20 genes have been found to encode proteins bearing the ArfGAP domain in the human genome: a number that is much larger than that of the Arf isoforms. Several Arf isoforms, such as Arf1 and Arf6, indeed have been shown to each employ multiple different ArfGAPs for their regulation and function. We have found that two ArfGAPs, namely AMAP1 and AMAP2, exhibit a novel biochemical property of directly and selectively binding to GTP-Arf6 without immediate GAPing activity, while they were previously shown to exhibit efficient catalytic GAPing activities to Arf isoforms except Arf6 *in vitro*. Such property of AMAPs appears to be important for AMAPs-mediated recruitment of auxiliary molecules, including paxillin, cortactin, amphiphysin, and intersectin, to sites of Arf6 activation. AMAPs thus appear to act as "effectors" rather than simple GAPs in some aspects of Arf6 function. This article presents methods and protocols developed for the functional characterization of AMAPs in Arf6 function. These methods may be applied to other types of ArfGAPs to further clarify the cellular functions of ArfGAPs as well as Arfs.

Introduction

Arf-family GTPases play pivotal roles in vesicle trafficking and membrane remodeling (Moss and Vaughan, 1998; Roth, 2000). There are six isoforms of Arf in mammals, and they are subclassified into three classes by their structural similarities: class I (Arf1–3), class II (Arf4 and 5), and class III (Arf6). Arf1 primarily functions at perinuclear areas, mediating transport between the Golgi, *cis*-Golgi network, and ER by employing coat protein complexes, such as COP I (Roth, 2000). On the other hand, Arf6 participates in the regulation of membrane recycling and actin-cytoskeletal remodeling at the cell periphery (Donaldson, 2003), and has also been implicated in higher order cellular functions, such as cell migration and cancer cell invasion (Hashimoto *et al.*, 2004a; Kondo *et al.*, 2000; Palacios

METHODS IN ENZYMOLOGY, VOL. 404
0076-6879/05 $35.00
DOI: 10.1016/S0076-6879(05)04021-8

et al., 2001; Sabe, 2003a,b; Tague *et al.*, 2004). Functions of the class II Arfs are largely unknown.

More than 20 genes have been found to encode proteins bearing the ArfGAP domain in the human genome. Although at least 10 different proteins bearing homology to members of the Arf family exist in mammals, there appears to be little or no cross reactivity of the ArfGAPs, and also ArfGEFs, with these Arf-like proteins (Kahn, 2003). Therefore, some Arf isoforms might employ multiple types of ArfGAPs, perhaps dependent on their different functions in different cellular contexts. Similar to GAPs for other Ras-superfamily small GTPases, the GAP activity of ArfGAPs can be measured *in vitro* by using GTP-loaded Arfs as substrates. Previous biochemical assays revealed that one of the ArfGAPs, AMAP2/PAG3/ Papα/KIAA0400, exhibits phosphatidylinositol 4,5-bisphosphate-dependent GAP activity for Arf1 and Arf5 but 10^2-fold to 10^3-fold less activity for Arf6 (Andreev *et al.*, 1999). There is a close isoform of AMAP2, namely AMAP1, in humans (Kondo *et al.*, 2000; Onodera *et al.*, 2005; Sabe, 2003a). ASAP1 (Brown *et al.*, 1998) and DEF-1 (King *et al.*, 1999) are murine and bovine orthologs of human AMAP1, respectively. The spectrum of biochemical GAP activities of ASAP1 towards the different Arf isoforms is essentially the same as that of AMAP1 (Brown *et al.*, 1998). DEF-1 has also been shown to exhibit significant GAP activity towards Arf1, but not Arf6, which was assessed by biochemically measuring the cellular levels of GTP-loaded Arfs in DEF-1-overexpressing cells (Furman *et al.*, 2002). In contrast to these biochemical results, our studies have indicated that AMAP2 and AMAP1 function in concert with Arf6 activity *in vivo*. We have shown that (1) overexpression of AMAP2 can suppress Arf6-dependent membrane protrusions, while it does not affect the subcellular localization of β-COP (Kondo *et al.*, 2000), which is a component of the COP-I coat and its localization to the Golgi is dependent on Arf1 activity; (2) Arf6 is known to play a role in the Fcγ receptor-mediated phagocytosis of macrophages (Zhang *et al.*, 1998), and we showed that AMAP2 colocalizes with Arf6 at Fcγ receptor phagocytic cups and that its overexpression blocks Fcγ receptor-mediated membrane extension and phagocytosis (Uchida *et al.*, 2001). This blockage can be restored by co-overexpression of Arf6, but not Arf1. In these experiments, we used a GAP-deficient mutant of AMAP2 as a control. We also employed another GAP, Git2-short, as another control, and showed that its overexpression affects the cellular localization of β-COP (Mazaki *et al.*, 2001) and does not block Fcγ receptor phagocytosis (Uchida *et al.*, 2001). Moreover, we have recently found that Arf6 and AMAP1 are both localized to invadopodia and play pivotal roles in the invasion of breast cancer cells (Hashimoto *et al.*, 2004a; Onodera *et al.*, 2005). Invadopodia are sites of plasma membrane extension and endocytosis (Coopman *et al.*, 1996), similar to Fcγ receptor phagocytic cups. Similar to the way AMAP2 plays a role in recruiting several endocytic proteins to sites of Arf6 activation for receptor internalization (Hashimoto *et al.*, 2004b; Uchida *et al.*, 2001; also see following), AMAP1 also appears to play a role in recruiting paxillin and cortactin, both of which are integral components of invadopodia (Bowden *et al.*, 1999), to sites of Arf6 activation for the formation of invadopodia (Onodera *et al.*, 2005).

The GEF domains of several Ras-superfamily GTPases have been shown to stably bind to the GDP-bound form of their cognate GTPases (Hart *et al.*, 1994; Hussain *et al.*, 2001). GAP proteins recognize the GTP-bound form of GTPases, but stable binding to the GTP-bound form of GTPases was an unexpected property. However, the primary structure of the ArfGAP domain is significantly different from other GAP domains. We hence examined whether AMAP2 binds stably and selectively to the GTP-bound form of Arf6 via its ArfGAP domain even in the presence of divalent cations, with the hope to solve the apparent discrepancy between the results of biochemical assays and cell biological assays of AMAP2 (Hashimoto *et al.*, 2004b; Sabe, 2003a). A noncatalytic model of the Arf-GAP1/Arf1 complex constructed by Goldberg (Goldberg, 1999) also supported our trial. Arf1 and Arf6 each employs different auxiliary proteins or coat proteins, closely related to the distinct cellular functions of Arf1 and Arf6. We have also analyzed which proteins, other than the Arfs, directly interact with AMAP2 to further assess its possible cofunction with Arf6. In addition, we examined the subcellular colocalization of AMAP2, its auxiliary proteins, and Arf6. We moreover conducted overexpression, as well as siRNA-mediated knock-down analyses of AMAP2 and compared their phenotypes with those seen upon manipulation of Arf6 activity and expression. In the following section, each protocol of these methods is described, most of which we used in an original study (Hashimoto *et al.*, 2004b). Protocols for the yeast two-hybrid system, which we used to identify proteins binding to the proline-rich domain (PRD) of AMAP, can be found elsewhere.

Binding Specificity of the AMAP ArfGAP Domains to GTP-Arf6

Preparation of GST-AMAP2 ArfGAP Domain

BL21(DE) bacterial cells transformed with pGEX-2TK-AMAP2 ArfGAP are cultured at 37° in 20 ml of Luria-Bertani (LB) medium containing 100 μg/ml ampicillin for 12 to 16 h to reach confluence. This preculture is then diluted 25 times with fresh LB medium containing 100 μg/ml ampicillin and further cultured at 37°. When the OD_{600} reaches

~0.6 (3 to 5 × 10⁸ cells/ml), isopropyl-β-D-thiogalactopyranoside (IPTG) is added to the culture to a final concentration of 0.1 mM, and cells are then grown at 25° for a further 2 h. Cells are aerated by vigorous shaking throughout the procedure (300 rpm by Brunswick Shaker). Use cotton plugs or loose aluminum cups to seal culture flasks. Plugs can even be taken off after IPTG is added. Cells are then quickly chilled and harvested by centrifugation for 15 min at 6000g at 4°. After washing once with ice-cold phosphate-buffered saline (PBS), bacterial pellets can be snap-frozen in liquid nitrogen and stored at –80° until use.

The cells are then suspended in 10 ml of ice-cold buffer A (1% Nonidet P-40, 100 mM EDTA, 1 mM PMSF, 1% aprotinin, 2 mg/ml leupeptin, and 3 μg/ml pepstatin A in PBS), and sonicated on ice 10 times for 30 sec with 30 sec intervals. The homogenate is then centrifuged for 15 min at 20,000g at 4°. The supernatant is collected and gently mixed with 500 μl bed volume of glutathione-Sepharose 4B beads (Amersham Pharmacia, Piscataway, NJ), and equilibrated with 5 ml of buffer A at 4° for 2 h. The mixture is then placed in a 10 ml chromatography column (731–1550, Bio-Rad, Hercules, CA) and the flow-through fraction collected. After the beads become packed, the flow-through fraction is reapplied twice to the column. The column is then washed with 50 ml of buffer A, then with 10 ml of buffer B (1 mM PMSF, 1% aprotinin, 2 μg/ml leupeptin, 3 μg/ml pepstatin A, and 0.02% NaN3 in PBS), and stored in buffer B at 4°. 0.5–1 mg of GST-ArfGAP is recovered from 500 ml culture. Protein concentration is determined using the Dc protein assay kit (Bio-Rad) using bovine serum albumin (BSA, fraction V; Sigma, St. Louis, MO) as a standard. The beads are diluted with fresh glutathione-Sepharose 4B beads to make GST fusion proteins bound to the beads to be 1 mg/ml of the bed volume. GST protein, encoded by the pGEX-2TK vector, is also prepared similarly and used as a control. The final preparation of proteins, separated by SDS-PAGE, is shown in Fig. 1A.

Binding of the GST-AMAP2 ArfGAP Domain with Arf In Vitro

COS-7 cells are maintained in Dulbecco's modified Eagle's medium (DMEM) with 10% fetal calf serum (FCS; Hyclone, Logan, UT). 5 × 10⁵ cells are seeded in a 100-mm culture dish. On the following day (16–24 h later), cells are transfected with 4 μg of HA-tagged wild type and mutant Arf cDNAs, each in the pcDNA3 vector (Invitrogen, Carlsbad, CA), using Poly-Fect (QIAGEN 40724 Hilden, Germany) according to the manufacturer's instructions. Twenty-four h later, cells are washed twice with ice-cold PBS and lysed in 300 μl of 1% NP-40 buffer (1% Nonidet P-40, 150 mM NaCl, 20 mM Tris-HCl (pH 7.4), 5 mM EDTA, 1 mM Na3VO4, 1 mM PMSF, 1% aprotinin, 2 μg/ml leupeptin, and 3 μg/ml pepstatin A) and incubated for 10 min at 4°

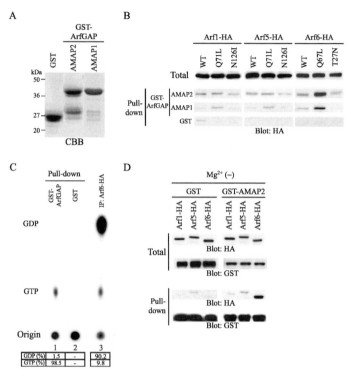

FIG. 1. Binding of AMAP2 with GTP-Arf6. (A) Protein preparations of GST-ArfGAP domains of AMAP1 and AMAP2, and GST are shown after being separated by SDS-PAGE (12%) and stained by Coomassie brilliant blue. (B) *In vitro* binding of GST-ArfGAP domains of AMAP1 and AMAP2 to wild type and mutant Arf proteins. WT, wild type; Q71L and Q67L, GTP hydrolysis-deficient mutants; and N126I and T27N, GDP dissociation-deficient mutants. Proteins coprecipitated were analyzed by immunoblotting using an anti-HA antibody. (C) Guanine nucleotide status of wild type Arf6 proteins bound to GST-AMAP2 ArfGAP domain. HA-Arf6, immunoprecipitated using an anti-HA antibody, was also analyzed. An autoradiograph of the thin layer chromatography is shown. (D) *In vivo* binding of wild type Arf6 to full length AMAP2. Immunoblots using anti-HA or anti-GST antibodies are shown. For (A–D), see text for details. Total: 10 μg of total cell lysates. Panels B–D are reproduced from Hashimoto *et al.* (2004b).

with occasional vortex mixing. Cleared lysates are then obtained by brief centrifugation (at 10,000g for 10 min at 4°); 250 μg of the lysate is then mixed with 5 μg of GST fusion protein bound to glutathione-beads, and incubated for 2 h at 4° with rotation. After washing four times with 1 μl of 1% NP-40 buffer, Arf proteins retained on the beads are subjected to SDS-PAGE, followed by immunoblotting analysis using an anti-HA antibody and visualization using an enzyme-linked chemiluminescence method (Fig. 1B). GST protein should be included as a negative control (Fig. 1B). The ArfGAP domain of AMAP1 exhibits the same properties of binding as the ArfGAP domain of AMAP2 (Fig. 1A and B).

Guanine Nucleotide Status of Wild Type Arf6 Bound to the AMAP2 ArfGAP Domain

Less than 10% of wild type Arf6 binds to GTP *in vivo* (Furman *et al.*, 2002; Hashimoto *et al.*, 2004b; also see Fig. 1C), which may explain why it is difficult to clearly detect the binding between wild type Arf6 and the AMAP2 ArfGAP domain in the above *in vitro* experiment. An assay to confirm the selective binding of the AMAP2 ArfGAP domain to the GTP-bound form of wild type Arf6 is performed as follows. In this assay, radiolabeling of cellular nucleotides and their analysis using thin layer chromatography are performed as described previously (Furman *et al.*, 2002; Vitale *et al.*, 2002).

5×10^5 COS-7 cells in a 100-mm dish are transfected with 1 μg of pcDNA3-Arf6-HA using PolyFect, as above. Twenty-four h later, cells are washed twice with phosphate-free DMEM (Gibco BRL, Grand Island, NY) and radiolabeled with 325 μCi/ml [^{32}P]orthophosphate (285.6 Ci/mg of phosphate, Perkin-Elmer Life Sciences, Boston, MA) for 16 h in phosphate-free DMEM and 10% FCS (total volume of the medium is 5ml), which is predialyzed against saline. Cell lysate is then prepared with a lysis buffer (20 mM Tris-HCl (pH 8.0), 100 mM NaCl, 1 mM MgCl$_2$, 1% Triton X-100, 0.05% cholate, 0.005% SDS, 1 mM dithiothreitol, 1 mM phenylmethyl-sulfonyl fluoride, 1 mM benzamidine, 5 μg/ml aprotinin, 5 μg/ml leupeptin, 1 mM NaF, and 1 mM Na$_3$VO$_4$). After clarification by centrifugation, the lysate is incubated with 5 μg of GST-ArfGAP bound to glutathione-beads for 2 h at 4°. After washing the beads four times with 1 ml of the above lysis buffer, the radioactive GTP- or GDP-bound Arf6 are eluted from the beads with 2 M formic acid at 70° for 3 min, and subjected to thin layer chroma-tography using polyethyleneimine-cellulose plates (Merck, Whitehouse Station, NJ) in a mixed solution of 1 M formic acid and 1 M LiCl. As a control, HA-tagged Arf6 is immunoprecipitated from the lysate using an anti-HA monoclonal antibody and analyzed similarly. GST-beads are also included as a control. As shown in Fig. 1C, GST-ArfGAP preferentially pulls down GTP-bound Arf6. No binding of GDP or GTP is detected with GST-beads.

Intracellular Binding of Full Length AMAP2 with Wild Type Arf6

COS-7 cells are transfected with 0.3 μg of pcDNA3-Arf-HA (encoding the HA-tagged wild type Arfs) and 4 μg of pEBG-AMAP2 (encoding the

GST-fusion form of the full length AMAP2) using PolyFect, as above. Twenty-four h later, cells are washed twice with ice-cold PBS and lysed in 300 μl of 1% NP-40 buffer. After clarification by centrifugation, the lysate containing 500 μg of protein (volume is adjusted to 300 μl by 1% NP-40 buffer) is incubated with 5-μl bed volume of glutathione-beads to pull-down the GST-AMAP2 proteins. After washing four times with 1 ml of 1% NP-40 buffer, proteins retained on the beads are subjected to immunoblotting analysis with an anti-HA antibody, as described above (Fig. 1D). The presence of 10 mM MgCl$_2$ in the lysis buffer does not affect the binding of GST-AMAP2 with Arf6-HA (Hashimoto et al., 2004b). As a negative control of GST expression, 2 μg of pcDNA-Arf-HA plus 2 μg of pEBG vector (for GST alone) are used instead of pEBG-AMAP2 (Fig. 1D).

Selective Binding of AMAP2 to Proteins Related to Arf6 Function

AMAP2 has several protein interaction modules including the PRD and src homology 3 (SH3) domain. We originally identified AMAP2 as a paxillin-binding protein (Kondo et al., 2000). Paxillin is a scaffold adaptor protein playing a role in mechanisms such as integrin signaling, and we showed that the SH3 domain of AMAP2 is responsible for its binding to paxillin. Consistent with our result that AMAP2 functions in Fcγ receptor phagocytosis (Uchida et al., 2001), paxillin is known to localize to Fcγ receptor phagocytic cups (Aderem and Underhill, 1999). To examine whether other proteins, with roles implicated in Arf6 function, also bind to AMAP2, we conducted a yeast two-hybrid screening using the AMAP2 PRD as a bait (Hashimoto et al., 2004b). Amphiphysin-IIm (Gold et al., 2000) and intersectin-Is (Hussain et al., 2001), which are both implicated in endocytosis, are identified as potential binding partners. Binding of AMAP2 with these proteins was confirmed in vitro and in vivo. Amphiphysin-IIm was originally identified from macrophages and shown to localize to early macrophage phagosomes as an integral component (Gold et al., 2000), again consistent with the notion that both Arf6 and AMAP2 are integral components of Fcγ receptor-mediated phagocytosis. Besides Fcγ receptor-mediated phagocytosis, Arf6 has been shown to play pivotal roles in other types of clathrin-independent endocytoses of cell surface receptors (Brown et al., 2001; Radhakrishna and Donaldson, 1997; Radhakrishna et al., 1996). Arf6 has also been proposed to be involved in some clathrin-dependent endocytoses (D'Souza-Schorey et al., 1995; Palacios et al., 2002).

Subcellular Colocalization of AMAP2 and its Auxiliary
Proteins with GTP-Arf6

Anaysis of the subcellular colocalization of AMAP2 and its auxiliary proteins with Arf6 is also necessary to assess when and where they function together with Arf6. We have shown the colocalization of AMAP2 with Arf6 at Fcγ receptor-mediated phagosomes of macrophage cells (Uchida *et al.*, 2001). In such cells, however, endomembrane structures are not easy to study using normal light microscopy. Instead, adherent cells such as HeLa and CHO have been frequently used for studying the subcellular localization of Arf proteins. In HeLa cells, the majority of a GTP hydrolysis-deficient mutant of Arf6, Arf6Q67L, has been shown to localize to the plasma membrane (Radhakrishna and Donaldson, 1997). In a subpopulation of cells, Arf6Q67L is also found to be enriched in large intracellular vacuoles, while in such cells localization of Arf6Q67L to the plasma membrane is barely observed (Brown *et al.*, 2001). On the other hand, the majority of wild type Arf6 and a GTP binding-deficient mutant of Arf6, Arf6T27N, are found localized to intracellular tubulovesicular structures, which probably represent recycling compartments (Radhakrishna and Donaldson, 1997). AMAP2 is also found localized to intracellular tubulovesicular structures in unstimulated HeLa cells (Hashimoto *et al.*, 2004b; see Fig. 2A).

To examine the colocalization of AMAP2, its auxiliary proteins, and Arf6, 1×10^5 HeLa cells, seeded in a 35-mm glass-bottomed dish (MatTek, Ashland, MA), are transfected with 1 μg of pEGFP-AMAP2, 1 μg of pcDNA3-Xpress-amphiphysin-IIm, and 0.5 μg of pcDNA3-Arf6-HA or its mutant using PolyFect reagent, as described previously. The pEGFP-AMAP2 plasmid is linearized beforehand by digestion with EagI, to minimize its expression. Twenty-four h later, cells are fixed and subjected to immunofluorescence analysis. Cell fixation is performed by the direct addition of 1/10 vol of prewarmed 40% paraformaldehyde in PBS to the culture medium and incubated for 10 min at 37°. Cells are then rinsed twice with prewarmed PBS, and then incubated with 0.1% BSA in PBS for 5 min at an ambient temperature. After rinsing once with PBS, cells are then subjected to incubation with primary antibodies at an ambient temperature for 1 h in the presence of 0.2% saponin (Sigma) and 0.1% BSA in PBS. After rinsing twice with PBS, cells are incubated with secondary antibodies (also in 0.2% saponin and 0.1% BSA in PBS) at an ambient temperature for 1 h. After rinsing twice with PBS, coverslips are mounted with 50% glycerol in PBS. Confocal images are acquired using a confocal laser scanning microscope (LSM 510 version 2.5; Carl Zeiss, 07745 Jena, Germany). Antibodies used are a rabbit polyclonal anti-HA antibody (Santa Cruz, Santa Cruz, CA),

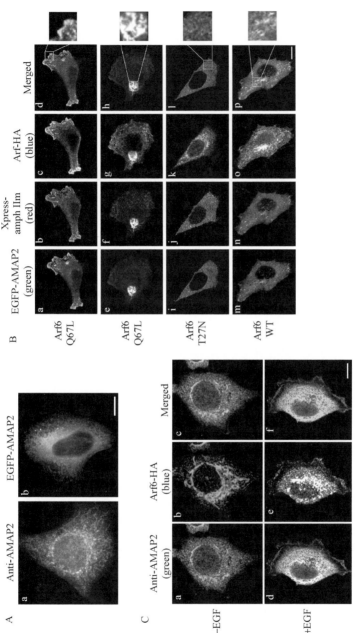

Fig. 2. Colocalization of AMAP2 with GTP-Arf6 in HeLa cells. (A) Subcellular localization of endogenous AMAP2 (a) and EGFP-AMAP2 (b). Endogenous AMAP2 was labeled using a rabbit anti-AMAP2 polyclonal antibody (Kondo *et al.*, 2000), coupled with a Cy2-conjugated anti-rabbit IgG antibody. EGFP-AMAP2 was visualized by autofluorescence from the tag. (B) Colocalization of AMAP2 with GTP-Arf6, but not GDP-Arf6. EGFP-AMAP2 colocalizes with Arf6Q67L at the plasma membrane (a–d) or at large intracellular vacuoles (e–h). No notable colocalization of AMAP2 with Arf6T27N (i–l), or with wild type Arf6 in unstimulated cells (m–p), was observed. (C) Colocalization of AMAP2 with wild type Arf6 at plasma membrane ruffles upon EGF stimulation. Cells expressing HA-tagged wild type Arf6 were stimulated briefly by EGF (+EGF) before fixation (d–f). Cells before the stimulation (-EGF) are also shown (a–c). Bars: 10 μm. Reproduced from Hashimoto *et al.* (2004b).

mouse monoclonal anti-Xpress antibody (Invitrogen, Carlsbad, CA), affinity purified Cy3-conjugated donkey polyclonal anti-rabbit IgG antibody (Jackson ImmunoResearch Laboratories, West Grove, PA), and affinity purified Cy5-conjugated donkey polyclonal anti-mouse IgG antibody (Jackson ImmunoResearch Laboratories). Enhanced green fluorescent protein (EGFP)-tagged AMAP2 was detected by its autofluorescence. BSA used should be of an IgG-free grade. Although high levels of AMAP2 overexpression act to block Arf6 function (more than 10–20-fold of the endogenous level: Kondo et al., 2000; Uchida et al., 2001), under the above conditions using linearized pEGFP-AMAP2 plasmid, the EGFP-AMAP2 protein is expressed at almost comparable levels to that of endogenous AMAP2 and exhibits much the same subcellular localization as seen with endogenous AMAP2. In cells expressing Arf6Q67L, EGFP-AMAP2 is predominantly redistributed to the Arf6Q67L-positive plasma membrane areas (Fig. 2B, a-d). Enrichment of EGFP-AMAP2 to the Arf6Q67L-positive intracellular vacuoles is also observed (Fig. 2B, e-h). On the other hand, much less EGFP-AMAP2 is colocalized with wild type Arf6-HA and Arf6T27N-HA (Fig. 2B, i-p). When wild type Arf6 is activated, such as by EGF stimulation of cells, Arf6 proteins are rapidly recruited to the plasma membrane (Boshans et al., 2000; Honda et al., 1999). Upon stimulation by EGF (10 ng/ml EGF for 7 min), AMAP2 is recruited to the wild type Arf6-positive plasma membrane areas (Fig. 2C). In contrast, the subcellular localization of AMAP2 is not notably correlated with that of Arf1 and its mutants (Hashimoto et al., 2004b).

Functional Relationship Between AMAP2 and Arf6 in Receptor Endocytosis

siRNA-mediated protein knock-down, as well as cDNA-mediated protein overexpression, function very well in HeLa cells (Elbashir et al., 2001). Arf6 has been shown to be involved in clathrin-independent endocytosis and recycling, such as for Tac and MHC class I molecules, but not in the clathrin-dependent endocytosis of transferrin (Tfn) in HeLa cells (Brown et al., 2001; Radhakrishna and Donaldson, 1997; Radhakrishna et al., 1996). To obtain further evidence for the involvement of AMAP2 in the cellular functions of Arf6, we examined the effects of overexpression, as well as siRNA-mediated downregulation, of AMAP2 on Arf6-mediated endocytosis in HeLa cells.

Small Interfering RNA (siRNA)-mediated Interference

Oligonucleotides used for human Arf6 siRNA are 5'-GCACCG-CAUUAUCAAUGACCGUU-3' and 5'-CGGUCAUUGAUAAUGCG

GUGCUU-3'; and for human AMAP2 siRNA, 5'-UUUAGGAAGU-GCGUUCCUGAAUU-3' and 5'-UUCAGGAACGCACUUCCUA-AAUU-3'. Duplex oligonucleotides with a similar length and irrelevant sequence, 5'-GCGCGCUUUGUAGGAUUCGdTdT-3' and 5'-CGAAU-CCUACAAAGCGCGCdTdT -3' (Dharmacon Research), are used as a control. 0.4×10^5 HeLa cells are seeded in a 35-mm glass-bottomed dish. Sixteen to 24 h later, cells are rinsed once with Opti-MEM (Gibco BRL) and transfected with 50 nM of siRNA duplexes in Opti-MEM using Oligofectamine (Invitrogen), according to the manufacturer's instructions. Four h later, a one-third volume of DMEM containing 30% FCS is added to the culture (Elbashir et al., 2001) and incubated for a further 32 h prior to analysis. To examine Tac uptake (see following), 12 h after the siRNA transfection, a cocktail containing 0.4 μg of pKCR-Tac, 0.1 μg of pEGFP-C1 (Clontech), and PolyFect (QIAGEN) in DMEM is added onto the cell culture and cells are incubated for a further 24 h prior to analysis. pEGFP-C1 is used to identify DNA transfected cells. Under these conditions of siRNA treatment, expression of endogenous Arf6 and AMAP2 proteins is suppressed to less than 10% of that in parental cells, which is assessed by immunoblotting of the cell lysates (Hashimoto et al., 2004b).

Overexpression of AMAP2 Proteins

To assess the effect of AMAP1 overexpression on Tac uptake (see following), 1×10^5 HeLa cells in a 35-mm glass-bottomed dish cells are cotransfected with 1.6 μg of pEGFP-AMAP2 and 0.4 μg of pKCR-Tac and incubated for 24 h.

Tac and Tfn Uptake

Ligand-independent autonomous internalization of Tac receptor molecules, expressed in HeLa cells by cDNA transfection, is measured using an anti-Tac antibody, as described previously (Donaldson and Radhakrishna, 2001). HeLa cells, manipulated as above, are rinsed once with ice-cold PBS and then incubated with 10 μg/ml of anti-Tac antibody (clone 7G7, Upstate Biotechnology) in PBS at 4° for 30 min. Cells are then washed twice with ice-cold DMEM containing 10% FCS, and returned to incubation at 37° with prewarmed DMEM containing 10% FCS. After 30 min of incubation, Tac antibodies remaining on the cell surface are removed by incubating cells with a low pH solution (0.5 M NaCl in 0.5% acetic acid (pH 3.0)) at an ambient temperature for 15 sec. Cells are rinsed once with DMEM (without FCS) before fixation in 4% paraformaldehyde, and then permeabilized with 0.2% saponin and labeled using

Cy3-conjugated anti-mouse IgG antibodies, as described previously, to detect internalized Tac antibodies. As a control, internalization of Tfn receptors may be assessed simultaneously. Tfn receptors internalize and recycle constantly regardless of whether or not they are being occupied by ligands. Using biotinylated human Tfn (Sigma), internalization of Tfn, and thereby Tfn receptors, can be measured according to a method described previously (Wigge *et al.*, 1997). Cells are rinsed twice with DMEM, incubated with DMEM (without FCS) for 1 h at 37°, and then incubated with 25 μg/ml Tfn for 1 h at 37° in DMEM containing 10% FCS. After rinsing twice with DMEM and once with a low pH solution, as above, cells were fixed, permeabilized, and labeled with Cy5-conjugated streptavidin to detect internalized biotinylated Tfn.

Amounts of internalized Tac or Tfn molecules are evaluated by quantifying the intensities of the fluorescent signals of each cell using computer software associated with the confocal laser scanning microscope. Focuses are adjusted about 3 μm above the surface of the glass chamber plate, across the center of the nucleus of the majority of cells. Transfection-positive cells are identified by immunolabeling of exogenous proteins by their tags or by the autofluorescence from cotransfected EGFP. Endocytosis blocked cells are defined as transfected cells in which Tac or Tfn uptake is less than 20% of that seen in untransfected cells in the same field or in control siRNA-treated cells, as described previously (Owen *et al.*, 1998; also see Fig. 3C and D). Endocytic blockage is then expressed as a percentage of that of the untransfected or control cells (Wigge *et al.*, 1997). In each assay, more than 50 transfected cells are examined in each of three independent experiments.

As shown in Fig. 3, overexpression of AMAP2 blocks the internalization of Tac, but not Tfn. Blockage of Tac uptake can be restored by the simultaneous overexpression of Arf6 (Hashimoto *et al.*, 2004b). As a control, blockage of both Tac and Tfn uptake should be confirmed by overexpression of the loss-of-function mutant of dynamin-2 (K44A: De Camilli *et al.*, 2001). siRNA-mediated knockdown of AMAP2 or Arf6 expression in HeLa cells each also significantly inhibits the uptake of Tac, while they do not notably affect the uptake of Tfn (Hashimoto *et al.*, 2004b). Since these siRNAs function very efficiently (more than 90% suppression is achieved), it may not be necessary to examine protein levels of endogenous AMAP2 and Arf6 in each cell to be examined. To show the specificity of these assays, we confirmed that siRNA-mediated knockdown of Git2/APAP2, which we have shown to act as a GAP for Arf1 *in vivo* (Mazaki *et al.*, 2001), as well as its overexpression does not affect the uptake of Tac and Tfn (Hashimoto *et al.*, 2004b).

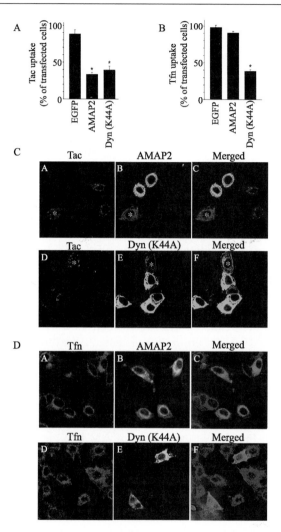

FIG. 3. Effects of AMAP2 overexpression on receptor internalizations in HeLa cells. Clathrin-independent endocytosis of Tac molecules (A and C) and clathrin-dependent endocytosis of Tfn molecules (B and D) were analyzed. See text for details. *, P<0.01 against values of the EGFP control in (A) and (B). In (C), cells marked by asterisks expressed detectable levels of EGFP-AMAP2 or dynamin2 K44A, but exhibited more than 20% of Tac uptake as compared to those of the untransfected cells in the same field. These cells were hence judged as cells in which Tac uptake was not blocked. Panels A an B are reproduced from Hashimoto *et al.* (2004b).

Future Studies

It is necessary to investigate the fate of the AMAP2/GTP-Arf6 complex. Our results suggest that further processing of this complex may be necessary for the recruitment of dynamin2 to sites of endocytosis (Hashimoto *et al.*, 2004b). Dynamin2 is thought to be necessary for the final fission process of endocytosis (De Camilli *et al.*, 2001). The mechanism regulating the processing of the AMAP2/GTP-Arf6 complex might hence be crucial for regulation of the onset of late stages of endocytosis. AMAP1, a close isoform of AMAP2, which functions in breast cancer invasion (Onodera *et al.*, 2005), also forms a stable complex with GTP-Arf6. Invadopodia are also sites of endocytosis. Moreover, both AMAP1 and AMAP2 are also highly enriched in the synaptic areas of neurons. Therefore, clarification of the fate of AMAP/GTP-Arf6 complexes is important for the further understanding of a variety of biological events, including endocytosis, cancer cell invasion, and synaptic transmission.

References

Aderem, A., and Underhill, D. M. (1999). Mechanisms of phagocytosis in macrophages. *Annu. Rev. Immunol.* **17,** 593–623.

Andreev, J., Simon, J. P., Sabatini, D. D., Kam, J., Plowman, G., Randazzo, P. A., and Schlessinger, J. (1999). Identification of a new Pyk2 target protein with Arf-GAP activity. *Mol. Cell. Biol.* **19,** 2338–2350.

Boshans, R. L., Szanto, S., van Aelst, L., and D'Souza-Schorey, C. (2000). ADP-ribosylation factor 6 regulates actin cytoskeleton remodeling in coordination with Rac1 and RhoA. *Mol. Cell. Biol.* **20,** 3685–3694.

Bowden, E. T., Barth, M., Thomas, D., Glazer, R. I., and Muller, S. C. (1999). An invasion-related complex of cortactin, paxillin and PKCμ associates with invadopodia at sites of extracellular matrix degradation. *Oncogene* **18,** 4440–4449.

Brown, F. D., Rozelle, A. L., Yin, H. L., Balla, T., and Donaldson, J. G. (2001). Phosphatidylinositol 4,5-bisphosphate and Arf6-regulated membrane traffic. *J. Cell Biol.* **154,** 1007–1017.

Brown, M. T., Andrade, J., Radhakrishna, H., Donaldson, J. G., Cooper, J. A., and Randazzo, P. A. (1998). ASAP1, a phospholipid-dependent arf GTPase-activating protein that associates with and is phosphorylated by Src. *Mol. Cell. Biol.* **18,** 7038–7051.

Coopman, P. J., Thomas, D. M., Gehlsen, K. R., and Mueller, S. C. (1996). Integrin $\alpha 3\beta 1$ participates in the phagocytosis of extracellular matrix molecules by human breast cancer cells. *Mol. Biol. Cell* **7,** 1789–1804.

De Camilli, P., Slepnev, V. I., Shupliakov, O., and Brodin, L. (2001). Synaptic vesicle endocytosis. *In* "Synapses" (W. M. Cowan, T. C. Südhof, and C. F. Stevens, eds.), pp. 217–274. Johns Hopkins University Press, Baltimore, Maryland.

Donaldson, J. G., and Radhakrishna, H. (2001). Expression and properties of ADP-ribosylation factor (ARF6) in endocytic pathways. *Methods Enzymol.* **329,** 247–256.

Donaldson, J. G. (2003). Multiple roles for Arf6: Sorting, structuring, and signaling at the plasma membrane. *J. Biol. Chem.* **278,** 41573–41576.

D'Souza-Schorey, C., Li, G., Colombo, M. I., and Stahl, P. D. (1995). A regulatory role for ARF6 in receptor-mediated endocytosis. *Science* **267,** 1175–1178.

Elbashir, S. M., Harborth, J., Lendeckel, W., Yalcin, A., Weber, K., and Tuschl, T. (2001). Duplexes of 21-nucleotide RNAs mediate RNA interference in cultured mammalian cells. *Nature* **411,** 494–498.

Furman, C., Short, S. M., Subramanian, R. R., Zetter, B. R., and Roberts, T. M. (2002). DEF-1/ASAP1 is a GTPase-activating protein (GAP) for ARF1 that enhances cell motility through a GAP-dependent mechanism. *J. Biol. Chem.* **277,** 7962–7969.

Gold, E. S., Morrissette, N. S., Underhill, D. M., Guo, J., Bassetti, M., and Aderem, A. (2000). Amphiphysin IIm, a novel amphiphysin II isoform, is required for macrophage phagocytosis. *Immunity* **12,** 285–292.

Goldberg, J. (1999). Structural and functional analysis of the ARF1-ARFGAP complex reveals a role for coatomer in GTP hydrolysis. *Cell* **96,** 893–902.

Hart, M. J., Eva, A., Zangrilli, D., Aaronson, S. A., Evans, T., Cerione, R. A., and Zheng, Y. (1994). Cellular transformation and guanine nucleotide exchange activity are catalyzed by a common domain on the dbl oncogene product. *J. Biol. Chem.* **269,** 62–65.

Hashimoto, S., Onodera, Y., Hashimoto, A., Tanaka, M., Hamaguchi, M., Yamada, A., and Sabe, H. (2004a). Requirement for Arf6 in breast cancer invasive activities. *Proc. Natl. Acad. Sci. USA* **101,** 6647–6652.

Hashimoto, S., Hashimoto, A., Yamada, A., Kojima, C., Yamamoto, H., Tsutsumi, T., Higashi, M., Mizoguchi, A., Yagi, R., and Sabe, H. (2004b). A novel mode of action of an ArfGAP, AMAP2/PAG3/Papα, in Arf6 function. *J. Biol. Chem.* **279,** 37677–37684.

Honda, A., Nogami, M., Yokozeki, T., Yamazaki, M., Nakamura, H., Watanabe, H., Kawamoto, K., Nakayama, K., Morris, A. J., Frohman, M. A., and Kanaho, Y. (1999). Phosphatidylinositol 4-phosphate 5-kinase α is a downstream effector of the small G protein ARF6 in membrane ruffle formation. *Cell* **99,** 521–532.

Hussain, N. K., Jenna, S., Glogauer, M., Quinn, C. C., Wasiak, S., Guipponi, M., Antonarakis, S. E., Kay, B. K., Stossel, T. P., Lamarche-Vane, N., and McPherson, P. S. (2001). Endocytic protein intersectin-l regulates actin assembly via Cdc42 and N-WASP. *Nat. Cell Biol.* **3,** 927–932.

Kahn, R. A. (2003). The Arf family. *In* "ARF Family GTPases" (R. Kahn, ed.), pp. vii–xxiii. Kluwer Academic Publishers, Dordrecht, Netherlands.

King, F. J., Hu, E., Harris, D. F., Sarraf, P., Spiegelman, B. M., and Roberts, T. M. (1999). DEF-1, a novel Src SH3 binding protein that promotes adipogenesis in fibroblastic cell lines. *Mol. Cell. Biol.* **19,** 2330–2337.

Kondo, A., Hashimoto, S., Yano, H., Nagayama, K., Mazaki, Y., and Sabe, H. (2000). A new paxillin-binding protein, PAG3/Papα/KIAA0400, bearing an ADP-ribosylation factor GTPase-activating protein activity, is involved in paxillin recruitment to focal adhesions and cell migration. *Mol. Biol. Cell* **11,** 1315–1327.

Mazaki, Y., Hashimoto, S., Okawa, K., Tsubouchi, A., Nakamura, K., Yagi, R., Yano, H., Kondo, A., Iwamatsu, A., Mizoguchi, A., and Sabe, H. (2001). An ADP-ribosylation factor GTPase-activating protein Git2-short/KIAA0148 is involved in subcellular localization of paxillin and actin cytoskeletal organization. *Mol. Biol. Cell* **12,** 645–662.

Moss, J., and Vaughan, M. (1998). Molecules in the ARF orbit. *J. Biol. Chem.* **273,** 21431–21434.

Onodera, Y., Hashimoto, S., Hashimoto, A., Morishige, M., Mazaki, Y., Yamada, A., Ogawa, E., Adachi, M., Sakurai, T., Manabe, T., Wada, H., Matsuura, N., and Sabe, H. (2005).

Expression of AMAP1, an ArfGAP, provides novel targets to inhibit breast cancer invasive activities. *EMBO J.* **24**, 963–973.

Owen, D. J., Wigge, P., Vallis, Y., Moore, J. D., Evans, P. R., and McMahon, H. T. (1998). Crystal structure of the amphiphysin-2 SH3 domain and its role in the prevention of dynamin ring formation. *EMBO J.* **17**, 5273–5285.

Palacios, F., Price, L., Schweitzer, J., Collard, J. G., and D'Souza-Schorey, C. (2001). An essential role for ARF6-regulated membrane traffic in adherens junction turnover and epithelial cell migration. *EMBO J.* **20**, 4973–4986.

Palacios, F., Schweitzer, J. K., Boshans, R. L., and D'Souza-Schorey, C. (2002). ARF6-GTP recruits Nm23-H1 to facilitate dynamin-mediated endocytosis during adherens junctions disassembly. *Nat. Cell Biol.* **4**, 929–936.

Radhakrishna, H., Klausner, R. D., and Donaldson, J. G. (1996). Aluminum fluoride stimulates surface protrusions in cells overexpressing the ARF6 GTPase. *J. Cell Biol.* **134**, 935–947.

Radhakrishna, H., and Donaldson, J. G. (1997). ADP-ribosylation factor 6 regulates a novel plasma membrane recycling pathway. *J. Cell Biol.* **139**, 49–61.

Roth, M. G. (2000). Arf. *In* "GTPases" (A. Hall, ed.), pp. 176–197. Oxford University Press, New York.

Sabe, H. (2003a). Paxillin-associated ARF GAPs. *In* "ARF Family GTPases" (R. Kahn, ed.), pp. 185–207. Kluwer Academic Publishers, Dordrecht, Netherlands.

Sabe, H. (2003b). Requirement for Arf6 in cell adhesion, migration, and cancer cell invasion. *J. Biochem. (Tokyo)* **134**, 485–489.

Tague, S. E., Muralidharan, V., and D'Souza-Schorey, C. (2004). ADP-ribosylation factor 6 regulates tumor cell invasion through the activation of the MEK/ERK signaling pathway. *Proc. Natl. Acad. Sci. USA* **101**, 9671–9676.

Uchida, H., Kondo, A., Yoshimura, Y., Mazaki, Y., and Sabe, H. (2001). PAG3/Papα/KIAA0400, a GTPase-activating protein for ADP-ribosylation factor (ARF), regulates ARF6 in Fcγ receptor-mediated phagocytosis of macrophages. *J. Exp. Med.* **193**, 955–966.

Vitale, N., Chasserot-Golaz, S., Bailly, Y., Morinaga, N., Frohman, M. A., and Bader, M. F. (2002). Calcium-regulated exocytosis of dense-core vesicles requires the activation of ADP-ribosylation factor (ARF) 6 by ARF nucleotide binding site opener at the plasma membrane. *J. Cell Biol.* **159**, 79–89.

Wigge, P., Vallis, Y., and McMahon, H. T. (1997). Inhibition of receptor-mediated endocytosis by the amphiphysin SH3 domain. *Curr. Biol.* **7**, 554–560.

Zhang, Q., Cox, D., Tseng, C. C., Donaldson, J. G., and Greenberg, S. (1998). A requirement for ARF6 in Fcγ receptor-mediated phagocytosis in macrophages. *J. Biol. Chem.* **273**, 19977–19981.

[22] Functional Assay of EFA6A, a Guanine Nucleotide Exchange Factor for ADP-Ribosylation Factor 6 (ARF6), in Dendritic Formation of Hippocampal Neurons

By Hiroyuki Sakagami, Akifumi Kamata,
Kohji Fukunaga, and Hisatake Kondo

Abstract

EFA6A is a guanine nucleotide exchange factor (GEF) that can activate ADP-ribosylation factor 6 (ARF6) *in vitro*, with prominent expression in the forebrain including the hippocampal formation. In this section, we describe the neuronal transfection method and show that the overexpression of a catalytically inactive mutant of EFA6A induces a prominent dendritic formation of the primary hippocampal neurons, suggesting the intimate involvement of EFA6A in the regulation of neuronal dendritic development. This reliable and consistent neuronal transfection method will also be applicable for the vector-based RNA interference method.

Introduction

Neuronal dendrites provide the sites to receive and integrate presynaptic inputs from axons. Their arborization patterns dramatically vary among neurons and define individual neuronal characters and specificities. The dendritic development includes several steps such as the initiation of the process extension, the dendritic outgrowth, branching, refinement, and spine formation (Scott and Luo, 2001). Possible mechanisms that govern the dendritic formation are thought to involve the regulation of the local cytoskeleton dynamics and the delivery of proteins and membrane to newly formed dendrites. In line with this notion, increasing evidence has accumulated that various small GTPases, especially the Rho family, are involved in the dendritic formation through the regulation of the cytoskeleton dynamics (Scott and Luo, 2001).

Like the Rho family, ADP ribosylation factor 6 (ARF6) has recently been in the spotlight because of its functional involvement in the regulation of cytoskeleton dynamics as well as endocytosis and membrane recycling based on the findings mainly from nonneuronal cell lines (Donaldson, 2003). The activity of ARF6 is strictly regulated by two factors: guanine nucleotide exchange factors (GEFs) and GTPase activating proteins

METHODS IN ENZYMOLOGY, VOL. 404
0076-6879/05 $35.00
DOI: 10.1016/S0076-6879(05)04022-X

(GAPs). GEFs activate ARF6 by facilitating the exchange of the bound GDP for GTP, while GAPs inactivate ARF6 by enhancing the intrinsic hydrolysis of the bound GTP to GDP. Possible downstream effectors of ARF6 include phosphatidylinositol 4-phosphate 5-kinase α, phospholipase D, Rac1, Arfaptin 2/POR1 (partner of Rac1), arfophilin, and a nucleoside diphosphate kinase (Nm23-H1) (Brown et al., 1993; Cockcroft et al., 1994; D'Souza-Schorey et al., 1997; Palacios et al., 2002; Santy and Casanova, 2001; Shin et al., 2001). Several GEFs including EFA6, cytohesin-1, ARNO, GRP1, and ARF-GEP100 have been identified to exhibit the ability to activate ARF6 (Derrien et al., 2002; Franco et al., 1999; Frank et al., 1998; Knorr et al., 2000; Langille et al., 1999; Someya et al., 2001). Among these, the EFA6 family is a new member of GEFs that exhibit preferential substrate specificity toward ARF6 in vitro. Recent cDNA cloning has revealed the existence of at least four closely related isoforms (termed as EFA6A–D) in the EFA6 family (Derrien et al., 2002; Franco et al., 1999). Our study has recently focused on possible functional involvement of the EFA6 family, especially EFA6A, in the dendritic formation of the hippocampal neurons because of its following attractive reasons (Sakagami et al., 2004; Suzuki et al., 2002). First, EFA6A mRNA is expressed most abundantly in the hippocampal neurons (Sakagami et al., 2004; Suzuki et al., 2002), which are known to exhibit a remarkable capacity for activity-dependent changes in synaptic structures and functions related to long-term potentiation and depression. In addition, the primary hippocampal culture is a relatively homogenous source of neuronal population and a very useful system to study the mechanisms of the neuronal morphogenesis including the establishment of polarity and the formation of axons and dendrites (Craig and Banker, 1994). Secondly, ARF6 and ARNO have recently been shown to regulate the dendritic development of hippocampal neurons (Hernandez-Deviez et al., 2002). Thirdly, the expression of EFA6A mRNA is dramatically increased in the hippocampal formation during early postnatal development, when the formation of the hippocampal dendrites is known to be most active (Pokorny and Yamamoto, 1981). Finally, EFA6A mRNA is localized at high levels not only in the cell bodies but also in the dendrites of hippocampal neurons (Sakagami et al., 2004). This characteristic somatodendritic localization of EFA6A mRNA enables EFA6A to be translated in the dendrites immediately in response to synaptic inputs, suggesting the prompt requirement of EFA6A for activity-dependent dendritic functions.

In this chapter, we describe a method to examine the involvement of EFA6A in the dendritic formation by transfecting its catalytically inactive mutant into the primary hippocampal neurons.

Methods

Construction of Expression Vectors

pCAGGS vector is a mammalian expression vector with a ubiquitously strong promoter activity based on a chicken β-actin promoter (Niwa *et al.*, 1991). To trace the exogenously expressed gene product, we constructed pCAGGS-FLAG vector that can express an amino-terminal FLAG epitope-tagged protein by inserting annealed oligonucletides containing the sequences of FLAG epitope, EcoRI, XhoI, and EcoRV restriction enzyme sites to EcoRI site of pCAGGS vector provided by Dr. J. Miyazaki (Osaka University) (Fig. 1A). The coding region of rat EFA6A was amplified by PCR with the rat cDNA cloned by us (Suzuki *et al.*, 2002) as a template and a combination of primers that create EcoRI sites (underlined) at both ends: sense 5'TGAATTCATGCCTCACTCTGGGCTCCTCAAGTC3' and antisense 5'AGAATTCTCAGGGCTTTGGCCTTGTACTGCCTG3'. The EcoRI-digested PCR fragment was subcloned to the EcoRI site of pCAGGS-FLAG. The catalytically inactive mutant of EFA6A (pCAGGS-FLAG-EFA6A(E246K)) was created by substituting glutamate for lysine at amino acid residue 246 in the Sec7 domain of rat EFA6A (Fig. 1B), which is equivalent to amino acid residue 242 of human EFA6A and is critical for the GEF activity (Franco *et al.*, 1999), by using GeneEditor *in vitro* site-directed mutagenesis system (Promega, Madison, WI, USA) with the primer 5'GGCTTTAATGGGTAAGACCCAGGAGCGGG3'. All the inserts were confirmed to be valid by sequencing. The loss of GEF activity of rat EFA6A(E246K) toward ARF6 was confirmed by ARF pull down assay with GGA1 (Golgi-localized, γ-ear-containing ARF-binding protein 1) (unpublished observation by Sakagami).

pCS2-Venus vector developed by Dr. A. Miyawaki (RIKEN, Saitama, Japan) carries a green fluorescent protein (GFP) variant, which is much more stable and enhanced fluorescent than the original GFP, under a cytomegalovirus promoter (Nagai *et al.*, 2002). This vector is cotransfected with a given expression vector to trace the overall dendritic morphology of transfected neurons.

All the expression vectors for neuronal transfection were purified by using EndoFree Plasmid Maxi kit (#12362, Qiagen, Germantown, MD, USA) and dissolved in sterile distilled water at a concentration of 1 $\mu g/\mu l$. Although the plasmid DNA for neuronal transfection was traditionally recommended to be purified by two rounds of CsCl gradients, this purification kit yields plasmids satisfactory for neuronal transfection without any problems such as neuronal toxicity.

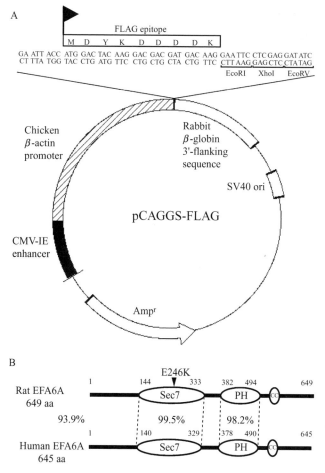

FIG. 1. Plasmid map of pCAGGS-FLAG (A) and the domain structure of rat and human EFA6A (B). (A) Plasmid map of pCAGGS-FLAG vector. The annealed oligonucleotides encoding FLAG epitope, EcoRI, XhoI, and EcoRV restriction enzyme sites were inserted to EcoRI site at the downstream of chicken β-actin promoter of pCAGGS vector. The coding region of rat EFA6A was subcloned to EcoRI site of pCAGGS-FLAG vector in the same reading frame as FLAG epitope. (B) Comparison of domain structure between rat and human EFA6A. The position of the amino acid mutated for a catalytically inactive mutant of rat EFA6A used in this study is shown by an arrowhead. Note that amino acid identities of EFA6A between rat and human are 93.9, 99.5, and 98.2% in the entire coding region, Sec7 domain, and PH domain, respectively. cc, coiled-coil motif.

Primary Hippocampal Neuronal Culture

Materials and Reagents

Cell Desk (Sumitomo Bakelite, Co., Ltd., Tokyo, Japan, #MS92132): Cell Desk is a 35 mm culture plastic dish containing a thin low autofluorescent plastic coverslip on which cells can be cultured. The viability and adherence of the cultured neurons on this coverslip is much better compared with those on the glass coverslip.

1 mg/ml laminin solution: Dissolve a vial of laminin (354232, 1mg, BD Biosciences, Bedford, MA, USA) with sterile water. Make aliquots (100 μl) of stock solution (1 mg/ml) and store them at $-80°$.

1 mg/ml poly-D-lysine solution (1 mg/ml): Dissolve a vial of poly-D-lysine solution hydrobromide (354210, 20 mg, BD Biosciences, Bedford, MA, USA) with 20 ml of sterile water. Make aliquots (1 ml) and store them at $-80°$.

Dissociation medium (DM), pH 7.4: 40.9 mM Na$_2$SO$_4$, 30 mM K$_2$SO$_4$, 5.8 mM MgCL$_2$, 0.252 mM CaCL$_2$, 1 mM HEPES, 20 mM glucose, 0.001% Phenol Red. Store at $4°$.

$10 \times$ Kynurenic acid/MgCL$_{12}$ solution ($10 \times$ Ky-Mg): 10 mM Kynurenic acid, 100 mM MgCL$_2$, 0.003% Phenol Red.

DM/Ky-Mg solution: Immediately before dissection, mix Ky-Mg to DM to a final of $\times$1.

Papain solution: Add 4.5 mg of L-cystein hydrochloride (Sigma, C788) to 10 ml of DM/Ky-Mg solution and adjust pH to 7.4 with 1 N NaOH. Add 150 unit of papain latex (Worthington, Freehold, NJ) and solubilize it completely in $37°$ water bath for approximately 5 min. Filter, sterilize, and keep it on ice until use.

Trypsin inhibitor solution: Add 100 mg of trypsin inhibitor (Sigma T9253, type II-O) to 10 ml of DM/Ky-Mg solution. Adjust pH to 7.4 with 1N NaOH. Filter, sterilize, and keep it on ice until use.

Neuronal maintenance medium: Neurobasal medium (GIBCO, Grand Island, NY), $1 \times$ B27 (GIBCO, Grand Island, NY), 1 mM glutamine (GIBCO, Grand Island, NY).

Neuronal plating medium: OptiMEM-1 reduced-serum medium (GIBCO, #31985), 20 mM glucose.

Procedure

1. One day before culture, coat culture dishes (Cell Desk) with 2 ml of laminin and poly-D-lysine solution diluted with sterile distilled water at a final concentration of 5 μg/ml and 50 μg/ml, respectively, overnight in

CO_2 incubator. After overnight incubation, wash the dishes three times with sterile distilled water and dry them in the hood. The coated plates can also be stored at $4°$ until use.

2. Under deep ether anesthesia, sacrifice a pregnant Wistar rat at gestational days 18–20 and remove the uterus. After dipping it in 70% ethanol, place it in a sterile 100-mm Petri dish and transfer it inside a culture hood.

3. Remove the brains from fetuses and collect them in DM/Ky-Mg solution in a 60-mm Petri dish. Under a dissecting microscope, carefully dissect out the hippocampus that lines the curved medial edge of the cerebral cortex. Remove the meninges and choroid plexes as much as possible. Collect the hippocampi to a 35-mm dish containing DM/Ky-Mg solution.

4. Transfer the hippocampi to a 50-ml falcon tube and rinse them three times with DM/Ky-Mg solution. After adding 5 ml of papain solution, incubate them at $37°$ for 15 min while gently mixing every 5 min. Remove the solution carefully with a disposable pipette, add another 5 ml of papain solution into the tube, and incubate them at $37°$ for 15 min.

5. Remove the papain solution and wash three times with 5 ml of DM/Ky-Mg solution.

6. Add 3 ml of trypsin inhibitor solution and incubate the specimens for 3 min at room temperature. Mix occasionally, let hippocampi settle down to the bottom of the tube, and remove the supernatant. Repeat this step two more times.

7. Wash the specimens three times with 5 ml of neuronal plating medium.

8. Triturate the hippocampi completely in 10 ml of neuronal plating medium by pipetting up and down approximately 30 times with a 10-ml disposable pipette. To remove chunks of tissues, pass the triturated solution through a 100-μm filter (Cell Strainer, #352360, BD Biosciences, Bedford, MA, USA) and determine the cell density with a hemocytomer.

9. Plate the cells onto a laminin- and poly-D-lysine-coated culture dish at a density of 2×10^6 cells in 2 ml of OptiMEM I/glucose solution. Allow the cells to adhere to the dish for 1–2 h in CO_2 incubator ($37°$, 5% CO_2). After 1–2 h, exchange the neuronal plating medium with 3 ml of the neuronal maintenance medium and incubate the plates in CO_2 incubator. In case the culture is maintained for more than one week, half of the medium is exchanged with the fresh culture medium every five days after plating.

Transfection with Lipofectamin 2000

To examine the role of EFA6A in the dendritic formation, especially dendritic outgrowth and branching, the transfection is usually carried out at 2 days *in vitro* (2 DIV) as follows.

1. Dilute 2 μg of pCAGGS-FLAG-EFA6A(E246K) or empty pCAGGS-FLAG together with 2 μg of pCS2-Venus in 100 μl of OptiMEM-1.
2. In a separate 1.5 ml-microtube, dilute 4 μl of Lipofectamine 2000 (LF2000) (Invitrogen, Carlsbad, CA, #11668, 1 mg/ml) in 100 μl of OptiMEM-1 by mixing it gently. Incubate for 5 min at room temperature.
3. Combine the diluted DNA and LF2000 solutions and incubate the mixture for 20 min at room temperature.
4. After 20-min incubation, add DNA-LF2000 complex drop-wise over the plate and distribute the complex evenly in the medium by moving the plate gently back and forth.

Immunostaining

Materials and Reagents

1M sodium phosphate buffer (PB, pH 7.4): Prepare 1M solution by adjusting pH to 7.4 with 1 M monobasic sodium phosphate and 1 M dibasic sodium phosphate.

1X phosphate buffered saline (PBS): 100 mM PB, 150 mM NaCl.

4% paraformaldehyde/0.1 M PB: Dissolve 8 g of paraformaldehyde in 100 ml of distilled water warmed at 50–60° by adding a few drops of 5N NaOH. Once in solution, add 100 ml of 0.2 M PB.

Procedure. Seven days after transfection (9 DIV), the cells were fixed in 4% paraformaldehyde in 0.1 M PB for 10 min at room temperature. They were washed five times with PBS and subsequently permeabilized in PBS containing 0.3% Triton × 100 for 15 min. After permeabilization, the cells were blocked in 5% normal goat serum in PBS containing 0.05% sodium azide for 30 min. They were incubated overnight at 4° with anti-FLAG monoclonal antibody (Sigma, M2, final concentration; 1 μg/ml) and rabbit polyclonal anti-GFP serum (1:5000). The antiserum against GFP was prepared by immunizing a rabbit with the glutathione S transferase (GST)–enhanced GFP fusion protein. After overnight incubation, the cells were washed six times with PBS for a total of 30 min and incubated with Alexa594-conjugated anti-mouse IgG (Molecular Probes, Inc., Eugene, OR; 1:2000) and Alexa 488-conjugated anti-rabbit IgG (Molecular Probes,

Inc. 1:2000) for 1 h at room temperature. The cells were washed with PBS and counterstained with DAPI (4',6-diamidino-2-phenylindole dihydrochloride) to visualize the nuclear morphology. The immunoreactive images were acquired by a confocal laser scanning microscopy (LSM5 PASCAL, Carl Zeiss, Jend, Germany). The dendritic morphology was analyzed with the aid of Neurocyte Image Analyzer Software (Kurabo, Osaka, Japan).

Results and Discussion

Various techniques for neuronal transfection have been reported including viral transfer, DNA/calcium-phosphate coprecipitation, electroporation, microinjection, biolistics, and liposomes. Although the ideal transfection method has not been developed, we describe a neuronal transfection method with LF2000, a commercially available cationic lipid reagent. This method yields consistent transfection efficiency without any time consumption, experienced culture technique, or cell toxicity.

The critical point of this lipofection is the reduced ratio of LF2000 to DNA by 1:1. According to the manufacturer's protocol, the ratio of LF2000 to DNA is recommended to be 2:1 to 3:1, which induces the neuronal cell death at a high level in our experience. Figure 2 shows the representative transfection efficiency of a vector encoding Venus, an enhanced variant of GFP, into the primary hippocampal neurons with LF2000 at various time points after plating. The maximal transfection efficiency was obtained in the primary hippocampal neurons at 2 DIV. The transfection efficiency ranges from 0.01% to 3% depending on plasmids. Although the transfection efficiency dramatically drops to $0.01 \sim 0.1\%$ after 5 DIV, it gives enough transfected cells to analyze the morphology of their dendrites even at 20 DIV.

By using this lipofection method, we transfected the plasmid encoding a catalytically inactive mutant of EFA6A (EFA6A(E246K)) or an empty vector together with pCS2-Venus into the primary hippocampal neurons at 2 DIV and examined its effect on the dendritic formation at 7 days after transfection (Fig. 3). In the control, the average number of the tips of the dendrites was approximately 15. In contrast, the overexpression of a catalytically inactive mutant of EFA6A increased the numbers of the tips of dendritic processes by 2-fold compared with those transfected with an empty vector. The enhancement of the dendritic formation by a catalytically inactive mutant of EFA6A is consistent with the previous finding that the overexpression of dominant negative mutants of ARF6 and ARNO induced prominent dendritic formation (Hernandez-Deviez et al., 2002), suggesting that ARF6 pathway may regulate the dendritic formation

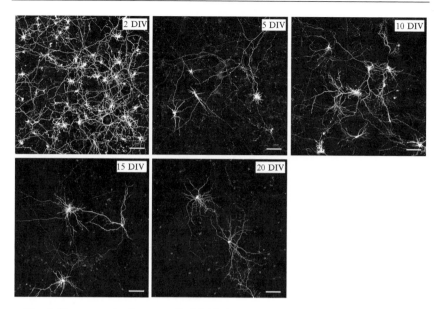

FIG. 2. Representative low-magnified (x10) figures showing the efficiency of transfection into the primary hippocampal neurons with LF2000 at various time points. The primary hippocampal neurons were transfected with pCS2-Venus (4 μg) by using LF2000 at 2, 5, 10, 15, and 20 DIV. The cells were fixed at 20–24 h after transfection and immunostained with antiGFP antibody. Note the maximal transfection efficiency at DIV2. Scale bars = 100 μm. (See color insert.)

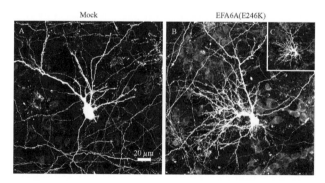

FIG. 3. Representative dendritic morphology of primary hippocampal neurons transfected with a catalytically inactive mutant of EFA6A. The hippocampal neurons were transfected with either an empty pCAGGS-FLAG (A) or pCAGGS-FLAGEFA6A(E246K) (B, C) together with pCS2-Venus at 2 DIV. At DIV 9, the cells were fixed and immunostained with anti-GFP (A, B) and anti-FLAG (C) antibodies. Note that the overexpression of a catalytically inactive mutant of EFA6A induced prominent dendritic formation compared with that of an empty vector (Mock). Scale bars = 20 μm. (See color insert.)

through EFA6A and ARNO. Although the mechanism by which a catalytically inactive mutant works as a dominant negative mutant remains to be characterized, it is possible that the inactive mutant can block EFA6A-ARF6 pathway by sequestering ARF6 or by competing the proper subcellular localization with intrinsic EFA6A. It should be noted that the carboxyl-terminal region of EFA6A containing a coiled-coil motif has been shown to exhibit the ability to regulate actin cytoskeleton probably through activation of Rac1 pathway independently of ARF6 pathway (Franco *et al.*, 1999). To make our findings conclusive, further studies are necessary by knocking down the intrinsic expression of EFA6A with RNA interference (RNAi) method. The neuronal transfection method that we describe here will be easily applicable for the vector-based RNAi methods.

Acknowledgments

The authors would like to thank Dr. J. Miyazaki (Osaka University Medical School) for kindly providing pCAGGS vector and Dr. A. Miyawaki (RIKEN) for pCS2-Venus vector. This work is supported by Grants-in-Aid for Young Scientists (#15700269) and for Scientific Research on Priority Areas (#16015215) to H.S. from the Ministry of Education, Science, Sports, Culture and Technology of Japan.

References

Brown, H. A., Gutowski, S., Moomaw, C. R., Slaughter, C., and Sternweis, P. C. (1993). ADPribosylation factor, a small GTPdependent regulatory protein, stimulates phospholipase D activity. *Cell* **75**, 1137–1144.

Cockcroft, S., Thomas, G. M., Fensome, A., Geny, B., Cunningham, E., Gout, I., Hiles, I., Totty, N. F., Truong, O., and Hsuan, J. J. (1994). Phospholipase D: A downstream effector of ARF in granulocytes. *Science* **263**, 523–526.

Craig, A. M., and Banker, G. (1994). Neuronal polarity. *Annu. Rev. Neurosci.* **17**, 267–310.

D'Souza-Schorey, C., Boshans, R. L., McDonough, M., Stahl, P. D., and Van Aelst, L. (1997). A role for POR1, a Rac1-interacting protein, in ARF6-mediated cytoskeletal rearrangements. *EMBO J.* **16**, 5445–5454.

Derrien, V., Couillault, C., Franco, M., Martineau, S., Montcourrier, P., Houlgatte, R., and Chavrier, P. (2002). A conserved C-terminal domain of EFA6-family ARF6-guanine nucleotide exchange factors induces lengthening of microvilli-like membrane protrusions. *J. Cell. Sci.* **115**, 2867–2879.

Donaldson, J. G. (2003). Multiple roles for Arf6: Sorting, structuring, and signaling at the plasma membrane. *J. Biol. Chem.* **278**, 41573–41576.

Franco, M., Peters, P. J., Boretto, J., van Donselaar, E., Neri, A., D'Souza-Schorey, C., and Chavrier, P. (1999). EFA6, a sec7 domain-containing exchange factor for ARF6, coordinates membrane recycling and actin cytoskeleton organization. *EMBO J.* **18**, 1480–1491.

Frank, S., Upender, S., Hansen, S. H., and Casanova, J. E. (1998). ARNO is a guanine nucleotide exchange factor for ADP-ribosylation factor 6. *J. Biol. Chem.* **273**, 23–27.

Hernandez-Deviez, D. J., Casanova, J. E., and Wilson, J. M. (2002). Regulation of dendritic development by the ARF exchange factor ARNO. *Nat. Neurosci.* **5**, 623–624.

Knorr, T., Nagel, W., and Kolanus, W. (2000). Phosphoinositides determine specificity of the guanine-nucleotide exchange activity of cytohesin-1 for ADP-ribosylation factors derived from a mammalian expression system. *Eur. J. Biochem.* **267**, 3784–3791.

Langille, S. E., Patki, V., Klarlund, J. K., Buxton, J. M., Holik, J. J., Chawla, A., Corvera, S., and Czech, M. P. (1999). ADP-ribosylation factor 6 as a target of guanine nucleotide exchange factor GRP1. *J. Biol. Chem.* **274**, 27099–27104.

Niwa, H., Yamamura, K., and Miyazaki, J. (1991). Efficient selection for high-expression transfectants with a novel eukaryotic vector. *Gene* **108**, 193–199.

Palacios, F., Schweitzer, J. K., Boshans, R. L., and D'Souza-Schorey, C. (2002). ARF6-GTP recruits Nm23-H1 to facilitate dynamin-mediated endocytosis during adherens junctions disassembly. *Nat. Cell Biol.* **4**, 929–936.

Pokorny, J., and Yamamoto, T. (1981). Postnatal ontogenesis of hippocampal CA1 area in rats. I. Development of dendritic arborisation in pyramidal neurons. *Brain Res. Bull.* **7**, 113–120.

Sakagami, H., Matsuya, S., Nishimura, H., Suzuki, R., and Kondo, H. (2004). Somatodendritic localization of the mRNA for EFA6A, a guanine nucleotide exchange protein for ARF6, in rat hippocampus and its involvement in dendritic formation. *Eur. J. Neurosci.* **19**, 863–870.

Santy, L. C., and Casanova, J. E. (2001). Activation of ARF6 by ARNO stimulates epithelial cell migration through downstream activation of both Rac1 and phospholipase D. *J. Cell. Biol.* **154**, 599–610.

Scott, E. K., and Luo, L. (2001). How do dendrites take their shape? *Nat. Neurosci.* **4**, 359–365.

Shin, O. H., Couvillon, A. D., and Exton, J. H. (2001). Arfophilin is a common target of both class II and class III ADP-ribosylation factors. *Biochemistry* **40**, 10846–10852.

Someya, A., Sata, M., Takeda, K., Pacheco Rodriguez, G., Ferrans, V. J., Moss, J., and Vaughan, M. (2001). ARF-GEP(100), a guanine nucleotide-exchange protein for ADP-ribosylation factor 6. *Proc. Natl. Acad. Sci. USA* **98**, 2413–2418.

Suzuki, I., Owada, Y., Suzuki, R., Yoshimoto, T., and Kondo, H. (2002). Localization of mRNAs for subfamily of guanine nucleotide-exchange proteins (GEP) for ARFs (ADP-ribosylation factors) in the brain of developing and mature rats under normal and postaxotomy conditions. *Brain Res. Mol. Brain Res.* **98**, 41–50.

[23] Functional Assay of ARNO and ARF6 in Neurite Elongation and Branching

By DELIA J. HERNÁNDEZ-DEVIEZ and JEAN M. WILSON

Abstract

During development of the nervous system, neurite outgrowth is necessary for the formation of connections between nerve cells. Neurons are highly polarized cells that send out distinct processes, axons, and dendrites; however, the molecular regulation of the differential growth of these processes remains incompletely understood. Primary cultures of rat hippocampal neurons have been used to study many aspects of neuronal cell

METHODS IN ENZYMOLOGY, VOL. 404
Copyright 2005, Elsevier Inc. All rights reserved.

0076-6879/05 $35.00
DOI: 10.1016/S0076-6879(05)04023-1

biology, including neurite extension, establishment of polarity, biogenesis of synapses, and membrane trafficking. After attachment to the substrate, hippocampal neurons begin sending out multiple processes by approximately 12 h after plating. The axonal process is derived from one of these processes, and is evident after 48 h in culture. Complete polarity of axons and dendrites is established after 7 days in culture. The establishment of these cultures and the ability to transfect them with potential regulatory genes allows the researcher to dissect out the pathways relevant to neurite extension. To study the role of small GTPases in neurite extension and branching, we describe methods for culture of hippocampal neurons, for transfection of these cells, and assessment of neurite extension and branching.

Introduction

Neurons have axonal and dendritic processes that define their unique morphology and function. These processes contain specialized sites to receive and send out information, which are critical for proper nervous system wiring. Controlled neurite elongation and branching are developmental events crucial for neurons to acquire their mature functional morphology. These events must be tightly regulated, and the molecular machinery involved in these processes is starting to be unraveled. An array of molecules has been shown to regulate neurite extension and branching, including cell surface receptors and ligands (Lin *et al.*, 2005; Whitford *et al.*, 2002), Rho GTPases and their effectors (Govek *et al.*, 2005; Luo, 2000), and lipid modifying enzymes (Hernandez-Deviez *et al.*, 2004; van Horck *et al.*, 2002; Yamazaki *et al.*, 2002). In addition, the Arf family of small GTPases has been implicated in membrane trafficking (Palacios *et al.*, 2001; Radhakrishna and Donaldson, 1997; Radhakrishna *et al.*, 1999) as well as reorganization of the cytoskeleton in non-neuronal cells (Boshans *et al.*, 2000; Frank *et al.*, 1998; Radhakrishna *et al.*, 1996), and we have shown that ARF6 and its exchange factor ARNO regulate dendrite branching and axonal elongation and branching in cultured rat hippocampal neurons (Hernandez-Deviez *et al.*, 2002, 2004).

The hippocampus is a source of a homogeneous population of pyramidal neurons. The sequence of developmental events that cultured hippocampal neurons follow to acquire their mature morphology has been previously described (Dotti *et al.*, 1988), and Banker and Goslin (1991) have described in detail the preparation of primary culture of rat hippocampal neurons. Here we present a description of the methods we employed to study the role of the small GTPase ARF6 and its exchange factor ARNO during neurite elongation and branching.

In our studies of neurite elongation and branching *in vitro*, hippocampi from embryonic day 17–18 rats are used as the source of neurons. These dissociated hippocampal cells were plated on coated coverslips, transfected with the plasmids of interest, and the effects upon neurite extension and branching quantified.

Culture of Rat Hippocampal Neurons

Successful culture of hippocampal neurons for transfection requires that the coverslip substrates be properly prepared to promote adhesion as well as for survival during the transfection process.

Coverslip Preparation

Neurons do not attach, mature, or survive properly when grown directly on plastic tissue culture dishes. In addition, immunofluorescence imaging and quantification can only be carried out easily on cells grown on coverslips. Therefore, we plate hippocampal neurons onto glass coverslips that have been cleaned, sterilized, and coated with poly-l-lysine to improved cell adhesion (Banker, 1991).

Coverslips are prepared as follows:

1. Place 12 mm round coverslips (Fisher Scientific, Pittsburg, PA: 12 mm circles, Cat. No. 12–545–80) into a 100 mm glass Petri dish.
2. Add 5 ml of concentrated nitric acid (HNO_3) and leave for 1–2 days.
3. Wash twice for 1 h and twice for 30 min with 15 ml of tissue culture grade water. Remove excess water and dry in an oven for 1 h at 60°.
4. Transfer the coverslips to 100 mm microbiological dishes and sterilize by UV irradiation for 30 min in a laminar flow hood. At this point coverslips can be stored under sterile conditions until the day before the dissection when they should be coated with poly-L-lysine.

We found that cells adhere better on freshly coated coverslips. The day before dissection, coat each coverslip with 1–2 drops of a sterile 1 mg/ml solution of poly-L-lysine (MW 30,000–70,000, Sigma Chemical Co., St. Louis, MO, Cat. No. P2636) dissolved in 0.1 M borate buffer, pH 8.5, and leave overnight. The dish should be sealed with parafilm to avoid evaporation. From this step onwards, avoid exposure to UV light, as cells plated on UV-irradiated poly-lysine do not develop well. Under sterile conditions, remove the poly-L-lysine and wash the coverslips two times for two h each with tissue culture grade water. Remove the final rinse and transfer the coverslips to 60 mm tissue culture dishes (about 10 coverslips per dish) and add 4 ml of Neurobasal media (Invitrogen Corporation, Carlsbad, CA,

GIBCO Cat. No. 21103–049) supplemented with B27-supplement (GIBCO Cat. No. 17504–044). These dishes are stored in the incubator at 37° under 5% CO_2, until seeding of cells.

Establishment of Hippocampal Cell Cultures

Hippocampal neurons can be grown in Neurobasal media with B27 supplement (as noted previously) or in glial-conditioned Neurobasal media. Glial cells are a source of neurotrophic factors important for neuronal cell differentiation (Banker, 1980; Booher and Sensenbrenner, 1972) and many groups have successfully cultured hippocampal neurons using glial support (Dotti *et al.*, 1988; Esch *et al.*, 2000; Jareb and Banker, 1997). Alternatively, commercially produced media are now available that provide sufficient neurotrophic factors for growth and differentiation. In our hands, we do not experience any differences in neuronal growth using either glial conditioned media or commercially available media. In fact, neuronal cultures became contaminated with bacteria more frequently when cocultured with glial cells or grown in glial conditioned media. In addition we found that Neurobasal media supplemented with B27 provides all the necessary factors for successfully culturing and transfecting hippocampal neurons, so herein we describe culture conditions in the absence of glial cell support.

In large part, we utilize the dissection and culturing techniques pioneered by Banker and colleagues (Banker and Goslin; Banker and Cowan, 1977), and it is worthwhile consulting their work before embarking upon the culture of hippocampal neurons.

The following protocol should provide 50 coverslips for transfection. To prepare primary cultures of hippocampal neurons:

1. Dissect hippocampi from 10 E17–18 timed pregnant Sprague-Dawley rats (Banker and Goslin, 1998). The resulting 20 hippocampi are transferred to a 15 ml tube with 4.5 ml of Ca^{++}/Mg^{++} free HBSS containing 10 mM Hepes buffer (pH 7.4), 10 mM glucose, penicillin (100 U/ml), and streptomycin (100 μg/ml).

2. Add 0.5 ml of 0.05% trypsin/0.53 mM EDTA and incubate for 15 min at 37°. Others have used 2.5% trypsin; however, we find that using a lower concentration of trypsin works as well and results in improved cell viability.

3. Remove the trypsin solution and wash twice for 5 min with 5 ml of Ca^{++}/Mg^{++} free HBSS. Bring the volume to 5 ml with Ca^{++}/Mg^{++} free HBSS and carefully dissociate the cells by passing them first through a regular Paster pipette (5 times) and then through a fire-polished Paster pipette (one half of the diameter) approximately 10 times.

4. Determine the cell density using a hemocytometer, and plate approximately 350,000 cells per 60 mm culture dish containing 10 coverslips in Neurobasal media plus B27 supplement.

5. These cells should be fed twice a week with Neurobasal media with B27 supplement. In higher density cultures (> 800,000 cells per dish), endogenous glial cells can proliferate and overwhelm the cultures making the visualization of individual neurons and neuronal processes difficult. To avoid this, proliferation can be inhibited with an antimitotic agent (0.005 mM cytosine arabinoside). However, we experience little glial growth and therefore do not include cytosine arabinoside in our cultures, since moderate glial proliferation does provide neurotrophic factors.

Transfection of Primary Neuronal Cultures with ARF6 and ARNO Plasmids

Primary culture cells have proven to be very difficult to transfect and hippocampal neurons do not escape this generalization. After attempting to transfect neuronal cell cultures using a variety of commercially available transfection reagents, as well as calcium phosphate precipitation, we found that we achieve consistent and reproducible transfection efficiencies (5–10%) using the nonliposomal lipid reagent Effectene (QIAGEN Inc., Valencia, CA). This reagent shows low cytotoxicity, can be used in the presence of serum, and requires less DNA than with other reagents that we tried. However, we also found that the quality of the plasmid is paramount when transfecting neuronal cell cultures. All plasmids are prepared using Qiagen mini-prep plasmid preparation kit. In addition, we find that further purification of the DNA preparation is crucial for efficient transfection. Therefore, to remove residual bacterial proteins we perform phenol:chloroform extraction of the DNA followed by ethanol precipitation. After purification, the plasmid is diluted to 1 μg/μl for all transfection experiments.

For a 60 mm dish containing 10 coverslips seeded with approximately 350,000 cells:

1. 1 μg of DNA is diluted with DNA-condensation buffer, Buffer EC, to a total volume of 150 μl.

2. 8 μl of Enhancer is added and the solution is vortexed for 1 sec, followed by incubation at room temperature for 5 min. If transfecting with more than one plasmid, the total DNA used should be 1 μg.

3. Next, 25 μl of Effectene transfection reagent is added to the DNA-Enhancer suspension, mixed by pipetting up and down 5 times and incubated an additional 10 min at room temperature.

4. The transfection mixture is then diluted into 1 ml of growth medium (Neurobasal media/B27) and mixed by pipetting up and down twice.

5. Prior to adding the DNA mix to the cells, the cells are washed once with 4 ml of PBS and then refed with 4 ml of fresh growth medium. The DNA mixture is added drop-wise to the cells and mixed by swirling the dish.

6. After 1 h incubation at $37°$ under 5% CO_2, the cells are rinsed twice with growth medium and allowed to express for 4–9 days before fixation and processing for analysis.

In general, we found that we have the highest transfection efficiency if the cells were transfected within 5 days of plating. However, transfection is still possible as late as 9 days after plating.

Single transfection with cDNAs encoding ARNO, ARNO-E156K, Rac1, and Rac1-N17, and coexpression of ARNO-E156K and Rac1 can be performed after 1 or 5 days *in vitro* with transfection efficiency of 5–10%. Transfections with either wild type or mutant ARF6 plasmids are 5–10% efficient only when performed in cells that are closer to a mature morphology. Therefore, ARF6, ARF6-T27N, ARF6-Q67L, and coexpression of ARNO-E156K and ARF6-Q67L plasmids are performed after 5 days *in vitro*. Regardless of the plasmid that the cells are transfected with, assessment of neurite extension is performed at 9 days *in vitro*. This is because there is sufficient opportunity for the cells to extend neurites and achieve complete polarity.

For many of these studies, it is possible to identify additional components of the regulatory pathway by performing cotransfection rescue experiments. For example, if the effects of ARNO-E156K are mediated by inhibition of GTP exchange on ARF6, cotransfection of ARNO-E156K plus active ARF6 should result in the "rescue" of the phenotype, and this was demonstrated (Hernandez-Deviez *et al.*, 2002, 2004). These "rescue" cotransfections are performed as described above, keeping the total amount of DNA in each transfection constant.

Morphological Analysis of Hippocampal Neurons Expressing ARNO, ARF6, and Mutant Forms of ARNO and ARF6

Because many small GTPases have been shown to affect different aspects of neuronal development, including axonal growth and dendritic complexity (Albertinazzi *et al.*, 2003; Li *et al.*, 2000; Luo, 2000; Nakayama *et al.*, 2000; Ng *et al.*, 2002), it is important when studying the effects of ARNO and ARF6 expression on neurite development to analyze the morphology of the transfected cells through immunofluorescent labeling and microscopy and to then quantify the differences upon neurite length and branching.

Fixation and Staining of Cells

After 9 days in culture, wash cells twice with phosphate-buffered saline (PBS, pH 7.4) to remove media and cell detritus. This immunofluorescence protocol is similar to many protocols used extensively in the scientific community, and variations would likely not affect the ability to analyze changes in neuronal morphology.

1. Fix cells in freshly prepared 4% paraformaldehyde in PBS for 20 min and then wash cells three times with PBS to remove any residual fixative.

2. Quench free aldehydes by incubating coverslips with 100 mM NH$_4$Cl in PBS for 15 min, then rinse briefly with PBS.

3. Cells are next blocked and permeabilized by incubating for 30 min with PBS containing 10% fetal bovine serum and 0.05% saponin (blocking buffer). We find manipulation of the coverslips is easiest by performing all labeling steps on parafilm on the bench top. Coverslips are then inverted onto 50–100 μl drops of blocking buffer into which primary antibody has been diluted and incubated for 30 min.

4. To differentiate between axons and dendrites, we double-label with a monoclonal antibody against the microtubule associate protein-2 (MAP2) (Sigma-Aldrich, St. Louis, MO; 1:200), to label the soma and dendrites. The transfected protein is labeled with either a polyclonal rabbit anti-*myc*, which recognizes the *myc* epitope tag on ARNO and the ARNO mutant (E156K) plasmids (Upstate Biotechnology, Lake Placid, NY; 1:100), or a polyclonal rabbit anti-HA (influenza hemagglutinin, HA), which recognizes the HA epitope tag on ARF6 and ARF6 mutants (ARF6-T27N and ARF6-Q67L) plasmids (Roche Diagnostics, Indianapolis, IN; 1:200). Alternatively, the proteins of interest can be expressed as a fusion with green fluorescent protein (GFP), eliminating the need to stain for the transfected gene. Coverslips are washed thoroughly by placing on drops of PBS on parafilm 3 times, 5 min each.

5. Place onto 50–100 μl drops of blocking buffer containing a mixture of secondary antibodies (Cy3-conjugated donkey anti-rabbit IgG and FITC-conjugated donkey anti-mouse IgG, Jackson Immunoresearch Laboratories, West Grove, PA) for 30 min. Optimal concentrations of secondary antibodies should range between 5–10 μg/ml of purified immunoglobulin G (IgG). Coverslips are then washed 3 times for 5 min with PBS, once with water, and mounted on a microscope slide with approximately 10 μl of Aqua-Poly/Mount (Polysciences, Inc., Warrington,

PA). All steps are carried out at room temperature, and secondary antibody incubations are carried out in the dark.

Imaging Parameters

For analysis of neurite extension and branching, it is critical to be able to image all the processes from a single neuron, and classify them as either an axon or dendrite. To obtain images that can be used for quantification, we find that both laser scanning confocal microscopy or deconvolution microscopy yield acceptable images. For imaging axons, it is necessary to use relatively low power magnification (usually 20X) to obtain images that include the entire neuron. For imaging dendritic branching, the 40X objective provides optimal magnification (Fig. 1). Transfected cells are identified and imaged at the appropriate wavelength and then the MAP2 labeling

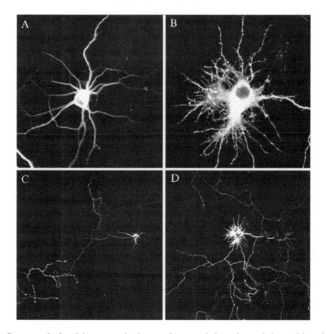

FIG. 1. Increased dendrite complexity and axonal length and branching in neurons expressing catalytically inactive ARNO. Rat hippocampal neurons in culture were transfected with GFP or GFP-ARNO-E156K one day after plating and visualized after 9 days. (A) Control cell expressing GFP showing dendrites originating from the cell body with several regular branches. (B) Cell expressing GFP-ARNO-E156K showing an extensively elaborated dendritic tree. Notice the numerous and heavily branched dendritic processes emanating from the cell body. (C) Control cell expressing GFP extending an axonal process with some collateral branches. (D) Cell expressing GFP-ARNO-E156K showing a very long axon with several collateral branches.

imaged to differentiate cell body and dendrites from the axonal process. These images are then merged prior to quantification.

Measurement of ARNO/ARF6 Effect on Neurite Elongation and Branching

To quantify the effects that ARNO/ARF6 and their mutants have on axonal length and axonal and dendritic complexity, we analyze the derived fluorescent images with a SimplePCI Image Analysis System (Compix Incorporated, Pittsburgh, PA). The software is calibrated to the microscope magnification. To enhance the visualization of dendrites and axons, the contrast is increased using the Compix System. Axons are identified based on their characteristic morphology (long, thin process with uniform diameter) or as MAP2-negative processes. Axonal length is determined by tracing the entire length of the process and the total length is calculated. Results are expressed as the mean of axonal length in microns. Axonal complexity is determined by counting the number of primary and secondary branch points along the axon. Dendrites are identified as MAP2-positive processes, and we assess dendritic complexity by counting branch points on main dendritic processes emanating from the neuronal cell body. Results of this quantification are expressed as the average number of branching points per dendritic arbour. However, other groups also have used quantification of the number of primary and secondary dendritic branches to assess effects upon branching (Ahnert-Hilger *et al.*, 2004; Gartner *et al.*, 2004; Lee *et al.*, 2005).

Because of the inherent variability in the length and branching neurites of control as well as experimental neurons, we use one-way analysis of variance (ANOVA) for statistical analysis of the data. Analysis of variance allows comparison of variability within as well as between samples, resulting in a more complete assessment of significant changes.

Acknowledgments

This work was supported by NIH DK43329 (JMW) and Consejo Nacional de Investigaciones Científicas y Tecnológicas, Venezuela and University of Los Andes, Mérida-Venezuela (DJH).

References

Ahnert-Hilger, G., Holtje, M., Grosse, G., Pickert, G., Mucke, C., Nixdorf-Bergweiler, B., Boquet, P., Hofmann, F., and Just, I. (2004). Differential effects of Rho GTPases on axonal and dendritic development in hippocampal neurones. *J. Neurochem.* **90,** 9–18.
Albertinazzi, C., Za, L., Paris, S., and De Curtis, I. (2003). ADP-ribosylation factor 6 and a functional PIX/p95-APP1 complex are required for Rac1B-mediated neurite outgrowth. *Mol. Biol. Cell* **14,** 1295–1307.

Banker, G., and Goslin, K (1998). Culturing Nerve Cells The MIT Press, Cambridge, Massachusetts.

Banker, G. A. (1980). Trophic interactions between astroglial cells and hippocampal neurons in culture. *Science* **209**, 809–810.

Banker, G. A., and Cowan, W. M. (1977). Rat hippocampal neurons in dispersed cell culture. *Brain Res.* **126**, 397–342.

Booher, J., and Sensenbrenner, M. (1972). Growth and cultivation of dissociated neurons and glial cells from embryonic chick, rat and human brain in flask cultures. *Neurobiology* **2**, 97–105.

Boshans, R. L., Szanto, S., van Aelst, L., and D'Souza-Schorey, C. (2000). ADP-ribosylation factor 6 regulates actin cytoskeleton remodeling in coordination with Rac1 and RhoA. *Mol. Cell Biol.* **20**, 3685–3694.

Dotti, C. G., Sullivan, J. M., and Banker, G. (1988). The establishment of polarity by hippocampal neurons in culture. *J. Neurosci.* **8**, 1454–1468.

Esch, T., Lemmon, V., and Banker, G. (2000). Differential effects of NgCAM and N-cadherin on the development of axons and dendrites by cultured hippocampal neurons. *J. Neurocytol.* **29**, 215–223.

Frank, S. R., Hatfield, J. C., and Casanova, J. E. (1998). Remodeling of the actin cytoskeleton is coordinately regulated by protein kinase C and the ADP-ribosylation factor nucleotide exchange factor ARNO. *Mol. Biol. Cell* **9**, 3133–3146.

Gartner, U., Alpar, A., Reimann, F., Seeger, G., Heumann, R., and Arendt, T. (2004). Constitutive Ras activity induces hippocampal hypertrophy and remodeling of pyramidal neurons in synRas mice. *J. Neurosci. Res.* **77**, 630–641.

Govek, E. E., Newey, S. E., and Van Aelst, L. (2005). The role of the Rho GTPases in neuronal development. *Genes Dev.* **19**, 1–49.

Hernandez-Deviez, D. J., Casanova, J. E., and Wilson, J. M. (2002). Regulation of dendritic development by the ARF exchange factor ARNO. *Nat. Neurosci.* **5**, 623–624.

Hernandez-Deviez, D. J., Roth, M. G., Casanova, J. E., and Wilson, J. M. (2004). ARNO and ARF6 regulate axonal elongation and branching through downstream activation of phosphatidylinositol 4-phosphate 5-kinase alpha. *Mol. Biol. Cell* **15**, 111–120.

Jareb, M., and Banker, G. (1997). Inhibition of axonal growth by brefeldin A in hippocampal neurons in culture. *J. Neurosci.* **17**, 8955–8963.

Lee, L. J., Lo, F. S., and Erzurumlu, R. S. (2005). NMDA receptor-dependent regulation of axonal and dendritic branching. *J. Neurosci.* **25**, 2304–2311.

Li, Z., Van Aelst, L., and Cline, H. T. (2000). Rho GTPases regulate distinct aspects of dendritic arbor growth in Xenopus central neurons *in vivo. Nat. Neurosci.* **3**, 217–225.

Lin, L., Rao, Y., and Isacson, O. (2005). Netrin-1 and slit-2 regulate and direct neurite growth of ventral midbrain dopaminergic neurons. *Mol. Cell. Neurosci.* **28**, 547–555.

Luo, L. (2000). Rho GTPases in neuronal morphogenesis. *Nat. Rev. Neurosci.* **1**, 173–180.

Nakayama, A. Y., Harms, M. B., and Luo, L. (2000). Small GTPases Rac and Rho in the maintenance of dendritic spines and branches in hippocampal pyramidal neurons. *J. Neurosci.* **20**, 5329–5338.

Ng, J., Nardine, T., Harms, M., Tzu, J., Goldstein, A., Sun, Y., Dietzl, G., Dickson, B. J., and Luo, L. (2002). Rac GTPases control axon growth, guidance and branching. *Nature* **416**, 442–447.

Palacios, F., Price, L., Schweitzer, J., Collard, J. G., and D'Souza-Schorey, C. (2001). An essential role for ARF6-regulated membrane traffic in adherens junction turnover and epithelial cell migration. *EMBO J.* **20**, 4973–4986.

Radhakrishna, H., Al-Awar, O., Khachikian, Z., and Donaldson, J. G. (1999). ARF6 requirement for Rac ruffling suggests a role for membrane trafficking in cortical actin rearrangements. *J. Cell Sci.* **112**, 855–866.

Radhakrishna, H., and Donaldson, J. G. (1997). ADP-ribosylation factor 6 regulates a novel plasma membrane recycling pathway. *J. Cell Biol.* **139,** 49–61.

Radhakrishna, H., Klausner, R. D., and Donaldson, J. G. (1996). Aluminum fluoride stimulates surface protrusions in cells overexpressing the ARF6 GTPase. *J. Cell. Biol.* **134,** 935–947.

van Horck, F. P., Lavazais, E., Eickholt, B. J., Moolenaar, W. H., and Divecha, N. (2002). Essential role of type I(alpha) phosphatidylinositol 4-phosphate 5-kinase in neurite remodeling. *Curr. Biol.* **12,** 241–245.

Whitford, K. L., Marillat, V., Stein, E., Goodman, C. S., Tessier-Lavigne, M., Chedotal, A., and Ghosh, A. (2002). Regulation of cortical dendrite development by Slit-Robo interactions. *Neuron* **33,** 47–61.

Yamazaki, M., Miyazaki, H., Watanabe, H., Sasaki, T., Maehama, T., Frohman, M. A., and Kanaho, Y. (2002). Phosphatidylinositol 4-phosphate 5-kinase is essential for ROCK-mediated neurite remodeling. *J. Biol. Chem.* **277,** 17226–17230.

[24] Analysis of the Interaction Between Cytohesin 2 and IPCEF1

By Kanamarlapudi Venkateswarlu

Abstract

IPCEF1 has been reported to interact with ADP-ribosylation factor GTP exchange factors of the cytohesin family and function by modulating the cytohesin 2 activity. This article describes methods used to study the interaction and activation of cytohesin GEFs by IPCEF1. The experimental approaches described here include physical and functional interaction assays by which the association of IPCEF1 with cytohesin 2 is explored both *in vitro* and *in vivo*. The methods used to analyze the physical association include GST-pull down and coimmunoprecipitation approaches. We also used yeast two-hybrid and colocalization assays to study the interaction between IPCEF1 and cytohesins. The functional relationship between IPCEF1 and cytohesin 2 was assessed by studying the effect of IPCEF1 on the *in vitro* and *in vivo* stimulation of ADP-ribosylation factor 6 GTP formation by cytohesin 2.

Introduction

The ADP-ribosylation factors (ARFs) family of small GTPases regulate intracellular membrane trafficking by cycling between an inactive GDP- and an active GTP-bound form (Moss and Vaughan, 1998). In mammalian cells, the ARF family consists of six ARF isoforms (ARF 1–6).

METHODS IN ENZYMOLOGY, VOL. 404
0076-6879/05 $35.00
DOI: 10.1016/S0076-6879(05)04024-3

ARF6, the most distantly related members of the ARF family, plays critical roles in cellular events such as receptor-mediated endocytosis, regulated exocytosis and cell spreading by regulating cytoskeleton reorganization, and membrane ruffling near the cell surface (Donaldson, 2003). The activation-inactivation cycle of ARFs is regulated by guanine-nucleotide exchange factors (GEFs), which promote GTP loading, and GTPase activating proteins (GAPs), which stimulate GTP hydrolysis (Randazzo et al., 2000). The family of cytohesin ARF GEFs (cytohesins 1–4) consist of a central catalytic Sec7 domain flanked by a C-terminal pleckstrin homology (PH) domain and an N-terminal coiled-coil (CC) domain (Jackson and Casanova, 2000). The PH domain of cytohesins 1–3 binds to phosphatidylinositol (3,4,5) trisphosphate (PIP3) in vitro (Cullen and Venkateswarlu, 1999). PIP_3 is a ubiquitous, mainly plasma-membrane-localized second messenger produced by agonist stimulated PI 3-kinase and plays an important role in cellular functions such as glucose uptake, membrane trafficking, cell adhesion, and cell secretion (Vanhaesebroeck et al., 2001).

Cytohesins 1–3 also bind PIP_3 in vivo through the PH domain and this interaction mediates translocation of the GFP-tagged cytohesins from the cytosol to the plasma membrane in agonist stimulated cells. Although they activate most of the ARFs in vitro by promoting GTP-binding, in vivo cytohesins 1–3 mainly activate ARF6 in a PIP_3-dependent manner (Hawadle et al., 2002). A few studies, however, have reported that cytohesins 1–3 localize to the Golgi and inhibit cell secretion and Golgi disassembly by activating ARF1 (Franco et al., 1998; Frank et al., 1998; Lee and Pohajdak, 2000; Monier et al., 1998). Together, these studies suggest that ARF specificity and regulation of cytohesins depends on their location. Approximately 50% of cytohesin 2 is found in the membrane fraction of some cell lines under nonstimulated conditions, suggesting that the localization and functions of this protein may be regulated by the binding partner proteins (Frank et al., 1998). In an effort to identify cytohesin 2 interacting proteins that may regulate its localization and activity, we have performed yeast two-hybrid screening of a rat brain cDNA library using cytohesin 2 as bait. This resulted in identification of a novel protein termed IPCEF1 (Interactor Protein for Cytohesin Exchange Factors 1). IPCEF1 contains a PH domain at the amino-terminal end and displays 79% amino acid identity to human KIAA0403, which was identified as a protein of unknown function in a brain gene-cloning project. IPCEF1 is widely distributed in rat tissues but abundant only in brain, spleen, lung, and testes. It colocalizes with cytohesin 2 to the cytosol and translocates with cytohesin 2 to the plasma membrane of cells stimulated with epidermal growth factor (EGF), where it modulates the ARF6 GEF activity of cytohesin 2 (Venkateswarlu, 2003). This article describes the experimental approaches by which the interaction with and activation of cytohesins by IPCEF1 are explored *in vitro* and *in vivo*.

Methods

Materials

Unless otherwise stated, all chemicals were purchased from Sigma (Poole, UK).

Yeast Two-Hybrid Analysis of Cytohesin 2 and IPCEF1 Interaction

The yeast two-hybrid system is a powerful genetic technique that has been used successfully to study protein-protein interactions *in vivo* from a wide variety of biological sources, including yeast, plant, and mammalian (Fields and Song, 1989). The two-hybrid system provides a sensitive method to detect relatively weak and transient protein-protein interactions that may not be biochemically detectable. The principle of this assay is based on the reconstitution of transcriptional factors that promotes the expression reporter genes, which then allow the proliferation of yeast under restrictive conditions. Using the two-hybrid system, we successfully isolated IPCEF1 and a novel kinesin motor proteins from a rat brain cDNA library as interactors of cytohesin 2 ARF GEF and centaurin-α1 ARF GAP, respectively (Venkateswarlu, 2003). The procedure used to map the binding domains of cytohesin 2 and IPCEF1, and to characterize the cytohesin 2 and IPCEF1 selectivity is outlined here.

To understand the structural requirements for the interaction between cytohesin 2 and IPCEF1, we used a deletion mutant approach to map the specific domains of cytohesin 2 and IPCEF1 required for their association. A series of cytohesin 2 deletion mutants were generated as LexA DNA binding domain (BD) fusion constructs, and we tested their ability to bind IPCEF1 fused to the activation domain of GAL4 (AD) in the yeast two-hybrid system (Fig. 1A). Full-length human cytohesin 2 and its deletion mutants cDNA sequences were amplified by PCR using cytohesin 2/ pEGFPC1 plasmid as a template (Venkateswarlu *et al.*, 1998a), Pfu DNA polymerase (Stratagene) and the following sets of primers containing *Eco*RI (sense, underlined) and *Sal*I (antisense, underlined) restriction sites. FL (the full length human cytohesin 2, 1–399 a.a.): sense primer 5'-CGC<u>GAATTC</u>ATGGCCAAGGAGCGGCGCAGG-3' and antisense primer 5'-CGC<u>GTCGAC</u>TCAGGGCTGCTCCTGCTTCTT-3', ΔPH (cytohesin 2 deletion mutant without the PH domain, 1–252 a.a.): FL sense primer and antisense primer 5'-CGC<u>GTCGAC</u>TCAGTCATTCCCGT-CATCCTCAGG-3', Sec7 (cytohesin 2 Sec7 domain, 61–252 a.a.): sense

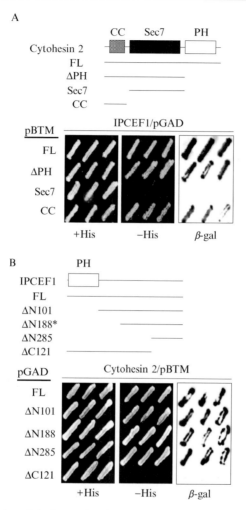

FIG. 1. Identification of the interacting domains of IPCEF1 and cytohesin 2 by yeast two-hybrid assay. (A) mapping of the IPCEF1 binding domain of cytohesin 2. Schematic representation of cytohesin 2 and its deletion mutants encoded by the BD-fusion cDNA constructs is shown in the top panel (FL = the full-length). (B) Mapping of the cytohesin 2 binding region within IPCEF1. Schematic representation of IPCEF1 and its deletion mutants encoded by the AD-fusion cDNA constructs is shown in the top panel (FL = the full-length). *The insert of the clone isolated by yeast two-hybrid screening using cytohesin 2 as bait. The BD-fusion (pBTM) constructs were cotransformed with the AD-fusion (pGAD) constructs into L40 yeast strain and tested for their ability to grow in the absence of histidine and to express β-galactosidase, which is analyzed by assaying the conversion of X-gal into a blue colored product. The bottom panel in both (A) and (B), left to right, shows 3 clones of each transformant grown on solid medium with (+His) and without (−His) histidine as well as filter β-galactosidase assays (β-gal). Reprinted from Venkateswarlu (2003), with permission.

primer 5′-CGCGAATTCATGCGGAACCGGAAGATGGCAATG-3′ and ΔPH antisense primer, and CC (the CC domain of cytohesin 2, 1–67 a.a.): FL sense primer and antisense primer 5′-CGCGTCGACTCACATTGC-CATCTTCCGGTTCCG-3′. The resulting cDNAs were digested with *Eco*RI and *Sal*I, and cloned into the same sites of a bait plasmid pBTM116 (29) for expression as LexA DNA binding domain (BD) fusion proteins.

To define the region of IPCEF1 involved in binding cytohesin 2, we generated a series of AD-fusion constructs containing various truncations of IPCEF1 and analyzed their interaction with BDcytohesin 2 in a yeast two-hybrid assay (Fig. 1B). IPCEF1 and its deletion mutant sequences were isolated by PCR using IPCEF1/pGEMT (Venkateswarlu, 2003) and the following sets of primers containing *Eco*RI (sense, underlined) and *Sal*I (antisense, underlined) sites, digested with *Eco*RI-*Sal*I and cloned into the same sites of pGAD424 (Clontech) for expression as GAL4 activation domain (AD) fusion proteins in yeast cells. IPCEF1 FL (IPCEF1 full-length, 1–406 a.a.): sense primer 5′-CGGAATTCATGAGTCGGAG AAGGATATCCTG-3′ and antisense primer 5′-CGCGTCGACAAGA-GAATTTTCCACACAGTCAG-3′, ΔN101 (IPCEF1, 102–406 a.a.): sense primer 5′-CGCGAATTCATGTGGCTAAACAAACTTGGATTTG-3′ and IPCEF1 FL antisense primer, ΔN188 (IPCEF1, 189–406 a.a.): sense primer 5′-CGCGAATTCATGAAAGAAAGACAGTCGTGGCTTG-3′ and IPCEF1 FL antisense primer, ΔN285 (IPCEF1, 286–406 a.a.): sense primer 5′-CGCGAATTCATGGAGAAACTGTACAAGTCATTG-3′; and IPCEF1 FL antisense primer, and ΔC121 (IPCEF1, 1–285 a.a.): IPCEF1 FL sense primer and antisense primer 5′-CGCGTCGACCT-CATCGTCTTCGGCGATTTTAT-3′.

Yeast two-hybrid screening was performed as described previously (Stephens and Banting, 1999). Yeast strain L40 (contains HIS3 and β-GAL reporters) was transformed with pBTM116 and pGAD10 constructs using the lithium acetate. Briefly, the parent yeast strain was grown over-night in 5 ml of YPD medium at 30° on a shaker at 200 rpm. The culture was diluted 50-fold and then allowed to grow for another 3 to 4 h. The yeast cells are harvested by centrifugation at 2000g for 5 min at room temperature, washed with 25 ml of sterile water and then with 1 ml of 0.1 *M* lithium acetate. The cells were resuspended in 0.5 ml of 0.1 *M* lithium acetate. 50 μl of cells were mixed with 240 μl of 50% polyethylene glycol (MW 3350), 36 μl of 1.0 *M* lithium acetate, 10 μl of salmon sperm DNA, and 1 μg of each plasmid, incubated at 30° for 30 min, and heat shocked at 42° for 20 min. The transformation mixture was centrifuged at 10,000g for 15 sec, the cell pellet washed with 1 ml of sterile water, and resuspended in 100 μl of sterile water, plated evenly on a 10-cm agar synthetic plate lacking tryptopan and leucine to select the transformants carrying the plasmids, and incubated at 30° for 2–3 days. The transformants were then assayed for growth on the histidine lacking media and for β-galactosidase activity. β-Galactosidase filter assays were carried out as described (Poullet and Tamanoi, 1995). The yeast transformants replica plated onto a 3 *MM* Whatmann filter paper were frozen in liquid nitrogen for 10 sec to break the cells. The filter was thawed at room temperature for 2 min, placed (cells facing up) in a 10-cm plate containing a 3 *MM* Whatmann filter paper soaked in 3.2 ml of Z buffer (60 m*M* Na_2HPO_4, 40 m*M* NaH_2PO_4, 10 m*M* KCl, 1 m*M* $MgSO_4$) containing 50 μl of 25 mg/ml of X-gal (5-bromo-4-chloro-3-indoyl-β-D-galactopyranoside) and 10 μl of β-mercaptoethanol, and incubated at 30° overnight.

Typical results are shown in Fig. 1. The results obtained by the histidine autotrophy assay correlate with those of the X-gal assay. In summary, the results demonstrated that only the fusion constructs containing the N-terminal CC domain of cytohesin 2 were capable of interacting with IP-CEF1, whereas the C-terminal PH domain and the central Sec7 domain were not required for the interaction. Similarly, the C-terminal 121 amino acids of IPCEF1 were found to be required for binding cytohesin 2, whereas the rest of the protein was not required for the interaction. Together these results demonstrate that the association of IPCEF1 with cytohesin 2 is mediated by the CC domain of cytohesin 2 and the C-terminal 121 amino acids of IPCEF1.

Detection of Cytohesin 2 and IPCEF1 Interaction Using GST-fusion Protein Pull-Down Assay

The principle of pull-down assay is based on the ability and specificity of a protein of interest, immobilized *via* its tag to an affinity matrix, to precipitate or pull-down its partner from the cell lysates. The glutathione S-transferase (GST) fusion protein pull-down approach is a commonly used method, in which the GST serves as a tag when fused to the protein of interest, and glutathione beads serve as the pull-down matrix. This section describes the use of GST pull-down assay to evaluate the interaction between cytohesin 2 and IPCEF1. This assay involves expression and purification of the GST-fusion protein in bacteria and expression of the interactor protein with an epitope-tag in mammalian cells, incubation of GST-fusion protein coupled to glutothione matrix with a lysate of cells expressing the epitope-tagged interactor protein, and analysis the protein complex by Western blotting.

The full-length human cytohesin 2 and its deletion mutants (ΔPH, Sec7, and CC) cDNA sequences were released from pBTM116 construct (see previous discussion) by *Eco*RI-*Sal*I digestion and subcloned into the same sites of pGEX-4T1 to express as GST-fusion proteins in bacteria. The pGEX plasmids containing the required insert were introduced into the *Escherichia coli* strain *BL21 (DE3)* using calcium chloride-based chemical transformation to express and purify the corresponding GST-fusion protein as follows (Venkateswarlu *et al.*, 1998b). 5 ml of LB broth containing 0.1 mg/ml ampicillin was inoculated with a single colony of the transformed strain and grown overnight at 37°. This was then diluted 1:100 in fresh LB broth containing 0.1 mg/ml ampicillin and grown at 37°, in an incubator with shaking at 200 rpm, until the cell density had reached a D_{600} of 0.5 (approximately 3 h). Protein expression was induced with 0.2 mM isopropyl β-D-thiogalactoside for 3 h at 37°, after which cells were collected by centrifugation. The resultant pellet was resuspended in 50 ml ice-cold buffer A (PBS containing 1 mM EDTA, 1 mM EGTA, 0.2 mM PMSF, and 1 mM β-mercaptoethanol). Cells were lysed by sonication (3 × 20 sec pulses, with 1 min at 4° between each sonication) and incubated at 4° with 1% Triton-X 100 for 1 h with gentle mixing prior to the removal of cell debris by centrifugation. A 50% slurry (1 ml) of glutathione–Sepharose 4B resin (Amersham Biosciences, Chalfont St. Giles, UK), prewashed with buffer A, was added to the resultant supernatant and incubated at 4° overnight with constant mixing. The resin was washed with 5 times with 10 ml of ice-cold buffer A and stored as a 25% slurry in buffer A containing 50% glycerol at −20°. The purity of protein bound to the resin was analyzed by SDS-polycarylamide gel electrophoresis (SDS-PAGE) and Coomassie Blue staining.

COS cells were plated into 10-cm dishes and allowed to grow to 70–80% confluency in Dulbecco's modified Eagle's medium supplemented with 10% fetal calf serum, penicillin (100 units/ml), streptomycin (0.1 mg/ml), and 2 mM glutamine in a humidified incubator with 5% CO2 at 37°. Then they were transfected with FLAG-tagged full-length IPCEF1 or its deletion mutants encoding plasmids (5 μg DNA) using the liposomal transfection reagent fuGENE6 (Roche Molecular Biochemicals, Lewes, UK) at a ratio of 4 μl of reagent/1 μg DNA according to the manufacturer's instructions. The full-length IPCEF1 (FL) and its deletion mutants (ΔC121 and ΔN285) cDNA sequences were released from pGAD424 construct (see previous) by *Eco*RI-*Sal*I digestion and subcloned into the same sites of pCMV-tag2 (Stratagene, Amsterdam, The Netherlands) for expression as FLAG-tagged proteins in mammalian cells. After 48 h of transfection, cells were washed twice with PBS and lysed in 1 ml of lysis buffer (50 mM Tris-HCl, pH 7.5, 150 mM NaCl, 1% Triton X-100, and 1% of protease inhibitor mixture). The lysates were clarified by centrifugation at 16,000g for 10 min at 4°. Ninety percent of the cell extract was incubated with 20 μg of GST or GST-fused protein coupled to glutathione beads for 1 h at 4°. The resin was washed three times with the lysis buffer and boiled in SDS-PAGE sample buffer. The samples were separated on SDS-PAGE gels, transferred to PVDF membranes (Whatman) and probed with anti-FLAG M2 monoclonal antibody. As shown in Fig. 2A, IPCEF1 interacted with GST-fusion proteins containing the CC domain of cytohesin 2, but not with the fusion proteins lacking the CC domain, or with GST alone. Similarly, GST-cytohesin 2 showed strong interaction with either IPCEF1 or the IPCEF1 deletion mutant containing the C-terminal 121 amino acids (ΔN285) but not with the deletion mutant lacking the C-terminal 121

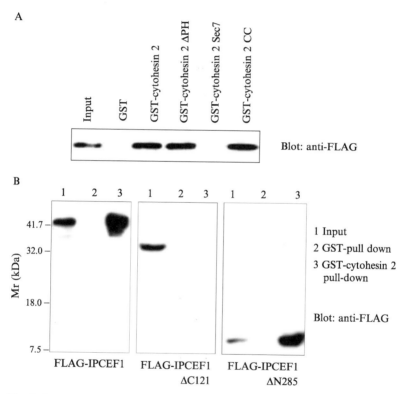

FIG. 2. Determination of IPCEF1 and cytohesin 2 binding sites by *in vitro* GST pull-down assay. (A) The lysate of COS cells expressing FLAG-IPCEF1 was incubated with either GST alone or GSTcytohesin 2 or GST-cytohesin 2 deletion mutants, all coupled to glutathione-beads. (B) Lysates of COS cells expressing FLAG-tagged IPCEF1 or its deletion mutants were incubated with either GST or GSTcytohesin 2 coupled to glutathione-beads. After washing the beads, the bound proteins were separated on SDS-PAGE, blotted onto PVDF membranes, and probed with a monoclonal anti-FLAG antibody (Blot). Positions of molecular mass standards, Mr (kDa) are shown. One-twentieth of the inputs (cell lysates) are also shown for each of the sets of experiments. Reprinted from Venkateswarlu (2003), with permission.

amino acids (ΔC121) (Fig. 2B). In agreement with the yeast two-hybrid data, these results reveal that the CC domain of cytohesin 2 and the C-terminal 121 amino acids of IPCEF1 are sufficient and necessary for the IPCEF1-cytohesin 2 interaction.

Coimmunoprecipitation of Cytohesin 2 and IPCEF1 in COS Cell Lysates

Coimmunoprecipitation is a powerful approach to analyze for physical interaction between proteins of interest in intact cells. The main advantages of this method, compared to *in vitro* GST pull-down assay, are that it is more physiologically relevant and does not require protein purification. The main disadvantages of this assay, however, are that it is relatively laborious and time-consuming and may not detect relatively weak and transient protein-protein interactions, which may be detectable by yeast two-hybrid assay. The basis of coimmunoprecipitation assay is to use a high-affinity antibody against protein of interest or the epitope-tag of protein of interest to precipitate protein complex under nondenaturing conditions, which preserve many of the protein-protein interactions that exist in intact cells. The protein complex then analyzed by Western blotting for the presence of interactor(s) of protein of interest. This section describes an application of this approach of the interaction of cytohesin 2 with IPCEF1.

COS cells were cotransfected with GFP or GFP-cytohesin 2 and FLAG or FLAG-tagged full-length IPCEF1 plasmids (1:1 ratio, 10 μg total DNA). After 2 days, COS cells were lysed as described above and the cell extracts were incubated with 5 μg of anti-GFP polyclonal antibody (Clontech) for 30 min at 4°. The immunocomplexes were incubated with 30 μl of protein A Sepharose for 2 h at 4° and washed 5 times with lysis buffer. The bound protein detected by immunoblotting using anti-FLAG M2 monoclonal antibody. As shown in Fig. 3, FLAG-IPCEF1 coimmunoprecipitated with GFP cytohesin 2 but not with GFP alone. This result clearly demonstrates that cytohesin 2 and IPCEF1 interact with each other *in vivo*.

IPCEF1 Modulates Cytohesin 2 GEF Activity In Vivo and In Vitro

We examined whether the IPCEF1-cytohesin 2 interaction is functionally relevant for the ARF GEF activity of cytohesin 2. Since cytohesin 2 has been shown to activate ARF6 both *in vitro* and *in vivo* (Frank *et al.*, 1998; Venkateswarlu and Cullen, 2000), we analyzed the effect of IPCEF1 on the ARF6 GEF activity of cytohesin 2 using both *in vitro* and *in vivo* binding assays. For *in vitro* studies, purified myristoylated ARF6-His$_6$ was incubated with [^{35}S]GTPγS in the presence and absence of cytohesin

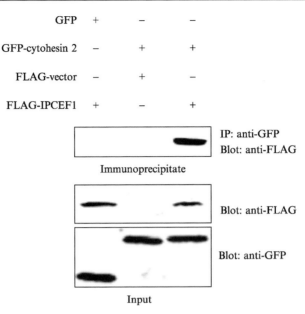

FIG. 3. *In vivo* interaction between cytohesin 2 and IPCEF1. Coimmunoprecipitation of FLAG-IPCEF1 with GFP-cytohesin 2. COS cells were transfected with the indicated expression vectors. After two days of transfection, cells were lysed and immunoprecipitated (IP) with an anti-GFP antibody. After washing, immunoprecipitates were resolved on SDS-PAGE, blotted onto PVDF membranes, and probed with a monoclonal anti-FLAG antibody (Blot) to detect FLAG-tagged IPCEF1. 5% of the input (cell lysates) was also immunoblotted with anti-GFP and anti-FLAG antibodies to ensure that GFP-cytohesin 2 and FLAG-IPCEF1, respectively, were expressed. Reprinted from Venkateswarlu (2003), with permission.

2 and/or IPCEF1, and the amount of $[^{35}S]GTP\gamma S$ bound to ARF6 was analyzed. ARF6 interacts specifically with downstream effectors such as metalothionine 2A (MT2A) when it is in the active GTP-bound form (Schweitzer and D'Souza-Schorey, 2002). Recently, Schweitzer and D'Souza-Schorey (2002) have made use of this observation and developed a GST-effector pull down assay to study ARF activation *in vivo*. To analyze the effect of IPCEF1 on the *in vivo* GEF activity of cytohesin 2, HA-tagged ARF6 was coexpressed with either GFP-cytohesin 2 or FLAG-IPCEF1, or both, in COS cells. Following serum starvation, cells were stimulated with EGF, the cells lysed, and the activated ARF6 purified from the cell lysates using GSTMT2A coupled to a glutathione-resin.

ARF6-HA/pXS construct, kindly provided by Dr J. Donaldson (National Institutes of Health, Bethesda, MD), was used as a template to isolate full-length ARF6 by PCR using forward (5'-CTAGCCATGGG GAAGGTGCTATCCAAAATC-3') and reverse (5'-CGGAATTCGAGA TTTGTAGTTAGAGGTTAACC-3') primers containing NcoI (underlined) and EcoRI (underlined) restriction sites respectively. The PCR product was digested with NcoI-EcoRI and cloned into the same sites of pTric-His2b (Novagen, Nottingham, UK) for expression of protein with a C-terminal His6 epitope-tag in E. coli. pBB131 plasmid encoding yeast N-myristoyltrans-ferase (obtained from Prof. J. I. Gordon, Washington University, St. Louis, MO) was used to myristoylate ARF6 in E. coli. Myristoylated ARF6 with His$_6$-tag at the C-terminus was expressed in E. coli BL21(DE3) strain by cotransforming with yeast N-myristoyltransferase plasmid (pBB131) and pur-ified using a Ni^{2+}-affinity column (Novagen) according to the procedure previously described (Glenn et al., 1998). GST-cytohesin 2 and GST-IPCEF1 recombinant proteins were prepared as described above. [^{35}S]GTPγS binding to recombinant myristoyled ARF6 was performed in triplicate using a rapid filtration procedure (Kliche et al., 2001). Briefly, 50 μl of the reaction buffer (50 mM HEPES, pH-7.5, 1 mM MgC$_{12}$, 0.1 M KCl, 1 mM dithiothreitol) containing 1 μM of ARF6, 4 μM [^{35}S]GTPγS (0.2 μCi) and 50 μg of liposomes (70% phosphatidylcholine, 30% phosphatidylglycerol, and 5% PIP3) was incubated with either GST (200 nM), GST-cytohesin 2 (50 nM), GST-IP-CEF1 (10–200 nM), GST-cytohesin 2 (50 nM) + GSTIPCEF1 (10–200 nM) at 37°. Reactions were stopped at the indicated times by adding 1.5 ml of ice cold washing buffer (20 mM HEPES, pH 7.5, 0.1 M NaCl, and 10 mM MgC$_{12}$) to each reaction tube. Samples were filtered through nitrocellulose membranes in a vacuum manifold (Millipore, Watford, UK). The filters were washed 3 times with 2 ml of ice-cold washing buffer and dried. Scintillation fluid was added to the filters and counted in a Beckman LS 5000 β-counter to quantify the amount of protein-[^{35}S]GTPγS bound.

As shown in Fig. 4, cytohesin 2 increased the binding of GTP to ARF6 in a time-dependent manner and the rate of binding was further increased by IPCEF1, which was ineffective when added in the absence of cytohesin 2. Similar results were also obtained using the deletion mutant of IPCEF1 that contains the cytohesin 2 binding site (ΔN285), but not the deletion mutant that lacks the cytohesin 2 binding site (ΔC121), in place of IPCEF1 (data not shown). Furthermore, IPCEF1 amplified cytohesin 2-mediated ARF6 activation in a concentration-dependent manner (data not shown), reaching saturation when the amount of IPCEF1 added was equal to that of cytohesin 2 (2.5 pmol).

In vivo ARF6 activation assay was performed as described (Schweitzer and D'Souza-Schorey, 2002). COS cells transfected with ARF6-HA, GFP, or GFP-cytohesin 2 and FLAG or FLAG-tagged full-length IPCEF1 plas-mids (3:1:1 ratio; 10 μg total DNA) were serum starved for 2 h and

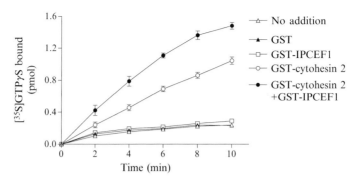

FIG. 4. Effect of IPCEF1 on the *in vitro* ARF6 GEF activity of cytohesin 2. ARF6 (50 pmol) was incubated alone or with GST (10 pmol), GST-cytohesin 2 (2.5 pmol), GST-IPCEF1 (1.25 pmol), or both GST-cytohesin 2 (2.5 pmol) and GST-IPCEF1 (1.25 pmol) for the indicated times at 37°. The bound GTP was then measured. Values are shown as the means of three independent assays performed in triplicate and expressed as the amount of GTPγS bound to ARF6. Error bars indicate standard error. Reprinted from Venkateswarlu (2003), with permission.

incubated for 5 min with or without 200 ng/ml EGF. ARF6-HA/pXS construct, kindly provided by Dr J. Donaldson (National Institutes of Health, Bethesda, MD), was used to express ARF6 with a C-terminal hemagglutinin (HA) epitope-tag in mammalian cells. The cells were lysed in 0.5 ml of lysis buffer B (50 mM Tris-HCl, pH 7.5, 10 mM MgC$_{l2}$, 0.5 M NaCl, 1% Triton X-100, 0.5% sodium deoxycholate, 0.1% SDS, and 0.1% protease inhibitor mixture [Sigma]). The lysates were clarified by centrifugation at 16,000g for 10 min at 4°. 90% of the cell extract was incubated with GST-MT2A coupled to glutathione beads in presence of 2 mM ZnC$_{l2}$. After 1 h mixing at 4°, the beads were washed 3 times with wash buffer (50 mM Tris-HCl, pH 7.5, 0.15 M NaCl, 10 mM MgC$_{l2}$, 1% Triton X-100, 2 mM ZnC$_{l2}$, and 0.1% protease inhibitors), boiled in SDS-PAGE sample buffer, and analyzed by immunoblotting using a monoclonal anti-HA antibody (Covance, Berkeley, CA). Immunoblots were scanned and the GTP-bound ARF6 precipitated with GST-MT2A was normalized to total ARF levels in the lysates to compare ARF6$_{GTP}$ levels in cells transformed with the indicated constructs. The full-length metalothionine 2A (MT2A) cDNA was amplified from MGC clone 12397 by PCR using sense (5′-CGC<u>GAATTC</u>ATGGACCCCAACTGCTCCTGCGCCGC-3′) and anti-sense (5′-CGC-<u>GTCGAC</u>GGCGCAGCAGCTGCACTTGTCCG-3′) primers having *Eco*RI (underlined) and *Sal*I (underlined) sites respectively. The cDNA was digested with *Eco*RI-*Sal*I and ligated to pGEX-4T1 for expression as GST-fusion protein in *E. coli*. The GST-MT2A fusion protein

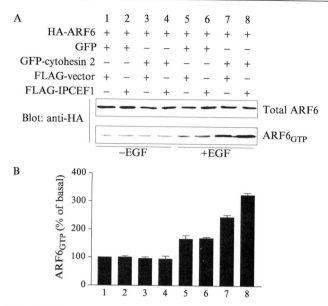

FIG. 5. Effect of IPCEF1 on the *in vivo* ARF6 GEF activity of cytohesin 2. (A) COS cells were transfected with the indicated expression plasmids. Two days later cells were serum starved for 2 h, incubated in the presence or absence of EGF (200 ng/ml), lysed, and ARF6$_{GTP}$ precipitated using GSTMT2A coupled to glutathione-beads. The precipitates were then immunoblotted (Blot) with anti-HA antibody. Before ARF6$_{GTP}$ precipitation, cell lysates were also immunoblotted (Blot) using an anti-HA antibody to determine total ARF6. (B) Quantification of data obtained from three similar experiments. The mean of three experiments is represented. Error bars indicate standard error. Reprinted from Venkateswarlu (2003), with permission.

was expressed in BL21(DE3) strain of *E. coli* and coupled to glutathione-beads as described above.

As shown in Fig. 5, ARF6 activation in EGF-treated cells was stimulated by cytohesin 2 and the activation further increased by IPCEF1. However, IPCEF1 was ineffective as a modulator for ARF6 activation in the absence of cytohesin 2. Although we used an equal amount of FLAG-IPCEF1 and GFP-cytohesin 2 plasmid DNA to transfect COS cells, the effect of IPCEF1 on the *in vivo* activation of ARF6 by cytohesin 2 was relatively small compared to its effect on the *in vitro* cytohesin 2 activity when the protein added to the assay was equal to that of cytohesin 2. This may be due to the low expression of IPCEF1 compared to that of cytohesin 2 (data not shown). In order to examine whether IPCEF1 affects the *in vivo* activation of ARF6 by cytohesin 2 in a concentration-dependent manner, we cotransfected COS cells with fixed amounts of DNA of

HA-ARF6 and GFP-cytohesin 2, and varying amounts of FLAG-IPCEF1 DNA and analyzed ARF6 activation in EGF stimulated cells using GST-effector pull-down assay. Consistent with the *in vitro* data, IPCEF1 modulated *in vivo* cytohesin 2 ARF6 GEF activity in a concentration-dependent manner, and its effect saturated when its expression levels were comparable to that of cytohesin 2 (data not shown). Together, these results suggest strongly that IPCEF1 potentiates ARF6 GEF activity of cytohesin 2 through direct interaction.

Conclusions

We described several methods to analyze cytohesin 2 GEF interaction with IPCEF1, including *in vitro* and *in vivo* binding assays, and functional studies. Using yeast two-hybrid screening, we analyzed IPCEF1 interaction with cytohesin 2 and the specific domains of each protein required for the interaction. We confirmed that IPCEF1 specifically binds cytohesins *in vitro* by GST pull-down assays and *in vivo* by immunoprecipitation assays. The cytohesin 2-IPCEF1 interaction is mediated by the CC domain of cytohesin 2 and the IPCEF1 C-terminal domain. Moreover, IPCEF1 potentiated ARF6 activation by cytohesin 2 *in vitro* as well as in EGF stimulated cells, suggesting that it modulates the ARF6 GEF activity of cytohesin 2. This implies a conformational change in the catalytic Sec7 domain upon IPCEF1 binding to the CC domain of cytohesin 2 that leads to an alteration in the catalytic activity of the Sec7 domain.

Acknowledgments

K.V. was recipient of a David Phillips Research Fellowship from the Biotechnology and Biological Science Research Council (BBSRC). This work was funded by the BBSRC and the Royal Society UK.

References

Cullen, P. J., and Venkateswarlu, K. (1999). Potential regulation of ADP-ribosylation factor 6 signalling by phosphatidylinositol 3,4,5-trisphosphate. *Biochem. Soc. Trans.* **27,** 683–689.

Donaldson, J. G. (2003). Multiple roles for Arf6: Sorting, structuring, and signaling at the plasma membrane. *J. Biol. Chem.* **278,** 41573–41576. Epub 2003 Aug 11.

Fields, S., and Song, O. (1989). A novel genetic system to detect protein–protein interactions. *Nature* **340,** 245–246.

Franco, M., Boretto, J., Robineau, S., Monier, S., Goud, B., Chardin, P., and Chavrier, P. (1998). ARNO3, a Sec7-domain guanine nucleotide exchange factor for ADP ribosylation factor 1, is involved in the control of Golgi structure and function. *Proc. Natl. Acad. Sci. USA* **95,** 9926–9931.

Frank, S., Upender, S., Hansen, S. H., and Casanova, J. E. (1998). ARNO is a guanine nucleotide exchange factor for ADP-ribosylation factor 6. *J. Biol. Chem.* **273**, 23–27.

Glenn, D. E., Thomas, G. M., O' Sullivan, A. J., and Burgoyne, R. D. (1998). Examination of the role of ADP-ribosylation factor and phospholipase D activation in regulated exocytosis in chromaffin and PC12 cells. *J. Neurochem.* **71**, 2023–2033.

Hawadle, M. A., Folarin, N., Martin, R., and Jackson, T. R. (2002). Cytohesins and centaurins control subcellular trafficking of macromolecular signaling complexes: Regulation by phosphoinositides and ADP-ribosylation factors. *Biol. Res.* **35**, 247–265.

Jackson, C. L., and Casanova, J. E. (2000). Turning on ARF: The Sec7 family of guanine-nucleotide-exchange factors. *Trends Cell. Biol.* **10**, 60–67.

Kliche, S., Nagel, W., Kremmer, E., Atzler, C., Ege, A., Knorr, T., Koszinowski, U., Kolanus, W., and Haas, J. (2001). Signaling by human herpesvirus 8 kaposin A through direct membrane recruitment of cytohesin-1. *Mol. Cell* **7**, 833–843.

Lee, S. Y., and Pohajdak, B. (2000). N-terminal targeting of guanine nucleotide exchange factors (GEF) for ADP ribosylation factors (ARF) to the Golgi. *J. Cell Sci.* **113**, 1883–1889.

Monier, S., Chardin, P., Robineau, S., and Goud, B. (1998). Overexpression of the ARF1 exchange factor ARNO inhibits the early secretory pathway and causes the disassembly of the Golgi complex. *J. Cell Sci.* **111**, 3427–3436.

Moss, J., and Vaughan, M. (1998). Molecules in the ARF orbit. *J. Biol. Chem.* **273**, 21431–21434.

Poullet, P., and Tamanoi, F. (1995). Use of yeast two-hybrid system to evaluate Ras interactions with neurofibromin-GTPase-activating protein. *Methods Enzymol.* **255**, 488–497.

Randazzo, P. A., Nie, Z., Miura, K., and Hsu, V. W. (2000). Molecular aspects of the cellular activities of ADP-ribosylation factors. *Sci. STKE* **2000**, RE1.

Schweitzer, J. K., and D'Souza-Schorey, C. (2002). Localization and activation of the ARF6 GTPase during cleavage furrow ingression and cytokinesis. *J. Biol. Chem.* **277**, 27210–27216.

Stephens, D. J., and Banting, G. (1999). Direct interaction of the trans-Golgi network membrane protein, TGN38, with the F-actin binding protein, neurabin. *J. Biol. Chem.* **274**, 30080–30086.

Vanhaesebroeck, B., Leevers, S. J., Ahmadi, K., Timms, J., Katso, R., Driscoll, P. C., Woscholski, R., Parker, P. J., and Waterfield, M. D. (2001). Synthesis and function of 3-phosphorylated inositol lipids. *Annu. Rev. Biochem.* **70**, 535–602.

Venkateswarlu, K. (2003). Interaction protein for cytohesin exchange factors 1 (IPCEF1) binds cytohesin 2 and modifies its activity. *J. Biol. Chem.* **278**, 43460–43469.

Venkateswarlu, K., and Cullen, P. J. (2000). Signaling via ADP-ribosylation factor 6 lies downstream of phosphatidylinositide 3-kinase. *Biochem. J.* **345**, 719–724.

Venkateswarlu, K., Gunn-Moore, F., Oatey, P. B., Tavare, J. M., and Cullen, P. J. (1998). Nerve growth factor- and epidermal growth factor-stimulated translocation of the ADP-ribosylation factor-exchange factor GRP1 to the plasma membrane of PC12 cells requires activation of phosphatidylinositol 3-kinase and the GRP1 pleckstrin homology domain. *Biochem. J.* **335**, 139–146.

Venkateswarlu, K., Oatey, P. B., Tavare, J. M., and Cullen, P. J. (1998). Insulin-dependent translocation of ARNO to the plasma membrane of adipocytes requires phosphatidyl-inositol 3-kinase. *Curr. Biol.* **8**, 463–466.

[25] Assay and Properties of the GIT1/p95-APP1 ARFGAP

By Ivan de Curtis and Simona Paris

Abstract

GIT1/p95-APP1 is an adaptor protein with an aminoterminal ARFGAP domain involved in the regulation of ARF6 function. GIT1/p95-APP1 forms stable complexes with a number of proteins including downstream effectors and exchanging factors for members of the Rho family of small GTPases. This protein can also interact with other adaptor proteins implicated in the regulation of cell adhesion and synapse formation. The stability of the endogenous and reconstituted complexes after cell lysis allows the biochemical identification and characterization of the GIT1 complexes that can be isolated from different cell types. This article presents methods for the identification of the endogenous and reconstituted GIT1 complexes that can be utilized for the biochemical and functional characterization of the complexes from different tissue and cell types.

Introduction

ARF (ADP-ribosylation factor) small GTPases are important regulators of membrane trafficking (Boman and Kahn, 1995). ARF6 is the unique member of the class III ADP-ribosylation factor (ARF) family of GTPases that has been implicated in the regulation of membrane trafficking between the endosomal compartment and the cell surface (Peters *et al.*, 1995; Radhakrishna *et al.*, 1996). Moreover, ARF6 appears to be functionally linked to the Rho family GTPase Rac1, since both Rac1 and ARF6 colocalize at the plasma membrane and on recycling endosomes, and Rac1-stimulated ruffling is blocked by the GTP binding-defective N27-ARF6 mutant (Radhakrishna *et al.*, 1999). While the details of the mechanisms driven by the ARF1 GTPase have been studied in detail (Spang, 2002), the knowledge on other members of the family is more limited. P95-APP1 is an ADP-ribosylation factor (ARF) GTPase-activating protein (GAP) of the GIT family (Di Cesare *et al.*, 2000), and is the avian orthologue of mammalian GIT1. There are two components of the GIT family: GIT1/p95-APP1/Cat1, and GIT2/PKL/p95-APP2/Cat2 (de Curtis, 2001; Donaldson and Jackson, 2000). The small GTPase ARF6 is a substrate

METHODS IN ENZYMOLOGY, VOL. 404
Copyright 2005, Elsevier Inc. All rights reserved.

in vitro for the members of this family of ARFGAPs (Vitale *et al.*, 2000). Moreover, a number of indications exist that GIT1/p95-APP1 may act as an ARF6 GAP *in vivo*, including its colocalization with ARF6 at the endocytic compartment of transfected fibroblasts and primary neurons (Albertinazzi *et al.*, 2003; Di Cesare *et al.*, 2000). GITs are composite proteins including different domains involved in the assembly of stable ubiquitous multimolecular complexes. GIT1/p95-APP1 is able to interact with PIX, a nucleotide exchanging factor for Rac and Cdc42 GTPases (Manser *et al.*, 1998), and with the focal adhesion protein paxillin (Turner *et al.*, 1999). By forming stable complexes with Pix and with the Rac/Cdc42 effector PAK that binds the Src homology 3 domain of PIX (Manser *et al.*, 1998), p95-APP1/GIT1 can indirectly interact with Rac GTPases in a GTP-dependent manner (Di Cesare *et al.*, 2000).

More recently, new interactions have been identified for the GIT proteins, including the adaptor proteins liprins (Ko *et al.*, 2003), and huntingtin (Goehler *et al.*, 2004). These findings, together with the functional studies performed so far on GIT1/p95-APP1, indicate that this is a ubiquitous adaptor protein that appears to be involved in a number of cellular functions including adhesion and migration, neurite extension, synaptogenesis, and receptor internalization and recycling (Claing *et al.*, 2001). Here we describe biochemical approaches for the identification and characterization of complexes including GIT1/p95-APP1, useful for the identification of new interactors of the complex, and for their functional analysis.

Isolation of the Endogenous GIT1/p95-APP1 Complex

The association of PIX to GIT1/p95-APP1 is quite stable, and allows the identification of the endogenous complex from cell and tissue extracts. Moreover, cotransfection experiments with plasmids coding for the different components of the complex allow the reconstitution of the complexes in the transfected cells. GIT1/p95-APP1 can be identified by pull-down experiments on columns of activated Rac GTP binding proteins, according to the protocol described later.

Expression and Purification of Rac GTPases

The Pix-p95-APP1 complex can be purified by affinity chromatography on Rac-GTP (Di Cesare *et al.*, 2000). Vertebrate Rac small GTP binding proteins include the ubiquitous Rac1, Rac2 that is specifically expressed in hematopoietic cells, and the neural-specific Rac3 (Cantrell, 2003; Corbetta *et al.*, 2005; Ridley, 2001). The open reading frames of individual Rac proteins are cloned in the expression vector pGEX-4T-1 vector (Pharmacia Biotech, Uppsala, Sweden). Competent cells are transformed with the plasmids, and sequences of the purified constructs are confirmed by automatic sequencing. Individual colonies of transformed *E. coli* strain BL21 (DE3) bearing the pLysS plasmid encoding T7 lysozyme are grown, and expression of the fusion protein is induced overnight at room temperature (22–24°) in the presence of 0,1 mM IPTG. Cell suspensions are sonicated on ice 10 times for 30 s each time, at maximum intensity with a Labsonic U Sonicator (Braun Biotech International, Melsungen, Germany). Sonicates are centrifuged at 20,000g for 10 min. Pellets and supernatants from bacterial lysates are then tested for the expression of the recombinant glutathione-S-transferase (GST) fusion proteins on 12% SDS-PAGE. GST-Rac proteins produced under these conditions are largely soluble. The recombinant GST-Rac proteins are purified from bacterial lysates on glutathione-Agarose (from Sigma-Aldrich). One ml of swollen Agarose beads prewashed with PBS are mixed with 10 ml of supernatant obtained from 150 ml of induced bacterial culture. The mix is incubated for 1 h with rotation at 4°. Beads are recovered by centrifugation for 5 min at 2,500g, 4°. Beads are washed four times after mixing with 14 ml of buffer A (50 mM NaCl, 5 mM MgCl2, 1 mM DTT, 50 mM Tris-Cl, pH 7.5). Beads are transferred to columns and washed with 20 ml of buffer A. Protein is eluted by buffer A with 10 mM glutathione (Sigma-Aldrich). One ml fractions are collected. Protein elution is checked by reading OD$_{280}$ against the elution buffer. Peak fractions are pooled and dialyzed against Exchange Buffer (50 mM NaCl, 1 mM EDTA, 1 mM DTT, 20 mM Hepes-NaOH, pH 7.5) to remove glutathione. The eluted, dialyzed protein is checked for purification by SDS-PAGE and Coomassie staining, and for total protein by Bio-Rad (Hercules, CA). Protein can be used immediately, or stored frozen after quick freezing for long-term storage at –20°.

Preparation of Nucleotide-Loaded GTPases

Both endogenous and overexpressed Pix and p95-APP1 proteins form stable complexes in cells (Di Cesare *et al.*, 2000). It is possible to identify the endogenous complexes from cell lines and tissues, or complexes reconstituted by overexpression of the components in cultured cells. One way to identify these complexes is by affinity chromatography using Rac-GTP-coated beads, by the following procedure.

Swollen glutathione-agarose beads are washed twice with 10 volumes of Exchange Buffer. The required amount of purified GST-GTPase is added to the washed beads (e.g., 250 μg of protein added to 25 μl of washed beads). After incubation for 1 h at 4° with rotation, the beads are collected by centrifugation for 3 min at 3,000g, 4°. The unbound material is used for protein assay (Biorad) to check for binding efficiency. Beads are washed twice with 10 volumes of Exchange Buffer, and used for loading with nucleotides.

The following is added to 50 μl of washed beads: 45 μl of Exchange Buffer, and 5 μl of 10 mM nucleotide (GDP-β-S or GTP-γ-S, from SigmaAldrich). Beads are resuspended and incubated for 10 min at 37° in waterbath, gently resuspending beads every 2 min. At the end of the incubation, 5 μl of 100 mM MgCl$_2$ are added (final $\approx$ 5 mM MgCl$_2$), and beads are resuspended and stored on ice.

Isolation of the Endogenous GIT1/p95-APP1 Complex

Add DTT (final 1 mM) and MgCl2 (final 5 mM) to tissue or cell lysate obtained with lysis buffer (1% Triton X-100, 150 mM NaCl, 20 mM Tris-HCl, pH 7.5) including anti-proteases (10 μM each of Chymostatin, Leupeptine, Antipain, and Pepstatin, from Sigma-Aldrich). Spin at 4° for 5 min at 20,000g in a refrigerated microfuge to remove aggregates. The lysate is added to the GTPase-coated beads loaded with nucleotides. For qualitative analysis, between 1 and 2 mg of protein are incubated with 25 μl of beads. For larger preparations, 10–20 mg of protein are incubated with 100 μl of beads. Beads are resuspended and incubated for 1 h at 4° with rotation. Beads are washed 3 times with lysis buffer containing 1 mM DTT and 5 mM MgCl$_2$. The final pellet can be used for analysis by SDS-PAGE and Silver staining or immunoblotting.

When Silver staining of gels is used, for better resolution after SDSPAGE, elution of the bound proteins from the beads is recommended, to reduce the background due to nonspecific binding. The elution is performed either by adding excess free glutathione, or by urea. For elution with glutathione, the beads are incubated with two volumes of 20 mM glutathione (Sigma-Aldrich) in lysis buffer with 1 mM DTT, 5 mM MgCl$_2$, and anti-proteases. After 15 min at 4° with rotation, beads are recovered by centrifugation, and the eluate is separated from beads. The procedure is repeated once, and the two eluates obtained from each sample are pooled.

For elution with urea, beads are incubated with 4 volumes of 8 M urea, 2% CHAPS, 65 mM DTT, at room temperature for 1 h. After separation of the eluate from beads, elution is repeated once, and the two eluates are pooled together. The elution with urea can be used for analysis of the bound proteins by two-dimensional electrophoresis.

The remaining beads are resuspended in two volumes of lysis buffer with 1 mM DTT, 5 mM MgCl2, and anti-proteases. Beads and eluates are then analyzed separately by silver staining after SDS-PAGE. Before SDS-PAGE, the pH of samples eluted with glutathione is neutralized with few μl of 0.5 M Tris-Cl, pH 6.8.

Analysis of the Eluted Fractions by Silver Staining

Qualitative analysis of the eluted fractions was carried out by silver staining. We have utilized a procedure that is optimized to allow analysis by MALDITOF microsequencing of the stained bands, if required (Shevchenko *et al.*, 1996). If the preparation has to be used for microsequencing, the isolation by affinity chromatography on Rac-GTP matrix should be scaled up by a factor of 5 to 10 (referred to as larger preparations in the previous paragraph).

For silver staining after SDS-PAGE, gels are fixed for 20–30 min with destaining solution containing methanol : acetic acid : double distilled (d.d.) water (45 : 5 : 45). All incubations are made in clean glass containers, under slow shaking conditions. For highest sensitivity, the gel is rinsed for 1 h or more. This helps to keep the background transparent during development. After rinsing for about 1 h with d.d. water, the gel is incubated for 2 min with 0.02% sodium thiosulfate. The solution is discarded and the gel is washed with two changes of d.d. water (1 min each). The gel is then incubated for 30 min with chilled (4°) 0.1% silver nitrate solution. The solution is discarded and the gel is washed with two changes of d.d. water (1 min each). The developing solution is then added (0.04% formalin, 2% sodium carbonate), and incubation with slow shaking continued until bands appear. If longer incubations are required, the developing solution may turn yellow, and should then be replaced by fresh developing solution. The development of bands is stopped by discarding the developing solution and by quickly adding 1% acetic acid. The gel can be stored in this solution at 4°.

Figure 1 shows an example of the analysis of the fractions obtained by affinity chromatography on GST-Rac1-coated beads. A clear band at 95 kDa corresponding to GIT1/p95-APP1 is observed in the eluate from GST-Rac1-GTP (arrow). The specificity of the eluted band is confirmed by the lack of a clear p95 band from eluates from GST-Rac1-GDP. Two more specific bands are observed in the eluate from GST-Rac1-GTP at 80 kDa (asterisk) and at 66 kDa (arrowhead). Analysis by either MALDI-TOFF or immunoblotting can be used to confirm that these bands correspond to β-PIX and PAK, respectively (data not shown).

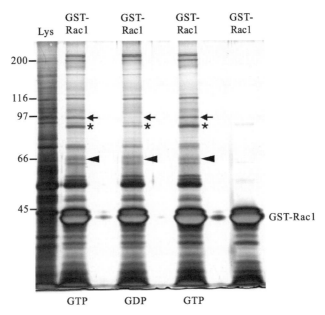

FIG. 1. Identification of the endogenous brain GIT1/p95-APP1 complex by affinity chromatography on Rac1-GTP columns. 100 μg of GST-Rac1 for each sample were adsorbed on 25 μl of glutathione-agarose beads. The adsorbed protein was loaded with GTPγS (GTP) or GDPβS (GDP), washed, and each sample was then incubated for 1 h with rotation at 4° with 2 mg of lysate from E13 avian brain. After washing, samples were eluted for 1 h at room temperature with 8 M urea, and one third of the eluted fractions were loaded on a 8% acrylamide gel. 6 μg of brain lysate (Lys), and eluate from beads loaded with GST-Rac1 and buffer only (last lane to the right) were included. After separation by SDS-PAGE, the gel was silver stained.

Characterization of the GIT1/p95-APP1 Complexes

Sucrose velocity gradients can be utilized to compare the size of protein complexes under different experimental conditions. Both GIT and PIX proteins can form homodimers (Kim *et al.*, 2001, 2003; Paris *et al.*, 2003), and velocity sucrose gradients and immunoprecipitation experiments have been used to determine the oligomeric state of these proteins.

Construction of Plasmids and Expression in Cells

The full-length avian GIT1/p95-APP1 and a fragment including amino acid residues 229–740, including both the PIX- and the paxillin-binding domains, are cloned into the pFLAG–CMV-2 vector (Kodak), to obtain the plasmids pFLAG–p95, and pFLAG–p95-C2, respectively. The pFLAG-p95-LZ monomeric mutant is obtained by mutating leucines 448 and 455 to prolines, by site-directed mutagenesis with the QuikChange™ site-directed mutagenesis kit (Stratagene GmbH, Heidelberg, Germany), using the oligonucleotides 5'-GTGAACAACAGCCCGAGCGATGAGC TGCG CCGGCCGCAGCGCGAGATC-3' and 5'-GATCTCGCGCTGC GGCCGGCG CAGCTCATCGCTCGGGCTGTTGTTCAC-3'. The pXJ 40-HA-βPIX-PG mutant (with tryptophans 43 and 44 changed into proline 43 and glycine 44, respectively) is prepared by site-directed mutagenesis (QuickChange™ kit) on the pXJ40-HA-βPIX plasmid (Manser *et al.*, 1998), with the primers PIXBIS5 (5'-GGAAGGAGGCCCGGGG-GAAGGCACAC-3') and PIXBIS3 (5'-GTGTGCCTTCCCCCGGGCCT CCTTCC-3'). The pXJ40-HA-βPIX-PGΔLZ plasmid is prepared by digesting pXJ40-HA-βPIX-PG with HindIII and BglII, and by ligation of the paired oligonucleotides deltaLZ1 (5'-AGCTTACTGCACAAGTG-CAAAGACGAGGCAGACCCTGAACTCAAGTTCACGCAAAGAG TCTGCTCCACAAGTGCCC GGGTAGA-3') and deltaLZ2 (5'-GA TCTCTACCCGGGCACTTGTGGAGCAGACTCTTTGCGTGAACT TGAGTTCAGGGTCTGCCTCGTCTTTGCACTTGTGCAGTA-3') to introduce a stop codon.

Chicken embryo fibroblasts (CEFs) are obtained from embryonic day-10 chicken embryos, and cultured as described (Albertinazzi *et al.*, 1998). Transient expression of proteins is achieved by transfection of CEFs by the calcium phosphate technique using 10 μg of plasmid DNA for each 6 cm diameter plate. 18–24 h after transfection, cells are extracted with lysis buffer (1% Triton X-100, 150 mM NaCl, 0.5 mg/ml phenyl methyl sulfonyl fluoride (PMSF), 20 mM Tris-Cl, pH 7.5). Cell lysates are clarified by differential centrifugation for 10 min at 18,000g in a refrigerated microcentrifuge, and utilized for analysis by sucrose velocity gradient centrifugation.

Sucrose Velocity Gradients

(a) Analysis of GIT1/p95-APP1. For analysis by sucrose velocity gradients, cell lysates are prepared from cell or tissues for the analysis of the endogenous proteins, or from transfected cells overexpressing dimeric or monomeric GIT1 mutants. Lysates are prepared by adding 9 volumes of lysis buffer (1% Triton X-100, 150 mM NaCl, 0.5 mM PMSF, 20mM Tris-Cl, pH 7.5; alternatively, 0.5–1% octylglucoside can be used instead of Triton X-100). Insoluble pellets are discarded by differential centrifugation. 300 μl aliquots of the cell or tissue lysates are applied to the top of 4.5 ml 5–20% sucrose (w/v, in lysis buffer) gradients, prepared with an

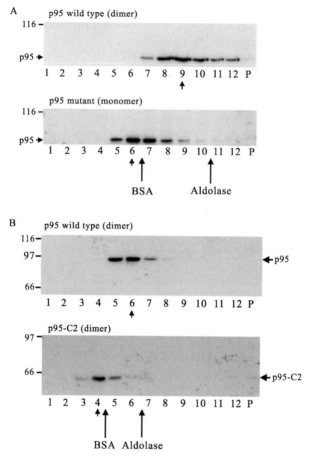

FIG. 2. Analysis of GIT1/p95-APP1 by velocity gradients. (A) 5–20% sucrose gradients (4,5 ml) were loaded with equal volumes of lysates (300 μl, corresponding to 700–800 μg of protein/gradient) from CEFs transfected with wild type FLAG-p95-APP1 (upper blot), or with the monomeric mutant FLAG-p95-LZ (lower blot). After 16 h centrifugation at 189,000g_{av}, 12 gradient fractions and the pellet were collected from the top (left), and analyzed by immunoblotting with anti-FLAG M5 mAb (SIGMA-Aldrich). Small arrows point to the peaks of distribution of the proteins on the gradients. Molecular mass markers are bovine serum albumin (BSA, 66 kDa) and aldolase (158 kDa). (B) 5–20% sucrose gradients were loaded with equal volumes of lysates (300 μl, corresponding to 360 μg of protein/gradient) from CEFs transfected with wild type FLAG-p95-APP1 (upper blot), or with the truncated mutant FLAG-p95-C2 (lower blot). After 9 h centrifugation at 189,000g_{av}, 12 fractions and the pellet were collected from the top (left), and analyzed by immunoblotting with anti-FLAG M5 mAb (SIGMA-Aldrich). One fourth of each fraction was loaded on the gel. Small arrows point to the peaks of distribution of the proteins on the gradients. Molecular mass markers are bovine serum albumin (BSA, 66 kDa) and aldolase (158 kDa).

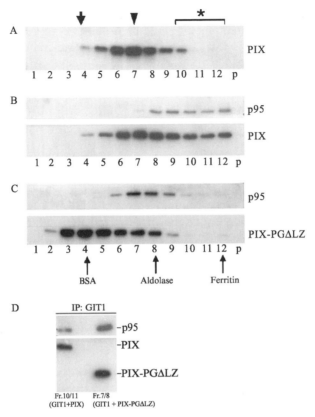

FIG. 3. PIX mediates the formation of large hetero-complexes with GIT1. 5–20% sucrose velocity gradients were loaded with equal amounts of lysates (240 μg protein/gradient) from CEFs transfected with wild-type βPIX (A), cotrasfected with wild-type GIT1 and βPIX (B), or wild-type GIT1 and the monomeric SH3 mutant βPIX-PGΔLZ (C). Gradients were centrifuged at 189,000g_{av} for 10 h. After centrifugation, 12 fractions and the pellet were collected from the top (left). The arrow, arrowhead, and asterisk indicate the position of the three peaks of protein distribution identified on the gradients. The distribution of protein standards loaded on a similar gradient showed peaks at fractions 5 (bovine serum albumin, 66 kD), 8 (aldolase, 158 kD), and 12 (ferritin, 440 kD). (D) Coprecipitation of monomeric PIX-PGΔLZ with GIT1. Fractions 10 and 11 from the gradient loaded with a lysate from CEFs transfected with GIT1 and PIX (experiment shown in panel B), and fractions 7 and 8 from the gradient loaded with a lysate from CEFs transfected with GIT1 and PIX-PGΔLZ (experiment shown in panel C) were immunoprecipitated with anti-GIT antibodies. Filters were blotted with anti-GIT (top) and anti-PIX (bottom) antibodies to detect the coprecipitating PIX polypeptides.

Autodensiflow gradient maker (Büchler, Labconco, Kansas City, MO). For a good separation of monomeric and dimeric full length GIT1, gradients are run for 16 h at 189,000g_{av} in a SW50.1 rotor (Beckman Coulter Inc., Fullerton, CA). For an indication of the molecular mass of the complexes, protein markers (0.4 mg of each of the following: bovine serum albumin, 66 kDa; transferrin, 76 kDa; aldolase, 158 kDa; catalase, 252 kDa; ferritin, 440 kDa) resuspended in 300 μl of lysis buffer are run on a separate gradient. Twelve 400-μl fractions are collected from the top of the gradients using the Autodensiflow gradient maker, and the pellets are resuspended in 400 μl of lysis buffer. About one fourth of each fraction is analyzed by SDS-PAGE and immunoblotting

Comparison of gradients loaded with cell extracts from cells transfected with either wild type or mutant forms of GIT1/p95-APP1 allows us to distinguish between the dimeric protein and the mutant monomeric protein (Fig. 2A), or between full-length dimers and dimeric fragments (Fig. 2B).

(b) Analysis of the Complexes between βPIX and GIT1/p95-APP1. On top of their ability to interact with each other, PIX and GIT proteins can form homodimers (Feng *et al.*, 2004; Kim *et al.*, 2001, 2003; Paris *et al.*, 2003). Therefore, larger complexes between the two proteins can be formed. Large endogenous complexes of GIT1 and PIX have been identified (Paris *et al.*, 2003). They could also be induced by overexpression of the two full-length proteins, as detected by velocity centrifugation on 5–20% sucrose gradients loaded with lysates from cotransfected cells (Fig. 3B), when compared to gradients loaded with extracts from cells overexpressing either PIX (Fig. 3A), or GIT1 alone (Paris *et al.*, 2003). Coexpression of the monomeric PIX-PGΔLZ mutant with wild type GIT1 prevented the formation of the larger complexes (Fig. 3C), although hetero-complexes between GIT and PIX-PGΔLZ are still formed (Fig. 3D). Under these conditions, the distribution of the PIX mutant extended to the upper part of the gradient, where the monomeric form of the overexpressed PIX-PGΔLZ migrated (Fig. 3C). Therefore, dimeric PIX may regulate the assembly of large oligomeric complexes with GIT1/p95-APP1. The use of velocity gradients represents a useful approach for the biochemical analysis of endogenous and reconstituted GIT1/p95-APP1 complexes.

Acknowledgments

This work was supported by Telethon-Italy (grant no. GGP02190), and by the "Progetto MIUR-CNR Genomica Funzionale-Legge 449/97."

References

Albertinazzi, C., Gilardelli, D., Paris, S., Longhi, R., and de Curtis, I. (1998). Overexpression of a neural-specific Rho family GTPase, cRac1B, selectively induces enhanced neuritogenesis and neurite branching in primary neurons. *J. Cell Biol.* **142,** 815–825.

Albertinazzi, C., Za, L., Paris, S., and de Curtis, I. (2003). ADP-ribosylation factor 6 and a functional PIX/p95-APP1 complex are required for Rac1B-mediated neurite outgrowth. *Mol. Biol. Cell.* **14,** 1295–1307.

Boman, A. L., and Kahn, R. A. (1995). Arf proteins: The membrane traffic police? *Trends Biochem. Sci.* **20,** 147–150.

Cantrell, D. A. (2003). GTPases and T cell activation. *Immunol. Rev.* **92,** 122–130.

Claing, A., Chen, W., Miller, W. E., Vitale, N., Moss, J., Premont, R. T., and Lefkowitz, R. J. (2001). Beta-arrestin-mediated ADP-ribosylation factor 6 activation and beta 2-adrenergic receptor endocytosis. *J. Biol. Chem.* **276,** 42509–42513.

Corbetta, S., Gualdoni, S., Albertinazzi, C., Paris, S., Croci, L., Consalez, G. G., and de Curtis, I. (2005). Generation and characterization of Rac3 knockout mice. *Mol. Cell. Biol.* **25,** 5763–5776.

de Curtis, I. (2001). Cell migration: GAPs between membrane traffic and the cytoskeleton. *EMBO Rep.* **2,** 277–281.

Di Cesare, A., Paris, S., Albertinazzi, C., Dariozzi, S., Andersen, J., Mann, M., Longhi, R., and de Curtis, I. (2000). P95-APP1 links membrane transport to Rac-mediated reorganization of actin. *Nat. Cell Biol.* **2,** 521–530.

Donaldson, J. G., and Jackson, C. L. (2000). Regulators and effectors of the ARF GTPases. *Curr. Opin. Cell Biol.* **12,** 475–482.

Feng, Q., Baird, D., and Cerione, R. A. (2004). Novel regulatory mechanisms for the Dbl family guanine nucleotide exchange factor Cool-2/alpha-Pix. *EMBO J.* **23,** 3492–3504.

Goehler, H., Lalowski, M., Stelzl, U., Waelter, S., Stroedicke, M., Worm, U., Droege, A., Lindenberg, K. S., Knoblich, M., Haenig, C., Herbst, M., Suopanki, J., Scherzinger, E., Abraham, C., Bauer, B., Hasenbank, R., Fritzsche, A., Ludewig, A. H., Buessow, K., Coleman, S. H., Gutekunst, C. A., Landwehrmeyer, B. G., Lehrach, H., and Wanker, E. E. (2004). A protein interaction network links GIT1, an enhancer of huntingtin aggregation, to Huntington's disease. *Mol. Cell* **15,** 853–865.

Kim, S., Lee, S. H., and Park, D. (2001). Leucine zipper-mediated homodimerization of the p21-activated kinase-interacting factor, βPix. Implication for a role in cytoskeletal reorganization. *J. Biol. Chem.* **276,** 10581–10584.

Kim, S., Ko, J., Shin, H., Lee, J. R., Lim, C., Han, J. H., Altrock, W. D., Garner, C. C., Gundelfinger, E. D., Premont, R. T., Kaang, B. K., and Kim, E. (2003). The GIT family of proteins forms multimers and associates with the presynaptic cytomatrix protein Piccolo. *J. Biol. Chem.* **278,** 6291–6300.

Ko, J., Kim, S., Valtschanoff, J. G., Shin, H., Lee, J. R., Sheng, M., Premont, R. T., Weinberg, R. J., and Kim, E. (2003). Interaction between liprin-alpha and GIT1 is required for AMPA receptor targeting. *J. Neurosci.* **23,** 1667–1677.

Manser, E., Loo, T. H., Koh, C. G., Zhao, Z. S., Chen, X. Q., Tan, L., Tan, I., Leung, T., and Lim, L. (1998). PAK kinases are directly coupled to the PIX family of nucleotide exchange factors. *Mol. Cell* **1,** 183–192.

Paris, S., Longhi, R., Santambrogio, P., and de Curtis, I. (2003). Leucine-zipper-mediated homo- and hetero-dimerization of GIT family p95-ARF GTPase-activating protein, PIX-, paxillin-interacting proteins 1 and 2. *Biochem. J.* **372,** 391–398.

Peters, P. J., Hsu, V. W., Ooi, C. E., Finazzi, D., Teal, S. B., Oorschot, V., Donaldson, J. G., and Klausner, R. D. (1995). Overexpression of wildtype and mutant ARF1 and ARF6: Distinct perturbations of nonoverlapping membrane compartments. *J. Cell Biol.* **128,** 1003–1017.

Radhakrishna, H., Klausner, R. D., and Donaldson, J. G. (1996). Aluminum fluoride stimulates surface protrusions in cells overexpressing the ARF6 GTPase. *J. Cell Biol.* **134,** 935–947.

Radhakrishna, H., Al-Awar, O., Khachikian, Z., and Donaldson, J. G. (1999). ARF6 requirement for Rac ruffling suggests a role for membrane trafficking in cortical actin rearrangements. *J. Cell Sci.* **112,** 855–866.

Ridley, A. J. (2001). Rho family proteins: Coordinating cell responses. *Trends Cell Biol.* **11,** 471–477.

Shevchenko, A., Wilm, M., Vorm, O., and Mann, M. (1996). Mass spectrometric sequencing of proteins from Silver stained polyacrilamide gels. *Anal. Chem.* **68,** 850–858.

Spang, A. (2002). ARF1 regulatory factors and COPI vesicle formation. *Curr. Opin. Cell Biol.* **14,** 423–427.

Turner, C. E., Brown, M. C., Perrotta, J. A., Riedy, M. C., Nikolopoulos, S. N., McDonald, A. R., Bagrodia, S., Thomas, S., and Leventhal, P. S. (1999). Paxillin LD4 motif binds PAK and PIX through a novel 95-kD ankyrin repeat, ARF-GAP protein: A role in cytoskeletal remodeling. *J. Cell Biol.* **145,** 851–863.

Vitale, N., Patton, W. A., Moss, J., Vaughan, M., Lefkowitz, R. J., and Premont, R. T. (2000). GIT proteins, A novel family of phosphatidylinositol 3,4,5-trisphosphate-stimulated GTPase-activating proteins for ARF6. *J. Biol. Chem.* **275,** 13901–13906.

[26]   Preparation and Characterization of Recombinant Golgin Tethers

By AYANO SATOH, MATTHEW BEARD, and GRAHAM WARREN

Abstract

Golgin tethers are integral or peripheral Golgi proteins with predicted coiled-coil domains and many are known to interact directly with small GTPases of the Ypt/Rab or Arl families. Here we describe the preparation of recombinant golgins: GM130, p115 (and truncations thereof), the N-terminal fragment of giantin, CASP, and golgin-84.

Introduction

In 1984, autoantibodies recognizing the Golgi apparatus were found in the serum of an autoimmune disease patient and this led to the cloning of the first golgin (Fritzler et al., 1993). Since that time a number of other Golgi autoantigens have been characterized though anti-Golgi autoantibodies have only been found in about 0.1% of all autoimmune diseases (Stinton et al., 2004). The term golgins now covers proteins that were not identified by autoantibodies and refers to Golgi proteins that contain extensive regions of predicted coiled-coil, implying that they form long, rod-like structures. They are either directly embedded in the Golgi membrane by spanning domains or are indirectly attached by proteins such as other golgins, GRASP, or small GTPases of the Ypt/Rab or Arl families. The exception is the SNARE family, which clearly belongs in a separate category (Barr and Short, 2003; Jackson, 2004; Shorter and Warren, 2002). Golgins have been implicated in the generation and maintenance of Golgi structure as well as in the control of the intra-Golgi flow of COPI vesicles (Barr and Short, 2003; Malsam et al., 2005; Shorter and Warren, 2002). This flow is thought to ensure the presentation of cargo to those enzymes that carry out post-translational modifications, in the correct sequence and for the appropriate length of time. Golgins may contribute to this process by helping the COPI vesicles to target to the correct compartment. They are thought to be part of the initial tethering process, which eventually leads to SNARE-mediated fusion (Pfeffer, 1999; Waters and Hughson, 2000).

The best-characterized golgin-tethering complex is that containing GM130, p115, and Giantin, which has been implicated in anterograde transport (Linstedt and Hauri, 1993; Nakamura et al., 1995; Shorter and Warren,

METHODS IN ENZYMOLOGY, VOL. 404
0076-6879/05 $35.00
DOI: 10.1016/S0076-6879(05)04026-7

2002; Waters *et al.*, 1992). GM130 is anchored to Golgi membranes via GRASP-65, a cisternal stacking protein (Barr *et al.*, 1997). Giantin is anchored in COPI vesicles via a spanning domain. p115 is a peripheral protein that can bind to GM130 and to Giantin, suggesting that it might form a bridge linking the vesicle to the membrane. Alternatively, or in addition, these interactions might stimulate the binding of p115 to Rab1, leading to p115 catalyzed assembly of those SNARE complexes needed for vesicle fusion (Puthenveedu and Linstedt, 2004; Shorter *et al.*, 2002).

More recently, a golgin-tethering complex containing golgin-84 and CASP has been described that has been implicated in the retrograde transport of Golgi enzymes in COPI vesicles (Malsam *et al.*, 2005). Both are spanning proteins (Bascom *et al.*, 1999; Gillingham *et al.*, 2002), with golgin-84 in the vesicle and CASP in the cisternal membrane (Malsam *et al.*, 2005). Rab1 is part of this complex, since it is known to bind to golgin-84 (Diao *et al.*, 2003; Satoh *et al.*, 2003). Other components remain to be characterized since it appears that golgin-84 and CASP do not interact directly.

Golgins are minor components of Golgi membranes so that preparation of sufficient amounts for biochemical analysis is difficult. Here we describe the preparation of recombinant forms of those golgins that constitute the tethering complexes just described.

Methods

Purification of Recombinant GM130

Plasmids: The open reading frame (ORF) (excluding the stop codon) for rat GM130 (Genbank accession number NM_022596) is amplified by PCR, incorporating NdeI and XhoI restriction sites into the primers. Primers are: 5'primer GM130 N-terminus: CCCCTCCCCGCCCCCA-TATGTCGGAAGAAACC (NdeI site underlined); 3'primer GM130-his AS: GCCGAAAGTCTCGAGTATAACCATGATTTTCACC (XhoI Site underlined). The PCR product is subcloned (in-frame) into pET23a (+) (Novagen, San Diego, CA) between the NdeI and XhoI sites. This plasmid allows expression, in bacteria expressing the T7 polymerase, of the encoded ORF as a fusion protein with a hexahistidine (his)-tag at the C-terminus. The recombinant plasmid (GM130-his/pET) should be verified by sequencing across the entire encoded ORF and transformed into BL21 (DE3) *E. coli* (Stratagene, La Jolla, CA) (which express T7 polymerase) for expression. A second expression plasmid is also constructed to encode GM130 fused to a C-terminal maltose binding protein (MBP) affinity-tag. The ORF for MBP is amplified by PCR from the plasmid pMALc2x (New

England Biolabs [NEB], Ipswich, MA), incorporating XhoI restriction sites into the primers. Primers are: 5'primer MBP Xho Sense: GACCA-TAG<u>CTCGAG</u>AAAATCGAAGAAGG (XhoI site underlined); 3'primer MBP Xho antisense: CGAATTCTGAAAT<u>CTCGAG</u>CTATTACCCG-AGGTTGTTG (XhoI Site underlined). The PCR product is subcloned into GM130-his/pET23a, in-frame and contiguous with the GM130 ORF. The recombinant plasmid (GM130-MBP/pET) should be verified by sequencing across the entire ORF and transformed into BL21(DE3) *E. coli* for expression. Note that GM130MBP, constructed by this strategy, retains an extreme C-terminal his-tag. If desired this could easily be deleted by QuikChange mutagenesis (Stratagene) prior to sequencing and transformation.

Protein expression: A single colony (or stab from a clonal glycerol stock) of BL21(DE3) *E. coli* transformed with GM130 expression vector is used to inoculate a starter culture (10 ml per 400 ml expression culture) of Luria-Bertani (LB) broth (1% tryptone, 0.5% yeast extract, 1% NaCl, pH ~7.5) supplemented with 0.1 mg/ml ampicillin. The starter culture is grown by shaking at 37° for 7–10 h then diluted into a larger, expression culture of LB broth supplemented with 0.1 mg/ml ampicillin. For MBP-tagged fusion proteins, glucose (2 mg/ml) is also added to the expression culture to repress amylase expression, as this enzyme degrades the amy-lose-affinity resin used in purification of MBP fusion proteins. The expression culture is grown with shaking at 27° for 12–16 h (overnight). Induction with isopropyl β-D-thiogalactoside (IPTG) is not necessary because BL21 (DE3) *E. coli* express sufficient T7 polymerase to allow low-level transcription from T7 promoters even without IPTG induction of T7 polymerase expression.

Protein purification: Bacteria are harvested by centrifugation (3000g, 4°, 20 min) and resuspended in lysis buffer (25 mM HEPES pH 7.4, 100 mM KCl, 2mM β-mercaptoethanol, EDTA free protease inhibitor cocktail [Roche, Indianapolis, IN]). The bacteria are lysed in an EmulsiFlex-C5 homogenizer (Avestin, Ottawa, Ontario, Canada) or by sonication and the lysate is cleared by centrifugation (17,600g, 4°, 20 min) to remove particulate matter. The cleared supernatant is adjusted (with lysis buffer) to a total volume of 50 ml for each 400 ml of expression culture that was harvested. The appropriate affinity resin, either amylose-resin (NEB) (for MBP-tagged proteins) or Ni-NTA-agarose (Qiagen, Valencia, CA) (for his-tagged proteins) is pre-equilibrated in lysis buffer, then added to the cleared supernatant (0.5 ml bed volume/400 ml expression culture). The cleared supernatant and the affinity resin mix are incubated in batch with end-over-end rotation at 4° for 1 h. The affinity resin is harvested by a brief centrifugation (2000g, 4°, 1 min) and washed three times (2 bed volumes

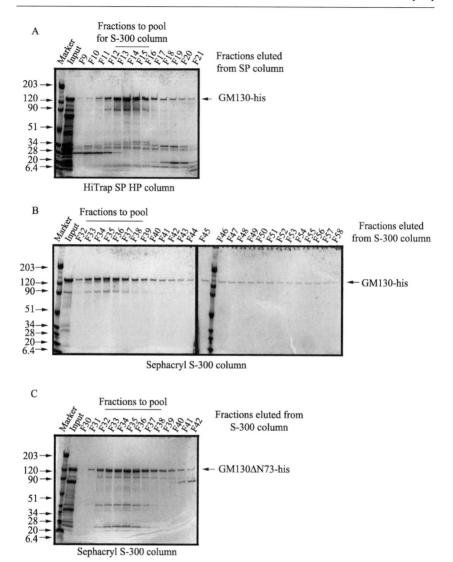

Fig. 1. Purification of recombinant GM130-his and GM130ΔN73-his. Recombinant GM130-his expressed in BL21(DE3) *E. coli* and concentrated by Ni-NTA-agarose affinity chromatography was (A) loaded onto a HiTrap SP HP cation-exchange column and eluted with a 30 ml 0.1–1.0 *M* KCl gradient. Eight-μl aliquots of the load and each 0.5 ml fraction were analyzed by SDS-PAGE with Coomassie staining (fractions 9–21 are shown). Peak fractions were pooled and (B) loaded onto a Hi-Prep 16/60 Sephacryl S-300 gel-filtration column. Eight-μl aliquots of the load and each 1 ml fraction were analyzed by SDS-PAGE with Coomassie staining (fractions 32–58 are shown). Peak fractions containing purified GM130-his were pooled and snap frozen as aliquots. (C) Recombinant GM130ΔN73-his

lysis buffer, 5 min on ice each). Bound protein is then eluted with the appropriate elution buffer: MBP-tag elution buffer (25 mM HEPES pH 7.4, 100 mM KCl, 2 mM β-mercaptoethanol, 0.05% Triton-X100, 10 mM maltose), or his-tag elution buffer (25 mM HEPES pH 7.4, 100 mM KCl, 2 mM β-mercaptoethanol, 0.05% TritonX-100, 350 mM imidazole). Elution is performed in batch, 3 elutions, 1 bed volume, incubated with end-over-end rotation for 10 min each at 4°. The eluted material should be pooled and desalted into buffer A (25 mM HEPES pH 7.4, 100 mM KCl, 2 mM β-mercaptoethanol, 0.05% Triton-X100) on a 10 DG desalting column (BioRad, Hercules, CA). The desalted eluate, which is not sufficiently pure to be used directly (Fig. 1A, Input), is filtered (0.45 μm) and loaded onto a 5 ml HiTrap SP HP cation-exchange column (Amersham Biosciences, Piscataway, NJ), equilibrated in lysis buffer, at a flow-rate of 2 ml/min. The column is washed with 30 ml buffer A at 2 ml/min then eluted at a flow-rate of 1 ml/min by a 30 ml linear gradient 0–100% buffer B (25 mM HEPES pH 7.4, 1 M KCl, 2 mM β-mercaptoethanol, 0.05% Triton-X100). Half ml fractions are collected and 8-μl aliquots are analyzed by SDS-PAGE with Coomassie staining to monitor the elution (an example is shown in Fig. 1A for his-tagged GM130). Fractions containing the protein of interest should be pooled and loaded onto a Hi-Prep 16/60 Sephacryl S-300 gel-filtration column (Amersham Biosciences) equilibrated in buffer A at a flow rate of 0.5 ml/min. One-ml fractions are collected and 8-μl aliquots are analyzed by SDS-PAGE with Coomassie staining to monitor the elution (an example is shown in Fig. 1B). Fractions containing the GM130-his or GM130-MBP should be pooled. Aliquots should be snap-frozen (liquid nitrogen) and stored at −80°. Protein concentration can be estimated by comparison with bovine serum albumin (BSA) standards on Coomassie stained gels. An example of pooled, purified GM130-his is shown in Fig. 2A. Typically, a 400-ml culture yielded 0.2 mg of GM130-his, which was judged 90% pure by Coomassie staining.

Purification of GM130ΔN73

Plasmids: The codons for the first 73 residues of GM130 (1–73) are deleted by QuikChange mutagenesis of GM130-his/pET to generate the plasmid GM130ΔN73-his/pET. Primers are: QM130deltaN73QCS:

---

expressed in BL21(DE3) *E. coli* and concentrated by Ni-NTA-agarose affinity chromatography was loaded onto a Hi-Prep 16/60 Sephacryl S-300 gel-filtration column. Eight-μl aliquots of the load and each 1 ml fraction were analyzed by SDS-PAGE with Coomassie staining (fractions 30–42 are shown). Peak fractions containing purified GM130ΔN73-his were pooled and snap frozen as aliquots.

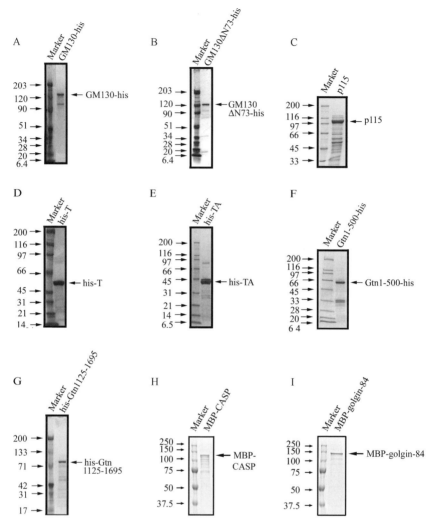

FIG. 2. Analysis of purified recombinant golgins. Coomassie stained SDS-PAGE gels showing purified (A) GM130-his, (B) GM130ΔN73-his, (C) p115, (D) his-T, (E) his-TA, (F) Giantin1-500-his, (G), his-Giantin1125-1695, (H) MBP-CASP, and (I) MBP-golgin-84. Marker = molecular weight markers.

GGAGATATACATATGACTATGTTTCTTGGTGTCGTCCC;   QM-
130deltaN73QCAS: GGGACGACACCAAGAAACATAGTCATATG-
TATATCTCC. The mutagenized plasmid should be verified by sequencing across the entire ORF and transformed into BL21(DE3) *E. coli* for
expression.

Protein expression: GM130ΔN73-his should be expressed as for full-
length GM130-his (above).

Protein purification: GM130ΔN73-his is purified as for full-length
GM130-his (above), except that the HiTrap SP HP cation-exchange chromatography step is omitted. This is because GM130ΔN73-his does not bind
to cation-exchange resin at neutral pH. Full-length GM130 binds to cation-
exchange resin because of the basic character of its N-terminal region
(which is deleted in GM130ΔN73-his). The cation-exchange step is retained in the purification protocol of full-length GM130 even though it
contributes little to the overall purification. This is because sequential
purification steps, first by a C-terminal affinity tag then by the basic character of the N-terminal domain, select for full-length molecules. An example of the purification achieved for GM130ΔN73-his by concentration on
Ni-NTA-agarose (Qiagen) followed directly by Sephacry S-300 gel-filtration is shown in Fig. 1C. An example of pooled, purified GM130ΔN73-his
is shown in Fig. 2B. Typically, a 400-ml culture yielded 0.2 mg of
GM130ΔN73-his, which was judged 90% pure by Coomassie staining.

Purification of Recombinant p115

Plasmids: The ORF (including the stop codon) for bovine p115 (Genbank accession number NM_174845) is amplified by PCR, incorporating
BamHI and NotI restriction sites into the primers. Primers are: 5′primer
Jo7: CCGCGGCTGAGC<u>GGATCC</u>ATGAATTTCCTCCG (BamHI site
underlined); 3′primer MB1: GTCTCTATCAT<u>GCGGCCGC</u>ATTT-
TAAAGC (NotI site underlined). The PCR product is subcloned (in-
frame) into pGEX-6P-1 (Amersham Biosciences) between the BamHI
and NotI sites. This plasmid allows bacterial expression of the encoded
ORF as a fusion protein with glutathione-S-transferase (GST) and short
linker sequence containing a PreScission protease (Amersham Biosciences) recognition site, at the N-terminus. The recombinant plasmid
(GST-p115/pGEX) is verified by sequencing across the encoded ORF
and transformed into BL21 *E. coli* for expression. A second expression
plasmid is also constructed to encode p115, fused to an N-terminal MBP-
tag. The PreScision protease site (from the linker sequence) together with
the p115 ORF is amplified by PCR incorporating EcoRI and XbaI restriction sites into the primers. Primers are: 5′primer MBP-PreScission p115

sense: CCTCCA<u>GAATTC</u>GATCTGGAAGTTCTGTTCC (EcoRI site underlined); 3′primer MBP-PreScission p115 antisense: CGCAT<u>TTCTA-GA</u>CTAGATATGACC (XbaI Site underlined). The PCR product is subcloned (inframe) into pMALc2x (NEB) between the EcoRI and XbaI sites. This plasmid allows bacterial expression of the encoded ORF as a fusion protein with MBP at the N-terminus. The recombinant plasmid (MBP-p115/pMAL) should be verified by sequencing across the encoded ORF and transformed into BL21 *E. coli* for expression.

Protein expression: A single colony (or stab from a clonal glycerol stock) of BL21 *E. coli* transformed with p115 expression vector is used to inoculate a starter culture (10 ml per 400 ml expression culture) of LB broth, supplemented with 0.1 mg/ml ampicillin. The starter culture is grown with shaking at 37° for 12–16 h (overnight), then diluted into a larger, expression culture of LB broth supplemented with 0.1 mg/ml ampicillin. For MBP-tagged fusion proteins 2 mg/ml glucose is also added to the expression culture. The expression culture is grown with shaking at 37° for 3–4 h until the OD at 600 nm reached 0.6–1.0. Expression is then induced by addition of IPTG (0.1 mM final concentration). Growth is continued for a further 4–6 h shaking at 27°.

Protein purification: Bacteria are harvested by centrifugation (3000g, 4°, 20 min) and resuspended in PreScission lysis buffer (50 mM Tris-HCl pH 7.0, 150 mM NaCl, 1 mM EDTA, 1 mM DTT), supplemented with EDTA free protease inhibitor cocktail. The bacteria are lysed in an EmulsiFlex-C5 homogenizer or by sonication, and the lysate is cleared by centrifugation (17,600g, 4°, 20 min) to remove particulate matter. The cleared supernatant is adjusted (with PreScission lysis buffer, supplemented with EDTA free protease inhibitor cocktail, and adjusted to 0.1% TritonX100) to a total volume of 50 ml for each 400 ml of expression culture that was harvested. The appropriate affinity resin, either amylose-resin (for MBP-tagged proteins) or glutathione-Sepharose 4B (Amersham Biosciences) (for GST-tagged proteins) is pre-equilibrated in PreScission buffer then added to the cleared supernatant (0.5 ml bed volume/400 ml expression culture). The cleared supernatant and the affinity resin mix are incubated as a batch, rotating end-over-end at 4° for 1 h. The affinity resin is harvested by brief centrifugation (2000g, 4°, 1 min) and washed three times with 2 bed volumes of PreScission proteolysis buffer (50 mM Tris-HCl pH 7.0, 150 mM NaCl, 1 mM EDTA, 1 mM DTT, 10% glycerol, 0.1% Triton-X100) for 5 min on ice each. Bound protein is then eluted by on-column cleavage with PreScission protease (Amersham Biosciences); the washed affinity resin is resuspended in 1 ml PreScission proteolysis buffer, supplemented with 40 U PreScission protease/400 ml expression culture. The on-column digestion is incubated with end-over-end rotation at 4° for

5 h. The affinity resin is pelleted by a brief centrifugation (10-sec pulse in a benchtop microcentrifuge) and the supernatant containing the released (untagged) p115 is collected. This supernatant is incubated with glutathione-Sepharose 4B (0.25 ml bed volume/400 ml expression culture) pre-equilibrated in PreScission buffer with end-over-end rotation at 4° for 1 h. This is to deplete the (GST-tagged) PreScission protease. The affinity resin is pelleted by a brief centrifugation (10-sec pulse in a benchtop microcentrifuge) and the supernatant (containing the p115) is collected and desalted into p115 buffer (25 mM HEPES pH 7.4, 100 mM KCl, 2 mM β-mercaptoethanol, 10% glycerol) on a 10 DG desalting column (BioRad). Fractions containing p115 should be pooled. Aliquots should be snap-frozen (liquid nitrogen) and stored at −80°. Protein concentration can be estimated by comparison with BSA standards on Coomassie stained gels. An example of pooled, purified p115 is shown in Fig. 2C. Typically, a 400-ml culture yielded 0.3 mg of p115, which was judged 80–90% pure by Coomassie staining.

Purification of his-TA and other Truncations within the p115 Tail-Acidic Domain

Plasmids: The codons for R5 and G6 of bovine p115 are mutagenized to a StuI restriction site (AGGCCT) by site-directed mutagenesis. A second round of mutagenesis is then performed to change the codons I651 and V652 to a PmlI site (CACGTG). The mutagenized plasmid is then digested with StuI and PmlI and religated to generate an ORF encoding the first five amino acids of p115 fused with V652 to the C-terminus. This region corresponds to the tail (T) and acidic (A) domains of p115. The construct is named TA. The ORF for TA is then subcloned (in-frame) into pQE9 (Qiagen). This plasmid allows expression of the encoded ORF as a fusion protein with an N-terminal his-tag (Dirac-Svejstrup et al., 2000). Further truncations and deletions within the TA region of p115 have also been constructed by Quikchange mutagenesis (Stratagene) of his-TA/pQE9. Recombinant plasmids should be verified by sequencing across the ORF and then transformed into XL1-Blue E. coli (Stratagene) for expression.

Protein expression: A single colony (or stab from a clonal glycerol stock) of XL1-Blue E. coli transformed with his-TA/pQE9 (or a derived expression vector) is used to inoculate a starter culture (10 ml per 400 ml expression culture) of LB broth, supplemented with 0.1 mg/ml ampicillin. The starter culture is grown with shaking at 37° for 12–16 h (overnight) then diluted into a larger, expression culture of LB broth supplemented with 0.1 mg/ml ampicillin. The expression culture is grown with shaking at 37° for 3–4 h until the OD at 600 nm reached 0.6–1.0. Expression is then induced by addition of IPTG (0.1 mM final concentration). Growth is continued for a further 3–4 h shaking at 27°.

Protein purification: Bacteria are harvested by centrifugation (3000g, 4°, 20 min) and resuspended in his-tag lysis buffer (50 mM NaH$_2$PO$_4$ pH 7.4, 500 mM NaCl, 2 mM β-mercaptoethanol, EDTA free protease inhibitor cocktail). The bacteria are lysed in an EmulsiFlexC5 homogenizer or by sonication and the lysate is cleared by centrifugation (17,600g, 4°, 20 min) to remove particulate matter. The cleared supernatant is adjusted (with lysis buffer) to a total volume of 50 ml for each 400 ml of expression culture that was harvested and adjusted to 10 mM imidazole. Ni-NTA-agarose (Qiagen) is pre-equilibrated in wash buffer (50 mM NaH$_2$PO$_4$ pH 7.4, 500 mM NaCl, 10 mM imidazole 2 mM β-mercaptoethanol, EDTA free protease inhibitor cocktail) then added to the cleared supernatant (0.5 ml bed volume/400 ml expression culture). The cleared supernatant and the affinity resin mix are incubated in batch with end-over-end rotation at 4° for 1 h. The affinity resin is harvested by a brief centrifugation (2000g, 4°, 1 min) and washed three times (2 bed volumes wash buffer, 5 min on ice each). Bound protein is then eluted with his-tag elution buffer (25 mM HEPES pH 7.4, 100 mM KCl, 2 mM β-mercaptoethanol, 0.05% TritonX100, 350 mM imidazole). Elution is performed in batch, 3 elutions of 1 bed volume, incubated with end-over-end rotation for 10 min each at 4°. The eluted material should be pooled and desalted into buffer A (25 mM HEPES pH 7.4, 100 mM KCl, 2 mM β-mercaptoethanol) on a 10 DG desalting column. Fractions containing his-TA are pooled. Aliquots should be snap-frozen (liquid nitrogen) and stored at −80°. Protein concentration can be estimated by comparison with BSA standards on Coomassie stained gels. A pooled and purified example of both his-T and his-TA is shown in Fig. 2D and E, respectively. Typically, a 400-ml culture yielded 4 mg of his-T, which was judged 90% pure by Coomassie staining. For his-TA the corresponding values are 4 mg and 90% purity.

Purification of Recombinant Giantin Fragments

Plasmids: The codons for residues 1–1197 in rat Giantin (GCP360) (Genbank accession number BAA05026) are amplified by PCR, incorporating NdeI and XhoI restriction sites into the primers. Primers are: (5′ primer MB3: GGAGCGCTCCTTCA*CATATG*CTGAGCCGATTATCG NdeI site underlined; 3′primer MB4: CGTGAGTGGGCTC*CTCGAGG*CCAGTGCCTGG XhoI site underlined. The PCR product is subcloned (in-frame) into pET23a(+) (Novagen) between the NdeI and XhoI sites. This plasmid allows expression, in bacteria expressing the T7 polymerase, of the encoded ORF as a fusion protein with a his-tag at the

C-terminus. The recombinant plasmid (Gtn1-1197-his/pET) should be verified by sequencing across the entire encoded ORF and transformed into BL21(DE3) *E. coli* (which express T7 polymerase) for expression. However, the encoded protein (although soluble and well expressed) appeared to precipitate onto the affinity resin during purification. Therefore a second expression plasmid was derived, encoding Giantin 1-500-his. The coding sequence between Giantin 500 and the start of the his-tag in Giantin 1-1197-his/pET is deleted by QuikChange mutagenesis (Stratagene) (5′primer QC1 Sense Gntn: AGCGACTCTTCTACTCTC-GAGCACCACCAC; 3′primer QC1 Antisense Gntn: GTGGTGGTGC-TCGAGAGTAGAAGAGTCGCT). Gtn1-500-his/pET should be verified by sequencing across the entire ORF and transformed into BL21(DE3) *E. coli* for expression. A third expression plasmid encoding the N-terminal region of Giantin fused to a C-terminal MBP-affinity tag is also constructed. The ORF for MBP is amplified by PCR from the plasmid pMALc2x, incorporating XhoI restriction sites into the primers. Primers are: 5′primer MBP Xho Sense: GACCATAGCTCGAGAAAATCGAA-GAAGG (XhoI Site underlined); 3′primer MBP Xho antisense: CGAATTCTGAAATCTCGAGCTATTACCCGAGGTTGTTG (XhoI Site underlined). The PCR product is subcloned into Gtn1-500-his/pET23a, in-frame and contiguous with the Giantin ORF. The recombinant plasmid (Gtn1-500-MBP/pET) is verified by sequencing across the entire ORF and transformed into BL21(DE3) *E. coli* for expression. Note that Gtn1-500-MBP, constructed by this strategy, retains an extreme C-terminal his-tag. If desired, this could easily be deleted by Quikchange mutagenesis.

The codons for residues 1967–2541 and for 1125–1695 in Giantin are excised by restriction digestion of the full-length Giantin cDNA, as approximately 1.7 kb KpnI-HindIII and HindIII fragments respectively. These are subcloned into pTrcHis (Invitrogen, Carlsbad, CA) to generate pTrc#1 and pTrc#2 respectively. This plasmid allows bacterial expression of the encoded ORF as a fusion protein with a his-tag at the N-terminus. Constructs should be verified by sequencing of the joins between the his-tag and Giantin fragments, and the following 300-500 bp at the 3′end (Lesa *et al.*, 2000).

Protein expression: A single colony (or stab from a clonal glycerol stock) of BL21(DE3) *E. coli* transformed with a Giantin fragment expression vector is used to inoculate a starter culture (10 ml per 400 ml expression culture) of LB broth, supplemented with 0.1 mg/ml ampicillin. The starter culture is grown with shaking at 37° for 12–16 h (overnight) then diluted into a larger, expression culture of LB broth supplemented with 0.1 mg/ml ampicillin. For MBP-tagged fusion proteins 2 mg/ml glucose is also added to the expression culture. The expression culture is grown with shaking at 37° for 3–4 h until the OD at 600 nm reached 0.6–1.0. Expression is then induced by addition of IPTG (0.1 mM final concentration). Growth is continued for a further 3–4 h shaking at 27°.

Protein purification: Bacteria are harvested by centrifugation (3000g, 4°, 20 min) and resuspended in either his-tag lysis buffer (50 mM NaH$_2$PO$_4$ pH 7.4, 500 mM NaCl, 2 mM β-mercaptoethanol, EDTA free protease inhibitor cocktail) (for his-tagged proteins), or MBP-tag lysis buffer (25 mM HEPES pH 7.4, 100 mM KCl, 2 mM β-mercaptoethanol, EDTA free protease inhibitor cocktail) (for MBP-tagged proteins). The bacteria are lysed in an EmulsiFlex-C5 homogenizer or by sonication and the lysate is cleared by centrifugation (17,600g, 4°, 20 min) to remove particulate matter. The cleared supernatant is adjusted (with lysis buffer) to a total volume of 50 ml for each 400 ml of expression culture that was harvested. For purification of his-tagged proteins the cleared supernatant is adjusted to 10 mM imidazole. The appropriate affinity resin, either amylose-resin (for MBP-tagged proteins) or Ni-NTA-agarose (for his-tagged proteins) is pre-equilibrated in lysis buffer, then added to the cleared supernatant (0.5 ml bed volume/400 ml expression culture). The cleared supernatant and the affinity resin mix are incubated in batch with end-over-end rotation at 4° for 1h. The affinity resin is harvested by a brief centrifugation (2000g, 4°, 1 min) and washed three times with either lysis buffer (for MBP fusion proteins) or wash buffer (50 mM NaH$_2$PO$_4$ pH 7.4, 500 mM NaCl, 10 mM imidazole 2 mM β-mercaptoethanol, EDTA free protease inhibitor cocktail) (for his-tagged proteins). Each wash is with 2 bed volumes buffer for 5 min on ice. Bound protein is then eluted with the appropriate elution buffer: MBP-tag elution buffer (25 mM HEPES pH 7.4, 100 mM KCl, 2 mM β-mercaptoethanol, 10 mM maltose), or his-tag elution buffer (25 mM HEPES pH 7.4, 100 mM KCl, 2 mM β-mercaptoethanol, 350 mM imidazole). Elution is performed in batch, 3 elutions of 1 bed volume, incubated with end-over-end rotation for 10 min each at 4°. The eluted material should be pooled and desalted into buffer A (25 mM HEPES pH 7.4, 100 mM KCl, 2 mM β-mercaptoethanol) on a 10 DG desalting column. Fractions containing the Giantin fragment should be pooled. Aliquots should be snap-frozen (liquid nitrogen) and stored at −80°. Protein concentration is estimated by comparison with BSA standards on Coomassie stained gels. A pooled and purified example of both Giantin1–500-his and his-Giantin1125-1695 is shown in Fig. 2F and G, respectively. Typically, a 400-ml culture yielded 2 mg of Giantin1–500-his, which was judged 90% pure by Coomassie staining. For his-Giantin1 125-1695 the corresponding values are 2 mg and 80–90% purity.

Coimmunoprecipitation of Recombinant GM130-his with Giantin1-500-MBP in the Presence of p115

A rapid functional test of recombinant p115, GM130, and Giantin1-500 can be performed by testing the ability of p115 to link recombinant GM130 to Giantin1-500 (Dirac-Svejstrup *et al.*, 2000). Equimolar concentrations (40 nM) of GM130-his and Giantin1-500-MBP are incubated with various concentrations of purified recombinant p115 in coimmunoprecipitation buffer (20 mM HEPES pH 7.4, 100 mM KCl, 5 mM MgCl$_2$, 1 mM DTT) with end-over-end rotation for 2 h at 4°. Anti-MBP antisera (NEB) are added (2 μl/400 μl reaction). Incubation is continued for a further 30 min, rotating at 4° before Protein A-Sepharose (Amersham Biosciences), pre-equilibrated in coimmunoprecipitation buffer, is added (100 μl bed volume/400 μl reaction). Incubation is continued for a further 1 h, rotating at 4°. Beads are washed 3 times in the same buffer, resuspended in 2X sample buffer (100 mM Tris pH 6.8, 3% SDS, 15% glycerol, 5% DTT, 0.01% bromophenol blue), and analyzed by SDS-PAGE and immunoblotting. An example is shown in Fig. 3A. Similar coimmunoprecipitation experiments can also be performed using a fixed concentration (400 nM) of truncations in p115 (purification described above) (Fig. 3B), and with the truncation GM130ΔN73-his substituted for full-length GM130-his (Fig. 3C). These truncations were tested because the binding sites for GM130 and Giantin both lie within the extreme C-terminal 25 amino acids of p115 (Dirac-Svejstrup *et al.*, 2000), and the binding site for p115 lies within the N-terminal 73 amino acids of GM130 (Dirac-Svejstrup *et al.*, 2000).

Purification of Recombinant CASP

Plasmids: The ORF including the stop codon and N-terminal myc-tag for human CASP is cut out from pCMV-GFP-Myc-CASP (kindly provided by S. Munro (Gillingham *et al.*, 2002)) and subcloned into pMALc2x (NEB) between the BamHI and XbaI sites. The recombinant plasmid is transformed into BL21(DE3) codon plus RIL (Stratagene) for expression.

Protein expression: A single colony (or stab from a clonal glycerol stock) of BL21 transformed with CASP expression vector is used to inoculate a starter culture (10 ml per 800 ml expression culture) of LB broth, supplemented with 0.05 mg/ml ampicillin. The starter culture is grown with shaking at 37° for 12–16 h (overnight) then diluted into a larger, expression culture of LB broth supplemented with 0.05 mg/ml ampicillin and 2 mg/ml glucose. The expression culture is grown with shaking at 37° for 2 h until the OD at 600 nm reached 0.6–1.0. Expression is induced by addition of IPTG (0.1 mM final concentration). Growth is continued for a further 4–5 h shaking at 30°.

A

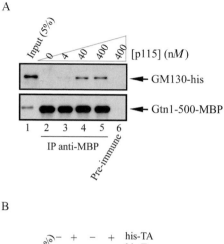

B

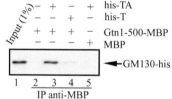

C

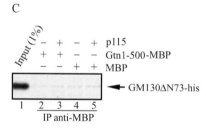

FIG. 3. p115 can directly link GM130 to Giantin. Purified, affinity tagged forms of GM130 and Giantin were incubated together with p115 or truncations thereof. Giantin1-500 (tagged with MBP) was immunoprecipitated by anti-MBP. Coimmunoprecipitating GM130 was analyzed by immunoblotting. (A) Full-length GM130-his, Gtn1-500-MBP and increasing concentrations of full-length p115. (B) Full-length GM130-his, Gtn1-500-MBP (or MBP alone), and a single concentration (400 nM) of either his-TA or his-T. (C) GM130ΔN73-his, Gtn1-500-MBP (or MBP alone), and a single concentration (400 nM) of full-length p115. Note that coimmunoprecipitation of GM130 by Giantin required the acidic domain of p115 and the N-terminal region of GM130.

Protein purification: Bacteria are harvested by centrifugation (3000*g*, 4°, 10 min) and resuspended in ~40 ml of a lysis buffer (20 m*M* HEPES pH 7.4, 200 m*M* KCl, 1 m*M* DTT or 1m*M* β-mercaptoethanol), supplemented with EDTA-free protease inhibitor cocktail and 1 m*M* PMSF. The bacteria are lysed in an EmulsiFlex-C5 homogenizer or by sonication and the lysate is cleared by centrifugation (14,600*g*, 4°, 30 min) to remove particulate matter. The cleared supplement is incubated with 0.8 ml of pre-equilibrated amylose-resin at 4° for 1 h with end-over-end rotation. The resin is collected by a brief centrifugation (1000*g*, 4°, 1 min) and washed three times (each with 10 bed volumes of lysis buffer). Bound protein is then eluted with the lysis buffer without protease inhibitor but containing 10 m*M* maltose. Elution is performed using a disposable column (BioRad). Fractions (0.4 ml) are collected and those containing CASP are pooled. Aliquots are snap-frozen (liquid nitrogen) and stored at −80°. Protein concentration can be estimated by comparison with BSA standards on Coomassie stained gels. An example of pooled, purified MBPCASP is shown in Fig. 2H. Typically, a 400-ml culture yielded 0.4 mg of MBP-CASP, which was judged 40–70% pure by Coomassie staining.

Purification of Recombinant Golgin-84

Plasmids: To remove 5′-untranslated region the ORF of human golgin-84 should be amplified by PCR from pBluescript-human golgin-84 (kindly provided by R. Nussbaum, (Bascom *et al.*, 1999) using the following primers; forward primer including EcoRI site: GGAATTC-TCTTGGTTTGTTGATC; reverse primer including BamHI site: CGGGATCCTCATTTGCCATATG. The PCR product is subcloned into pGEM-T easy (Promega, Madison, WI). The ORF including EcoRI and BamHI sites is cut out from the pGEM vector and subcloned into pMALc2x (NEB) between the EcoRI and BamHI sites. The recombinant plasmid should be verified by sequencing across the entire encoded ORF and transformed into BL21(DE3) codon plus RIL (Stratagene) for expression. A second expression plasmid is constructed to encode his-tagged rat golgin-84. The ORF for rat golgin-84 is amplified by PCR from rat liver cDNA library (Origene, Rockville, MD) using the following reagents and conditions; 1 μl of the cDNA library; 1 μl of 100 p*M* each of forward primer: GGTCCTCGCCGCCCTTCCACGTCT; reverse primer: CCTAGTCTTAGTGCTGATCTTTCTG; Advantage II DNA polymerase mixture (Clontech); 35 cycles of denaturing at 95° for 30 sec and annealing and extension at 68° for 3 min. The reaction volume is 50 μl. To generate a his-tag at the N-terminus, and a stop codon before its transmembrane domain, the nested PCR should be performed using the forward primer:

CATCATCATCATCATCATTCTTGGTTTGCTGATCTTGCTG;   reverse primer: CTATCGTGCTATGGGGTATCTTCTG, with Pfu Turbo
DNA polymerase (Stratagene) followed by the addition of Taq DNA
polymerase (Roche) at 72° for 30 min. The PCR product is gel-purified
and directly subcloned into pTrcHis2-TOPO vector (Invitrogen). The recombinant plasmid (his-rat golgin-84 Δ transmembrane domain [TMD]) is
transformed into BL21(DE3) codon plus RIL (Stratagene) for expression.

Protein expression: A single colony (or stab from a clonal glycerol
stock) of BL21 transformed with golgin-84 expression vector is used to
inoculate a starter culture (10 ml per 800 ml expression culture) of LB
broth, supplemented with 0.05 mg/ml ampicillin. The starter culture is
grown with shaking at 37° for 12–16 h (overnight) then diluted into a larger,
expression culture of LB broth supplemented with 0.05 mg/ml ampicillin.
For MBP-tagged fusion proteins 2 mg/ml glucose is also added to the
expression culture. The expression culture is grown with shaking at 37°
for 2 h until the OD at 600 nm reaches 0.6–1.0. Expression is induced by
addition of IPTG (0.1 mM final concentration). Growth is continued for a
further 4–5 h shaking at 30°.

Protein purification: Bacteria are harvested by centrifugation (3000g,
4°, 10 min) and resuspended in ~40 ml of a lysis buffer (20 mM HEPES pH
7.4, 200 mM KCl, 1 mM DTT or 1 mM β-mercaptoethanol), supplemented
with EDTA free protease inhibitor cocktail and 1 mM PMSF. The bacteria
are lysed in an EmulsiFlex-C5 homogenizer, or by sonication, and the
lysate is cleared by centrifugation (14,600g, 4°, 30 min) to remove particulate matter. The cleared supplement is incubated with 0.8 ml of pre-equilibrated appropriate affinity resin, either amylose-resin or Ni-NTA-agarose
at 4° for 1 h with end-over-end rotation. For his-tagged protein, the lysis
buffer is supplemented with 10 mM imidazole to reduce nonspecific binding. The resin is collected by brief centrifugation (1000g, 4°, 1 min) and
washed three times (each with 10 bed volumes of lysis buffer). Bound
protein is then eluted with the lysis buffer without protease inhibitor but
containing either 10 mM maltose or 350 mM imidazole. Elution is performed using a disposable column (BioRad). Fractions (0.4 ml) are collected and containing golgin-84 should be pooled. Aliquots should be snap-
frozen (liquid nitrogen) and stored at −80°. Protein concentration can be
estimated by comparison with BSA standards on Coomassie stained gels.
The eluted material containing 350 mM imidazole should be desalted into
buffer A (25 mM HEPES pH 7.4, 100 mM KCl, 2 mM β-mercaptoethanol)
on a 10 DG desalting column before snap-freezing. An example of pooled,
purified MBP-golgin-84 is shown in Fig. 2I. Typically, a 400-ml culture
yielded 0.8 mg of MBP-golgin-84, which was judged 70–90% pure by
Coomassie staining.

Acknowledgment

We thank all members of the Warren, Mellman, and Toomre laboratories for helpful comments and discussions, and Lesa G., Lowe M., Barr F., Nakamura N., Nussbaum R., and Munro S. for generous provision of plasmids. This work was supported by the National Institutes of Health and the Ludwig Institute for Cancer Research. A.S. was supported by the American Heart Association.

References

Barr, F. A., Puype, M., Vandekerckhove, J., and Warren, G. (1997). GRASP65, a protein involved in the stacking of Golgi cisternae. *Cell* **91**, 253–262.

Barr, F. A., and Short, B. (2003). Golgins in the structure and dynamics of the Golgi apparatus. *Curr. Opin. Cell. Biol.* **15**, 405–413.

Bascom, R. A., Srinivasan, S., and Nussbaum, R. L. (1999). Identification and characterization of golgin-84, a novel Golgi integral membrane protein with a cytoplasmic coiled-coil domain. *J. Biol. Chem.* **274**, 2953–2962.

Diao, A., Rahman, D., Pappin, D. J., Lucocq, J., and Lowe, M. (2003). The coiled-coil membrane protein golgin-84 is a novel rab effector required for Golgi ribbon formation. *J. Cell Biol.* **160**, 201–212.

Dirac-Svejstrup, A. B., Shorter, J., Waters, M. G., and Warren, G. (2000). Phosphorylation of the vesicle-tethering protein p115 by a casein kinase II-like enzyme is required for Golgi reassembly from isolated mitotic fragments. *J. Cell Biol.* **150**, 475–488.

Fritzler, M. J., Hamel, J. C., Ochs, R. L., and Chan, E. K. (1993). Molecular characterization of two human autoantigens: Unique cDNAs encoding 95- and 160-kD proteins of a putative family in the Golgi complex. *J. Exp. Med.* **178**, 49–62.

Gillingham, A. K., Pfeifer, A. C., and Munro, S. (2002). CASP, the alternatively spliced product of the gene encoding the CCAAT-displacement protein transcription factor, is a Golgi membrane protein related to giantin. *Mol. Biol. Cell* **13**, 3761–3774.

Jackson, C. L. (2004). N-terminal acetylation targets GTPases to membranes. *Nat. Cell Biol.* **6**, 379–380.

Lesa, G. M., Seemann, J., Shorter, J., Vandekerckhove, J., and Warren, G. (2000). The amino-terminal domain of the golgi protein giantin interacts directly with the vesicle-tethering protein p115. *J. Biol. Chem.* **275**, 2831–2836.

Linstedt, A. D., and Hauri, H. P. (1993). Giantin, a novel conserved Golgi membrane protein containing a cytoplasmic domain of at least 350 kDa. *Mol. Biol. Cell* **4**, 679–693.

Malsam, J., Satoh, A., Pelletier, L., and Warren, G. (2005). Golgin tethers define subpopulations of COPI vesicles. *Science* **307**, 1095–1098.

Nakamura, N., Rabouille, C., Watson, R., Nilsson, T., Hui, N., Slusarewicz, P., Kreis, T. E., and Warren, G. (1995). Characterization of a *cis*-Golgi matrix protein, GM130. *J. Cell Biol.* **131**, 1715–1726.

Pfeffer, S. R. (1999). Transport-vesicle targeting: Tethers before SNAREs. *Nat. Cell Biol.* **1**, E17–E22.

Puthenveedu, M. A., and Linstedt, A. D. (2004). Gene replacement reveals that p115/SNARE interactions are essential for Golgi biogenesis. *Proc. Natl. Acad. Sci. USA* **101**, 1253–1256.

Satoh, A., Wang, Y., Malsam, J., Beard, M. B., and Warren, G. (2003). Golgin-84 is a rab1 binding partner involved in Golgi structure. *Traffic* **4**, 153–161.

Shorter, J., Beard, M. B., Seemann, J., Dirac-Svejstrup, A. B., and Warren, G. (2002). Sequential tethering of Golgins and catalysis of SNAREpin assembly by the vesicle-tethering protein p115. *J. Cell Biol.* **157**, 45–62.

Shorter, J., and Warren, G. (2002). Golgi architecture and inheritance. *Annu. Rev. Cell Dev. Biol.* **18**, 379–420.

Stinton, L. M., Eystathioy, T., Selak, S., Chan, E. K., and Fritzler, M. J. (2004). Autoantibodies to protein transport and messenger RNA processing pathways: Endosomes, lysosomes, Golgi complex, proteasomes, assemblyosomes, exosomes, and GW bodies. *Clin. Immunol.* **110**, 30–44.

Waters, M. G., Clary, D. O., and Rothman, J. E. (1992). A novel 115-kD peripheral membrane protein is required for intercisternal transport in the Golgi stack. *J. Cell Biol.* **118**, 1015–1026.

Waters, M. G., and Hughson, F. M. (2000). Membrane tethering and fusion in the secretory and endocytic pathways. *Traffic* **1**, 588–597.

[27] Purification and Functional Properties of the Membrane Fissioning Protein CtBP3/BARS

By Carmen Valente, Stefania Spanò, Alberto Luini, and Daniela Corda

Abstract

The fissioning protein CtBP3/BARS is a member of the CtBP transcription corepressor family of proteins. The characterization of this fissioning activity of CtBP3/BARS in both isolated Golgi membranes and in intact cells has indicated that the CtBP family includes multifunctional proteins that can act both in the nucleus and in the cytoplasm. The fissiogenic activity of CtBP3/BARS has a role in the fragmentation of the Golgi complex during mitosis and during intracellular membrane transport. This was demonstrated using a number of approaches and reagents, which are discussed in the following text, and which include recombinant proteins and mutants, antibodies, protein overexpression, RNA interference, antisense oligonucleotides, cell permeabilization, and electron miscroscopy, together with biochemical assays such as that for ADP-ribosylation.

Introduction

CtBP3/BARS (BARS) is a member of the CtBP family of proteins (Chinnadurai, 2002), and it was initially identified as the 50-kDa substrate of brefeldin A (BFA)-dependent ADP-ribosylation (De Matteis *et al.*, 1994; Di Girolamo *et al.*, 1995). This protein is involved in the disassembly of the Golgi complex induced by BFA, suggesting that it has a role in the maintenance of the structure and/or function of this organelle (Mironov *et al.*, 1997). This was later confirmed as the fissiogenic activity of BARS (Weigert

METHODS IN ENZYMOLOGY, VOL. 404
0076-6879/05 $35.00
DOI: 10.1016/S0076-6879(05)04027-9

et al., 1999) is essential for specific steps of intracellular membrane transport (Bonazzi *et al.*, 2005), and for the fragmentation of the Golgi complex during mitosis (Hidalgo Carcedo *et al.*, 2004). BFA-dependent ADP ribosylation of BARS has also been shown to inhibit its functions (Weigert *et al.*, 1999).

The initial protein characterization indicated that the ADP-ribosylation of BARS is modulated by the G-protein $\beta\gamma$ subunit (Di Girolamo *et al.*, 1995). BARS is not a classical GTP-binding protein (as later confirmed also by sequence analysis [Spanò *et al.*, 1999]), and there is no indication that it may directly interact with $\beta\gamma$ (Di Girolamo *et al.*, 1995; Spanò *et al.*, 1999).

After its purification and cloning (see Methods), rat BARS was revealed to have 97% and 79% identity to CtBP1 and CtBP2, respectively (the transcription corepressor proteins, C-terminus-binding protein 1 and 2, that have been cloned from human and mouse; Criqui-Filipe *et al.*, 1999; Schaeper *et al.*, 1995; Turner and Crossley, 1998) (see also Fig. 1; [Spanò *et al.*, 1999, 2006]). CtBP1 and CtBP2 can themselves be considered as "BARS" proteins as they are also substrates of the ADP-ribosylation induced by BFA (Fig. 2). Whether this protein modification can interfere with the corepression activities of CtBP1 and CtBP2 has not been directly assessed, and it is also not yet known if these proteins can induce fission.

The crystal structure of the binary complex of rat BARS and human CtBP1 with NAD(H) has been resolved (Kumar *et al.*, 2002; Nardini *et al.*, 2003) (Fig. 3). Structural modeling and binding studies have shown that BARS is able to bind NAD(H) and long-chain acyl-CoAs through the

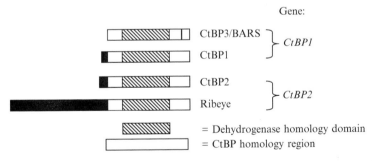

FIG. 1. Schematic representation of the sequences of the CtBP family members. The scheme includes: CtBP3/BARS and CtBP1, which are alternative splice variants of the *CtBP1* gene (Spanò *et al.*, 2006); and CtBP2 and Ribeye (Schmitz *et al.*, 2000), which are alternative splice variants of the *CtBP2* gene. The region conserved in all of the CtBPs is shown in white. This includes a dehydrogenase homology region (striped), with a weak, but significant, similarity with the D-stereoisomer-specific 2-hydroxyacid NAD-dependent dehydrogenases (Nardini *et al.*, 2003). All the other regions with no recognizable homology with other known proteins are shown in black. CtBP1 and CtBP3/BARS are also now referred to as CtBP1-L and CtBP1-S, respectively, as they represent the long and short forms of the *CtBP1* gene product.

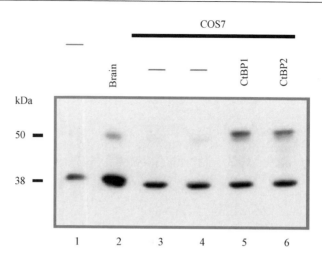

FIG. 2. BFA-dependent ADP-ribosylation of CtBPs in COS7 cells. The 50-kDa ADP-ribosylated CtBPs are shown in lanes 2 to 4, whereas overexpressed CtBP1 and CtBP2 are shown in lanes 5 and 6, respectively. The 38-kDa ADP-ribosylated glyceraldehyde-3-phosphate dehydrogenase is also shown (De Matteis *et al.*, 1994). Samples are: control buffer (lane 1), 30 µg cytosol from rat brain (lane 2), 10 µg (lane 3), or 30 µg (lane 4) cytosol from nontransfected COS7 cells, 10 µg cytosol from T7-CtBP1-transfected COS7 cells (lane 5), and 10 µg cytosol from CtBP2-transfected COS7 cells (lane 6), all [^{32}P]-ADP-ribosylated (see text for details), analyzed by SDS-PAGE and revealed by gas ionization (Instant Imager, Packard Instrument, CT, USA). The T7-tagged human CtBP1 cDNA construct was kindly provided by Dr. Chinnadurai (Schaeper *et al.*, 1995); mouse CtBP2 cDNA was obtained as an EST database clone (accession number AA433525) and cloned in pcDNA3 (Spanò *et al.*, 2002).

same site. This structural analysis suggests that NAD(H) binding promotes BARS dimerization, whereas acyl-CoA binding induces the monomeric conformation (Nardini *et al.*, 2003). This mechanism could represent the "functional" molecular switch between "transcriptional activity" and "fission activity" of the CtBP proteins.

Experimental Approaches to Study CtBP3/BARS Function

The activity of BARS (an enriched preparation from rat brain, as described by Spanò *et al.*, 1999) was originally investigated in *in vitro* experiments using isolated Golgi membranes (prepared according to Cluett and Brown, 1992). These were examined by both conventional electron microscopy (EM) (Mironov *et al.*, 1997) and by negative-staining EM (see Methods) after exposure to cytosol, BARS, or BARS and cofactors (Weigert *et al.*, 1999). In isolated Golgi complex preparations, both cisternae and tubular structures can be visualized by EM (Weigert *et al.*,

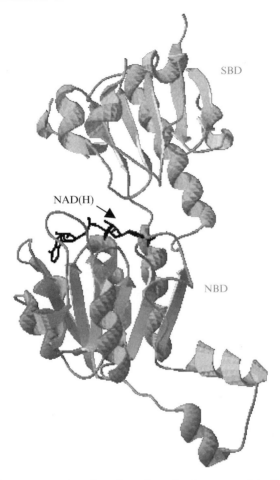

FIG. 3. Ribbon representation of the three-dimensional structure of the BARS monomer. The SBD is depicted in cyan, and the NBD in green. The NAD(H) that was co-crystallized with the BARS is shown in black, bound in its interdomain cleft (Nardini *et al.*, 2003).

1999), and the latter undergo a rapid disruption into fragments of irregular sizes upon exposure to cytosol. This process is BARS dependent, and it is associated with the formation of "fission intermediates," that is, tubules presenting constricted areas that are thought to represent the actual fission sites, and which precede the formation of transport intermediates (Weigert *et al.*, 1999). Isolated Golgi cisternae are not affected by this treatment. As the ADP-ribosylated protein (see Methods) is inactive in these assays, this indicates that BFA-induced ADP-ribosylation is a modification that can lead to loss of function (Weigert *et al.*, 1999).

The BARS-induced fissioning activity requires the presence of acyl-CoAs as cytosolic cofactors. Both BARS and endophilin, a second, independent fissiogenic protein, are associated with an acyltransferase activity that is selective for lysophosphatidic acid and that increases the membrane content of phosphatidic acid, an effect that has been related to membrane fission (Schmidt et al., 1999; Weigert et al., 1999). This activity of BARS was studied by in vitro acyltransferase assays using the recombinant protein and the acceptor lysolipid (Weigert et al., 1999). As this activity of BARS (and endophilin) is very slow (Schmidt et al., 1999; Weigert et al., 1999), this raised concerns regarding its specificity and its functional significance in the fission reaction. The acyltransferase activity seems to be specific since it is not expressed in a large number of control (fission-unrelated) recombinant proteins (our unpublished results); moreover, it is inhibited by antibodies against BARS and by the ADP-ribosylation of BARS, both treatments that block the effect of BARS on membrane fission (Weigert et al., 1999). However, more recently, we have reported that the acyltransferase activity is not essential for the BARS-dependent fragmentation of the Golgi complex during mitosis, and for basolateral transport from the Golgi to the plasma membrane (Bonazzi et al., 2005; Hidalgo Carcedo et al., 2004; Spanò et al., 2006). Thus, while the specific rearrangements of the membrane bilayer required to achieve fission involve changes in their lipid composition, structure, and dynamics (Burger, 2000; Corda et al., 2002; Schmidt et al., 1999), these modifications might be supported by other mechanisms, such as the recruitment of specific lipids by protein components of the fissioning machinery, or the activity of lipid-modifying enzymes recruited to the fission sites (Corda et al., 2002; Huttner and Schmidt, 2000).

Methods Involved in the In Vitro Studies

Protein Purification

The ADP-ribosylated form of BARS was purified from rat brain cytosol through four chromatographic steps (achieving an 800-fold enrichment), and then run on two-dimensional gel electrophoresis, microsequenced, and cloned, as detailed by Spanò et al. (1999). A full-length open reading frame (ORF, GenBank Accession Number AF067795) that codes for a 430-amino acid protein with a predicted mass of 47 kDa was isolated, and it was confirmed to code for BARS by overexpression and antibody recognition (see following) (Spanò et al., 1999) (Fig. 1). Based on its three-dimensional (3D) structure that indicates a strong structural similarity between BARS and the family of the D-stereoisomer-specific

2-hydroxyacid NAD dehydrogenases, two domains were identified within BARS that were originally described for these NAD dehydrogenases. These are known as the nucleotide-binding (NBD) and substrate-binding (SBD) domains (Nardini *et al.*, 2003; see also Fig. 3). Based on this observation, two mutant forms of the protein, NBD (residues 112–309) and SBD (residues 1–113, 307–430), were expressed, used in the BARS-dependent assays, and found to either act as dominant negatives (Hidalgo Carcedo *et al.*, 2004) or to mimic the action of BARS (Bonazzi *et al.*, 2005; Spanò *et al.*, 2006). Additional mutants were generated by combining different domains of the protein: the N-terminal portion (NTP, consisting of the N-terminal subdomain of SBD and NBD, aa 1–350), the C-terminal portion (CTP, consisting of the C-terminal subdomain of SBD and NBD, aa 112–430), and the C-terminal domain (CTD, representing the extreme C-terminal subdomain, aa 306–430) (Yang *et al.*, 2006). Two point mutants of BARS have also been expressed: G172E, in which a residue in the cofactor-binding site is mutated (Nardini *et al.*, 2003), and D355A, where the mutated D belongs to a charged cluster that is thought to be required for fission (our unpublished results).

Expression and Purification of Recombinant BARS

GST-BARS and His-tagged BARS and their mutant forms (BARS-G172E, BARS-SBD, BARS-NBD, BARS-D355A, BARS-NTP, BARS-CTP, and BARS-CTD) have been subcloned for bacterial expression, as described previously (Bonazzi *et al.*, 2005; Hidalgo Carcedo *et al.*, 2004; Nardini *et al.*, 2003; Weigert *et al.*, 1999; Yang *et al.*, 2006), and expressed, as detailed in the following material.

Purification of GST-BARS and Mutants

Solutions

LB: 1% Tryptone Peptone, 0.5% yeast extract, 1% NaCl; autoclaved for 15 min at 121°;

GST lysis buffer: 20 mM Tris-HCl, pH 8.0, 100 mM NaCl, 1 mM ethylenediaminetetraacetic acid (EDTA);

Protease inhibitors: 2.0 μg/ml aprotinin, 0.5 μg/ml leupeptin, 2 μM pepstatin, 0.5 mM 1,10-phenanthroline, 1 mM phenylmethylsulfonyl fluoride (PMSF);

PBS: phosphate-buffered saline;

GST elution buffer: 100 mM Tris-HCl, pH 8.0, 20 mM glutathione, 5 mM dithiothreitol (DTT).

The following procedure was used to purify GST-BARS, GST-BARS-G172E, GST-BARS-NBD, GST-BARS-SBD, and it leads to 2–5 mg of the recombinant proteins. *Escherichia coli* XL1-Blue transformed with pGEX4T1-BARS (Weigert *et al.*, 1999) are inoculated in 4 ml of LB containing 60 μg/ml ampicillin and grown at 37° under continuous shaking (200 rpm) for 4–8 h. Two ml of this culture is inoculated into 50 ml of the LB medium and incubated under the same conditions overnight. The overnight culture is diluted 1:20 in 1 L of the same medium at 37° under continuous shaking (200 rpm), and the OD_{600} is monitored until it reaches 0.4. Then the bacteria are induced with the addition of 0.1 mM isopropyl β-D-thiogalactopyranoside (IPTG) for 2 h at 37°. After this, the culture is cooled on ice and centrifuged at 4,000g in a JA10 rotor for 10 min at 4°. After discarding the supernatant, the pellet is resuspended in 50 ml GST lysis buffer containing the cocktail of protease inhibitors, and lysozyme and Triton X-100 are added to final concentrations of 1 mg/ml and 1%, respectively. The samples are then frozen by immersion in liquid nitrogen. The lysate can be stored at −80° overnight or for a few days. The suspension is thawed by transferring it to a 4° bath, and after adding the protease inhibitors and lysozyme again, it is incubated with gentle agitation at 4° for 30 min, followed by sonication in ice for 6 × 15 s and centrifugation at 20,000g in a JA20 rotor for 20 min at 4°. The supernatant is recovered and added to 2 ml of glutathione Sepharose 4B matrix (Amersham Pharmacia Biotech, NJ, USA) that has been previously equilibrated in GST lysis buffer. The suspension is incubated with gentle agitation at 4° for 30 min, and then centrifuged at 700g for 5 min, to sediment the matrix. The matrix is washed 5× with 50 ml PBS (centrifuging at 700g for 5 min). After this washing, the matrix is packed into a 10-ml chromatography column (Bio-Rad Laboratories, UK) and drained, and the bottom cap of the column is replaced. The protein is eluted by adding 2 ml GST elution buffer and incubated for 5 min at room temperature; the bottom cap is then removed and the eluate is collected in 0.5 ml fractions. The elution and collection steps are repeated at least 4×. GST-BARS generally peaks after around 2 ml of elution. The fractions containing at least 0.2 mg/ml of protein are pooled, dialysed twice against 1,000× volume PBS, and stored at −80°.

Purification of His-tagged BARS and Mutants

Solutions

LB: 1% Tryptone Peptone, 0.5% yeast extract, 1% NaCl; autoclaved for 15 min at 121°;

His lysis buffer: 50 mM sodium phosphate buffer, pH 8.0, 300 mM
NaCl, 10 mM imidazole;
Protease inhibitors: 2.0 μg/ml aprotinin, 0.5 μg/ml leupeptin, 2 μM
pepstatin, 1 mM PMSF
His wash buffer: 50 mM sodium phosphate buffer, pH 8.0, 300 mM
NaCl, 20 mM imidazole;
His elution buffer: 50 mM sodium phosphate buffer, pH 8.0, 300 mM
NaCl, 250 mM imidazole.

The following procedure was used to purify His-BARS, His-BARS-
NTP, His-BARS-G172E and His-BARS-D355A, and it leads to about 2 mg
of the recombinant proteins. $E.$ $coli$ BL21pLysS transformed with pET11d-
His-BARS (Nardini et $al.$, 2003) are inoculated in 4 ml LB containing 60
μg/ml ampicillin and 10 μg/ml chloramphenicol and incubated overnight at
37° under continuous shaking (200 rpm). The culture is diluted 1:20 in 80
ml of the same medium, and the OD$_{600}$ is monitored until it reaches 0.6.
The bacteria are then induced with the addition of 0.4 mM IPTG for 2 h at
37°. Finally, the culture is cooled in ice and centrifuged at 6,200g in a JA14
rotor for 10 min at 4°. After discarding the supernatant, the pellet is
resuspended in 4 ml His lysis buffer containing the cocktail of protease
inhibitors, and frozen by immersion in liquid nitrogen. The lysate can be
stored at −80° overnight or for a few days. The suspension is thawed by
transferring it to a 4° bath, and the protease inhibitors are added again,
with lysozyme added to a final concentration of 1 mg/ml. The lysate is
incubated with gentle agitation at 4° for 30 min, before being sonicated in
ice for 8 × 15 s. Then 10 mM MgCl$_2$ and 10 μg/ml DNAse I are added, and
the lysate is incubated for 15 min in ice, and then centrifuged at 25,000g in a
JA20 rotor for 20 min at 4°. The supernatant is recovered and added to 0.5
ml Ni-NTA beads (Qiagen, CA, USA) that have been previously equilibrated in His lysis buffer. The suspension is incubated with gentle agitation at 4° for 1 h and then packed into a 10-ml chromatography column
(Bio-Rad Laboratories, UK). The column is washed 5× with 10 ml His
wash buffer, and the protein is eluted by adding 0.5 ml His elution buffer
and collecting it in a clean tube in which there is EDTA to a 1 mM final
concentration. The elution and collection steps are repeated at least 5×.
His-BARS usually peaks in the first fraction. The fractions containing at
least 0.2 mg/ml of protein are pooled, dialyzed twice against 1000× volume
PBS, and stored at −80°.

A similar procedure was used to purify His-BARS-NBD, His-BARS-
SBD, His-BARS-CTP, and His-BARS-CTD, with the only change being
that the IPTG induction was at 0.1 mM and incubated overnight at 25°,
rather than for 2 h at 37°.

ADP-ribosylation

Solutions

PFB: 50 mM potassium phosphate buffer, pH 7.5, 1.25 mM MgCl$_2$, 0.5 mM adenosine 5'-triphosphate (ATP), 0.5 mM guanosine 5'-triphosphate (GTP), 5 mM thymidine.

The ADP-ribosylation assay is usually carried out with 100 ng recombinant protein in the presence of 1.5 mg/ml rat brain membranes (as the enzyme source), 60 μg/ml BFA (Sigma Aldrich, USA; dissolved at 10 mg/ml in DMSO, aliquoted, and stored at $-20°$), 30 μM β-nicotinamide adenine dinucleotide (NAD$^+$; Sigma Aldrich, USA) and 3 μCi [^{32}P]-NAD$^+$ (specific activity, 800 Ci/mmol; Amersham Pharmacia Biotech, NJ, USA, or NEN-Dupont, UK) in 50 μl 1× PFB and 5 mM DTT, using the following procedure. A first mixture (40 μl/sample) is prepared by mixing 2× PFB, 10 mg/ml BFA, 1 M DTT, and H$_2$O, and then adding the membranes and recombinant protein at the end (e.g., 20 μl 2× PFB, 0.3 μl 10 mg/ml BFA, 0.25 μl 1 M DTT, 8 μl H$_2$O, 9.4 μl 8 mg/ml rat brain membranes, and 2 μl 0.05 μg/μl recombinant protein). A second mixture (10 μl/sample) is prepared by mixing 2× PFB, 3 mM NAD$^+$, [^{32}P]-NAD$^{+,}$ and H$_2$O (e.g., 5 μl 2× PFB, 0.5 μl 3 mM NAD$^+$, 0.6 μl 5.3 μCi/μl [^{32}P]-NAD$^+$, and 3.9 μl H$_2$O). The second mixture is immediately added to the first, and the incubation is carried out for 2 h at 37°. The sample is then centrifuged at 18,000g for 10 min, the supernatant is recovered, and the ADP-ribosylated proteins are analyzed by 10% SDS-PAGE. The PAGE is followed by autoradiography and quantified using gas ionization counting (Instant Imager, Packard Instrument, CT, USA) (Fig. 2).

The ADP-ribosylation of endogenous BARS was carried out as described above, with the incubation of rat brain cytosol (as the substrate source) at a final concentration of 5 mg/ml, with 250 μM NAD$^+$ and 480 μCi/ml [^{32}P]-NAD$^+$ (Spanò *et al.*, 1999).

Acyl-CoA and NAD Binding Experiments

Five μg (100 pmoles) of recombinant His-BARS are spotted onto nitrocellulose filters (PROTRAN, PerkinElmer Life Sciences Boston, MA, USA) using a slot-blot apparatus (Bio-Rad Laboratories, UK). To analyze the binding of NAD, the filters are incubated at room temperature for 2 h in 1.5 ml 20 mM Hepes, pH 7.4, 0.1% Triton X-100 in the presence of different concentrations of NAD$^+$ (specific activity: 5 μCi/μmol). After this, the filters are washed for 10 min each with 3× 2 ml 20 mM Hepes, pH 7.4, 0.1% Triton X-100, dried, and quantified by gas ionization counting

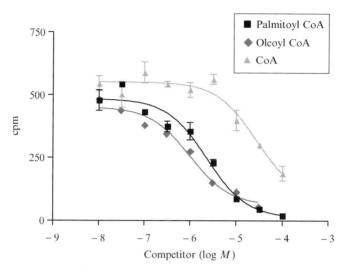

Fig. 4. Acyl-CoA binding assays. BARS was incubated with [^{14}C]-palmitoyl-CoA and increasing concentrations of unlabeled acyl-CoAs or CoA, as indicated. The amount of radiolabeled palmitoyl-CoA that remained bound to the protein was then quantified by gas ionization (Instant Imager, Packard Instrument, CT, USA) and the data were analyzed and plotted with the help of Prism (Graphpad Software). Representative experiment performed in duplicate.

(Instant Imager) (Nardini *et al.*, 2003). To investigate the binding of acyl-CoA, the filters are incubated with 6×10^4 cpm of [^{14}C]-acyl-CoA (palmitoyl-CoA or oleoyl-CoA) and different concentrations of unlabeled acyl-CoA, in 20 m*M* Hepes, pH 7.4, 0.01% Triton X-100 (Fig. 4). The filters are then washed for 10 min each with 3×2 ml of the same buffer and dried, and the amount of labeled acyl-CoA on the nitrocellulose is quantified by gas ionization counting (Instant Imager) (Yang *et al.*, 2006).

Acyltransferase Activity Assay

Solutions

Acyltransferase buffer: 20 m*M* Hepes, pH 7.4, 50 m*M* KCl, 40 m*M* sucrose, 3 m*M* MgCl$_2$, 5 m*M* DTT, 1 m*M* ATP, 10 U/ml creatine phosphokinase, 5 m*M* creatine phosphate.

This activity is measured in an incubation of 30 min at 37° that contains 0.1 mg/ml recombinant protein with 5 μM [^{14}C]-palmitoyl-CoA (specific activity, 60 mCi/mmol; PerkinElmer), 90 μM 18:1 lysophosphatidic acid, 1 m*M* GTP, and 5× acyltransferase buffer, diluted with H$_2$O to a final volume of 12.5 μl. The reaction is stopped on ice by adding 5 μl cold methanol/ 1 M hydrochloric acid (1:1), vortexed, and then the lipids are extracted in 50 μl chloroform/ methanol (2:1, v/v). The lower organic phase is loaded onto an oxalate-pretreated TLC plate (plates from Kiesegel 60 F_{254}, 20 × 20 cm, Merck, were pretreated by dipping for 80 s in a solution of 5 g potassium oxalate, 146 mg EDTA, 250 ml ethanol, 250 ml water, and then dried). Lipids were separated by running the TLC plates with chloroform/methanol/33% ammonium hydroxide/water (54:42:2.9:9.1). The radiolabeled spots are quantified by gas ionization counting (Instant Imager). [^{14}C]-palmitoyl-CoA, [^{14}C]-phosphatidic acid, and [^{3}H]-lysophosphatidic acid are used as standards.

Depletion of BARS

Solutions

KHM: 25 mM Hepes, pH 7.2, 125 mM potassium acetate, 2.5 mM magnesium acetate.

To deplete 25 μl of mitotic cytosol (at 12 mg/ml) of BARS, 10 μl of the p50–2 antiserum is coupled to 10 μl of protein A-Sepharose beads (Amersham Pharmacia Biotech, NJ, USA) in PBS for 1 h at 4°. To deplete 25 μl of rat brain cytosol (at 12 mg/ml) of BARS, 20 μl of the p50–2 antiserum is coupled to 20 μl of protein A-Sepharose beads. The antibody-protein A-Sepharose complex is washed (3×) with KHM permeabilization buffer and incubated with mitotic or rat brain cytosol. After 30 min, the beads are pelleted and the cytosol is incubated with fresh antibody-protein A complex for another 30 min. Mock incubations are performed by binding protein-A Sepharose beads to pre-immune IgGs. Thirty-five μg of total protein is run on a 10% SDS-PAGE and the depletion efficiency is monitored by immuno-blotting (Hidalgo Carcedo *et al.*, 2004).

Negative Staining

Preparation of Formvar-coated Grids for EM. Coated grids for EM are prepared as follows: formvar is solubilized in 1% chloroform at its final concentration and filtered through paper (Whatman paper n°3). Glass slides (76 × 26 mm) that have been cleaned with paper towels to remove any dust are soaked for 20 s in the formvar solution, rapidly removed, touched against a sheet of filter paper with one edge, and left to dry in air. The slides covered with the formvar film are exposed to a lamp light for 20 min and then the film is cut into rectangles (50 × 15 mm), removed from the slide, and floated on deionized water. Ten grids (3 mm in diameter,

200 mesh, in nickel or copper; Electron Microscopy Sciences, USA) are placed on each film, covered with a parafilm strip, and allowed to dry. The grids are stored in a covered petri dish to avoid dust contamination, and cleaned under a gentle nitrogen flow prior to use (Weigert *et al.*, 1999).

Preparation of Samples

Solutions

Incubation buffer: 25 m*M* Hepes, pH 7.4, 50 m*M* KCl, 200 m*M* sucrose, 1 m*M* ATP, 1 m*M* GTP, 5 m*M* DTT, 2.5 m*M* MgCl$_2$, 10 U/ml creatine phosphokinase, 10 m*M* creatine phosphate.

Two μl of Golgi membranes (at 0.1 mg/ml) are incubated with 2 mg/ml cytosol at 37° for 20 min in incubation buffer in a final volume of 10 μl. The reactions are stopped by the addition of 2 μl 2% glutaraldehyde in PBS, pH 7.4, at room temperature, and placed on a 1% formvar-coated grid for 5 min. The excess fluid is removed, and the grid is gently washed with 20 m*M* Hepes, pH 6.8. Ten mg/ml BSA in PBS, pH 7.4, is applied for 1 min, and 3 washing steps are performed with 20 m*M* Hepes, pH 6.8. Finally, the contraster (NanoW 2018; Nanoprobes, USA) is applied for 2× 20 s; the excess is removed with a strip of filter paper, and the grid is left to dry slowly (Weigert *et al.*, 1999).

Cellular Studies

Two types of approaches have been undertaken to evaluate the function of BARS in cells. First, the role of BARS in the fragmentation of the Golgi complex during mitosis has been assessed by using permeabilized normal rat kidney (NRK; epithelial) cells incubated with mitotic cytosol, an assay that mimics the fragmentation and dispersion of the Golgi complex membranes observed at the onset of mitosis (as detailed by Acharya *et al.*, 1998; Colanzi *et al.*, 2000).

The reagents that are essential for this study include mitotic cytosol, prepared according to Acharya *et al.* (1998), and the depletion of BARS by either an antibody or by antisense oligonucleotides, or its inhibition by a dominant-negative form (as detailed in the following). BARS is necessary for this fragmentation since BARS-depleted mitotic cytosol induces the formation of tubules that are not fragmented (Hidalgo Carcedo *et al.*, 2004). It is important to point out that in these experiments, as in other assays involving isolated membranes or permeabilized cells to which the recombinant protein (or a dominant negative) is to be added, the Golgi membranes need to be salt-washed. This is probably because the residual proteins that otherwise remain associated to the membranes are sufficient to sustain membrane fission.

The second line of experiments was focused on the role of BARS in the formation of intracellular transport carriers in intact cells. These have indicated that this protein controls basolateral transport from the Golgi complex to the plasma membrane and fluid-phase endocytosis (Bonazzi *et al.*, 2005), two traffic steps that are independent of dynamin, the second fission machinery protein that is known to operate in mammalian cells (Altschuler *et al.*, 1998; Damke *et al.*, 1994; Guha *et al.*, 2003; Kasai *et al.*, 1999). These results are compatible with the cellular localization of BARS that has been assessed in permeabilized cells using affinity-purified anti-BARS antibodies, which labeled both the nucleus and the cytoplasm, with a signal on the Golgi complex and the plasma membrane (see also Bonazzi *et al.*, 2005; Fig. 5). The morphological techniques used in these studies have been detailed elsewhere, and they include fluorescence and confocal microscopy (Polishchuk *et al.*, 2000, 2003), correlative light-EM (Polishchuk *et al.*, 2000), EM and cryo-immuno-gold labeling (Liou *et al.*, 1996; Mironov *et al.*, 2001).

Methods Involved in the *In Vivo* Studies

Cell Transfection

To express BARS in mammalian cells, a BARS-coding sequence was subcloned in pCDNA3 (Spanò *et al.*, 1999). COS7, Hela, NRK, and MDCK cells can be transfected with the lipofectamine-plus reagent (GIBCO BRL, UK). The cells are seeded on glass coverslips in 24-well plates in normal culture medium at a concentration suitable for 50–70% confluence at transfection. The fibronectin is layered onto the coverslip before the seeding of the cells, by adding to each well 200 μl 10 μg/ml fibronectin (Sigma Aldrich, WI, USA) in PBS containing 0.9 mM $CaCl_2$ and 0.5 mM $MgCl_2$, incubating for 30 min at room temperature, removing the solution, and allowing the fibronectin to dry for a few min. Twenty to twenty-four hours after seeding, a transfection mixture is prepared by first diluting the DNA and lipofectamine in separate polypropylene tubes, and then mixing the two solutions by pipetting. For each well, 400 ng DNA and 4 μl Plus reagent are diluted in 25 μl OPTI-MEM (GIBCO BRL, UK), and 1 μl lipofectamine is diluted in 25 μl OPTI-MEM. After 15 min, the two solutions are mixed and incubated for 15 min at room temperature. In the meantime, the cells are washed with OPTI-MEM medium. After this, the transfection mixture is brought to 0.25 ml and added to the cells. After 4 h of incubation under normal growth conditions, the transfection mixture is

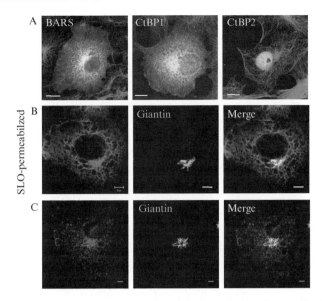

FIG. 5. Subcellular localization of BARS. (A) To examine the localization of the individual CtBPs, COS7 cells were transfected with BARS (CtBP3/BARS), CtBP1, or CtBP2, and 24 h after transfection they were fixed with 4% paraformaldehyde for 10 min, and double stained with 1 μg/ml affinity purified SN1 antibody (green) and an anti-tubulin antibody (red). (B) (C) COS7 cells were transfected with BARS, and 24 h after the transfection they were fixed with 4% paraformaldehyde (B), or first permeabilized with SLO, incubated at 37° for 5 min, and fixed with 4% paraformaldehyde (C), for 10 min at room temperature. The cells were double-labeled with 1 μg/ml affinity-purified SN1 antibody (red) and a monoclonal anti-giantin antibody (green); the merged images are also shown. The Golgi and plasma membrane localization of BARS are seen after SLO-permeabilization, which removes the soluble cytosolic pool of BARS.

replaced by complete culture medium and the cells are incubated under normal growth conditions for an additional 24–72 h.

Human fibroblasts can be transfected by electroporation. For transfection, 4×10^6 cells are resuspended in 500 μl PBS containing 20 μg DNA. The suspension is placed in an electroporation cuvette (Bio-Rad Laboratories, UK) with a gap of 0.4 cm, and kept in ice for 10 min before electroporation. For the electroporation, a Gene-Pulser Electroporator (Bio-Rad Laboratories, UK) can be used with the following working parameters: voltage, 380 V; capacitance, 850 μF. After the pulse, the cells are resuspended in normal culture medium and plated onto coverslips in 24-well plates at a density of 1.5×10^5 cells/well (Bonazzi et al., 2005).

Antisense Oligonucleotides

A 21-bp oligonucleotide sequence (5'-TGA TGC CCA AGG TCT CTC CAC-3'; MWG-Biotech, Italy) was designed based on the sequence of BARS, in order to inhibit the synthesis of both CtBP1 and BARS. A sequence (5'-GAG CCT TTC TCA GCC ACC TAG-3'; MWG-Biotech, Italy) obtained by randomly scrambling the above nucleotides was used as control. The NRK cells are plated at 20% confluence in normal growth medium. After 12–18 h, and every 24 h thereafter for the following 5 days, the oligonucleotides are added directly to culture medium at a final 8 μM concentration. The efficiency of the treatment is monitored by SDS-PAGE and Western blotting (Hidalgo Carcedo *et al.*, 2004).

SiRNAs Transfection

COS7 cells at 50% confluence in 24-well plates can be transfected with a Smart Pool of siRNA sequences (CCGUCAAGCAGAUGAGACAUU; GGAUAGAGACCACGCCAGUUU; GCUCGCACUUGCUCAACA-AUU; GAGCAGGCAUCCAUCGAGAUU; Dharmacon, USA) directed against CtBP1 and BARS using oligofectamine (Invitrogen, USA), according to the manufacturer's instructions. Briefly, the transfection mixture is prepared by first diluting the siRNAs and oligofectamine in separate polypropylene tubes, and then mixing the two solutions by pipetting. For each well, 50 ng siRNAs are diluted in 40 μl OPTI-MEM (GIBCO BRL, UK), and 2 μl oligofectamine is diluted in 7.5 μl OPTI-MEM. After 5 min, the two solutions are mixed and incubated for 20 min at room temperature. In the meantime, the COS7 and HeLa cells are washed with OPTI-MEM medium. After this, the transfection mixture is brought to 0.25 ml and added to the cells. After 4 h of incubation under normal growth conditions, growth medium containing 3× the normal concentration of serum is added to the cells, without removing the transfection mixture. The cells are incubated for an additional 48 h and the level of knock-down is evaluated by SDS-PAGE and Western blotting (Bonazzi *et al.*, 2005).

Microinjection

Solutions

Microinjection buffer: 10 mM sodium phosphate buffer, pH 7.4, 70 mM KCl.

COS7 cells are injected with the purified BARS protein at concentrations of 2–3 mg/ml in microinjection buffer. This should increase the intracellular

BARS concentration by some 5-fold to 15-fold, based on the calculations that the intracellular concentration is around 20 μg/ml and on the assumption that 5–10% of the cell volume is injected. Prior to injection, the protein is mixed with 0.4 mg/ml fluorescein isothiocyanate (FITC)- or tetramethyl-rhodamine B isothiocyanate (TRITC)-labeled dextran (Molecular Probes) as a marker for the microinjected cells (Bonazzi et al., 2005). To give an example, in studies of basolateral and apical transport (using the vesicular stomatitis virus glycoprotein and p75, respectively), the proteins were microinjected 1 h after the beginning of the 20° incubation in the transport assay. After injection, the cells were allowed to recover for 1 h before proceeding with the experimental protocol (Bonazzi et al., 2005). The BARS (p50–2) and dynamin (DYN2) antibodies were injected at 4.5 mg/ml, 3 h before further experimental procedures.

Cell Permeabilization

Solutions

Streptolysin O (SLO; Sclavo, Italy): 25 U/ml SLO is dissolved in sterile water;

Buffer A: 20 mM Hepes, pH 7.2, 110 mM potassium acetate, 2 mM magnesium acetate, 1 mM DTT (added just before the experiment);

Buffer B: 25 mM Hepes, pH 7.2, 75 mM potassium acetate, 2.5 mM magnesium acetate, 5 mM ethylene glycol-bis(β-aminoethyl ether) N,N,N′,N′,-tetracetic acid (EGTA), 1.8 mM CaCl$_2$.

The SLO is thawed in a 37° bath and 1 vol. 2× buffer A is added. This solution is incubated 5 min at 37° to activate the SLO. After this incubation, the SLO is diluted to 2 U/ml with 1× buffer A and kept in ice before being used within 30 min. The buffer A, the buffer B, and a 24-well plate with 70–90%-confluent cells are put on ice. The cells are washed twice with 500 μl ice-cold 1× buffer A, and 250 μl 2 U/ml SLO are added. After a 10-min incubation in ice, the cells are gently washed with 500 μl ice-cold 1× buffer A. Following the addition of 500 μl ice-cold buffer B, the cells are transferred to a 37° bath. After a 5-min incubation, the cells are fixed and processed for immuno-fluorescence (Bonazzi et al., 2005).

Preparation of Antibodies

An affinity-purified antibody raised against full-length BARS (p50–2) was used for Western blotting, immuno-fluorescence and immuno-electro-microscopy experiments, and it was able to recognize endogenous BARS using all of these techniques. Characterization of this affinity-purified antibody indicated that it recognizes all three of the CtBPs expressed in mammalian tissues (Spanò *et al.*, 2002), indicating that the endogenous BARS pool stained by this antibody will also include the other CtBPs. An affinity-purified antibody raised against a BARS-derived peptide (SN1) also recognizes all three of the CtBPs (Spanò *et al.*, 2002), although it cannot detect the endogenous levels of the protein either by Western blotting or by immuno-fluorescence. The affinity-purified SN1 antibody was thus used in the course of the studies to detect the overexpression of BARS, CtBP1, and CtBP2 by Western blotting, immuno-fluorescence, and immuno-electromicroscopy. These antibodies that were raised against BARS were obtained through the immunization of rabbits. One mg of the SN1 peptide (aa 147 to 162 of BARS) or 0.5 mg of recombinant GST-BARS were resuspended in 2 ml PBS. Two ml of complete Freund's adjuvant were added, and this mixture was used to immunize New Zealand rabbits. The rabbits were boosted after 21 and 42 days with 1 mg (or 0.5 mg, respectively) of antigen containing the same volume of incomplete Freund's adjuvant (Sigma Aldrich, WI, USA). Blood was taken from the rabbits, and it was allowed to clot at $37°$ for 60 min, and then kept overnight at $4°$ to allow the clot to contract. The clot was removed from the serum, along with other insoluble material, by centrifugation at $10,000g$ for 10 min at $4°$, and afterwards it was stored at $-80°$. The affinity-purified antibodies were obtained by passing purified IgGs through a column of protein-A-Sepharose that was covalently linked to the antigen. The total IgGs were purified by protein-A-Sepharose beads; 500 mg of this was suspended in 5 ml distilled water, packed into a column, and washed with 100 ml distilled water under a constant flow. The packed beads were washed with 20 ml PBS, and 2 ml antiserum was loaded onto the column at 0.5 ml/min using a fast protein liquid chromatography (FPLC) system (Pharmacia Bio-Tec, UK). The beads were washed with 30 ml PBS, and the retained IgGs were detached with 15 ml 0.1 *M* glycine, pH 2.5. Fractions of 0.5 ml were collected, and their protein content quantified by spectrophotometric analysis using a protein assay kit (Bio-Rad). The six fractions containing the highest concentrations of protein were pooled and neutralized with 1 *M* Tris, pH 11. These total IgGs were passed through a column to which the recombinant His-BARS or the SN1 peptide were covalently coupled via a Hi-trap N-hydroxysuccinimide (NHS)-activated matrix (Amersham Pharmacia Biotech, NJ, USA), according to the manufacturer's instructions. The BARS antibodies were purified through the column with a procedure analogous to that given for the total IgGs.

Further Technical Considerations

BARS Overexpression

The use of overexpression or knock-down of BARS requires some comments. Different effects can be observed, depending on the experimental conditions that are followed: while acute administration of BARS stimulates BARS-dependent processes, prolonged exposure to BARS inhibits them. To give an example, stimulatory effects of BARS on the formation of VSVG-containing post-Golgi transport carriers were observed if the microinjection of purified recombinant BARS was followed by a recovery period of 1 h before the analysis of transport-carrier formation (Bonazzi *et al.*, 2005). In contrast, an inhibition of post-Golgi transport-carrier formation was seen when BARS was administered to cells through DNA transfection and the cells were analyzed after 18 h. Similar results were obtained when the cells were microinjected with the p50–2 anti-BARS antibody and analyzed 3 h after this treatment. The mechanism through which prolonged exposure to BARS results in a blocking effect on post-Golgi transport is currently under investigation, and it may be related to the ability of BARS to interfere with lipid metabolism and composition at the Golgi membranes (Weigert *et al.*, 1999).

BARS Localization

Another practical aspect of BARS that deserves comment is its localization in different cell types. The localization of endogenous BARS at the Golgi complex can be detected in human fibroblasts after a 10-min fixation with 4% paraformaldehyde and staining with 10 μg/ml affinity-purifed p50–2 antibody (Bonazzi *et al.*, 2005). In other cell types, especially when overexpressed, BARS is largely cytosolic and becomes clearly detectable at the Golgi complex only after permeabilization with SLO prior to fixation (see preceding, and Fig. 5). The SLO permeabilization procedure removes the excess cytosolic BARS that is present in cells after transfection without affecting the pool of BARS that is associated with the Golgi complex, the plasma membrane, and the nuclear envelope.

As recently reported, a well-defined Golgi localization of BARS has been observed in COS7 cells after the microinjection of low doses of the p50–2 anti-BARS antibody before fixation, suggesting that fixation decreases the BARS staining on the Golgi complex (Bonazzi *et al.*, 2005). In addition, in human fibroblasts immuno-EM analysis has shown BARS localization at the *cis-* and *trans-*sides of the Golgi stacks (Bonazzi *et al.*, 2005).

Acknowledgments

We wish to thank C. Cericola and G. Turacchio for invaluable technical assistance during these studies, A. Colanzi, C. Hidalgo Carcedo, and M. Bonazzi for discussion, G. Chinnadurai for kindly providing cDNAs, C. P. Berrie for editorial assistance, R. Le Donne and E. Fontana for preparation of the figures, and the Italian Association for Cancer Research (AIRC, Milan, Italy), the Italian Foundation for Cancer Research (FIRC, Milan, Italy), Telethon (Italy), and the MIUR for financial support. CV is supported by an FIRC fellowship.

References

Acharya, U., Mallabiabarrena, A., Acharya, J. K., and Malhotra, V. (1998). Signaling via mitogen-activated protein kinase kinase (MEK1) is required for Golgi fragmentation during mitosis. *Cell* **92,** 183–192.

Altschuler, Y., Barbas, S. M., Terlecky, L. J., Tang, K., Hardy, S., Mostov, K. E., and Schmid, S. L. (1998). Redundant and distinct functions for dynamin-1 and dynamin-2 isoforms. *J. Cell Biol.* **143,** 1871–1881.

Bonazzi, M., Spanò, S., Turacchio, G., Cericola, C., Valente, C., Colanzi, A., Kweon, H. S., Hsu, V. W., Polishchuck, E. V., Polishchuk, R. S., Sallese, M., Pulvirenti, T., Corda, D., and Luini, A. (2005). CtBP3/BARS drives membrane fission in dynamin-independent transport pathways. *Nat. Cell Biol.* **7,** 570–580.

Burger, K. N. (2000). Greasing membrane fusion and fission machineries. *Traffic* **1,** 605–613.

Chinnadurai, G. (2002). CtBP, an unconventional transcriptional corepressor in development and oncogenesis. *Mol. Cell* **9,** 213–224.

Cluett, E. B., and Brown, W. J. (1992). Adhesion of Golgi cisternae by proteinaceous interactions: Intercisternal bridges as putative adhesive structures. *J. Cell Sci.* **103,** 773–784.

Colanzi, A., Deerinck, T. J., Ellisman, M. H., and Malhotra, V. (2000). A specific activation of the mitogen-activated protein kinase kinase 1 (MEK1) is required for Golgi fragmentation during mitosis. *J. Cell Biol.* **149,** 331–339.

Corda, D., Hidalgo Carcedo, C., Bonazzi, M., Luini, A., and Spanò, S. (2002). Molecular aspects of membrane fission in the secretory pathway. *Cell. Mol. Life. Sci.* **59,** 1819–1832.

Criqui-Filipe, P., Ducret, C., Maira, S. M., and Wasylyk, B. (1999). Net, a negative Ras-switchable TCF, contains a second inhibition domain, the CID, that mediates repression through interactions with CtBP and de-acetylation. *EMBO J.* **18,** 3392–3403.

Damke, H., Baba, T., Warnock, D. E., and Schmid, S. L. (1994). Induction of mutant dynamin specifically blocks endocytic coated vesicle formation. *J. Cell Biol.* **127,** 915–934.

De Matteis, M. A., Di Girolamo, M., Colanzi, A., Pallas, M., Di Tullio, G., McDonald, L. J., Moss, J., Santini, G., Bannykh, S., Corda, D., and Luini, A. (1994). Stimulation of endogenous ADP-ribosylation by brefeldin A. *Proc. Natl. Acad. Sci. USA* **91,** 1114–1118.

Di Girolamo, M., Silletta, M. G., De Matteis, M. A., Braca, A., Colanzi, A., Pawlak, D., Rasenick, M. M., Luini, A., and Corda, D. (1995). Evidence that the 50-kDa substrate of brefeldin A-dependent ADP-ribosylation binds GTP and is modulated by the G-protein beta gamma subunit complex. *Proc. Natl. Acad. Sci. USA* **92,** 7065–7069.

Guha, A., Sriram, V., Krishnan, K. S., and Mayor, S. (2003). Shibire mutations reveal distinct dynamin-independent and -dependent endocytic pathways in primary cultures of Drosophila hemocytes. *J. Cell Sci.* **116,** 3373–3386.

Hidalgo Carcedo, C., Bonazzi, M., Spanò, S., Turacchio, G., Colanzi, A., Luini, A., and Corda, D. (2004). Mitotic Golgi partitioning is driven by the membrane-fissioning protein CtBP3/BARS. *Science* **305,** 93–96.

Huttner, W. B., and Schmidt, A. (2000). Lipids, lipid modification and lipid-protein interaction in membrane budding and fission—insights from the roles of endophilin A1 and synaptophysin in synaptic vesicle endocytosis. *Curr. Opin. Neurobiol.* **10**, 543–551.

Kasai, K., Shin, H. W., Shinotsuka, C., Murakami, K., and Nakayama, K. (1999). Dynamin II is involved in endocytosis but not in the formation of transport vesicles from the trans-Golgi network. *J. Biochem. (Tokyo)* **125**, 780–789.

Kumar, V., Carlson, J. E., Ohgi, K. A., Edwards, T. A., Rose, D. W., Escalante, C. R., Rosenfeld, M. G., and Aggarwal, A. K. (2002). Transcription corepressor CtBP is an NAD (+)-regulated dehydrogenase. *Mol. Cell* **10**, 857–869.

Liou, W., Geuze, H. J., and Slot, J. W. (1996). Improving structural integrity of cryosections for immunogold labeling. *Histochem. Cell Biol.* **106**, 41–58.

Mironov, A., Colanzi, A., Silletta, M. G., Fiucci, G., Flati, S., Fusella, A., Polishchuk, R., Mironov, A., Jr., Di Tullio, G., Weigert, R., Malhotra, V., Corda, D., De Matteis, M. A., and Luini, A. (1997). Role of NAD+ and ADP-ribosylation in the maintenance of the Golgi structure. *J. Cell Biol.* **139**, 1109–1118.

Mironov, A. A., Beznoussenko, G. V., Nicoziani, P., Martella, O., Trucco, A., Kweon, H. S., Di Giandomenico, D., Polishchuk, R. S., Fusella, A., Lupetti, P., Berger, E. G., Geerts, W. J., Koster, A. J., Burger, K. N., and Luini, A. (2001). Small cargo proteins and large aggregates can traverse the Golgi by a common mechanism without leaving the lumen of cisternae. *J. Cell Biol.* **155**, 1225–1238.

Nardini, M., Spanò, S., Cericola, C., Pesce, A., Massaro, A., Millo, E., Luini, A., Corda, D., and Bolognesi, M. (2003). CtBP/BARS: A dual-function protein involved in transcription co-repression and Golgi membrane fission. *EMBO J.* **22**, 3122–3130.

Polishchuk, E. V., Di Pentima, A., Luini, A., and Polishchuk, R. S. (2003). Mechanism of constitutive export from the Golgi: Bulk flow via the formation, protrusion, and en bloc cleavage of large trans-golgi network tubular domains. *Mol. Biol. Cell* **14**, 4470–4485.

Polishchuk, R. S., Polishchuk, E. V., Marra, P., Alberti, S., Buccione, R., Luini, A., and Mironov, A. A. (2000). Correlative light-electron microscopy reveals the tubular-saccular ultrastructure of carriers operating between Golgi apparatus and plasma membrane. *J. Cell Biol.* **148**, 45–58.

Schaeper, U., Boyd, J. M., Verma, S., Uhlmann, E., Subramanian, T., and Chinnadurai, G. (1995). Molecular cloning and characterization of a cellular phosphoprotein that interacts with a conserved C-terminal domain of adenovirus E1A involved in negative modulation of oncogenic transformation. *Proc. Natl. Acad. Sci. USA* **92**, 10467–10471.

Schmidt, A., Wolde, M., Thiele, C., Fest, W., Kratzin, H., Podtelejnikov, A. V., Witke, W., Huttner, W. B., and Soling, H. D. (1999). Endophilin I mediates synaptic vesicle formation by transfer of arachidonate to lysophosphatidic acid. *Nature* **401**, 133–141.

Schmitz, F., Konigstorfer, A., and Sudhof, T. C. (2000). RIBEYE, a component of synaptic ribbons: A protein's journey through evolution provides insight into synaptic ribbon function. *Neuron* **28**, 857–872.

Spanò, S. (2002). Functional and molecular characterization of CtBP3/BARS, a protein involved in the control of the Golgi complex. *PhD thesis.*.

Spanò S, Hidalgo Carcedo C, Luini A, and Corda D. (2006). CtBP3/BARS and membrane fission. *In* "CtBP Family Proteins" (G. Chinnadurai, Ed.), Georgetown, TX: Landes Bioscience.

Spanò, S., Silletta, M. G., Colanzi, A., Alberti, S., Fiucci, G., Valente, C., Fusella, A., Salmona, M., Mironov, A., Luini, A., and Corda, D. (1999). Molecular cloning and functional characterization of brefeldin A-ADP-ribosylated substrate. A novel protein involved in the maintenance of the Golgi structure. *J. Biol. Chem.* **274,** 17705–17710.

Turner, J., and Crossley, M. (1998). Cloning and characterization of mCtBP2, a co-repressor that associates with basic Kruppel-like factor and other mammalian transcriptional regulators. *EMBO J.* **17,** 5129–5140.

Weigert, R., Silletta, M. G., Spanò, S., Turacchio, G., Cericola, C., Colanzi, A., Senatore, S., Mancini, R., Polishchuk, E. V., Salmona, M., Facchiano, F., Burger, K. N., Mironov, A., Luini, A., and Corda, D. (1999). CtBP/BARS induces fission of Golgi membranes by acylating lysophosphatidic acid. *Nature* **402,** 429–433.

Yang, J.-S., Lee, S. Y., Spanò, S., Gad, H., Zhang, L., Bonazzi, M., DeBono, C. A., Branch Moody, D., Barr, F. A., Corda, D., Luini, A., and Hsu, V. W. (2006). BARS participates in the fission of COPI vesicles from Golgi membrane. *EMBO J.*

[28]   *In Vitro* Assays of Arf1 Interaction with GGA Proteins

By Hye-Young Yoon, Juan S. Bonifacino, and
Paul A. Randazzo

Abstract

ADP-ribosylation factor 1 (Arf1) is a GTP-binding protein that regulates membrane traffic. This function of Arf1 is, at least in part, mediated by Arf1•GTP binding to coat proteins such as coatomer, clathrin adaptor protein (AP) complexes 1 and 3, and γ-adaptin homology-Golgi associated Arf-binding (GGA) proteins. Binding to Arf1•GTP recruits these coat proteins to membranes, leading to the formation of transport vesicles. Whereas coatomer and the AP complexes are hetero-oligomers, GGAs are single polypeptide chains. Therefore, working with recombinant GGAs is straightforward compared to the other Arf1 effectors. Consequently, the GGAs have been used as a model for studying Arf1 interactions with effectors and as reagents to determine Arf1•GTP levels in cells. In this chapter, we describe *in vitro* assays for analysis of GGA interaction with Arf1•GTP and for determining intracellular Arf1•GTP levels.

Introduction

Arfs are members of a family of Ras-like small GTP-binding proteins (Moss and Vaughan, 1998; Randazzo *et al.*, 2000). They are ubiquitously expressed in eukaryotic cells and are highly conserved. The six mammalian

METHODS IN ENZYMOLOGY, VOL. 404   0076-6879/05 $35.00
                                          DOI: 10.1016/S0076-6879(05)04028-0

Arf proteins are grouped into class I (Arf1, 2, and 3), class II (Arf4 and 5), and class III (Arf6) based on sequence homology. Arfs were originally named for their function as cofactors for ADP-ribosylation of heterotrimeric G proteins catalyzed by cholera toxin. Subsequent studies, however, have shown that their main physiologic function is regulation of membrane traffic.

Arf regulation of membrane traffic depends on their interaction with a subset of coat proteins that are critical components of the membrane traffic machinery (Bonifacino and Glick, 2004; Bonifacino and Lippincott-Schwartz, 2003; Kirchhausen, 2002; Owen *et al.*, 2004; Robinson and Bonifacino, 2001). Among the coat proteins that interact with Arf are a heteroheptameric complex named coatomer, which polymerizes to form COPI coats, and the heterotetrameric adaptors AP-1 and AP-3, which are incorporated into clathrin coats. The structurally related non-clathrin adaptor AP-4 also interacts with Arf, as do the single polypeptide GGA clathrin adaptors. Although all Arfs bind to these proteins to some extent, class I Arfs such as Arf1 and Arf3 are the most active for coat protein recruitment. Therefore, we focus our discussion and methods on Arf1. In the current paradigm (Bonifacino and Glick, 2004; Bonifacino and Lippincott-Schwartz, 2003; Nie *et al.*, 2003; Randazzo *et al.*, 2000; Spang, 2002; Springer *et al.*, 1999). Arf1•GDP exchanges nucleotide to form Arf1•GTP. Arf1•GTP binds tightly to membranes via its myristoylated N-terminal α-helix and to the protomer form of the coat proteins via its switch 1 and switch 2 regions. This results in recruitment of the coat protomers to the cytosolic surface of membranes. The coat protomers can then bind to and trap transmembrane cargo molecules, as well as polymerize into a vesicle coat that drives deformation of the membrane and budding of a transport vesicle.

Three GGA proteins exist in humans (i.e., GGA1, GGA2, and GGA3) and 1–3 in most other eukaryotes (Bonifacino, 2004; Ghosh and Kornfeld, 2004). The GGAs are comprised of four domains, from N- to C-terminus, VHS, GAT, hinge, and GAE (Bonifacino, 2004; Boman *et al.*, 2000; Dell'Angelica *et al.*, 2000; Hirst *et al.*, 2000; Nakayama and Wakatsuki, 2003). The VHS domain of the mammalian GGAs binds to acidic cluster dileucine or DXXLL sorting motifs in intracellular transport receptors such as the cation-independent and cation-dependent mannose 6-phosphate receptors (Ghosh *et al.*, 2003; Nielsen *et al.*, 2001; Puertollano *et al.*, 2001a; Takatsu *et al.*, 2001). The GAT domain binds to Arf1•GTP, Rabaptin-5, ubiquitin, and TSG101 (Bilodeau *et al.*, 2004; Dell'Angelica *et al.*, 2000; Mattera *et al.*, 2003, 2004; Puertollano *et al.*, 2001b; Scott *et al.*, 2004; Shiba *et al.*, 2004). The unstructured hinge region binds to clathrin (Puertollano *et al.*, 2001b). The GAE domain interacts with accessory proteins including γ-synergin, p56, Rabaptin-5, enthoprotin, and aftiphilin (Lui *et al.*, 2003; Mattera *et al.*, 2003).

Given that GGAs comprise a single modular polypeptide, expression of the recombinant domains has been relatively straightforward. The domains expressed in bacteria are soluble and have the same activities, including cargo, Arf1, clathrin, and accessory protein binding, as in the full-length proteins expressed in mammalian cells. Because of these properties, GGA has been extensively studied with significant progress in understanding structure–function relationships. The crystal structures of the VHS (Misra *et al.*, 2001; Zhu *et al.*, 2003a), GAT (Collins *et al.*, 2003b; Miller *et al.*, 2003; Zhai *et al.*, 2003; Zhu *et al.*, 2003b, 2004) and GAE (Collins *et al.*, 2003a; Miller *et al.*, 2003; Nogi *et al.*, 2002) domains, alone or in complexes with their binding partners, have been determined. The domain that interacts with Arf1, GAT, is an elongated, all α-helical fold that forms two subdomains. There is an N-terminal "hook" subdomain consisting of a short α-helix folding with the N-terminal portion of a longer α-helical segment, with a loop separating the two α-helices. This structure interacts with the switch 1 and switch 2 regions (parts of Arf that are sensitive to nucleotide) of Arf1•GTP. The C-terminal subdomain consists of a three α-helix bundle and is involved in binding ubiquitin, Rabaptin 5, and TSG101. The putative binding sites are far apart so that both binding sites can be occupied simultaneously.

Because of their simpler structure, the GGAs have been used as a model for studying the interactions of Arf1 with effectors. Much work has been done using yeast two-hybrid and mutagenesis (see for example Kuai and Kahn, 2000; Kuai *et al.*, 2000; Puertollano *et al.*, 2001b). Using *in vitro* assays, GGA binding to Arf1 has been further characterized in respects that could not be done by two-hybrid assays. In addition to quantifying the relative effects of switch 1 and switch 2 mutants, the *in vitro* assays have allowed examination of the effects of cargo, acid phospholipids, and domains adjacent to the GAT domain, on Arf1-GGA interaction (Hirsch *et al.*, 2003; Jacques *et al.*, 2002). Here, we describe several approaches we have used for studying GGA interactions with Arf1 in solution, and a method using GGA for the determination of Arf1•GTP levels *in vivo*.

Methods

Preparation of Recombinant GGA Domains

For the assays described in this chapter, constructs comprised of the GAT domain of GGA proteins, and additional domains as necessary for the question being addressed, are expressed in bacteria. The proteins are fused to tags to aid in purification. GST-fusion proteins work well for all the methods described (Dell'Angelica *et al.*, 2000; Puertollano *et al.*, 2001b). GST-VHSGAT$_{GGA3}$ (residues 1–313 of GGA3) and GST-GAT$_{GGA3}$ (residues 147–313) were generated using the plasmid pGEX-5X-1 (Amersham Biosciences, Piscataway, NJ). The open reading frames were amplified with *Eco*RI and *Not*I restriction sites at the 5′ and 3′ ends and were subcloned into pGEX-5X-1 by standard DNA recombinant procedures. His$_{10}$-tagged proteins are also useful and have the added benefit that they do not dimerize as readily. His$_{10}$-VHSGATGGA1 (residues 1–315 of GGA1) and His$_{10}$-GAT$_{GGA1}$ (residues 148–315 of GGA1) were generated using the plasmid pET19 (Novagen, Madison, WI) (Hirsch *et al.*, 2003; Jacques *et al.*, 2002; Puertollano *et al.*, 2001b). The open reading frame was amplified by PCR incorporating *Nde*I and *Bam*HI restriction sites that were used to subclone into the plasmid.

GST-fusion proteins and His$_{10}$-tagged proteins are expressed in *E. coli* BL21(DE3) bacteria using the same protocol. Transformed bacteria are selected with ampicillin. A single colony is grown in 200 ml Luria-Bertani (LB) medium containing 100 μg ampicillin per ml until OD$_{600}$ = 0.6 at 37°. The bacteria are cooled to 4° and refrigerated overnight. The next day, the bacteria are collected by centrifugation and resuspended in 1–2 l of LB medium with 100 μg/ml ampicillin. The culture is maintained at 37° until it reaches OD$_{600}$ = 0.6 and then isopropyl thio-β-D-galactylpyranoside (IPTG) is added to a final concentration of 1 mM. After IPTG induction, the bacteria are grown for an additional 3 h at 37° and then harvested by centrifugation at 1500–2500g for 20 min at 4°.

To purify the bacterially expressed proteins, the cell pellets from 250–500 ml of cell culture are suspended in 10 ml of phosphate-buffered saline (PBS) containing a Complete® protease inhibitor tablet (Roche, Indianapolis, IN) and 0.1% (w/v) Triton X-100 and lysed with a French press operated at 12,000 psi (double the volume of PBS if using 1–2l of cell culture). For the His10-tagged protein, an EDTA-free protease inhibitor cocktail is used (also available from Roche). The soluble material is clarified by centrifugation at 100,000g for 60 min at 4°. Both GST-fusion proteins and His$_{10}$-tagged proteins can be purified by batch adsorption to and elution from glutathione-Sepharose 4 B (Amersham Biosciences) or a metal-chelating resin (e.g., Talon® from Clontech or Ni-NTA from Qiagen, Valencia, CA), respectively, using methods described by the manufacturer of the resin. We prefer to use columns. For GST-fusion proteins, the clarified cell lysate is loaded onto 300 μl of glutathione-Sepharose 4 B packed in a Poly-Prep chromatography column (Bio-Rad) equilibrated with PBS. The column is washed with 1–2 ml of ice-cold PBS and proteins eluted with 10 mM glutathione, 50 mM Tris-HCl, pH 8.0, and 100 mM

NaCl in 5 fractions of 300 μl. 5 μl samples of the 5 fractions are analyzed by SDS-PAGE and the 1 or 2 fractions that together contain more than 90% of the proteins are taken. For His10-tagged proteins, we use a HisTrap™ column (Amersham Biosciences). The His10-tagged protein is adsorbed to the column equilibrated with 20 mM Tris-HCl, pH 8.0, 500 mM NaCl, and 10 mM imidazole, pH 7.0. The protein is then eluted with a gradient from 10 to 500 mM imidazole, pH 7.0, in 500 mM NaCl. Both GST-fusion and His10-tagged proteins are desalted by using a PD-10 column (Amersham Biosciences), equilibrated, and run with ice-cold PBS or 20 mM Tris-HCl, pH 7.5, 100 mM NaCl, and 1 mM dithiothreitol (DTT). The 0.5–1.5 ml sample is applied to a PD-10 column and 1 ml fractions are collected. The 1 or 2 fractions containing greater than 90% of the protein, determined using the Bio-Rad dye-binding protein assay, are pooled.

Preparation of Other Recombinant Proteins

The preparation of His$_{10}$[325-724]ASAP1 is described in Randazzo *et al.* (2000). The preparation of myristoylated Arf1 is described in Chapter 16 of this volume (Preparation of Myristoylated Arf1 and Arf6). The preparation of non-myristoylated Arf1, is described in Randazzo *et al.* (1992). The same method is used for the purification of [L8K]Arf1, which is described in Yoon *et al.* (2004).

Loading Arf1 with [^{35}S]GTPγS and [α^{32}P]GTP

For the methods we describe here, Arf1•GTP is used at a concentration that is much lower than the measured dissociation constant for the GGA•Arf1•GTP complex (K_d) or the Michaelis constant (i.e., the concentration of Arf1•GTP that gives half maximal velocity of GAP-induced GTP hydrolysis, K_m) for Arf GAP. Under this condition, the equations that are derived for the analysis of the data are simple hyperbolics rather than quadratics (see following).

For two assays, Arf1•[^{35}S]GTPγS is prepared by incubating 1–5 μM Arf1 with 15 μM [^{35}S]GTPγS (specific activity $\approx$ 50,000 cpm/pmol) in 20 mM Tris-HCl, pH 7.5, 100 mM NaCl, 1 mM EDTA, 0.5 mM MgCl$_2$, 1 mM DTT, and 0.1% (w/v) Triton X-100 for 1 h at 30°. [α^{32}P]GTP•Arf1 is prepared in a similar way. However, in this case, steps are taken to ensure that contaminating nucleotidases (very difficult to completely remove from Arf1) do not significantly degrade GTP before it is able to bind to Arf1. Two approaches have worked for us. One way is to incubate 1–5 μM Arf1 with 10 μM [α^{32}P]GTP (specific activity $\approx$ 50,000 cpm/pmol) in 20 mM Tris-HCl, pH 7.5, 1 mM EDTA, 0.5 mM MgCl$_2$, 1–2 mM ATP, 1 mM DTT, and 0.1% (w/v) Triton X-100 for 30–60 min at 30°. High concentrations of

ATP inhibit nonspecific nucleotidases. Another way to prepare $[\alpha^{32}P]$ GTP•Arf1 for the GAP assay is to incubate Arf1 with 25 mM HEPES, pH 7.4, 100 mM NaCl, 3.5 mM MgCl$_2$, 1 mM EDTA, 1 mM ATP, 1 μM $[\alpha^{32}P]$GTP (specific activity = 50,000–250,000 cpm/pmol), 25 mM KCl, 1.25 U/ml pyruvate kinase, and 3 mM phosphoenolpyruvate. This buffer contains a GTP regenerating system. If using Arf1 that has not been myristoylated, include 0.1% (w/v) Triton X-100. For myristoylated Arf1, use either micelles of 3 mM dimyristoylphosphatidylcholine and 0.1% cholate, pH 7.4 or use vesicles prepared by extrusion or sonication (see Chapter 15 of this volume, Assay and Properties of the Arf GAPs AGAP1, ASAP1, and ArfGAP1).

Three Assays for GGA•Arf1 Interactions

Direct Determination of Binding. With a K_d of greater than 200 nM for Arf1•GTP•GGA interactions, conventional pull-down assays are not quantitative due to rapid dissociation during the washes. Nonetheless, these assays are useful for establishing that there is a specific interaction between GGA and Arf1•GTP. We show an example of data from a direct binding assay in Fig. 1. In a typical experiment, 10 μg of GST or GST-VHSGATGGA3 are added to 450 μl of a cell lysate, for example, bovine brain lysate, containing 20 mM Tris-HCl, pH 8.0, 100 mM NaCl, 0.1% (w/v)

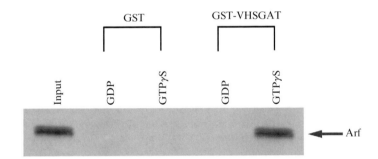

FIG. 1. Pull-down assay for Arf•GTP association with GST-VHSGAT$_{GGA3}$. GSTVHSGAT$_{GGA3}$ (10 μg) or GST was incubated with a bovine brain lysate containing 0.1% (w/v) Triton X-100 and 100 μM GDP or GTPγS for 30 min at 30° and then chilled to 4° for 5 min. GST and GST-VHSGAT$_{GGA3}$ were precipitated with glutathione-agarose. The precipitate was washed 3 times with ice-cold PBS containing 0.1% (w/v) Triton X100 and the associated proteins were fractionated by SDS-PAGE and transferred to nitrocellulose. Arf was detected by immunoblotting with a mouse monoclonal antibody 1D9 from Affinity BioReagents used at a dilution of 1:500, a goat anti-mouse IgG-HRP conjugate (Bio-Rad) used at a dilution of 1:10,000 and ECL plus Western blotting detection reagent (Amersham Biosciences).

Triton X-100, and 20 μM GTPγS. The mixture is incubated at 30° for 30–60 min and then chilled. Glutathione-agarose beads, 25 μl, are added and the mixture is incubated an additional 30 min at 4°. The beads are collected by a brief centrifugation (13,000g in a table-top refrigerated microcentrifuge for 30 sec) and washed 3 times with ice-cold PBS containing 0.1% (w/v) Triton X-100. Arf1 can be detected in the pellet by immunoblotting using a commercially available antibody, such as monoclonal mouse anti-Arf (1D9) from Affinity Bioreagents (Golden, CO) (Fig. 1). This approach can be exploited for measuring Arf1•GTP levels *in vivo* as described in the following.

For quantitative analysis, we use a method that is a variation of dialysis binding assays (Jacques *et al.*, 2002). In this case, instead of using a dialysis membrane to separate two volumes, one with the "receptor" (in this case, a GGA construct such as GST-VHSGAT, abbreviated "GGA" in the equations that follow) and one excluding the "receptor," we generate two de facto compartments by immobilizing GST-GGA protein on glutathione-agarose beads. After a brief incubation with Arf1•GTPγS, two volumes are generated by a brief centrifugation, maintaining the temperature of the assay, and separated into two scintillation vials by pipetting. The volume with the beads contains Arf1•GTPγS•GGA and free Arf1•GTPγS, whereas the volume excluding the beads contains only free Arf1•GTPγS. Assuming the volume containing the beads is 20% of the total reaction volume, then,

$$(\text{Arf1} \bullet \text{GTP}\gamma\text{S})_{pellot} = 0.2(\text{Arf1} \bullet \text{GTP}\gamma\text{S})_{total} + 0.8\frac{B_{max}[\text{GGA}]}{[\text{GGA}] + K_d} \quad (1)$$

where K_d is the Arf•GGA dissociation constant (also abbreviated K_{GGA}) and B_{max} is the maximum binding. To perform this assay, Arf1 is loaded with [^{35}S]GTPγS in one reaction. Varying amounts of a GGA fragment fused to GST (e.g., GST-VHSGAT) are immobilized on glutathione-agarose beads such that 10 μl of the beads added to a 50–100 μl reaction will yield a GGA protein concentration of between 0 and 5 μM. Arf1•GTPγS is then added to the GGA protein immobilized on 10 μl of beads in a reaction that contains 20 mM Tris-HCl, pH 7.5, 100 mM NaCl, 1 mM MgCl$_2$, 1 mM DTT, 0.1% (w/v) Triton X-100, and other additions such as phospholipids in a total volume of 100 μl. The mixture is incubated for 5–10 min at 30°. The beads are separated from the bulk solution by a 5–10 sec centrifugation at 13,000g in a tabletop microcentrifuge. [^{35}S]GTPγS in 80 μl of the supernatant and in the 20 μl containing the beads are quantified by scintillation spectrometry. The fraction of [^{35}S]GTPγS in the pellet is plotted against the concentration of GGA protein in the pellet and the data are fit to Eq. (1), using a nonlinear least squares algorithm, to

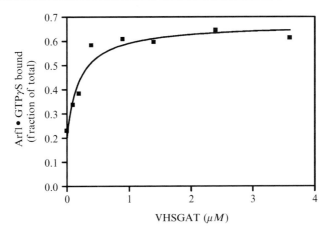

FIG. 2. Results from a direct binding assay. The fraction of total Arf1•GTPγS that was associated with GST-VHSGAT$_{GGA3}$ immobilized on glutathione beads is plotted against the concentration of GST-VHSGAT$_{GGA3}$ in the assay. The data were fit to Eq. (1).

determine the K_d. For fitting the data, we use a program called GraphPad Prism (GraphPad Software, San Diego, CA). Other scientific graphics programs also have suitable curve fitting capabilities. The data are entered with the amount of Arf1•GTPγs in the pellet as "y" and the concentration of GGA as "x." The process of analyzing the data is menu driven and also well explained in the software's documentation. Example data are shown in Fig. 2.

We have also used this approach for characterizing the binding of Arf1•GTP to the Arf GAP, AGAP1, in which case the determined K_d fit well with that calculated using other approaches. This approach has an advantage over surface plasmon resonance or isothermal titrating calorimetry in that (i) it does not require chemical concentrations of Arf1•GTP, which are sometimes difficult to achieve and (ii) the results are not skewed by differences in efficiency of GTP binding that might occur when using Arf1 mutants.

Binding Determined by Inhibition of Arf GAP Activity

GGA proteins bind Arf1 through the switch 1 and switch 2, which overlap the GAP binding site. Therefore, GGA binding to Arf1•GTP inhibits GAP activity (see Fig. 3). If the concentrations of both Arf1•GTP and Arf GAP (in this case, we use ASAP1) are less than the K_m for the GAP, and the concentration of Arf1•GTP is significantly less than

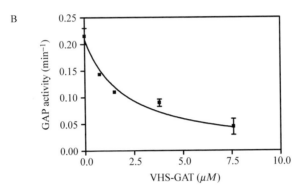

FIG. 3. Determination of GGA-Arf binding by inhibition of GAP activity. (A) Schematic of reaction. (B) Data from example experiment. GST-VHSGAT$_{GGA3}$, Arf1, and [325–724] ASAP1 were used as described in the text. Data were fit to Eq. (2) and GAP activity was calculated using Eq. (3).

the K_d for the GGA•Arf1•GTP complex, then we can derive Eq. (2) describing the relationship of GAP activity and GGA concentration. In the equation V_{obs} is the rate of GTP hydrolysis observed in the presence of a given concentration of GGA and V_{noGGA} is the rate in the absence of GGA. The identical equation is obtained under equilibrium and steady state assumptions. Based on this equation, the concentration of GGA that gives half maximal inhibition is the K_d for the GGA•Arf1•GTP complex (Hirsch *et al.*, 2003; Jacques *et al.*, 2002; Puertollano *et al.*, 2001b).

$$V_{obs} = \frac{V_{noGGA}}{1 + \frac{GGA}{K_d}} \tag{2}$$

For this assay, we use ASAP1, a robust Arf GAP with a turnover number (k_{cat}) of approximately 30/sec, and a K_m of approximately 5 μM (Che *et al.*, 2005). With these parameters, nanomolar concentrations of Arf1 and the GAP can be used with an excellent signal to noise ratio for a

3–5 min assay. Arf1 is loaded with $[\alpha^{32}P]GTP$ using one of the two methods described above. In separate tubes kept at 4°, add between 0.2 and 1 nM [325–724]ASAP1[1] in 20 mM HEPES, pH 7.4, 100 mM NaCl, 2 mM MgCl$_2$, 1 mM GTP, 1 mM DTT, 360 μM phosphatidic acid, and 90 μM phosphatidylinositol 4, 5bisphosphate in 0.1% Triton X-100 and varying amounts of the GGA being examined in a total volume of 22.5 μl. Initiate the GAP reaction by the addition of 2.5 μl of the mixture containing Arf1•GTP and simultaneously shifting the reaction mixture to 30°. Always, as described in Chapter 15, Assay and Properties of the Arf GAPs AGAP1, ASAP1, and Arf GAP1), include no GAP control to correct for GDP that binds to Arf1 during loading. The reaction is stopped after 3–5 min by dilution into 2 ml of ice-cold 20 mM Tris-HCl, pH 8.0, 100 mM NaCl, 10 mM MgCl$_2$, 1 mM DTT. Arf1 is then trapped on nitrocellulose filters. Nucleotide is eluted from the filters using 2 M formic acid and a sample of this eluate is separated on a PEI (polyethylenimine)-cellulose TLC plate developed in 1 M CHOOH: 1 M LiCl. To extend the useful range of the assay, we use a mathematical transform of the data, Equation (3) in which V is the velocity (expressed as a first order rate constant), $(Arf1 \bullet GTP)_0$ is the concentration of Arf1•GTP at time 0 (or in the blank), and $(Arf1 \bullet GTP)$ is the concentration of Arf1•GTP at time t in the presence of GAP, as described in Randazzo *et al.* (2001). Further details of this assay are given in Chapter 15 (Assay and Properties of the Arf GAPs AGAP1, ASAP1, and Arf GAP1) of this volume and Randazzo *et al.* (2001).

$$V = \frac{\ln \frac{(Arf1 \bullet GTP)_0}{(Arf1 \bullet GTP)}}{t} \tag{3}$$

A sample set of data is shown in Fig. 3B. We use a nonlinear least squares algorithm to fit the data to Eq. (2). The K_d determined is nearly identical to the value obtained by other methods (Jacques *et al.*, 2002).

One disadvantage of this approach is that it is difficult to determine the role of phospholipids because the Arf GAP interaction with Arf1 is also dependent on phospholipids. However, the GAP does not have to be under optimal conditions: simply add more GAP if conditions are less than optimal. Also, other GAPs can be used that do not have as restrictive phospholipids requirements. For instance, Arf GAP1, which does not require phosphatidylinositol 4, 5-bisphosphate, also appears to bind a site on Arf1 that overlaps the binding site for GGA (Jacques *et al.*, 2002) and, therefore, can be used in the assay.

---

[1] When diluting a highly purified GAP to nanomolar concentrations, the protein is stabilized by including a carrier protein such as bovine serum albumin at a concentration of 100 μg/ml.

Binding Determined by Slowing GTPγS Dissociation

This method is based on an assay developed by Herrmann *et al.* (1995) for the determination of binding affinities of Ras effectors for Ras•GTP. Arf1•GTP and Arf1•GTPγS dissociate at a rate determined by phospholipids, Mg^{2+} concentration and, as illustrated in Fig. 4, associated proteins. Effectors like GGA slow the dissociation rate. We can measure the dissociation rate by first loading Arf1 with a $[^{35}S]$GTPγS of high specific activity and then incubating the Arf1•$[^{35}S]$GTPγS in a reaction mixture containing a high concentration of GTPγS or GDP. As the $[^{35}S]$GTPγS dissociates, the unlabeled nucleotide competes for rebinding. The dissociation rate can be measured as the rate of loss of protein-associated ^{35}S. The dissociation rate in the absence of effector is k_{-1} and in the presence of effector is k_{-2}. The total dissociation rate is

$$-d(\text{Arf1} \bullet \text{GTP}\gamma\text{S})/_{dt} = k_{-1}[\text{Arf1} \bullet \text{GTP}\gamma\text{S}] + k_{-2}[\text{GGA} \bullet \text{Arf1} \bullet \text{GTP}\gamma\text{S}] \tag{4}$$

If $k_{-2} \ll k_{-1}$, this reduces to

$$-d(\text{Arf1} \bullet GTP\gamma S)/_{dt} = k_{-1}[\text{Arf1} \bullet GTP\gamma S] \tag{5}$$

We assume that Arf1 • GTPγS $< K_d$ for the GGA • Arf1 • GTP complex. Under this condition, Arf1•GTPγS that is not bound to GGA is

$$[\text{Arf1} \bullet \text{GTP}\gamma\text{S}] = \frac{[\text{Arf1} \bullet \text{GTP}\gamma\text{S}]_{total} K_d}{[\text{GGA}] + K_d} \tag{6}$$

substituting Eq. (6) into (5) gives

$$-d(\text{Arf1} \bullet \text{GTP}\gamma\text{S})/_{dt} = \frac{k_{-1}}{1 + \dfrac{[\text{GGA}]}{K_d}}[\text{Arf1} \bullet \text{GTP}\gamma\text{S}]_{total} \tag{7}$$

From this expression, the observed dissociation rate at a given concentration of GGA fragment, the k_{obs}, is related to the k_{-1} as described by Eq. (8).

$$k_{obs} = \frac{k_{-1}}{1 + \dfrac{[\text{GGA}]}{K_d}} \tag{8}$$

The concentration of GGA that slows dissociation by 50%, the K_d, can be determined from a plot kobs against the concentration of GGA.

To be consistent with our assumptions for the derivation, we use a low concentration of Arf1•$[^{35}S]$GTPγS, prepared as described above. This assay benefits from using Arf1 that is not myristoylated because the

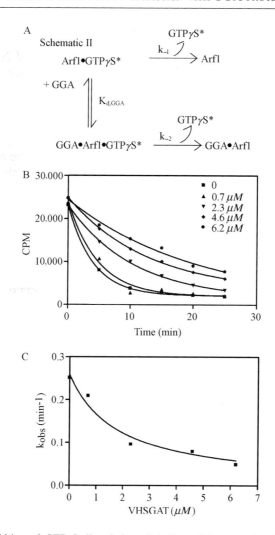

FIG. 4. Inhibition of GTPγS dissociation of Arf1 used in assay for GGA binding to Arf1. (A) Schematic of reactions. (B) Effect of GGA on GTPγS dissociation from Arf1. Nonmyristoylated Arf1•[^{35}S]GTPγS was incubated with the indicated amount of GST-VHSGAT$_{GGA3}$. After the indicated period of time, samples were removed and protein-bound [^{35}S]GTPγS was determined by filter binding followed by scintillation spectrometry. (C) Plot for determining K$_{d,GGA3}$. The dissociation data, presented in Fig. 4B, were fit to single exponential decay equation (if examine dissociation over linear range, use linear least squares fitting) to determine the dissociation rate. The observed dissociation rates were plotted against the concentration of GGA3 and fit to Eq. (8) to determine the K$_d$.

dissociation rates in the presence of lipid are greater and easier to measure than those for myristoylated Arf1. Mg^{2+} is buffered to approximately 1 μM to maximize uncatalyzed dissociation of the Arf1•GTPγS complex. The reaction mixture contains 20 mM Tris-HCl, pH 7.5, 100 mM NaCl, 0.5 mM $MgCl_2$, 1 mM EDTA, 0.1% (w/v) Triton X-100, and the GGA protein fragment, for example, VHSGAT, as well as any other additions such as phospholipids in a total volume of 100 μl. Samples of the reaction are removed at 6–9 time points ranging from 0 to 60 min and quenched by dilution into ice-cold 10 mM Tris-HCl, pH 8.0, 100 mM NaCl, 10 mM $MgCl_2$, and 1 mM DTT. Protein-bound nucleotide is trapped on nitrocellulose filter disks. Scintillation spectrometry is used to quantify ^{35}S. The progress curves are fit to a single exponential decay equation (cpm = $cpm_0 \cdot e^{-k_{obs} \cdot t}$, e.g., in Fig. 4B, all curves can be fit to this equation) or a line if dissociation is less than 15%) to determine k_{obs}. This estimated value of k_{obs} is then plotted against GGA concentration (Fig. 4C) and fit to Eq. (8) using a scientific graphics program to determine the affinity.

One drawback of this method is that it is dependent on the dissociation rates for the Arf1•GTPγS complex. Some mutants of Arf1, such as Δ17Arf1, do not have a rapid GTP dissociation rate so signal to noise may be problematic. On the other hand, this method has worked for many switch 1 and switch 2 mutants of Arf1 and is also useful for comparing different proteins derived from GGA. We expect that it will be useful for other Arf effectors but have not yet tested them.

Use of GGA for Determining Cellular Levels of Arf1•GTP

GGA binds Arf1•GTP in preference to Arf1•GDP. This difference in binding has been exploited in an assay to measure intracellular Arf•GTP levels (Santy and Casanova, 2001). The rationale is identical to that for assays developed for Ras (deRooij and Bos, 1997; Franke et al., 1997) and Rho (Sander et al., 1998) family proteins. Because the presence of the VHS domain of GGA3 improves the affinity of the GAT domain for Arf1 • GTP, we use a fusion protein of GST-VHSGAT$_{GGA3}$ expressed and purified as described previously. This interaction is low affinity; therefore, temperature control is critical. In a typical experiment, cells grown on 35 mm well plates are transfected with expression vectors for Arf1-HA or mutants and any other proteins of interest, such as GAPs or GEFs. After 18–24 h, cells are lysed in 0.5 ml of 50 mM Tris-HCl, pH 7.5, 100 mM NaCl, 2 mM $MgCl_2$, 1% (w/v) Triton X-100, and protease inhibitors at 4°. The lysates are cleared by addition of 20 μl of Sepharose CL-4B beads, mixing and separating the beads from the lysate by centrifugation at 16,000g for 20 sec at 4°. Fifty μg of GST-VHSGAT$_{GGA3}$ immobilized on 20 μl of glutathione-Sepharose

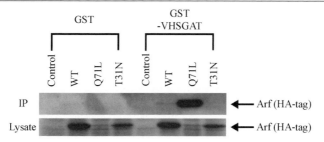

FIG. 5. Use of GST-VHSGAT$_{GGA3}$ as a reagent for determining Arf1•GTP levels *in vivo*. HEK 293 cells were transfected with plasmids directing expression of epitope tagged Arf1, [T31N]Arf1 or [Q71L]Arf1. Eighteen hours later, the cells were lysed. The lysates were incubated with either GST or GST-VHSGAT$_{GGA3}$ and Arf1 associating with GST-VHSGAT$_{GGA3}$ was determined as described in the text. As anticipated, the constitutively active mutant of Arf1, [Q71L], gave a greater signal than wild type protein whereas no signal was detected with the dominant negative protein.

CL-4B beads are added to the cleared lysates and the mixture is incubated at 4° for 1 h. The beads are collected by a brief centrifugation (30 sec at 13,000g in a microcentrifuge).

The beads are washed three times with 50 mM Tris-HCl, pH 7.5, 100 mM NaCl, 2 mM MgCl$_2$, and 1% (w/v) Triton X-100 at 4°. To analyze the bound proteins, 60 μl sample buffer is added to the beads and the mixture is heated at 95° for 5 min. The beads are removed by centrifugation and samples of the supernatant are fractionated by SDSPAGE. Proteins are eluted from the beads by boiling in SDS-PAGE sample buffer, electrophoresed on a 15% SDS-PAGE gel, and transferred to an Immobilon P membrane (Milipore, Bedford, MA). The blot is incubated sequentially with monoclonal mouse anti-HA antibody (Roche, Indianapolis, IN) (1:3000) and with goat anti-mouse IgG-HRP conjugate (1:10,000, Bio-Rad, Hercules, CA). The IgG HRP conjugate is detected using ECL plus Western blotting reagent (Amersham Biosciences) (Fig. 5).

This method is reasonably robust when assaying Arf1 and Arf5. Our laboratory and others following the outlined protocol have obtained interpretable data with both Arf isoforms. Assay of Arf6 appears to be more variable. This may be related to cell differences as well as the solubility properties and stability of Arf6 relative to Arf1.

References

Bilodeau, P. S., Winistorfer, S. C., Allaman, M. M., Surendhran, K., Kearney, W. R., Robertson, A. D., and Piper, R. C. (2004). The GAT domains of clathrin-associated GGA proteins have two ubiquitin binding motifs. *J. Biol. Chem.* **279,** 54808–54816.

Boman, A. L., Zhang, C. J., Zhu, X. J., and Kahn, R. A. (2000). A family of ADP-ribosylation factor effectors that can alter membrane transport through the *trans*-Golgi. *Mol. Biol. Cell* **11,** 1241–1255.

Bonifacino, J. S. (2004). The GGA proteins: Adaptors on the move. *Nat. Rev. Mol. Cell Biol.* **5,** 23–32.

Bonifacino, J. S., and Glick, B. S. (2004). The mechanisms of vesicle budding and fusion. *Cell* **116,** 153–166.

Bonifacino, J. S., and Lippincott-Schwartz, J. (2003). Opinion – coat proteins: Shaping membrane transport. *Nat. Rev. Mol. Cell Biol.* **4,** 409–414.

Che, M. M., Boja, E. S., Yoon, H.-Y., Gruschus, J., Jaffe, H., Stauffer, S., Schuck, P., Fales, H. M., and Randazzo, P. A. (2005). Regulation of ASAP1 by phospholipids is dependent on the interface between the PH and Arf GAP domains. *Cell. Signal.* **17,** 1276–1288.

Collins, B. M., Praefcke, G. J. K., Robinson, M. S., and Owen, D. J. (2003a). Structural basis for binding of accessory proteins by the appendage domain of GGAs. *Nat. Struct. Biol.* **10,** 607–613.

Collins, B. M., Watson, P. J., and Owen, D. J. (2003b). The structure of the GGA1-GAT domain reveals the molecular basis for ARF binding and membrane association of GGAs. *Dev. Cell* **4,** 321–332.

Dell'Angelica, E. C., Puertollano, R., Mullins, C., Aguilar, R. C., Vargas, J. D., Hartnell, L. M., and Bonifacino, J. S. (2000). GGAs: A family of ADP ribosylation factor-binding proteins related to adaptors and associated with the Golgi complex. *J. Cell Biol.* **149,** 81–93.

deRooij, J., and Bos, J. L. (1997). Minimal Ras-binding domain of Raf1 can be used as an activation-specific probe for Ras. *Oncogene* **14,** 623–625.

Franke, B., Akkerman, J. W. N., and Bos, J. L. (1997). Rapid Ca^{2+}-mediated activation of Rap1 in human platelets. *EMBO J.* **16,** 252–259.

Ghosh, P., Griffith, J., Geuze, H. J., and Kornfeld, S. (2003). Mammalian GGAs act together to sort mannose 6-phosphate receptors. *J. Cell Biol.* **163,** 755–766.

Ghosh, P., and Kornfeld, S. (2004). The GGA proteins: Key players in protein sorting at the *trans*-Golgi network. *Eur. J. Cell Biol.* **83,** 257–262.

Herrmann, C., Martin, G. A., and Wittinghofer, A. (1995). Quantitative-analysis of the complex between P21(Ras) and the Ras-binding domain of the human Raf-1 protein-kinase. *J. Biol. Chem.* **270,** 2901–2905.

Hirsch, D. S., Stanley, K. T., Chen, L. X., Jacques, K. M., Puertollano, R., and Randazzo, P. A. (2003). Arf regulates interaction of GGA with mannose-6-phosphate receptor. *Traffic* **4,** 26–35.

Hirst, J., Lui, W. W. Y., Bright, N. A., Totty, N., Seaman, M. N. J., and Robinson, M. S. (2000). A family of proteins with gamma-adaptin and VHS domains that facilitate trafficking between the *trans*-Golgi network and the vacuole/lysosome. *J. Cell Biol.* **149,** 67–79.

Jacques, K. M., Nie, Z. Z., Stauffer, S., Hirsch, D. S., Chen, L. X., Stanley, K. T., and Randazzo, P. A. (2002). Arf1 dissociates from the clathrin adaptor GGA prior to being inactivated by arf GTPase-activating proteins. *J. Biol. Chem.* **277,** 47235–47241.

Kirchhausen, T. (2002). Clathrin adaptors really adapt. *Cell* **109,** 413–416.

Kuai, J., Boman, A. L., Arnold, R. S., Zhu, X. J., and Kahn, R. A. (2000). Effects of activated ADP-ribosylation factors on Golgi morphology require neither activation of phospholipase D1 nor recruitment of coatomer. *J. Biol. Chem.* **275,** 4022–4032.

Kuai, J., and Kahn, R. A. (2000). Residues forming a hydrophobic pocket in ARF3 are determinants of GDP dissociation and effector interactions. *FEBS Lett.* **487,** 252–256.

Lui, W. W. Y., Collins, B. M., Hirst, J., Motley, A., Millar, C., Schu, P., Owen, D. J., and Robinson, M. S. (2003). Binding partners for the COOH-terminal appendage domains of the GGAs and gamma-adaptin. *Mol. Biol. Cell* **14,** 2385–2398.

Mattera, R., Arighi, C. N., Lodge, R., Zerial, M., and Bonifacino, J. S. (2003). Divalent interaction of the GGAs with the Rabaptin-5-Rabex-5 complex. *EMBO J.* **22**, 78–88.

Mattera, R., Puertollano, R., Smith, W. J., and Bonifacino, J. S. (2004). The trihelical bundle subdomain of the GGA proteins interacts with multiple partners through overlapping but distinct sites. *J. Biol. Chem.* **279**, 31409–31418.

Miller, G. J., Mattera, R., Bonifacino, J. S., and Hurley, J. H. (2003). Recognition of accessory protein motifs by the gamma-adaptin ear domain of GGA3. *Nat. Struct. Biol.* **10**, 599–606.

Misra, S., Puertollano, R., Bonifacino, J. S., and Hurley, J. H. (2001). Structural basis of acidic/ dileucine-motif-based sorting by GGA proteins. *Mol. Biol. Cell* **12**, 85A.

Moss, J., and Vaughan, M. (1998). Molecules in the ARF orbit. *J. Biol. Chem.* **273**, 21431–21434.

Nakayama, K., and Wakatsuki, S. (2003). The structures and function of GGAs, the traffic controllers at the TGN sorting crossroads. *Cell Struct. Funct.* **28**, 431–442.

Nie, Z. Z., Hirsch, D. S., and Randazzo, P. A. (2003). Arf and its many interactors. *Curr. Opin. Cell Biol.* **15**, 396–404.

Nielsen, M. S., Madsen, P., Christensen, E. I., Nykjaer, A., Gliemann, J., Kasper, D., Pohlmann, R., and Petersen, C. M. (2001). The sortilin cytoplasmic tail conveys Golgi-endosome transport and binds the VHS domain of the GGA2 sorting protein. *EMBO J.* **20**, 2180–2190.

Nogi, T., Shiba, Y., Kawasaki, M., Shiba, T., Matsugaki, N., Igarashi, N., Suzuki, M., Kato, R., Takatsu, H., Nakayama, K., and Wakatsuki, S. (2002). Structural basis for the accessory protein recruitment by the gamma-adaptin ear domain. *Nat. Struct. Biol.* **9**, 527–531.

Owen, D. J., Collins, B. M., and Evans, P. R. (2004). Adaptors for clathrin coats: Structure and function. *Ann. Rev. Cell Dev. Biol.* **20**, 153–191.

Puertollano, R., Aguilar, R. C., Gorshkova, I., Crouch, R. J., and Bonifacino, J. S. (2001a). Sorting of mannose 6-phosphate receptors mediated by the GGAs. *Science* **292**, 1712–1716.

Puertollano, R., Randazzo, P. A., Presley, J. F., Hartnell, L. M., and Bonifacino, J. S. (2001b). The GGAs promote ARF-dependent recruitment of clathrin to the TGN. *Cell* **105**, 93–102.

Randazzo, P. A., Nie, Z., Miura, K., and Hsu, V. (2000). Molecular aspects of the cellular activities of ADP-ribosylation factors. *Sci. STKE* **59**, RE1.

Randazzo, P. A., Miura, K., and Jackson, T. R. (2001). Assay and purification of phosphoinositide-dependent ADP-ribosylation factor (ARF) GTPase activating proteins. *Methods Enzymol.* **329**, 343–354.

Robinson, M. S., and Bonifacino, J. S. (2001). Adaptor-related proteins. *Curr. Opin. Cell Biol.* **13**, 444–453.

Sander, E. E., van Delft, S., ten Klooster, J. P., Reid, T., van der Kammen, R. A., Michiels, F., and Collard, J. G. (1998). Matrix-dependent Tiam1/Rac signaling in epithelial cells promotes either cell-cell adhesion or cell migration and is regulated by phosphatidylino-sitol 3-kinase. *J. Cell Biol.* **143**, 1385–1398.

Santy, L. C., and Casanova, J. E. (2001). Activation of ARF6 by ARNO stimulates epithelial cell migration through downstream activation of both Rac1 and phospholipase D. *J. Cell Biol.* **154**, 599–610.

Scott, P. M., Bilodeau, P. S., Zhdankina, O., Winistorfer, S. C., Hauglund, M. J., Allaman, M. M., Kearney, W. R., Robertson, A. D., Boman, A. L., and Piper, R. C. (2004). GGA proteins bind ubiquitin to facilitate sorting at the *trans*-Golgi network. *Nat. Cell Biol.* **6**, 252–259.

Shiba, Y., Katoh, Y., Shiba, T., Yoshino, K., Takatsu, H., Kobayashi, H., Shin, H. W., Wakatsuki, S., and Nakayama, K. (2004). GAT (GGA and Tom1) domain responsible for ubiquitin binding and ubiquitination. *J. Biol. Chem.* **279**, 7105–7111.

Spang, A. (2002). Arf1 regulatory factors and COPI vesicle formation. *Curr. Opin. Cell Biol.* **14,** 423–427.

Springer, S., Spang, A., and Schekman, R. (1999). A primer on vesicle budding. *Cell* **97,** 145–148.

Takatsu, H., Katoh, Y., Shiba, Y., and Nakayama, K. (2001). Golgi-localizing, gammaadaptin ear homology domain, ADP-ribosylation factor-binding (GGA) proteins interact with acidic dileucine sequences within the cytoplasmic domains of sorting receptors through their Vps27p/Hrs/STAM (VHS) domains. *J. Biol. Chem.* **276,** 28541–28545.

Yoon, H.-Y., Jacques, K., Nealon, B., Stauffer, S., Premont, R. T., and Randazzo, P. A. (2004). Differences between AGAP1, ASAP1 and Arf GAP1 in substrate recognition: Interaction with the N-terminus of Arf1. *Cell. Signal.* **16,** 1033–1044.

Zhai, P., He, X. Y., Liu, J. A., Wakeham, N., Zhu, G. Y., Li, G. P., Tang, J., and Zhang, X. J. C. (2003). The interaction of the human GGA1 GAT domain with Rabaptin-5 is mediated by residues on its three-helix bundle. *Biochemistry* **42,** 13901–13908.

Zhu, G. Y., He, X. Y., Zhai, P., Terzyan, S., Tang, J., and Zhang, X. J. C. (2003a). Crystal structure of GGA2 VHS domain and its implication in plasticity in the ligand binding pocket. *FEBS Lett.* **537,** 171–176.

Zhu, G. Y., Zhai, P., He, X. Y., Terzyan, S., Zhang, R. G., Joachimiak, A., Tang, J., and Zhang, X. J. C. (2003b). Crystal structure of the human GGA1 GAT domain. *Biochemistry* **42,** 6392–6399.

Zhu, G. Y., Zhai, P., He, X. Y., Wakeham, N., Rodgers, K., Li, G. P., Tang, J., and Zhang, X. J. C. (2004). Crystal structure of human GGA1 GAT domain complexed with the GAT-binding domain of Rabaptin5. *EMBO J.* **23,** 3909–3917.

[29] The Role of EFA6, Exchange Factor for Arf6, for Tight Junction Assembly, Functions, and Interaction with the Actin Cytoskeleton

Abstract

In polarized epithelial cells, the tight junction has been ascribed several functions including the regulation of the paracellular permeability, an impediment to the diffusion of molecules between the apical and basolateral domains, a site of delivery of transport vesicles for basolateral proteins, and a scaffold for structural and signaling molecules. The tight junction is anchored physically into the apical actin cytoskeleton circumscribing the cell, which is known as the perijunctional actomyosin ring. This connection was first suggested by experiments using the actin depolymerizing drug cytochalasin, which was also found to disrupt the transepithelial permeability. Since then a large number of studies have reported the effects of drugs, molecular tools, or physiological and pathological conditions that alter coordinately actin organization and the tight junction. In support of this model, proteins of the tight junction, such as the members of

METHODS IN ENZYMOLOGY, VOL. 404
0076-6879/05 $35.00
DOI: 10.1016/S0076-6879(05)04029-2

the ZO family and occludin, have been shown to bind to actin. However, very little is known regarding the molecular mechanisms by which the actin cytoskeleton modulates tight junction functions. We have studied the role of the Exchange Factor for Arf6, EFA6, in tight junction assembly. By combining a large panel of methods, including morphological, physiological, and biochemical, described in detail hereafter we demonstrated that EFA6 plays a role in the physical association of the tight junction to the perijunctional actomyosin ring.

Introduction

Polarized epithelial cells are characterized by two distinct plasma membrane domains: the free apical domain exposed to the lumen and the basolateral domain facing the neighboring cells and the basal lamina. These two membrane domains display distinct lipid and protein compositions and are separated by a junctional complex, the tight junction (TJ), that forms a diffusion barrier that prevents mixing of the surfaces and provides a tight seal between cells. Tight junctions were first described morphologically and functionally as the Zonula Occludens, referred to as an apical fusion point that prevents the free diffusion of protein markers or ionic lanthanum salts between the plasma membranes of two contacting cells (Farquhar and Palade, 1963; Goodenough and Revel, 1970). The prevention of paracellular diffusion is known as the barrier function of the TJ, whereas preventing the free diffusion of lipids and proteins between the apical and lateral domains of the plasma membrane is known as the fence function of the TJ. (Schneeberger and Lynch, 1992). A laborious biochemical approach allowed for the enrichment of a massive preparation of TJ and led to the identification of the first integral membrane protein called occludin (Furuse et al., 1993). After years of debate around the importance of the role of occludin in the TJ assembly and functions, knock-out mice for occludin definitively demonstrated that occludin is not essential, and thus suggested that other proteins must be important in the composition of the TJ (Saitou et al., 1998). In a second round of TJ purification by the same group, a whole new family of proteins called claudins was identified (Furuse et al., 1998). The members of this family are believed to represent the basic unit of the TJ and were shown to directly govern the barrier function (Tsukita et al., 2001). The role of occludin is still under debate and may be required at special stages of TJ assembly and/or in the regulation of their dynamics. Numerous cytosolic molecules associate with the TJ, among which are the proteins ZO-1 and the closely related ZO-2 and ZO-3 (Schneeberger and Lynch, 2004; Stevenson et al., 1986).

These molecules contain 3 PDZ-domains (PSD-95, Dlg, ZO-1 homology domain), one SH3 domain, and one guanylyl kinase-like domain and are believed to act as molecular scaffolds. ZO-1 binds to many proteins including ZO-2 and ZO-3, occludin, claudins, ZONAB (ZO-1 associated nucleic acid-binding protein), and α-catenin. In addition, the C-terminal domain of ZO-1 binds directly to actin filaments and serves as a bridge between the TJ and the underlying actin cytoskeleton (Mitic and Anderson, 1998). A large selection of antibodies directed against the different components of the TJ is available for the analysis at the morphological and biochemical levels of the TJ. An important characteristic of the TJ is its association with the dense apical actomyosin ring. Several functions have been ascribed to this distinctive actin cytoskeleton, which are not only important in fully polarized cells, but are also active during the development of cell polarity. These include acting as an anchoring structure for the TJ with contractile activity, a platform for structural and signaling molecules where they organize as functional complexes, and a sorting apparatus by selective retention of proteins at the cell surface and exclusion from nascent endocytic vesicles (Drubin and Nelson, 1996; Turner, 2000). While the small G proteins of the Rho family have been extensively studied for organizing the actin cytoskeleton in nonpolarized cells, little is known about their role in mammalian polarized epithelium and much less in cell polarity development. Original studies on RhoA, Rac1, and Cdc42 in polarized MDCKII cells indicated their involvement in regulating TJ functions (Jou *et al.*, 1998; Rojas *et al.*, 2001). However, the expression of the mutants had multiple effects on the actin cytoskeleton, cell polarity, and TJ assembly, probably as a result of altering numerous signaling pathways. Our studies focus on the small G protein Arf6 that has been implicated in both intracellular transport and actin rearrangement in fibroblastic cell lines (Chavrier and Goud, 1999). We have been especially interested in its specific exchange factor, EFA6. Similar to the other exchange factors, it is a multidomain protein that comprises the catalytic exchange domain (Sec7), a Pleckstrin Homology domain responsible for its membrane localization, and a C-terminal region involved in actin cytoskeleton rearrangement (Franco *et al.*, 1999). The following sections will describe the methods we used to analyze the role of EFA6 on the above-mentioned properties of the TJ.

Cells

Our laboratory has worked almost exclusively on MDCKII cells and the methodological conditions described below have been calibrated for this particular cell line. However, the methods can be applied to all cell lines by adjusting the experimental conditions. EFA6 is expressed under the control of the tetracycline responding element using the Tet-off system (Luton *et al.*, 2004). Routinely, cells are passaged in 10 cm dishes in MEM supplemented with 5% decomplemented FCS, penicillin-streptomycin, and in the absence or presence of 20 ng/ml doxycycline. The presence of doxycycline suppresses the expression of EFA6 and allows for it to be expressed when removed. To obtain a fully polarized monolayer, the cells are plated at confluency and grown for at least 3 days on 12 mm or 24 mm, 0.4-μm pore size Transwell™ polycarbonate filters (Corning Costar Corp., Cambridge, MA).

Confocal Immunofluorescence of Polarized Cells

Electron microscopy of samples prepared by freeze fracture is the only technique that enables direct observations on the intimate structure of the TJ such as the number of strands, their reticulation, and the position and the width occupied by the TJ along the lateral membrane. In contrast, confocal immunofluorescence analyses provide information at a more macroscopic level. Using specific probes, it is possible to obtain useful information regarding the general morphology of the monolayer, the actin cytoskeleton or microtubular network, and the localization of markers for the apical or basal domains of the plasma membrane or the junctional complexes. Immunofluorescence procedures generally begin by a fixation step. The method of fixation depends largely on the epitope analyzed. I will describe the most commonly used, which is the fixation by paraformalde-hyde (PFA), that preserves most of the proteins and intracellular struc-tures. Other methods include a modified PFA fixation protocol or use glutaraldehyde (Apodaca *et al.*, 1994), methanol or methanol/acetone as fixative agents (Balda *et al.*, 1996; Wong and Gumbiner, 1997). Cells grown on 12 mm filters are rinsed three times in ice-cold PBS-CM (calcium 1 mM, magnesium 0.5 mM). The cells are then fixed on ice with 3.7% PFA (1 ml basal and 0.5 ml apical) for 30 min. The PFA is rapidly removed by three quick washes in PBS-CM and quenched 10 min at room temperature by adding 1 ml apical and basal of the quenching solution (75 mM NH4Cl and 20 mM glycine in PBS). The cells are then washed twice in PBS-CM before permeabilization at 37° for 15–60 min in 0.025% saponin, 0.7% fish skin gelatin (blocking agent) prepared in PBS-CM (PFS). The permeabilization solution is aspirated and 120 μl of antibody solution diluted in PFS added apically. The filter is then gently touched at an angle onto a piece of paper to adsorb the liquid left underneath and laid down on a 30 μl drop of antibody solution deposited on a piece of parafilm taking care to avoid bubbles. The incubation period is performed in a humid chamber at 37° for a minimum of 1 h. Then, the filters are washed four times 5 min in PFS and further processed for the labeling with the secondary antibodies following the same protocol. After four washes of 5 min in PFS and two more in PBS-CM we postfix the samples in 3.7% PFA for 15 min at room temperature. For mounting, a square of nail polish of the size of the coverslip is drawn on a clean slide and a 30 μl drop of mounting media containing antifade is placed in the middle. The filter is carefully cut out from the support and placed cells facing up into the mounting media. A coverslip is placed on top of the filter in-line, with nail polish applying a gentle but steady pressure and sealed using additional nail polish. It is important to handle the sample very carefully to avoid wrinkling of the filter and trapping bubbles. The slides can be stored at $-20°$ for several months. Because TJs are detergent insoluble we often perform a Triton X-100 extraction either before or after fixation. Cells are extracted with 0.5% Triton X-100, 300 mM sucrose, 10 mM PIPES pH 6.8, 3 mM MgCl2 for 13 min, rapidly but gently rinsed in PBS-CM, and processed as described above.

Paracellular Permeability and Barrier Function

Paracellular transport of ions and larger molecules is restricted by the intercellular TJ. However, it is now well appreciated that the TJ does not constitute a passive filtering seal isolated from the rest of the cell. It is instead a highly regulated and dynamic structure physically associated to an acto-myosin cytoskeleton and functionally linked to a wide variety of signaling pathways that can contribute to modulation of TJ permeability (Cereijido et al., 1998; Schneeberger and Lynch, 2004; Tsukita et al., 2001). Because the mechanisms of regulation are still largely unknown one has to be very prudent when interpreting experiments designed to measure the paracellular permeability of ionic and nonionic molecules (Schneeberger and Lynch, 2004; Tsukita and Furuse, 2000; Tsukita et al., 2001). Therefore, the two measures must be used in conjunction and subjected to independent morphological and biochemical studies.

Transepithelial Electrical Resistance (TER)

The flow of ionic molecules is estimated by measuring the resistance of the cell monolayer subjected to an alternating current using a pair of electrodes designed to convey the current and measure the voltage deflection. Although the procedure is simple, particular precautions must be taken to minimize variability. We calibrate the instrument (EVOM, World Precision Instruments, Sarasota, FL) by bathing the electrodes for 2 h at 37° in the "TER medium," which is our cell culture medium (MEM) without serum and buffered with an Hepes solution to a 10 mM final concentration pH 7.4. Before taking the measure, the culture medium from the well and the insert is completely aspirated and replaced precisely by 0.75 ml and 1.5 ml of prewarmed "TER medium" in the top and bottom reservoirs, respectively. The value obtained is the result of the additive resistance of the filter and the cell monolayer. Thus, the resistance of a control filter without cells is measured and subtracted from the value obtained with the sample. As the resistance is inversely proportional to the surface area of the sample, the value is further divided by the surface area and expressed as omh. cm^2. The variation from one filter to another is such that it is necessary to measure triplicates for each sample. We also do not recommend measuring TER on the same filters at close time intervals.

Paracellular Diffusion of Nonionic Hydrophilic Tracers

Several tracers can be used to measure the permeability of the cell monolayer to nonionic molecules. We have used the small radioactive molecule ^{14}C-mannitol (2.5 μCi/filter; 50 mCi/mmol), but ^{3}H-inulin can been used as well, and larger molecules such as different polymers of fluorescently coupled dextran molecules ranging from 3 K to 40 K (Rhodamine B isothiocyanate-conjugated dextran, RITC). The concentration of tracer used depends on its size and ranges from 3 to 0.2 mg/ml. The principle of the method consists in adding a given amount of tracer on one side of the cells, usually apically, and measure over time its free diffusion to the other side. In order to alleviate a differential in pressure, we add accurately 200 μl of the tracer solution diluted in MEM on top of a 12 mm filter unit and 800 μl of MEM in the bottom compartment so that their levels are coincident. At the end of the assay, the total amount of RITC-dextran and ^{14}C-mannitol in the basal medium are quantitated using a spectrofluorimeter and a γ-counter, respectively. The results are reported as the total amount of tracer found in the basal compartment in mole/hr.cm^2. Our standard procedure includes the measurement of triplicates for each sample. Within an experiment where cells are submitted to various conditions, the positive control for minimum permeability is the untreated fully polarized wild-type cell monolayer, whereas, an empty filter unit is used as the negative control for maximum cell permeability.

Calcium Switch

The search for the role of calcium in biological functions eventually led to the appreciation of its absolute requirement for the establishment of cell-cell contacts and the subsequent TJ assembly and development of cell polarity epithelial cells (Schneeberger and Lynch, 1992). It was later shown to participate in the homotypic interaction of the E-cadherin molecules forming the adherens junction (Gumbiner and Simons, 1986). Cell biologists took advantage of this calcium dependency to study the morphological and signaling events controlling cell polarity by depleting and/or repleting calcium from the cell culture medium (Cereijido *et al.*, 1998; Denker and Nigam, 1998). Thus, either TJ assembly or disassembly can be followed morphologically with immunofluorescence, functionally by monitoring the TER and paracellular permeability, or biochemically (see following). To synchronously induce cell-cell contacts and subsequent TJ formation, the cells must be present as a confluent monolayer before the addition of calcium. One method is to first form a fully polarized monolayer and then remove the calcium containing media for at least 4 h. In another method, "contact naive" cells grown in large dishes are trypsinized and plated as an instant monolayer in the absence of calcium. The preformed monolayer ensures a homogenous coverage of the entire surface of the filter and the even distribution of the cells as a monolayer, whereas the use of "contact naive" cells may lead to either empty areas in the monolayer or multilayering, which is especially detrimental to the accurate measurement of paracellular permeability. However, when analyzing the role of a protein, it is important to ensure that the rapid removal of calcium does not affect its given expression levels. Another major concern is that the repertoire of proteins present at the cell surface is altered due to the trypsinization of the "contact naive" cells. We have used both methods and found no difference for our proteins and use routinely the more convenient first procedure of preforming confluent monolayers. Nevertheless, I will describe both methods. In the first method, the fully polarized confluent monolayer is washed three times quickly in PBS and three times 10 min under agitation in PBS-EGTA 2 mM. At t = 0 cells are incubated in MEM containing 5 μM calcium, low calcium medium (LCM), and when necessary with or without doxycycline, which controls the exogenous expression of EFA6. It is recommended not to leave the cells without any calcium, which would be toxic to many intracellular functions, and 5 μM calcium is insufficient to promote cadherin homotypic interactions. Normally, MDCK cells require 4 h low calcium to result in the loss of cell contacts, after which the "calcium switch" is performed by replacing the medium with normal calcium containing MEM with or without doxycycline. Because MDCK cells overexpressing EFA6 were found to have stabilized TJ, it was necessary to incubate the cells 6 h in LCM to disrupt all cell-cell contacts as monitored by TER, paracellular permeability, and morphology. Therefore, before doing a "calcium switch" assay we advise performing a time course following calcium depletion to verify that the conditions used are appropriate to reach full disassembly of the TJ. To obtain "contact naive" cells, the cells are trypsinized on day one, physically separated by iterative pipetting, counted, and plated at 1.5×10^6 cells in a 10 cm dish. The procedure is repeated on day 2. On day 3, trypsinized cells are plated as an "instant confluent monolayer" at 0.62×10^6 cells for a 12 mm filter or 2.5×10^6 cells for a 24 mm filter in LCM. We let the cells sit for a minimum of 3 h before starting the "calcium switch."

EFA6 Effects on TJ Permeability and Actin Cytoskeleton

Through analyses by confocal immunofluorescence and permeability assays, we did not detect any difference in TJ function between fully polarized cells expressing EFA6 and control cells. However, using the calcium switch assay, we observed that EFA6 expression accelerated the assembly of the TJ and the polarized rearrangement of the actin cytoskeleton. We monitored this assembly morphologically using confocal microscopy looking at occludin, claudin 2, and ZO-1. They were all found to concentrate into nascent cell-cell contacts and to migrate apically coordinately. We also found that the gain of permeability functions (TER and dextran-RITC paracellular permeability) correlated in a temporal manner with the morphological assembly of the TJ described above. On the other hand, upon calcium removal, EFA6 delayed the disassembly of the TJ and loss of permeability functions. EFA6 was first described for its potent impact on the actin cytoskeleton not only by activating Arf6 but also independently through its C-terminal domain. Hence, we suspected that EFA6 could stabilize the apical actin ring onto which the TJ is anchored. To test this hypothesis, we examined the sensitivity of the actin cytoskeleton to latrunculin B by confocal immunofluorescence in cells expressing or not EFA6. The mode of action of this drug makes it a very useful tool to analyze the stability of actin cytoskeleton structures. Indeed, it does not depolymerize actin, but instead by forming a 1:1 complex with free G-actin, it sequesters the actin monomers released during turnover of actin structures, thus depleting the cells of de novo actin filament polymerization (Gronewold et al., 1999; Spector et al., 1983). It appears that the basal stress fibers and the bundles of lateral cortical actin are very sensitive to latrunculin treatment and become undetectable to phalloidin after only 5 min exposure to 5 μM latrunculin. In contrast, the apical microvilli and even more so the apical actin ring of the TJ are resistant for up to 30 min and 60 min, respectively (Luton et al. [2004] and our unpublished observations). Altogether, these results prompted us to look at the effects of EFA6 expression on the stability of the TJ at the cell surface.

Biochemical Analyses

Similar to any other cell surface proteins, occludin and claudin molecules follow a life cycle. After protein synthesis, this includes cell surface delivery, presence at the plasma membrane, endocytosis–recycling, and finally degradation. In the following sections I describe two experiments that address these different steps. These assays are based on the possibility to biotinylate the lysine residue(s) of the extracellular domain of occludin and claudin molecules. One lysine residue is present in the first extracellular loop of human, murine, canine, and Xenopus occludin. In rat kangaroo and chicken, the lysine is replaced by an asparagine. As occludin was first cloned from chicken, it explains why the tentatives to biotinylate occludin were unfruitful. In human, all the claudins, except claudin 18, contain at least one (claudins 1, 15, 23) to five (claudins 10 and 12) available lysine residues. In other species, including monkey, mouse, rat, dog, bovine, and fish among the sequences of claudin present in the databases only the murine claudin 18 does not contain a lysine residue.

Endocytosis Assay

The rate of endocytosis of TJ proteins is a balance between their incorporation within endocytic vesicles and their retention at the plasma membrane. Retention at the cell surface is due to their presence within a multiprotein complex and their tight association with the apical ring of actin. Thus the rate of endocytosis reflects the shift in equilibrium between cell surface TJ proteins that are incorporated within the TJ and the TJ proteins that are not incorporated within the TJ. The assay is performed on cells grown on 12 mm filters. The cells are washed three times in ice-cold PBS-CM and incubated twice for 30 min in PBS-CM containing 1 mg/ml of the cleavable NHS-SS-biotin. The amount of biotin is a compromise between the detection level of occludin and claudin 1 which possess only one available lysine and the toxic effect of a massive biotinylation of the cell surface proteins. The cells are successively washed in ice-cold PBS-CM to eliminate the excess of biotin and washed in MEM-BSA (MEM, 0.6% BSA, 20 mM Hepes pH7.4, penicillin/streptomycin) supplemented with 50 mM NH4Cl to quench the remaining biotin. Endocytosis is then started by incubating the cells at 37° for different periods of time in MEM-BSA. The cells are then rapidly cooled down by three quick washes in PBS-CM. Control samples are processed for immunoblot quantitation. These samples will be used to indicate the total amount of biotinylated occludin, claudin, or control proteins. The other cells are treated with the reducing agent glutathione (50 mM glutathione, 50 mM TAPS-NaOH pH 8.60, 100 mM NaCl, 0.2 % BSA) two times 1 h at 4° to remove the remaining biotin from the cell surface. We found this step to be the most critical and at the origin of the greatest variability. The glutathione solution needs to be prepared extemporary and its pH adjusted exactly to 8.60 with a strong buffer as glutathione is very acidic. Although Tris can be used, TAPS, which has a higher buffer capacity at pH 8.60 is better to compensate for the drastic acidification of the glutathione solution. Following three quick washes in cold PBSCM the cells are solubilized in lysis buffer (1% Triton-X100, 20 mM Hepes pH 7.4, 2 mM EDTA, 125 mM NaCl, 0.2 mM PMSF, and protease inhibitors) for 20 min at 4° and spun 30 min at 15,000g in a microfuge. The endocytosed material found in the supernatant is precleared on sepharose beads and the biotinylated proteins recovered after a 1 h incubation with streptavidin-agarose beads. The bound proteins are eluted by vortexing the beads 10 min at 60° in equal volume of 100 mM DTT and then prepared for SDS-PAGE and immunoblot analysis. By recovering all the biotinylated proteins, we can analyze by Western blot within the same sample occludin, claudin 2, and control proteins such as the EGF-R or Tf-R. Performing this experiment on cells expressing EFA6, we found that EFA6 expression decreased specifically the endocytosis of TJ proteins (Luton *et al.*, 2004). This effect could either reflect a direct impact on plasma membrane trafficking or cell surface retention by increased anchoring to the apical actin cytoskeleton. To discriminate between these two possibilities we directly measured the half-life of the TJ proteins at the cell surface.

Cell Surface Delivery and Half-Life Assay

The experiment is designed to determine the time course of the different steps of the journey of the TJ proteins from the Golgi to the plasma membrane before intracellular degradation. The assay is based on a regular pulse-chase experiment coupled to cell surface biotinylation as described above. Cells grown on 12 mm filters are rinsed twice and starved for 20 min at 37° in MEM without cysteine/methionine (MEM-Cys-Met), 5% dialyzed FCS. The pulse labeling is performed by placing the filter on a 30 μl drop of MEM-Cys-Met containing a mix of ^{35}S-Cys/Met (66 μCi/filter at 1175 Ci/mmol) for a short period of 15 min. After four quick washes in MEM-BSA, the cells are then incubated in MEM-BSA at 37° for different periods of time to allow for the intracellular trafficking of the TJ proteins and to monitor over time their rate of transport to the cell surface, their half-life at the plasma membrane, and their disappearance from the cell surface by endocytosis. At the end of each time point, the cells are rinsed three times quickly in ice-cold PBS-CM and similarly to the endocytosis assay, the cells are submitted to biotinylation at 4° and then solubilized for protein analysis. After preclearing of the supernatants, the proteins are immuno-precipitated using specific antibodies coupled to protein A or protein G sepharose beads, and the immunoprecipitates are washed five times in lysis buffer, and twice in cold distilled water. The immunocomplexes are dissociated by two incubations at 75° in 1.6% SDS, 8.3% glycerol, 167 mM Tris pH 6.8 under mild vortexing for 5 min. The pooled eluates are transferred to a new tube containing 1 ml lysis buffer and 10 μl streptavidin-agarose beads and further incubated under head-to-head rotation for 1 h at 4° to recover the biotinylated proteins. After five washes in large volumes of lysis buffer and two washes in cold distilled water, two elutions are performed at 60° in 10 μl DTT 100 mM under mild vortexing for 5 min. The pooled eluates containing the metabolically labeled and biotinylated proteins are resolved by SDS-PAGE, visualized, and analyzed using a PhosphorImager (Fujifilm BAS-1500, Fuji Film, France). An example of this experiment is shown in Fig. 1. We compared occludin to the transferrin receptor in control and EFA6 expressing cells. To interpret the result it is important to remember that a small pool of molecules is synchronously traveling through the cell as a single cohort. The first part of the curve is linear and reflects the rate of delivery to the plasma membrane, which was identical in the two cell lines. The slope of the curves inflects before reaching a maximum. The peak reflects the maximum number of molecules that were present at the same time at the cell surface. Given an equivalent delivery rate to the cell surface, molecules that remain at the cell surface will result in a higher and later peak of radiolabeled biotinylated proteins.

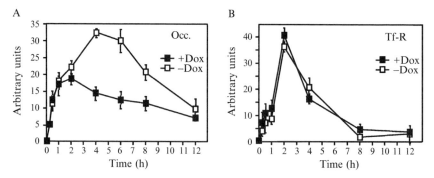

Fig. 1. EFA6 expression promotes the cell surface retention of occludin. The quantity of occludin (A) and transferrin receptor (B) present at the plasma membrane of control and EFA6-expressing MDCK cells was determined after pulse-chase and cell-surface biotinylation. The amounts of protein were quantified by densitometry and plotted against time.

Both parameters, the maximum intensity and time of the peak, were comparatively increased in EFA6 expressing cells demonstrating that EFA6 had a retention effect on occludin at the cell surface. In the control experiment, EFA6 had no effect on transferrin receptor cell surface delivery or half-life (Luton *et al.*, 2004). While the calcium switch allows one to analyze the TJ assembly and disassembly, this method tests the stability of the TJ at steady state without the disturbance of the calcium switch.

Detergent Partitioning

Partitioning into the detergent-insoluble fraction is used to monitor the amount of proteins associated with the cortical actin cytoskeleton and thought to be incorporated into TJ. We have refined the basic procedure to optimize the recovery of Tx-100 insoluble material from MDCK cells grown on 24 mm filters. To let the TJ reach full maturity the cells are grown on filters for a minimum of 5 days changing the culture medium every day. On the day of the experiment, the cells are washed three times in ice-cold PBS-CM. The filters are carefully excised from their support and slide cells down into a 24-well dish containing 0.4 ml of extraction buffer (0.5% Triton X-100, 120 mM NaCl, 10 mM Tris pH 7.5, 25 mM KCl, 1.8 mM CaCl2, 0.1 mM DTT, 0.25 mM PMSF, 0.1 mg/ml DNAseI, 0.1 mg/ml RNAse) and agitated on ice for 15 min. The cellular material remaining on the filter membranes is scraped using a rubber policeman and the recovered material is spun at 15,000g at 4° for 20 min. The supernatant (detergent-soluble fraction) is saved, while the pellet (detergent-insoluble fraction) is resuspended by energic pipeting in 40 μl of SDS buffer (1% SDS, 10 mM Tris pH 7.5, 2 mM EDTA) and boiled for 5 min. The final volume is completed to 0.4 ml with extraction buffer. Equal amounts of proteins of each sample are then analyzed by SDSPAGE and immunoblot. When coupled to a calcium switch, the assembly of the TJ can be visualized by the redistribution of the TJ proteins from the soluble to the insoluble fraction. Another indication of the incorporation of occludin into the TJ is its phosphorylation whereby occludin appears over time as a doublet in the insoluble fraction. We had shown by independent experiments that EFA6 could stabilize the TJ and the apical actin ring, thus we asked whether the TJ proteins were stabilized by an increased association to the apical actin cytoskeleton. In EFA6 expressing cells, we found that the detergent-insoluble fraction of all the TJ proteins, including ZO-1, claudin 2, and occludin, but not that of E-cadherin or the transferrin receptor, was enhanced while their total cellular amount was unchanged (Luton *et al.*, 2004). This result allowed us to couple the effects of EFA6 on the increased actin cytoskeleton stability to the TJ retention at the cell surface.

Acknowledgments

The author thanks Stéphanie Klein, Jean-Paul Chauvin, Mariagrazzia Partisani, Drs. Michel Franco, Sylvain Bourgoin, André Le Bivic, and Pierre Chardin for various contributions to the work presented in this chapter. Gratitude is also expressed to Drs. Karen L. Singer and Eva Faurobert for critical reading of the manuscript. The work described herein was made possible by grants from the Centre National de la Recherche Scientifique, the Association pour la Recherche sur le Cancer, and salary support from the Institut National de la Santé et de la Recherche Médicale.

References

Apodaca, G., Katz, L. A., and Mostov, K. E. (1994). Receptor-mediated transcytosis of IgA in MDCK cells via apical recycling endosomes. *J. Cell Biol.* **125**, 67–86.

Balda, M. S., Whitney, J. A., Flores, C., Gonzalez, S., Cereijido, M., and Matter, K. (1996). Functional dissociation of paracellular permeability and transepithelial electrical resistance and disruption of the apical-basolateral intramembrane diffusion barrier by expression of a mutant tight junction membrane protein. *J. Cell Biol.* **134**, 1031–1049.

Cereijido, M., Valdés, J., Shoshani, L., and Contreras, R. G. (1998). Role of tight junctions in establishing and maintaining cell polarity. *Annu. Rev. Physiol.* **60**, 161–177.

Chavrier, P., and Goud, B. (1999). The role of ARF and Rab GTPases in membrane transport. *Curr. Opin. Cell Biol.* **11**, 466–475.

Denker, B. M., and Nigam, S. K. (1998). Molecular structure and assembly of the tight junction. *Am. J. Physiol.* **274**, F1–F9.

Drubin, D. G., and Nelson, W. J. (1996). Origins of cell polarity. *Cell* **84**, 335–344.

Farquhar, M. G., and Palade, G. E. (1963). Junctional complexes in various epithelia. *J. Cell Biol.* **17**, 375–412.

Franco, M., Peters, P. J., Boretto, J., van Donselaar, E., Neri, A., D'Souza-Schorey, C., and Chavrier, P. (1999). EFA6, a sec7 domain-containing exchange factor for ARF6, coordinates membrane recycling and actin cytoskeleton organization. *EMBO J.* **18**, 1480–1491.

Furuse, M., Fujita, K., Hiiragi, T., Fujimoto, K., and Tsukita, S. (1998). Claudin-1 and -2: Novel integral membrane proteins localizing at tight junctions with no sequence similarity to occludin. *J. Cell Biol.* **141**, 1539–1550.

Furuse, M., Hirase, T., Itoh, M., Nagafuchi, A., Yonemura, S., Tsukita, S., and Tsukita, S. (1993). Occludin: A novel integral membrane protein localizing at tight junctions. *J. Cell Biol.* **123**, 1777–1788.

Goodenough, D. A., and Revel, J. P. (1970). A fine structural analysis of intercellular junctions in the mouse liver. *J. Cell Biol.* **45**, 272–290.

Gronewold, T. M., Sasse, F., Lunsdorf, H., and Reichenbach, H. (1999). Effects of rhizopodin and latrunculin B on the morphology and on the actin cytoskeleton of mammalian cells. *Cell Tissue Res.* **295**, 121–129.

Gumbiner, B., and Simons, K. (1986). A functional assay for proteins involved in establishing an epithelial occluding barrier: Identification of a uvomorulin-like polypeptide. *J. Cell Biol.* **102**, 457–468.

Jou, T.-S., Schneeberger, E. E., and Nelson, W. J. (1998). Structural and functional regulation of tight junctions by RhoA and Rac1 small GTPases. *J. Cell Biol.* **142**, 101–115.

Luton, F., Klein, S., Chauvin, J. P., Le Bivic, A., Bourgoin, S., Franco, M., and Chardin, P. (2004). EFA6, exchange factor for ARF6, regulates the actin cytoskeleton and associated tight junction in response to E-cadherin engagement. *Mol. Biol. Cell* **15**, 1134–1145.

Mitic, L. L., and Anderson, J. M. (1998). Molecular architecture of tight junctions. *Annu. Rev. Physiol.* **60**, 121–142.

Rojas, R., Ruiz, W. G., Leung, S.-M., Jou, T.-S., and Apodaca, G. (2001). Cdc42-dependent modulation of tight junctions and membrane protein traffic in polarized Madin-Darbey canine kidney cells. *Mol. Biol. Cell* **12**, 2257–2274.

Saitou, M., Fujimoto, K., Doi, Y., Itoh, M., Fujimoto, T., Furuse, M., Takano, H., Noda, T., and Tsukita, S. (1998). Occludin-deficient embryonic stem cells can differentiate into polarized epithelial cells bearing tight junctions. *J. Cell Biol.* **1998**, 397–408.

Schneeberger, E. E., and Lynch, R. D. (1992). Structure, function, and regulation of cellular tight junctions. *Am. J. Physiol.* **262**, L6647–L6661.

Schneeberger, E. E., and Lynch, R. D. (2004). The tight junction: A multifunctional complex. *Am. J. Physiol. Cell. Physiol.* **286**, C1213–C1228.

Spector, I., Shochet, N. R., Kashman, Y., and Groweiss, A. (1983). Latrunculins: Novel marine toxins that disrupt microfilament organization in cultured cells. *Science* **219**, 493–495.

Stevenson, B. R., Siliciano, J. D., Mooseker, M. S., and Goodenough, D. A. (1986). Identification of ZO-1: A high molecular weight polypeptide associated with the tight junction (zonula occludens) in a variety of epithelia. *J. Cell Biol.* **103**, 755–766.

Tsukita, S., and Furuse, M. (2000). Pores in the wall: Claudins constitute tight junction strands containing aqueous pores. *J. Cell Biol.* **149**, 13–16.

Tsukita, S., Furuse, M., and Itoh, M. (2001). Multifunctional strands in tight junctions. *Nat. Mol. Cell Biol.* **2**, 285–293.

Turner, J. R. (2000). "Putting the squeeze" on the tight junction: Understanding cytoskeletal regulation. *Sem. Dev. Biol.* **11**, 301–308.

Wong, V., and Gumbiner, B. M. (1997). A synthetic peptide corresponding to the extracellular domain of occludin perturbs the tight junction permeability barrier. *J. Cell Biol.* **136**, 399–409.

[30] *In Vitro* Reconstitution of ARF-Regulated Cytoskeletal Dynamics on Golgi Membranes

By Ji-Long Chen, Weidong Xu, and Mark Stamnes

Abstract

Function of the secretory pathway is intimately connected to the cytoskeleton. Cytoskeletal dynamics and molecular motors are involved in organelle morphology and positioning, as well as the formation and translocation of trafficking intermediates such as vesicles. At least three classes of small GTPases, the ADP-ribosylation factor (ARF), Rho, and Rab families, have been implicated in regulating cytoskeleletal dynamics and molecular motor function within the secretory pathway. We have used the reconstitution of transport vesicle formation on isolated Golgi membranes to characterize mechanisms of ARF1 regulated actin polymerization. ARF1 affects cytoskeletal function in part by recruiting a complex between

METHODS IN ENZYMOLOGY, VOL. 404   0076-6879/05 $35.00
          DOI: 10.1016/S0076-6879(05)04030-9

the vesicle-coat protein, coatomer, and the Rho-related GTPase, Cdc42, to the Golgi apparatus. Cdc42 can activate actin polymerization on Golgi membranes through an Arp2/3-dependent mechanism. Coatomer-bound Cdc42 plays a further role in regulating vesicle motility via the motor protein, dynein. Future studies elucidating the molecular mechanisms connecting vesicular transport with actin dynamics will provide important clues to the overall contribution of the cytoskeleton and molecular motors to protein transport. This article describes methods and reagents for characterizing cytoskeletal regulation at the Golgi apparatus through the cell-free reconstitution of vesicle formation.

Introduction

The structural and dynamic properties of the cytoskeleton contribute to the ability of eukaryotic cells to carry out numerous general and compartmentalized functions. Because the cytoskeleton contributes simultaneously to so many different processes, it is often challenging to characterize its contribution to a specific task such as vesicular transport. Experiments utilizing whole cells can be difficult to interpret since direct and indirect effects of disrupting the cytoskeleton are hard to separate. An additional complication is that some cellular processes adapt to cytoskeletal disruption. In higher eukaryotic cells, for example, protein transport through the secretory pathway is inhibited by acute disruption of microtubules with nocodazole. Remarkably, after several hours of nocodazole treatment, the cells adapt and resume trafficking in a microtubule-independent manner (Thyberg and Moskalewski, 1999). Thus, whether one concludes that microtubules are dispensable or indispensable for transport depends on the protocol used with nocodazole. In part because of these issues, the exact contribution of the cytoskeleton and molecular motors to transport through the secretory pathway remains unclear. Cell-free reconstitution approaches can provide an opportunity to study the role of the cytoskeleton in isolation and thus avoid some of the pitfalls of whole-cell experiments.

We have used the reconstitution of COPI transport vesicle formation on isolated rat-liver Golgi membranes as one approach to characterize the role of the cytoskeleton during protein transport (Chen et al., 2004, 2005; Fucini et al., 2000, 2002). The cell-free reconstitution of intra-Golgi transport and COPI vesicle assembly was developed in the laboratory of J. Rothman during the 1980s (Balch et al., 1984; Malhotra et al., 1989; Serafini et al., 1991). These studies helped establish that GTP-bound ARF1 initiates COPI vesicle assembly by triggering the assembly of the coat protein, coatomer. Current studies stem from the observation that activation of ARF1-dependent COPI vesicle assembly with the non-hydrolyzable GTP-analog, GTPγS, leads to a concomitant increase in actin levels on the Golgi membrane (Fucini *et al.*, 2000; Godi *et al.*, 1998).

We have characterized details of ARF1-dependent actin polymerization on Golgi membranes. ARF1 activates actin polymerization by recruiting a second GTPase, Cdc42, through a binding interaction with coatomer (Chen *et al.*, 2004; Fucini *et al.*, 2002; Luna *et al.*, 2002; Stamnes, 2002; Wu *et al.*, 2000). The binding interaction between Cdc42 and coatomer is disrupted by a putative cargo-receptor protein, p23. Thus, both vesicle coat proteins and cargo proteins have the potential to influence actin dynamics on Golgi membranes. Actin polymerization on Golgi membranes also requires the Arp2/3 complex (Chen *et al.*, 2004; Matas *et al.*, 2004). Both Arp2/3 and its activator, the Cdc42 effector Wiskott-Aldrich syndrome protein (WASP), can bind Golgi membranes when ARF1 is activated.

Characterization of actin dynamics using this cell-free system can also yield insight into its role during trafficking. Our recent studies have revealed that coatomer-bound Cdc42 affects the recruitment of the microtubule-based motor protein, dynein, to COPI vesicles (Chen *et al.*, 2005). Cdc42 regulates dynein recruitment through its ability to affect actin dynamics. Previous studies have provided one step toward a full understanding of the regulation and the role of the cytoskeleton during trafficking. Future progress will rely on combined approaches utilizing genetics, whole-cell characterization, and cell-free reconstitution systems. Here we outline methods we have used for characterizing cytoskeletal regulation at the Golgi. These approaches or modifications thereof should be useful for addressing many aspects of cytoskeletal regulation on organelles.

Incubation Conditions for Golgi-Membrane Binding and Vesicle Budding Reactions

We have modified assays used to elucidate the molecular mechanisms of intra-Golgi transport and vesicle formation *in vitro* (Serafini and Rothman, 1992) in order to characterize regulatory mechanisms connecting cytoskeletal dynamics to the secretory machinery. In short, isolated Golgi membranes are incubated with cytosol as a source for vesicle coat proteins, cytoskeletal proteins, and regulatory small GTPases. The membranes or newly formed transport vesicles are then reisolated and their association with cytoskeletal proteins are assayed. Activation or inhibition of the various small GTPases provides an opportunity to determine their specific roles in connecting cytoskeletal dynamics to the secretory machinery.

Rat-Liver Golgi Membranes

Rat-liver Golgi membranes are prepared using the protocol described for rabbit liver (Serafini and Rothman, 1992) with only minor modifications. We have obtained comparable results using membranes prepared from freshly dissected rat liver or frozen rat-liver tissue obtained from a commercial source (Pel-Freez, Rogers, AR). Currently, we use frozen liver almost exclusively because of its convenience and lower cost. We find it convenient to prepare the postnuclear supernatant from livers (Serafini and Rothman, 1992) that is then frozen in 50 ml aliquots and stored at $-80°$. The postnuclear supernatant can be thawed and fractionated as need for Golgi membranes arises. The protein concentration of the final rat-liver Golgi membrane preparation as extracted from the gradient is typically about 1 mg/ml and the sucrose concentration is about 0.8 M.

Cytosol

Bovine-brain cytosol is prepared as described previously (Serafini and Rothman, 1992). The brains are obtained from a local slaughterhouse. We have tried to use commercially available frozen bovine and rat brains for cytosol preparations without success. Cytosol is stored in small aliquots (0.5–1.0 ml), that are frozen in liquid nitrogen and stored at $-80°$. The protein concentration of typical bovine-brain cytosol preparations is approximately 20 mg/ml.

Incubation Conditions

Reaction conditions for the reconstitution of regulated cytoskeletal dynamics on Golgi membranes were modeled on the conditions shown previously to reconstitute intra-Golgi-transport and Golgi vesicle formation. The only significant modification is that higher ATP concentrations are required for consistent ARF1-dependent actin polymerization. The final reaction conditions are as follows: 25 mM HEPES (pH 7.2), 2.5 mM Mg(Acetate)$_2$, 15 mM KCl, Golgi membranes (0.2 mg/ml), and bovine-brain cytosol (1.0 mg/ml), and an ATP regenerating system (1.0 mM creatine phosphate, 80 units/ml creatine phosphokinase, 250 μM UTP, 500 μM ATP). The sucrose is adjusted to a final concentration of 0.2 M, including that derived from the membrane preparation.

Vesicle coat assembly and actin polymerization can be initiated by including the nonhydrolyzable GTP analog, GTPγS, (20 μM) in the reaction. The reactions are normally incubated at 37° for 20 min. Typical reaction volumes are 0.5–1.0 ml for Golgi-binding assays and 2–10 ml when vesicles will be isolated.

*The Use of Two-Stage Reactions to Characterize the Involvement of
Specific GTPases*

Constitutive activation of GTP-binding proteins with nonhydrolyzable
GTP analogs (i.e., GTPγS) is a powerful tool for analyzing their downstream
effects. However, since GTPγS activates many GTPases, it is not sufficient for
defining the role of a specific GTPase in cytosol. In some cases, we have used
two-stage reactions to define the involvement of GTPases more precisely. In
this case, the Golgi membranes are first incubated with one or more recombinant GTPases, that is, ARF1 plus GTPγS. The membranes are then reisolated by sedimentation in a microfuge. The ARF1-bound membranes can
then be resuspended in a new reaction containing cytosol (but not GTPγS) to
examine the consequences on cytoskeletal dynamics. We used this approach,
to first determine that ARF1 was a necessary, but not a sufficient, GTPase for
actin polymerization on Golgi membranes (Fucini *et al.*, 2002).

Following one- or two-stage incubations, the Golgi membranes or
Golgi-derived vesicles can be isolated and characterized as elaborated
below.

Rapidly Reisolating Golgi Membranes by Flotation

Vesicle coat assembly was characterized using Golgi-binding assays in
which the membranes were reisolated from the reconstitution reaction by
sedimentation (Donaldson *et al.*, 1992; Palmer *et al.*, 1993; Stamnes and
Rothman, 1993; Traub *et al.*, 1993). We found that this approach was
unsatisfactory for characterizing cytoskeletal dynamics on the Golgi complex since F-actin and microtubules are found in a pellet fraction after
centrifugation regardless of whether they are associated with the membranes. To overcome this problem, we have used small (200 μL) sucrose
gradients that allow Golgi membranes to be rapidly reisolated from binding reactions by flotation (Fucini *et al.*, 2000) (Fig. 1).

Solutions

45% (by weight) sucrose, 25 mM HEPES (pH 7.2), 25 mM KCl
35% (by weight) sucrose, 25 mM HEPES (pH 7.2), 25 mM KCl
15% (by weight) sucrose, 25 mM HEPES (pH 7.2), 25 mM KCl

Procedure. Golgi membranes are sedimented from the reaction by
microcentrifugation (15,000g) for 20 min. The membrane pellet is then
resuspended with 50 μL of the 45% sucrose solution. The sample is
placed in a 7 × 20 mm thickwall ultracentrifugation tube appropriate for a
TLA-100 rotor (Beckman-Coulter, Fullerton, CA). The sample is overlaid
with 125 μL 35% sucrose solution, then 25 μL of the 15% sucrose solution.

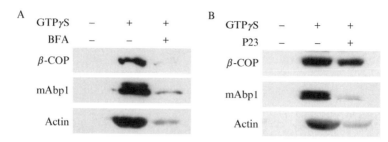

Fig. 1. Activating GTP-binding proteins with GTPγS stimulates actin polymerization on Golgi membranes. Shown are Golgi-binding assays in which the membranes were reisolated by flotation following the incubation with cytosol. Western blots were probed with antibodies against the coatomer subunit (β-COP), the actin binding protein (mAbp1), and actin. (A) Brefeldin A (BFA) was used to inhibit ARF activation as indicated. (B) A peptide corresponding to C-terminal coatomer binding motif of the p23 protein (P23) was included in the incubation to disrupt Cdc42 binding. Note that brefeldin A inhibits both coatomer binding and actin polymerization. Disrupting the coatomer/Cdc42 binding interaction with the peptide affects actin but not vesicle coat assembly.

The samples are centrifuged at 100,000 rpm (386,000g) for 40 min. The TLA-100 rotor allows 20 samples to be characterized simultaneously. After centrifugation, the Golgi membranes are recovered from the 35%/15% sucrose interface using a micropipette. Occasionally, some of the membrane forms a "pellet" on the side of the tube at the interface. We find that this pellet is easily displaced and recovered with the pipette tip. With some care, consistent membrane recovery is obtained from all samples.

For most analyses, the recovered membranes are diluted and precipitated with 10% trichloroacetic acid. The presence of Golgi-membrane-bound cytoskeletal proteins and vesicle coat proteins can be ascertained using SDS-PAGE and Western blotting (Fig. 1).

Vesicle Budding Reactions and the Extraction of Golgi-Derived Vesicles

While binding to whole Golgi membranes is informative, some aspects of cytoskeletal function in the secretory pathway are better ascertained by characterizing the composition of *in vitro* generated Golgi-derived transport vesicles.

Solutions

High-salt stripping buffer: 250 mM KCl, 25 mM HEPES (pH 7.2), 2.5 mM Mg acetate, 0.2 M sucrose

Low-salt stripping buffer: 50 mM KCl, 25 mM HEPES (pH 7.2), 2.5 mM Mg acetate, 0.2 M sucrose

Procedure. The reaction conditions for vesicle budding reactions are the same as for binding experiments except that palmitoyl-CoA (5 μM) is also included (Pfanner *et al.*, 1989). Both COPI vesicles and clathrin/AP1-coated vesicles are formed when Golgi membranes are incubated with cytosol and activated ARF1. Irreversibly coated COPI vesicles are generated when ARF1 is constitutively activated in the reaction either using GTPγS or recombinant constitutively active ARF1(Q71L). Following the incubation, the membranes are washed by resuspension in low-salt-stripping buffer. The vesicles are extracted from the membranes by incubating with high-salt-stripping buffer. After a final sedimentation step 15,000g for 20 min, the vesicles are present in the high-salt supernatant.

Constitutively coated COPI vesicles can be enriched from the high-salt supernatant by exploiting the fact that their buoyant density is greater than the contaminating Golgi fragments. We have used both a rapid sedimentation procedure (Fig 2A) and flotation through an isopycnic sucrose gradient (Fig. 2B) in order to purify vesicles.

Enriching Vesicles Using Isopycnic Gradient Centrifugation

COPI vesicles can be isolated by flotation through an isopycnic sucrose gradient as described by Ostermann *et al.* (1993).

Solutions

70%, 45%, 40%, 35%, and 30% (by weight) sucrose solutions each with 250 mM KCl, 25 mM HEPES (pH 7.2), and 2.5 mM Mg acetate.

Procedures. The high-salt supernatant (0.5 ml) from the budding reaction is mixed with 1.0 ml 70% sucrose solution and placed at the bottom of a 13 × 51 mm tube for the SW-55 rotor (Beckman-Coulter). This is then overlaid sequentially with 0.5 ml 45%-, 1 ml 40%-, 1 ml 35%-, and 1 ml 30%-sucrose solutions. The samples are then centrifuged at 368,000g for 16 h. The gradients can be fractionated from the top with a micropipettor. The sucrose concentration of each fraction can be determined with a refractometer. COPI vesicles should fractionate with a buoyant density equivalent to 42% sucrose. Each fraction is diluted with 1 volume of water and centrifuged at 436,000g in a TLA-120.2 rotor (Beckman-Coulter) for 20 min. The membrane pellets are resuspended in Laemmli buffer and characterized by SDS-PAGE (Fig. 2B).

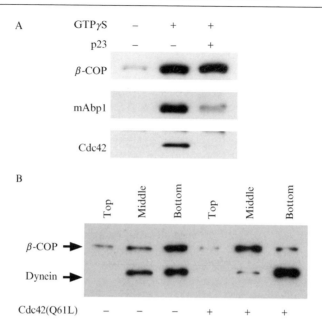

FIG. 2. Cytoskeletal proteins are found in COPI vesicle-enriched fractions. Shown are Western blots of Golgi-budding assays. (A) Following the incubation, vesicles were extracted from the membranes and enriched by sedimentation through a 35% sucrose cushion. Note that the presence of the p23 C-terminal peptide (p23) prevents Cdc42 and mAbp1 from appearing with the vesicles. (B) A vesicle extract from budding reactions carried out in presence or absence of recombinant constitutively-active Cdc42, Cdc42(Q61L), was fractionated by flotation on a continuous sucrose gradient. Following the centrifugation, three fractions were obtained corresponding to the top, middle, and bottom (load) of the gradient. COPI vesicles fractionate in the middle of the gradient. Note that the presence of active Cdc42 decreased the amount of the motor protein, dynein, present in the middle vesicle fraction.

A Golgi-Vesicle Enriched Fraction Obtained by Sedimentation

Vesicles can alternatively be enriched from the high-salt supernatant by sedimentation through a 35% sucrose cushion (Fig. 2A). This approach has the advantages that it is rapid, it can be done on a small scale, and many samples can be accommodated. However, some care must be taken to control for non-membrane-associated cytoskeletal filaments that could sediment together with the vesicles.

Procedures

A 50 μL 35% sucrose cushion is added to the bottom of a 7 × 20 mm centrifuge tube for a TLA-100 rotor. The high-salt supernatant (150 μL) is layered on top of the cushion and the sample is centrifuged for 30 min at

436,000*g*. The supernatant is carefully removed from the tube and the vesicle-enriched pellet is characterized by SDS-PAGE (Fig. 2A).

Using Liposomes to Study Membrane-Associated Cytoskeletal Dynamics

Reconstitution of transport vesicle assembly on liposomes devoid of membrane proteins has been useful for defining the contribution of soluble proteins or the specific contribution of the cytosolic domains of Golgi membrane proteins (Bremser *et al.*, 1999; Spang *et al.*, 1998). We have shown that ARF1-dependent actin polymerization is reconstituted on liposomes generated from a rat-liver lipid extract concurrently with COPI vesicle assembly (Chen *et al.*, 2004) (Fig. 3). Actin polymerized on liposomal membranes in response to ARF1 activation shared important properties with ARF-dependent actin characterized previously on rat-liver Golgi membranes. We have used the reconstitution of ARF1-dependent actin

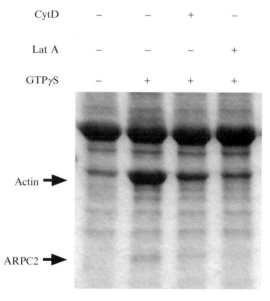

FIG. 3. Arp2/3 binds to liposomes in an actin-dependent manner. Shown is a Coomassie-blue stained gel of sedimented liposomes following incubation with cytosol. Activation of GTPases with GTPγS causes the appearance of additional protein bands on the gel. Protein sequencing and mass spectrometry were used to identify several of these proteins including actin and the Arp2/3 subunit, ARPC2. The actin toxins cytochalasin D (CytD) and latrunculin A (Lat A) reduce the levels of both actin and the ARPC2. The experiment indicates that actin plays an important role in the binding of Arp2/3 complex to the liposomal membranes.

polymerization on liposomes in order to characterize the specific contribution of cytosolic proteins to the reaction. Replacing Golgi membranes with liposomes also provides a greatly simplified system that facilitates the isolation and identification of cytosolic proteins that bind membranes and contribute to cytoskeletal regulation. In this way, we identified Arp2/3 as an ARF1-dependent membrane binding protein (Chen *et al.*, 2004) (Fig. 3). Liposomes with a defined phospholipid composition could ostensibly be used in place of the liver-lipid extract in order to define the contribution of phospholipids to cytoskeletetal dynamics. Conditions and procedures that we have used successfully for reconstitution on liposomes are outlined below.

Additional Materials and Solutions

> Rat-liver lipid extract (Avanti Polar Lipids, Alabaster, AL)
>
> Extruder with 400 nm pore diameter polycarbonate membrane (Avestin, Ottawa, Canada)
>
> Resuspension buffer 20 mM HEPES (pH 7.2), 100 mM K acetate, 250 mM sucrose
>
> Wash buffer 25 mM HEPES (pH 7.2), 25 mM KCl

Procedures. The rat liver lipid extract in chloroform is dried by evaporation in a glass test tube. Alternatively, synthetic phospholipids could be combined at this step to generate liposomes with a defined lipid composition. A test tube evaporator such as the Evap-O-Rac System (Cole-Parmer, Vernon Hills, IL) facilitates the drying step. Resuspension buffer is added to the glass tube in order to obtain a final lipid concentration of 2.3 mg/ml. The phospholipids are resuspended by vigorous mixing with a vortex mixer. In order to maintain conditions similar to those used previously to study vesicle assembly and membrane fusion, we have processed the resuspended liposomes to generate unilamellar membranes with a consistent size. This was done by subjecting the liposomes to 10 freeze/thaw cycles. The membranes were then extruded through the polycarbonate membrane. We have used 4–5 passages through a membrane with a 400 nm pore size for our experiments. The liposomes can be stored frozen at $-80°$ in small aliquots.

Reaction Conditions. The reaction conditions are identical to those described above with the exception that the rat-liver Golgi membranes are replaced by the liposomes at a final phospholipid concentration of 0.23 mg/ml.

Isolating Liposomes from the Reaction. The liposomes can be reisolated from the reaction by sedimentation using a microcentrifuge 15,000g for 30 min. The liposomes are washed twice with the wash buffer. The membranes can then be resuspended with Laemmli sample buffer for characterization by SDS-PAGE (Fig. 3). We have not been successful using flotation to reisolate liposomes. Thus, there is the potential for nonmembrane bound cytoskeletal polymers to sediment with the liposomes. Incubations carried out in the absence of liposomes should be done to control for this possibility.

Inhibiting Small GTPases Involved in Cytoskeletal Dynamics at the Golgi

Several inhibitors are useful for defining the requirements for small GTP-binding proteins in regulating cytoskeletal dynamics at the Golgi complex. A requirement for ARF can be demonstrated using brefeldin A to block its nucleotide exchange reaction (Fig. 1A). General inhibition of all Rho-family GTPases can by accomplished using *Clostridium difficile* toxin B (Fucini *et al.*, 2002). The Cdc42/coatomer binding interaction can be specifically disrupted using a peptide corresponding to the C-terminus of the p23 putative cargo receptor (Fucini *et al.*, 2002; Wu *et al.*, 2000) (Figs. 1B and 2A).

Materials and Solutions

10 mM brefeldin A (Sigma, St. Louis, MO), in methanol
p23 C-terminal peptide, YLRRFFKAKKLIE (custom synthesized)
40 μg/ml *Clostridium difficile* Toxin B (Calbiochem, San Diege, CA)
10 mM UDP-glucose (Calbiochem)

Procedures. Brefeldin A inhibits ARF activation at concentrations greater than 200 μM. Since brefeldin A inhibits the nucleotide exchange reaction rather than downstream effector interactions, it is usually necessary to preincubate the samples with brefeldin A for 5 min at 37° before adding a nonhydrolyzable GTP analog. Inhibition of ARF1 with brefeldin A should prevent coatomer binding, vesicle budding, and actin polymerization even in the presence of GTPγS.

Toxin B inhibits Rho-related GTPases by glucosylating them. Therefore, 100 μM UDP-glucose must be added to the reaction as a substrate in addition to the toxin. Actin polymerization on Golgi membranes should be significantly inhibited at Toxin B concentrations greater than 1 μg/ml.

The cytoplasmic tail domain of p23 disrupts the binding interaction between coatomer and Cdc42 (Fucini *et al.*, 2002; Wu *et al.*, 2000). The synthetic peptide containing the dibasic coatomer-binding motif can be used to examine signaling events specifically downstream of coatomer/ Cdc42. Cdc42 binding to Golgi membranes should be completely inhibited at peptide concentrations greater than 250 μM. We have found that in this concentration range, the p23 peptide inhibits the recruitment of the actin-binding protein mAbp1 to the membrane (Fucini *et al.*, 2002) (Fig. 1B). In addition, disrupting Cdc42 signaling stimulates dynein recruitment to COPI vesicles (Chen *et al.*, 2005) (Fig. 2B).

Detecting and Identifying Cytoskeletal Proteins

We have characterized the recruitment and function of regulatory and cytoskeletal proteins to Golgi membranes, vesicles, or liposomes using fractionation by SDS-PAGE. The binding of specific proteins was ascertained by Western blotting and probing with appropriate antibodies. Commercially available antibodies that have been used successfully in our studies are listed below.

Antibodies

Anti-Cdc42 (Zymed Laboratories, Carlsbad, CA)
Anti-Rac (Cytoskeleton)
Anti-Rho (Cytoskeleton)
Anti-actin (Sigma)
Anti-α-tubulin (Sigma)
Anti-dynein IC 74.1 (Covance, Princeton, NJ)
Anti-β-COP (Sigma)

Identification of Novel Membrane-Bound Cytoskeletal Proteins

This approach can also be used to identify novel or unexpected proteins that contribute to cytoskeletal function on membranes. We have successfully isolated and identified individual protein bands from the SDS-PAGE gels and identified them using mass spectrometry (Chen *et al.*, 2004). This has been particularly feasible with the vesicle fractions and liposome binding assays where the background from resident Golgi proteins is greatly reduced or absent (Fig. 3).

Conclusions

The protocols outlined in this article were used to characterize the regulation of cytoskeletal dynamics associated with COPI vesicle formation on Golgi membranes. With some modification it seems likely that related approaches could be used to elucidate the role of the cytoskeleton with other vesicle types or on other organelles. Combining these cell-free approaches with genetic and whole-cell studies will ultimately provide a much clearer picture of the interface between vesicular transport and the cytoskeleton.

Acknowledgments

The authors thank J. Ahluwalia, R. Fucini, A. Navarrete, J. Topp, and C. Vadakkan for contributing to the development of these assays. This work was supported by an NIH grant (RO1 GM068674) to M.S.

References

Balch, W. E., Dunphy, W. G., Braell, W. A., and Rothman, J. E. (1984). Reconstitution of the transport of protein between successive compartments of the Golgi measured by the coupled incorporation of N-acetylglucosamine. *Cell* **39,** 405–416.

Bremser, M., Nickel, W., Schweikert, M., Ravazzola, M., Amherdt, M., Hughes, C. A., Sollner, T. H., Rothman, J. E., and Wieland, F. T. (1999). Coupling of coat assembly and vesicle budding to packaging of putative cargo receptors. *Cell* **96,** 495–506.

Chen, J. L., Fucini, R. V., Lacomis, L., Erdjument-Bromage, H., Tempst, P., and Stamnes, M. (2005). Coatomer-bound Cdc42 regulates dynein recruitment to COPI vesicles. *J. Cell Biol.* **169,** 383–389.

Chen, J. L., Lacomis, L., Erdjument-Bromage, H., Tempst, P., and Stamnes, M. (2004). Cytosol-derived proteins are sufficient for Arp2/3 recruitment and ARF/coatomer-dependent actin polymerization on Golgi membranes. *FEBS Lett.* **566,** 281–286.

Donaldson, J. G., Cassel, D., Kahn, R. A., and Klausner, R. D. (1992). ADP-ribosylation factor, a small GTP-binding protein, is required for binding of the coatomer protein beta-COP to Golgi membranes. *Proc. Natl. Acad. Sci. USA* **89,** 6408–6412.

Fucini, R. V., Chen, J. L., Sharma, C., Kessels, M. M., and Stamnes, M. (2002). Golgi vesicle proteins are linked to the assembly of an actin complex defined by mAbp1. *Mol. Biol. Cell* **13,** 621–631.

Fucini, R. V., Navarrete, A., Vadakkan, C., Lacomis, L., Erdjument-Bromage, H., Tempst, P., and Stamnes, M. (2000). Activated ADP-ribosylation factor assembles distinct pools of actin on Golgi membranes. *J. Biol. Chem.* **275,** 18824–18829.

Godi, A., Santone, I., Pertile, P., Devarajan, P., Stabach, P. R., Morrow, J. S., Di Tullio, G., Polishchuk, R., Petrucci, T. C., Luini, A., and De Matteis, M. A. (1998). ADP ribosylation factor regulates spectrin binding to the Golgi complex. *Proc. Natl. Acad. Sci. USA* **95,** 8607–8612.

Luna, A., Matas, O. B., Martinez-Menarguez, J. A., Mato, E., Duran, J. M., Ballesta, J., Way, M., and Egea, G. (2002). Regulation of protein transport from the Golgi complex to the endoplasmic reticulum by CDC42 and N-WASP. *Mol. Biol. Cell* **13,** 866–879.

Malhotra, V., Serafini, T., Orci, L., Shepherd, J. C., and Rothman, J. E. (1989). Purification of a novel class of coated vesicles mediating biosynthetic protein transport through the Golgi stack. *Cell* **58,** 329–336.

Matas, O. B., Martinez-Menarguez, J., and Egea, G. (2004). Association of Cdc42/N-WASP/Arp2/3 signaling pathway with Golgi membranes. *Traffic* **5,** 838–846.

Ostermann, J., Orci, L., Tani, K., Amherdt, M., Ravazzola, M., Elazar, Z., and Rothman, J. E. (1993). Stepwise assembly of functionally active transport vesicles. *Cell* **75,** 1015–1025.

Palmer, D. J., Helms, J. B., Beckers, C. J., Orci, L., and Rothman, J. E. (1993). Binding of coatomer to Golgi membranes requires ADP-ribosylation factor. *J. Biol. Chem.* **268,** 12083–12089.

Pfanner, N., Orci, L., Glick, B. S., Amherdt, M., Arden, S. R., Malhotra, V., and Rothman, J. E. (1989). Fatty acyl-coenzyme A is required for budding of transport vesicles from Golgi cisternae. *Cell* **59,** 95–102.

Serafini, T., Orci, L., Amherdt, M., Brunner, M., Kahn, R. A., and Rothman, J. E. (1991). ADP-ribosylation factor is a subunit of the coat of Golgi-derived COP-coated vesicles: A novel role for a GTP-binding protein. *Cell* **67,** 239–253.

Serafini, T., and Rothman, J. E. (1992). Purification of Golgi cisternae-derived non-clathrin-coated vesicles. *Methods Enzymol.* **219,** 286–299.

Spang, A., Matsuoka, K., Hamamoto, S., Schekman, R., and Orci, L. (1998). Coatomer, Arf1p, and nucleotide are required to bud coat protein complex I-coated vesicles from large synthetic liposomes. *Proc. Natl. Acad. Sci. USA* **95,** 11199–11204.

Stamnes, M. (2002). Regulating the actin cytoskeleton during vesicular transport. *Curr. Opin. Cell. Biol.* **14,** 428–433.

Stamnes, M. A., and Rothman, J. E. (1993). The binding of AP-1 clathrin adaptor particles to Golgi membranes requires ADP-ribosylation factor, a small GTP-binding protein. *Cell* **73,** 999–1005.

Thyberg, J., and Moskalewski, S. (1999). Role of microtubules in the organization of the Golgi complex. *Exp. Cell Res.* **246,** 263–279.

Traub, L. M., Ostrom, J. A., and Kornfeld, S. (1993). Biochemical dissection of AP-1 recruitment onto Golgi membranes. *J. Cell Biol.* **123,** 561–573.

Wu, W. J., Erickson, J. W., Lin, R., and Cerione, R. A. (2000). The gamma-subunit of the coatomer complex binds Cdc42 to mediate transformation. *Nature* **405,** 800–804.

[31]   Assays and Properties of Arfaptin 2 Binding to Rac1 and ADP-Ribosylation Factors (Arfs)

By OK-HO SHIN and JOHN H. EXTON

Abstract

Arfaptin 1 and 2 were identified as targets for GTP bound ADP-ribosylation factors (Arfs). Arfaptin 1 had no significant effects on guanine nucleotide binding to Arfs, nor enzymatic activities of guanine nucleotide exchange factor (GEF) and GTPase activating protein (GAP) acting on Arfs. However, arfaptin 1 inhibited Arf activation of cholera toxin and phospholipase D (PLD) in a dose-dependent manner. Only GTP-bound forms of Arf1, 5, and 6 interacted with arfaptin 1 and 2, but GTP-Arf1 showed the strongest binding to the arfaptins. In contrast to the binding of Arfs to arfaptins, GDP-Rac1 or dominant negative Rac1-N17N bound to arfaptin 2, whereas GTP-Rac1 or dominant active Rac1-Q61L did not bind to arfaptin 2. Neither GTP-Rac1 nor GDP-Rac1 bound to arfaptin 1. Based on our observation, we propose that arfaptin 2 is a target for GDP-Rac1 and for GTP-Arf1, and is involved in interactions between the Rac1 and Arfs signaling pathways. This chapter describes methods for investigating the interactions of arfaptins 1 and 2 with GTP- or GDP-liganded Arfs and Rac1.

Identification of Arfaptin 1 and 2 as Binding Proteins for GTP-ARFs

Arfaptin 1 and 2 were initially identified as binding proteins for GTP-ARF3 by yeast two-hybrid screening (Kanoh *et al.*, 1997). Messenger RNAs from both arfaptin 1 ($\sim$3.4 kb) and arfaptin 2 ($\sim$1.8 kb) were ubiquitously expressed in various human tissues. Endogenous arfaptin 1 was found in the cytosol in HL60 cells and translocated to the isolated Golgi membranes in the presence of GTP-Arf3 (Kanoh *et al.*, 1997). Arfaptin 2 shows 60% identity and 81% homology to arfaptin 1. Arfaptins do not have any known domains for GEF or GAP for ARFs. The effects of arfaptin 1 on GTPγS or GDPβS binding to ARFs were minimal, and arfaptin 1 did not interfere with the action of a GEF from rat spleen (Tsai *et al.*, 1998). Arfaptin 1 inhibited ARF-dependent activation of both cholera toxin and PLD activities in a dose-dependent manner (Tsai *et al.*, 1998). The crystal structure of arfaptin 2 has been deduced and reveals an elongated and crescent-shaped dimer of three helix coiled coils (Tarricone *et al.*, 2001).

METHODS IN ENZYMOLOGY, VOL. 404
0076-6879/05 $35.00
DOI: 10.1016/S0076-6879(05)04031-0

Yeast Two-Hybrid Assay of Arfaptin 2 Interactions with Arfs

Preparation of Plasmids

GTP-bound forms of Arf 1 and 5 were generated by Q71L substitution in Arf1 and 5, and by Q67L in Arf6. For GDP-bound forms, T31N substitution was introduced in Arf1 and 5, and T27N in Arf6. Yeast overexpressing Arf5-Q71L protein at high level, detectable by Western blotting, did not grow (Shin *et al.*, 1999). For this reason, we employed yeast vectors (pGBT9, pGAD424, and pGAD10) inducing very low levels of protein expression.

Procedure for Yeast-Two Hybrid Binding Assay

SFY526 yeast was cotransformed by both pGBT9-Arfs and pGAD10-arfaptin 2 plasmids. For controls, we employed either pGAD424 vector alone or pGAD424-α-actin. After 3-day cultivation, 50–300 yeast colonies grown on a tryptophan- and leucine-deficient agar plate were transferred onto 75 mm VWR grade 410 paper filters and permeabilized in liquid nitrogen. Each filter was placed on another filter paper that had been presoaked in X-gal buffer (60 mM Na$_2$HPO$_4$, 40 mM β-mercaptoethanol, and 0.33 mg/ml 5-bromo-4-chloro-3-indolyl-β-D-galatoside, pH 7), and incubated at 30° until color development.

Interactions Between Arfaptin 2 and Arfs

Table I shows that arfaptin 2 interacted with only dominant active forms of Arf1 and 5 by yeast two-hybrid assay. We could not determine interaction between Arf6 and arfaptin 2 by this assay because Arf6-Q67L fused with GAL4 DNA binding domain was autoactivating. In addition, yeast did not express Arf6-Q67L fused with the GAL4 activating domain.

Yeast Two-Hybrid Interaction of Arfaptin 2 with Rac1

Preparation of Rac1 Plasmids

Q61 has been identified as a key residue involved in GTP hydrolysis of Rac1 (Hirshberg *et al.*, 1997). Several studies (Dorseuil *et al.*, 1996; Xu *et al.*, 1994) have indicated that the Q61L mutation in Rac1 produces greater activation compared to the G12V mutation, which was previously employed to determine interaction between Rac1 and arfaptin 2 lacking N-terminal 38 amino acids (Van Aelst *et al.*, 1996). Rac1-Q61L or Rac1-T17N

TABLE I
YEAST TWO-HYBRID INTERACTION ASSAY BETWEEN ARFs AND ARFAPTIN 2

	GAL4 activating	
GAL4 DNA binding (pGBT9)	pGAD424-α-actin	pGAD10-arfaptin 2
Arf1-Q71L	-[a]	Blue[b]
Arf1	-	-
Arf1-T31N	-	-
Arf5-Q71L	-	Blue[b]
Arf5	-	-
Arf5-T31N	-	-
Arf6-Q67L	Blue[c]	Blue[b]
Arf6	-	-
Arf6-T27N	-	-

β-Galactosidase activity was determined by a colony-lift filter assay for the 3-day-old SFY526 yeast transformants containing the indicated plasmids.
[a] Color development was not observed after 24 h incubation at 30°.
[b] Blue color started to develop within 30 min of incubation.
[c] pGAD424 vector alone also showed blue color when cotransformed with pGBT9-ARF6-Q67L.

mutation was generated by a PCR mutation method using *pfu* polymerase (Stratagene, La Jolla, CA), and ligated into GAL4 DNA binding domain vectors, pGBT9 or pAS2-1.

Determination of Rac1 Mutant Expression in Yeast

SFY526 yeast cells were transformed by both pACT2-arfaptin 2 and pAS2-1-Rac1 mutants. After 3-day cultivation on both tryptophan- and leucine-deficient selection plates, several 2–5 mm size colonies were combined and cultured overnight in a selection liquid medium. Overnight cultures were centrifuged (1000g for 5 min), and yeast cells were resuspended in complete YPD medium and further cultivated for another 5 h. Yeast protein was extracted using 8 M urea and 5% SDS, and protein samples equivalent to 0.75 OD_{600} units of cells were determined for Rac1 expression by Western blotting using GAL4 DNA binding domain monoclonal antibody (CLONTECH, Palo Alto, CA).

Procedure for Yeast Two-Hybrid Interaction Assay Between Rac1 and Arfaptins

We have employed a low-level expression vector for Rac1 because yeast did not express Rac1-Q61 protein from pAS2-1 vector. SFY526 yeast was

TABLE II
Yeast Two-Hybrid Interaction Assay Between Rac1 Mutants and Arfaptins

GAL4 DNA binding (pGBT9)	GAL4 activating			
	pACT2 alone	pGAD-GH-PAK3	pACT2-arfaptin 2	pACT2-arfaptin 1
Rac1-Q61L	-[a]	Blue[b]	-	-
Rac1	-	-	Blue[c]	-
Rac1-T17N	-	-	Blue[c]	-

β-Galactosidase activity was determined by a colony-lift filter assay for the 3-day-old SFY526 yeast transformants containing the indicated plasmids.
[a] Color development was not observed after 24 h incubation at 30°.
[b] 10–50% of colonies showed blue color after 24 h incubation.
[c] Blue color started to develop after more than 2 h incubation.

cotransformed with both pGBT9-Rac1 mutants and pACT2-arfaptin 2 plasmid. For controls we have employed either pACT2 vector alone, pGAD GH-PAK3, or pACT2-arfaptin 1. After 3-day cultivation, we determined interactions using the colony lift filter assay described previously.

Interactions Between Arfaptins and Rac1 Mutants

Yeast was found to express wild type or T17N mutant form of Rac1 protein, but did not express Rac1-Q61L protein from a high-level expression pAS2-1 vector (Shin et al., 2001). 10–50% yeast colonies grown on tryptophan- and leucine-deficient agar plates after cotransformation with both pGBT9-Rac1-Q61L and pGAD-GH-PAK3 showed blue color, but no colonies from yeast cotransformed with both pGBT9-Rac1-Q61L and pACT2-arfaptin 2 showed blue color suggesting no interaction between Rac1-Q61L and arfaptin 2 (Table II). On the other hand, both wild type and T17N form of Rac1 showed blue color, when cotransformed with arfaptin 2. None of Rac1 mutants showed interactions with arfaptin 1 (Table II).

GST Pull-Downs with Active Rac1 and ARFs

Preparation of Recombinant Rac1 and ARFs in CHO-K1 Cells

Actively growing 9.5×10^5 CHO-K1 cells were plated in a 100 mm plate using Dulbecco's modified Eagle medium (DMEM) containing 10% fetal bovine serum and proline (0.034 g/liter) and were cultivated for 24 h.

6 ml of Opti-MEM medium (Invitrogen) containing 6 μg of eukaryotic expression plasmids and 30 μl of LipofectAMINE reagent (Life Technologies, Inc.) was added to the plates washed three times with 6 ml of Opti-MEM. After 5 h incubation, 6 ml of DMEM containing 10% fetal bovine serum and proline (0.034 g/liter) was added and the cells were cultivated for another 19 h. The cells were washed three times using phosphate-buffered saline and 0.5 ml of binding buffer (50 mM HEPES pH 7.2, 100 mM KCl, 5 mM NaCl, 1 mM MgCl$_2$, 0.5 mM EGTA, 1 mM EDTA, 2.1 μg/ml aprotinin, 2.5 μg/ml leupeptin, 1 mM dithiothreitol, and 1 mM phenylmethylsulfonyl fluoride) containing 0.5% Triton X-100 was added. After scraping, the cell suspension was incubated for 1 h at 4° with rocking. The clear supernatant containing recombinant Rac1 or Arf proteins was collected after centrifugation at 16,000g for 10 min using an Eppendorf microcentrifuge and kept at $-80°$ until used. The protein concentration of CHO-K1 lysate was between 1.3 to 2.0 mg/ml and recombinant proteins represented 0.2–0.3% of total protein as determined by densitometric analyses using recombinant proteins as standards.

Preparation of GST-Fusion Proteins

E. coli DH5α or BL21(DE3) cells were transformed with pGEX4T2, pGEX4T-arfaptin 2, or pGEX-PBD-PAK3 plasmid. Transformed cells were grown at 37° to $A_{600} = 0.8$, and protein expression was induced with 1 mM isopropyl-1-thio-β-D-galactopyranoside for 4 h at 37°. The cells were washed with buffer (10 mM Tris pH 8, 150 mM NaCl, and 1 mM EDTA) and kept at $-80°$ until used. GST-fusion proteins were extracted using 1.5% sodium N-lauryl sarcosine (Sigma, St. Louis, MO) and 2% Triton X-100 (Frangioni *et al.*, 1993), and then affinity-purified using glutathione-Sepharose 4B (Amersham Pharmacia Biotech, Piscataway, NJ).

Procedure for GST Pull-Downs

CHO-K1 lysate containing 0.5% Triton X-100 was diluted with the same volume of binding buffer to decrease Triton X-100 to 0.25% from 0.5%. 200 μM GTPγS was added to lysates and incubated for 30 min at 30°. 17 μg of GST or GST-fusion proteins attached to 25 μl of glutathione-Sepharose beads were washed two times with binding buffer containing 0.1% Triton X-100 and incubated with CHO-K1 lysates preincubated with GTPγS. The binding mixtures were incubated for 30 min at 4° with rocking, then beads washed 5 times using 1 ml binding buffer containing 0.1% Triton X-100. Rac1 or Arf bindings was determined by Western blotting after 14% Tris-Glycine SDS-polyacrylamide gel electrophoresis (PAGE).

Arfaptin 2 Pulls Down GTP-Arfs, but not GTP-Rac1

Arfaptin 2 did not bind to the GTP-bound mutant form of Rac1 (Rac1-Q61L). Recombinant Rac1-Q61L was shown to be functional since it strongly bound to the binding domain of PAK3 (Fig. 1A). GTP-bound mutant forms of Arf1, 5, and 6 showed binding to arfaptin 2, but Arf1 binding was much stronger than that of Arf5 or 6 (Fig. 1B).

Guanine Nucleotide Effects on Pull-Downs of Rac1 and Arfs

Procedure for GTPγS or GDPβS Treatment

Wild type Rac1 or Arf proteins were generated in CHO-K1 cells and each lysate was pretreated with either 200 μM GTPγS or 1 mM GDPβS for 30 min at 30°, and GST pull-downs were performed as described earlier.

Differential Binding of Arfaptin 2 to Arfs and Rac1

Rac1 treated with GTPγS showed slight binding to arfaptin 2, which was much less than that observed with GDPβS treatment. In contrast,

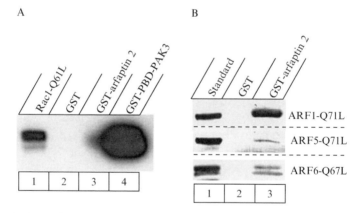

Fig. 1. *In vitro* interaction of GST-arfaptin 2 with GTP-bound mutant forms of Arfs or Rac1. Each Arf or Rac1 mutant was expressed in CHO-K1 cells, and total 0.5% Triton X-100 lysate was used for GST pull-downs. 17 μg of GST, GST-arfaptin 2, or GST-PBD-PAK3 proteins immobilized on glutathione-Sepharose beads was incubated with 0.5 ml of CHO-K1 lysate over-expressing the GTP-bound form of Rac1 (A) or Arf1, 5, 6 (B) in the presence of 10 μM GTPγS for 30 min at 4°. After washing, 50 μl of 2× SDS sample buffer was added to each sample and boiled for 10 min. Each 20 μl of sample preparation was used for 14% SDS-PAGE, and Arf or Rac1 association was determined by Western blotting. For standards, 60 ng of Rac1-Q61L, 40 ng of Arf1-Q71L, 50 ng of Arf5-Q71L, and 45 ng of Arf6-Q67L (upper band, nonmyristoylated form; lower band, myristoylated form) were used in this specific experiment. Another independent pull-down showed similar results.

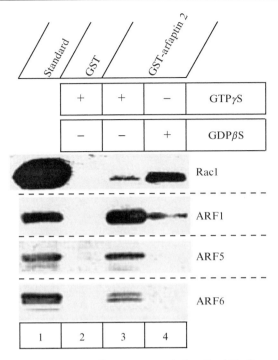

FIG. 2. Effects of GTPγS or GDPβS treatment on Arfs or Rac1 binding to GST-arfaptin 2. Wild type forms of Arfs or Rac1 were expressed in CHO-K1 cells, and total 0.5% Triton X-100 lysate was used for GST pull-downs. Each lysate overexpressing Arfs or Rac1 was incubated with 200 μM GTPγS or 1 mM GDPβS for 30 min at 30° and used for pull-downs. For this specific experiment, 15 μg of GST or GST-arfaptin 2 protein immobilized on glutathione-Sepharose beads was incubated with 0.5 ml CHO-K1 cell lysates overexpressing Arfs or Rac1 for 30 min at 4°. After washing, 50 μl of 2× SDS sample buffer was added to each sample and boiled for 10 min. Each 20 μl of sample preparation was used for 14% SDS-PAGE, and Arf or Rac1 association was determined by Western blotting. For standards, 67 ng of Rac1-Q61L, 57 ng of ARF1-Q71L, 58 ng of ARF5-Q71L, and 66 ng of ARF6-Q67L (upper band, nonmyristoylated form; lower band, myristoylated form) were used in this specific experiment. Another independent pull-down showed similar results.

Arf1, 5, and 6 bound to arfaptin 2 when treated with GTPγS, whereas little or no binding was observed in the presence of GDPβS (Fig. 2).

Discussion

Wild type Rac1 did not stably bind to GTP in the yeast nuclei as we demonstrated by cotransformation of SFY526 yeast with wild type Rac1 and PAK3 (Table II; also see Fig. 2 from Shin *et al.*, 2001). Similarly wild type Arfs did not interact with any known binding protein for GTP-Arfs in yeast two-hybrid assays (Shin *et al.*, 2001; Van Valkenburgh *et al.*, 2001). This may be due to the lack of GEF activity or high GAP activity in the yeast nuclei where binding was actually measured. We had to employ low expression vectors in the yeast two-hybrid interaction assay to overcome the cytotoxic effect of Rac1-Q61L protein. For GST pull-downs, we generated recombinant Rac1 or Arfs in CHO-K1 cells because Arfs generated in *E. coli* were 50–100 times less effective in activation of cholera toxin compared to purified natural Arfs (Tsai *et al.*, 1998). This is probably because these small G proteins are post-translationally modified and this is important for binding to the target molecules.

We have demonstrated that arfaptin 2 interacts with either GDP-Rac1 or GTP-Arfs, but preferentially with GTP-Arf1. Arfaptin 2 also interacts with GTP-Arl1 (Lu *et al.*, 2001; Van Valkenburgh *et al.*, 2001). Both Arf1-Q71L and Arl1-Q71L are located in Golgi and trigger expansion and vesiculation of Golgi, but the Golgi abnormality induced by Arl1-Q71L was less dramatic than that caused by Arf1-Q71L (Van Valkenburgh *et al.*, 2001). Recently, the arfaptin 2 C-terminal coiled-coil domain was identified as a BAR domain, which either detects or induces curvature in membranes (Peter *et al.*, 2004).

In summary, our experiments demonstrate the importance of examining the interactions of proteins with small G proteins in both the active GTP-bound and inactive GDP-bound states. This is illustrated by the fact that arfaptin 2 interacts with the GDP-liganded form of Rac1, but not the GTP-bound form. Conversely, arfaptins 1 and 2 interact with the GTP-bound forms of Arf1, 5, and 6, but not the GDP-bound forms. These reciprocal interactions of arfaptin 2 with GDP- or GTP-bound forms of Rac1 and Arfs suggest an interesting form of cross-talk in the actions of this arfaptin *in vivo*.

Acknowledgments

We thank Dr. Joel Moss (National Institutes of Health) for supplying Arf cDNA and antibodies, Dr. Richard A. Cerione (Cornell University) for supplying the pGEX-PBD-PAK3 plasmid, and Dr. Linda Van Aelst (Cold Spring Harbor Laboratory) for supplying the pGAD GH-PAK3 plasmid.

References

Dorseuil, O., Reibel, L., Bokoch, G. M., Camonis, J., and Gacon, G. (1996). The Rac target NADPH oxidase p67[phox] interacts preferentially with Rac2 rather than Rac1. *J. Biol. Chem.* **271**, 83–88.
Frangioni, J. V., and Neel, B. G. (1993). Solubilization and purification of enzymatically active glutathione S-transferase (pGEX) fusion proteins. *Anal. Biochem.* **210**, 179–187.

Hirshberg, M., Stockley, R. W., Dodson, G., and Webb, M. W. (1997). The crystal structure of human rac1, a member of the rho-family complexed with a GTP analogue. *Nat. Struct. Biol.* **4,** 147–152.

Kanoh, H., Williger, B. T., and Exton, J. H. (1997). Arfaptin 1, a putative cytosolic target protein of ADP-ribosylation factor, is recruited to Golgi membranes. *J. Biol. Chem.* **272,** 5421–5429.

Lu, L., Horstmann, H., Ng, C., and Hong, W. (2001). Regulation of Golgi structure and function by ARF-like protein 1 (Arl1). *J. Cell Sci.* **114,** 4543–4555.

Peter, B. J., Kent, H. M., Mills, I. G., Vallis, Y., Butler, P. J. G., Evans, P. R., and McMahon, H. T. (2004). BAR domains as sensors of membrane curvature: The amphiphysin BAR structure. *Science* **303,** 495–499.

Shin, O. H., Ross, A. H., Mihai, I., and Exton, J. H. (1999). Identification of arfophilin, a target protein for GTP-bound class II ADP-ribosylation factors. *J. Biol. Chem.* **274,** 36609–36615.

Shin, O. H., and Exton, J. H. (2001). Differential binding of arfaptin 2/POR1 to ADP-ribosylation factors and Rac1. *Biochem. Biophys. Res. Commun.* **285,** 1267–1273.

Tarricone, C., Xiao, B., Justin, N., Walker, P. A., Rittinger, K., Gamblin, S. J., and Smerdon, S. J. (2001). The structural basis of arfaptin-mediated cross-talk between Rac and Arf signalling pathways. *Nature* **411,** 215–219.

Tsai, S. C., Adamik, R., Hong, J. X., Moss, J., Vaughan, M., Kanoh, H., and Exton, J. H. (1998). Effects of arfaptin 1 on guanine nucleotide-dependent activation of phospholipase D and cholera toxin by ADP-ribosylation factor. *J. Biol. Chem.* **273,** 20697–20701.

Van Aelst, L., Joneson, T., and Bar-Sagi, D. (1996). Identification of a novel Rac1-interacting protein involved in membrane ruffling. *EMBO J.* **15,** 3778–3786.

Van Valkenburgh, H., Shern, J. F., Sharer, J. D., Zhu, X., and Kahn, R. A. (2001). ADP-ribosylation factors (ARFs) and ARF-like 1 (ARL1) have both specific and shared effectors. *J. Biol. Chem.* **276,** 22826–22837.

Xu, X., Barry, D. C., Settleman, J., Schwartz, M. A., and Bokoch, G. M. (1994). Differing structural requirements for GTPase-activating protein responsiveness and NADPH oxidase activation by Rac. *J. Biol. Chem.* **269,** 23569–23574.

[32] Analysis of Arf Interaction with GGAs *In Vitro* and *In Vivo*

By KAZUHISA NAKAYAMA and HIROYUKI TAKATSU

Abstract

Small GTPases of the ADP-ribosylation factor (Arf) family regulate membrane traffic and dynamics in eukaryotic cells. GGAs (Golgi-localizing, γ-adaptin ear homology domain, Arf-binding proteins) are a family of monomeric clathrin adaptor proteins that were originally identified as proteins interacting with Arfs and found to associate mainly with membranes of the *trans*-Golgi network (TGN). Like other adaptor and coat proteins,

METHODS IN ENZYMOLOGY, VOL. 404   0076-6879/05 $35.00
DOI: 10.1016/S0076-6879(05)04032-2

membrane association of GGAs is regulated by Arfs in a GTP-dependent manner. Together with or independent of the adaptor protein complex AP-1, GGAs mediate sorting of transmembrane proteins, including mannose 6-phosphate receptors, between the TGN and endosomes by cla-thrin-coated vesicles. This chapter describes methods to examine the inter-action between Arfs and GGAs and to analyze the cellular function of GGAs regulated by Arfs.

Introduction

ADP-ribosylation factors are a family of small GTPases that regulate membrane trafficking and dynamics in eukaryotic cells. There are six Arfs (Arf1–Arf6) in mammals, even though human do not have Arf2. These Arfs are grouped into three classes on the basis of sequence similarity: class I, Arf1–Arf3; class II, Arf4 and Arf5; and class III, Arf6 (Moss and Vaughan, 1995; Welsh et al., 1994). Among them, relatively well character-ized to date are Arf1 and Arf6. Arf1 is involved in the formation of coated carrier vesicles by promoting membrane recruitment of coat proteins, such as the COPI complex and the heterotetrameric AP-1 clathrin adaptor complex (Moss and Vaughan, 1995; Nie et al., 2003; Shin and Nakayama, 2004). On the other hand, Arf6 regulates endocytic and recycling processes, and remodeling of actin cytoskeleton (Donaldson, 2003; Nie et al., 2003).

In 2000, a novel family of monomeric clathrin adaptor proteins, re-ferred to as GGAs (for Golgi-localizing, γ-adaptin ear homology domain, Arf-binding proteins), was identified (Bonifacino, 2004; Nakayama and Wakatsuki, 2003). There are three GGAs (GGA1–GGA3) in mammals and two (Gga1p and Gga2p) in yeasts. These proteins have a common structural organization (Fig. 1A). The N-terminal VHS (Vps27p/Hrs/Stam) domain binds cargo proteins, such as mannose 6-phosphate receptors (MPRs), by recognizing acidic amino acid cluster-dileucine motifs. The following GAT (GGA and Tom1) domain interacts with Arfs in a GTP-dependent manner and is responsible for association of GGAs with mem-branes of the trans-Golgi network (TGN). The C-terminal GAE (γ-adaptin ear homology) domain interacts with a variety of accessory proteins that may regulate formation and tethering–docking of carrier vesicles. The Pro-rich hinge region of variable lengths connecting the GAT and GAE domains resembles those of adaptins and is responsible for recruitment of clathrin. In mammalian cells, GGAs were found to regulate selective transport of transmembrane proteins, including MPRs, between the TGN and endosomes via clathrin-coated vesicles.

In this chapter, we describe methods to examine the interaction of GGAs with Arfs through their GAT domains and to analyze membrane association of GGAs under the regulation of Arfs (Takatsu et al., 2002).

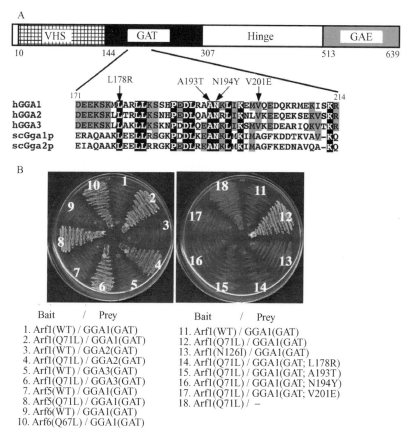

FIG. 1. Two-hybrid analyses of interactions between Arfs and the GAT domains. (A) Schematic representation of the domain organization of human GGA1 and alignment of the GAT domain sequences of human (h) and *S. cerevisiae* (sc) GGAs. Residues conserved in all GGAs are shown as white letters on a black background, and those conserved in at least three members are shaded in gray. Positions of mutated residues identified by the reverse two-hybrid screening are indicated. (B) Growth of yeast cells harboring indicated combinations of a bait vector for Arf and a prey vector for the GAT domain on plates lacking tryptophan, leucine, and histidine. Reprinted from *Biochemical Journal* (Takatsu *et al.*, 2002) with the permission of The Biochemical Society.

Two-Hybrid Analysis and Pull-Down Assay of Interaction between Arf and GGA

Plasmid Construction

Q71L and Q67L mutations of Arf1 and Arf5, and Arf6, respectively, and an N126I mutation of Arf1 were introduced into their cDNAs of mouse origin by a PCR-based strategy (Takatsu *et al.*, 2002; Toda *et al.*,

1999). cDNA fragments covering the VHS + GAT (aa 1–327), VHS (aa 1–147), GAT (aa 141–327), and GAE (aa 515–639) domains of human GGA1, and the GAT domains of human GGA2 (aa 157–342) and GGA3S (aa 107–286) were amplified by PCR of their respective full-length cDNA fragments (Takatsu et al., 2000). For two-hybrid analyses, the cDNA fragments for Arfs, in which the N-terminal helix regions were deleted (Δ1–17 for Arf1 and Arf5, and Δ1–13 for Arf6), and the fragments for GGAs were subcloned into the pGBT9 bait vector and the pGAD10 or pGAD424 prey vector (Clontech, Mountain View, CA), respectively. For expression in *Escherichia coli* as proteins fused to glutathione *S*-transferase (GST), the cDNA fragments for the GAT domains were separately subcloned into the pGEX-4T vector (Amersham Biosciences, Piscataway, NJ). Expression vectors for C-terminally HA-tagged Arfs were described previously (Hosaka et al., 1996).

Note: Although, in their two-hybrid screening and analyses for GGAs, Kahn and colleagues (Boman et al., 2000; Kuai et al., 2000) used modified vectors for Arfs, which were designed to place the Arf polypeptide N-terminal to the GAL4 DNA-binding domain, we used the conventional two-hybrid vectors available from Clontech. However, in our two-hybrid constructs, the N-terminal helix regions of Arfs were deleted.

Two-Hybrid Analysis

Yeast two-hybrid analyses were performed essentially according to the Clontech's instructions. Briefly, yeast Y190 cells were cotransformed with a pGBT9-based bait vector and pGAD-based prey vector, and were grown on synthetic medium lacking tryptophan and leucine. Colonies were picked up, and streaked on the same medium for a filter assay for β-galactosidase activity or on medium containing 25 mM 3-aminotriazole and lacking tryptophan, leucine, and histidine for a growth assay under histidine-deficient conditions.

Pull-Down Assay

1. The GAT domain of each GGA protein fused to the C-terminus of GST is expressed in *E. coli* BL21(DE3) cells and purified using glutathione-Sepharose 4B beads (Amersham Biosciences) as described in the manufacturer's instruction. The buffers used are as follows: GST-fusion protein binding buffer: PBS, pH 7.4, 5 mM β-mercaptoethanol, 5 mg/ml DNase I, 5 mg/ml RNase A, and a protease inhibitor mixture (Complete™–EDTA free, Roche Diagnostics, Indianapolis, IN); GST-fusion protein washing buffer: PBS, pH 7.4, 5 mM β-mercaptoethanol, Complete™–EDTA free.

2. HEK-293 cells grown on a 10-cm dish are transfected with an expression vector for Arf-HA using a FuGENE6 transfection reagent (Roche Diagnostics) and incubated for 24 h.

3. The cells are lysed in 0.65 ml of pull-down binding buffer (50 mM Tris-Cl, pH 7.5, 100 mM NaCl, 2 mM MgCl$_2$, 0.5% sodium deoxycholate, 1% Triton X-100 10% glycerol, Complete™–EDTA free) for 20 min on ice with constant mixing, and centrifuged at maximum speed (15,000g) for 20 min at 4° in a microcentrifuge to remove unbroken cells.

4. The lysate (containing ∼60 μg protein) is precleared by mixing with glutathione-Sepharose 4B beads for 20 min at 4°, followed by centrifugation at maximum speed for 2 min at 4° in a microcentrifuge to remove the beads.

5. The supernatant is preincubated with 200 μM GDP or GTPγS for 30 min at room temperature, then with 10 μg of the GST-fusion protein prebound to glutathione-Sepharose beads for 1 hr at 4° with gentle shaking, and centrifuged at 2000g for 5 min at 4°.

6. The beads are washed three times with pull-down washing buffer (50 mM Tris-Cl, pH 7.5, 100 mM NaCl, 2 mM MgCl$_2$, 1% NP-40, 10% glycerol, Complete™–EDTA free).

7. The bound proteins are eluted from the beads by boiling in an SDS-PAGE sample buffer, electrophoresed on a 15% SDS-polyacrylamide gel, and electroblotted on to a PVDF membrane.

8. The blot is incubated sequentially with monoclonal rat anti-HA antibody, 3F10 (Roche Diagnostics), and with peroxidase-conjugated anti-rat IgG (Jackson ImmunoResearch Laboratories West Grove, PA), and detected using a Renaissance Chemiluminescence reagent *Plus* (Perkin-Elmer Life Science, Boston, MA).

We analyzed interaction between Arf and the GAT domain of GGA by yeast two-hybrid and GST-pull-down assays. As shown in Fig. 1B, the yeast two-hybrid analyses showed that the GGA1-GAT domain interacts with a GTP-bound mutant of Arf1 (Arf1(Q71L)) but not with wild-type Arf1 or its nucleotide-free mutant (Arf1(N126I)) (streaks 11–13), and the GAT domains from all human GGAs (streaks 1–6) interact with Arf1(Q71L) but not wild-type Arf1. Furthermore, the GTP-bound mutants of all classes of Arfs (Arf1, Arf5, and Arf6) interact with the GGA1-GAT domain (streaks 1, 2, and 7–10).

The two-hybrid data were confirmed by pull-down assays. As shown in Fig. 2A, upper panel, the GST-fusion proteins of the GAT domains of all GGAs efficiently pulled down wild-type Arf1 in the presence of GTPγS, but with very low efficiency in the presence of GDP. In contrast to wild-type Arf1, the pull-down efficiencies of Arf1(Q71L) were much higher,

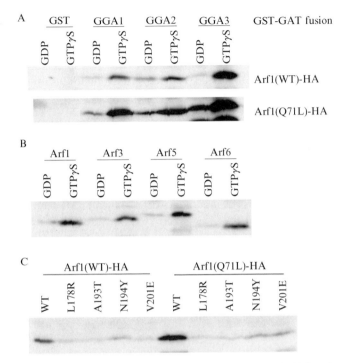

FIG. 2. Pull-down assays for interactions between Arfs and the GAT domains. (A) Arf1
(WT)-HA or Arf1(Q71L)-HA in the lysates of HEK293 cells was subjected to pull down with
GST fused to the GAT domain of GGA1, GGA2, or GGA3 in the presence of GDP or
GTPγS. (B) C-terminally HA-tagged Arf1, Arf3, Arf5, or Arf6 in HEK293 cell lysates was
subjected to pull down with the GST fusion of the GGA1-GAT domain in the presence of
GDP or GTPγS. (C) Arf1(WT)-HA or Arf1(Q71L)-HA in the lysates of HEK293 cells was
subjected to pull down with the GST fusion of the GGA1-GAT domain with an indicated
amino acid substitution in the presence of GTPγS. Reprinted from *Biochemical Journal*
(Takatsu *et al.*, 2002) with the permission of The Biochemical Society.

both in the presence of GDP and in the presence of GTPγS (Fig. 2A,
lower panel). These results indicate that the GTP-bound active, but not the
GDP-bound inactive, form of Arf1 binds the GAT domain.

We also examined the specificity of the GGA1-GAT domain to Arf
isoforms. As shown in Fig. 2B, the GGA1-GAT domain efficiently pulled
down Arfs of all classes (class I, Arf1 and Arf3; class II, Arf5; and class III,
Arf6) in the presence of GTPγS.

These data obtained by two-hybrid and pull-down experiments consistently indicate that the GAT domains from all GGAs interact with GTP-
bound Arfs, and that the GAT domain interacts with all classes of Arfs
with comparable efficiencies.

The Arf pull-down assay is analogous to the assays for selective detection of GTP-bound Rac and Cdc42 with the CRIB domain of PAK and for that of GTP-bound Rho with the Rho-binding domain of Rhotekin (Benard and Bokoch, 2002; Ren *et al.*, 1999), and is now widely used for detection of Arf activation in the cells (Santy and Casanova, 2001; Shinotsuka *et al.*, 2002; Shin *et al.*, 2004, 2005).

Isolation of GAT Domain Mutants Defective in Binding to Arfs

Error-Prone PCR and Reverse Two-Hybrid Screening

We isolated GAT domain mutants defective in Arf binding by random mutagenesis of the cDNA fragment for the GGA1-GAT domain by error-prone PCR and subsequent reverse two-hybrid screening of GAT clones that cannot interact with Arfs.

1. A cDNA fragment for the GAT domain of GGA1 is subjected to PCR with *Taq* DNA polymerase in the presence of 50–100 μM MnCl$_2$.

2. A pool of the randomly mutagenized cDNA fragments is subcloned into the pGAD10 vector and transformed into Y190 yeast cells harboring the pGBT9 vector for Arf1(Q71L).

3. The transformed cells are plated on medium lacking tryptophan and leucine, and subjected to a filter assay for β-galactosidase activity.

4. Colonies that develop pale blue color or do not develop blue color by 18-h incubation with 5-bromo-4-chloro-3-indolyl-β-D-galactopyranoside, are picked up and subjected to colony PCR and sequence analysis.

Note 1: Increasing the concentration of MnCl$_2$ in the reaction mixture reduces the fidelity of PCR; namely, increases the incidence of mutation (Cadwell and Joyce, 1992; Yahara *et al.*, 2001).

Note 2: Although the AH109 yeast strain is generally used as a host for two-hybrid screening, we routinely use the Y190 strain for reverse two-hybrid screening, because the latter strain develops brighter blue color than the former does.

Whereas two-hybrid screening is utilized to identify interactions between two polypeptide molecules, reverse two-hybrid screening is to detect loss of the interactions. By this screening, it is possible to determine amino acid residues of given polypeptides that are crucial for particular protein-protein interactions. To isolate GAT domain mutants that lack interaction with Arf, we took advantage of reverse two-hybrid screening by a procedure as described above. Sequence analysis of plasmid clones isolated from colonies that developed pale blue color or did not develop blue color revealed that the insert cDNAs have frame shift mutations, single missense mutations, or double or triple mutations. The single missense mutations (L178R, A193T, N194Y, and V201E) thus identified (Fig. 1B, streaks 14–17) were found to be restricted in a subregion of the GGA1-GAT domain, where residues are highly conserved in human and yeast GGAs; among the four mutations, three (L178R, A193T, and N194Y) are at residues conserved in all the human and yeast GGAs. The two-hybrid data were confirmed by pull-down assays (Fig. 2C); even in the presence of GTPγS, Arf1 was pulled down at extremely low efficiencies with the mutants of the GGA1-GAT domain. These results indicate that these residues play a crucial role in Arf binding. A following X-ray crystallographic study revealed a mode of the interaction between the GAT domain and Arf, and demonstrated the importance of these residues (Shiba *et al.*, 2003).

Correlation Between Arf Binding and Golgi Association of GGAs

Plasmid Construction

For expression of N-terminally HA-tagged GGA and the GAT domain, the cDNA fragments were separately subcloned into the pcDNA3-HAN vector (Shin *et al.*, 1997). An expression vector for C-terminally Myc-tagged Arf1(Q71L) was described previously (Takatsu *et al.*, 2002).

DNA Transfection and Immunofluorescence Analysis

HeLa cells grown in wells of 8-well Lab-Tek-II chamber slides (Nunc A/S, Roskilde, Denmark) are transfected with the expression vector for HA-GGA alone or in combination with that for Arf1(Q71L)-Myc using the FuGENE6 reagent, incubated for 10–16 h, and processed for indirect immunofluorescence analysis. The fixed cells are incubated sequentially with a combination of monoclonal rat anti-HA antibody and either monoclonal mouse anti-γ-adaptin antibody (100.3; Sigma, St. Louis, MO) or monoclonal mouse anti-Myc antibody (9E10; Roche Diagnostics), and with a combination of Cy3-conjugated anti-rat IgG (Jackson ImmunoResearch Laboratories, West Grove, PA) and Alexa488-conjugated anti-mouse IgG (Molecular Probes, Eugene, OR).

Arfs have been shown to be responsible for membrane recruitment of coat protein complexes, including the COPI and AP-1 complexes (Nie *et al.*, 2003; Shin and Nakayama, 2004). To examine whether interaction of Arf and the GAT domain is required for recruitment of GGAs onto TGN membranes, we performed the following experiments. When

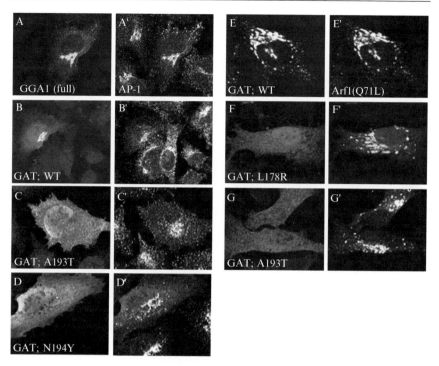

FIG. 3. Localization of GGA1 and its GAT domain, and GAT domain mutants defective in Arf binding. (A–D) HeLa cells expressing HA-tagged full-length GGA1, the wild-type GGA1-GAT domain, and its A193T and N194Y mutants were double-stained with antibodies to HA (A–D) and γ-adaptin (A′–D′), a subunit of the AP-1 clathrin adaptor complex. (E–G) HeLa cells expressing the wild-type GGA1-GAT domain tagged with HA, or its L178R or A193T mutant, in combination with Arf1(Q71L)-Myc were double-stained with anti-HA and anti-Myc antibodies. Reprinted from *Biochemical Journal* (Takatsu *et al.*, 2002) with the permission of The Biochemical Society.

expressed in HeLa cells, full-length GGA1 (Fig. 3A), GGA2, and GGA3 (not shown), were localized on perinuclear structures positive for AP-1, indicating that they associate with TGN membranes. Furthermore, the GAT domains alone of GGA1 (Fig. 3B and E), GGA2, and GGA3 (not shown) are sufficient for association with the TGN. In striking contrast, the GAT domain mutants defective in Arf binding were largely cytoplasmic (Fig. 3C and D) even when Arf1(Q71L) was coexpressed (Fig. 3F and G). These observations indicate that the GAT domains are responsible for the recruitment of GGAs onto TGN membranes through interacting with GTP-bound Arf.

Acknowledgments

We thank Hye-Won Shin for critical reading of the manuscript. This work was supported in part by grants from the Japan Society for Promotion of Science, from the Ministry of Education, Culture, Sports, Science and Technology of Japan, from the Protein 3000 Project, from the Naito Foundation, and from the Takeda Science Foundation.

References

Benard, V., and Bokoch, G. M. (2002). Assay of Cdc42, Rac, and Rho GTPase activation by affinity methods. *Methods Enzymol.* **345,** 349–359.

Boman, A. L., Zhang, C.-J., Zhu, X., and Kahn, R. A. (2000). A family of ADP-ribosylation factor effectors that can alter membrane transport through the *trans*-Golgi. *Mol. Biol. Cell* **11,** 1241–1255.

Bonifacino, J. S. (2004). The GGA proteins: Adaptors on the move. *Nat. Rev. Mol. Cell Biol.* **5,** 23–32.

Cadwell, R. C., and Joyce, G. F. (1992). Randomization of genes by PCR mutagenesis. *PCR Methods Appl.* **2,** 28–33.

Donaldson, J. G. (2003). Multiple roles for Arf6: Sorting, structuring, and signaling at the plasma membrane. *J. Biol. Chem.* **278,** 41573–41576.

Hosaka, M., Toda, K., Takatsu, H., Torii, S., Murakami, K., and Nakayama, K. (1996). Structure and intracellular localization of mouse ADP-ribosylation factors type 1 to type 6 (ARF1–ARF6). *J. Biochem.* **120,** 813–819.

Kuai, J., Boman, A. L., Arnold, R. S., Zhu, X., and Kahn, R. A. (2000). Effects of activated ADP ribosylation factors on Golgi morphology require neither activation of phospholipase D1 nor recruitment of coatomer. *J. Biol. Chem.* **275,** 4022–4032.

Moss, J., and Vaughan, M. (1995). Structure and function of ARF proteins: Activators of cholera toxin and critical components of intracellular vesiclar transport processes. *J. Biol. Chem.* **270,** 12327–12330.

Nakayama, K., and Wakatsuki, S. (2003). The structure and function of GGAs, the traffic controllers at the TGN sorting crossroads. *Cell Struct. Funct.* **28,** 431–442.

Nie, Z., Hirsch, D. S., and Randazzo, P. A. (2003). Arf and its many interactors. *Curr. Opin. Cell Biol.* **15,** 396–404.

Ren, X.-D., Kiosses, W. B., and Schwartz, M. A. (1999). Regulation of the small GTP-binding protein Rho by cell adhesion and cytoskeleton. *EMBO J.* **18,** 578–585.

Santy, L. C., and Casanova, J. E. (2001). Activation of ARF6 by ARNO stimulates epithelial cell migration through downstream acticavion of both Rac1 and phospholipase D. *J. Cell Biol.* **154,** 599–610.

Shiba, T., Kawasaki, M., Takatsu, H., Nogi, T., Matsugaki, N., Igarashi, N., Suzuki, M., Kato, R., Nakayama, K., and Wakatsuki, S. (2003). Molecular mechanism of membrane recruitment of GGA by ARF in lysosomal protein transport. *Nat. Struct. Biol.* **10,** 386–393.

Shin, H.-W., Morinaga, N., Noda, M., and Nakayama, K. (2004). BIG2, a guanine nucleotide exchange factor for ADP-ribosylation factors: Its localization to recycling endosomes and implication in the endosome integrity. *Mol. Biol. Cell* **15,** 5283–5294.

Shin, H.-W., and Nakayama, K. (2004). Guanine nucleotide exchange factors for Arf GTPases: Their diverse functions in membrane traffic. *J. Biochem.* **136,** 761–767.

Shin, H.-W., Shinotsuka, C., and Nakayama, K. (2005). Expression of BIG2 and analysis of its function in mammalian cells. *Methods Enzymol.* **404,** 206–215.

Shin, H.-W., Shinotsuka, C., Torii, S., Murakami, K., and Nakayama, K. (1997). Identification and subcellular localization of a novel mammalian dynamin-related protein homologous to yeast Vps1p and Dnm1p. *J. Biochem.* **122,** 525–530.

Shinotsuka, C., Yoshida, Y., Kawamoto, K., Takatsu, H., and Nakayama, K. (2002). Overexpression of an ADP-ribosylation factor-guanine nucleotide exchange factor, BIG2, uncouples brefeldin A-induced adaptor protein-1 coat dissociation and membrane tubulation. *J. Biol. Chem.* **277,** 9468–9473.

Takatsu, H., Yoshino, K., and Nakayama, K. (2000). Adaptor γ ear homology domain conserved in γ-adaptin and GGA proteins that interact with γ-synergin. *Biochem. Biophys. Res. Commun.* **271,** 719–725.

Takatsu, H., Yoshino, K., Toda, K., and Nakayama, K. (2002). GGA proteins associate with Golgi membranes through interaction between their GGAH domains and ADP-ribosylation factors. *Biochem. J.* **365,** 369–378.

Toda, K., Nogami, M., Murakami, K., Kanaho, Y., and Nakayama, K. (1999). Colocalization of phospholipase D1 and GTP-binding-defective mutant of ADP-ribosylation factor 6 to endosomes and lysosomes. *FEBS Lett.* **442,** 221–225.

Welsh, C. F., Moss, J., and Vaughan, M. (1994). ADP-ribosylation factors: A family of ~20-kDa guanine nucleotide-binding proteins that activate cholera toxin. *Mol. Cell. Biochem.* **138,** 157–166.

Yahara, N., Ueda, T., Sato, K., and Nakano, A. (2001). Multiple roles of Arf1 GTPase in the yeast exocytic and endocytic pathways. *Mol. Biol. Cell* **12,** 221–238.

[33]   Arf6 Modulates the β-Actin Specific Capping Protein, βcap73

By Alice Y. Welch,* Kathleen N. Riley,*
Crislyn D'Souza-Schorey, and Ira M. Herman[†]

Abstract

Recent work from our laboratory has revealed that isoactin cytoskeletal and membrane dynamics are coordinately regulated. In this chapter, we review some of the recent and relevant scientific literature focusing on key aspects of cytoskeletal and membrane-mediated signal transduction. Additionally, we highlight some of the strategic molecular, biochemical, and cell-based methodologies that we have either developed or implemented in our efforts aimed at revealing the pivotal role(s) that the actin isoforms play in controlling cell shape and motility during developmental and/or disease-associated events. Furthermore, we address the central position of β-actin and its barbed end-specific capping protein, βcap73, in modulating

*These authors contributed equally.
[†]To whom correspondence should be addressed: ira.herman@tufts.edu.

METHODS IN ENZYMOLOGY, VOL. 404   0076-6879/05 $35.00

nonmuscle cell membrane dynamics and cell migration. In studying the molecular mechanisms mediating these cytoskeletal protein interactions, we have recently recognized that cell motility and β-actin dynamics are controlled by the direct association of βcap73 with the plasma membrane- and endosome-associated protein, ADP-ribosylation factor 6 (Arf6).

Introduction

Revealing the molecular regulators and signaling pathways that coordinate cell shape and motility, whether during developmental or human disease-related phenomena, continues to be a central focus in biomedical research today. At the same time, we are interested in understanding the mechanisms regulating actin cytoskeletal remodeling, seeking to divulge the manner in which membrane dynamics might modulate actin-mediated motility. Indeed, recent work in the field and work carried out in our laboratory indicate that membrane and actin cytoskeletal dynamics are exquisitely synchronized to control the timely delivery and subcellular targeting of isoactin-specific and membrane-associated protein regulators, facilitating such developmentally diverse phenomena as cell cycle events during tissue morphogenesis or cancer cell invasion during tumor metastasis.

Of the isoactins, nonmuscle β-actin is the one that is targeted to and retained at the plasma membrane when cells undergo morphological changes. Disrupting β-actin localization or increasing the cellular γ-actin:β-actin ratio decreases spreading and inhibits the formation of motile cell structures. It is the regulated β-actin assembly at the plasma membrane that contributes to cell motility. Although many actin binding proteins (ABPs) provide spatial and temporal control of actin dynamics in cells by regulating every aspect of actin dynamics, including polymerization–depolymerization rates, nucleotide exchange, sequestration of monomers, filament end blocking, filament cross-linking, and membrane associations (Borisy and Svitkina, 2000; Cooper and Schafer, 2000; Pollard and Borisy, 2003), to date, βcap73 is the only ABP identified that binds specifically to β-actin.

Originally isolated from microvascular pericytes (Shuster, 1995; Shuster et al., 1996), which produce elaborate membrane structures, βcap73 binds directly and specifically to β-actin filaments. In vitro, either bacterially expressed and purified βcap73, or the endogenous protein isolated from mammalian cells, regulates actin assembly in an isoform-specific fashion. Indeed, when recombinant βcap73 is added to isoactin assembly reactions, βcap73 clearly inhibits β-actin assembly whereas α-actin assembly is unaffected (Fig. 1). Control experiments using gelsolin and CapZ show that α-actin assembly can be inhibited by these non-isoactin-specific capping proteins, but not by β-actin-specific βcap73. Additionally, βcap73 directly interacts with the cytoskeleton-membrane linker protein, ezrin,

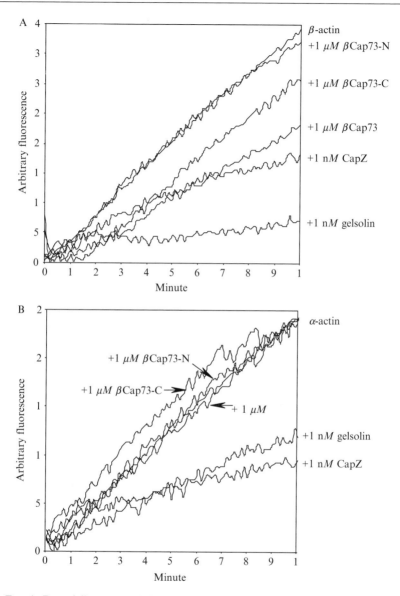

FIG. 1. Bacterially expressed βcap73 inhibits β-actin assembly. (A) β-actin assembly reactions; (B) α-actin assembly reactions; βcap73-N, N-terminus. βcap73-C, C-terminus, which contains the β-actin-binding domain). Purified gelsolin and CapZ are positive controls for capping activity. (1 mM actin [50% labeled] is seeded with sheared F-actin of the appropriate isoform. Assembly is initiated with 2 mM MgCl₂ and 100 mM KCl. βCAP73 and seeds are added just before initiating the reactions at 25°.) Interestingly, bacterially expressed βcap73 inhibits β-actin assembly less effectively than βcap73, which is purified to homogeneity from living, mammalian cells and used in comparable isoactin capping assays (Welch and Herman, 2002).

and colocalizes with ezrin in the leading edge of motile cells. The simultaneous binding of βcap73 to both ezrin and β-actin barbed ends ultimately sequesters β-actin at the plasma membrane; and, because βcap73 is a barbed end binding protein, it is well positioned to control β-actin-specific assembly. Indeed, overexpression of βcap73 in bovine retinal pericytes causes these cells to become rounder and smaller, indicating that excessive capping of β-actin barbed ends limits cell spreading. Tightly controlled β-actin assembly is one requirement for membrane morphology to change appropriately in response to extracellular signals.

Cells receive motility cues from a variety of sources, such as the extracellular matrix, soluble growth factors, or loss of cell-cell contact (Lauffenburger, 1996; Stossel, 1993). Signals received at the cell surface are transduced through the plasma membrane to the cytoplasm, where they are amplified via specific kinases and their downstream effectors. A large body of research shows that the Rho family of GTPases, which includes Rho, Rac, and CDC42, relay motility signals to the actin cytoskeleton. Of key interest for this signaling is activation of the Rho GTPases (reviewed in Aspenstrom, 1999b; Shuster and Herman, 1998). Different classes of cell surface receptors signal to the Rho GTPases, including tyrosine kinase receptors, G-protein-coupled receptors, and adhesion receptors (Kjoller and Hall, 1999). The effects of Rho family GTPases on formation of actin-based structures has been examined primarily by overexpression, where overexpression of constitutively activated Rho, Rac, and CDC42 stimulate the formation of stress fibers, ruffles, and filopodia, respectively (Ridley et al., 1999). The difference in the structures generated by the different Rho members indicates that these proteins signal to effectors that speak to different subsets of ABPs in order to assemble and organize actin into these structures (Aspenstrom, 1999a). Perhaps most interesting, recent work from our laboratory has revealed that Rho GTPase signals through the actin cytoskeleton in an isoform-specific fashion (Kolyada et al., 2003).

An important downstream effector of Rho GTPases is PI-5 kinase, which synthesizes PIP_2 from PI-4-phosphate. This PIP_2 generation regulates the actin binding activities of profilin (Lassing and Lindberg, 1985), capping protein (Schafer et al., 1996), and gelsolin (Weeds and Maciver, 1993), leading to changes in actin dynamics and network architecture. Interestingly, a more recently identified downstream effector of the Rho GTPases also contributes to actin regulation via PIP2 generation; this downstream effector is the ADP-ribosylation factor, Arf6.

An important molecular link between the actin cytoskeleton and the intracellular membrane pool is Arf6, a small GTPase of the Ras superfamily. During cell migration in response to injury, Arf6 helps to regulate the delivery of early and recycling endosomes to the leading edge (Riley

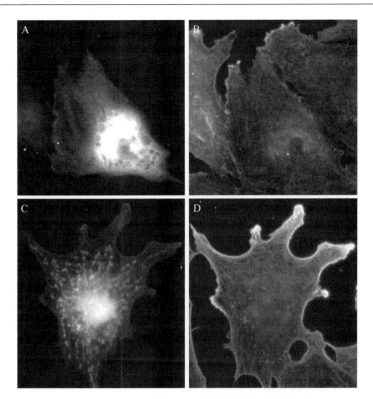

FIG. 2. β-actin and HA-Arf6 localization. Localization of HA-Arf6 (A) and (C) and endogenous β-actin (B) and (D) in capillary endothelial cells expressing HA-Arf6-Q67L (A) and (B) or HA-Arf6-T22N (C) and (D), crawling in response to injury *in vitro*.

et al., 2003). In our studies of capillary endothelial cells, we have found that transfection of the constitutively active form, Arf6-Q67L, increases the rate of wound closure in a scratch assay as well as the magnitude of individual cell spreading. This point mutant is especially concentrated at the leading edge in migrating cells. The inactive form, Arf6-T22N, causes a reduction in cell spreading and impaired leading edge formation (Fig. 2). Additionally, we have shown via co-immunoprecipitation that Arf6 interacts directly with βcap73 (Fig. 3).

Isoactin Assembly Assays

1. Skeletal muscle α-actin is purified from chicken muscle acetone powder, prepared as previously described (Herman and Pollard, 1979; Spudich and Watt, 1971).

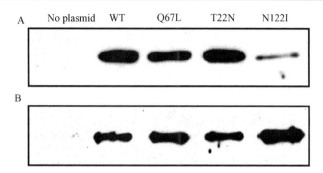

FIG. 3. Co-immunoprecipitation of βcap73 with HA-Arf6. Western blot of anti-HA immunoprecipitation from EC transfected with the indicated HA-Arf6 plasmid, probed with anti-HA (A), and anti-βcap73 (B).

2. Erythrocyte β-actin is purified from erythrocyte acetone powder prepared as previously described (Puszkin *et al.*, 1978; Shuster *et al.*, 1996) with some modifications, including:

a. Acetone powder is extracted twice on ice at 4° G buffer (2 mM Tris, pH 8.0, 0.2 mM CaCl$_2$, 0.2 mM DTT, 0.2 mM ATP, 0.02% NaN$_3$).

b. Extracts are pooled and concentrated to 1/10 the starting volume before polymerization for 1 h at room temperature (RT).

c. Following polymerization, the actin solution is brought up to 800 mM KCl and incubated 1 h at RT.

d. Polymerized actin is pelleted by centrifugation at 100,000g for 2 h at 4°, depolymerized by dialysis in G buffer, and then gel-filtered in G buffer (Sephadex G-150, Sigma, St. Louis, MO).

e. Fractions are collected and samples are analyzed for purity by 10% SDS-PAGE followed by Coomassie blue staining.

3. Actins are pyrene-labeled based on the method of Kouyama and Mihashi (Kouyama and Mihashi, 1981).

a. Pyrene-iodoacetamide (Molecular Probes, Eugene, OR) is added to the labeling reaction at a molar ratio of 1:1.125 (actin:pyrene) in the following order and final concentrations: 40 μM isoactin, 2 mM MgCl$_2$, 45 μM pyrene-iodoacetamide, 100 mM KCl. The reaction is polymerized–labeled for 90 min at RT in the dark.

b. Polymerized actin is centrifuged at 100,000g for 2 h at 10°. Pelleted pyrene-actin is depolymerized by dialysis in G buffer at 4°, then centrifuged again before being frozen as aliquots in liquid nitrogen and stored at −80°. Any actin nuclei resulting from freeze–thaw cycle are removed by centrifugation at 100,000g prior to use in actin assembly assays.

4. Barbed-end capping assays are performed as previously described (Weber *et al.*, 1987) with modifications. 1 μM (50% pyrene labeled) α- or β-actin in G buffer is seeded with 0.05–0.1 nM sheared F-actin seeds (isoactins polymerized at 30 μM are sheared through a 50 μL Hamilton syringe just prior to addition to the reaction).

   a. To test capping activity, the 1 μM isoactin solution is pre-incubated with 0.2 mM MgCl$_2$ for 3 min at 25° to allow for Ca^{+2}/Mg^{+2} exchange.

   b. At time = 0, isoactin seeds, various amounts of βcap73, 2 mM MgCl$_2$, and 100 mM KCl are added in rapid succession, mixed, and placed in a 0.5 cm square fluorimetry cuvette.

   c. Fluorescence intensity is detected on a Perkin-Elmer Luminescence Spectrometer LS 50B set at excitation = 365 nM and emission = 407 nM. The reactions polymerized at 25° in 300 μl reaction volumes.

Expression and Purification of GST-βcap73

1. GST-βcap73 plasmids are transformed into DH5α bacteria (subcloning efficiency), plated for ampicillin selection, and grown overnight at 37°.

2. Colony selection and overnight growth at 37°, while shaking, is then followed by dilution (1:100) and large-scale culture, the following day. The large-scale culture is grown in 2XYT medium at 28° to an OD$_{600}$ ~ 1.5 before induction with 0.1 mM IPTG for 2 h at 28°. After induction, the cells are pelleted and stored at −80°.

3. To purify βcap73 domains, bacterial cell pellets are resuspended in ice-cold PBS containing protease inhibitors and then lysed by sonication on ice.

   a. The particulate fraction is removed by centrifugation at 20,000g for 20 min at 4°.

   b. Glutathione-sepharose (Pharmacia) is pre-equilibrated in cold phosphate buffered saline (PBS), added as a slurry to the supernatants (1 ml of bed/100 ml of starting culture), and mixed on a rotator for 30 min at RT.

   c. The mixtures are poured into columns and washed with 30 column volumes of PBS. The protein-bound matrices are then equilibrated into Factor Xa cleavage buffer (50 mM Tris, pH 7.5, 150 mM NaCl, and 1 mM CaCl$_2$) and moved into tubes.

   d. The matrix-bound GST-βcap73 fusion proteins are treated with Factor ×a for 2 h at RT, rotating.

e. βcap73 domains relieved of the GST moiety are then collected as flow-through when the matrices were replaced into columns.

Analysis of Endothelial Cell Motility Following Injury *In Vitro*

1. Capillary endothelial cells from retina are isolated by physical separation of the retina from the posterior pole of bovine eyes.

2. Sterile transfer of pooled retinae into PBS (phosphate buffered saline) is followed by mincing and sieving through a series of Nitex membrane filters, (100, 40 μM).

3. Endothelial isolates, derived from 3–4 cell endothelial capillary islands are then subcloned, with sterile cloning rings, in DMEM (Gibco BRL, Gaithersburg, MD) supplemented with 5% bovine calf serum (Hyclone) and 0.5% each L-glutamine and penicillin-streptomycin-fungizone (Gibco).

4. Capillary-derived retinal endothelial cells are then cultured, expanded, and stored, as previously published (Herman and D'Amore, 1985; Riley *et al.*, 2003).

5. Square 1 cm^2 coverslips are placed in 24-well tissue culture plates and sterilized by UV irradiation for 30 min. Retinal endothelial cells are then plated at 100 K cells per well and cultured for 24 h in serum-containing media. Alternatively, cells are plated at 30 K cells/well in a 96-well tissue culture plate and cultured as above.

6. Transfection is carried out with the Qiagen Effectene reagent kit, according to manufacturer's instructions (0.2 μg of DNA per well in 24-well plates). The following plasmids for expression of Arf6 proteins with HA tags were used: wild-type Arf6; a dominant-negative, GTP binding-defective mutant, T27N; and a constitutively active, GTP hydrolysis-defective mutant, Q67L. Another HA-tagged Arf6 mutant, which is completely incapable of binding nucleotide, N122I, was a generous gift of Dr. Ralph Isberg (Tufts University). After 24 h in DNA/transfection reagent, cells are washed with PBS and fed with fresh media. Coverslips are used for live wound healing assays or prepared for indirect immunofluorescence.

7. Monolayer injury is created on the coverslips with a fire-polished Pasteur pipette or a single-edged razor blade. The coverslip is mounted in a specially designed culture chamber, covered with fresh media, and sealed with a clean glass coverslip. The chamber is mounted on the stage of a Zeiss Axiovert fluorescence microscope workstation (40× objective, NA = 0.75), which includes a Nevtek warm air curtain, Hamamatsu cooled-CCD camera, Sutter DG-4 fluorescence excitation source, and a PC running Universal Imaging Systems Metamorph 5.0. A target field along the wound edge is selected, and phase contrast images are captured every 5

min and combined into time-lapse video files. (Hoock *et al.*, 1991; Shuster, 1995; Young and Herman, 1985). For 96-well plates, monolayer injury is produced with a plastic 96-pin array (Yarrow *et al.*, 2004), the plate is mounted as described above, and viewed with a 10x objective.

8. Rate of cell motility is determined by analyzing multiple images from a time-lapse series using Adobe Photoshop 5.5. In each image, the space devoid of cells is measured in pixels, then converted to square microns. The difference in empty area between each pair of images is calculated, then divided by the elapsed time to give a rate of movement.

$$\frac{[(\text{empty area, image B}) - (\text{empty area, image A})]}{5 \text{ min}} = \mu\text{m}^2/\text{min}$$

Individual cell spreading is also measured in Adobe Photoshop 5.5, by selecting each cell around its border and converting this area measurement into square microns. Spreading is expressed as a percentage of the size of that cell at the first time point.

9. Two h after monolayer injury, coverslips are removed from the chamber or the culture plate, fixed with 4% formaldehyde in DMEM, and permeablized for 90 sec at room temperature with buffered Triton-X100 as described (Hoock *et al.*, 1991). Cells are then incubated in the specified primary antibodies diluted in PBS/azide, followed by the corresponding secondary antibodies. Slides are imaged using the microscope setup described above for time-lapse imaging. Primary antibodies for immuno-fluorescence are anti-βcap73 (murine monoclonal IgM) and anti-β-actin (affinity-selected, rabbit polyclonal IgG), rat monoclonal anti-HA IgG (Roche Molecular Biochemicals, Indianapolis, IN). Secondary antibodies are Alexa 488- and 546-conjugated antibodies from Molecular Probes (Eugene, OR). With the anti-βcap73 IgM, the secondary antibody is rabbit anti-mouse IgM and Alexa-labeled goat anti-rabbit IgG is the tertiary antibody.

Analysis of Protein Interaction by Co-Immunoprecipitation

1. Retinal endothelial cells are plated in 6-well tissue culture plates, 600 K cells per well.

2. Transfection is performed as described above (0.4 μg of DNA per well).

3. Cells are lysed, as previously described (Welch and Herman, 2002; Witczak *et al.*, 1999), in RIPA buffer containing 150 mM NaCl, 30 mM Tris-HCl, pH 8.0, 0.1% SDS, 0.5% sodium deoxycholate, and 1% Nonidet P-40, then incubated 10 min at room temperature in the presence of

protease inhibitors. For each sample, 10 μg of anti-HA primary antibody is incubated with 10 μl of packed protein A/Sepharose beads (Pharmacia, Piscataway, NJ) for 1 h at room temperature with gentle rotation. Meanwhile, 250 μl of lysate ($\sim$400 μg protein) is precleared with 10 μl of protein A/Sepharose for 1 h at room temperature. The precleared lysate is then incubated with the antibody-bead complex overnight at 4° with gentle rotation.

4. After overnight incubation, the beads are washed five times in RIPA buffer and one time in buffer containing 30 mM Tris-HCl, pH 8.0, and 50 mM NaCl. The beads are then boiled in 50 ml of 1× sample buffer for 3 min, and the supernatant collected for SDS-polyacrylamide gel electrophoresis (PAGE).

5. The samples are separated by electrophoresis on 1.5 mm thick, 10% acrylamide gels and transferred to nitrocellulose (Schleicher and Schuell, Keene, NH) overnight at 200 mA in a Tris-buffered methanol SDS solution. Western blotting is performed as described by Amersham. Briefly, blots are blocked with 5% nonfat dry milk in TBST (20 mM Tris-HCl, pH 7.5, 150 mM NaCl, 0.05% Tween-20) for at least 1 h at room temperature. For antigen detection, the blots are incubated with primary antibody at $\sim$5 μg/ml for 2 h at room temperature, then incubated 1 h in $\sim$0.4 μg/ml HRP-conjugated secondary antibody. Detection is performed with Supersignal Western detection reagents (Pierce, Rockford, IL). The signal is recorded using Kodak X-OMAT Blue XB-1 scientific imaging film. Films are imaged using an AGFA StudioStar flatbed scanner and Adobe Photoshop 5.0.

References

Aspenstrom, P. (1999a). Effectors for the Rho GTPases. *Curr. Opin. Cell Biol.* **11,** 95–102.
Aspenstrom, P. (1999b). The Rho GTPases have multiple effects on the actin cytoskeleton. *Exp. Cell Res.* **246,** 20–25.
Borisy, G. G., and Svitkina, T. M. (2000). Actin machinery: Pushing the envelope. *Curr. Opin. Cell Biol.* **12,** 104–112.
Cooper, J. A., and Schafer, D. A. (2000). Control of actin assembly and disassembly at filament ends. *Curr. Opin. Cell Biol.* **12,** 97–103.
Herman, I. M., and Pollard, T. D. (1979). Comparison of purified anti-actin and fluorescent-heavy meromyosin staining patterns in dividing cells. *J. Cell Biol.* **80,** 509–520.
Herman, I. M., and D'Amore, P. A. (1985). Microvascular pericytes contain muscle and nonmuscle actins. *J. Cell Biol.* **101,** 43–52.
Hoock, T. C., Newcomb, P. M., and Herman, I. M. (1991). Beta actin and its mRNA are localized at the plasma membrane and the regions of moving cytoplasm during cellular response to injury. *J. Cell Biol.* **112,** 653–664.
Kjoller, L., and Hall, A. (1999). Signaling to Rho GTPases. *Exp. Cell Res.* **253,** 166–179.

Kolyada, A. Y., Riley, K. N., and Herman, I. M. (2003). Rho GTPase signaling modulates cell shape and contractile phenotype in an isoactin-specific manner. *Am. J. Physiol. Cell Physiol.* **285,** C1116–C1121.

Kouyama, T., and Mihashi, K. (1981). Fluorimetry study of N-(1-pyrenyl)iodoacetamide-labelled F-actin: Local structural change of actin protomer both on polymerization and on binding of heavy meromyosin. *Eur. J. Biochem.* **114,** 33–38.

Lassing, I., and Lindberg, U. (1985). Specific interaction between phosphatidylinositol 4,5-bisphosphate and profilactin. *Nature* **314,** 472–474.

Lauffenburger, D. A. (1996). Cell motility. Making connections count. *Nature* **383,** 390–391.

Pollard, T. D., and Borisy, G. G. (2003). Cellular motility driven by assembly and disassembly of actin filaments. *Cell* **112,** 453–465.

Puszkin, S., Maimon, J., and Puszkin, E. (1978). Erythrocyte actin and spectrin: Interactions with muscle contractile and regulatory proteins. *Biochim. Biophys. Acta* **513,** 205–220.

Ridley, A. J., Allen, W. E., Peppelenbosch, M., and Jones, G. E. (1999). Rho family proteins and cell migration. *Biochem. Soc. Symp.* **65,** 111–123.

Riley, K. N., Maldonado, A. E., Tellier, P., D'Souza-Schorey, C., and Herman, I. M. (2003). βCap73-ARF6 interactions modulate cell shape and motility after injury *in vitro. Mol. Biol. Cell* **14,** 4155–4161.

Schafer, D. A., Jennings, P. B., and Cooper, J. A. (1996). Dynamics of capping protein and actin assembly *in vitro*: Uncapping barbed ends by polyphosphoinositides. *J. Cell Biol.* **135,** 169–179.

Shuster, C. B., and Herman, I. M. (1998). The mechanics of vascular cell motility. *Microcirculation* **5,** 239–257.

Shuster, C. B., Lin, A. Y., Nayak, R., and Herman, I. M. (1996). Beta cap73: A novel beta actin-specific binding protein. *Cell Motil. Cytoskeleton* **35,** 175–187.

Shuster, C. B., and Herman, I. M. (1995). Indirect association of ezrin with F-actin: Isoform specificity and calcium sensitivity. *J. Cell Biol.* **128,** 837–848.

Spudich, J. A., and Watt, S. (1971). The regulation of rabbit skeletal muscle contraction. I. Biochemical studies of the interaction of the tropomyosin-troponin complex with actin and the proteolytic fragments of myosin. *J. Biol. Chem.* **246,** 4866–4871.

Stossel, T. P. (1993). On the crawling of animal cells. *Science* **260,** 1086–1094.

Weber, A., Northrop, J., Bishop, M. F., Ferrone, F. A., and Mooseker, M. S. (1987). Kinetics of actin elongation and depolymerization at the pointed end. *Biochemistry* **26,** 2537–2544.

Weeds, A., and Maciver, S. (1993). F-actin capping proteins. *Curr. Opin. Cell Biol.* **5,** 63–69.

Welch, A. Y., and Herman, I. M. (2002). Cloning and characterization of betaCAP73, a novel regulator of beta-actin assembly. *Int. J. Biochem. Cell Biol.* **34,** 864–881.

Witczak, O., Skalhegg, B. S., Keryer, G., Bornens, M., Tasken, K., Jahnsen, T., and Orstavik, S. (1999). Cloning and characterization of a cDNA encoding an A-kinase anchoring protein located in the centrosome, AKAP450. *EMBO J.* **18,** 1858–1868.

Yarrow, J. C., Perlman, Z. E., Westwood, N. J., and Mitchison, T. J. (2004). A high-throughput cell migration assay using scratch wound healing, a comparison of image-based readout methods. *BMC Biotechnol.* **4,** 21.

Young, W. C., and Herman, I. M. (1985). Extracellular matrix modulation of endothelial cell shape and motility following injury *in vitro. J. Cell Sci.* **73,** 19–32.

[34]   Functional Assay of Effectors of ADP Ribosylation Factor 6 During Clathrin/AP-2 Coat Recruitment to Membranes

By Michael Krauss and Volker Haucke

Abstract

Arf proteins play pivotal roles in membrane traffic, cell signaling, and actin cytoskeletal rearrangements. We describe here methods to functionally analyze interacting partner proteins of recombinantly produced N-myristoylated Arf6. Combined evidence from affinity purification and chemical crosslinking experiments, *in vitro* recruitment assays, and the analysis of lipid kinase activities indicates that Arf6-GTP facilitates clathrin/AP-2 recruitment to synaptic membranes by direct binding and activation of the brain-specific phosphatidylinositol 4-phosphate 5-kinase type Iγ (PIPKIγ). These methods shall help to mechanistically dissect the role of Arf6 in regulating exo-endocytic vesicle cycling at synapses and in related membrane trafficking events.

Introduction

The ADP-ribosylation factors (Arfs) constitute a family of Ras-related low-molecular weight GTP-binding proteins that are found in virtually all eukaryotic cells. Six mammalian Arfs and a number of Arf-like proteins have been identified, with the most extensively studied ones being Arf1 and Arf6. Like all small GTPases Arf proteins cycle between a GTP-bound active and a GDP-bound inactive state. The association of activated Arfs with effector molecules is believed to mediate a number of physiological functions, including the recruitment of COPI, GGA1-3, AP-1, and AP_3 coat proteins onto membranes, the activation of lipid-modifying enzymes, and changes in cell morphology by modulation of the actin cytoskeleton (Donaldson and Jackson, 2000; Nie *et al.*, 2003; Randazzo *et al.*, 2000). Arf proteins have been implicated in budding events from the Golgi complex and from peripheral membranes, based on their ability to directly interact with coat proteins or by stimulating phospholipase D activity (Donaldson and Jackson, 2000; Nie *et al.*, 2003; Randazzo *et al.*, 2000). In addition, Arf proteins have been shown to associate with and activate phosphatidylinositol (PI) kinases, thereby stimulating the synthesis of PI 4-phosphate (PIP) and PI 4,5-bisphosphate (PIP$_2$) (De Matteis and Godi, 2004). In turn, the

METHODS IN ENZYMOLOGY, VOL. 404   0076-6879/05 $35.00
DOI: 10.1016/S0076-6879(05)04034-6

presence of PIP has been shown to be essential for the recruitment of AP-1 to the Golgi membrane (Wang *et al.*, 2003), and PIP_2 has been found to trigger association of AP-2, epsin, CALM/AP180, HIP1, Dab2, and dynamin with the plasma membrane (McMahon and Mills, 2004; Owen *et al.*, 2004; Wenk and De Camilli, 2004). Whereas a role of Arf1 in coat recruitment to the Golgi complex and the *trans*-Golgi network (TGN) is clearly established, the involvement of an Arf GTPase during clathrin/AP-2 coat recruitment to the plasma membrane has been less clear. Morphometric and biochemical studies in which GTPγS, a nonhydrolyzable analogue of GTP, potently stimulated the formation of clathrin/AP-2-coated structures on synaptic membranes (Takei *et al.*, 1996) have implicated a function for a yet unidentified GTPase during this process. The only Arf protein that has been shown to localize to the plasma membrane is Arf6 (Donaldson, 2003; Randazzo *et al.*, 2000). Furthermore, Arf6 has been suggested to regulate clathrin-mediated endocytosis from the apical (Altschuler *et al.*, 1999) and the basolateral (Palacios *et al.*, 2002) membrane surfaces of MDCK cells. In addition, a synaptic Arf6-specific guanine exchange factor (GEF), mSec7, has been shown to increase the rate of presynaptic vesicle cycling (Ashery *et al.*, 1999). Together, these findings prompted us to investigate a potential role for Arf6 during clathrin/AP-2 coat recruitment. Here we describe the development of a cell-free system to analyze the function and downstream effectors of recombinant N-myristoylated Arf6. Active recombinant N-myristoylated Arf proteins fused to a carboxyl terminal His_6-tag can be purified by a simple, one-step procedure from *E. coli* coexpressing N-myristoyl transferase (Fig. 1A). Recombinant myr-Arf1 (Q71L) (not shown) or myr-Arf6 (Q67L) are both active with regard to their ability to bind to the GAT domain of GGA proteins (Fig. 1B) or to AP-1 adaptor complexes (Fig. 3B). When incubated with lysed synaptosomal membranes, constitutively active myr-Arf6 (Q67L)-GTP stimulates membrane recruitment of clathrin, AP-2, and AP180 *in vitro* (Fig. 2). Affinity purification (Fig. 3) in combination with chemical cross-linking of immobilized Arf6-GTP-bound proteins has allowed us to track down the brain-specific isoform of type I PI 4-phosphate 5-kinase (PIPKIγ) (Ishihara *et al.*, 1998) as an Arf6 effector protein. This kinase has been shown to be essential for the generation of PIP_2 in brain (Wenk *et al.*, 2001) and to regulate the rates of both exo- (Milosevic *et al.*, 2005) and endocytosis (Di Paolo *et al.*, 2004). In contrast to membrane traffic at the Golgi, clathrin or plasmalemmal endocytic adaptor proteins do not appear to bind to activated Arf6. In order to further characterize the interaction between Arf6 and PIPKIγ, we have employed biochemical methods as well as enzymatic lipid kinase assays described in Krauss *et al.* (2003). The availability of recombinant myristoylated Arf proteins and mutants thereof will hopefully provide a useful tool

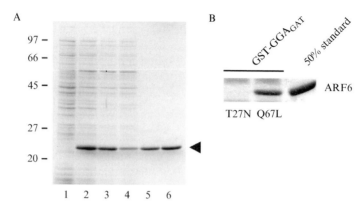

FIG. 1. Purification of recombinant myristoylated ARF6. (A) Samples of each purification step were analyzed by Coomassie blue staining after SDS-PAGE. Lane 1, total cell extract of uninduced cells; lane 2, total cell extract of cells induced with IPTG; lane 3, detergent-extract of lysed bacteria; lane 4, unbound proteins after adsorption to Ni-NTA-agarose; lane 5, proteins adsorbed to Ni-NTA-agarose; lane 6, proteins eluted from Ni-NTA-agarose by treatment with imidazole. Molecular weight markers (in kD) are shown on the left. The arrowhead indicates the molecular weight of recombinant ARF6-His6. (B) Purified ARF6 mutants were monitored for their ability to associate with a GST-GGA$_{GAT}$ affinity matrix. Proteins retained by the matrix were analyzed by SDS-PAGE and Coomassie blue staining.

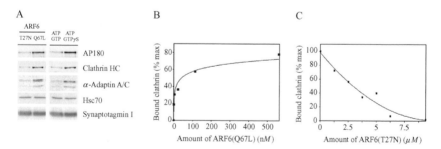

FIG. 2. Activated ARF6 interacts with PIPKIγ in brain. (A) PIPKIγ specifically interacts with ARF6(Q67L). Immunoblot analysis of PIPKIγ affinity-purified with myristoylated His6tagged ARF6(Q67L), ARF1(Q71L), or Arfaptin 2. Std., 5% of the extract used for affinity purification. (B) PIPKIγ can be cross-linked to ARF6(Q67L) during recruitment of clathrin/AP-2 to synaptic membranes. Cross-linked proteins were analyzed by immunoblotting. *Top,* Immunoblot analysis. *Bottom,* Coomassie stained gel demonstrating that equal amounts of ARF6 have been recovered in each sample. Reproduced from *The Journal of Cell Biology,* 2003, 113–124, by copyright permission of The Rockefeller University Press.

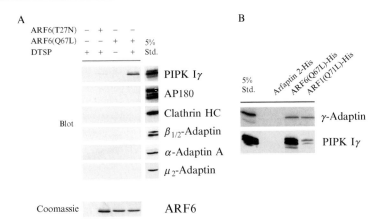

FIG. 3. ARF6-GTP stimulates clathrin/ AP-2 recruitment to synaptic membranes. (A) Coat recruitment to LP2 membranes was performed in presence of 200 μM GTP or GTPγS as indicated. All samples were analyzed by quantitative immunoblotting. (B) Dose-dependence of the stimulatory effect of ARF6(Q67L) on clathrin recruitment to membranes as shown in (A). Values were normalized to the amount of clathrin bound in presence of ATP and GTPγS (100%). (C) Dose-dependence of the inhibitory effect of ARF6(T27N) on clathrin recruitment to membranes as shown in (A). Values were normalized to the amount of clathrin bound in presence of ATP and GTPγS (100%). Reproduced from *The Journal of Cell Biology*, 2003, 113–124, by copyright permission of The Rockefeller University Press.

for the further analysis of Arf-dependent membrane traffic and signaling pathways *in vitro*, and to unravel the cascades that regulate the Arf GTPase cycle at synapses and elsewhere.

Methods

Purification of recombinant myristoylated Arf1 and Arf6 mutants

Human Arf6 cDNA (kind gift of Julie Donaldson), or human Arf1 cDNA (kind gift of Dennis Shields) were subcloned into a pET21 vector (Novagen, Madison, WI) encoding a hexa-histidine (His)-tag fused to the carboxyl terminus of Arf1 or Arf6. To allow for amino terminal myristoylation of the recombinant Arf proteins *E. coli* (BL21) cells were cotransformed with pET21b-Arf1/6 (Amp^R) and pBB131 encoding yeast myristoyl transferase (Kan^R) (Duronio *et al.*, 1990). Cells were grown in 1 l of 2×YT-ampicillin/kanamycin medium at 30° to an optical density of 0.8. Then, the medium was supplemented with 200 μM of myristic acid (dissolved as an 80 mM stock in ethanol) and cells were grown for additional 15 min. The expression of recombinant proteins was induced by addition of 0.5 mM isopropylthiogalactoside (IPTG) and subsequent incubation for additional 6 h at 25°. The bacteria were harvested by centrifugation for 15 min at 2500g and resuspended in 40 ml of ice-cold PBS containing 1 mM PMSF and 2 mM MgCl$_2$ (PBS-MgCl$_2$). Cells were lysed by incubation with a tip of spatula of lysozyme for 15 min on ice followed by sonification (30 sec, 50% duty cycle, 70% power; Bandelin, Germany) and subsequent treatment with 1% Triton-X-100 for 10 min on ice. The lysate was centrifuged for 15 min at 30,000g and the supernatant was rotated for 2–3 h with 600 μl of a 1:1 slurry of Ni-NTA-agarose (Quiagen, Valencia, CA) prewashed in PBS at 4°. To increase the specificity of binding, the mixture was supplemented with 10 mM imidazole. After adsorption the resin was washed three times for 10 min with PBS-MgCl$_2$ containing 10 mM imidazole and once for 5 min with PBS-MgCl$_2$ containing 20 mM imidazole. Resin-bound recombinant Arf1 or 6 were used immediately after purification as affinity matrix to identify binding partners from rat brain extracts (see following). For elution of recombinant myristoylated Arf1 or 6 beads were resuspended in PBS-MgCl$_2$ supplemented with 250 mM imidazole, 5 mM MgCl$_2$, 5 mM GDP or GTP, and 50% glycerol and incubated for 45 min. The eluates were cleared by centrifugation for 15 min at 30,000g and frozen in aliquots in liquid N$_2$ and stored at −80° until use for recruitment experiments. As exemplified for the purification of Arf6(Q67L)-His$_6$ (Fig. 1A) mutant Arf proteins are expressed in high amounts in bacteria and can be prepared to about 90% purity (as estimated by Coomassie blue staining of protein on SDS-PAGE gels). Typically, 4–15 mg of eluted recombinant Arf can be gained from 1 l of bacterial culture. The eluted myr-Arf1 or myr-Arf6 proteins (GDP-, or GTP-locked mutants) were then assayed for their ability to bind to the GST-fused GAT domain of mouse GGA1 (kind gift of Peter Schu, University of Göttingen, Freiburg, Germany). To this aim, the GAT domain of GGA1 was subcloned into pGEX5T-1 (Amersham Biosciences), and the GST-fusion protein was purified according to the manufacturer's instructions. 100 μg of resin-bound GST-GGA$_{GAT}$ was incubated with 30 μg of myr-Arf1 or myr-Arf6 mutants for 1.5 h at 4° in 100 mM KCl, 2 mM MgCl$_2$, 1% Triton-X-100, and 10% glycerol in 20 mM HEPES pH 7.4. The resin was then recovered by centrifugation for 5 min at 2500g and subsequently washed four times for 5 min at 4°. Finally, the beads were washed once in the same buffer and extracted by boiling in 120 μl Laemmli buffer. Thirty μl of the extract were separated by SDS-PAGE and the eluted proteins were analyzed by Coomassie blue staining. As illustrated in Fig. 1B, about 25% of the constitutively active Arf6 or Arf1 mutant proteins added (data not shown) bound to the GST-GGA$_{GAT}$ resin. By contrast, the inactive mutants did not associate with the GAT domain.

Preparation of Rat Brain Cytosol and LP2 Membranes for Recruitment Experiments

For preparation of brain cytosol, 10 rat brains were homogenized in 100 ml of breaking buffer (500 mM KCl, 10 mM MgCl$_2$, 250 mM sucrose, 25 mM Tris-HCl, pH 8, 1 mM DTT, 2 mM EGTA, and 1 mM PMSF) using a glass-Teflon homogenizer. The homogenate was centrifuged at 9000g for 60 min. The resulting supernatant was further centrifuged for 2 h at 184,000g. The high-speed supernatant was dialyzed against 50 mM KCl, 25 mM Tris-HCl pH 8.0, 1 mM DTT (dialysis buffer) to remove endogenous nucleotides. The dialyzed material was centrifuged (2 h, 184,000g) and the supernatant subsequently precipitated in 60% ammonium sulfate. The precipitate was resuspended in 10 ml dialysis buffer using a glass-Teflon homogenizer and dialyzed once more as described above. Finally, the dialyzed solution was clarified by centrifugation (1.5 h, 100,000g). The supernatant constituting rat brain cytosol, which had a typical protein concentration of 10 mg/ml, was shock-frozen in liquid nitrogen and stored in aliquots at $-80°$ until use.

LP2 membranes were prepared from lysed synaptosomes as described previously (Huttner *et al.*, 1983). Briefly, 14 rat brains were homogenized with a glass-Teflon homogenizer in 150 ml of 320 mM sucrose buffered with 4 mM HEPES, pH 7.4. The homogenate was centrifuged for 10 min at 1000g. The resulting supernatant was centrifuged further at 9000g. The pellet containing synaptosomes was washed by resuspension in 120 ml of buffered sucrose and subsequent centrifugation for 15 min at 10,000g. The synaptosomal pellet was resuspended in 13 ml of buffered sucrose, rapidly diluted into a 10-fold excess of ice-cold water, and homogenized in a glass-Teflon homogenizer (five strokes at 2000 rpm). The homogenate was immediately supplemented with 20 mM HEPES-NaOH, pH 7.4 and left on ice for 30 min. The suspension was centrifuged for 20 min at 25,000g. The supernatant was collected and subjected to ultracentrifugation (for 2 h at 165,000g). The pellet was resuspended in 6 ml of cytosolic buffer (25 mM HEPES-KOH, pH 7.2, 25 mM KCl, 2.5 mM magnesium acetate, and 150 mM potassium glutamate), homogenized in a glass-Teflon homogenizer (ten strokes at 1000 rpm) and forced five times through a 27-gauge needle. The resulting suspension constituting LP2 membranes typically had a protein content of 2 mg/ml. We observed that significant amounts of the coat components AP-2, AP180, and clathrin were associated with purified LP2 membranes. In addition, these membranes contained small, but significant amounts of Arf6 (Krauss *et al.*, 2003). To be able to monitor *de novo* recruitment of cytosolic proteins and to remove prebound Arf6, LP2 membranes were washed by resupending them in 0.1 M Na$_2$CO$_3$, pH 9.5

at a concentration of 0.5 mg/ml with a glass-Teflon homogenizer (three strokes by hand) in a total volume of 1 ml. The membranes were incubated on ice for 15 min and subsequently recovered by centrifugation at 90,000g in a Beckman TLA 120.2 rotor. The pellet was rinsed once carefully with 500 μl of cytosolic buffer and finally resuspended with a glass-Teflon homogenizer (three strokes by hand) in 500 μl of cytosolic buffer. We confirmed by Western blotting that this washing procedure efficiently removed coat proteins and Arf6 (Krauss *et al.*, 2003).

Arf6-Mediated Recruitment of Clathrin/AP-2 Coats onto LP2 Membranes

Morphometric studies had demonstrated that the generation of clathrin/AP-2-coated buds on native synaptic membranes is potently stimulated in the presence of ATP and GTPγS (Takei *et al.*, 1996). We have established an *in vitro* assay that allowed us to analyze the recruitment of clathrin/AP-2 coat proteins onto LP2 membranes biochemically and to investigate directly if Arf6 played a regulatory role in this process. Recruitment experiments were performed in 1× cytosolic buffer in a total volume of 400 μl. Each sample contained 35 μg LP2 membranes and 50 μg of rat brain cytosol. In addition, all samples were supplemented with 2 mM ATP and 200 μM GTPγS or GTP, as indicated. After incubation at 37° for 15 min, the samples were cooled on ice and then carefully loaded on top of a cushion of 0.5 M sucrose in cytosolic buffer. After centrifugation at 150,000g for 1 h at 4° in a Beckman TLA100.2 rotor, the resulting pellets were rinsed with 500 μl of cytosolic buffer to remove loosely associated cytosolic components and recentrifuged at 175,000g for 15 min. The final pellet was resuspended in 160 μl of pre-heated Laemmli sample buffer. 20–30 μl of each sample were subjected to SDS-PAGE and analyzed by Western blotting using antibodies specifically recognizing different coat components (α-adaptin A/C: clone AC1-M11, Affinity Bioreagents Inc., Golden, CO; monoclonal antibodies recognizing AP180 and clathrin heavy chain were a kind gift of Pietro de Camilli). Antibodies directed against Hsc70 (clone 3A3, Affinity Bioreagents Inc.) or synaptotagmin I (Cl41.1; kind gift of Reinhard Jahn) were used to demonstrate that equal amounts of cytosol and LP2 membranes had been present in each sample. Signals from Western blots were quantified after decoration with a rabbit anti-mouse IgG (Jackson ImmunoResearch, New Baltimore, PA) using ^{125}I-protein A (Amersham Biosciences, Freiburg, Germany) for detection and phosphoimage analysis (Image Reader 3000; Fuji, Düsseldorf, Germany). As illustrated in Fig. 2A, the concomitant presence of ATP and GTPγS, but not GTP, strongly enhances membrane recruitment of AP-2, AP180, and clathrin to synaptic LP2 membranes. These data indicate that activation of a GTPase is involved in clathrin/AP-2 coat formation at synaptic membranes. To investigate directly whether Arf6 contributes to AP-2/clathrin coat recruitment, we performed the same experiment in the presence of 1 μM recombinant myristoylated Arf6(Q67L), a mutant of Arf6 lacking GTPase activity, which is therefore locked in a constitutively active GTP-bound state. Arf6(Q67L) could stimulate recruitment of coat proteins in the presence of 200 μM GTP. The extent of stimulation seen with Arf6 (Q67L)-GTP was similar to that observed after addition of GTPγS in the absence of exogenously added recombinant protein (Fig. 2A). Three lines of evidence confirm that this effect is indeed specifically mediated by Arf6: first, a dominant negative Arf6(T27N) mutant, which is unable to bind GTP and is therefore locked in the GDP-bound inactive state, does not stimulate coat recruitment (Fig. 2A). Second, stimulation of coat recruitment by Arf6(Q67L) is dose-dependent, with half-maximal stimulation observed at about 120 nM Arf6 (Fig. 2B). Third, in the presence of ATP and GTPγS, an excess of recombinant Arf6(T27N) attenuates membrane association of clathrin/AP-2 in a concentration-dependent manner, indicating that this mutant inhibits activiation of endogeneous Arf6, presumably by blocking the Arf-GEFs (Fig. 2C).

Identification of Binding Partners of Recombinant Myristoylated Arf6 by Chemical Cross-Linking

A cross-linking approach was chosen to identify binding partners of recombinant myr-Arf6 during coat recruitment to synaptic membranes. To this aim, recruitment experiments were performed in the presence of recombinant Arf6-His$_6$ mutants as described previously. Thereafter, the samples were treated with an amine-reactive homo-bifunctional, cleavable reagent to covalently cross-link potential interactors to recombinant Arf6. The recombinant protein was then recovered from the mixture by virtue of its His$_6$-tag. In a total volume of 1.5 ml cytosolic buffer, 1 mg of washed LP2 membranes was incubated with 5 mg rat brain cytosol, 2 mM ATP, 200 μg Arf6(T27N)-His$_6$ or Arf6(Q67L)-His$_6$ and 200 μM GDP or GTP, respectively, for 15 min at 37°. The mixture was chilled on ice and 0.5 mM of the cross-linker 3,3'-dithio-bis(proprionic acid N-hydroxysuccinimide ester (DTSP, dissolved as a 500 mM stock solution in DMSO) was added. Cross-linking was allowed to take place during a 1 h incubation at 4°. Then, the samples were supplemented with 1% Triton-X-100 and incubated on ice for a further 10 min under repeated vortexing. The solution was cleared by a 15 min centrifugation at 21,000g. To the supernatant 6 ml of hot 8 M guanidinium hydrochloride were added and the mixture was supplemented with 10 mM imidazole. The samples were denatured by shaking them for 1 h at room temperature. His$_6$-tagged proteins were extracted twice by incubating the solution with 25 μl of washed Ni^{2+}-NTA-agarose for 1 h at room temperature. To allow for efficient recovery of the resin during centrifugation, 30 μl of a 1:1 slurry of uncoupled sepharose beads were added at the end of the first incubation. The beads were washed twice in 1 ml of PBS containing 6 M guanidinium hydrochloride, 10 mM imidazole, and 1% Triton-X-100, then once in the same buffer lacking detergent and finally once in PBS containing 20 mM imidazole. Beads were ultimately extracted twice by boiling in 70 μl of Laemmli sample buffer. 45 μl of the extract were separated by SDS-PAGE and analyzed by Western blotting using antibodies directed against subunits of AP-2 (α-adaptin A, $\beta_{1/2}$-adaptin, and μ2-adaptin: BD Transduction Labs, Heidelberg, Germany), against AP180, the heavy chain of clathrin, or PIPK Iγ (BD Transduction Labs). As demonstrated in Fig. 3A, no cross-link of either Arf6 mutant was observed with the individual subunits of AP-2, with AP180, or with clathrin. Arf6(Q67L), but not the inactive mutant Arf6(T27N), could be efficiently cross-linked to PIPK Iγ. In addition, no cross-link between PIPK Iγ and Arf6(Q67L)-His$_6$ was detected if the cross-linker had been omitted prior to denaturation. 15 μl of each sample was separated by SDS-PAGE and analyzed by Coomassie blue staining to evaluate the amount of Arf6-His$_6$ recovered by Ni-NTA agarose. Equal amounts of Arf6 were detected in each sample, constituting 30–40% of the total recombinant protein initially added to the assay.

Pull-Down Experiments

Recombinant myristoylated His$_6$-tagged Arf bound to Ni^{2+}-NTA-agarose was used as an affinity matrix for incubations with a detergent extract from rat brain to verify the interaction of Arf6(Q67L) with PIPK Iγ and to evaluate its specificity. For this purpose, one brain was homogenized in 10 ml of 320 mM sucrose, 4 mM HEPES pH 7.4 containing 1 mM PMSF and a mammalian tissue protease inhibitor cocktail (Sigma, St. Louis, MO) using a glass-Teflon homogenizer. The homogenate was centrifuged at 1000g for 10 min. The postnuclear supernatant was supplemented with 1% Triton-X-100, 15 mM imidazole, 20 mM HEPES, pH 7.4, 5 mM EDTA, 2 mM GTP, and 10 mM MgCl$_2$ (buffer A), left on ice for 15 min, and cleared by consecutive centrifugations for 15 min at 30,000g and then for 15 min at 150,000g. The resulting supernatant was used at a concentration of 2.5 mg/ml for incubation with 0.15 mg/ml myristoylated Arf6(Q67L)-His$_6$, Arf1(Q71L)-His$_6$, or His$_6$-tagged arfaptin 2 bound to Ni^{2+}-NTA agarose (1 h at 4°). Samples were washed four times in buffer A (the last step contained 20 mM imidazole) and once in buffer A lacking detergent and glycerol. Finally, proteins associated with the resin were extracted twice by boiling in 120 μl Laemmli buffer. 25 μl of each sample were separated by SDS-gel electrophoresis and analyzed by Western blotting using antibodies recognizing one of the large subunits of AP-1 (γ-adaptin: BD Transduction labs) or PIPK Iγ. As indicated in Fig. 3B both Arf1(Q71L) and Arf6 (Q67L) can pull down AP-1 from a rat brain extract to a similar extent, consistent with previous cross-linking experiments. By contrast, Arf6 (Q67L) is much more efficient in associating with PIPK Iγ than Arf1(71L).

Conclusions

The *in vitro* system described here allows for the biochemical analysis of the activities of recombinant myristoylated Arf proteins and mutants thereof. In combination with other biochemical interaction assays, including affinity chromatography and cross-linking, these methods offer the possibility to track down regulators and effectors of Arf6-dependent trafficking and signaling pathways. By extension these methods should also be applicable to other small GTPases or lipid-modified proteins in general. While the systems described here may allow the functional analysis of Arf-mediated protein-protein interactions, it is clear that data obtained *in vitro* must be complemented by additional cell biological and genetic tools to establish their physiological significance in living cells and *in vivo*.

References

Ashery, U., Koch, H., Scheuss, V., Brose, N., and Rettig, J. (1999). A presynaptic role for the ADP ribosylation factor (ARF)-specific GDP/GTP exchange factor msec7-1. *Proc. Natl. Acad. Sci. USA* **96,** 1094–1099.

De Matteis, M. A., and Godi, A. (2004). PI-loting membrane traffic. *Nat. Cell. Biol.* **6,** 487–492.

Di Paolo, G., Moskowitz, H. S., Gipson, K., Wenk, M. R., Voronov, S., Obayashi, M., Flavell, R., Fitzsimonds, R. M., Ryan, T. A., and De Camilli, P. (2004). Impaired PtdIns(4,5)P2 synthesis in nerve terminals produces defects in synaptic vesicle trafficking. *Nature* **431,** 415–422.

Donaldson, J. G. (2003). Multiple roles for Arf6: Sorting, structuring, and signaling at the plasma membrane. *J. Biol. Chem.* **278,** 41573–41576.

Donaldson, J. G., and Jackson, C. L. (2000). Regulators and effectors of the ARF GTPases. *Curr. Opin. Cell. Biol.* **12,** 475–482.

Huttner, W. B., Schiebler, W., Greengard, P., and De Camilli, P. (1983). Synapsin I (protein I), a nerve terminal-specific phosphoprotein. III. Its association with synaptic vesicles studied in a highly purified synaptic vesicle preparation. *J. Cell. Biol.* **96,** 1374–1388.

Ishihara, H., Shibasaki, Y., Kizuki, N., Wada, T., Yazaki, Y., Asano, T., and Oka, Y. (1998). Type I phosphatidylinositol-4-phosphate 5-kinases. Cloning of the third isoform and deletion/substitution analysis of members of this novel lipid kinase family. *J. Biol. Chem.* **273**, 8741–8748.

Krauss, M., Kinuta, M., Wenk, M. R., De Camilli, P., Takei, K., and Haucke, V. (2003). ARF6 stimulates clathrin/AP-2 recruitment to synaptic membranes by activating phosphatidyl-inositol phosphate kinase type Igamma. *J. Cell. Biol.* **162**, 113–124.

McMahon, H. T., and Mills, I. G. (2004). COP and clathrin-coated vesicle budding: Different pathways, common approaches. *Curr. Opin. Cell Biol.* **16**, 379–391.

Milosevic, I., Sorensen, J. B., Lang, T., Krauss, M., Nagy, G., Haucke, V., Jahn, R., and Neher, E. (2005). Plasmalemmal phosphatidylinositol-4,5-bisphosphate level regulates the releasable vesicle pool size in chromaffin cells. *J. Neurosci.* **25**, 2557–2565.

Nie, Z., Hirsch, D. S., and Randazzo, P. A. (2003). Arf and its many interactors. *Curr. Opin. Cell Biol.* **15**, 396–404.

Owen, D. J., Collins, B. M., and Evans, P. R. (2004). Adaptors for clathrin coats: Structure and function. *Annu. Rev. Cell. Dev. Biol.* **20**, 153–191.

Randazzo, P. A., Nie, Z., Miura, K., and Hsu, V. W. (2000). Molecular aspects of the cellular activities of ADP-ribosylation factors. *Sci. STKE* **2000**, RE1.

Takei, K., Mundigl, O., Daniell, L., and De Camilli, P. (1996). The synaptic vesicle cycle: A single vesicle budding step involving clathrin and dynamin. *J. Cell. Biol.* **133**, 1237–1250.

Wang, Y. J., Wang, J., Sun, H. Q., Martinez, M., Sun, Y. X., Macia, E., Kirchhausen, T., Albanesi, J. P., Roth, M. G., and Yin, H. L. (2003). Phosphatidylinositol 4 phosphate regulates targeting of clathrin adaptor AP-1 complexes to the Golgi. *Cell* **114**, 299–310.

Wenk, M. R., and De Camilli, P. (2004). Protein-lipid interactions and phosphoinositide-metabolism in membrane traffic: Insights from vesicle recycling in nerve terminals. *Proc. Natl. Acad. Sci. USA* **101**, 8262–8269.

Wenk, M. R., Pellegrini, L., Klenchin, V. A., Di Paolo, G., Chang, S., Daniell, L., Arioka, M., Martin, T. F., and De Camilli, P. (2001). PIP kinase Igamma is the major PI(4,5)P(2) synthesizing enzyme at the synapse. *Neuron* **32**, 79–88.

[35] Assays to Study Phospholipase D Regulation by Inositol Phospholipids and ADP-Ribosylation Factor 6

By Dale J. Powner, Trevor R. Pettitt, and Michael J. O. Wakelam

Abstract

Phospholipase D (PLD) is an enzyme implicated in the regulation of both exocytic and endocytic vesicle trafficking as well as many other processes. Consistent with this, the small GTPase Arf6 and regulated changes in inositol phospholipids levels are two factors that regulate both PLD and vesicle trafficking. Here we describe three methodologies through which the activation of PLD by Arf6 and inositol phospholipids may be investigated. The first method described is an *in vitro* protocol that

METHODS IN ENZYMOLOGY, VOL. 404
0076-6879/05 $35.00
DOI: 10.1016/S0076-6879(05)04035-8

allows the analysis of purified proteins or cell lysates. Furthermore, this protocol can be used to analyze the effects of different inositol phospholipids by changing the composition of the substrate vesicle. The major advantage of this protocol lies in the ability to analyze the effects of direct interactions on PLD activation by using pure proteins and lipids. The other two methods are *in vivo* protocols for the analysis of PLD activation in response to extracellular stimuli. Modification of cellular composition using overexpression/deletion or knockout of specific genes can be utilized with these protocols to characterize PLD activation pathways. The first of these methods uses the detection of radiolabeled PLD products and can be used for most cell types whereas the second of these two protocols is used to measure PLD products when radiolabeling of cells is not possible, such as freshly isolated cells that will not survive long enough to attain radiochemical equilibrium.

Introduction

The hydrolysis of the most abundant plasma membrane phospholipid, phosphatidylcholine (PtdCho), into phosphatidate (PtdOH) and choline is catalyzed by members of the PLD enzyme family of which there are two isoforms each comprising two splice variants (PLD1a, PLD1b, PLD2a, and PLD2b) (for detailed reviews see McDermott *et al.*, 2004; Powner and Wakelam, 2002). Much evidence has demonstrated that the PtdOH released via this reaction functions as a second messenger while the same is not true for the released choline. It is possible that under some circumstances the signaling properties of this PtdOH are due to its conversion into either 1, 2 diacylglycerol (DAG) by a lipid phosphate phosphatase (LPP) activity or lysophosphatidic acid (LPA) by a phospholipase A_2 (PLA$_2$) activity (Fig. 1). However, evidence is growing that PLD-generated PtdOH primarily acts as a messenger by direct interaction with a target protein, for example, some cAMP-specific phosphodiesterase 4 family members (Baillie *et al.*, 2002; Nemoz *et al.*, 1997), the mammalian target of rapamycin (mTOR) (Fang *et al.*, 2001), phosphatidylinositol 4-phosphate 5kinase type I (PIPkinI) (α, β and γ) family members (Ishihara *et al.*, 1998), and Raf kinase (Rizzo *et al.*, 1999). PLD activation regulates diverse cellular processes including protein translation (Fang *et al.*, 2001), actin cytoskeletal rearrangements (Cross *et al.*, 1996), adhesion, spreading (both Powner and Wakelam, unpublished observations), migration (Santy and Casanova, 2001), exocytosis (Brown *et al.*, 1998), endocytosis (Shen *et al.*, 2001), and the oxidative burst of neutrophils (McPhail *et al.*, 1995). The targets of PLD-generated PtdOH in many of these processes are largely undefined, although some may be inferred from the targets identified so far. As well as

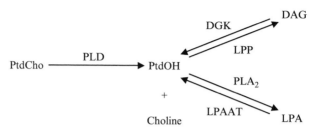

FIG. 1. Ptd OH metabolism.

being generated by PLD, PtdOH can be generated by the phosphorylation of DAG by DAG kinase (DGK) and the acylation of LPA by LPA acyl transferase (LPAAT) (Fig. 1). However, there is evidence to show that the fatty acyl chain species of these PtdOHs may differ from those generated by PLD and as such, these lipids may have different signaling targets/properties (Pettitt *et al.*, 1997).

Currently, selective chemical inhibitors of PLD1 and PLD2 are not available; however, PLD-generated PtdOH-specific signaling events can be defined and separated from other PtdOH signaling events in the presence of short chain primary aliphatic alcohols such as ethanol, propan-1-ol, and butan-1-ol. This is because the PLDs alone catalyze a trans-phosphatidylation reaction utilizing such alcohols; these primary alcohols are stronger nucleophilic acceptors than the water utilized in the normal hydrolysis reaction and a phosphatidylalcohol (e.g., phosphatidylbutan-1-ol [PtdBuOH]) is formed rather than PtdOH (Yang *et al.*, 1967). This phosphatidylalcohol is both unable to interact with the targets of PtdOH, thus inhibiting the PLD-dependent signaling, and is relatively stable in comparison to PtdOH and may therefore be isolated and quantified as a measurement of PLD activity. In our experiments we use butanol at concentrations between 0.1–0.5% (v/v) (10–50 mM) above which non-PLD-dependent affects of the alcohol become apparent, though the extent of this varies between cell types. The employment of primary butanol allows the use of either the secondary or *tert*-butanol as controls for the nonspecific effects of an alcohol on the cells, as these do not act as nucleophilic acceptors in the transphos-phatidylation reaction. These controls are not available for each of the other alcohols.

In the resting cell, PLD1 is generally found upon secretory vesicles/granules, the Golgi complex, and/or recycling endosomes (Brown *et al.*, 1998; Freyberg *et al.*, 2001; Hughes and Parker, 2001), while PLD2 is predominantly found at the plasma membrane (Colley *et al.*, 1997).

Following cell stimulation, PLD1 can translocate to the plasma membrane where it becomes activated (Brown *et al.*, 1998; Powner *et al.*, 2002) and in some cases may become internalized via an endocytic pathway (Hughes and Parker, 2001); whether or not this is PLD-dependent may depend upon the cell type and/or the mechanism of stimulation. In contrast, PLD2 generally remains at the plasma membrane, but may also be found along the endocytic pathway (Powner and Wakelam, unpublished observations). Consistent with this, phosphatidylinositol 3-phosphate, a lipid known to regulate the endocytic pathway, has been shown to bind to full-length PLD1 (Hodgkin *et al.*, 2000) and both PLD1 and PLD2 contain PX domains that are predicted to have both phosphatidylinositol 3-phosphate and PtdOH binding sites, although binding of either of these lipids to a PLD PX domain has yet to be shown. PLD1 and PLD2 also contain PH domains and in the case of PLD1 its plasma membrane localization and activation requires phosphatidylinositol 4,5-bisphosphate ($PtdIns(4,5)P_2$) binding to this domain (Hodgkin *et al.*, 2000) but also to a further site (Sciorra *et al.*, 1999). While the presence of $PtdIns(4,5)P_2$ is sufficient to activate PLD2 and its availability is a major factor in the regulation of this isoform (Divecha *et al.*, 2000; Lopez *et al.*, 1998), considerable evidence suggests that this enzyme may also be subjected to negative regulation by specific proteins (e.g., tubulin, α-actinin, actin, α-synuclein) (Chae *et al.*, 2005; Jenco *et al.*, 1998; Lee *et al.*, 2001; Park *et al.*, 2000). In contrast, $PtdIns(4,5)P_2$ alone is not sufficient to activate PLD1. This requires the phosphoinositide 3-kinase-dependent binding and activation of Arf GTPase family members, Rho family GTPase members, or PKC family members containing a C2 domain, either alone or in combination (Hodgkin *et al.*, 1999; Powner *et al.*, 2002). There is some evidence that PLD2 activation may also be regulated by Arf family GTPases, although this activation appears far less potent than for PLD1 (Lopez *et al.*, 1998).

The small GTPase, Arf6, regulates both actin cytoskeletal rearrangements and endocytosis (reviewed by [Donaldson, 2003]), two processes in which PLD1 and 2 have also been implicated. GTPases, such as Arf6, cycle between the signal perpetuating, "active," GTP-bound state and the "inactive," GDP-bound state. In the active GTP-bound state, conformational changes allow this protein to interact with downstream effectors such as PLD1. GTP exchange factors (GEFs) regulate the transition from "inactive" to the "active" state, whereas GTPase activating proteins (GAPs) promote the GTPase activity of the protein and therefore conversion from the "active" to the "inactive" state. It is possible that overexpression or deletion of Arf6-specific GEFs and GAPs may be useful in characterizing PLD regulation by Arf6. Mutations within Arf6 can "force" the protein to remain within either the "active" state (Q67 to L), or the "inactive," dominant negative state (T27 to N). However, it seems that the activation of PLD1 by Arf6 requires the GTPase to cycle between the "active" and "inactive" states, as neither of these mutants is able to stimulate PLD1 when overexpressed (Powner et al., 2002). Therefore when analyzing the regulation of PLD by Arf6, overexpression of the mutation T157 to A may perhaps be of more use, as this mutation allows Arf6 to cycle faster between the "active" and "inactive" states, potentially enhancing PLD activation (Santy, 2002).

In Vitro PLD Assay

This technique provides the most controllable method for the analysis of PLD regulation by direct interaction with Arf6 and inositol phospholipids and measures the release of radiolabeled choline from PtdCho. Generation of PLD and Arf6 proteins (and their derivatives) is best done using a baculovirus expression system followed by affinity purification making use of appropriate tags such as His or GST incorporated into the expression plasmid. More crudely, membranes containing PLD and Arf6 can be prepared from cell lysates although generally transfection/siRNA depletion of these proteins and their derivatives/regulators makes these experiments more informative. Immunoprecipitation of overexpressed, tagged versions of PLD and Arf6 may be used to enrich for these proteins and may also allow their copurification with other cofactors.

Protocol

Sample Preparation (on ice)

1. In at least triplicate for each condition, prepare 200 ng of purified PLD protein (or derivatives) diluted in Solution A in glass tubes (Fisher Scientific [Loughborough, UK], 10 × 75 mm Borosilicate Glass Tubes, Cat. Code FB59523), on ice, to the appropriate volume considering point 2.

2. Add to 1 μg of purified Arf6 protein (or derivatives) diluted in Solution A to the appropriate concentration to step 1., such that the total volume of step 1 + step 2 does not exceed 20 μl. Other activator/inhibitor proteins/solutions may also be added, for example, cell cytosol, Arf1, Rac, PKCα, actin, α-actinin. Transfection/siRNA depletion of PLD/Arf6/Arf6 regulators/inositol phospholipids regulators may also be used to prepare samples for analysis by this technique. One well of a 12-well cell culture plate at 80–90% confluence should give enough protein, ∼5 μg. Prepare samples by scraping the cells in to 25 μl of ice cold Lysis Buffer and add 20 μl of this to a glass tube on ice. All samples should exclude detergents as these will collapse substrate vesicles.

3. Add 5 μl of either 10× Mg/Ca/GTP or 10× Mg/Ca (i.e., ± GTP) and 5 μl 10× NaCl, maintain on ice. GTP is added to allow the induction of the "active" conformation of the GTPase and samples in the absence of GTP are therefore used to control for PLD activity in the absence of GTPase stimulation. Nonhydrolyzable GTPγS may be substituted for GTP so as to maintain the GTPases in their "active" conformations.

Substrate Vesicle Preparation (Room Temperature)

4. Mix together enough stock lipids (generally solubilized in CHCl$_3$) to give 6.85 nmoles phosphatidylethanolamine (PtdEtn), 0.6 nmoles PtdIns (4,5)P$_2$, 0.43 nmoles PtdCho (16:1.4:1) per reaction (+5–10% to account for errors), in one glass tube and keep at room temperature. These amounts have been optimized for PLD activity using dipalmitoyl lipids (diC16) (Avanti Polar Lipids). However, investigation of the effects of inositol phospholipids on PLD activity should proceed with the replacement of equimolar levels of PtdIns(4,5)P$_2$ with the lipid to be tested.

5. Add 0.05 μCi ^{3}[H]-PtdCho per reaction to this tube keeping at room temperature. Other head-group radiolabeled lipids may be substituted for this PtdCho to test for PLD activity towards other substrates; however, different conditions will be required to capture the released head-group.

6. Dry under a stream of N$_2$.

7. Add 20 μl of 2.5× Sonication Buffer per reaction.

8. Probe sonicate for 10–15 sec at 10 rms with the probe just below the surface. This should remove the dried lipid from the bottom of the tube and generate a misty/metallic/grey translucent solution due to the formation of the vesicles. Resonicate if required. Keep at room temperature—otherwise the vesicles will collapse.

9. Transfer 20 μl vesicles to a scintillation vial and add 12 ml of scintillation fluid; determine the radioactivity by scintillation counting for 1 min. This should give upwards of 50,000 d.p.m., if not resonicate. Save the vial and recount for 10 min together with the test samples after the assay to give total counts.

Assay (37°)

10. Warm samples to 37° for 5 min.

11. Add 20 μl of vesicles to each sample tube.

12. Leave reaction for between 30–60 min at 37°.

13. Stop reaction with the addition of 750 μl of Stop Solution (ice-cold).

14. Phase-separate by adding 250 μl CHCl$_3$ and 400 μl water (this produces a mixture comprising CHCl$_3$:methanol(MeOH):H$_2$O [1:1:0.9] that is essential for phase separation).

15. Centrifuge for 5 min at 350g. If necessary at this stage samples can be left overnight at 4°.

16. Load 600 μl of the aqueous upper phase to cation-exchange columns (to capture the released [3]$[H]$-choline) containing a reservoir of 9 ml of water. Start columns running and discard flow through—a slight pressure applied to the head of the column may be required to start flow.

17. Wash columns once with 9 ml of water. Discard.

18. Elute the [3]$[H]$-choline with 8 ml of 1 M KCl directly into scintillation vials.

19. Add 12 ml of scintillation fluid (must be more scintillant than sample to ensure an optically clear mix).

20. Mix and determine radioactivity by scintillation counting (with Total −9.) for 10 min.

Solutions

Solution A: 137 mM NaCl, 8.1 mM Na_2HPO_4, 2.7 mM KCl, 1.15 mM KH_2PO_4 pH 7.5, 2.5 mM EDTA. Lysis Buffer: (prepared on day of use) Solution A containing 1.56 g/l benzamidine, 1.54 g/l dithiolthreitol, 200 μM PMSF (from a 200 mM stock in propan-2-ol). 10× NaCl: 4 M NaCl. 10× Mg/Ca: 30 mM $MgCl_2$, 20 mM $CaCl_2$. 10× Mg/Ca/GTP: 30 mM $MgCl_2$, 20 mM $CaCl_2$, 300 μM GTP (nonhydrolyzable GTPγS may be substituted to maintain all GTPases in the "active" conformation). Sonication Buffer: 125 mM HEPES pH 7.5, 200 mM KCl, 7.5 mM EGTA, 2.5 mM dithiothreitol. [3]$[H]$-PtdCho: 1 μCi/μl 1,3 phosphatidyl (N-methyl [3]$[H]$) choline, dipalmitoyl (GE Healthcare).

Cation-Exchange Column: Dowex 50W×8 100–200, H-form (SUPELCO, Cat. Code 50X8200500G). Using a stirrer, make a 50% v/v solution of the Dowex in deionized water. Block the narrow part of a glass Pasteur pipette with some glass wool. Add 1 ml Dowex solution to the Pasteur pipette, this results in ∼3 cm on top of the glass wool once the water has drained through. Attach these to 20 ml plastic syringe barrels (to be used as a reservoir) using flexible tubing and position such that eluants may be collected. Stop Solution: $CHCl_3$:MeOH (1:2 v/v)

Whole Cell PLD Assay

This technique allows the characterization of PLD activation in whole cells in response to extracellular stimuli by measuring the accumulation of radiolabeled phosphatidylbutan-1-ol (PtdBuOH). Transfection or siRNA depletion of cells with PLD and Arf6 derivatives/regulators can be used in these experiments. Similarly transfection of cells with inositol phospholipid-binding protein domains, such as the PtdIns(4,5)P$_2$-specific PH domain from PLCδ1, or transfection/siRNA depletion of inositol phospholipid generating/modulating proteins such as PtdIns(4,5)P$_2$-generating phosphatidylinositol 4-phosphate 5-kinase Iα, may facilitate the determination of the effects of these lipids on PLD activity in the cell. Overexpression and siRNA techniques require high percentage levels of transfection in order to maximize the signal to background ratio (e.g., using adenoviral/retroviral infection). However, overexpression should be optimized such that the levels of overexpressed protein per cell are as low as is possible to generate a detectable response. Cell-permeant inositol phospholipids (with fatty acyl chains of no more than 8 carbons) (Avanti Polar Lipids) may also be used to investigate the regulation of PLD activation in this whole cell assay. These should be prepared as a 1–5 mM stock by first drying down in a glass vial under a stream of N$_2$ to remove organic solvent, followed by sonication into a solution containing 10 mM HEPES pH 7.4, 10 mM Glucose, 125 mM NaCl, and 100 ng/ml fatty acid-free BSA. As a guide, for both cell-permeant PtdOH and PtdIns(4,5)P$_2$ a concentration of around 100 mM when added directly to culture medium gives maximal effects in RBL-2H3 cells (mass spectroscopy shows that ~1% of this lipid is taken up within 5 min of addition to achieve these effects).

Protocol

1. In at least triplicate for each condition, radiolabel cells of 50–60% confluence in 12-well cell culture plates for 16 h in 5 μCi/ml ^{3}[H]-palmitate. This radiolabeled lipid should be dried down in a glass vial under a stream of N$_2$ and resuspended in the appropriate amount of culture medium prior to adding to cells.

2. After the radiolabeling period, aspirate the medium and wash the cells three times with serum-free cell culture medium containing 10 mM HEPES pH 7.4 and 5 mM NaHCO$_3$. This allows cells to be manipulated on the bench although the temperature should still be maintained at 37°. Incubate the cells in a minimum volume of this medium at 37°.

3. Treat the cells as desired, for example, with the appropriate agonist for the desired length of time—these conditions will require optimization, but in general PLD activation is rapid and detectable within 1 min. For agonist stimulated cells, butanol is generally added at a final concentration of 0.3–0.5% for 5–10 min prior to the addition of the agonist and reaction is generally stopped between 5–15 mins later. For measurement of PLD activity in the absence of agonist stimulation, butan-1-ol is generally added to cells for between 10–15 mins before the reaction is stopped.

4. At the point of stopping a particular reaction condition, carefully aspirate the media from the cells and add 1 volume of ice-cold methanol. For cells in suspension, it is best to centrifuge them and aspirate off their media. However, if not possible, 1 volume of MeOH may be added directly to culture medium where this is equal to 0.9 volumes—see point 8.

5. Keeping everything ice-cold, remove the cells to a glass vial sized at least 3 volumes. (Fisher Scientific, 46× 12.5 mm Glass Vials with screw lids, ~3 ml volume, Cat. Code TUL-520-007H.) For adherent cells this will require them to be scraped off the culture surface.

6. Add 1 volume of $CHCl_3$ + 10μg PtdBuOH (Avanti Polar Lipids).

7. Vortex and leave at room temperature for 20 min.

8. Add 0.9 volumes 0.88% KCl and vortex—unless MeOH was added to 0.9 volumes of culture medium in step 4. Under all circumstances in order to isolate the lipids from the rest of the cell debris the ratio of $CHCl_3$:MeOH:aqueous should be 1:1:0.9. 0.88% KCl provides the best aqueous solution for separation; however, any aqueous solution, such as culture media, should also work.

9. Centrifuge at 800g for 5 min to promote phase separation.

10. Remove the top phase to radioactive waste.

11. Dry the lower phase under a stream of N_2 or in a vacuum centrifuge.

12. Resuspend the lipid extract in 50 μl of chloroform.

13. Remove 10 μl of step 12. to a scintillation vial; this gives a measure of the Total Lipid in the sample-multiply the contents here by 4 in the final calculation.

14. Apply the remainder to the loading area, but 1 cm from the bottom, of a silica thin-layer chromatography (TLC) plate that has been activated in an oven at 110° for 15 min before use (Fisher Scientific, Whatman Silica gel 150 Å TLC plates, LK5DF, 250 μm, 20 × 20 cm, Cat. Code TLC-240-020H). Separation can be further improved by prerunning blank TLC plates in the separating solvent before drying them in a fume cupboard prior to activation at 110°.

15. Place the TLC plate into a TLC tank with a depth of 0.5 cm of separating solvent, seal the top, and run for ~1.5 h or until the solvent front is about 2–3 cm from the top of the plate.

16. Air-dry the plates in a fume cupboard.

17. Place the TLC plate in an iodine atmosphere for 5–10 min to stain for the added PtdBuOH and mark this area with a pencil. The Retention factor (Rf) should be approximately 0.3–0.4.

18. Allow the iodine to sublime from the plate surface in a fume cupboard (exposure to steam may be required to speed this up for heavily

stained plates) and then remove the marked area to a scintillation vial containing 0.5 ml of methanol.

19. Add scintillation fluid to steps 13. and 18., vortex, and determine the radioactivity in a scintillation counter for 10 min.

20. The percentage PtdBuOH formed in a sample is calculated as a measure of PLD activity. This is determined by the d.p.m. from step 18. divided by the d.p.m. from [step 13. × 4] multiplied by 100.

Solutions

Separating solvent: Mix 50 ml trimethylpentane, 110 ml ethyl acetate, 20 ml glacial acetic acid, 100 ml water in a separating funnel, leave the phases to separate, and pour the lower phase to waste. The upper phase is the TLC solvent. The level added to the TLC tanks should be to just below the level of the sample loaded on to each plate (~0.5 cm).

Determination of PLD Activity in Nonradiolabeled Cells

It is not possible to radiolabel some cells; in particular, freshly isolated cells such as platelets cannot retain their viability during the period necessary to achieve radiochemical equilibrium. Analysis of PLD activity in these cells makes use of the transphosphatidylation property of the enzyme; however, rather than isolate a radiolabeled PtdBuOH, mass spectroscopy is used to identify and quantify unlabeled PtdBuOH following separation by HPLC. It is unlikely that all laboratories wishing to analyze PLD activity will have access to the appropriate mass spectrometer; however, there are now a number of laboratories with which it should be possible to interact. Use of specific mouse gene knockouts/knockins of PLD/Arf6/inositol lipid regulating enzymes or cell permeant lipids may aid these investigations.

Protocol

Extraction procedure

1. Sufficient lipid for a phosphatidylalcohol analysis is normally available from approximately 5×10^6 mammalian cells. Each condition should be analyzed in at least triplicate.

2.–5. Follow steps 2 to 5 from whole cell PLD assay.

6. Add 1 volume of $CHCl_3$ and internal standards.

7.–11. Follow steps 7 to 11 from Assay whole cell PLD assay.

12. Resuspend the lipid extract in 50 μl of $MeOH:CHCl_3:H_2O$ (5:5:1).

13. Transfer into a silanized glass, mass spectrometer–autosampler vial insert (Alltech [Cannforth, Lancs, UK], Hi-temp HMDS silane-treated 100 μl glass LV Cat. Code 952018).

14. Rinse glass vial from step 11. with a further 50 μl of MeOH:CHCl$_3$: H$_2$O (5:5:1) and combine with step 13.

15. Dry again in a vacuum centrifuge.

16. Resuspend in 15 μl of CHCl$_3$.

LC-MS Analysis

17. Inject 0.5 μl lipid extract from step 15. onto a Luna silica column (3 μm, 1.0 × 150 mm; Phenomenex).

18. Separate using a gradient of 100% Solvent 1 changing to 100% Solvent 2 over 20 min at a flow rate of 100 μl/min.

19. Detection of phosphatidylalcohol species (eluting at about 3 min) is by negative electrospray ionization (ESI). We use a Shimadzu QP8000a with a probe voltage of -4 kV, a sheath nitrogen flow of 4 l/min, and a desolvation line temperature of 300°. The major phosphatidylalcohol species from cultured mammalian cells is usually 16:0/18:1.

Solutions

Internal Standards: Suitable phosphatidylalcohol internal standards are not commercially available. Those that are available have diC16 and diC18:1 acyl profiles that will mask endogenously derived phosphatidylalcohols of the same structure. As such, quantification has to be related to an internal standard of a different lipid class. We normally add a range of diC12 species that are not found in any biological samples we have examined (500 ng each of diC12-DAG, diC12-PtdOH, diC12PtdCho, diC12-PtdEtn, diC12-phosphatidylglycerol, and diC12phosphatidylserine-Avanti Polar Lipids). Solvent 1: 100% CHCl$_3$/MeOH/H$_2$O (90:9.5:05) containing 7.5 mM ethylamine. Solvent 2: Acetonitrile/CHCl$_3$/MeOH/H$_2$O (30:30:35:5) containing 10 mM ethylamine.

Acknowledgment

We thank the Wellcome Trust and the Medical Research Council for financial support.

References

Baillie, G. S., Huston, S., Scotland, G., Hodgkin, M. N., Gall, I., Peden, A. H., MacKenzie, C., Houslay, E. S., Currie, R. A., Pettitt, T. R., Walmsley, A. R., Wakelam, M. J. O., Warwicker, J., and Houslay, M. D. (2002). TAPAS-1, a novel microdomain within the unique N-terminal region of the PDE4A1 cAMP phosphodiesterase that allows rapid

Ca^{2+} triggered membrane association with selectivity for interaction with phosphatidic acid. *J. Biol. Chem.* **277,** 28298–28309.

Brown, F. D., Thompson, N. T., Saqib, K. M., Clark, J. M., Powner, D., Thompson, N. T., Solari, R., and Wakelam, M. J. O. (1998). Phospholipase D1 localises to secretory granules and lysosomes and is plasma-membrane translocated on cellular stimulation. *Cur. Biol.* **8,** 835–838.

Chae, Y. C., Lee, S., Lee, H. Y., Heo, K., Kim, J. H., Kim, J. H., Suh, P. G., and Ryu, S. H. (2005). Inhibition of muscarinic receptor-linked phospholipase D activation by association with tubulin. *J. Biol. Chem.* **280,** 3723–3730.

Colley, W. C., Sung, T.-C., Roll, R., Jenco, J., Hammond, S. M., Altshuller, Y., Bar-Sagi, D., Morris, A. J., and Frohmam, M. A. (1997). Phospholipase D2, a distinct phospholipase D isoform with novel regulatory properties that provokes cytoskeletal reorganisation. *Cur. Biol.* **7,** 191–201.

Cross, M. J., Roberts, S., Ridley, A. J., Hodgkin, M. N., Stewart, A., Claesson-Welsh, L., and Wakelam, M. J. O. (1996). Stimulation of actin stress fibre formation mediated by activation of phospholipase D. *Cur. Biol.* **6,** 588–597.

Divecha, N., Roefs, M., Halstead, J. R., D'Andrea, S., Fernandez-Borga, M., Oomen, L., Saquib, K. M., Wakelam, M. J. O., and D'Santos, C. (2000). Interaction of the Type Iα PIPkinase with phospholipase D: A role for the local generation of phosphatidylinositol 4,5-bisphosphate in the regulation of PLD2 activity. *EMBO J.* **19,** 5440–5449.

Donaldson, J. G. (2003). Multiple roles for Arf6: Sorting, structuring, and signaling at the plasma membrane. *J. Biol. Chem.* **278,** 41573–41576.

Fang, Y., Vilella-Bach, M., Bachman, R., Flanigan, A., and Chen, J. (2001). Phosphatidic acid-mediated mitogenic activation of mTOR signaling. *Science* **294,** 1942–1945.

Freyberg, Z., Sweeney, D., Siddhanta, A., Bourgoin, S., Frohman, M. A., and Shields, D. (2001). Intracellular localization of phospholipase D1 in mammalian cells. *Mol. Biol. Cell* **12,** 943–955.

Hodgkin, M. N., Clark, J. M., Rose, S., Saqib, K. M., and Wakelam, M. J. O. (1999). Characterisation of protein kinase C and small G-protein regulated phospholipase D activity in the detergent-insoluble fraction of HL60 cells. *Biochem. J.* **339,** 87–93.

Hodgkin, M. N., Masson, M. R., Powner, D. P., Saqib, K. M., Ponting, C. P., and Wakelam, M. J. O. (2000). Phospholipase D regulation and localisation is dependent upon a phosphatidylinositol 4,5-bisphosphate-specific PH domain. *Cur. Biol.* **10,** 43–46.

Hughes, W. E., and Parker, P. J. (2001). Endosomal localization of phospholipase D 1a and 1b is defined by the C-terminal of the proteins, and is independent of activity. *Biochem. J.* **356,** 727–736.

Ishihara, H., Shibasaki, Y., Kizuki, N., Katagiri, H., Yazaki, Y., Asano, T., and Oka, Y. (1998). Type 1 phosphatidylinositol-4-phosphate 5 kinases. Cloning of the third isoforms and deletion/substitution analysis of members of this novel lipid kinase family. *J. Biol. Chem.* **273,** 8741–8748.

Jenco, J. M., Rawlingson, A., Daniels, B., and Morris, A. J. (1998). Regulation of phospholipase D2: Selective inhibition of mammalian phospholipase D isoenzymes by α- and β-synucleins. *Biochem.* **37,** 4901–4909.

Lee, S., Park, J. B., Kim, J. H., Kim, Y., Kim, J. H., Shin, K. J., Lee, J. S., Ha, S. H., Suh, P. G., and Ryu, S. H. (2001). Actin directly interacts with phospholipase D, inhibiting its activity. *J. Biol. Chem.* **276,** 28252–28260.

Lopez, I., Arnold, R. S., and Lambeth, J. D. (1998). Cloning and initial characterization of a human phospholipase D2 (hPLD2). *J. Biol. Chem.* **273,** 12846–12852.

McDermott, M., Wakelam, M. J. O., and Morris, A. J. (2004). Phospholipase D. *Biochem. Cell Biol.* **82,** 225–253.

McPhail, L. C., Qualliotinemann, D., and Waite, K. A. (1995). Cell-free activation of neutrophil NADPH oxidase by a phosphatidic acid-regulated protein-kinase. *PNAS* **92,** 7931–7935.

Nemoz, G., Sette, C., and Conti, M. (1997). Selective activation of rolipram-sensitive, cAMP-specific phosphodiesterase isoforms by phosphatidic acid. *Mol. Pharm.* **51,** 242–249.

Park, J. B., Kim, J. H., Kim, Y., Ha, S. H., Kim, H. H., Yoo, J. S., Du, G., Frohman, M. A., Suh, P. G., and Ryu, S. H. (2000). Cardiac phospholipase D2 localizes to sarcolemmal membranes and is inhibited by α-actinin in an ADP-ribosylation factor-reversible manner. *J. Biol. Chem.* **275,** 21295–21301.

Pettitt, T. R., Martin, A., Horton, T., Liossis, C., Lord, J. M., and Wakelam, M. J. O. (1997). Diacylglycerol and phosphatidate generated by phospholipases C and D respectively have distinct fatty acid compositions and functions: Phospholipase D-derived diacylglycerol does not activate protein kinase C in PAE cells. *J. Biol. Chem.* **272,** 17354–17359.

Powner, D. J., Hodgkin, M. N., and Wakelam, M. J. O. (2002). Antigen-stimulated activation of phospholipase D1b by Rac1, ARF6 and PKCα in RBL2H3 cells. *Mol. Biol. Cell* **13,** 1252–1262.

Powner, D. J., and Wakelam, M. J. O. (2002). The regulation of phospholipase D by inositol phospholipids and small GTPases. *FEBS Lett.* **531,** 62–64.

Rizzo, M. A., Shome, K., Vasudevan, C., Stolz, D. B., Sung, T. C., Frohman, M. A., Watkins, S. C., and Romero, G. (1999). Phospholipase D and its product, phosphatidic acid, mediate agonist-dependent Raf-1 translocation to the plasma membrane and the activation of the mitogen-activated protein kinase pathway. *J. Biol. Chem.* **274,** 1131–1139.

Santy, L. C. (2002). Characterisation of a fast cycling ADP-ribosylation factor 6 mutant. *J. Biol. Chem.* **277,** 40185–40188.

Santy, L. C., and Casanova, J. E. (2001). Activation of ARF6 by ARNO stimulates epithelial cell migration through downstream activation of both Rac1 and phospholipase D. *J. Cell Biol.* **154,** 599–610.

Sciorra, V. A., Rudge, S. A., Prestwich, G. D., Frohman, M. A., Engebrecht, J. A., and Morris, A. J. (1999). Identification of phosphoinositide binding motif that mediates activation of mammalian and yeast phospholipase D isoenzymes. *EMBO J.* **20,** 5911–5921.

Shen, Y., Xu, L., and Foster, D. A. (2001). Role for phospholipase D in receptor-mediated endocytosis. *Mol. Cell. Biol.* **21,** 595–602.

Yang, S. F., Freer, S., and Benson, A. A. (1967). Transphosphatidylation of phospholipase D. *J. Biol. Chem.* **242,** 477–484.

[36] Assay and Functional Properties of the Tyrosine Kinase Pyk2 in Regulation of Arf1 Through ASAP1 Phosphorylation

By ANAMARIJA KRULJAC-LETUNIC and ANDREE BLAUKAT

Abstract

The Arf-GTPase-activating protein (GAP) ASAP1 has been identified in a yeast-two-hybrid screen as a prominent binder of the proline-rich tyrosine kinase 2 (Pyk2) via SH3 domain-mediated interaction. Following binding, Pyk2 directly phosphorylates ASAP1 on tyrosine residues 308 and 782 *in vitro* and in cells. To understand the functional impact of this interaction and subsequent phosphorylation, nonphosphorylated and Pyk2-phosphorylated ASAP1 have been generated. This material can be used for lipid-protein overlay assays and fluorimetric Arf-GTPase tests to show that the Pyk2-mediated tyrosine phosphorylation of ASAP1 modulates its GAP activity towards Arf1 *in vitro*. These studies provide the first evidence for a regulation of ASAP1 and hence Arf1 activity by tyrosine phosphorylation and suggest a functional link between tyrosine kinases and Arf GTPases. Furthermore, an assay to study cellular Arf activation and thus GAP as well as guanine nucleotide exchange factor (GEF) activities and their regulation is described.

Introduction

ADP-ribosylation factors (Arfs) belong to the family of monomeric GTP-binding proteins and are involved in intracellular membrane traffic, modulation of actin cytoskeleton, and activation of phospholipase D (Boman and Kahn, 1995; Exton, 2002; Kirchhausen, 2000; Randazzo *et al.*, 2000; Rothman and Wieland, 1996). They act as molecular switches that flip between the inactive GDP-bound and active GTP-bound state. When GTP is bound, the best-characterized family member Arf1 translocates to membranes of the Golgi apparatus and recruits vesicle coat proteins. At physiological concentration of magnesium, Arfs tightly bind GDP, making the spontaneous exchange of GDP for GTP very slow. For their activation Arfs need the action of guanine nucleotide exchange factors (GEFs) that catalyze the release of GDP and allow rapid exchange with ready to bind GTP. Due to their low intrinsic GTPase activity, subsequent hydrolysis of bound GTP

METHODS IN ENZYMOLOGY, VOL. 404
Copyright 2005, Elsevier Inc. All rights reserved.

by Arfs must be assisted by GTPase activating proteins (GAPs) (Donaldson and Jackson, 2000; Randazzo and Hirsch, 2004). Therefore, any protein that physically interacts with an ArfGEFs or ArfGAPs could potentially regulate Arf activity and Arf-dependent cellular events. ASAP1 is an ArfGAP that displays *in vitro* activity towards Arf1 and Arf5 (Brown *et al.*, 1998). As a number of other GAPs and GEFs, ASAP1 contains a pleckstrin homology (PH) domain that is a phosphoinositide-binding motif responsible for membrane targeting (Jackson *et al.*, 2000). Through its PH domain, ASAP1 interacts with different phosphoinositides, but primarily binding to phosphatidylinositol 4,5-bisphosphate (PI4,5P2) seems to be critical for the catalytic activity of ASAP1. It appears that binding to PI4,5P2 is not only important for targeting of ASAP1 to membranes (where Arf-GTP is localized) but also to induce a conformational change in the ASAP1 catalytic domain necessary to promote ASAP1 activation (Kam *et al.*, 2000). In a yeast two-hybrid screen, ASAP1 has been identified as an interaction partner of the tyrosine kinase Pyk2 and subsequent experiments showed that Pyk2 directly phosphorylates ASAP1 (Kruljac-Letunic *et al.*, 2003). Pyk2 has been implicated in the activation of mitogen activated protein (MAP) kinase cascades upon stimulation of cells with different ligands that increase intracellular calcium or during cellular stress, such as radiation or changes in extracellular osmolarity (Avraham *et al.*, 2000). Interestingly, tyrosine 308 that was identified as a major ASAP1 phosphorylation site *in vitro* and in intact cells is in proximity to the PH domain (spanning amino acids 340–431 of human ASAP1) and therefore, the introduction of a negative charge by Pyk2-mediated tyrosine phosphorylation might effect binding to PI4,5P2 and thereby influence ASAP1 GAP activity. Here we describe approaches that are used to analyze effects of Pyk2-mediated phosphorylation on ASAP1 binding to phospholipids and the potency of ASAP1 to induce GTP hydrolysis on Arf1 *in vitro*. In addition we suggest an assay that can be employed to follow levels of activated Arf1 in intact cells.

Buffers

> 2× BBS: 50 mM HEPES, 280 mM NaCl, 1.5 mM Na2HPO4, adjusted to pH 6.92 (it is suggested to prepare several solutions with slightly different pH values and choose the optimal one based on test transfections with reporter genes, such as GFP or β-glactosidase).
> HGNT: 20 mM HEPES, pH 7.4, 10% glycerol, 150 mM NaCl, 1% Triton X-100, 2 mM. EDTA, 25 mM NaF TBS: 0.15 M NaCl, 0.05 M Tris, adjusted to pH 7.5.

TBST: 0.05% Tween 20, 0.15 M NaCl, 0.05 M Tris, adjusted to pH 7.5.
Buffer A: 20 mM Tris, pH 8.0, 1 mM DTT, 200 μM GDP.
Buffer B: 20 mM Tris, pH 8.0, 1 mM DTT, 5 μM GDP, 1 mM MgCl2.
Buffer D: 10 mM MES, pH 5.7, 1 mM MgCl2, 1 mM DTT, 5 μM GDP.
Hydration buffer: 25 mM HEPES, pH 7.4, 150 mM KCl.
Reaction buffer: 25 mM HEPES, pH 7.4, 150 mM KCl, 2 mM MgCl2.
Kinase buffer: 10 mM Tris, pH 7.5, 10 mM MgCl2 and 100 μM ATP:
    3× SDS sample buffer: 150 mM Tris-HCl, pH 6.8, 6% SDS, 30%
    glycerol, 0.2% bromophenol blue, 150 mM DTT.

Reagents

Protease inhibitor cocktail: 100 mM AEBSF, 80 μM Aprotinin, 5 mM
Bestatin, 1.5 mM E-64, 2 mM Leupeptin, 1 mM Pepstatin A (#539134,
Calbiochem, Schwalbach, Germany).
    Phosphatase inhibitor cocktail: 200 mM Imidazole, 100 mM NaF, 115
    mM Sodium molybdate, 100 mM Sodium orthovanadate, 400 mM
    Sodium tartarate dihydrate (#524625, Calbiochem) anti-Flag M2
    affinity matrix (#A2220, Sigma-Aldrich, St. Louis, MO).
    3× Flag peptide (#F4799, Sigma-Aldrich).
    Phospholipid array (#P-6001, Echelon Bioscience, Salt Lake City, UT).
    Lipids: PI(4,5)P2 (#524644, Calbiochem), phosphatidylcholin, Phos-
    phatidylethanolamine, phosphatidylserine, phosphatidylinositol,
    sphingomyelin, cholesterol (Avanti Polar Lipids, Alabaster, AL).
    Antibodies: anti-ASAP1 (#sc-11539, Santa Cruz, Santa Cruz, CA),
    anti-phosphotyrosine pY99 (#sc-7020, Santa Cruz), anti-Flag (#F3165,
    Sigma), anti-HA (#1666606, Roche Applied Science), antiPyk2 (#694,
    kind gift from I. Dikic, University of Frankfurt, Germany), anti-mouse-
    horseradish peroxidase conjugate (#NA931, Amersham Biosciences,
    Buckinghamshire, UK).
Enhanced chemiluminescence substrate (#2015200, Roche Applied Science,
    Mannheim, Germany).

Expression and Purification of Flag-ASAP1 from Transfected
    HEK293T Cells

HEK293T cells, grown until 60–80% confluence in 10-cm cell culture
dishes, are transiently transfected using a modified calcium phosphate
method (similar to the commercial MBS kit from Stratagene, La Jolla,
CA). For transfection of each dish, a total of 15 μg of plasmid DNA (4 μg

Flag-tagged ASAP1 in pcDNA3 + 11 μg "empty" pcDNA3) is diluted in 450 μl of water and supplemented with 50 μl 2.5 M CaCl2 and 550 μl of 2× BBS. This solution is thoroughly mixed and the DNA is allowed to form precipitate for 20 min at room temperature before it is drop-wise added onto the cells under constant swirling. Cells are incubated with the DNA precipitate for 3 h in DMEM medium containing 10% calf serum (*Note*: it is essential *not* to use fetal serum for this procedure). Thereafter fresh medium containing 10% fetal bovine serum is added and cells are incubated for 48 h at 37°, 5% CO2 before they are harvested and lysed. For cell lysis 1 ml/10 cm-dish of ice-cold HGNT lysis buffer supplemented with protease and phosphatase inhibitor cocktails is added. Lysates are incubated under gentle rocking for 45 min at 4° and subsequently centrifuged at 4° for 20 min at 6000g to remove cellular debris. Supernatants obtained from ten 10-cm dishes are pooled and incubated with 400 μl of anti-Flag M2 affinity matrix for 3 h on a rotating platform at 4°. Thereafter, the matrix is washed three times with 10 ml of HGNT lysis buffer and once with 10 ml of TBS. Bound Flag-ASAP1 is eluted at 4° by incubation with TBS containing 150 μg/ml 3× Flag peptide (Sigma-Aldrich). Eluates can be divided into aliquots and stored shock-frozen at −80° without apparent loss of activity for several weeks. The purity of Flag-ASAP1 and its phosphorylation status are analyzed by Coomassie blue staining and Western blotting with corresponding antibodies, respectively (Fig. 1). Though not necessary for the described assays, the obtained material may be further purified by additional chromatography steps.

Nonphosphorylated Flag-ASAP1 (note that "nonphosphorylated" here only relates to tyrosine phosphorylation; in addition ASAP1 is constitutively phosphorylated on several serine residues [Kruljac-Letunic *et al.*, 2003]) is purified from cells transiently transfected only with plasmid DNA encoding Flag-ASAP1, while tyrosine phosphorylated Flag-ASAP1 can be isolated from cells transfected with plasmid DNAs encoding both the active tyrosine kinase Pyk2 and Flag-ASAP1. Tyrosine phosphorylation is monitored by Western blotting using antiphosphotyrosine antibodies (Fig. 1B). The second tyrosine phosphorylated protein seen in these blots represents Pyk2 that associates with ASAP1 via a SH3 domain interaction and therefore copurifies with ASAP1 (Kruljac-Letunic *et al.*, 2003). Pyk2 itself does not affect the subsequent assays alone (not shown) but can be employed to further increase the tyrosine phosphorylation of ASAP1 by performing an *in vitro* kinase reaction (see following). In a similar way, the effect of other tyrosine kinases, such as Src and FAK, both shown to interact with ASAP1 can be analyzed (Brown *et al.*, 1998; Liu *et al.*, 2002).

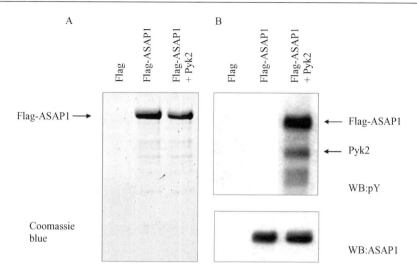

FIG. 1. Affinity purification of phosphorylated and nonphosphorylated Flag-ASAP1 from transfected HEK293T cells. HEK293T cells transfected with the indicated constructs are lysed and incubated with anti-Flag affinity matrix. Following elution with 3×Flag peptide, purity of ASAP1 and its phosphorylation status are analyzed by Coomassie blue staining (A) 10 μl of purified sample per lane and Western blotting with anti-phosphotyrosine (pY) or anti-ASAP1 antibodies (B) 5 μl of purified sample per lane.

Overlay Assay to Analyze the Effect of ASAP1 Tyrosine Phosphorylation on Phospholipid Binding

The PH domain of ASAP1 has been shown to be critical for phospho-inositide binding and allosteric activation of its GAP activity (Kam *et al.*, 2000). Due to the proximity of the major ASAP1 tyrosine phosphorylation site mapped by 2D phosphopeptide analysis (Kruljac-Letunic *et al.*, 2003), an influence of this additional negative charge on phosphoinositide binding has been hypothesized. A protein-lipid overlay assay with purified Flag-ASAP1 is used to analyze the effect of tyrosine phosphorylation on phospholipids binding (Dowler *et al.*, 2002). Initially, a nitrocellulose membrane array of 15 immobilized phospholipids (100 pmol/spot) and a blank spot are blocked with 3% fatty acid-free BSA in TBST for 1 h at room temperature. Thereafter membranes are incubated for 1 h at room temperature in the same solution supplemented with 0.5 μg/ml of purified phosphorylated or nonphosphorylated Flag-ASAP1 (cf. Fig. 1). Following washing with TBST for 30 min, membranes are incubated with a monoclonal anti-Flag antibody (1 μg/ml, in blocking solution for 1 h at room temperature) and an anti-mouse-horseradish peroxidase conjugate (1:5000 in 5% fat-free milk powder in TBS for 1 h at room temperature). Finally, Flag-ASAP1 bound to phospholipids is detected using an enhanced chemiluminescence substrate.

A representative experiment demonstrates reduced binding of tyrosine phosphorylated ASAP1 to PI(3,4)P2, PI(4,5)P2, PI(3,4,5)P3, PI(3)P, and PS while interactions with PI(3,5)P2, PI(4)P, and PI(5)P are virtually unchanged (Fig. 2). Interestingly, in particular PI(4,5)P2 has been shown to be an important regulator or ASAP1 GAP activity (Kam *et al.*, 2000; Kruljac-Letunic *et al.*, 2003). For control experiments, Pyk2 alone (the "kinase contamination"), an unrelated Flag-tagged construct and/or Flag-ASAP1 mutants with a disturbed PH domain or lacking corresponding phosphorylation sites are used ([Kruljac-Letunic *et al.*, 2003]) and not shown). The same method may be applied to get a first hint on PH domain binding specificity and on its modulation by posttranslational modifications. The obtained luminescence signals can be quantitatively measured with an appropriate camera system (e.g., VersaDoc 5000 from BioRad, Hercules, CA) and apparent affinities and the role of individual phosphorylation sites

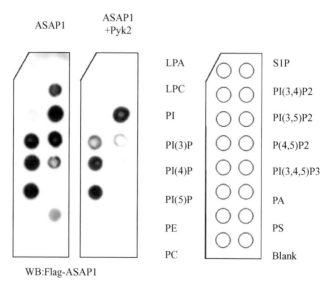

FIG. 2. Protein-lipid overlay assay to demonstrate the effect of ASAP1 tyrosine phosphorylation on phospholipid binding. Protein-lipid overlay assays are performed with immobilized phospholipids (100 pmol/spot) incubated with 0.5 μg/ml nonphosphorylated (left panel) or phosphorylated ASAP1 (middle panel). Binding of Flag-ASAP1 to immobilized phospholipids is detected by Western blotting with an anti-Flag antibody. Positions of the individual lipids on the array are shown on the right panel.

may be further analyzed using different concentrations of ASAP1 or mutants lacking individual phosphorylation sites.

In Vitro Arf GTPase Assay to Study the Modulation of ASAP1 GAP Activity by Tyrosine Phosphorylation

Expression and Purification of myr-Arf1 from E. coli

Human Arf1 is expressed in *E. coli* BL21 (DE3) together with N-myristoyl transferase and purified in the myristoylated form in three steps: (i) precipitation with 35% ammonium sulfate saturation; (ii) DEAE-Sepharose chromatography; (iii) MonoS chromatography as described (Franco *et al.*, 1995), including slight modifications. Briefly, a bacterial preculture is diluted in 1 l LB medium containing antibiotics, grown until the OD600 reached 0.2–0.4, and supplemented with 50 μM myristic acid. The culture is further grown until an OD600 of 0.6 and protein expression is induced with 1 mM IPTG for 4 h at 27°. Following lysis by high pressure using a cell disrupter (e.g., Avestin EmulsiFlex-C5) in buffer A and two centrifugation steps (10,000g for 15 min and 100,000g for 1 h), soluble proteins in the supernatant are precipitated with ammonium sulfate at 35% saturation at 4°. Under these conditions myristoylated Arf1 precipitates while its nonmyristoylated form remains in the soluble phase. The precipitate is separated by centrifugation, resuspended in buffer A, and dialyzed against buffer B. The clear solution is loaded on a DEAE anion exchange matrix (Amersham Bioscience, Buckinghamshire, UK) and eluted with a linear NaCl gradient (0–1 M). Eluted fractions containing Arf1 are pooled, dialyzed against buffer D, and applied on a Mono S cation exchange matrix (Amersham Bioscience). Bound Arf1 protein is eluted with a linear NaCl gradient (0–0.5 M), divided into aliquots, shock-frozen, and can be stored without loss of activity for several weeks at −80°.

Liposomes Preparation

Liposomes are prepared in a pear-shaped flask in a rotating evaporator at 37° as previously described (Matsuoka *et al.*, 1998). Briefly, 100 μl of a 3 M lipid stock mixture resembling the Golgi lipid composition (43 mol% phosphatidylcholine (PC), 19 mol% phosphatidylethanolamine (PE), 5 mol % phosphatidylserine (PS), 10 mol% phosphatidylinositol (PI), 7 mol% sphingomyelin (SM), 16 mol% cholesterol (CL)) are diluted in 2 ml chloroform and evaporated under an argon stream. The obtained lipid film is hydrated in 900 μl of hydration buffer. Lipid suspensions are subjected to

10 freezing/thawing cycles and extruded 21 times through a polycarbonate filter with a pore size of 0.8 μm.

Fluorimetric In Vitro Arf1-GTPase Assay

Arf1-GTPase activity can be followed with a fluorimetric assay based on the change of the Arf1 intrinsic tryptophan fluorescence during its transition from the GTP- to the GDP-bound state (Antonny et al., 1997). The assay is performed with a fluorimeter (e.g., Jasco FP6500) in a cylindrical cuvette at 37° with permanent stirring. In a typical experiment, the cuvette initially contains 60 μl of a liposome suspension with 1 mol% PI (4,5)P2 in 530 μl reaction buffer. Once the base line of the fluorescence is stable (within about 1 min), Arf1-GDP (4 μg) and GTP (10 μM) or the nonhydrolyzable analogue GTPγS are injected and the GDP/GTP exchange is accelerated by addition of 4 mM EDTA (Paris et al., 1997). GTP-loaded Arf1 is stabilized by addition of 4 mM MgCl2 and the Arf1-GTPase activity is followed after injection of purified nonphosphorylated or phosphorylated Flag-ASAP1 (about 2 μg). Tryptophan fluorescence is recorded at 340 nm (bandwidth 20 nm) upon excitation at 297.5 nm (bandwidth 5 nm). Data are collected (e.g., using the Spectra Managertm software, Jasco, Easton, MD) and fitted to exponential curves using an appropriate software package (e.g., Origin, OriginLab Corp., Northampton, MA).

Before injection into the cuvette, the stoichiometry of ASAP1 phosphorylation in purified samples (with or without copurified Pyk2 kinase), can be increased by an in vitro kinase reaction for 30 min at 30° in kinase buffer (Fig. 3A). The effect of the increase in stoichiometry of phosphorylation can be observed by comparison of ASAP1 GAP activity of samples in which an in vitro kinase reaction is performed with (+ATP) and without ATP (-ATP). A representative experiment demonstrates the inhibitory effect of Pyk2-mediated tyrosine phosphorylation on ASAP1 GAP activity towards Arf1 (Fig. 3B). In a more detailed study we have mapped phosphorylated tyrosines 308 and 782 in ASAP1 being responsible for reduced PI(4,5)P2 binding and GAP activity (Kruljac-Letunic et al., 2003). The same procedure may be used to determine the effect of other kinase, such as Src and FAK on ASAP1 GAP activity.

Cellular Arf1-GTPase Assay

Cellular levels of activated, GTP-bound Arf1 can be monitored with a recently described pull-down assay using a GST-GGA fusion protein as a probe (Lorraine et al., 2001). This fusion protein is composed of the VHS and GAT domains of human GGA3 (Golgi-associated, γ-adaptin homologous, Arf-interacting protein 3), an ubiquitously expressed coat protein localized in

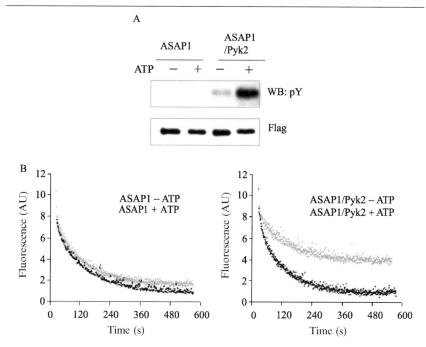

FIG. 3. *In vitro* Arf-GTPas assay to measure the GAP activity of ASAP1. (A) The stoichiometry of ASAP1 tyrosine phosphorylation can be increased by *in vitro* kinase reaction performed with purified Flag-ASAP1/Pyk2 complexes. This increase is followed by Western blotting with an anti-phosphotyrosine antibody (pY) and levels of Flag-ASAP1 are controlled with an anti-Flag antibody. (B) Arf-GTPase activities of nonphosphorylated (left panel) and phosphorylated ASAP1 (right panel) are monitored in a fluorimetric Arf1-GTPase assay as a decrease of the intrinsic Arf1 tryptophan fluorescence at 340 nm (in arbitrary units; AU) upon excitation at 297.5 nm. Black dots indicate sample prior and grey dots after *in vitro* kinase reaction.

coats of vesicles budding from the *trans*-Golgi network (TGN). The GAT domain of GGA3 specifically interacts with Arf1-GTP (Puertollano *et al.*, 2001) and can therefore be used to selectively precipitate activated Arf1. The specificity of GST-GGA3 binding to activated Arf1 can be shown in HEK293T cells transiently transfected with HA-tagged wild-type, constitutively activated (Arf1Q71L), and dominant-negative inactive (Arf1T31N) Arf1 constructs with a modified calcium phosphate method (see previous). 48 h after transfection, cells are lysed in HGNT buffer and incubated at 4° for 2 h on a rotating platform with equal amounts of GST-GGA fusion proteins (about 5 μg) bound to glutathion sepharose beads. GST-GGA can be expressed in *E. coli* and isolated by incubation of

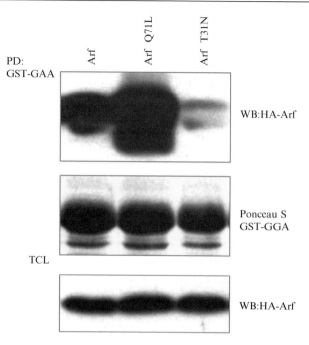

FIG. 4. Pull-down assay to monitor cellular levels of activated Arf1. HEK293T cells, transfected with indicated HA-tagged Arf1 constructs, are lysed and subjected to pull-down assays with GST-GGA as a probe. Bound HA-Arf1 is detected with a monoclonal anti-HA antibody (upper panel). Ponceau S staining of membranes after protein transfer confirms equal amounts of GST-GGA probes (middle panel). Comparable levels of HA-Arf1 variants in lysates used for pull-downs are verified by Western blotting with an anti-HA antibody (lower panel).

bacterial lysates with glutathion sepharose according to manufacturer's instructions (Amersham Biosciences). Following incubation, beads are washed three times with HGNT lysis buffer and associated HA-Arf1-GTP is released in SDS sample buffer (25 μl) during 5 min incubation at 98°. Thereafter, samples are subjected to 12% SDS-PAGE and the amount of HA-Arf1-GTP is analyzed by Western blotting with an anti-HA antibody (Fig. 4).

This pull-down assay should be useful for future studies on the effects of Arf GAPs and GEFs on cellular levels of Arf-GTP and the modulation of their activities, for instance, by tyrosine phosphorylation. However, we have not yet optimized the experimental conditions to follow the impact of ASAP1 on cellular Arf1-GTP.

Acknowledgments

The authors would like to thank Joerg Moelleken and Felix Wieland for support and sharing reagents as well as equipment. Furthermore, we acknowledge Ivan Dikic (anti-Pyk2 antibody #694), Paulo Randazzo (ASAP1 cDNA), Wilhelm Just (HA-tagged Arf variants cDNA), and Juan S. Bonifacino (GST-GGA cDNA) for providing reagents.

References

Antonny, B., Beraud-Dufour, S., Chardin, P., and Chabre, M. (1997). N-terminal hydrophobic residues of the G-protein ADP-ribosylation factor-1 insert into membrane phospholipids upon GDP to GTP exchange. *Biochemistry* **36**, 4675–4684.

Avraham, H., Park, S. Y., Schinkmann, K., and Avraham, S. (2000). RAFTK/Pyk2-mediated cellular signalling. *Cell. Signal* **12**, 123–133.

Boman, A. L., and Kahn, R. A. (1995). Arf proteins: The membrane traffic police? *Trends Biochem. Sci.* **20**, 147–150.

Brown, M. T., Andrade, J., Radhakrishna, H., Donaldson, J. G., Cooper, J. A., and Randazzo, P. A. (1998). ASAP1, a phospholipid-dependent arf GTPase-activating protein that associates with and is phosphorylated by Src. *Mol. Cell. Biol.* **18**, 7038–7051.

Donaldson, J. G., and Jackson, C. L. (2000). Regulators and effectors of the ARF GTPases. *Curr. Opin. Cell. Biol.* **12**, 475–482.

Dowler, S., Kular, G., and Alessi, D. R. (2002). Protein lipid overlay assay. *Sci. STKE* **2002**, PL6.

Exton, J. H. (2002). Regulation of phospholipase D. *FEBS Lett.* **531**, 58–61.

Franco, M., Chardin, P., Chabre, M., and Paris, S. (1995). Myristoylation of ADP-ribosylation factor 1 facilitates nucleotide exchange at physiological Mg2+ levels. *J. Biol. Chem.* **270**, 1337–1341.

Jackson, T. R., Kearns, B. G., and Theibert, A. B. (2000). Cytohesins and centaurins: Mediators of PI 3-kinase-regulated Arf signaling. *Trends Biochem. Sci.* **25**, 489–495.

Kam, J. L., Miura, K., Jackson, T. R., Gruschus, J., Roller, P., Stauffer, S., Clark, J., Aneja, R., and Randazzo, P. A. (2000). Phosphoinositide-dependent activation of the ADP-ribosylation factor GTPase-activating protein ASAP1. Evidence for the pleckstrin homology domain functioning as an allosteric site. *J. Biol. Chem.* **275**, 9653–9663.

Kirchhausen, T. (2000). Three ways to make a vesicle. *Nat. Rev. Mol. Cell. Biol.* **1**, 187–198.

Kruljac-Letunic, A., Moelleken, J., Kallin, A., Wieland, F., and Blaukat, A. (2003). The tyrosine kinase Pyk2 regulates Arf1 activity by phosphorylation and inhibition of the Arf-GTPase-activating protein ASAP1. *J. Biol. Chem.* **278**, 29560–29570.

Liu, Y., Loijens, J. C., Martin, K. H., Karginov, A. V., and Parsons, J. T. (2002). The association of ASAP1, an ADP ribosylation factor-GTPase activating protein, with focal adhesion kinase contributes to the process of focal adhesion assembly. *Mol. Biol. Cell* **13**, 2147–2156.

Lorraine, C., Casanova, S., and Casanova, J. E. (2001). Activation of Arf6 by ARNO stimulates epithelial cell migration through downstream activation of both Rac1 and phospholipase D. *J. Cell Biol.* **154**, 599–610.

Matsuoka, K., Orci, L., Amherdt, M., Bednarek, S. Y., Hamamoto, S., Schekman, R., and Yeung, T. (1998). COPII-coated vesicle formation reconstituted with purified coat proteins and chemically defined liposomes. *Cell* **93**, 263–275.

Paris, S., Beraud-Dufour, S., Robineau, S., Bigay, J., Antonny, B., Chabre, M., and Chardin, P. (1997). Role of protein-phospholipid interactions in the activation of ARF1 by the guanine nucleotide exchange factor Arno. *J. Biol. Chem.* **272,** 22221–22226.

Puertollano, R., Randazzo, P. A., Presley, J. F., Hartnell, L. M., and Bonifacino, J. S. (2001). The GGAs promote Arf-dependent recruitment of clathrin to the TGN. *Cell* **105,** 93–102.

Randazzo, P. A., Andrade, J., Miura, K., Brown, M. T., Long, Y. Q., Stauffer, S., Roller, P., and Cooper, J. A. (2000). The Arf GTPase-activating protein ASAP1 regulates the actin cytoskeleton. *Proc. Natl. Acad. Sci. USA* **97,** 4011–4016.

Randazzo, P. A., and Hirsch, D. S. (2004). Arf GAPs: Multifunctional proteins that regulate membrane traffic and actin remodelling. *Cell Signal.* **16,** 401–413.

Rothman, J. E., and Wieland, F. T. (1996). Protein sorting by transport vesicles. *Science* **272,** 227–234.

[37] ADP-Ribosylation Factor 6 Regulation of Phosphatidylinositol-4,5-Bisphosphate Synthesis, Endocytosis, and Exocytosis

By Yoshikatsu Aikawa and Thomas F. J. Martin

Abstract

Unlike other members of the ADP-ribosylation factor (ARF) family, Arf6 is localized to the plasma membrane and endosomes, and regulates membrane traffic from and into the plasma membrane. Arf6 regulates a clathrin-independent endocytic membrane recycling pathway in non-polarized cells and clathrin-dependent endocytosis in polarized cells. It also regulates recycling endosome traffic back to the plasma membrane as well as dense-core vesicle exocytosis in neuroendocrine cells. A key effector for Arf6 is phosphatidylinositol 4-monophosphate 5-kinase, which catalyzes plasma membrane synthesis of phosphatidylinositol-4,5-bis-phosphate (PIP_2), a common required cofactor for several endocytic and exocytic membrane trafficking pathways. Long-term expression of a constitutively active Arf6 mutant in cells can lead to the depletion of PIP_2 from the plasma membrane, its accumulation in intracellular vacuoles, and the inhibition of PIP_2-dependent membrane trafficking at the plasma membrane.

Background on Arf Family

ADP-ribosylation factors (Arfs) are low molecular mass (20 kDa) GTPases in the family of Ras-related GTP-binding proteins. There are six mammalian Arfs (Arf1-6) that are ubiquitously expressed in eukaryotic

METHODS IN ENZYMOLOGY, VOL. 404
0076-6879/05 $35.00
DOI: 10.1016/S0076-6879(05)04037-1

cells. By structural similarity, Arfs are subclassified into class I (Arf1–3), class II (Arf4, 5), and class III (Arf6) proteins. Like other GTPases, the Arf proteins are molecular switches that cycle between GDP-bound inactive and GTP-bound active states. In their active GTP-bound state, Arfs interact with effector proteins to execute their function, which is to control membrane trafficking in eukaryotic cells. Arfs promote the membrane recruitment of peripheral proteins and activate lipid-modifying enzymes that participate in membrane transitions such as vesicle formation. Studies have focused on the roles of Arf1 and Arf6, the most dissimilar of the Arfs in sequence (66% amino acid identity). Arf1 is Golgi-localized and plays an essential role in vesicle formation (Randazzo *et al.*, 2000). Arf6 is localized to the plasma membrane and regulates endosome-plasma membrane trafficking (Donaldson, 2003). This chapter will focus on functional studies of Arf6 directed at understanding its role in membrane trafficking into and from the plasma membrane.

Studies Utilizing Arf6 Mutants

With the precedent from studies of Ras-related GTPases, studies of Arf6 (D'Souza-Schorey *et al.*, 1995; Peters *et al.*, 1995) utilized the overexpression of point mutants designed to alter the GTPase cycle of the protein. Roles of Arf6 are inferred from the cellular phenotypes generated by expressing proteins defective in GTP hydrolysis or in GTP binding. The Q67L mutant is predicted to be defective in GTPase activity, which would "lock" it into the active GTP-bound state. T27N and N122I mutants are predicted to be defective in GTP binding and would remain in the inactive GDP-bound state. While mutant overexpression studies have provided most of our present knowledge about the cell biology of Arf6, there are some caveats associated with this approach. First, because wild type Arf6 proteins function cyclically in rapid cycles of GTP exchange, long-term expression of GTP-locked or GDP-bound Arf6 may cause cumulative effects of long-term blockade. Second, because nucleotide exchange rates of the mutants have rarely been characterized in a cellular context (Macia *et al.*, 2004; Vitale *et al.*, 2002), mutant Arf6 proteins may sequester accessory factors such as Arf6-GEFs and have unanticipated secondary consequences.

Two approaches have been used to reduce secondary consequences of Arf6 mutant overexpression. The first is to reduce expression time. Whereas Arf6Q67L expression accelerated clathrin-independent endocytosis at early times following transfection, it inhibited this pathway at longer expression times (Naslavsky *et al.*, 2004). Expressing Arf6 mutants with an inducible promoter is also useful for distinguishing immediate primary from cumulative secondary effects (Paleotti *et al.*, 2005). A second approach has been to redesign and characterize new Arf6 mutants. Because the constitutively active Arf6Q67L cannot accurately recapitulate the cycle of activated wild type Arf6 due to its lack of GTP turnover, Santy (2002) designed and employed a fast cycling mutant (T157A), which promoted actin cytoskeletal rearrangements anticipated for an activated Arf6. Macia *et al.* (2004) found that Arf6T27N readily dissociated bound nucleotide and was prone to aggregation. A redesigned Arf6T44N mutant was shown to be mostly GDP-bound in cells and to localize to the plasma membrane, whereas Arf6T27N localized to an endosomal compartment.

In another approach, Arf6 mutant overexpression was bypassed entirely through the use of siRNA-mediated down-regulation (Hashimoto *et al.*, 2004; Houndolo *et al.*, 2005), which confirmed essential roles for Arf6 in endocytosis. Because Arf6 null cells may also exhibit cumulative and secondary effects of Arf6 loss-of-function, studies to conditionally induce siRNA production and to acutely re-express Arf6 in downregulated cells will provide the strongest approaches to identifying primary cellular phenotypes of Arf6 dysfunction.

Arf6 Regulates Endocytic Membrane Trafficking

One of the first reports on Arf6 function (D'Souza-Schorey *et al.*, 1995) showed that transferrin uptake by a clathrin-dependent endocytic pathway was blocked in CHO cells expressing Arf6Q67L, which suggested a role for Arf6 in the clathrin-dependent pathway. However, subsequent studies (Altschuler *et al.*, 1999; Palacios *et al.*, 2002) in polarized MDCK cells found that expression of the Arf6Q67L mutant stimulated transferrin uptake. In recent studies, GTP-bound Arf6 was shown to recruit AP-2 and clathrin to membranes (Krauss *et al.*, 2003; Paleotti *et al.*, 2005) by direct interactions between GTP-bound Arf6 and AP-2. However, transferrin uptake was inhibited by expression of the constitutively active Arf6Q67L mutant. Studies of β-adrenergic receptor internalization via clathrin-dependent endocytosis showed that ligand-dependent internalization was inhibited by Arf6Q67L expression (Claing, 2004) or by siRNA-mediated down regulation of Arf6 (Houndolo *et al.*, 2005). However, downregulation of Arf6 did not affect constitutive internalization of transferrin (Hashimoto *et al.*, 2004; Houndolo *et al.*, 2005). To date, results on the role of Arf6 in clathrin-dependent endocytosis do not provide a coherent picture. There appear to be cell type differences with stimulation by Arf6Q67L observed in polarized epithelial but not in nonpolarized cells. Neither study that utilized siRNA-mediated downregulation of Arf6 in nonpolarized cells (Hashimoto *et al.*, 2004; Houndolo *et al.*, 2005) found an effect on transferrin uptake. It is possible that inhibition of transferrin uptake observed with long-term Arf6Q67L expression is a secondary consequence (see following).

A consistent set of studies indicate that Arf6 is a regulatory component in a clathrin-independent endocytic pathway of membrane recycling. Proteins such as MHC1, integrins, E cadherin, interleukin-2 receptor α, CD59, M2 muscarinic receptor, and SNAP-25, which lack signals for clathrin-dependent endocytosis, are internalized by a pathway that is clathrin- and dynamin-independent and regulated by Arf6 (Aikawa and Martin, 2003; Brown *et al.*, 2001; Delaney *et al.*, 2002; Donaldson, 2003; Naslavsky *et al.*, 2003, 2004; Radhakrishna and Donaldson, 1997). Arf6Q67L expression promotes plasma membrane invaginations (D'Souza-Schorey *et al.*, 1998; Franco *et al.*, 1999; Peters *et al.*, 1995) and an acceleration of uptake via this pathway (Aikawa and Martin, 2003; Aikawa *et al.*, submitted; Brown *et al.*, 2001; Naslavsky *et al.*, 2004). siRNA-mediated silencing of Arf6 was found to abolish uptake via the clathrin-independent pathway without altering transferrin uptake (Hashimoto *et al.*, 2004).

A potential problem (or advantage depending upon the circumstances) with using the Arf6Q67L mutant, especially in long-term or high-level expression studies, is that it induces the formation of a vacuolar compartment of endocytosed membrane and leads to the depletion of recycling components from the plasma membrane that are trapped in the vacuole (Aikawa and Martin, 2003; Aikawa *et al.*, submitted; Brown *et al.*, 2001). Because, among other components, PIP$_2$ depletion from the plasma membrane occurs under these conditions (Fig. 1A) (Aikawa and Martin, 2003; Brown *et al.*, 2001), PIP$_2$-requiring events at the plasma membrane can be inhibited. The reported inhibition of transferrin uptake by Arf6Q67L expression in nonpolarized cells (D'Souza-Schorey *et al.*, 1995, 1998; Franco *et al.*, 1999; Paleotti *et al.*, 2005) might be secondary to the depletion of plasma membrane PIP$_2$, which is essential for clathrin-dependent endocytosis (Jost *et al.*, 1998). Recent studies revealed another indirect effect of Arf6Q67L expression. SNAP-25, which is required for trafficking of early endosomes into the recycling endosome in neuroendocrine cells, normally cycles between the plasma membrane and recycling endosomes by an Arf6-regulated pathway. Arf6Q67L expression leads to the trapping of this protein in a vacuolar compartment, which indirectly leads to an inhibition of transferrin uptake into the recycling endosome (Aikawa *et al.*, submitted).

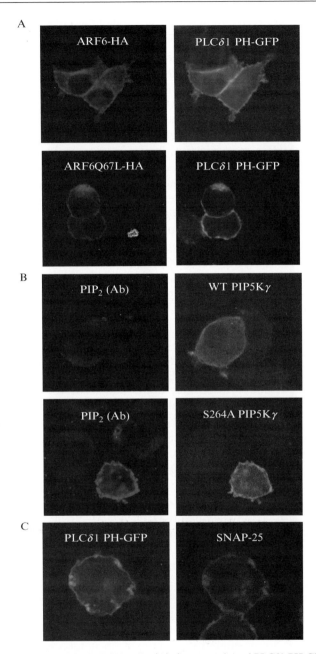

FIG. 1. Detection of PIP$_2$ in PC12 cells. (A) Co-expression of PLCδ1 PH-GFP was used to detect PIP$_2$ (right panels) in cells expressing HA-tagged wild type ARF6 (upper row) or HA-tagged ARF6Q67L (lower row). ARF6 proteins were detected with HA antibody.

Arf6 Regulates Exocytic Membrane Trafficking

Arf6 is thought to regulate the delivery of recycling endosomal vesicles back to the plasma membrane (D'Souza-Schorey, 1998; Prigent *et al.*, 2003; Radhakrishna and Donaldson, 1997) although this exocytic pathway has not been fully characterized. Several studies have found that Arf6 regulates the exocytosis of dense-core vesicles in neuroendocrine cells (Aikawa and Martin, 2003; Lawrence and Birnbaum, 2003; Vitale *et al.*, 2002). It was reported that Arf6 is present on secretory granules in chromaffin cells (see Vitale *et al.*, 2002), but this was not the case in neuroendocrine PC12 cells, where Arf6 localized to the plasma membrane and to endosomes as in other cell types (Aikawa and Martin, 2003). Overexpression of Arf6Q67L in PC12 cells trapped recycling plasma membrane constituents such as PIP$_2$, phosphatidylinositol 4-phosphate 5-kinase (PI(4)P 5-kinase), and SNAP-25 in an intracellular vacuolar compartment (Aikawa and Martin, 2003; Aikawa *et al.*, submitted). Because PIP$_2$ is essential for dense-core vesicle exocytosis (Hay *et al.*, 1995), its depletion from the plasma membrane inhibited stimulus-evoked vesicle exocytosis (Aikawa and Martin, 2003). Arf6 was also found to participate in the acute Ca^{2+}-stimulated synthesis of PIP$_2$ and enhancement of vesicle exocytosis (Aikawa and Martin, 2003). In MIN6 insulinoma cells, expression of the Arf6T27N mutant inhibited stimulus-evoked insulin secretion, which was attributed to a role for Arf6 in maintaining plasma membrane PIP$_2$ required for vesicle exocytosis (Lawrence and Birnbaum, 2003).

Arf6 Regulates Plasma Membrane PIP 5-Kinase and PIP$_2$ Synthesis

Arf1 functions in Golgi trafficking by recruiting cytosolic coat proteins as well as by activating lipid-modifying enzymes (Godi *et al.*, 1999; Jones *et al.*, 2000; Randazzo *et al.*, 2000). For clathrin-independent endocytosis, Arf6 stimulation of PIP$_2$ synthesis and F actin assembly may mediate the invagination and pinching off of the plasma membrane

---

ARF6Q67L-expressing cells (two days post-transfection) contain an intracellular vacuole enriched in PIP$_2$ and a plasma membrane depleted in PIP$_2$. (B) PIP$_2$ antibody was used in immunocytochemistry (left panels) to detect PIP$_2$ in cells expressing wild type HA-tagged PI(4)P 5-kinase Iγ (upper row) or a phosphorylation site mutant (S264) of HA-tagged PI(4)P 5-kinase Iγ (lower row) that is likely constitutively active (Aikawa and Martin, 2003). (C) Expression of PLCδ1 PH-GFP (left panel) was used to detect PIP$_2$ in cells treated with latrunculin. SNAP-25 antibody was used to detect SNAP-25 (right panel). SNAP-25 is recycled by an ARF6-regulated pathway. Inhibition of F actin assembly by latrunculin leads to membrane invaginations that fail to pinch off and internalize, which contain PIP$_2$ and SNAP-25 (Aikawa *et al.*, submitted).

(Aikawa *et al.*, submitted; Brown *et al.*, 2001). A key effector identified for Arf6 is PI(4)P 5-kinase. While class I and class III Arfs activate PI(4)P 5-kinase (Aikawa and Martin, 2003; Brown *et al.*, 2001; Godi *et al.*, 1999; Honda *et al.*, 1999; Jones *et al.*, 2000; Krauss *et al.*, 2003; Oude Weernink *et al.*, 2004), a major role for Arf6 emerged because it uniquely colocalizes with PI(4)P 5-kinase at the plasma membrane (Honda *et al.*, 1999).

Arf6Q67L expression results in a persistent activation of PIP_2 synthesis at the plasma membrane followed by dramatically enhanced plasma membrane invagination and internalization with the formation of a PIP_2-containing, actin-coated vacuole (Brown *et al.*, 2001). In PC12 cells, Arf6Q67L expression results in extensive recruitment of PI(4)P 5-kinase to endocytosing membrane and the depletion of PIP_2 from the plasma membrane (Fig. 1A) (Aikawa and Martin, 2003). The persistent stimulation of PI(4)P 5-kinase by Arf6Q67L and resulting extensive PIP_2 synthesis at the plasma membrane appear to drive actin-mediated membrane involution and internalization because treatment with latrunculin results in membrane invaginations containing PIP_2 and Arf6Q67L that fail to internalize (Aikawa *et al.*, submitted).

PI(4)P 5-kinase is an important effector for Arf6 but other potential effectors have been identified, which include PLD, the Sec10 subunit of the exocyst complex, the NDP kinase Nm23-H1, AP-2, arfaptin 2/POR, and arfophilin (see Donaldson, 2003; Donaldson and Jackson, 2000; Palacios *et al.*, 2002; Paleotti *et al.*, 2005; Prigent *et al.*, 2003). Surprisingly for an effector, the association of Arf6 with PI(4)P 5-kinase *in vivo* did not depend upon the nucleotide status of Arf6 as assessed by co-immunoisolation from detergent extracts of PC12 cells (Aikawa and Martin, 2003), which is similar to Rho GTPase interactions with PI(4)P 5-kinase (Oude Weernink *et al.*, 2004). Instead, the phosphorylation state of PI(4)P 5-kinase Iγ was a strong determinant for Arf6 association (Aikawa and Martin, 2003). *In vitro* studies, in contrast, indicated that a recombinant Arf6Q67L protein, but not a recombinant Arf6T27N protein, interacted with PI(4)P 5-kinase Iγ (Krauss *et al.*, 2003). In any case, the activation of PI(4)P 5-kinase by Arf6 is dependent upon GTP (or GTPγS) (Honda *et al.*, 1999; Jones *et al.*, 2000; Krauss *et al.*, 2003; Oude Weernink *et al.*, 2004).

Detection of Cellular PIP_2 by Immunocytochemistry

Expression of Arf6Q67L stimulates PIP_2 synthesis in cells but, as a consequence, also promotes extensive plasma membrane internalization, which can lead to intracellular PIP_2 accumulation and plasma membrane

PIP$_2$ depletion (Aikawa and Martin, 2003; Brown *et al.*, 2001). Intracellular accumulation of PIP$_2$ and its depletion from the plasma membrane have been detected using three methods. In the first, the PLCδ1 PH domain GFP fusion protein is co-expressed in cells (Fig. 1A and C) as detailed elsewhere (Balla and Varnai, 2002). This method has the advantage of detecting dynamic changes in PIP$_2$ localization in living cells (see Brown *et al.*, 2001). Because there is some uncertainty about non-PIP$_2$ cofactors needed for PLCδ1 PH localization (Varnai *et al.*, 2002), it is desirable to utilize a second method for PIP$_2$ localization.

Immunocytochemistry using PIP$_2$ monoclonal antibodies from several sources (K. Fukami, University of Tokyo for KT10; M. Umeda, Tokyo Metropolitan Institute of Medical Science for AM212; PerSeptive Biosystems Inc. for kt3g) work quite well for this purpose (Fig. 1B). PC12 cells are washed, fixed with 4% formaldehyde (w/v) for 30 min, permeabilized with 0.2% saponin for 30 min, and blocked for 1 h (in 0.1% bovine serum albumin/1% fetal bovine serum) prior to overnight antibody incubation. Typically high dilutions of PIP$_2$ monoclonal antibody (1:1000–1:10,000 of ascites) are used to minimize nonspecific background staining. PIP$_2$ can also be localized in mechanically permeabilized cells (Aikawa *et al.*, submitted), which has the advantage of not employing detergents. PC12 were permeabilized by passage through a ball homogenizer (Hay and Martin, 1992) and washed extensively. For incubation with PIP$_2$ antibody, washing, and incubation with secondary antibodies, permeable cells were tethered onto glutaraldehyde-activated, poly-L-lysine-coated glass coverslips (Weidman *et al.*, 1993). Incubation times with antibodies are minimized (∼60 min on ice) so that the unfixed antibody-decorated cells can be quickly imaged by microscopy.

References

Aikawa, Y., and Martin, T. F. J. (2003). ARF6 regulates a plasma membrane pool of phosphatidylinositol(4,5)bisphosphate required for regulated exocytosis. *J. Cell Biol.* **162,** 647–659.

Altschuler, Y., Liu, S.-H., Katz, L., Tang, K., Hardy, S., Brodsky, F., Apodaca, G., and Mostov, K. (1999). ADP-ribosylation factor 6 and endocytosis at the apical surface of Madin-Darby canine kidney cells. *J. Cell Biol.* **147,** 7–12.

Balla, T., and Varnai, P. (2002). Visualizing cellular phosphoinositide pools with GFP-fused protein modules. *Sci. STKE* **125,** PL3.

Brown, F. D., Rozelle, A. L., Yin, H. L., Balla, T., and Donaldson, J. G. (2001). Phosphatidylinositol 4,5-bisphosphate and Arf6-regulated membrane traffic. *J. Cell Biol.* **154,** 1007–1017.

Claing, A. (2004). Regulation of G protein-coupled receptor endocytosis by ARF6 GTP-binding protein. *Biochem. Cell Biol.* **82**, 610–617.

Delaney, K. A., Murph, M. M., Brown, L. M., and Radhakrishna, H. (2002). Transfer of M2 muscarinic acetylcholine receptors to clathrin-derived early endosomes following clathrin-independent endocytosis. *J. Biol. Chem.* **277**, 33439–33446.

Donaldson, J. G. (2003). Multiple roles for ARF6: Sorting, structuring, and signaling at the plasma membrane. *J. Biol. Chem.* **278**, 41573–41576.

Donaldson, J. G., and Jackson, C. L. (2000). Regulators and effectors of the ARF GTPases. *Curr. Opin. Cell Biol.* **12**, 475–482.

D'Souza-Schorey, C., Li, G., Colombo, M. I., and Stahl, P. D. (1995). A regulatory role for ARF6 in receptor-mediated endocytosis. *Science* **267**, 1175–1178.

D'Souza-Schorey, C., van Donselaar, E., Hsu, V. W., Yang, C., Stahl, P. D., and Peters, P. J. (1998). ARF6 targets recycling vesicles to the plasma membrane: Insights from an ultrastructural investigation. *J. Cell Biol.* **140**, 603–616.

Franco, M., Peters, P. J., Boretto, J., van Donselaar, E., Neri, A., D'Souza-Schorey, C., and Chavrier, P. (1999). EFA6, a sec 7 domain-containing exchange factor for ARF6, co-ordinates membane recycling and actin cytoskeleton organization. *EMBO J.* **18**, 1480–1491.

Godi, A., Pertile, P., Meyers, R., Marra, P., Di Tullio, G., Iurisci, C., Luini, A., Corda, D., and De Matteis, M. A. (1999). ARF mediates recruitment of phosphatidylinositol 4-kinase and stimulates synthesis of PI(4,5)P2 on the Golgi complex. *Nature Cell Biol.* **1**, 280–287.

Hashimoto, S., Hashimoto, A., Yamada, A., Kojima, C., Yamamoto, H., Tsutsumi, T., Higashi, M., Mizoguchi, A., Yagi, R., and Sabe, H. (2004). A novel mode of action of an ArfGAP, AMAP2/PAG3/Papα, in Arf6 function. *J. Biol. Chem.* **279**, 37677–37684.

Hay, J. C., and Martin, T. F. (1992). Resolution of regulated secretion into sequential MgATP-dependent and calcium-dependent stages mediated by distinct cytosolic proteins. *J. Cell Biol.* **119**, 139–151.

Hay, J. C., Fisette, P. L., Jenkins, G. H., Fukami, K., Takenawa, T., Anderson, R. A., and Martin, T. F. (1995). ATP-dependent inositide phosphorylation required for Ca^{2+} -activated secretion. *Nature* **374**, 173–177.

Honda, A., Nogami, M., Yokozeki, T., Yamazaki, M., Nakamura, H., Watanabe, H., Kawamoto, K., Nakayama, K., Morris, A. J., Frohman, M. A., and Kanaho, Y. (1999). Phosphatidylinositol 4-phosphate 5-kinase a is a downstream effector of the small G protein ARF6 in membrane ruffle formation. *Cell* **99**, 521–532.

Houndolo, T., Boulay, P., and Claing, A. (2005). G protein-coupled receptor endocytosis in ADP-ribosylation factor 6-depleted cells. *J. Biol. Chem.* **280**, 5598–5604.

Jones, D. H., Morris, J. B., Morgan, C. P., Kondo, H., Irvine, R. F., and Cockcroft, S. (2000). Type I phosphatidylinositol 4-phosphate 5-kinase directly interacts with ADP-ribosylation factor 1 and is responsible for phosphatidylinositol 4,5-bisphosphate synthesis in the Golgi compartment. *J. Biol. Chem.* **275**, 13962–13966.

Jost, M., Simpson, F., Kavran, J. M., Lemmon, M. A., and Schmid, S. L. (1998). Phosphatidylinositol-4,5-bisphosphate is required for endocytic coated vesicle formation. *Curr. Biol.* **8**, 1399–1402.

Krauss, M., Kinuta, M., Wenk, M. R., De Camilli, P., Takei, K., and Haucke, V. (2003). ARF6 stimulates clathrin/AP-2 recruitment to synaptic membranes by activating phosphatidyl-inositol phosphate kinase type Ig. *J. Cell Biol.* **162**, 113–124.

Lawrence, J. T. R., and Birnbaum, M. J. (2003). ADP-ribosylation factor 6 regulates insulin secretion through plasma membrane phosphatidylinositol 4,5-bisphosphate. *Proc. Natl. Acad. Sci. USA* **100,** 13320–13325.

Macia, E., Luton, F., Partisani, M., Cherfils, J., Chardin, P., and Franco, M. (2004). The GDP-bound form of Arf6 is located at the plasma membrane. *J. Cell Sci.* **117,** 2389–2398.

Naslavsky, N., Weigert, R., and Donaldson, J. G. (2003). Convergence of non-clathrin and clathrin-derived endosomes involves Arf6 inactivation and changes in phosphoinositides. *Mol. Biol. Cell* **14,** 417–431.

Naslavsky, N., Weigert, R., and Donaldson, J. G. (2004). Characterization of a nonclathrin endocytic pathway: Membrane cargo and lipid requirements. *Mol. Biol. Cell* **15,** 3542–3552.

Oude Weernink, P. A., Schmidt, M., and Jakobs, K. H. (2004). Regulation and cellular roles of phosphoinositide 5-kinases. *Eur. J. Pharm.* **500,** 87–99.

Palacios, F., Schweitzer, J. K., Boshans, R. L., and D'Souza-Schorey, C. (2002). ARF6-GTP recruits Nm23-H1 to facilitate dynamin-mediated endocytosis during adherins junctions assembly. *Nat. Cell Biol.* **4,** 929–935.

Paleotti, O., Macia, E., Luton, F., Klein, S., Partisani, M., Chardin, P., Kirchhausen, T., and Franco, M. (2005). The small G-protein Arf6-GTP recruits the AP-2 adaptor complex to membranes. *J. Biol. Chem.* **280,** 21661–21666.

Peters, P. J., Hsu, V. W., Ooi, C. E., Finazzi, D., Teal, S. B., Oorschot, V., Donaldson, J. G., and Klausner, R. D. (1995). Overexpression of wild-type and mutant ARF1 and ARF6: Distinct perturbations of nonoverlapping membrane compartments. *J. Cell Biol.* **128,** 1003–1017.

Prigent, M., Dubois, T., Raposo, G., Derrien, V., Tenza, D., Rosse, C., Camonis, J., and Chavrier, P. (2003). ARF6 controls post-endocytic recycling through its downstream exocyst complex effector. *J. Cell Biol.* **163,** 1111–1121.

Radhakrishna, H., and Donaldson, J. G. (1997). ADP-ribosylation factor 6 regulates a novel plasma membrane recycling pathway. *J. Cell Biol.* **139,** 49–61.

Randazzo, P. A., Nie, Z., Miura, K., and Hsu, V. W. (2000). Molecular aspects of the cellular activities of ADP-ribosylation factors *http://www.stke.org/cgi/content/full/OC_sig-trans;2000/59/re1.*

Santy, L. C. (2002). Characterization of a fast cycling ADP-ribosylation factor 6 mutant. *J. Biol. Chem.* **277,** 40185–40188.

Varnai, P., Lin, X., Lee, S. B., Tuymetova, G., Bondeva, T., Spat, A., Rhee, S. G., Hajnoczky, G., and Balla, T. (2002). Inositol lipid binding and membrane localization of isolated pleckstrin homology (PH) domains: Studies on the PH domains of phospholipase C delta 1 and p130. *J. Biol. Chem.* **277,** 27412–27422.

Vitale, N., Chasserot-Golaz, S., Bailly, Y., Morinaga, N., Frohman, M. A., and Bader, M.-F. (2002). Calcium-regulated exocytosis of dense-core vesicles requires the activation of ADP-ribosylation factor 6 by ARF nucleotide binding site opener at the plasma membrane. *J. Cell Biol.* **159,** 79–89.

Weidman, P., Roth, R., and Heuser, J. (1993). Golgi membrane dynamics imaged by freeze-etch electron microscopy: Views of different membrane coatings involved in tubulation versus vesiculation. *Cell* **75,** 123–133.

[38]   Interaction of Arl1 GTPase with the GRIP Domain of Golgin-245 as Assessed by GST (Glutathione-S-transferase) Pull-Down Experiments

By Lei Lu, Guihua Tai, and Wanjin Hong

Abstract

Arl1 is a member of the Arf/Arl family of Ras-like GTPase superfamily. Arl1 is enriched in the *trans*-Golgi network (TGN). We have recently shown that Arl1 regulates TGN recruitment of GRIP domain-containing Golgin-97 and Golgin-245 by interacting with the conserved GRIP domain present in their carboxyl (C)-termini. We describe here methods for the analysis of the interaction between Arl1(GTP) and the GRIP domain of Golgin-245 using *in vitro* GST pull-down experiments. GST-Arl1(GTP) can recover endogenous Golgin-245 from HeLa cell cytosol. Furthermore, GST-GRIP domain of Golgin-245 can efficiently retain endogenous active Arl1. A pull-down assay is developed to quantify the relative level of active Arl1.

Introduction

Arf/Arl family of small GTPases are key regulators of membrane trafficking in eukaryotic cells. Members of this family are classified into Arf, Arl, and Sar subfamilies. Arfs and Sar1 are relatively well-characterized and their functions include vesicle coat recruitment (Arf1–3 and Sar1) and regulation of phospholipids metabolism in membrane trafficking (Arf1 and Arf6) (Takai *et al.*, 2001). Arls are different from Arfs in three experimental activities: 1. Arls lack the cofactor activity in cholera toxin-catalyzed ADP ribosylation of $G\alpha s$; 2. Arls do not activate phospholipase D (PLD); and 3. Arls do not rescue the Arf1–Arf2 double deletion yeast mutant (Moss and Vaughan, 1998). The Arl subfamily contains a growing number of members, including Arl1–11, Arl4L, ArfRP1, and Ard1 (Burd *et al.*, 2004). Arls are implicated in diverse cellular functions as they have diverse cellular localizations, such as the Golgi apparatus (Arl1 and ArfRP1), nucleus (Arl4, Arl4L, and Arl7), mitochondria (Arl2), endoplasmic reticulum (ER) (Arl6), cilia (Arl6), and microtubule network (Arl2 and Arl3) (Burd *et al.*, 2004; Fan *et al.*, 2004). As the first member of Arls, Arl1 is one of the most characterized. Recently, we and others have shown

METHODS IN ENZYMOLOGY, VOL. 404   0076-6879/05 $35.00

that Arl1 is a key regulator of structure and function of the Golgi apparatus (Lu *et al.*, 2001; Van Valkenburgh *et al.*, 2001). To understand the molecular mechanism governing the cellular function of Arl1, yeast two-hybrid interaction screens have identified that golgin-97 and golgin-245 are candidate partners of Arl1(GTP) (Lu *et al.*, 2001; Van Valkenburgh *et al.*, 2001). Golgin-97 and golgin-245 are characterized by a conserved C-terminal GRIP domain of about 50 residues, and the GRIP domain is sufficient for Golgi targeting through a saturable process (Barr, 1999; Kjer-Nielsen *et al.*, 1999; Munro and Nichols, 1999). Our results suggest that these two golgins are effectors of Arl1 (Lu and Hong, 2003). The interaction of Arl1 (GTP) with the GRIP domain is the molecular mechanism responsible for Golgi recruitment of golgin-97 and golgin-245 (Lu and Hong, 2003). Using GST pull-down experiments, the following methods characterize the interaction of golgin-245 with Arl1 in a guanine nucleotide dependent manner, the hallmark of interaction between a small GTPase and its effectors.

The Retention of Cytosolic Golgin-245 by Immobilized GST-Arl1(GTP)

Expression of GST-Arl1 Fusion Protein

The coding region for full-length rat Arl1 (Gene Bank No.: U12402) is ligated into *Eco*RI/*Xho*I sites of bacteria expression vector pGEX-KG (Amersham, Buckinghamshire, UK). The resulting construct, Arl1-pGEX-KG, expresses recombinant fusion protein with GST fused to the N-terminus of Arl1. DH5α *E. coli* cells are transformed with this construct. Ampicillin-resistant colonies are then screened in miniscale for IPTG (isopropyl β-D-thiogalactopyranoside) (Invitrogen, Carlsbad, CA) induced-expression of GST-Arl1 (~50 kd). The bacterial clone expressing control GST is selected likewise by transforming bacteria cells with pGEX-KG vector.

For purification of GST or GST-Arl1, 500 ml of bacterial culture is grown to an OD600 of 0.5–1 at 37° followed by overnight induction with 0.25 mM IPTG (final concentration) at room temperature with shaking. Bacteria are pelleted by a spin using J6B swing bucket centrifuge (Beckmann, Fullerton, CA) and resuspended in 30 ml of lysis buffer (50 mM Tris-HCl, pH 8.0, 0.1% Triton X-100, 0.5 mM MgCl$_2$, 1 mg/ml lysozyme, 5 mM dithiothreitol [DTT], 0.5 mM phenylmethylsulfonyl fluoride [PMSF], and Complete Protease Inhibitors [EDTA-free] [Roche, Basel, Switzerland]). After 15 min of incubation on ice, the lysis of bacteria is aided by sonication 3 times (1 min each) on ice. The resulting lysate is cleared by centrifugation at 14 krpm for 30 min with a Sorvall SS34 rotor (Thermo Electron, Asheville, NC). To purify GST or GST-Arl1, 400 μl (drained volume) of glutathione sepharose 4B beads (Amersham) (about 500 μl slurry) equilibrated with GST washing buffer (50 mM Tris-HCl, pH 8.0, 0.1% Triton X-100, 0.5 mM MgCl$_2$) is added to the lysate. The mixture is rotated at 4° for 4 h, after which the sepharose beads are subjected to 3 times of washing with cold GST washing buffer. The beads are collected by centrifugation after each washing. The recombinant GST or GST-Arl1 attached onto the beads is then kept at 4° in GST washing buffer. Typically, this preparation yields about 1.5 mg of GST-Arl1 and more than 5 mg of GST.

In Vitro *Guanine Nucleotide Exchange Reactions of GST-Arl1*

A small GTPase, such as Arf1, has two different states of guanine nucleotide binding with either GTP or GDP. The conversion between these two states is catalyzed *in vivo* by GTPase activating proteins (GAPs) (from GTP to GDP) and guanine nucleotide exchange factors (GEFs) (from GDP to GTP). The guanine nucleotide exchange for Arf1 could be achieved *in vitro* in the presence of ethylenediaminetetraacetic acid (EDTA), detergent and a high concentration of GTP or GDP. A previous study reported that recombinant Arf1 purified from bacteria is predominantly in its GDP-bound form (Weiss *et al.*, 1989); thus, *in vitro* guanine nucleotide exchange is needed to obtain its active GTP-bound form. For Arf1, the exchange of guanine nucleotide is induced by chelating the bound Mg^{2+} and the addition of phospholipid and detergent. The exchange reaction can be stopped by high Mg^{2+} concentration (Kahn and Gilman, 1986).

Recombinant GST-Arl1 purified from bacteria is likely in its GDP-bound form. The *in vitro* exchange reactions for loading GTPγS or GDP onto GST-Arl1 are carried out according to previously described methods for Arf1 and Rab5 (Christoforidis and Zerial, 2000; Randazzo *et al.*, 1995) with some modifications. Briefly, GST-Arl1 bound to sepharose slurry is transferred to two tubes in equal amount. One is subjected to exchange for loading of GTPγS, the other to exchange for loading of GDP. For GTPγS exchange, the beads are first equilibrated in nucleotide exchange (NE) buffer (20 mM Hepes, pH 7.5, 100 mM NaCl, 10 mM EDTA, 5 mM MgCl$_2$, and 1 mM DTT) supplemented with 0.1% (w/v) sodium cholate and 10 μM GTPγS (Roche). Subsequently, beads are incubated with NE buffer containing 1 mM GTPγS, 0.1% sodium cholate and 3 mM L-α-dimyristoylphosphatidylcholine (DMPC) (Sigma, St. Louis, MO) (30 mM DMPC stock is freshly prepared by sonication in NE buffer) for 1.5 h at room temperature. The beads are then washed twice and equilibrated with nucleotide stabilizing (NS) buffer (20 mM Hepes, pH 7.5, 100 mM NaCl, 5 mM MgCl2, 1 mM DTT) supplemented with 10 μM GTPγS. The GST-Arl1(GTPγS) beads are stored at 4° in about 1 ml of NS buffer containing

10 μM GTPγS. The procedure is exactly the same for the GDP exchange to produce GST-Arl1(GDP) except that GTPγS is substituted by GDP (Roche) at the same concentrations. The recombinant GST beads are also subjected to the same GTPγS exchange reaction for mock control.

Preparation of HeLa Cell Cytosol

HeLa cells are grown on five 14-cm petri dishes to confluence in RPMI medium supplemented with 10% fetal bovine serum (FBS). The cells are first washed with cold NS buffer and then scraped off from the dish after adding 700 μl of cold NS buffer supplemented with 1 mM DTT and 1 mM PMSF. The scraped cells are combined (about 5 ml) and homogenized in a Dounce homogenizer. The resulting lysate is subjected to centrifugation in a table-top centrifuge to remove nuclei and unbroken cells. The cytosol is obtained by centrifugation at 55 krpm for 1.5 h with a TLA100.2 rotor (Beckman) to remove membrane debris from the lysate. The cytosol is first precleared by incubating with >1 mg of GST immobilized on sepharose beads. Bovine serum albumin (BSA, Sigma) is then added to the cleared cytosol at a final concentration of 4 mg/ml as a blocking reagent.

Immobilized GST-Arl1 Pulls-Down Endogenous Golgin-245 from HeLa Cytosol

Around 60 μg of GST fusion protein, GST-Arl1(GTPγS), GST-Arl1 (GDP), or GST, immobilized on the beads is first blocked for nonspecific binding by incubating with NS buffer containing 4 mg/ml BSA at 4° for 2 h. The beads are then incubated overnight with 500 μl of precleared HeLa cytosol at 4°. After washing the beads 3 times with cold NS buffer supplemented with 1 mM DTT, proteins retained on the beads are eluted by boiling in 2 × SDS sample buffer and resolved by 6% SDS-PAGE. After transferring to a filter, the amounts of golgin-245 retained by the beads are detected by Western blot using a monoclonal antibody against golgin-245 (BD Bioscience, Franklin Lakes, NJ). As shown in Fig. 1, a significant amount of golgin-245 is specifically retained by GST-Arl1(GTPγS), but not GST-Arl1(GDP) or control GST.

GST-GRIP Domain of Golgin-245 Pulls-down Active Arl1

Cloning and Expression of GST-GRIP

The multiple cloning site of pGEX-KG vector is modified by digestion with *Bam*HI and *Eco*RI, and subsequently ligated with the annealed DNA duplex of two oligonucleotides (5′-GAT CAG AAT CGC TA GCG

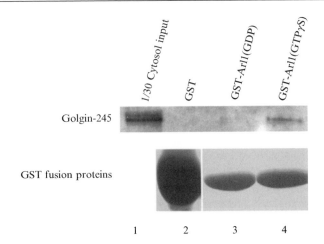

FIG. 1. GST-Arl1(GTP) pulls-down endogenous golgin-245 from HeLa cytosol. Upper panel, Western blot analysis of proteins retained by immobilized GST, GST-Arl1(GDP), or GST-Arl1(GTPγS) resolved by 6% SDS-PAGE. The filter is probed with a monoclonal antibody against golgin-245. Lower panel, Coomassie blue stained gel showing the loading of each GST fusion protein. Lane 1 is 3% of cytosol input used for the pull-down experiments. Golgin-245 is pulled-down by GST-Arl1(GTPγS) (lane 4), but not by GST-Arl1(GDP) (lane 3) or mock GTPγS treated GST (lane 2). This figure is adapted and modified from Lu and Hong, 2003; with the copyright permission of *Mol. Biol. Cell* of the American Society for Cell Biology.

GAT CCT TAA C-3' and 5'-AAT TGT TAA GGA TCC GCT AGC GAA TTC T-3'). The resulting vector has *Eco*RI site at the 5' of *Bam*HI site in its multiple cloning sites. The coding region for the extreme C-terminal 106 amino acid region of human golgin-245 (Gene Bank No.: NP_002069), which comprises of its GRIP domain, is cloned into *Eco*RI/*Bam*HI sites of the modified vector. The GST-GRIP immobilized on the glutathione sepharose beads is produced similarly as described above for GST-Arl1.

GST-GRIP Selectively Pulls-Down Cytosolic Active Arl1

GST-GRIP on beads is first blocked by NS buffer containing 4 mg/ml BSA. About 60 μg of GST-GRIP or GST beads is incubated at 4° overnight with 500 μl of HeLa cytosol with the addition of 0.1% Triton X-100 and 100 μM GTPγS (final concentrations). The beads are washed 3 times with cold NS buffer containing 0.1% Triton X-100. Proteins retained on the beads are then eluted by boiling in 2 × SDS sample buffer and resolved by 12% SDS-PAGE. After transferring to a filter, Arl1 is detected by Western blot probed with rabbit polyclonal antibody against Arl1. As shown in Fig. 2, cytosolic Arl1 is retained by immobilized GST-GRIP but not GST. The binding is restricted to active GTP-bound Arl1, as GST-GRIP does not pull-down transiently expressed Arl1T31N (the GDP-restricted mutant of Arl1), whereas transiently expressed Arl1Q71L (the GTP-restricted mutant of Arl1) is efficiently pulled-down by immobilized GST-GRIP (data not shown). As evident in Fig. 2, Arl1 is very efficiently retained by GST-GRIP beads and this observation suggests that this pull-down experiment can be developed to quantify the relative amount of GTP-bound Arl1 in the cell.

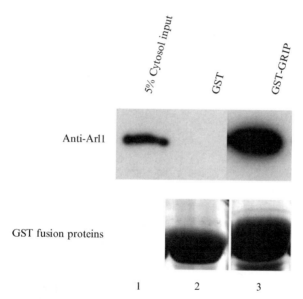

FIG. 2. GST-GRIP of golgin-245 pulls-down endogenous Arl1 in the presence of GTPγS. Upper panel, Western blot of proteins retained by immobilized GST or GST-GRIP after being resolved by 12% SDS-PAGE. The filter is probed with a rabbit polyclonal antibody against Arl1. Lower panel, Coomassie blue stained gel showing the loading of each GST fusion protein. Lane 1 is the 5% of cytosol input used for the pull-down experiments. Arl1 is pulled-down by GST-GRIP (lane 3), but not by GST (lane 2). This figure is adapted and modified from Lu and Hong, 2003; with the copyright permission of *Mol. Biol. Cell* of the American Society for Cell Biology.

A Pull-down Assay Measuring the Relative Amount of Cellular
  Active Arl1

In eukaryotes, the balance between GTP- and GDP-bound forms of a small GTPase is controlled by its GAPs and GEFs, which, in turn, are regulated by upstream signaling events. Knowing the content of the active form of a small GTPase can shed light on understanding its cellular regulation in a complex signaling network. Methods have been developed for measuring the relative levels of active forms of small GTPases, and they all utilize GTPase-binding regions of their effectors, which specifically bind to GTP-bound forms of GTPases. For examples, Ras binding domain (RBD) of Raf is used to assay the level of active Ras (de Rooij and Bos, 1997); p21 binding domain (PBD) of PAK is used to measure relative amount of active Rho (Benard and Bokoch, 2002); and the GAT domain of GGA1 is used to pull-down active Arf1 (Santy and Casanova, 2001; Shinotsuka et al., 2002). The identification of the GRIP domain of golgin-97 and golgin-245 as the binding domain for GTP-bound Arl1 offers a strategy to develop an assay for measuring the relative amount of active Arl1 in the cell. Based on our finding that active Arl1 is retained efficiently by immobilized GST-GRIP and the conditions described in other assays (Santy and Casanova, 2001; Shinotsuka et al., 2002), the following protocol is developed and used to assay the level of active Arl1 in cultured mammalian cells. The change of levels of active Arl1 is observed when cells are treated with Brefeldin A (BFA) and during recovery after treatment.

Methods

NRK cells are grown on 9-cm petri dishes to about 50% confluence in DMEM medium supplemented with 10% FBS. Cells are treated at 37° with 10 μg/ml of BFA (Epicenter, Madison, WI) for 0 min, 5 min, 10 min, 30 min, 60 min, or 60 min followed by recover incubation in the absence of BFA for 60 min, respectively. The culture medium is subsequently changed to cold phosphate buffered saline (PBS) and the culture dishes are chilled on ice. After removing cold PBS, 450 μl chilled cell lysis buffer (50 mM Tris, pH 7.5, 100 mM NaCl, 2 mM MgCl$_2$, 0.1% SDS, 0.5% sodium deoxycholate, 1% Triton X100, 10% glycerol) supplemented with Complete Protease Inhibitors (EDTA-free) is applied to each dish followed by scraping on ice to collect the cell lysate. The lysate is briefly vortexed and centrifuged at 4° Triton X-100 to remove cell debris. Then 50 μl of glutathione sepharose 4B beads pre-equilibrated with cell lysis buffer is used to preclear the lysate at 4° for 10 min. An aliquot (5% of the lysate, for example) of the precleared lysate is set aside as the loading control. 400

μl of each precleared lysate is then added to a tube containing ~30 μg of GST-GRIP immobilized on glutathione sepharose 4B beads prewashed with cell lysis buffer. After rotating the mixture at 4° for 30 min, the beads are subsequently washed 3 times with cold cell lysis buffer. The bound Arl1 is eluted by boiling in 2 × SDS sample buffer and resolved by 12% SDS-PAGE. After transferring to a filter, Western blot is performed using rabbit polyclonal antibody against Arl1. A representative result is shown in Fig. 3.

Results

In NRK cells without BFA treatment, around 20% of Arl1 is active as measured by this GST-GRIP pull-down assay (compare lane 1 to the loading control of lane 8 in Fig. 3). The level of active Arl1 is greatly reduced after 5 min of BFA treatment (lane 2) and further reduced when BFA treatment proceeds to 10 min (lane 3). After 10 min of BFA treatment, the level of active Arl1 is reduced to a plateau, which is ~2% of total cellular Arl1. In contrast to Arf1 and its effectors, which dissociate from the Golgi apparatus in less than 2 min, Arl1 dissociates from the Golgi apparatus only after 5 min of BFA treatment (Lowe *et al.*, 1996; Lu *et al.*, 2003). The kinetics of Arl1 inactivation during BFA treatment as measured

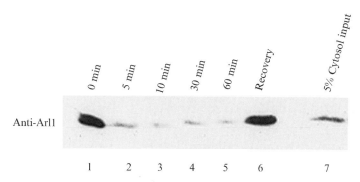

FIG. 3. An assay for measuring the relative amount of active Arl1 in control and BFA-treated NRK cells. NRK cells are treated with 10 μg/ml of BFA for 0 min (lane 1), 5 min (lane 2), 10 min (lane 3), 30 min (lane 4), 60 min (lane 5), or 60 min followed by recovery incubation in the absence of BFA for 60 min (lane 6). The cell lysates are incubated with immobilized GST-GRIP and the retained proteins are resolved by 12% SDS-PAGE and transferred to a filter, which is probed with rabbit antibody against Arl1. 5% of cytosol input is used as a loading control (lane 7).

here is consistent with the conclusion that its dissociation from the Golgi membrane is due to conversion of its GTP-bound to GDP-bound form. Typically, around 20% of total cellular Arl1 is retained by immobilized GST-GRIP, implying that around 20% of total Arl1 in NRK cells is in the GTP-bound active form. A similar result is also obtained in 293T cells (data not shown). There are factors that could complicate the interpretation of this data. For example, it is possible that the active Arl1 content in the cell is actually higher due to incomplete pull-down of Arl1(GTP) from the cell lysate by GST-GRIP. On the other hand, as reported for POR1 (Zhu et al., 2000), the binding to Arl1's effector GRIP domain could potentially enhance the conversion of Arl1 to its active form, which would result in higher estimation. Although the lack of other experiments precludes an absolute measurement of the percentage of active Arl1, the data is in good agreement with the finding that, generally, 20~30% of Ras GTPase is in active form in mammalian cells (Burgering et al., 1991; Osterop et al., 1992). This method is useful to assay the relative amount of cellular GTP-bound Arl1 in response to various cellular events (such as BFA treatment as described previously).

Comments

In the pull-down experiments using recombinant protein, the GRIP domain of golgin-245 interacts selectively with GTP-bound Arl1, thus supporting the proposal that golgin-245 is an effector of Arl1 *in vivo*. This interaction is direct as shown recently by our X-ray crystal structure of the GRIP domain complexed with Arl1(GTP) (Wu et al., 2004).

References

Barr, F. A. (1999). A novel Rab6-interacting domain defines a family of Golgi-targeted coiled-coil proteins. *Curr. Bio.* **9,** 381–384.

Benard, V., and Bokoch, G. M. (2002). Assay of Cdc42, Rac, and Rho GTPase activation by affinity methods. *Methods Enzymol.* **345,** 349–359.

Burd, C. G., Strochlic, T. I., and Gangi Setty, S. R. (2004). Arf-like GTPases: Not so Arf-like after all. *Trends Cell Biol.* **12,** 687–694.

Burgering, B. M., Medema, R. H., Maassen, J. A., van de Wetering, M. L., van der Eb, A. J., McCormick, F., and Bos, J. L. (1991). Insulin stimulation of gene expression mediated by p21ras activation. *EMBO J.* **10,** 1103–1109.

Christoforidis, S., and Zerial, M. (2000). Purification and identification of novel Rab effectors using affinity chromatography. *Methods* **20,** 403–410.

de Rooij, J., and Bos, J. L. (1997). Minimal Ras-binding domain of Raf1 can be used as an activation-specific probe for Ras. *Oncogene* **14,** 623–625.

Fan, Y., Esmail, M. A., Ansley, S. J., Blacque, O. E., Boroevich, K., Ross, A. J., Moore, S. J., Badano, J. L., May-Simera, H., Compton, D. S., Green, J. S., Lewis, R. A., van Haelst, M. M., Parfrey, P. S., Baillie, D. L., Beales, P. L., Katsanis, N., Davidson, W. S., and Leroux, M. R. (2004). Mutations in a member of the Ras superfamily of small GTP-binding proteins causes Bardet-Biedl syndrome. *Nat. Genet.* **36,** 989–993.

Kahn, R. A., and Gilman, A. G. (1986). The protein cofactor necessary for ADPribo-sylation of Gs by cholera toxin is itself a GTP binding protein. *J. Biol. Chem.* **261,** 7906–7911.

Kjer-Nielsen, L., Teasdale, R. D., van Vliet, C., and Gleeson, P. A. (1999). A novel Golgi-localisation domain shared by a class of coiled-coil peripheral membrane proteins. *Curr. Biol.* **9,** 385–388.

Lowe, S. L., Wong, S. H., and Hong, W. J. (1996). The mammalian ARF-like protein 1 (Arl1) is associated with the Golgi complex. *J. Cell Sci.* **109,** 209–220.

Lu, L., and Hong, W. (2003). Interaction of Arl1-GTP with GRIP domains recruits autoantigens golgin-97 and golgin-245/p230 onto the Golgi. *Mol. Biol. Cell* **14,** 3767–3781.

Lu, L., Horstmann, H., Ng, C., and Hong, W. J. (2001). Regulation of Golgi structure and function by ARF-like protein 1 (Arl1). *J. Cell Sci.* **114,** 4543–4555.

Moss, J., and Vaughan, M. (1998). Molecules in the ARF orbit. *J. Biol. Chem.* **273,** 21431–21434.

Munro, S., and Nichols, B. J. (1999). The GRIP domain—a novel Golgi-targeting domain found in several coiled-coil proteins. *Curr. Biol.* **9,** 377–380.

Osterop, A. P., Medema, R. H., Bos, J. L., vd Zon, G. C., Moller, D. E., Flier, J. S., Moller, W., and Maassen, J. A. (1992). Relation between the insulin receptor number in cells, autophosphorylation and insulin-stimulated Ras.GTP formation. *J. Biol. Chem.* **267,** 14647–14653.

Randazzo, P. A., Weiss, O., and Kahn, R. A. (1995). Preparation of recombinant ADP-ribosylation factor. *Methods Enzymol.* **257,** 128–135.

Santy, L. C., and Casanova, J. E. (2001). Activation of ARF6 by ARNO stimulates epithelial cell migration through downstream activation of both Rac1 and phospholipase D. *J. Cell Biol.* **154,** 599–610.

Shinotsuka, C., Yoshida, Y., Kawamoto, K., Takatsu, H., and Nakayama, K. (2002). Overexpression of an ADP-ribosylation factor-guanine nucleotide exchange factor, BIG2, uncouples brefeldin A-induced adaptor protein-1 coat dissociation and membrane tubulation. *J. Biol. Chem.* **277,** 9468–9473.

Takai, Y., Sasaki, T., and Matozaki, T. (2001). Small GTP-binding proteins. *Physiol. Rev.* **81,** 153–208.

Van Valkenburgh, H., Shern, J. F., Sharer, J. D., Zhu, X., and Kahn, R. A. (2001). ADP-ribosylation factors (ARFs) and ARF-like 1 (ARL1) have both specific and shared effectors: Characterizing ARL1-binding proteins. *J. Biol. Chem.* **276,** 22826–22837.

Weiss, O., Holden, J., Rulka, C., and Kahn, R. A. (1989). Nucleotide binding and cofactor activities of purified bovine brain and bacterially expressed ADP-ribosylation factor. *J. Biol. Chem.* **264,** 21066–21072.

Wu, M., Lu, L., Hong, W., and Song, H. (2004). Structural basis for recruitment of GRIP domain golgin-245 by small GTPase Arl1. *Nat. Struct. Mol. Biol.* **11,** 86–94.

Zhu, X., Boman, A. L., Kuai, J., Cieplak, W., and Kahn, R. A. (2000). Effectors increase the affinity of ADP-ribosylation factor for GTP to increase binding. *J. Biol. Chem.* **275,** 13465–13475.

[39] Functional Analysis of Arl1 and Golgin-97 in Endosome-to-TGN Transport Using Recombinant Shiga Toxin B Fragment

By GUIHUA TAI, LEI LU, LUDGER JOHANNES, and WANJIN HONG

Abstract

A direct transport route from early/recycling endosome (EE/RE) to the *trans*-Golgi network (TGN) is exploited by Shiga toxin (Mallard *et al.*, 1998) and TGN38 (Ghosh *et al.*, 1998). To facilitate the study of this pathway, both *in vivo* and *in vitro* transport assays using recombinant Shiga toxin B fragments (STxB) as protein cargos have facilitated the analysis of this transport event (Johannes *et al.*, 1997; Mallard *et al.*, 1998, 2002; Tai *et al.*, 2004). We describe here the application of these assays to study the role of a small GTPase Arl1 and its effector golgin-97 in this transport process.

Introduction

Exogenously added protein toxins such as Shiga toxin and Cholera toxin as well as endogenous proteins such as TGN38 are transported directly from the EE/RE to the TGN, bypassing late endosomes (Ghosh *et al.*, 1998; Mallard *et al.*, 1998; Sandvig and van Deurs, 2002). Shiga toxin is composed of two subunits. The A-subunit is a RNA N-glycosidase and inhibits protein synthesis by inactivating 28S RNA of the 60S ribosomal subunit in the cytosol. The B-subunit is a homo pentamer of five B-fragments, and it is responsible for the binding of Shiga toxin to cell surface receptor glycosphingolipid Gb3, internalization and subsequent retrograde trafficking from plasma membrane, via endosomes and then the Golgi apparatus, to the endoplasmic reticulum (Sandvig and van Deurs, 2002). Using recombinant Shiga toxin B fragment as a protein cargo, both *in vivo* and *in vitro* transport assays have been used to analyze the role played by several proteins in EE/RE-to-TGN transport (Johannes *et al.*, 1997; Mallard *et al.*, 1998, 2002; Tai *et al.*, 2004).

Arl1 is an Arf1-like small GTPase and is implicated in the regulation of Golgi structure and function (Lu *et al.*, 2001). Morphological, biochemical, and yeast two-hybrid analyses have shown that GTP-bound Arl1 interacts with the GRIP domain of golgin-97 and that the interaction is responsible

METHODS IN ENZYMOLOGY, VOL. 404
0076-6879/05 $35.00
DOI: 10.1016/S0076-6879(05)04039-5

for recruitment of golgin-97 to the TGN (Lu and Hong, 2003). Mutation of Y697 conserved in all GRIP domains to Alanine, which abolished the interaction with Arl1, abrogates the TGN-targeting property of the GRIP domain.

The function of Arl1 and golgin-97 in EE/RE-to-TGN transport has been investigated using the *in vivo* and *in vitro* STxB transport assays (Lu *et al.*, 2004). Both Arl1 and golgin-97 are involved in STxB transport from the EE/RE to the TGN. Golgin-97 may function as a tethering factor in this trafficking event. The GRIP domain of golgin-97 and its interaction with Arl1 are important for the function of golgin-97. We summarize three assays used to study the role of Arl1 and golgin-97.

Inhibition of *In Vitro* STxB Transport by Recombinant GRIP
    Domain of Golgin-97

Expression of GST-GRIP Domain of Golgin-97

The coding region for the extreme C-terminal 108 amino acids, which comprises GRIP domain of human golgin-97 (GenBank Accession No.: AI803928), is cloned into *Eco*RI/*Bam*HI site of a modified pGEX-KG vector (see Chapter 38 of this volume). The Y697 to A mutation is introduced by using standard PCR mutagenesis and cloned similarly. The resulting constructs express fusion proteins with GST located at the N-terminus of the GRIP domain (GST-GRIP or GST-GRIP/Y697A, respectively). GST-GRIP and GST-GRIP/Y697A fusion proteins are purified as described in Chapter 38 in this volume. Fusion proteins on glutathione sepharose 4 Fast Flow beads (Amersham, Buckinghamshire, UK) are washed with 50 mM Tris-HCl, pH 8.0, 0.5 mM MgCl$_2$ and eluted with the same buffer containing 10 mM reduced glutathione at room temperature. Eluted proteins are concentrated and dialysed extensively against cytosol buffer (25 mM Hepes, pH 7.3, and 125 mM KOAc).

In Vitro *STxB Transport Assay*

Media and buffers

  Complete MEM medium (for STxB binding and internalization): sulfate-free minimum essential medium (Sigma-Aldrich, St. Louis, MO) containing 5% fetal bovine serum (FBS), 1 mM CaCl$_2$, and 10 mM Hepes, pH 7.3.
  HKM buffer (for SL-O perforation): 25 mM Hepes, pH 7.3, 125 mM KOAc, 2.5 mM Mg(OAc)$_2$, and 1 mM DTT.

HES buffer (for semi-intact cells): 25 mM Hepes, pH 7.3, 1 mM EDTA, and 0.25 M sucrose.

RB buffer (reaction buffer): 250 mM Hepes, pH 7.3, 250 mM KCl, and 15 mM Mg(OAc)$_2$.

ATP mix (for ATP regeneration): 10 mM ATP, pH 7.0 (Roche, Basel, Switzerland), 150 mM creatine phosphate (Roche), 210 units/ml creatine phosphokinase (Roche), and 20 mM MgCl$_2$.

Cytosol buffer (for cytosol, antibody, antigen, and recombinant proteins): 25 mM Hepes, pH 7.3, and 125 mM KOAc.

10T80 Buffer (for STxB preparation): 10 mM Tris-HCl, pH 8.0.

20T80 Buffer (for STxB preparation): 20 mM Tris-HCl, pH 8.0.

20T75 Buffer (for STxB preparation): 20 mM Tris-HCl, pH 7.5.

Preparation of STxB

The plasmid, pSTxB(sulf)2, for expressing recombinant Shiga toxin B fragment with two sulfation sites (STxB) was described previously (Johannes *et al.*, 1997) and was transformed into *E. coli* DH5α cells. The STxB gene is under the regulation of heatshock inducible promoter. The protein is secreted to periplasm.

Extraction of Periplasm

LB medium (125 ml) containing 100 μg/ml ampicillin (LBamp) is inoculated with DH5α cells transformed with pSTxB(sulf)2. Cells are grown overnight at 30°. The culture is diluted 4 times with LBamp medium preheated to 50° and then incubated at 42° for 4 h. Cells are pelleted by centrifugation at 6000 rpm for 10 min with a GS3 rotor. After washing twice with 100 ml of 10T80 buffer, the cells are resuspended in 33 ml of 10T80 buffer containing 25% sucrose and 1mM EDTA and incubated at room temperature for 10 min. After spinning down the cells (7000 rpm for 20 min with a GS3 rotor), cells are resuspended in 33 ml cold water containing complete protease inhibitors (Roche). The periplasm is extracted by incubation on ice for 10 min with gentle agitation from time to time. After centrifugation at 7000 rpm for 10 min with a GS3 rotor, the supernatant containing the periplasm is harvested. After the addition of 2 M Tris-HCl, pH 8.0 to a final concentration of 20 mM, the periplasmic extract can be frozen in liquid nitrogen and stored at $-70°$.

Purification of STxB by Q-sepharose

Q-sepharose beads (Amersham) equilibrated in 20T80 buffer are added to periplasmic extract at 100 ml extract per ml beads. The extract is incubated with the beads in a cold room for 3 h with rotation. The beads are recovered by brief spin (100g, for 5 min) and subsequently transferred to an empty column. After washing with 5× bed volumes of 20T80 buffer, the bound proteins are eluted with 20T75 buffer containing 0.5 M NaCl. Fractions are collected at 1 ml per fraction. An aliquot of each fraction is analyzed for protein contents, and fractions with high concentrations are combined.

Purification on Mono Q Column

The combined fractions derived from Q-sepharose are diluted 5–10 times with 20T75 buffer and loaded onto a Mono Q HR 5/5 column. The column is eluted at a flow rate of 1 ml/min with 15 ml of 20T75 buffer followed by a salt gradient from 0 to 1 M NaCl in 20T75 buffer within 46 min. Fractions (1 ml each) are collected. STxB is generally eluted at about 0.4 M NaCl as a single peak. Aliquots of fractions can be analyzed by SDS-PAGE for the presence of STxB and for the purity of STxB.

Preparation of Rat Liver Cytosol

Two-month old rats (Wistar) are starved for 1 day and then sacrificed by CO_2 and neck dislocation. The head of the rat is severed to drain away the blood from the heart. The liver is then removed as quickly as possible and washed 3 times with ice-cold cytosol buffer. After trimming away fat tissues, the liver is cut into small pieces in about 25 ml of ice-cold cytosol buffer. The liver pieces are homogenized in 3 volume of cytosol buffer using a Dunce homogenizer. The homogenate is centrifuged for 30 min at 12,000 rpm with a SS34 rotor at 4°. The supernatant is carefully removed, filtered through cheese cloths, and then centrifuged again for 2 h at 40,000 rpm with a SW41 rotor at 4°.

After centrifugation, there will be 3 phases: top phase (white) is the fat, middle phase (red) is the cytosol, and the bottom phase (deep red) contains the red blood cells and debris. The cytosol is collected using a needle attached to a syringe and aliquoted. The aliquots are snap-frozen in liquid nitrogen and stored at −70°.

Preparation of Semi-Intact HeLa Cells

Sulfate Starvation

HeLa cells are grown on 10 cm tissue culture dishes for 2 days to 90% confluence in RPMI medium containing 10 mM Hepes, pH 7.3, and 10% FBS. After rinsing cells once with warm complete MEM medium, 10 ml of warm complete MEM medium is added and the dishes are incubated for 1 h at 37° in a CO_2 incubator.

STxB Binding and Internalization

Culture medium is removed and the dishes are cooled on ice. After the addition of 5 ml ice-cold complete MEM medium containing 1 μM of STxB, the dishes are incubated in an 18° water bath for 60 min with gentle rotations at 15 min intervals (STxB is accumulated in the EE/RE at 18°). The STxB-containing medium is aspirated. The dishes are rinsed once and then incubated in 10 ml fresh complete MEM medium pre-incubated at 18° for a further 30 min at 18° (to chase the internalized STxB in the EE/RE).

Cell Perforation by Streptolysin O (SL-O)

Culture medium is removed. The cells are rinsed with ice-cold HKM buffer and then incubated on ice for 10 min with 5 ml HKM buffer containing 2 $\mu g/ml$ of SL-O (SL-O binds to cell surface at 4°). SL-O is purchased from Prof. Dr. Sucharit Bhakdi, Johannes Gutenberg-Universitat Mainz, Germany, and dissolved in HKM buffer at 0.25 mg /ml. The SL-O solution is aspirated and the cells are rinsed once with ice-cold HKM buffer. Fresh ice-cold HKM buffer is then added and the dish is incubated at 18° for 30 min (cells perforate at 18°). After rinsing once with ice-cold HKM buffer, 10 ml of ice-cold HKM buffer is added, and the dish is incubated on ice for 1 h to leak out the cytosol.

Collection of Semi-Intact Cells

After aspirating the HKM buffer, 5 ml of ice-cold HES buffer is added to the dish. The cells are scraped off with a rubber cell-lifter and the resulting cell suspension is transferred into a 15-ml Falcon tube (BD Biosciences, Franklin Lakes, NJ). This process is repeated once to ensure maximum recovery of the scraped cells. Subsequently, the cells are recovered by centrifugation at 750g for 5 min at 4° and resuspended in 150 μl of ice cold HES buffer by pipetting up and down 10 times using a blue tip. Typically, 180–200 μl of semi-intact cells with a protein concentration of 1.5–4 $\mu g/\mu l$ is obtained. The semi-intact cells are stored at –70° if not used immediately.

Re-Constitution of EE/RE-to-TGN Transport

The transport reaction mixture is assembled on ice in 1.5-ml micro-centrifuge tubes (nonautoclaved) in the following order (tapping the bottom of the tube after each addition to ensure good assembly): First, add 20 μl of cytosol buffer (positive control) or cytosol buffer containing 10 μg

of GST-GRIP, 10 μg of GST-GRIP/Y697A, or 10 μg of heat-inactivated GST-GRIP; second, add 5 μl of rat liver cytosol (final concentration of around 1.2 mg/ml) or cytosol buffer (negative control); third, add 20 μl of premix (4 μl RB buffer, 6 μl ATP mix, 0.15 μl of 100 mM GTP) (the final concentration is 0.2 mM GTP, GTP is from Roche), 0.15 μl of 0.5 M DTT (the final concentration is 1 mM DTT, DTT is from Sigma-Aldrich), and 9.7 μl H$_2$O; fourth, add 10 μl of semi-intact cells; and finally, add 20 μl of ^{35}S-sulfate (the final concentration is 1 mCi ^{35}S/ml, ^{35}S-sulfate is purchased from Amersham).

The mixture is pre-incubated on ice for 1 h followed by incubation in a 37° water bath for 90 min (transport reaction). Membrane pellet is recovered by centrifugation at 14,000g for 30 min at 4° and careful removal of the supernatant. The pellet is dissolved in 30 μl Laemmli sample buffer by brief vortex first and then heated at 95° for 5 min and finally vortexed thoroughly.

Analysis of ^{35}S-Sulfated STxB

Proteins are separated on 15% modified Laemmli peptide gels, and transferred onto nitrocellulose filters. (*Note:* Do not let the sample front run into the running buffer as it contains significant amounts of free ^{35}S-sulfate. Cut the front off before transfer.) Dry filters are assembled in a phosphoimaging cassette and exposed overnight or for longer for a good signal to noise ratio, although the bands are visible in one hour. Scanning and quantification are carried out using a Molecular Imager FX (Bio-Rad, Hercules, CA).

Cytosol-dependent transport is obtained by subtracting the value of negative control (STxB transport without cytosol) from each assay and used for the calculation thereafter. The extent of transport is presented as percentage of STxB transport relative to the positive control (complete transport reaction without extra reagents), which is defined as 100%.

Result and Discussion

Hela cells are sulfate-starved and incubated with STxB at 18° to accumulate STxB in the EE/RE (synchronization). Then the plasma membrane is selectively perforated with SL-O and semi-intact cells are harvested. EE/RE-TGN transport is reconstituted in test tubes with the semi-intact cells supplemented with rat liver cytosol, ATP regeneration system, ^{35}S-sulfate, and the reagents to be tested at 37°. STxB is specially engineered to contain two tyrosine sulfation sites. Therefore, the arrival of STxB to the TGN is monitored by its acquisition of ^{35}S incorporation, as tyrosine sulfation is a

TGN specific event. ^{35}S-labeled STxB is quantified after SDS-PAGE and blotting to nitrocellulose filters.

The *in vitro* transport assay is routinely characterized to ensure that the assay has indeed faithfully reconstituted the EE/RE-to-TGN transport *in vitro* (Tai *et al.*, 2004). First, the assay must be cytosol-dependent. The basal level of transport in the absence of cytosol is usually 5–10% of that in the presence of 1.2 mg/ml of rat liver cytosol. Second, the assay is tested with some commonly used inhibitors such as Brefeldin A and guanosine 5'-O-(3-thio) triphosphate (GTPγS), as their inhibitory role in EE/RE-to-TGN traffic is well documented. Finally, the assay is tested using antibodies (such as antibody against syntaxin16) or recombinant proteins (such as GST-GRIP of golgin-97) that have been shown to inhibit *in vitro* EE/RE-to-TGN transport of STxB.

To investigate the function of golgin-97 in STxB transport from the EE/RE to the TGN, two GST fusion proteins, GST-GRIP containing intact golgin-97 GRIP domain, and GST-GRIP/Y697A containing a mutation in the GRIP domain, are prepared and tested in the *in vitro* transport assay.

Lane 1 and lane 4 in Fig. 1 are the positive controls (i.e., the complete transport reaction), which are set as 100%. The addition of 133 μg/μl of GST-GRIP reduces transport to 30% of the control (lane 2), suggesting that golgin-97 GRIP domain inhibits the STxB transport. The transport is essentially unaffected when the same amount of heat-denatured GST-GRIP is added to the reaction (lane 3), indicating that proper folding of the GRIP domain is necessary for the inhibition. On the other hand, GST-GRIP/Y697A does not inhibit STxB transport (lane 5), suggesting that residue Y697 is required for the GRIP domain to inhibit *in vitro* transport. Furthermore, the lack of inhibition by GST-GRIP/Y697A suggests that GST moiety dose not inhibit STxB transport. Together, these data suggest that the GRIP domain of golgin-97 inhibits *in vitro* STxB transport and the conserved residue Y697, which was earlier shown to be important for Golgi targeting and interaction with Arl1, is essential for its function.

Effect of Arl1 Knockdown on *In Vivo* STxB Transport as Revealed by a Biochemical Approach

siRNA Treatment

siRNA for human Arl1 (5'-AAG AAG AGC UGA GAA AAG CCA-3') and control siRNA with a random sequence (5'-AAG UCG GUG UGC UCU UGU UGG-3') are synthesized at Dharmacon Research Inc. (Lafayette, CO). Knockdown experiments are typically conducted in HeLa cells seeded in 24-well plates. After reaching ~30% confluence, the cells are

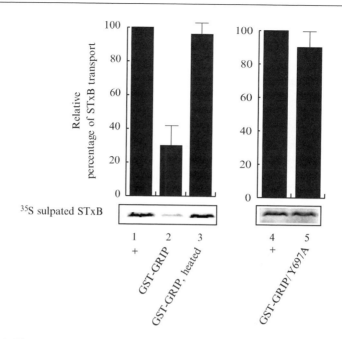

FIG. 1. The recombinant GRIP domain of golgin-97 inhibits *in vitro* EE/RE-TGN transport. STxB transport assays are carried out in the absence (lanes 1 and 4) or presence of 133 μg/ml of GST-GRIP domain proteins (lane 2, GST-GRIP; lane 3, heat-denatured GST-GRIP; lane 5, GST-GRIP/Y697A). Extent of transport relative to controls (set as 100%) is calculated from two to three independent experiments. Error bars represent standard deviations. Bottom panel shows representative phosphor-imaging data. Reprinted from *Molecular Biology of the Cell* (*Mol. Biol. Cell* (2004). **15,** 4426–4443.) with the permission of The American Society for Cell Biology.

transfected with either Arl1 siRNA or control siRNA using Oligofectamine transfection reagent (Invitrogen, Grand Island, NY) according to protocol provided by Dharmacon Research Inc. The optimal knockdown is observed after 48 h of incubation with the siRNA mixture and typically ~80% of endogenous Arl1 is knocked down as revealed by Western blot analysis (Fig. 2A). It is at this stage that the *in vivo* STxB transport assay is performed.

In Vivo *STxB Transport Assay—Biochemical Analysis*

Cells are washed twice with complete MEM medium (see *in vitro* transport assay) and then sulphate-starved in the same medium for 1 h at 37° in a CO_2 incubator. The cells are incubated with complete MEM medium containing 1 μM of STxB and 0.5 mCi/ml ^{35}S-sulfate (Amersham) for 90 min at 37° in a CO_2 incubator. The medium is aspirated carefully.

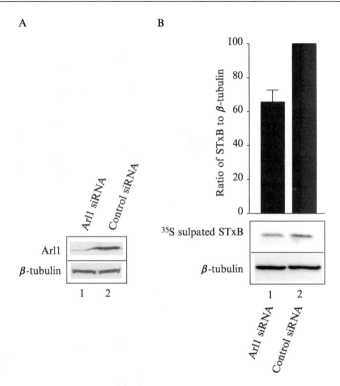

FIG. 2. Knockdown of Arl1 reduces *in vivo* trafficking of STxB to the TGN. HeLa cells grown on 24-well plates are transfected with Arl1 or control siRNA for 48 h followed by (A) Western blot analysis with antibodies against Arl1 and β-tubulin or (B) analysis of the levels of ³⁵S-labeled STxB (assessed by *in vivo* transport) and β-tubulin (assessed by Western blot) after the *in vivo* STxB transport assay. Adapted from *Molecular Biology of the Cell* (*Mol. Biol. Cell* (2004). **15**, 4426–4443) with the permission of The American Society for Cell Biology.

The cells are washed three times with ice-cold PBS containing 1 m*M* MgCl$_2$ and 1 m*M* CaCl$_2$ and solubilized in Laemmli sample buffer. Proteins are separated on 15% modified Laemmli peptide gels and transferred onto nitrocellulose filters. The filters are scanned and ³⁵S-labeled STxB bands are quantified as described in the *in vitro* transport assay.

Result and Discussion

The levels of Arl1 and β-tubulin in cells treated with Arl1 or control siRNA are examined by Western blotting analysis. Figure 2A shows that Arl1 (lane 1) is reduced to <30% in Arl1 siRNA transfected cells as compared to cells transfected with control siRNA (lane 2). The arrival of STxB to the TGN from plasma membrane is analyzed by its acquisition of ^{35}S-incorporation. Figure 2B shows that STxB transport is reduced to 66% in Arl1 siRNA-treated cells (Fig. 2B, lane 1) relative to cells transfected with control siRNA (Fig. 2B, lane 2). Different from the *in vitro* STxB biochemical transport assay described previously, this *in vivo* STxB transport experiment actually assays the transport of the STxB from plasma membrane to the TGN in Arl1-depleted HeLa cells. These results suggest that endogenous Arl1 is involved in STxB transport to the TGN.

Effect of Arl1-Q71L Overexpression on *In Vivo* STxB Transport as Revealed by a Morphological Approach

Rat Arl1 coding sequence (Gene Bank No.: U12402) is subjected to PCR mutagenesis to introduce Q71L mutation and subsequently cloned into *Eco*RI/*Bam*HI sites of tetracycline-inducible mammalian expression vector pSTAR (Lu *et al.*, 2001; Zeng *et al.*, 1998). HeLa cells are cultured in RPMI medium supplemented with 10% FBS at 37°. Transient transfection of Arl1-Q71L in pSTAR is carried out using Effectene transfection reagent (QIAGEN, Hilden, Germany) according to the manufacturer's protocol. 12 μg/ml of doxycyclin is added to induce the expression of Arl1. After about 20 h of incubation with the DNA mixture, transfected cells are incubated in growing medium (RPMI with 10% FBS) containing 1 μM STxB for 3 h at 37° in a CO_2 incubator. Cells are washed three times with PBS containing 1 mM $MgCl_2$ and 1 mM $CaCl_2$ and then processed for indirect immunofluorescent microscopy as described elsewhere (Lu *et al.*, 2001). Overexpressed Arl1-Q71L can be detected by a limited amount of Arl1 rabbit polyclonal antibody. Endogenous golgin-97 and internalized STxB are analyzed by golgin-97 monoclonal antibody CDF4 (Molecular Probes, Eugene, OR) and STxB monoclonal antibody 13C4 (the hybridoma is available from American Type Culture Collection, Manassas, VA), respectively.

It is relatively easy to identify Arl1-Q71L overexpressed cells from nontransfected cells under immunofluorescent microscopy study because the overexpression of Arl1-Q71L causes the significant expansion of Golgi apparatus (Lu *et al.*, 2001). Cells are transfected with dominant active Arl1 (Arl1-Q71L) followed by continuous internalization of STxB for 3 h. As expected, STxB is accumulated in the TGN as it colocalizes well with endogenous golgin-97 in nontransfected cells (Fig. 3D–F). In contrast, STxB remains in peripheral punctuates in Arl1-Q71L overexpressed cells (Fig. 3A–C). This result supports a role of Arl1 in traffic of STxB to the TGN.

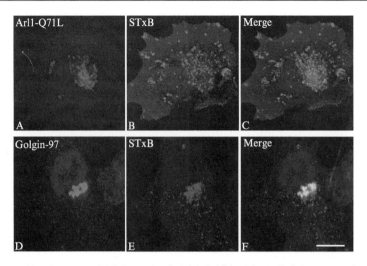

FIG. 3. The transport of STxB to the Golgi is inhibited in Arl1-Q71L expressing cells. HeLa cells are transiently transfected to express Arl1-Q71L and then allowed to continuously internalize STxB for 3 h. The cover slips are then processed for immunofluorescence microscopy to reveal overexpressed Arl1-Q71L (by a limited amount of Arl1 antibody) (A) Golgin-97 (in nontransfected cells; D), and STxB (B and E). Bar, 10 μm. Reprinted from *Molecular Biology of the Cell* (*Mol. Biol. Cell* (2004). **15**, 4426–4443) with the permission of The American Society for Cell Biology.

Concluding Remarks

Small GTPases regulate diverse intracellular processes, and we have shown that activated Arl1 interacts with the GRIP domain of golgin-97 and golgin-245, leading to their recruitment onto the *trans*-Golgi network. The biological relevance of this event is established by the *in vitro* transport assay measuring transport of STxB from the EE/RE to the TGN. The conclusion reached by *in vitro* assay is supported by *in vivo* experiments, suggesting that Arl1-golgin-97 cascade emerges as a regulatory loop for EE/RE-TGN transport. The assays described here will facilitate the identification of other players of the molecular machinery regulating this trafficking event.

References

Ghosh, R. N., Mallet, W. G., Soe, T. T., McGraw, T. E., and Maxfield, F. R. (1998). An endocytosed TGN38 chimeric protein is delivered to the TGN after trafficking through the endocytic recycling compartment in CHO cells. *J. Cell Biol.* **142**, 923–936.

Johannes, L., Tenza, D., Antony, C., and Goud, B. (1997). Retrograde transport of KDEL-bearing B-fragment of Shiga toxin. *J. Biol. Chem.* **272**, 19554–19561.

Lu, L., Horstmann, H., Ng, C., and Hong, W. (2001). Regulation of Golgi structure and function by ARF-like protein 1 (Arl1). *J. Cell. Sci.* **114**, 4543–4555.

Lu, L., and Hong, W. (2003). Interaction of Arl1-GTP with GRIP domains recruits autoantigens Golgin-97 and Golgin-245/p230 on to the Golgi. *Mol. Biol. Cell* **14**, 3767–3781.

Lu, L., Tai, G., and Hong, W. (2004). Autoantigen Golgin-97, an effector of Arl1 GTPase, participates in traffic from the endosome to the *trans*-Golgi network. *Mol. Biol. Cell* **15**, 4426–4443.

Mallard, F., Antony, C., Tenza, D., Salamero, J., Goud, B., and Johannes, L. (1998). Direct pathway from early/recycling endosomes to the Golgi apparatus revealed through the study of Shiga toxin B-fragment transport. *J. Cell Biol.* **143**, 973–990.

Mallard, F., Tang, B. L., Galli, T., Tenza, D., Saint-Pol, A., Xu, Y., Antony, C., Hong, W., Goud, B., and Johannes, L. (2002). Early/recycling endosomes-to-TGN transport involves two SNARE complexes and a Rab6 isoform. *J. Cell Biol.* **156**, 653–664.

Sandvig, K., and van Deurs, B. (2002). Membrane traffic exploited by protein toxins. *Annu. Rev. Cell Dev. Biol.* **18**, 1–24.

Tai, G., Lu, L., and Hong, W. (2004). Participation of the syntaxin 5/Ykt6/GS28/GS15 SNARE complex in transport from the early-recycling endosome to the *trans*-Golgi network. *Mol. Biol. Cell* **15**, 4011–4022.

Zeng, Q., Tan, Y. H., and Hong, W. (1998). A single plasmid vector (pSTAR) mediating efficient tetracycline-induced gene expression. *Anal. Biochem.* **259**, 187–194.

[40]   Assays Used in the Analysis of Arl2 and Its Binding Partners

By J. Bradford Bowzard, J. Daniel Sharer, and Richard A. Kahn

Abstract

Arl2 is a ~20 kDa GTPase in the ADP-ribosylation factor (Arf) family within the Ras superfamily with roles in microtubule dynamics that impact the cytoskeleton, cell division, and cytokinesis. Arl2 has been implicated as a regulator of the pathway responsible for formation of properly folded tubulin heterodimers and in adenine nucleotide transport in mitochondria. The identification and characterization of Arl2 binding partners and regulators of Arl2 activities are critical steps in the further dissection of these and likely other Arl2-dependent functions. This chapter describes methods for preparing recombinant Arl2, loading different radiolabeled guanine nucleotides onto the GTPase, identifying high-affinity Arl2 binding proteins, and assaying Arl2 GTPase activating

METHODS IN ENZYMOLOGY, VOL. 404
0076-6879/05 $35.00
DOI: 10.1016/S0076-6879(05)04040-1

proteins (GAPs). These methods may also prove useful for studies of other Arls or other GTPases.

Introduction

The ADP-ribosylation factor family of small GTPases contains six Arf, 14 Arf-like (Arl), and two Sar proteins in mammals (Logsdon and Kahn, 2003; Li *et al.*, 2004). The Arf proteins have been extensively studied and are best known for their roles in membrane traffic as regulators of coat protein recruitment (Boman and Nilsson, 2003) and lipid metabolism (Randazzo *et al.*, 2003). Arls are more divergent in sequence and function (Schurmann and Joost, 2003), with different Arls implicated in membrane traffic (Lu *et al.*, 2004), tubulin folding (Bhamidipati *et al.*, 2000), germ cell development (Lin *et al.*, 2000), primary cilia function (Chiang *et al.*, 2004; Fan *et al.*, 2004), and cytokinesis (Okai *et al.*, 2004).

Arl2 has been the most extensively studied of the Arls because it has emerged from numerous genetic studies in model organisms as a regulator of microtubule dynamics. Loss of function mutations in the Arl2 ortholog causes altered cell polarity in *S. pombe* (Radcliffe *et al.*, 2000) and chromosome instability and supersensitivity to microtubule destabilizing agents in *S. cerevisiae* (Hoyt *et al.*, 1990; Stearns *et al.*, 1990). Mutant Arl2 disrupts microtubule arrays in *Arabidopsis* embryos and leads to developmental arrest (Mayer *et al.*, 1999), while loss of Arl2 function in *C. elegans* results in loss of cytoskeletal integrity and chromosome segregation defects (Antoshechkin and Han, 2002; Li *et al.*, 2004). Biochemical studies in mammalian cells suggest that Arl2 influences microtubules by interacting with cofactor D, a component of the tubulin folding pathway (Bhamidipati *et al.*, 2000; Shern *et al.*, 2003). In addition to its role in tubulin folding in the cytoplasm, Arl2 is also thought to play a role in mitochondria through its interaction with the binder of Arl2 (BART) (Sharer and Kahn, 1999) and the adenine nucleotide translocase (ANT1) (Sharer *et al.*, 2002). The nature of this role is not yet clear but Arl2 may represent a link between the metabolic health of the cell and the microtubule-based cytoskeleton and chromosome segregation machinery.

Further insights into the mechanisms of Arl2 action will require a greater understanding of the regulators and effectors of Arl2. The most fruitful approach to working out the signaling network of regulatory GTPases has proven to be the biochemical purification and characterization of proteins that bind directly to the GTPase, particularly those that bind in a nucleotide-dependent fashion. However, to date, no guanine nucleotide exchange factors or GTPase activating proteins (GAPs) and only a handful of potential effectors of Arl2 have been identified (Bhamidipati *et al.*, 2000; Hanzal-Bayer *et al.*, 2002; Kolonin and Finley, 2000; Sharer and Kahn, 1999; Shern *et al.*, 2003; Van Valkenburgh *et al.*, 2001). Our laboratory has been engaged in the identification of Arl2 binding partners, using a variety of techniques, for the last several years (Sharer and Kahn, 1999; Shern *et al.*, 2003; Van Valkenburgh *et al.*, 2001).

Toward this end we have developed a number of biochemical assays particularly useful for the detection of Arl2 interactors. Here we briefly describe these assays, including the gel-overlay procedure for identification of high-affinity Arl2 interactors, and two Arl2 GAP assays. These assays were derived from those used earlier to study Arfs or other GTPases but have been adapted to take advantage of the nucleotide binding and exchange properties of Arl2.

Purification of Recombinant Human Arl2 and BART

The procedures for producing recombinant Arl2 and BART have been previously published (Clark *et al.*, 1993; Sharer and Kahn, 1999) and will be briefly outlined here. The full-length open reading frame of each cDNA was subcloned into pET-3c or a related vector. While Arl2 is expressed to high levels in bacteria and is both stable and readily purified from bacterial lysates, we have found that epitope tagging (e.g., His6) of BART both stabilizes the protein against degradation in bacteria and facilitates its purification. The following procedure is the same for Arl2 and BART except where noted. Each preparation requires only two steps of chromatography with the second being gel filtration.

Plasmids are transformed into the BL21(DE3) strain of *E. coli*, and the following day a single fresh colony is picked into each of two 2l baffled flasks containing 500 ml of LB plus 100 μg/ml ampicillin. After growing at 37° to an $OD_{600} = 0.8$, recombinant protein is induced by the addition of IPTG to a final concentration of 1 mM. The cultures are grown for an additional 3 h, harvested by centrifugation for 20 min at 6000g, resuspended in Buffer A (25 mM HEPES pH 7.4, 10 mM NaCl, 1 mM DTT), and lysed by two passages through a French pressure cell. The cell lysate is spun at 100,000g for 1 h to remove insoluble material.

The Arl2 lysate is loaded onto a 2.6 × 40 cm column containing 100 ml MacroPrep DEAE (BioRad, Hercules, PA) resin previously equilibrated with Buffer A. After washing, the bound proteins are eluted with a linear gradient of 0–50% Buffer B (25 mM HEPES pH 7.4, 1 M NaCl, 1 mM DTT) over 250 ml. Arl2 elutes at 200–300 mM NaCl and can be monitored by running aliquots of each fraction on 15% SDS-PAGE gels and staining with Coomassie blue. Immunoblotting with Arl2-specific antisera or monitoring GTP binding of fractions can also be used for confirmation if desired. Because the recombinant BART is expressed with a C-terminal His6 tag, the S100 is loaded onto a 5 ml HiTrap chelating column (GE Healthcare, Piscataway, NJ) charged with NiCl$_2$. Elution of bound proteins is achieved with a gradient of 0–100% Buffer I (Buffer A + 400 mM imidazole). The BART elution profile is monitored as above by Coomassie staining or western blotting.

Fractions containing Arl2 (or BART) are combined, concentrated by ultrafiltration to a volume less than ~6 ml and a total protein concentration of ~10mg/ml, and loaded onto a 2.6 × 60 cm Sephacryl S-100 gel filtration column (GE Healthcare). Elution profiles are monitored as above. Fractions containing Arl2 or BART are pooled, concentrated to at least 1 mg/ml, flash frozen in liquid N$_2$, and stored at −80° until needed. Yields are typically in excess of 5 mg protein/l of bacterial culture.

Comments

1. The purification of Arl2 should be completed as quickly as possible once the bacterial cells are lysed because the protein is rapidly proteolyzed (visible as a doublet by Coomassie staining).

2. Ease of purification is in large part a consequence of high protein expression. If the desired protein is not the most abundant protein in bacterial lysate S100, we suggest investing some time in optimizing protein expression (lower temperature of induction, longer times of induction, compare different transformants).

Nucleotide Binding to Arl2

The nucleotide binding properties of Arl2 differ substantially from those of Arfs and must be considered when designing an assay to detect Arl2 or its regulators (e.g., effectors or GAPs). The most important differences are that nucleotide binding to Arl2 is rapid (e.g., Fig. 1A) and to high stoichiometry, binding of GTP does not require lipids or detergents, and high (millimolar) concentrations of Mg^{++} do not stabilize nucleotide bound to Arl2 (e.g., Fig. 1B) (Clark *et al.*, 1993). Thus, Arl2 is relatively easy to load with radiolabeled nucleotides but difficult to stabilize once bound. Functional consequences of these properties include difficulties in accurate quantitation in some (e.g., GAP) assays as the substrate (Arl2•GTP) is unstable and its concentration is changing even over a short

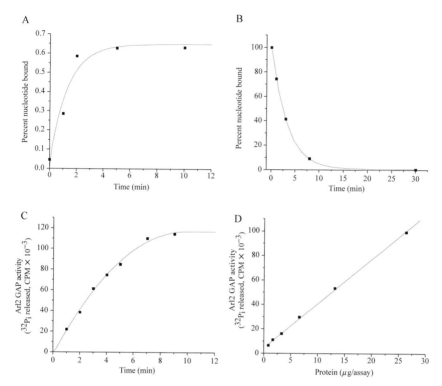

FIG. 1. Properties of nucleotide binding to Arl2 and Arl2 GAP assays. (A) Binding of
$[\gamma^{32}P]$-GTP to Arl2 is rapid and efficient. High specific activity Arl2•GTP is achieved using
the nucleotide loading protocol described in the text and is suitable for gel overlay or Arl2
GAP assays. (B) The GTP bound to Arl2 is in rapid equilibrium with free nucleotides and
cannot rebind when excess unlabeled nucleotides are present. This graph shows the amount of
protein bound $[^{32}P]$GTP as a function of time after dilution into buffer containing excess
unlabeled nucleotides and incubated at $30°$, prior to determining the amount of protein bound
counts by filtration on nitrocellulose membranes, as described in the text. (C) The rate of
hydrolysis of the gamma phosphate on Arl2-bound GTP in the presence of the GAP activity is
linear with time for at least 5 min, as shown here using the charcoal-based Arl2 GAP assay.
(D) The (charcoal) Arl2 GAP assay yields a linear concentration dependence when a single
sample is diluted and assayed over a >20-fold range of protein concentrations.

time course. In contrast, the high stoichiometry of binding has allowed the
successful use of assays, for example, the gel overlay assay, that lack
sufficient sensitivity to identify binding partners when Arf was used as
the probe. We provide a brief description of an assay for loading nucleotide
onto Arl2. With minor changes this can be used to quantify guanine
nucleotide binding sites on purified Arl2, to assay for the presence of

Arl2 during purification, or to load the GTPase prior to use in the gel overlay or GAP assays.

Arl2 is purified from bacteria as described above and is used in the binding reaction at a concentration of 1 μM. Nucleotide binding is performed in Binding Buffer (25 mM HEPES, pH 7.4, 100 mM NaCl, 2.5 mM MgCl$_2$, 1 mM EDTA, 0.1% Lubrol PX, 1 mM DTT, and (optional) 0.1 mg/ml bovine serum albumin [BSA]). A 2× stock buffer can be made up and stored frozen for use over several months. The nature, concentration, and specific activity of the guanine nucleotide will depend upon whether binding is being performed to quantify nucleotide-binding sites or to preload the Arl2 for use in another assay. To quantify binding sites, any radiolabeled GDP or GTP (or GTPγS) can be used and the concentration of nucleotide in the assay is typically 5–10 μM (5–10-fold excess over the GTPase). A specific activity of 10–30,000 cpm/pmol is used as this provides a favorable signal-to-noise ratio and sufficient bound counts to find even low amounts of Arl2 in column fractions during purification.

Binding is initiated by addition of the Arl2 and is performed at 30°. Under these conditions, binding is rapid and usually reaches steady state within 10 min. The reaction is stopped by taking an aliquot (typically 10 μl) and diluting into 2 ml of ice-cold Stop Buffer (25 mM HEPES, pH 7.4, 100 mM NaCl, 10 mM MgCl$_2$, 1 mM DTT). Bound and free ligand are then separated by filtration on 25 mm nitrocellulose disks (BA85, Schleicher and Schuell, Florham Park, NJ), with at least 4×2 ml washes, as described previously for Arfs (Randazzo $et\ al.$, 1995). Filters are added to scintillation cocktail and the bound counts are determined by liquid scintillation counting. Drying of filters is not usually required for ^{35}S or ^{32}P labeled nucleotides, but is useful when using ^{3}H-labeled ligands as the water on the filter can cause substantial quenching. A parallel sample in which no GTPase was added is run to determine background, attributed to nonspecific binding of nucleotides to the filter, which is subtracted from the experimental values to determine specific binding. The specific activity of the ligand in the binding reaction is determined by spotting some of it directly onto a filter and counting without washing, using the equation: CPM on the filter/ (volume cocktail spotted × concentration of nucleotide in the binding cocktail spotted) = cpm/pmol nucleotide.

When the highest specific activity radiolabeled Arl2 is the goal, rather than quantitation of binding sites, it is necessary only to vary the guanine nucleotide composition in the reaction.

Arl2 is loaded to high specific activity with [α-^{32}P]GTP for the gel overlay or the TLC-based Arl2 GAP assay and with [γ-^{32}P]GTP for the charcoal-based GAP assay. For example, to load Arl2 with radiolabeled GTP for GAP or gel overlay assays, we typically use 0.3 μCi [^{32}P]GTP

(6000 Ci/mmol) per pmol Arl2 in the loading reaction. Under these conditions, greater than 50% of the ligand is bound to Arl2 (Fig. 1A), thus we typically achieve $>3 \times 10^6$ bound cpm in a 10 μl aliquot.

Gel Overlay Assay

The gel overlay assay is basically the same as a Western blot but instead of probing the solid-phased proteins with an antibody, a radiolabeled GTPase is used to detect protein binding partners (see Fig. 2). This assay (Sharer and Kahn, 1999) was developed by modifying the method of Lounsbury (1994). Because this technique requires that the binding partner, or at least the GTPase binding domain, bind to nitrocellulose

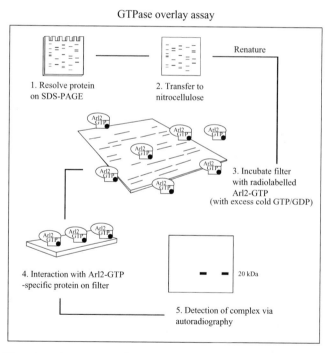

FIG. 2. The gel overlay assay can detect high-affinity Arl2 binding partners. As described in the text, proteins from a sample of interest are resolved by SDS-PAGE (1) and electrophoretically transferred to nitrocellulose (2). After renaturation of the transferred proteins and blocking (3), the filter is incubated with pre-loaded Arl2•[α-^{32}P]GTP (4). Three quick washes at 4° are followed by autoradiography to reveal the location of proteins that retain the Arl2•[α-^{32}P]GTP (5). This cartoon shows binding to BART, a 20-kDa protein.

filters and refold after electrophoresis in SDS, many binding partners will not be detected with this technique. Advantages to this methodology include the ability to assay the interaction in a crude mixture (e.g., a whole cell lysate), test specificity by comparing binding of GTP-loaded GTPase to GDP-loaded GTPase, and learn immediately the size of the binding partner. This assay was used to identify and purify the first Arl2 effector, BART (Sharer and Kahn, 1999), and was later used to identify a binding site for the Arl2•GTP•BART complex on the adenine nucleotide translocase (ANT1) (Sharer et al., 2002).

Proteins to be assayed are prepared in Laemmli sample buffer and loaded onto a single lane of an SDS-PAGE gel (up to 25 μg protein loaded per lane), resolved by standard electrophoresis at room temperature, and electrophoretically transferred to nitrocellulose filters using the method of Towbin (1979). The only differences between this method and standard SDS-PAGE conditions are that the samples are not boiled before loading and some proteins survive better if the sample buffer contains only as much SDS as the running buffer (0.1%), instead of the usual 2% SDS. Filters are then incubated for at least 1 h at 4° in Renaturation Buffer (10 mM MOPS, pH 7.1, 100 mM KOAc, 5 mM MgOAc, 0.25% Tween 20, 5 mM DTT, and 0.5% BSA). The addition of a mixture of 3 mM dimyristoyl phosphatidyl-choline and 0.1% sodium cholate to the refolding step was found to increase activity in the assay, presumably by aiding the refolding of proteins bound to the filter (Sharer and Kahn, 1999). The filter is incubated with shaking in 15 ml Blocking Buffer (20 mM MOPS pH 7.1, 100 mM KOAc, 5 mM MgOAc, 0.1% Triton X-100, 0.5% BSA, 5 mM DTT, 50 mM GTP, and 50 mM GDP) for at least 30 min at room temperature. High specific activity Arl2•[α-^{32}P]GTP is then incubated with the filter in 15 ml of blocking buffer for 15 min at room temperature with gentle shaking. Note that the blocking buffer (also used for washes) contains 50 mM GTP and GDP to lower the specific activity of free nucleotides to such a point that binding of radionucleotides by filter-bound nucleotide-binding proteins is insignificant. The filter is then quickly washed 3 times with binding buffer and exposed to X-ray film (Eastman Kodak Co., XAR, Rochester, NY) or phosphor screens (Molecular Dynamics, Piscataway, NJ) for analysis. Specificity of the interaction is checked by parallel incubation of samples with Arl2•[α-^{32}P]GDP and [α-^{32}P]GTP in the absence of Arl2.

Comments

1. This assay can be done in a quantitative fashion to determine an apparent binding constant for the protein interaction (Sharer and Kahn, 1999).

2. As noted above, guanine nucleotides dissociate rapidly from Arl2, even in the presence of excess Mg^{2+} so performing the washing steps at 4° and as quickly as possible is important. Binding of effectors to Arl2•GTP often slows the dissociation of GTP and thus helps maintain the sensitivity of the assay when searching for binding partners of the GTPase or for binders to the GTPase•effector complex.

Arl2 GAP Assays

Two different protocols are described here for assaying Arl2 GAP activity in crude cell or tissue lysates and for use in protein purification. In each case the Arl2 is preloaded with $[^{32}P]$GTP and the amount of Arl2-dependent GTP hydrolysis in a single round is determined. In one case the amount of Arl2-bound GTP and GDP is resolved and quantified by thin layer chromatography, making $[\alpha\text{-}^{32}P]$GTP the required radionucleotide. In the other case, the Arl2 and GAP-dependent release of free ^{32}Pi is determined by precipitation of all radionucleotides with charcoal, leaving the ^{32}Pi behind in solution, and making $[\gamma\text{-}^{32}P]$GTP the required radionucleotide. The advantage of the former assay is increased confidence that you are measuring hydrolysis of GTP that is bound to Arl2 as protein-bound nucleotides are first resolved from free nucleotides by filtration on nitrocellulose (as described above for the nucleotide binding assay) and then bound GDP and GTP are resolved by TLC. The disadvantages of this assay are that it is quite labor intensive, is less sensitive, and allows for a more limited number of assays to be processed per day. The latter assay is more sensitive, allows far greater throughput, is more amenable to assaying fractions off columns during purification, and is more readily quantified.

Accurate determination of Arl2 GAP activity is hampered by the relatively rapid dissociation of GTP from Arl2 (Fig. 1B), yielding a constantly decreasing concentration of substrate (Arl2•GTP) in the assay. Addition of cold GTP is necessary to reduce the background of GTP hydrolysis due to other GTP-binding proteins, nucleotidases, or nucleotide diphosphate kinases that are present in tissue lysates and purified protein preparations. The high concentration of free nucleotides also prevents the rebinding of radionucleotides to Arl2. Therefore, the Arl2 GAP reaction times are kept to a minimum, typically 5 min or less, to balance formation of detectable products of the GAP reaction with the loss of substrate from nucleotide dissociation. The net result is that some sensitivity is sacrificed for a greater assurance of specificity.

Preparation of Arl2 GAP Samples

Although bovine brain S100 is an abundant source of Arl2, no Arl2 GAP activity was detectable from this tissue fraction. The finding that a pool of cellular Arl2 is in mitochondria (Sharer and Kahn, 1999) suggested this organelle as a potential location for an Arl2 GAP. Indeed, we were able to detect Arl2 GAP activity in detergent extracts of partially purified bovine brain mitochondria using either of the assays described below (J. B. Bowzard, J. D. Sharer, J. Shern, and R. A. Kahn, manuscript in preparation).

Frozen bovine brains are purchased from Pel-Freez Biologicals and stored at $-80°$ until used. One brain is thawed overnight at $4°$. The brain stem and cerebellum are removed and the remaining tissue washed with Buffer A and minced with a spatula. The minced tissue is then homogenized with a Kinematica PT 2000 polytron fitted with a 20 mm generator for 3 min in a total of 1.5l Buffer A. The sample is transferred to 1l centrifuge bottles and spun at $1000g$ for 1 h. The supernatant is carefully removed, transferred to 250 ml centrifuge bottles, and spun at $25,000g$ for 25 min. The S25 is removed and discarded and the P25 is resuspended by pipetting in 300 ml MIB (25 mM HEPES pH 7.4, 70 mM sucrose, 200 mM D-mannitol). The resuspended P25 is homogenized by 10 strokes of the loose pestle in a glass dounce. This mixture is then spun at $1000g$ for 25 min and the S1 spun at $25,000g$ for 25 min. After resuspension in 100 ml MIB, the pellet is dounced as before. Following a third $1000/25,000g$ set of spins, the P25 is resuspended in 20–40 ml MIB to yield protein concentrations of 5–10 mg/ml. This crude mitochondrial membrane preparation can be frozen at $-80°$ for future use.

The Arl2 GAP activity is solubilized by bringing the crude mitochondrial membranes to a final concentration of a 2% CHAPS at a detergent: protein ratio of 2.5 detergent:1 total protein (w/w) and incubating for 30 min on ice. The sample is then spun at $100,000g$ for 1 h and the supernatant is recovered and stored at $4°$.

TLC-Based Assay

This assay has been extensively used for monitoring GAP activity for other GTPases, including Arfs (Randazzo and Kahn, 1994), and will only be briefly described here. The basis for this procedure is that a GAP acting on Arl2•GTP will stimulate the hydrolysis of the bound nucleotide, resulting in the formation of Arl2•GDP. If nucleotides bound to the GTPases can be captured efficiently, the ratio of GDP/GTP bound will indicate the extent of hydrolysis. No differences in the rate of dissociation of GDP and

GTP have been noted, so this ratio is maintained, though sensitivity of the assay is compromised by any delay in processing samples. After stopping the reaction, the Arl2•GXP is captured by filter binding, the bound nucleotide is eluted with formic acid, and GDP is resolved from GTP using thin layer chromatography. GAP activity is measured as the percent conversion of GTP to GDP as compared to a no GAP control.

Arl2 is loaded as described above using $[\alpha\text{-}^{32}P]GTP$ (3000 Ci/mmol) as the isotope. The assay is performed in 50 μl reactions containing 12.5 μl 4× GAP buffer (100 mM HEPES pH 7.4, 400 mM NaCl, 10 mM MgCl$_2$, 4 mM GTP, 4 mM DTT), 5 μl 25 mg/ml phosphatidyglycerol, and 27.5 μl of the sample to be tested. The reactants are combined and held on ice until the reaction is started by the addition of 5 pmol Arl2•$[\alpha\text{-}^{32}P]GTP$ and transfer to 30°. The reactions are stopped at appropriate time points (typically <10 min), by dilution into cold Stop Buffer on ice and immediate filtration on nitrocellulose filters, as described above. Retained (protein-bound) nucleotides are eluted from the filter by incubating 1 min in 0.4 ml 0.8 M formic acid and are then resolved by spotting 10 μl on PEI cellulose TLC plates. The TLC plates are developed with a solution of 0.5 M formic acid and 0.5 M LiCl$_2$, dried, exposed to a phosphorimager screen overnight, and quantitated on a Molecular Dynamics Storm phosphorimager. Data are typically expressed as the percentage of GTP converted to GDP, or better as GDP/(GDP + GTP). This allows for variation in the recovery of protein-bound nucleotides on filters and in spotting of TLC plates. Note that sensitivity is further compromised by the fact that only 10 μl of the 400 μl formic acid eluate is analyzed.

Arl2 GAP Assay Using Charcoal-Based Separation

The GAP assay described here is modified from the procedure of Higashijima et al. (1987). It also measures GAP-dependent hydrolysis of GTP bound to Arl2 in a single round reaction but the product, ^{32}Pi, is resolved from substrate by precipitation of all nucleotides in solution with activated charcoal. Because samples contain many other activities that can yield ^{32}Pi, it is important to perform adequate controls to ensure that the Arl2- and Arl2 GAP-dependent hydrolysis is what is being measured. Thus, each sample is used in two parallel reactions. One (experimental) uses the Arl2•$[\gamma\text{-}^{32}P]GTP$ as substrate and the other uses equivalent amounts of free $[\gamma\text{-}^{32}P]GTP$ from a "mock load." The amount of GTP hydrolysis in the mock reaction is subtracted from that in the parallel reaction to obtain the Arl2-dependent hydrolysis. This background hydrolysis can be substantial, particularly with crude samples, and compromises accurate measurements of Arl2 GAP activities in these samples. However, this usually becomes less problematic after one or two steps of purification.

The activated charcoal is prepared as a 5% (w/v) suspension in 50 mM NaH$_2$PO$_4$, pH 7.4 by washing 5× by centrifugation at 1000g for 10 min, followed by removal of the supernatant, and resuspension of the pellet in the original volume of the phosphate buffer. Arl2 is loaded as described above using [γ-^{32}P]GTP as the source of labeled nucleotide. For example, Arl2 (5 pmol) is loaded in a volume of 5 μl for each GAP reaction to be performed. Thus, to run 20 Arl2 GAP reactions, 100 pmol of Arl2 ($\sim$2 μg) is added to 50 μl of 2× Load Buffer (50 mM HEPES pH 7.4, 200 mM NaCl, 5 mM MgCl$_2$, 2 mM EDTA), plus [γ-^{32}P]GTP (30 μCi at 6000 Ci/mmol), and enough water is added to bring the total volume to 100 μl. The loading reaction is started by the addition of the [γ-^{32}P]GTP, incubated at 30° for 15–30 min, and stored on ice until needed. In the "mock load" reaction Arl2 is replaced by buffer.

For each sample to be analyzed the GAP reaction is performed as in the TLC-based assay but the reaction is stopped by dilution into 15 volumes of ice-cold activated charcoal suspension and maintained at 4° thereafter. After the reactions are stopped, the samples are spun at 1000g for 10 min and supernatant equal to one-half the total volume is removed (with care to avoid the charcoal pellet), mixed with scintillation fluid, and counted. This assay is most conveniently performed in 12 × 75 mm polypropylene tubes.

When analyzing large numbers of samples, a single time point is used (typically 4 min) and the entire 50 μl reaction is stopped by the addition of 750 μl of phosphate-buffered charcoal.

An aliquot (400 μl) of the supernatant is removed after centrifugation for scintillation counting. Larger numbers of samples may necessitate longer times of incubation if only one person is running the assay, but care should be taken to remain in the linear range of the assay. We have found that nucleotide dissociation from Arl2 in the assay becomes limiting for quantitation within $\sim$8 min in the GAP reaction (Fig. 1C).

Several sources of background counts are present in this assay and need to be considered. One is the free ^{32}Pi in the commercially obtained isotope preparation. This background steadily increases as the nucleotide preparation ages due to spontaneous hydrolysis of the GTP and radiation damage. Another is caused by the presence of nucleotidases and nucleoside diphosphate kinases in the samples being tested and in the Arl2 preparation. All of these nonspecific counts can be accounted for by including two control reactions in each assay: the mock reaction, in which free $[\gamma\text{-}^{32}P]GTP$ replaces the $Arl2\bullet[\gamma\text{-}^{32}P]GTP$, is run in parallel for each experimental sample assayed, and a "no GAP" reaction, in which buffer replaces the experimental sample to be assayed. The former will yield information on the extent of hydrolysis that is occurring on the free nucleotide in the reaction (remember that in this assay we do not separate bound and free nucleotide from the loading reaction before addition into the GAP reaction). The latter control gives an estimate of the amount of hydrolysis that results solely from incubating Arl2 with the GTP. This takes into account any basal GTPase activity of the GTPase as well as any contaminating nucleotidases or nucleoside diphosphate kinases in the Arl2 preparation. (*Note*: we have never seen a recombinant protein preparation that did not have some contaminating activity.) To calculate the Arl2-specific GAP activity, the following formula is used: (experimental sample$_{Arl2 \; load}$-experimental sample$_{mock \; load}$)–(no GAP$_{Arl2 \; load}$-no GAP$_{mock \; load.}$)

Comments

1. It is likely that this assay uses substrate at concentrations below the K_D so we are underestimating the activity. The loss of substrate in the reaction due to nucleotide dissociation also leads to underestimation of Arl2 GAP activity. Despite these issues we have found the charcoal assay to be specific, reproducible, reasonably quantitative (Fig. 1D), and rapid enough to assay fractions from chromatography columns to allow purification of the activity.

2. Both the assay and the Arl2 GAP itself are sensitive to a wide variety of buffer components typically used in chromatographic procedures. For example, high salt interferes with the assay so chromatography fractions from the end of a salt gradient must be diluted before being assayed.

Acknowledgments

The authors would like to thank Jack Shern for valuable input in the development of these assays and other members of the Kahn Lab, past and present, for insightful discussions. This work was supported by National Institutes of Health grants R01GM068029 (RAK) and F32GM067465 (JBB).

References

Antoshechkin, I., and Han, M. (2002). The *C. elegans* evl-20 gene is a homolog of the small GTPase ARL2 and regulates cytoskeleton dynamics during cytokinesis and morphogenesis. *Dev. Cell* **2**, 579–591.

Bhamidipati, A., Lewis, S. A., and Cowan, N. J. (2000). ADP ribosylation factor-like protein 2 (Arl2) regulates the interaction of tubulin-folding cofactor D with native tubulin. *J. Cell Biol.* **149**, 1087–1096.

Boman, A., and Nilsson, T. (2003). Coat proteins. *In* "ARF Family GTPases" (R. A. Kahn, ed.), Vol. 1, pp. 241–257. Kluwer Academic Publishers, Dordrecht.

Chiang, A. P., Nishimura, D., Searby, C., Elbedour, K., Carmi, R., Ferguson, A. L., Secrist, J., Braun, T., Casavant, T., Stone, E. M., and Sheffield, V. C. (2004). Comparative genomic analysis identifies an ADP-ribosylation factor-like gene as the cause of Bardet-Biedl syndrome (BBS3). *Am. J. Hum. Genet.* **75**, 475–484.

Clark, J., Moore, L., Krasinskas, A., Way, J., Battey, J., Tamkun, J., and Kahn, R. A. (1993). Selective amplification of additional members of the ADP-ribosylation factor (ARF) family: Cloning of additional human and Drosophila ARF-like genes. *Proc. Natl. Acad. Sci. USA* **90**, 8952–8956.

Fan, Y., Esmail, M. A., Ansley, S. J., Blacque, O. E., Boroevich, K., Ross, A. J., Moore, S. J., Badano, J. L., May-Simera, H., Compton, D. S., Green, J. S., Lewis, R. A., van Haelst, M. M., Parfrey, P. S., Baillie, D. L., Beales, P. L., Katsanis, N., Davidson, W. S., and Leroux, M. R. (2004). Mutations in a member of the Ras superfamily of small GTP-binding proteins causes Bardet-Biedl syndrome. *Nat. Genet.* **36**, 989–993.

Hanzal-Bayer, M., Renault, L., Roversi, P., Wittinghofer, A., and Hillig, R. C. (2002). The complex of Arl2-GTP and PDE delta: From structure to function. *EMBO J.* **21**, 2095–2106.

Higashijima, T., Ferguson, K. M., Smigel, M. D., and Gilman, A. G. (1987). The effect of GTP and Mg2+ on the GTPase activity and the fluorescent properties of Go. *J. Biol. Chem.* **262**, 757–761.

Hoyt, M. A., Stearns, T., and Botstein, D. (1990). Chromosome instability mutants of *Saccharomyces cerevisiae* that are defective in microtubule-mediated processes. *Mol. Cell. Biol.* **10**, 223–234.

Kolonin, M. G., and Finley, R. L., Jr. (2000). A role for cyclin J in the rapid nuclear division cycles of early Drosophila embryogenesis. *Dev. Biol.* **227**, 661–672.

Li, Y., Kelly, W. G., Logsdon, J. M., Jr., Schurko, A. M., Harfe, B. D., Hill-Harfe, K. L., and Kahn, R. A. (2004). Functional genomic analysis of the ADP-ribosylation factor family of GTPases: Phylogeny among diverse eukaryotes and function in *C. elegans. Faseb. J.* **18**, 1834–1850.

Lin, C. Y., Huang, P. H., Liao, W. L., Cheng, H. J., Huang, C. F., Kuo, J. C., Patton, W. A., Massenburg, D., Moss, J., and Lee, F. J. (2000). ARL4, an ARF-like protein that is developmentally regulated and localized to nuclei and nucleoli. *J. Biol. Chem.* **275**, 37815–37823.

Logsdon, J. M., Jr., and Kahn, R. A. (2003). The Arf family tree. *In* "ARF Family GTPases" (R. A. Kahn, ed.), Vol. 1, pp. 1–21. Kluwer Academic Publishers, Dordrecht.

Lounsbury, K. M., Beddow, A. L., and Macara, I. G. (1994). A family of proteins that stabilize the Ran/TC4 GTPase in its GTP-bound conformation. *J. Biol. Chem.* **269**, 11285–11290.

Lu, L., Tai, G., and Hong, W. (2004). Autoantigen golgin-97, an effector of Arl1 GTPase, participates in traffic from the endosome to the trans-golgi network. *Mol. Biol. Cell* **15,** 4426–4443.

Mayer, U., Herzog, U., Berger, F., Inze, D., and Jurgens, G. (1999). Mutations in the pilz group genes disrupt the microtubule cytoskeleton and uncouple cell cycle progression from cell division in Arabidopsis embryo and endosperm. *Eur. J. Cell Biol.* **78,** 100–108.

Okai, T., Araki, Y., Tada, M., Tateno, T., Kontani, K., and Katada, T. (2004). Novel small GTPase subfamily capable of associating with tubulin is required for chromosome segregation. *J. Cell Sci.* **117,** 4705–4715.

Radcliffe, P. A., Vardy, L., and Toda, T. (2000). A conserved small GTP-binding protein Alp41 is essential for the cofactor-dependent biogenesis of microtubules in fission yeast. *FEBS Lett.* **468,** 84–88.

Randazzo, P. A., and Kahn, R. A. (1994). GTP hydrolysis by ADP-ribosylation factor is dependent on both an ADP-ribosylation factor GTPase-activating protein and acid phospholipids. *J. Biol. Chem.* **269,** 10758–10763.

Randazzo, P. A., Nie, Z., and Hirsch, D. S. (2003). Arf and phospholipids. *In* "ARF Family GTPases" (R. A. Kahn, ed.), Vol. 1, pp. 49–69. Kluwer Academic Publishers, Dordrecht.

Randazzo, P. A., Weiss, O., and Kahn, R. A. (1995). Preparation of recombinant ADP-ribosylation factor. *Methods Enzymol* **257,** 128–135.

Schurmann, A., and Joost, H.-G. (2003). Arf-like proteins. *In* "ARF Family GTPases" (R. A. Kahn, ed.), Vol. 1, pp. 325–350. Kluwer Academic Publishers, Dordrecht.

Sharer, J. D., and Kahn, R. A. (1999). The ARF-like 2 (ARL2)-binding protein, BART. Purification, cloning, and initial characterization. *J. Biol. Chem.* **274,** 27553–27561.

Sharer, J. D., Shern, J. F., Van Valkenburgh, H., Wallace, D. C., and Kahn, R. A. (2002). ARL2 and BART enter mitochondria and bind the adenine nucleotide transporter. *Mol. Biol. Cell* **13,** 71–83.

Shern, J. F., Sharer, J. D., Pallas, D. C., Bartolini, F., Cowan, N. J., Reed, M. S., Pohl, J., and Kahn, R. A. (2003). Cytosolic Arl2 is complexed with cofactor D and protein phosphatase 2A. *J. Biol. Chem.* **278,** 40829–40836.

Stearns, T., Hoyt, M. A., and Botstein, D. (1990). Yeast mutants sensitive to antimicrotubule drugs define three genes that affect microtubule function. *Genetics* **124,** 251–262.

Towbin, H., Staehelin, T., and Gordon, J. (1979). Electrophoretic transfer of proteins from polyacrylamide gels to nitrocellulose sheets: Procedure and some applications. *Proc. Natl. Acad. Sci. USA* **76,** 4350–4354.

Van Valkenburgh, H., Shern, J. F., Sharer, J. D., Zhu, X., and Kahn, R. A. (2001). ADP-ribosylation factors (ARFs) and ARF-like 1 (ARL1) have both specific and shared effectors: Characterizing ARL1-binding proteins. *J. Biol. Chem.* **276,** 22826–22837.

[41] Assay and Functional Analysis of the ARL3 Effector RP2 Involved in X-Linked Retinitis Pigmentosa

By R. Jane Evans, J. Paul Chapple, Celene Grayson, Alison J. Hardcastle, and Michael E. Cheetham

Abstract

Mutations in *RP2* cause X-linked retinitis pigmentosa and also macular and peripapillary atrophy. RP2 is a functional homologue of the tubulin folding cofactor, cofactor C, as it can replace the β tubulin GTPase stimulating activity of cofactor C in an *in vitro* assay. An important difference between RP2 and cofactor C is their subcellular localization. RP2 is targeted to the cytoplasmic face of the plasma membrane by dual N-terminal acylation, and this post-translational modification is important for protein function. The activity of tubulin folding cofactors is modulated by certain ADP ribosylation factor-like (Arl) proteins. It has been shown that RP2 can interact directly with Arl3. Here we describe the methodologies that we have developed to analyze the interaction of RP2 with Arl3 and to investigate the effect of RP2 post-translational modifications on its subcellular and tissue localization.

Introduction

X-linked retinitis pigmentosa (XLRP) is a severe form of retinal degeneration. Patients in the early stages of disease suffer from night blindness and constriction of visual fields as a result of peripheral photoreceptor degeneration. As disease progresses, impairment of central vision occurs, resulting in loss of visual acuity and blindness. Mutations in the retinitis pigmentosa 2 (*RP2*) gene have been shown to account for up to 15% of XLRP (Breuer *et al.*, 2002; Hardcastle *et al.*, 1999; Schwahn *et al.*, 1998). Furthermore, mutations in *RP2* can also cause macular and peripapillary atrophy (Dandekar *et al.*, 2004), indicating phenotypic heterogeneity associated with *RP2* mutations such that *RP2* may account for a larger proportion of retinal disease. The gene product, RP2, is a ubiquitously expressed 350 amino acid protein (Chapple *et al.*, 2000; Schwahn *et al.*, 1998). To date, we know little of the function of this important disease protein. Our major clues to RP2 function are its homology to the tubulin folding cofactor, cofactor C; its interaction with the small GTPase Arl3; and its post-translational modification dependent subcellular targeting.

METHODS IN ENZYMOLOGY, VOL. 404   0076-6879/05 $35.00
          DOI: 10.1016/S0076-6879(05)04041-3

The predicted amino acid sequence of RP2 is similar to cofactor C (>30% identity over 151 amino acids). In the current model of tubulin folding, cofactor C together with cofactors D and E facilitate the assembly of the α/β tubulin heterodimer and act as a GTPase activating protein (GAP) by stimulating GTP hydrolysis by β tubulin (Tian et al., 1996, 1997). Furthermore, cofactors C and D together with cofactor E have been shown to stimulate the GTPase activity of native tubulin (Tian et al., 1999), suggesting that they could also regulate microtubules. RP2 is a functional homologue of cofactor C. In the presence of cofactor D, RP2 also stimulates the GTPase activity of tubulin (Bartolini et al., 2002). Evidence for a shared structural element between the two proteins comes from the observation that the pathogenic mutation R118H in RP2 (Hardcastle et al., 1999; Schwahn et al., 1998), at a residue conserved with cofactor C, abolishes the tubulin-GTPase stimulation (tubulin-GAP) activity in both RP2 and cofactor C (Bartolini et al., 2002). These data suggest that this residue may act as an "arginine finger" to trigger the tubulin-GAP activity and that the R118H mutation in RP2 may cause retinitis pigmentosa due to this loss of tubulin-GAP activity. It should be noted, however, that although RP2 in conjunction with cofactor D can replace cofactor C to function as a tubulin-GAP, RP2 cannot replace cofactor C in the heterodimerization reaction (Bartolini et al., 2002).

Arl2 binds to cofactor D and modulates the tubulin-GAP activity of cofactors C and D (Bhamidipati et al., 2000; Shern et al., 2003). Given the association of Arl2 with cofactor D, Bartolini (2002) screened Arl2, Arl3, and Arl4 for an interaction with RP2. Arl3, not Arl2, bound specifically to RP2, but did not affect the tubulin-GAP activity of RP2 and cofactor D. Arl3 is targeted to the connecting cilium in photoreceptors, decorates microtubule structures within cells, and cosediments with microtubules suggesting that Arl3 is a microtubule-associated protein (MAP) (Grayson et al., 2002). These data, coupled with the functional homology with cofactor C, suggest that Arl3 and RP2 cooperate to regulate the cytoskeleton.

One of the key differences between RP2 and cofactor C is membrane association mediated via post-translational modification. Dual N-terminal acyl modification by myristoylation and palmitoylation target RP2 to the cytoplasmic face of the plasma membrane in cultured cells (Chapple et al., 2000) and in cells throughout the retina (Grayson et al., 2002). A pathogenic mutation ΔS6 (Rosenberg et al., 1999; Schwahn et al., 1998) in RP2 prevents the plasma membrane targeting of the protein (Chapple et al., 2000, 2002; Schwahn et al., 2001), illustrating that this post-translational modification is vital for the function of the protein in the retina. RP2 is not partitioned to either the apical or basolateral domains of polarized epithelial cells. A population of RP2 does, however, partition into detergent-resistant domains (DRMs) within the plasma membrane, suggesting that RP2 could be associated with membrane microdomains (Chapple *et al.*, 2003).

Here, we describe methods that we have used and developed to investigate the interaction of RP2 with Arl3 and the post-translational modifications of RP2 coupled to the consequences of these modifications for the subcellular localization of RP2.

RP2 cDNA Amplification

RP2 is a relatively low abundance, ubiquitous mRNA (Schwahn *et al.*, 1998) and protein (Chapple *et al.*, 2000). We amplified full length RP2 cDNA by PCR from a human brain cDNA library (Clontech, Palo Alto, CA) and cloned into the pGEM-T vector, using oligonucleotide primers based on the untranslated regions of the mRNA (Forward 5′-GGAAGTGCCTGAGCTAGTGAG-3′ Reverse 5′-CTATATTCA-CAAGTTGGGAAAGGC-3′). Due to the low abundance of the message this required 40 cycles of amplification. The cDNA sequence was confirmed and this construct was used as a template for further rounds of PCR to introduce restriction endonuclease recognition sites appropriate for subcloning and/or site-directed mutagenesis.

Using Yeast-2-Hybrid to Study the Arl3-RP2 Interaction

The interaction of RP2 with Arl3 was demonstrated by Bartolini (2002) using purified recombinant proteins and *in vitro* translated products coupled with GST pull downs. These analyses showed that RP2 did not appear to interact with Arl2 or Arl4, but only Arl3. Furthermore, the interaction was strongest between nonmyristoylated RP2 and Arl3(Q71L), the GTP locked form of Arl3 (Bartolini *et al.*, 2002). In order to investigate this interaction further we have developed a version of the CytoTrap yeast-2-hybrid system (Stratagene, Amsterdam, The Netherlands) expressing RP2 and Arl3. The CytoTrap system uses a fusion of a bait protein (in this case RP2) with human Sos (hSos) in a *cdc25* defective yeast strain to identify protein:protein interactions in the cytoplasm on the basis of the prey (Arl3), which is myristoylated, bringing the hSos fusion to the membrane (Fig. 1). Mutant *cdc25* yeast grow normally at the permissive temperature of 25° (Fig. 1A) but at 37° the temperature-sensitive *cdc25* can no longer stimulate nucleotide exchange by Ras and activate the Ras pathway, therefore the yeast cannot grow at 37° (Fig. 1B). When the hSos-RP2 fusion protein is brought to the membrane by its interaction with myristoylated

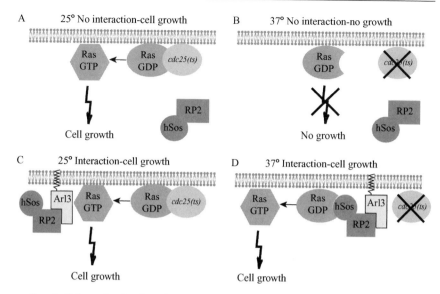

FIG. 1. Schematic showing the use of the CytoTrap system to study the RP2-Arl3 interaction.

Arl3, the hSos moiety in the fusion protein stimulates Ras nucleotide exchange and permits growth at the nonpermissive temperature of 37° (Fig. 1C, D; Fig. 2). We have produced a hSos-RP2 fusion plasmid (pSos-RP2), which is a fusion of the C-terminus of hSos with the N-terminus of RP2. This fusion silences the RP2 dual acylation motif (see following) and the protein is not recruited to the membrane. The fusion protein is expressed at high levels, is soluble, and does not transactivate when expressed alone at the nonpermissive temperature 37° (Fig. 2A). Human Arl3 was cloned into the prey plasmid pMyr-Arl3, which produced a hybrid form of Arl3 with a myristoylation motif at its N-terminus. This system that confirms the interaction of RP2 with Arl3 will allow the delineation of the Arl3-RP2 binding sites and the determination of the effect of mutations in RP2 and Arl3 on the protein:protein interactions.

Cloning RP2 and Arl3

Human RP2 (Chapple *et al.*, 2000) was subcloned into the pSos vector by double restriction enzyme digestion with *Not*I and *Nco*I, which produced an in-frame hSos fusion. Human Arl3 was cloned into the "prey" plasmid pMyr from IMAGE clone 3610040. Arl3 was engineered to contain *Eco*R1 and *Sal*1 restriction sites by PCR (Forward 5′-CGGAATT

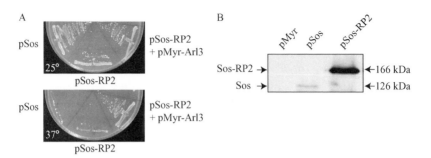

Fig. 2. (A) Yeast Cytotrap strains transformed with pSos, pSos-RP2, and pSos-RP2 + pMyr-Arl3 as indicated grown at 25° and 37°. Note only pSos-RP2 + pMyr-Arl3 grows at the nonpermissive temperature of 37° (B) Western blot of yeast cell lysates showing production of soluble, stable full length Sos-RP2 (probed with anti-Sos1; BD Biosciences 610096; at 1:250).

CATGGGCTTGCTCTCAATTTTG-3′ Reverse 5′-CGGTCGACTCCCG CAGCTCCTGCATC-3′) for in frame cloning into pMyr. The resulting clones were sequence verified using vector-specific primers.

Yeast Strain Phenotype Verification

The yeast strain cdc25Hα phenotype was checked to ensure there was no aberrant growth on inappropriate selective media. This strain is *ura3-52, his3-200, ade2-101, lys2-801, trp1-901, leu2-3, 112, cdc25-2 Gal+* and should not grow on plates lacking tryptophan, leucine, histidine, or uracil. This yeast strain should also not grow at 37°. The phenotype was checked by streaking out onto SD/glucose plates (0.17% (w/v) yeast nitrogen base without amino acids, 0.5% (w/v) ammonium sulfate, 2% (w/v) dextrose, 1.5% (w/v) agar, with appropriate 1× dropout solution added after cooling to 55°) each lacking one of the 4 amino acids and incubated at 25°. Yeast cells were also streaked out onto YPAD plates (1% (w/v) yeast extract, 2% (w/v) bacto-peptone, 2% (w/v) dextrose, 0.004% (w/v) adenine sulfate, 1.5% (w/v) agar) and incubated at 37° to ensure the phenotype had not reverted to allow growth at the restrictive temperature.

Yeast Transformation

The "bait" plasmid (pSos-RP2) was transformed into the yeast strain cdc25Hα to check for transactivation and expression. Transformation of the yeast strain was performed using a lithium acetate based method as follows. A single colony of cdc25Hα from a YPAD plate was used to inoculate a 10 ml culture of YPAD and grown overnight at 25° with shaking until saturation. This culture was then used to inoculate 50 ml of fresh YPAD to an OD_{600} of 0.2 and grown until an OD_{600} of 0.75–1.0. Cells were centrifuged at 2000g for 2 min and supernatant removed. The yeast were resuspended in 10 ml sterile water and centrifuged again and the supernatant removed. Cells were resuspended in 5 ml LiOAc/TE (10 mM LiOAC, 10 mM Tris-HCl (pH 7.5), 1 mM EDTA (pH 8.0)) and centrifuged again at 2000g for 2 min. The supernatant was removed and the yeast was resuspended in 250 μl of LiOAc/TE and incubated at 25° for 15 min. For each transformation 50 μl of cells were mixed with 1 μg of plasmid DNA, 50 μg of sonicated DNA salmon sperm (Stratagene), and 10% (v/v) DMSO. 300 μl of 40% (w/v) PEG3350 in LiOAc/TE was added to each reaction and mixed by gentle trituration. The cells were incubated for 90 min at 25° followed by a heat shock at 42° for 10 min. The cells were then centrifuged at 2000g for 2 min, supernatant removed, washed in 1 ml sterile water, and pelleted again. 750 μl of the supernatant was removed and the cells were resuspended in the remaining 250 μl and plated onto appropriate selective media by spreading with approximately 10–15 4-mm sterile glass beads. The plates were incubated at 25° until colonies appeared, usually after about 3–4 days. To ensure the bait (pSos-RP2) or pMyr-Arl3 did not cause transactivation of the system, single yeast colonies transformed with the respective plasmid constructs were resuspended in 25 μl of sterile water, and 2.5 μl was spotted onto an SD/glucose (-L) plate. This was repeated for various control transformations using SD/galactose (-UL) plates (0.17% (w/v) yeast nitrogen base without amino acids, 0.5% (w/v) ammonium sulfate, 2% (w/v) galactose, 1% (w/v) raffinose, 1.5% (w/v) agar, appropriate 1× dropout solution added after cooling to 55°. These plates were then incubated at 37° for 5 days and growth was compared to the positive and negative controls.

Verification of Bait Protein Expression

To ensure expression of the bait fusion protein (hSos-RP2) yeast cell total protein was extracted and the expression level determined using SDS-PAGE and immunoblotting (Fig. 2B). A saturated yeast culture was harvested at 2000g for 5 min. Cells were resuspended in 1 ml cold distilled water and lysed by addition of 150 μl freshly made NaOH/β-ME buffer (1.85 M NaOH, 7.5% (v/v) β-mercaptoethanol) on ice for 15 min. The cells were vortexed briefly and incubated with 150 μl of 55% (w/v) TCA in water on ice for a further 10 min. The protein extract was then pelleted by centrifugation at 12,000g for 10 min at 4°. The supernatant was removed and the protein extract resuspended in 300 μl SU buffer (5% (w/v) SDS,

125 mM Tris-HCl (pH 6.8), 0.1 mM EDTA, 0.005% (w/v) bromophenol blue, 8 M urea, 15 mg/ml DTT). 4 μl SDS-PAGE sample buffer was added to a 16 μl aliquot of SU buffer suspension and heated for 3 min at 65° before SDS-PAGE and immunoblotting. Sos fusion proteins were visualized by anti-Sos1 at 1:250 (BD Biosciences, Oxford, UK), goat anti-mouse-HRP (Pierce, Tattenhall, UK), and enhanced chemiluminescence ECL+.

Protein Interactions

Yeast cells were cotransformed with a total of 1.5 μg plasmid DNA (0.75 μg pMyr-Arl3 and 0.75 μg pSos-RP2) as described previously. A positive control of 0.75 μg pSos-RP2 with 0.75 μg pMyr-SB (the SB interacts with the hSos moiety), and a negative control of pSos-RP2 with pMyr-LaminC, were also cotransformed. Transformations were grown for 36–40 h at 25°. Single transformant colonies were picked and streaked out onto SD/galactose plates in duplicate and the plates were incubated at 37° for up to 7 days and observed daily for growth (Fig. 2A).

Investigating Post-Translational Modifications and Subcellular Localization of RP2

Antisera developed against recombinant RP2 showed that in cultured cells the protein appeared to localize to the plasma membrane (Chapple et al., 2000). Bioinformatic analyses revealed that this could be a consequence of a dual acylation motif at the N-terminus of the protein. Therefore we designed a series of experiments to test the significance of this potential motif for myristoylation and palmitoylation.

Inhibition of Myristoylation

The importance of myristoylation for RP2 targeting to the plasma membrane was tested by inhibition of myristoylation with 2-bromopalmitate. This was tested in the neuroblastoma cell line SH-SY5Y, which was maintained in Dulbecco's modified Eagle's medium-F12 with 10% (v/v) fetal calf serum. Cells were seeded at 1×10^5 cells/ml onto chamber slides. After 24 h, cells were transferred to media containing 2.5% (v/v) fetal calf serum, with or without 50 μM 2-bromopalmitate (Aldrich, Poole, UK). After a further 16 h, cells were fixed in ice-cold methanol and processed for immunofluorescence with affinity-purified antisera S974, as we have described (Chapple et al., 2000). Fluorescence was detected with an LSM510 laser scanning confocal microscope (Carl Zeiss, Oberkochen, Germany).

Construction of Plasmids

Full-length RP2 cDNA was subcloned into the *Bam*HI site of the expression vector pBK-CMV to create an untagged RP2 expression vector. For mutagenesis of the coding region, RP2 was PCR amplified from the pGEM-T RP2 construct using 5′ primers where sequence alterations had been introduced to produce the required amino acid sequence changes G2A, C3S, G2A/C3S, and ΔS6. An RP2-GFP fusion was created by cloning the RP2 open reading frame into the *Bam*HI/*Age*I site of pEGFP-N1. Cloning into the *Bam*HI-*Age*I site of pEGFP-N1 was also used to generate certain chimeras of N-terminal regions of RP2 fused to GFP. Oligonucleotides with a *Bam*HI site in the 5′ end and an *Age*I site at the 3′ end were annealed, phosphorylated, and cloned into the GFP vector. The N-terminal 15 amino acid sequence of RP2 was cloned into the *Bam*HI/*Age*I site of pEGFP-N1 using annealed phosphorylated oligonucleotides. This stratagem was also used to introduce a range of mutations at the N-terminus (Chapple *et al.*, 2000; Chapple *et al.*, 2002). All constructs were confirmed by sequencing.

Basic Subcellular Fractionation

Subcellular fractionation experiments were performed either on endogenous RP2 expressed in SH-SY5Y cells, untagged human RP2 transfected into CHO cells, or RP2-GFP transfected into a range of cells. We exploited the primate specificity of antiserum S974 to study untagged RP2 in CHO cells and GFP fluorescence or anti-GFP antibodies to study the RP2-GFP. Cells were washed twice and scraped into PBS. They were pelleted by centrifugation at 1000*g* and resuspended in breaking buffer (250 m*M* sucrose, 10 m*M* triethanolamine, 10 m*M* acetic acid 1 m*M* EDTA, pH 7.45, Sigma protease inhibitor cocktail, Poole, UK). The cells were then allowed to swell for 10 min prior to breaking by passing 30 times through a steel ball bearing cell homogenizer with a clearance of 12 μm (HGM, Heidelberg, Germany). After breakage, cells were centrifuged at 1850*g*, and the supernatant fraction removed. The pellet was resuspended in an equal volume of buffer prior to comparison of pellet and supernatants by immunoblotting analysis.

Sucrose Gradient Centrifugation

Supernatant fractions from the basic subcellular fractionation were further separated on a linear sucrose gradient. A 14 ml 10–40% (w/w) sucrose gradient in 50 m*M* HEPES pH 7.2, 90 m*M* KCL was prepared with a 65% (w/v) sucrose cushion. Supernatants (1 ml) were loaded on to the sucrose gradients and centrifuged using a swing out rotor at 78,800 average g for 16 h at 4°. In order to test membrane association dependent fractionation, detergent (NP-40) was added to the supernatant fraction to a final concentration of 0.5% (v/v), 10 min prior to loading onto the gradient. After centrifugation 1 ml fractions were collected starting from the bottom of the tube (65% (w/v) sucrose). Fractions were stored at $-70°$ until immunoblotting analysis.

Detergent Resistant Membrane (DRM) Isolation

As dual acylated proteins have been previously shown to be associated with membrane microdomains at the plasma membrane, we investigated their association with detergent resistant membranes DRMs (Chapple et al., 2003). DRMs were isolated using a variation of the method of Madore (1999). SH-SY5Y cells pellets were resuspended in 10 mM Tris-HCl pH 8.2 containing a protease inhibitor cocktail (Sigma). Triton X-100 was added to SH-SY5Y samples to a final concentration of 1% (v/v) and the samples were then incubated for 30 min at 4°. The gradient consisted of the sample in a 40% (w/v) sucrose cushion with 30% and 5% sucrose steps overlaid. 1% (v/v) Triton X-100 was present throughout the gradient. Following centrifugation (200,000g for 18 h at 4°) fractions of the gradient were withdrawn and equal fraction volumes analyzed by immunoblotting. RP2 was compared with a range of other membrane associated proteins known to be DRM associated (e.g., Lyn, Fyn) or not DRM associated (e.g., Rho A, Transferrin receptor). The DRM marker GM-1 was detected by dot blotting fractions and binding of biotinylated cholera toxin β-subunit (Sigma).

Defining the Localization of RP2 in the Retina

An essential aspect of understanding the role of RP2 in the pathogenesis of XLRP was to localize the expression of the protein in the human retina. We developed a method to characterize the retinal expression RP2 based on immunofluorescent labeling of "thick" Vibratome sections and optical sectioning using a confocal microscope. Importantly, the retinae studied were fresh surgical samples that had been lightly fixed. This coupled with use of Vibratome sections ensured good morphological and antigen preservation. Using these techniques we were able to localize RP2 to the plasma membrane of the rod and cone photoreceptors in the adult human retina (Fig. 3). RP2 is also present on the plasma membrane of every other cell type within the retina. This localization was confirmed by dual labeling with a range of subcellular markers (Grayson et al., 2002).

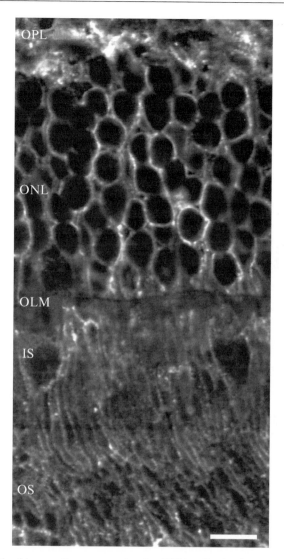

FIG. 3. Confocal immunofluorescence showing RP2 localization in photoreceptors of the human retina. An 80-μm Vibratome section of human retina stained with rabbit hRP2-337-350 antiserum was detected with Cy3 conjugated donkey anti-rabbit and visualized by optical sectioning using a Zeiss LSM510 confocal microscope. OPL outer plexiform layer; ONL outer nuclear layer; OLM outer limiting membrane; IS inner segment; OS outer segment. Scale bar is 10 μm.

Immunofluorescence Labeling and Confocal Scanning Microscopy

Adult human retinae were fixed with 4% (v/v) paraformaldehyde in PBS pH 7.3, within 2 min of enucleation for at least 2 h. After thorough rinsing in PBS, the retina tissue was embedded in 5% (w/v) low melting point agarose and cut into 80-μm sections using a Vibratome. The sections were blocked with 5% (v/v) normal donkey serum (Jackson Immunoresearch, Luton, UK) and 2% (w/v) BSA overnight at 4° prior to antibody incubations. The hRP2-337-350 antiserum was used at 1:100 in blocking buffer at 4° overnight. The sections were washed extensively in PBS, with at least 5 washes of at least 30 min duration at 4°. The primary antibody was detected with Cy3 conjugated donkey anti-rabbit at 1:100 (Jackson Immunoresearch), in blocking buffer at 4° overnight. Nuclei were labeled with DAPI (0.5 μg/ml) in the final PBS wash and sections were mounted in mounting media containing 15 mM sodium azide (Dako, Ely, UK). With retinal staining it is extremely important to verify the specificity of the immunostaining. This should be done by staining sections with no primary antiserum, pre-immune serum, and pre-absorbed antisera (e.g., hRP2-337-350 was absorbed with the peptide epitope at 20 μg/μl). For double labeling experiments, it is important to incubate rabbit primary antibodies with mouse secondaries and vice versa to ensure there is no cross-reactivity between antisera in the double labeling procedures. The labeled retinal sections were visualized with a Zeiss LSM 510 laser scanning confocal microscope.

Future Directions

The data suggest that the XLRP protein RP2 and Arl3 cooperate to regulate the cytoskeleton in some way. Arl3 in itself appears to act as a MAP, as it decorates microtubule structures including the photoreceptor axoneme (Grayson *et al.*, 2002). This association of Arl3 with specialized microtubule structures and cilia in particular may be conserved as a *Leishmania* homologue of Arl3 has been shown to be important for flagellum integrity (Cuvillier *et al.*, 2000). The association of RP2 with membrane microdomains also warrants further investigation, as it suggests a potential link to vesicular transport and/or signaling processes. As GTP-Arl3 interacts with unmyristoylated RP2 preferentially, important questions are where in the cell the interaction takes place and what regulates the nucleotide binding state of Arl3. The major challenges ahead are to clarify the consequences of the RP2:Arl3 interaction and determine the role of these proteins in the retina. Given the strong association between

RP2 and Arl3 we believe it is likely that Arl3 will also be important for retinal function. Clearly, little is known about RP2 and the related family of tubulin folding cofactors, but their relevance to human disease is becoming clear and they are likely to be important cellular factors.

Acknowledgments

We are grateful to our colleagues and collaborators who have helped develop these methodologies and provided reagents. This work was supported by the Medical Research Council (MRC), Wellcome Trust, and Fight for Sight.

References

Bartolini, F., Bhamidipati, A., Thomas, S., Schwahn, U., Lewis, S. A., and Cowan, N. J. (2002). Functional overlap between retinitis pigmentosa 2 protein and the tubulin-specific chaperone cofactor C. *J. Biol. Chem.* **277,** 14629–14634.

Bhamidipati, A., Lewis, S. A., and Cowan, N. J. (2000). ADP ribosylation factor-like protein 2 (Arl2) regulates the interaction of tubulin-folding cofactor D with native tubulin. *J. Cell Biol.* **149,** 1087–1096.

Breuer, D. K., Yashar, B. M., Filippova, E., Hiriyanna, S., Lyons, R. H., Mears, A. J., Asaye, B., Acar, C., Vervoort, R., Wright, A. F., Musarella, M. A., Wheeler, P., MacDonald, I., Iannaccone, A., Birch, D., Hoffman, D. R., Fishman, G. A., Heckenlively, J. R., Jacobson, S. G., Sieving, P. A., and Swaroop, A. (2002). A comprehensive mutation analysis of RP2 and RPGR in a North American cohort of families with X-linked retinitis pigmentosa. *Am. J. Hum. Genet.* **70,** 1545–1554.

Chapple, J. P., Hardcastle, A. J., Grayson, C., Spackman, L. A., Willison, K. R., and Cheetham, M. E. (2000). Mutations in the N-terminus of the X-linked retinitis pigmentosa protein RP2 interfere with the normal targeting of the protein to the plasma membrane. *Hum. Mol. Genet.* **9,** 1919–1926.

Chapple, J. P., Hardcastle, A. J., Grayson, C., Willison, K. R., and Cheetham, M. E. (2002). Delineation of the plasma membrane targeting domain of the X-linked retinitis pigmentosa protein RP2. *Invest. Ophthalmol. Vis. Sci.* **43,** 2015–2020.

Chapple, J. P., Grayson, C., Hardcastle, A. J., Bailey, T. A., Matter, K., Adamson, P., Graham, C. H., Willison, K. R., and Cheetham, M. E. (2003). Organization on the plasma membrane of the retinitis pigmentosa protein RP2: Investigation of association with detergent-resistant membranes and polarized sorting. *Biochem. J.* **372,** 427–433.

Cuvillier, A., Redon, F., Antoine, J. C., Chardin, P., DeVos, T., and Merlin, G. (2000). LdARL-3A, a Leishmania promastigote-specific ADP-ribosylation factor-like protein, is essential for flagellum integrity. *J. Cell Sci.* **113,** 2065–2074.

Dandekar, S. S., Ebenezer, N. D., Grayson, C., Chapple, J. P., Egan, C. A., Holder, G. E., Jenkins, S. A., Fitzke, F. W., Cheetham, M. E., Webster, A. R., and Hardcastle, A. J. (2004). An atypical phenotype of macular and peripapillary retinal atrophy caused by a mutation in the RP2 gene. *Br. J. Ophthalmol.* **88,** 528–532.

Grayson, C., Bartolini, F., Chapple, J. P., Willison, K. R., Bhamidipati, A., Lewis, S. A., Luthert, P. J., Hardcastle, A. J., Cowan, N. J., and Cheetham, M. E. (2002). Localization in the human retina of the X-linked retinitis pigmentosa protein RP2, its homologue cofactor C and the RP2 interacting protein Arl3. *Hum. Mol. Genet.* **11,** 3065–3074.

Hardcastle, A. J., Thiselton, D. L., Van Maldergem, L., Saha, B. K., Jay, M., Plant, C., Taylor, R., Bird, A. C., and Bhattacharya, S. (1999). Mutations in the RP2 gene cause disease in 10% of families with familial X-linked retinitis pigmentosa assessed in this study. *Am. J. Hum. Genet.* **64,** 1210–1215.

Madore, N., Smith, K. L., Graham, C. H., Jen, A., Brady, K., Hall, S., and Morris, R. (1999). Functionally different GPI proteins are organized in different domains on the neuronal surface. *EMBO J.* **18,** 6917–6926.

Rosenberg, T., Schwahn, U., Feil, S., and Berger, W. (1999). Genotype-phenotype correlation in X-linked retinitis pigmentosa 2 (RP2). *Ophthalmic Genet.* **20,** 161–172.

Schwahn, U., Lenzner, S., Dong, J., Feil, S., Hinzmann, B., van Duijnhoven, G., Kirschner, R., Hemberger, M., Bergen, A. A., Rosenberg, T., Pinckers, A. J., Fundele, R., Rosenthal, A., Cremers, F. P., Ropers, H. H., and Berger, W. (1998). Positional cloning of the gene for X-linked retinitis pigmentosa 2. *Nature Genet.* **19,** 327–332.

Schwahn, U., Paland, N., Techritz, S., Lenzner, S., and Berger, W. (2001). Mutations in the X-linked RP2 gene cause intracellular misrouting and loss of the protein. *Hum. Mol. Genet.* **10,** 1177–1183.

Shern, J. F., Sharer, J. D., Pallas, D. C., Bartolini, F., Cowan, N. J., Reed, M. S., Pohl, J., and Kahn, R. A. (2003). Cytosolic Arl2 is complexed with cofactor D and protein phosphatase 2A. *J. Biol. Chem.* **278,** 40829–40836.

Tian, G., Huang, Y., Rommelaere, H., Vandekerckhove, J., Ampe, C., and Cowan, N. J. (1996). Pathway leading to correctly folded beta-tubulin. *Cell* **86,** 287–296.

Tian, G., Lewis, S. A., Feierbach, B., Stearns, T., Rommelaere, H., Ampe, C., and Cowan, N. J. (1997). Tubulin subunits exist in an activated conformational state generated and maintained by protein cofactors. *J. Cell Biol.* **138,** 821–832.

Tian, G., Bhamidipati, A., Cowan, N. J., and Lewis, S. A. (1999). Tubulin folding cofactors as GTPase-activating proteins: GTP hydrolysis and the assembly of the alpha/beta-tubulin heterodimer. *J. Biol. Chem.* **274,** 24054–24058.

[42] Reconstitution of Transport to Recycling Endosomes *In Vitro*

By RENÉ BARTZ, CORINNE BENZING, AND OLIVER ULLRICH

Abstract

Using an *in vitro* assay instead of an approach *in vivo* greatly facilitates the analysis of complex intracellular mechanisms. This becomes particularly important for studying vesicular trafficking in both the endocytic and exocytic pathways with multiple transport routes connecting cellular organelles. Our chapter describes a novel cell-free assay that reconstitutes endosomal transport to recycling endosomes. The method measures transport of transferrin, a marker for endocytosis/recycling, from an endosome-enriched donor fraction to immunoisolated Rab11-positive acceptor recycling endosomes. Transfer of acridinium-labeled transferrin is detected by a highly sensitive chemiluminescence reaction using a luminometer.

METHODS IN ENZYMOLOGY, VOL. 404
0076-6879/05 $35.00
DOI: 10.1016/S0076-6879(05)04042-5

Introduction

Vesicular transport of molecules and lipids occurs between different compartments of mammalian cells, a process that needs to be tightly regulated. Part of this regulation machinery are Rab proteins, small GTPases of the Ras superfamily (Pfeffer and Aivazian, 2004; Seabra and Wasmeier, 2004; Segev, 2001; Zerial and McBride, 2001). The number of more than 60 different Rab proteins in mammals and their specific localization to subcellular compartments reflects the complexity of vesicular transport and the involvement of these proteins in distinct transport steps. Since their first discovery in the 1980s, it has been shown that Rab proteins are involved in many aspects of vesicular transport such as vesicle budding as well as tethering and fusion. They also play a role in recruiting molecular motor proteins to specific organelles and are subsequently involved in organelle movement along the cytoskeleton (Nielsen *et al.*, 1999). Rab proteins also seem to promote the remodeling of membranes in order to establish microdomains.

We are particularly interested in understanding the function of Rab proteins in the endocytic recycling pathway, where Rab11 seems to play a key role (Green *et al.*, 1997; Ren *et al.*, 1998; Ullrich *et al.*, 1996). This chapter describes a novel *in vitro* assay, reconstituting transport of transferrin to recycling endosomes (Bartz *et al.*, 2003). The assay is based on our previous biochemical approach to analyze immunoisolated Rab11-positive recycling endosomes and Rab5-positive sorting endosomes (Trischler *et al.*, 1999).

General Considerations

To investigate vesicular trafficking in a cell-free transport assay (e.g., from endosomal membranes to recycling endosomes) four prerequisites are essential: first, a physiologically relevant marker by which the vesicular transport step can be followed and quantified; second, defined donor membranes; third, defined acceptor membranes; and fourth, appropriate transport conditions.

A well-established marker for the endocytotic/recycling pathway is transferrin bound to its specific receptor (Lemmon and Traub, 2000). In our assay (Fig. 1), transferrin is covalently coupled to a nonradioactive, low molecular weight acridinium-ester (Bartz *et al.*, 2003). The resulting acridinium-transferrin (Ac-Tfn) can be detected with high sensitivity by adding trigger solution to the compound through injection by a luminometer. Cleavage of acridinium leads to a rapid chemiluminescence reaction emitting light at a maximum of 430 nm, which is measured by the same apparatus.

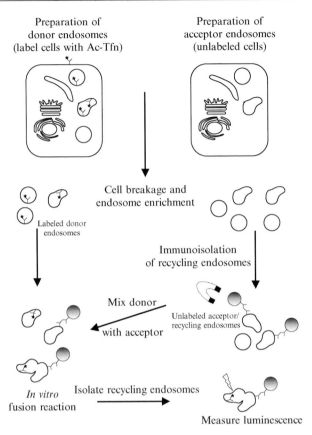

FIG. 1. Principle of the *in vitro* fusion assay. After labeling the endocytic pathway with acridinium-transferrin (Ac-Tfn) labeled endosomes (donor) are enriched by density centrifugation. Recycling endosomes (acceptor) are purified from unlabeled cells first by density centrifugation followed by immunoisolation using anti-Rab11 coated magnetic beads. Transport of Ac-Tfn from the donor endosomes to acceptor recycling endosomes is performed at 37° and in the presence of cytosol and ATP. After washing the beads-bound membranes, transferred Ac-Tfn is measured by detecting acridinium cleavage in a luminometer.

Donor endosomal membranes are prepared from cells where the endocytic/recycling pathway has been labeled with Ac-Tfn. After binding to its receptor, iron-loaded Ac-Tfn is internalized via clathrin-mediated endocytosis and sequentially moves through sorting and recycling endosomes that specifically contain the endosomal marker proteins Rab5 and Rab11, respectively, as previously shown (Trischler *et al.*, 1999). Depending on the internalization time, Tfn is found after 3 min in Rab5-positive sorting endosomes. After 3 min internalization followed by a 10 min chase, it has moved to Rab11-positive recycling endosomes. After approximately 30 min, the Tfn-cycle is complete and the receptor releases its iron-free ligand into the extracellular space (Ullrich *et al.*, 1996). Therefore, by varying the internalization protocol, sorting endosomes or recycling endosomes can be preferentially labeled as donor membranes with Ac-Tfn. However, for the experiments described in this chapter, cells are incubated continuously with Ac-Tfn for 30 min, thus the complete endocytic/recycling pathway is labeled. To prepare an endosome-enriched fraction containing both labeled sorting and recycling endosomes, cells are homogenized and then fractionated by flotation in a sucrose step gradient (Gorvel *et al.*, 1991; Trischler *et al.*, 1999).

Acceptor recycling endosomes are purified using a combined protocol of subcellular fractionation and immunoisolation. First, an endosome-enriched fraction is prepared from unlabeled cells by subcellular fractionation as described previously. Second, the endosome-enriched fraction is used as starting material for immunoisolation of recycling endosomes with anti-Rab11 antibodies. Rab11 is specifically present on the surface of these organelles (Green *et al.*, 1997; Ren *et al.*, 1998; Trischler *et al.*, 1999; Ullrich *et al.*, 1996). In general, Rab proteins can be used as subcellular marker proteins as they show a very specific distribution on distinct organelles and are ideal targets for isolating specific organelles (Pfeffer and Aivazian, 2004; Seabra and Wasmeier, 2004; Segev, 2001; Zerial and McBride, 2001). However, Rab proteins cycle between their distinct membranes and the cytosol in a GTP/GDP-dependent manner. Therefore, an immunoisolation directly out of cell lysates is difficult since cytosolic Rabs compete with the membrane-bound form of the protein. The cytosolic pool of Rabs can be removed by the subcellular fractionation step prior to immunoisolation.

Anti-Rab11 antibodies used for immunoisolation are bound to paramagnetic beads (Dynabeads). Magnetic beads have been used extensively to enrich for certain cell populations; however, more recently they are also used for isolating organelles (Bartz *et al.*, 2003; Espenshade *et al.*, 2002; Gagescu *et al.*, 2000; Harsay and Schekman, 2002; Luers *et al.*, 1998; Trischler *et al.*, 1999). Beads with bound organelles can either be precipitated by low-speed centrifugation or by using a magnetic device.

Cells and Cell Culture

For our experiments we use CHO (Chinese hamster ovary) cells, kindly provided by M. Zerial (Max Planck Institute for Molecular Cell Biology and Genetics, Dresden, Germany). Cells are cultured on 10-cm tissue culture dishes in Ham's F12 containing 10% (v/v) heat-inactivated fetal bovine serum (FBS), 100 U/ml penicillin, 100 μg/ml streptomycin, and 2 mM L-glutamine at 37° in 5% CO_2 in a standard tissue culture incubator. Media and reagents for cell culture are obtained from Invitrogen (Karlsruhe, Germany). For maintenance of CHO cells they are split 1:5 two to three times a week. The day before the experiment confluent cells are split 1:2 to reach a confluency of about 60%.

Labeling of Transferrin with Acridinium Ester

40 mg lyophilized human holo-transferrin (Sigma, Munich, Germany) are reconstituted in 7 ml labeling buffer (100 mM bicarbonate buffer, pH 8.5) and incubated for 10 min at room temperature (RT). The solution is transferred into a small glass beaker and stirred gently. 1.9 mg acridinium C_2 NHS ester (Biotrend, Cologne, Germany) are dissolved in 750 μl dimethylformamide (DMF; Pierce) and added dropwise into the stirring transferrin solution. Moderate mixing continues for an additional 30 min at RT. To quench remaining active groups of the ester, 80 μl of a 1 M glycine solution (in labeling buffer, pH 8.5) are added and the solution is stirred slowly for 15 min at RT. The solution is then applied to a PD-10 column (Pharmacia, Freiburg, Germany) pre-equilibrated with column buffer (50 mM Tris-HCl, pH 8.0) to separate coupled from free acridinium ester. Labeled transferrin is eluted with 13 ml of column buffer and 500-μl fractions are collected. Peak protein fractions are determined according to Bradford (Bradford, 1976) using BioRad (Munich, Germany) reagent. In parallel, peak fractions containing acridinium-coupled transferrin are detected by diluting 1 μl of each fraction in 1 ml H_2O and measuring relative light units (RLUs) using a luminometer (Berthold, model LB 9501, Bad Wildbad, Germany). First, acridinium trigger solution 1 (Biotrend, Cologne, Germany) is injected into the sample followed with a delay of 1.2 s by the injection of trigger solution 2 and a read-out time of 2 s. Fractions with highest protein concentration and highest RLU read-out are pooled and dialysed (MWCO: 10,000 Daltons) for 4 h (with one buffer change after 2 h) at 4° against iron-saturation buffer (1 mM NaHCO$_3$, 2 mM Na$_2$NTA, 250 mM Tris-HCl; first adjust pH to 8.2 with HCl, then add 250 mM FeCl$_3$) to ensure iron-loading of transferrin (Stoorvogel et al., 1987). Acridinium-transferrin is then dialysed for 12 h (with buffer changes after 4 and 8 h) against dialysis buffer (20 mM Hepes, 150 mM NaCl, pH 7.4). After dialysis, the sample is aliquoted and frozen at –20° until use.

Internalization of Acridinium-Transferrin (Ac-Tfn) into Endosomes

CHO cells grown on 10-cm tissue culture dishes to 60% confluency are washed twice with phosphate-buffered saline (PBS) at 37°. Next, serum transferrin is depleted from cells by an incubation in serum-free Ham's F12 for 1 h at 37° in a CO_2 incubator. Internalization of Ac-Tfn occurs by incubating cells with serum-free Ham's F12 containing 20 μg/ml Ac-Tfn at 37° and 5% CO_2. Depending on the incubation time, different compartments along the endocytic pathway can be labeled preferentially. For instance, a 3 min pulse labels mainly sorting endosomes while a 3 min pulse followed by washing cells twice with serum-free media and further incubation (chase) with serum-free Ham's F12 for 10 min labels preferentially recycling endosomes (Trischler *et al.*, 1999). Experiments described in this chapter are performed with 30 min internalization time, labeling both sorting and recycling endosomes. After incubation, cells are immediately placed on ice and the incubation media is exchanged to PBS (4°). Cells are washed twice with PBS at 4°, followed by one wash with PBS containing 5 mg/ml bovine serum albumin (BSA) for 20 min at 4°. BSA is finally removed by washing cells twice with PBS (4°). Cells are now ready for preparation of endosome-enriched fractions.

Preparation of Endosome-Enriched Fractions

The procedure is performed by a slightly modified protocol that was first described by Gorvel *et al.* (Gorvel *et al.*, 1991). Labeled or unlabeled CHO cells at 60% confluency are washed twice with PBS at 4° and are then detached gently from the dishes using a cell scraper. Cells of 6 dishes are pooled into one 15 ml tube and are spun for 5 min at 500g at 4°. After removing the supernatant cells are resuspendend in 600 μl homogenization buffer (250 mM sucrose, 3 mM imidazole, pH 7.4) and then broken using a 22G½ needle. Breaking efficiency is determined by phase contrast microscopy. Unbroken cells and nuclei are removed by spinning for 10 min at 1000g at 4°. The pooled postnuclear supernatants (PNS) of 30 dishes are transferred to the bottom of a Kontron TST 28.38 centrifugation tube (38 ml) and adjusted to a sucrose concentration of 40.6% using a stock solution of 62% sucrose/3 mM imidazole, pH 7.4. The solution is then carefully overlaid with 12 ml 35% sucrose/3 mM imidazole, pH 7.4, followed by 8 ml 25% sucrose/3 mM imidazole, pH 7.4 and filled up to the top with homogenization buffer. Step-flotation gradients are centrifuged at 108,000g for 3 h at 4°. A visible 35/25% interphase enriched for endosomes (early and recycling endosomes) is collected using a fractionator (Auto Densi-Flow, Labconco, Kansas City, USA). Fractions are used directly or snap-frozen in liquid nitrogen and stored for up to 2 months at –80°.

Coupling of Anti-Rab11 Antibodies to Magnetic Beads
  for Immunoadsorption

10 ml tosylactivated Dynabeads (Dynal, Norway) M-450 (4×10^8 beads/
ml; approx. 30 mg/ml) are resuspended thoroughly by vortexing and transferred into a 15 ml tube. Beads are spun down at $1000g$ for 7 min, the
supernatant is removed, and magnetic beads are washed once (resuspended
and centrifuged) in 5 ml 100 mM borate buffer (pH 9.5). For coupling beads
with secondary anti-rabbit antibodies, beads are resuspended in 5 ml borate
buffer, then 1.5 mg affinity purified goat anti-rabbit IgG (Fc-specific)
(Dianova, Hamburg, Germany) are added to the suspension. Coupling occurs
on a rotating wheel at $37°$ for 24 h. After incubation, beads are pulled down by
centrifugation at $500g$ for 3 min at RT. Beads are washed twice with 50 mM
Tris/100 mM NaCl/0.1% BSA, pH 8.5. Remaining uncoupled sites are
blocked by incubation in this wash buffer for 20 h at RT under slow rotation.
Finally, magnetic beads are centrifuged at $500g$ for 3 min and washed twice
with 5 ml PBS/0.1% BSA/0.1% Tween 20 and twice with 5 ml PBS/0.1% BSA
to remove Tween. The coated Dynabeads are resuspended in 5 ml PBS/0.1%
BSA containing 0.02% (w/v) sodium azide as a bacteriostatic agent and can
be stored for several months at $4°$. Binding efficiency of antibodies to beads
is calculated by determing free protein (antibody) concentrations before and
after the coupling reaction (usually 70–80% coupling efficiency).

For a single point of the fusion assay, 50 μl (3 mg) of the goat anti-rabbit
coupled M-450 bead-suspension are added into a 1.5 ml reaction tube and
suspended in 1 ml PBS/0.5% BSA. Beads are spun down at $500g$ for 2 min,
supernatant is removed, and beads are washed two times in 1 ml PBS/0.5%
BSA. For primary antibody binding, beads are resuspended in 1.4 ml PBS/
0.5% BSA and 5 μg affinity purified anti-Rab11 antibodies (rabbit polyclonal) are added to the suspension. Polyclonal rabbit Rab11-antiserum
was raised against full-length recombinant His_6-Rab11 expressed in
Escherichia coli and affinity purified essentially as previously described
(Bartz *et al.*, 2003). Binding of antibodies to beads occurs at $4°$ under slow
rotation for 12 h. Subsequently, unbound antibodies are removed by washing beads three times with PBS/0.5% BSA. Beads are finally resuspended
in 1 ml PBS/0.5% BSA, transferred into a new 1.5 ml reaction tube, spun
down again at $500g$ for 2 min, and supernatant is removed.

Immunoadsorption of Recycling Endosomes

For specific isolation of recyling endosomes, 100 μl of unlabeled endosome-enriched fraction (from the step-flotation gradient) diluted with ice-
cold 750 μl PBS/0.1% BSA are added to 3 mg of anti-Rab11-coupled

magnetic beads and gently resuspended. The mixture of beads and endosomes is then incubated at 4° under slow rotation for 4 h. After incubation, beads are collected with a magnetic particle collector (MPC, Dynal, Norway) placed on ice. Supernatant is removed and beads are washed once with 1 ml ice-cold fusion buffer (12 mM Hepes, 75 mM potassium acetate, and 1.5 mM magnesium acetate, pH 7.3) containing 0.1% BSA. Beads coated with Rab11-positive acceptor recycling endosomes are collected by the MPC and are now ready for the fusion reaction.

In Vitro Fusion Assay

The combined methods of preparing an endosome-enriched fraction labeled with acridinium-transferrin and specific immunoadsorption of unlabeled Rab11-positive recycling endosomes provide the basis of our *in vitro* fusion assay. Using this assay, transport of the physiological marker transferrin from labeled donor endosomes to immunoisolated acceptor recycling endosomes can be followed by measuring arrival of acridinium-transferrin to beads-bound endosomes as detected by lashlight-luminescence in a luminometer.

For each point of the *in vitro* fusion assay, 3 mg magnetic beads with immunoisolated recycling endosomes are resuspended on ice in a total volume of 150 μl ice-cold reation mix, containing fusion buffer (12 mM Hepes, 75 mM potassium acetate, 1.5 mM magnesium acetate, pH 7.3, final concentration), 50 μl rat liver cytosol (10 mg/ml), and 50 μl donor endosome-enriched fraction from cells incubated with Ac-Tfn for 30 min. Fusion is measured in the presence of an ATP-regenerating system (1 mM ATP, 8 mM creatine phosphate, and 40 μg/ml creatine phosphokinase) or an ATP-depletion system (for each assay point 15 μl of hexokinase (NH_4)$_2SO_4$-precipitate suspension (1500 U/ml) is centrifuged and the pellet is resuspended in 10 μl 0.5 mM D-glucose). All components for the ATP-regenerating and depletion system were purchased from Roche Applied Science (Mannheim, Germany).

To prepare rat liver cytosol, freshly removed rat livers are minced and rinsed in homogenization buffer (250 mM sucrose, 3 mM imidazole, pH 7.4) at 4°. The minced tissue is transferred into a Potter–Elvehjem and homogenized in ice-cold homogenization buffer (5 ml/g tissue), followed by centrifugation of the homogenate at 15,000g for 10 min at 4°. The supernatant is passed through sterile gauze and the filtrate centrifuged at 200,000g at 4° for 1 h. The supernatant is diluted to 10 mg/ml with homogenization buffer and aliquots are frozen at –80°.

The *in vitro* transport reaction is started by transferring the samples to a 37° waterbath for different periods of time (for up to 60 min) with

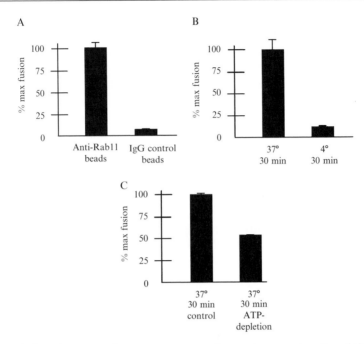

FIG. 2. Requirements of transport to recycling endosomes *in vitro*. (A) After immunoisolation of acceptor recycling endosomes using beads coated with anti-Rab11-antibodies (anti-Rab11) or with anti-rabbit antibodies (IgG control), beads were incubated with donor endosomes for 30 min at 37°. Only for anti-Rab11-isolated membranes efficient transfer of Ac-Tfn was detected, thus excluding unspecific binding of labeled endosomes to magnetic beads. (B) Incubation of anti-Rab11 isolated acceptor membranes with donor endosomes for 30 min at 37° or 4° demonstrates the temperature-dependency of transport to recycling endosomes. (C) Acceptor membranes immunoisolated with anti-Rab11 were incubated for 30 min at 37° with donor endosomes in the presence of an ATP-regenerating system (control) or an ATP-depletion system (ATP-depletion). The approximately 50% inhibition after depletion of cytosolic ATP reflects the general requirement of ATP for membrane transport. Error bars represent the standard deviation of three experiments (reprinted from Biochemical and Biophysical Research Communications, Vol. 3 [Bartz *et al.*, 2003]. © 2003, with permission from Elsevier).

occasional manual shaking to prevent magnetic beads from settling. After incubation, beads are collected on ice with the magnetic particle collector to stop the fusion reaction. Following collection, beads are washed at 4° with 1.5 ml 300 mM KCl/0.1% BSA by rotating beads on a wheel for 2 min. The suspension is transferred into a new reaction tube and beads are collected using the magnetic particle collector. Then, magnetic particles are washed once with PBS followed by an incubation with 100 μl 0.1% Triton X-100 for 5 min on ice in order to solubilize the membranes. Finally, beads are collected by the magnet and acridinium-transferrin in the supernatant is detected by measuring flashlight chemiluminescence in acridinium trigger solution (Biotrend, Cologne, Germany) for 2 s using a Berthold luminometer LB 9501.

Testing the General Requirements for Transport to Recycling Endosomes *In Vitro*

To study the requirements of the *in vitro* transport reaction to recycling endosomes, we performed several control experiments (Fig. 2). First, we tested for nonspecific binding of labeled endosomes to beads. Anti-Rab11 antibodies or nonspecific IgG antibodies were bound to magnetic beads and used for immunoisolation of unlabeled acceptor endosomes. Both samples were mixed with Ac-Tfn-labeled endosomes for a 30 min fusion reaction. Magnetic beads with control antibodies showed only little nonspecific binding of labeled endosomes (Fig. 2A). Next, temperature- and ATP-dependency of *in vitro* transport were tested. As shown in Fig. 2B, *in vitro* transport of Ac-Tfn was only observed at 37° but not at 4°, consistent with the inhibition of transport in living cells at low temperatures. We then depleted cytosolic ATP using an ATP-depletion system in order to investigate the ATP-dependence of transport. Our results reflect the requirement of ATP for membrane transport since transport was found to be inhibited by approximately 50% after depletion of cytosolic ATP (Fig. 2C).

Acknowledgments

The authors are very grateful Dr. John K. Zehmer for critical reading of the manuscript. This work was supported by a grant from the Deutsche Forschungsgemeinschaft (to O.U.). R.B. was supported by a graduate scholarship of the University of Mainz (LGFG).

References

Bartz, R., Benzing, C., and Ullrich, O. (2003). Reconstitution of vesicular transport to Rab11-positive recycling endosomes *in vitro*. *Biochem. Biophys. Res. Commun.* **312,** 663–669.

Bradford, M. M. (1976). A rapid and sensitive method for the quantitation of microgram quantities of protein utilizing the principle of protein-dye binding. *Anal. Biochem.* **72,** 248–254.

Espenshade, P. J., Li, W. P., and Yabe, D. (2002). Sterols block binding of COPII proteins to SCAP, thereby controlling SCAP sorting in ER. *Proc. Natl. Acad. Sci. USA* **99,** 11694–11699.

Gagescu, R., Demaurex, N., Parton, R. G., Hunziker, W., Huber, L. A., and Gruenberg, J. (2000). The recycling endosome of Madin-Darby canine kidney cells is a mildly acidic compartment rich in raft components. *Mol. Biol. Cell* **11,** 2775–2791.

Gorvel, J. P., Chavrier, P., Zerial, M., and Gruenberg, J. (1991). rab5 controls early endosome fusion *in vitro*. *Cell* **64,** 915–925.

Green, E. G., Ramm, E., Riley, N. M., Spiro, D. J., Goldenring, J. R., and Wessling-Resnick, M. (1997). Rab11 is associated with transferrin-containing recycling compartments in K562 cells. *Biochem. Biophys. Res. Commun.* **239,** 612–616.

Harsay, E., and Schekman, R. (2002). A subset of yeast vacuolar protein sorting mutants is blocked in one branch of the exocytic pathway. *J. Cell Biol.* **156,** 271–285.

Lemmon, S. K., and Traub, L. M. (2000). Sorting in the endosomal system in yeast and animal cells. *Curr. Opin. Cell Biol.* **12,** 457–466.

Luers, G. H., Hartig, R., Mohr, H., Hausmann, M., Fahimi, H. D., Cremer, C., and Volkl, A. (1998). Immuno-isolation of highly purified peroxisomes using magnetic beads and continuous immunomagnetic sorting. *Electrophoresis* **19,** 1205–1210.

Nielsen, E., Severin, F., Backer, J. M., Hyman, A. A., and Zerial, M. (1999). Rab5 regulates motility of early endosomes on microtubules. *Nat. Cell Biol.* **1,** 376–382.

Pfeffer, S., and Aivazian, D. (2004). Targeting Rab GTPases to distinct membrane compartments. *Nat. Rev. Mol. Cell. Biol.* **5,** 886–896.

Ren, M., Xu, G., Zeng, J., De Lemos-Chiarandini, C., Adesnik, M., and Sabatini, D. D. (1998). Hydrolysis of GTP on rab11 is required for the direct delivery of transferrin from the pericentriolar recycling compartment to the cell surface but not from sorting endosomes. *Proc. Natl. Acad. Sci. USA* **95,** 6187–6192.

Seabra, M. C., and Wasmeier, C. (2004). Controlling the location and activation of Rab GTPases. *Curr. Opin. Cell Biol.* **16,** 451–457.

Segev, N. (2001). Ypt/Rab GTPases: Regulators of protein trafficking. *Sci STKE* **2001,** RE11.

Stoorvogel, W., Geuze, H. J., and Strous, G. J. (1987). Sorting of endocytosed transferrin and asialoglycoprotein occurs immediately after internalization in HepG2 cells. *J. Cell Biol.* **104,** 1261–1268.

Trischler, M., Stoorvogel, W., and Ullrich, O. (1999). Biochemical analysis of distinct Rab5- and Rab11-positive endosomes along the transferrin pathway. *J. Cell Sci.* **112**(Pt. 24), 4773–4783.

Ullrich, O., Reinsch, S., Urbe, S., Zerial, M., and Parton, R. G. (1996). Rab11 regulates recycling through the pericentriolar recycling endosome. *J. Cell Biol.* **135,** 913–924.

Zerial, M., and McBride, H. (2001). Rab proteins as membrane organizers. *Nat. Rev. Mol. Cell Biol.* **2,** 107–117.

[43] Robust Colorimetric Assays for Dynamin's Basal and Stimulated GTPase Activities

By MARILYN LEONARD, BYEONG DOO SONG,
RAJESH RAMACHANDRAN, and SANDRA L. SCHMID

Abstract

Dynamin, unlike many GTPase superfamily members, exhibits a relatively rapid basal rate of GTP hydrolysis that is not rate-limited by GTP binding or GDP dissociation. Also unique to dynamin GTPase family

METHODS IN ENZYMOLOGY, VOL. 404   0076-6879/05 $35.00
Copyright 2005, Elsevier Inc. All rights reserved.   DOI: 10.1016/S0076-6879(05)04043-7

members is their ability to self-assemble into rings and helical stacks of rings either in solution or onto lipid templates. Self-assembly stimulates dynamin's GTPase activity by >100-fold. Given these robust rates of GTP hydrolysis compared to most GTPases, GTP hydrolysis by dynamin can be easily measured using a simple colorimetic assay to detect released phosphate. We describe this assay and report variations in assay conditions that have contributed to the wide range of reported values for dynamin's basal and assembly-stimulated rates of GTP hydrolysis.

Introduction

Dynamin, a ~100 kD multidomain GTPase, is required in higher eukaryotes for multiple modes of endocytosis (Conner and Schmid, 2003; Hinshaw, 2000; McNiven *et al.*, 2000). Dynamin self-assembles *in vitro* into helical arrays onto lipid templates, or in solution in the presence of nonhydrolyzable GTP analogues (Carr and Hinshaw, 1997; Hinshaw and Schmid, 1995; Sweitzer and Hinshaw, 1998; Takei *et al.*, 1995). Self-assembly stimulates dynamin GTPase activity by as much as ~100-fold, although reported values vary considerably (Barylko *et al.*, 1998; Stowell *et al.*, 1999; Warnock *et al.*, 1996). The dynamin and dynamin-related GTPases are unique in that they encode their own GAP (GTPase activating protein) activity within the C-terminally located GED (GTPase effector domain) that also functions in self-assembly (Marks *et al.*, 2001; Sever *et al.*, 1999; Song *et al.*, 2004a). GED has been proposed to function as an assembly-dependent GAP in mediating dynamin's rapid assembly-stimulated GTPase activity (Sever *et al.*, 1999). More recently, GED has also been shown to play a role in the basal GTPase activity of unassembled dynamin (Narayanan *et al.*, 2005; Song, *et al.*, 2004a).

Dynamin's function *in vivo* is best studied in the context of clathrin-mediated endocytosis and synaptic vesicle recycling, where it is targeted to nascent coated pits, and subsequently self-assembles into collar-like structures at the necks of deeply invaginated coated pits (Conner and Schmid, 2003). Dynamin self-assembly and assembly-stimulated GTPase activity are thought to be directly involved in late stages of coated vesicle formation and perhaps directly in membrane fission (Hinshaw, 2000; Praefcke and McMahon, 2004; Song and Schmid, 2003). Dynamin also has a measurable basal rate of GTP hydrolysis, whose function has been implicated in early stages of clathrin-coated vesicle formation (Narayanan *et al.*, 2005; Song *et al.*, 2004b). However, as for assembly-stimulated rates of GTP hydrolysis, there are significant variations in the reported rates for dynamin's basal GTPase activity, ranging from ~0.1–0.3 min^{-1} (Binns *et al.*, 1999; Marks *et al.*, 2001) to 1–2 min^1 (Barylko *et al.*, 2001; Sever *et al.*, 1999;

Shpetner and Vallee, 1992; Warnock *et al.*, 1996). We believe that these reported ranges of dynamin's GTPase activity reflect, in part, differences in assay conditions and assembly templates.

Here we describe a simple, colorimetric assay to measure GTP hydrolysis by dynamin and discuss some of the variables that can affect measurements of dynamin's basal and assembly-stimulated GTPase activity.

Materials and Reagents

Dynamin and GED Purification and Storage

Recombinant full length dynamin is expressed using a baculovirus expression system in Tn5 insect cells grown in suspension, as described in detail elsewhere (Damke *et al.*, 2001). Dynamin is purified to near homogeneity by affinity chromatography using GST-amph2SH3 domain immobilized on Glutathione-Sepharose, as originally described by Marks and colleagues (2001). The eluted protein is dialyzed overnight into 20 mM Hepes-KOH pH 7.5, 150 mM KCl, 1 mM EDTA, 1mM EGTA, 0.5 mM DTT dispensed in 100 μl aliquots, snap frozen in liquid nitrogen, and stored at –80°. Purified dynamin can be stored without loss of activity for several months under these conditions. However, we have observed that storage in divalent cation-containing buffers, or in the absence of DTT, results in enhanced aggregation and loss of activity. For assays, aliquots are thawed rapidly at room temperature and centrifuged for 3 min at 20,000g rpm in a microfuge to remove any aggregated protein. In general, thawed dynamin is used the same day.

The isolated GTPase effector domain (GED, corresponding to aa 618–752 of human dynamin-1) is expressed as a GST fusion protein in *E. coli* and purified by adsorption onto Glutathione-Sepharose (GSH-Seph). The GST-GED is eluted with 15 mM glutathione in 100 mM Tris-HCl pH 8.0, 150 mM KCl and digested at 4° for 20 h with thrombin (Amersham, Piscataway, NJ) at a ratio of 10 U/mg fusion protein to cleave the GST moiety. The digested material is dialyzed into 20 mM Tris-HCl pH 8.0, 100 mM NaCl, 1 mM DTT, and rechromatographed on GSH-Seph to remove the cleaved GST. The flow-through fractions containing GED are loaded onto Fast Flow Q Sepharose and step eluted sequentially with 150 mM, 250 mM, 300 mM, 350 mM, 450 mM, and 500 mM NaCl in 20 mM Tris-HCl pH 8.0, 1 mM DTT. This step removed the thrombin and any contaminating GST. Purified GED elutes at ∼300 mM NaCl. Isolated GED is dialyzed into 20 mM Hepes-KOH pH 7.0, 150 mM KCl, 1 mM EDTA, 1 mM EGTA, 1 mM DTT and stored in aliquots at –80°. The concentration of GED is determined by Abs at 280 nm using the extinction coefficient $0.27_{1\ mg/ml}$. The presence of 1 mM DTT in the buffers is essential as we have found that GED can be deactivated by oxidation. For assays, aliquots are thawed rapidly at room temperature and centrifuged for 3 min at 20,000g rpm in a microfuge to remove any aggregated protein.

Liposome Preparation and Storage

All phospholipids are purchased from Avanti Polar Lipids, Inc (Alabaster, AL). Cholesterol is from Calbiochem (San Diego, CA). Stock solutions of 1,2-dioleoyl-sn-glycero-3-phosphocholine (DOPC, 10 mg/ml), porcine brain L-α-phosphatidylinositol-4,5-bisphosphate, triammonium salt (PI4,5P$_2$, 1 mg/ml), and cholesterol (1 mg/ml) are prepared in chloroform and stored at –20°. Typically, we use liposomes containing DOPC, PI4,5P$_2$, and cholesterol in a final ratio of 80:15:5 mol%, respectively, which parallels the composition of lipid nanotubules (see following). Liposomes with lower concentrations of PI4,5P$_2$ were less effective templates for assembly-stimulated GTPase activity. Lipids and cholesterol stocks are mixed in a glass tube to yield the desired molar ratios. Typically, 48 μl of 10 mg/ml DOPC, 125 μl of 1 mg/ml PI4,5P$_2$, 15 μl of 1 mg/ml cholesterol to yield a molar ratio of 80:15:5 is dried under a stream of N$_2$ at room temperature, and further dried under vacuum and low heat (37°) for at least 2 h. The dried lipid mixture is resuspended in 300 μl of 20 mM Hepes, pH 7.5, 100 mM KCl, to yield a final concentration of 2.5 mM total lipids, warmed to 37°, and vortexed thoroughly to ensure complete hydration. The resuspended lipid mixture is then subjected to 3 freeze-and-thaw cycles followed by passage/extrusion, using an Avanti Mini-extruder, through a 0.4 μm polycarbonate membrane 21 times to yield unilamellar liposomes. The liposome suspension is stored on ice and used within one week of production.

Lipid Nanotubule Preparation and Storage

Lipid nanotubules are generated from a mixture of C24: 1 β-D-Galactosyl Ceramide (GalCer), DOPC, PI4,5P$_2$, and cholesterol in an approximate ratio of 40:40:15:5 mol%. The GalCer is first resuspended in a small volume of methanol and stored at –20° as a 10 mg/ml stock solution in methanol:chloroform (1:19). Lipids and cholesterol stocks are added in the following proportions: DOPC (4 μl), GalCer (4 μl), cholesterol (3 μl), and PI4,5P$_2$ (25 μl) to a glass vial, mixed by vortexing well and the solvent gently evaporated under a stream of nitrogen gas. After evaporation of all the solvent, the dried lipids are resuspended in 100 μl of 20 mM Hepes-KOH, pH 7.5, and 100 mM NaCl by vortexing well. Lipid nanotubules are then generated by sonication for 2 min at room temperature in a water bath sonicator (Branson Ultrasonic, Danburg, CT, Model 200), by holding the tube in the position that generates maximum turbulence in the lipid suspension. The final concentration of total lipids is ~1.3 mM and ~250 μM for PI4,5P$_2$.

Buffers and Reagents

GTPase assay buffer: 20 mM HEPES-KOH pH 7.5, 150 mM KCl, 2 mM MgCl$_2$, 1 mM DTT, Malachite Green Stock Solution (1 mM Malachite Green, 10 mM ammonium molybdate in 1N HCl): Dissolve 34 mg Malachite Green Carbinol base (Aldrich Chemicals, Milwaukee, WI) in 40 mL 1N HCl. Dissolve 1 g ammonium molybdate tetrahydrate (Sigma Chemical Co.) in 14 mL 4N HCl. Mix the two solutions together and bring to 100 mL with distilled water. Filter through a 0.45 μ Millipore filter and store in the dark at 4°.

GTP Stock: GTP (lithium salt, Sigma Chemical Co.) is resuspended in 20 mM Hepes, pH 7.4) to ≥100 mM. Nucleotide concentration is determined by UV adsorption at 252 nm (E$_{1\ mM}$ = 13.7) and adjusted to 100 mM. GTP stock is dispensed in 10 μl aliquots, snap frozen, and stored at –70° for several months.

    0.5 M EDTA, pH 8.0
    0.5 M KH$_2$PO$_4$ HCB50: 20 mM Hepes-KOH pH 7.0, 2 mM EGTA,
      2 mM MgCl$_2$, 0.1 mM DTT, 50 mM
    NaCl
    HCB150: same as HCB50, except containing 150 mM NaCl

Methods

Basal GTPase Activity

For basal GTPase assays, dynamin is diluted to 2× final concentration (typically 0.5–1 μM) in GTPase assay buffer. GTP is also diluted to 2× final concentration (typically over a range of 0.025–1.2 mM GTP for K$_m$ determinations or 0.5–1 mM GTP, for routine GTPase assays). To initiate the reaction, PCR tubes containing 90 μl of the dynamin stock are transferred to 37°. An equal volume of the GTP stock is added and mixed rapidly by up and down pipetting. 20 μl aliquots are removed after increasing times of incubation (typically over a 5–90 min time course) and transferred to microtiter wells containing 5 μl of 0.5 M EDTA. The reaction is immediately halted in the presence of 100 mM EDTA, which chelates the Mg^{2+} required for GTP hydrolysis. A 0 min time point is taken for each sample by adding 10 μl dynamin and 10 μl GTP directly into a microtiter well containing EDTA. When the time course is completed, 150 μl of the Malachite Green Stock solution is added to each well and the absorbance at 650 nm is determined using a microplate reader. A standard curve from 10–100 μM Pi is generated for each experiment and read in parallel. Samples with buffer only ± GTP are used as blank controls. Malachite Green detection of free phosphate is linear over a range of 1–100 μM (Fig. 1A). The use of PCR tubes and 96-well templates allows us to conveniently perform these assays using a multi-channel pipette.

We and others have previously described dynamin GTPase assays that followed the hydrolysis of α^{32}P-GTP using thin-layer chromatography

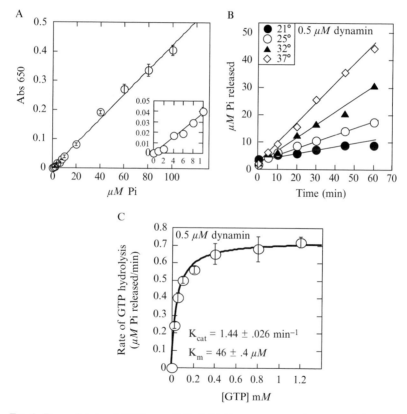

FIG. 1. Dynamin's basal GTPase activity is highly temperature-dependent (A) Standard curve for Malachite Green detection of free phosphate. Inset shows enlargement of curve at low concentrations of free phosphate. Data shown avg ± std. dev. values obtained from triplicate samples. (B) Time course of GTP hydrolysis by 0.5 μM dynamin measured under basal GTPase conditions at the indicated temperatures. (C) The basal rate of GTP hydrolysis of dynamin, measured at 37°, is plotted against the initial concentration of GTP to obtain the Michaelis-Menten constants k_{cat} and K_m for dynamin's basal GTPase activity. Data shown avg ± std. dev. values obtained from triplicate samples performed during single experiment.

(Barylko *et al.*, 2001; Damke *et al.*, 2001). The ratio of [GDP]/([GDP] + [GTP]) is determined for each sample to obtain an accurate measure of GTP hydrolysis that is independent of possible variations in the sample volumes spotted on TLC. This method, however, is prone to error if nonreacting, radiolabeled substances comigrate with GTP. Indeed, by allowing reactions to go to completion, we found that some ^{32}P GTP batches had materials that did not react but had the same retention time as GTP. The presence of these nonreacting materials lowered measured values for GTP hydrolysis up to 50% for basal GTPase rates when compared to those obtained using the colorimetric method.

Representative time courses for the hydrolysis of GTP measured using the colorimetric assay are shown in Fig. 1B. We find that the k_{cat} for GTPase activity is independent of the [dynamin] between 0.2 and 1 μM (data not shown) indicating that, under these assay conditions, the rate of GTP hydrolysis does not depend on intermolecular dynamin-dynamin interactions. However, positive cooperativity is detected at concentrations of dynamin >1 μM (Song *et al.*, 2004a) and at lower [dynamin] when assayed in low salt buffers (Warnock *et al.*, 1996). Importantly, the basal rate of GTP hydrolysis by dynamin is strongly temperature dependent (Table I). This temperature dependence can account for the discrepancies in reported basal rates of GTP hydrolysis as in some cases assays were performed at room temperature (Binns *et al.*, 1999; Marks *et al.*, 2001), and in others assays were performed at 37° (Sever *et al.*, 1999; Song *et al.*, 2004b).

Michaelis-Menten kinetic parameters, K_m and k_{cat}, can be measured by determining the initial rates of GTP hydrolysis at various concentrations of GTP (Fig. 1C). We typically measure GTPase rates at 25, 50, 100, 200, 400, 800, and 1200 μM GTP and determine K_m and k_{cat} by plotting the initial rate of GTP hydrolysis (v), determined during the linear phase of the time course, versus [GTP] and fitting the curve to the Michaelis-Menten equation: $v = V_{max}[GTP]/(K_m + [GTP])$. The Michaelis-Menten constant is a complex term reflecting the rates of GTP association (k_{on}) and dissociation (k_{off}) and of GTP hydrolysis (k_{cat}), thus $K_m = (k_{off} + k_{cat})/k_{on}$. The basal rate of GTP hydrolysis ($\sim$0.02 s^{-1}) is very slow compared to rate of GTP dissociation (2 s^{-1}, [Binns *et al.*, 1999]); therefore, under these conditions, $k_{off} + k_{cat} \approx k_{off}$ and the K_m for GTP is a close approximation of GTP binding affinity ($K_d = k_{off}/k_{on}$). Interestingly, the K_m for GTP is also significantly affected by temperature (Table I). Nonetheless, when measured at room temperature, the K_m for GTP (3.4 $\pm$ 1.7 μM) is indeed in good agreement to the value for K_d (2.5 μM) measured at room temperature using MANT-GTP by stop-flow kinetics also (Binns *et al.*, 1999).

Finally, we find that while the values we determine for k_{cat} are fairly constant between experiments and between preparations of dynamin; the values we obtain for K_m under basal conditions (ranging from ~30 μM to ~100 μM) are more variable (note large standard deviation in Table I and difference between these values and those obtained in the experiment shown in Fig. 1C). While we do not understand the nature of this variability, it may reflect varying degrees of micro-aggregation in the dynamin samples, as self-assembly can alter the K_m for GTP. The variation is reduced, but not eliminated, when dynamin is centrifuged to remove large aggregates before use (see previous).

Liposome-Stimulated GTPase Assays

In the absence of GTP, or in the presence of nonhydrolyzable analogues of GTP, dynamin will self-assemble onto and tubulate liposomes. This activity has been observed with liposomes of varied composition, including those formed solely using dioleoylphosphatidylserine (Sweitzer and Hinshaw, 1998), or Folch fraction I, a brain lipid extract (Takei et al., 2001). Correspondingly, dynamin's GTPase activity is stimulated 10–100-fold when measured in the presence of liposomes. We refer to this as dynamin's liposome-stimulated GTPase activity.

Liposome-stimulated GTPase assays are performed as described for basal except that the final concentration of dynamin used in the assay is lower (typically 0.1–0.5 μM). Liposomes are added from a 20$\times$ stock solution to the dynamin in assay buffer just prior to mixing this 2$\times$ dynamin-liposome stock with an equal volume of the 2$\times$ GTP stock in assay buffer to initiate the incubation. The remainder of the assay is as described above except that time points are taken more frequently and typically over a 0–15 min time course. The data in Fig. 2A shows that dynamin's GTPase activity could be stimulated >100-fold upon assembly onto a liposome template composed of DOPC:PI4,5P$_2$:cholesterol (80:15:5 mol%), with maximum stimulation occurring at >150 μM lipid (Fig. 2A). As previously described (Tuma and Collins, 1994), the concentration dependence for dynamin was sigmoidal, indicating positive cooperativity at low concentrations of dynamin (Fig. 2B).

The degree of stimulation of dynamin's GTPase activity depends on the composition of the liposomes. The data in Fig. 3A shows that dynamin's GTPase activity is stimulated ~100-fold when assayed in the presence of liposomes composed of DOPS, but only ~10-fold in the presence of liposomes composed of DOPC:PI4,5P$_2$ (90:10 mol%), although dynamin was able to tubulate liposomes of both compositions (Fig. 3B). While we have

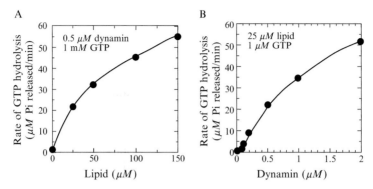

FIG. 2. Dynamin's GTPase activity is stimulated by assembly onto PI4,5P$_2$-containing liposomes. (A) Dependence of the rate of GTP hydrolysis by dynamin on the concentration of liposomes, assayed at fixed concentrations of dynamin (0.5 μM) and GTP (1 mM). (B) Dependence of the rate of GTP hydrolysis by dynamin on the concentration of dynamin in the assayed at fixed concentrations of liposomes (25 μM) and GTP (1 mM). Note the sigmoidal shape of the curve indicating cooperativity.

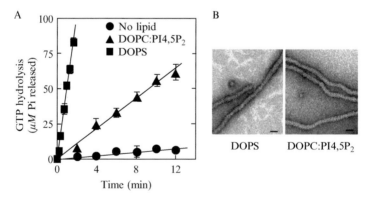

FIG. 3. Dynamin's liposome-stimulated GTPase activity varies with liposome composition. (A) Dynamin's GTPase activity is greater when assayed in the presence of liposomes composed exclusively of DOPS, compared to liposomes prepared from DOPC:PI4,5P$_2$ (90:10 mol%). Liposomes composed of a higher mol% PI4,5P$_2$ (see Fig. 2) are significantly more effective templates. (B) Negative-stain electron micrographs of dynamin-generated lipid tubules formed on DOPS and DOPC:PI4,5P$_2$ liposomes are indistinguishable, despite the observed differences in rates of GTP hydrolysis. Scale bar = 50 nm.

not systematically varied liposome size, we find that dynamin GTPase activity is equally stimulated by liposomes extruded using either a 0.4 μM or a 0.1 μm polycarbonate filter.

Lipid Nanotubule-Stimulated GTPase Activity

Lipid mixtures containing a high proportion of unsaturated galactocer-amides spontaneously form elongated nanotubules rather than more spherical liposomes (Goldstein *et al.*, 1997). Lipid nanotubules (LT), which are ~20 nm in diameter, resemble the dimensions of the constricted necks of clathrin coated pits, and are highly efficient templates for eliciting dynamin's assembly-stimulated GTPase activity (Stowell *et al.*, 1999). LT-stimulated GTPase assays are performed exactly as described for liposome-stimulated GTPase assays. We find that it is important to add the lipid nanotubes (final [lipids] 100–150 μM) to the diluted dynamin solution, as adding dynamin directly to the lipid nanotubules before dilution can decrease lipid stimulation rates by 50%. As for basal GTPase activity, LT-stimulated GTP hydrolysis is also highly temperature sensitive (Fig. 4A; Table I). The K_m for GTP measured in the presence of lipid nanotubules is lower than that determined for basal GTPase activity and exhibits less variability between experiments (Fig. 4B, Table I).

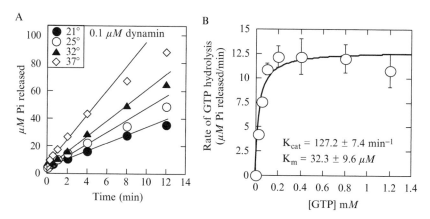

FIG. 4. Dynamin's lipid nanotubule-stimulated GTPase activity is temperature dependent. (A) Time course of GTP hydrolysis by 0.1 μM dynamin measured in the presence of 400 μM PI4,5P$_2$-containing lipid nanotubules (LT) at the indicated temperatures. (B) The LT-stimulated rate of GTP hydrolysis of dynamin, measured at 37°, is plotted against the initial concentration of GTP to obtain the Michaelis-Menten constants k_{cat} and K_m for dynamin's LT-stimulated GTPase activity. Data shown avg. ± std. dev. values obtained from triplicate samples performed during single experiment.

TABLE I

TEMPERATURE DEPENDENCE OF MICHAELIS-MENTEN PARAMETERS FOR DYNAMIN'S BASAL AND
LIPID NANOTUBULE (LT)-STIMULATED GTPASE ACTIVITY. VALUES SHOWN ARE AVERAGE ±
STANDARD DEVIATION (N ≥ 3 INDEPENDENT EXPERIMENTS)

	Basal kcat	LT-stimulated kcat	Basal Km	LT-stimulated Km
22° (RT)	0.19 ± 0.03 min-1	7.6 ± 0.2 min-1	3.4 ± 1.7 μM	5.3 ± 2.5 μM
37°	2.6 ± 0.98 min-1	105 ± 47 min-1	102 ± 35 μM	37 ± 18 μM

GED-Stimulated GTPase Activity

We have reported that GED, expressed independently in *E. coli*, has GTPase activating protein (GAP) activity for intact dynamin (Sever *et al.*, 1999). Based on this and other observations, we proposed that GED functions as an assembly-dependent GAP responsible for dynamin's assembly-stimulated GTPase activity. Recent data suggests that GED might function in a manner similar to RGS-type GAPs that interact with switch regions of the GTPase domain to stabilize transition states and enhance the orientation/functionality of catalytic residues located within the GTPase domain (Narayanan *et al.*, 2005; Song *et al.*, 2004a).

To assay GED's GAP activity, GED (1–10 μM) and dynamin (0.15 μM) are mixed in a final volume of 175 μl HCB150. The mixture is dialyzed overnight at 4° against HCB50. We perform this microdialysis in the caps derived from 1.5 ml microfuge tubes as follows: the top portion of 1.5 ml microfuge tubes (Sarstedt, Numbrecht, Germany) are cut with a razor blade generating caps with attached rings. The 175 μl sample is transferred into a cap (each cap having been prelabeled using a Sharpie). A small square piece of dry Pierce Snake Skin® dialysis membrane MWCO 10,000 is placed over the cap and sealed down tightly by closing the microfuge tube-derived ring over the cap. The samples (up to 24 individual caps) are placed into 1 L of cold HCB50 buffer in a glass beaker and dialyzed overnight with sufficient stirring to keep the caps in suspension. After dialysis the samples are collected by puncturing the dialysis tubing with a pipette tip and removing the sample. We find this configuration works best to maintain constant volume between samples. While stimulated GTPase activity can be measured after simply preincubating GED and dynamin in low salt on ice (Sever *et al.*, 1999), we find that overnight dialysis yields more reproducible and more robust levels of stimulation. While we do not know the mechanism of GED-stimulation, this requirement is reminiscent of refolding experiments. Therefore, we speculate that it might reflect a requirement for a slow, induced fit between the isolated, recombinant

GED and intact dynamin that is facilitated by gradually lowering the ionic strength.

To initiate GTPase assays, 175 μl of the dynamin/GED mixtures in PCR tubes are transferred to 37° and 25 μl of 4 mM GTP, 5mM MgCl$_2$ in H$_2$0 is added and rapidly mixed. A "0 min time" point is taken while the mixtures are still on ice by transferring 17.5 μl of dynamin/GED and 2.5 μl of GTP stock to a microtiter well containing 5 μl EDTA. 20-μl aliquots are removed for each subsequent time point and the colorimetric determination of Pi released is performed as described above. Additional controls include dynamin alone and GED alone, subjected to the same dialysis (Fig. 5A).

Note that the GED-stimulated GTPase assays are performed in HCB50, which is both lower ionic strength and lower pH than our normal GTPase assay buffer. Both parameters were important for efficient GED-stimulation of dynamin GTPase activity. Figure 5B shows that the pH optimum for GED stimulated GTPase activity of dynamin is pH 6.8–7.0. Although variable, under optimum conditions, isolated GED routinely stimulates dynamin's GTPase activity 10–20-fold.

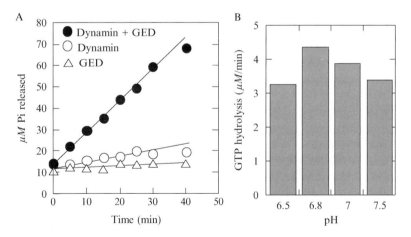

FIG. 5. Isolated, recombinant GED exhibits GTPase activating protein (GAP)-activity towards dynamin. (A) Time course of GTP hydrolysis by 0.15 μM dynamin (○), 5 μM GED (Δ), or both (●) after individual components or the mixture was dialyzed overnight against HCB50. (B) GTP hydrolysis was determined at the indicated pH for mixtures of dynamin (0.15 μM) and GED (5 μM) after dialysis in buffer also at the indicated pH.

References

Barylko, B., Binns, D., Lin, K. M., Atkinson, M. A., Jameson, D. M., Yin, H. L., and Albanesi, J. P. (1998). Synergistic activation of dynamin GTPase by Grb2 and phosphoinositides. *J. Biol. Chem.* **273**, 3791–3797.

Barylko, B., Binns, D. D., and Albanesi, J. P. (2001). Activation of dynamin GTPase activity by phosphoinositides and SH3 domain-containing proteins. *Meth. Enzymol.* **329**, 486–496.

Binns, D. D., Barylko, B., Grichine, N., Adkinson, A. L., Helms, M. K., Jameson, D. M., Eccleston, J. F., and Albanesi, J. P. (1999). Correlation between self-association modes and GTPase activation of dynamin. *J. Protein Chem.* **18**, 277–290.

Carr, J. F., and Hinshaw, J. E. (1997). Dynamin assembles into spirals under physiological salt conditions upon the addition of GDP and gamma-phosphate analogues. *J Biol. Chem.* **272**, 28030–28035.

Conner, S. D., and Schmid, S. L. (2003). Regulated portals of entry into the cell. *Nature* **422**, 37–44.

Damke, H., Muhlberg, A. B., Sever, S., Sholly, S., Warnock, D. E., and Schmid, S. L. (2001). Expression, purification, and functional assays for self-association of dynamin-1. *Meth. Enzymol.* **329**, 447–457.

Goldstein, A. S., Likyanov, A. N., Carlson, P. A., Yager, P., and Gelb, M. H. (1997). Formation of high-axial-ratio-microstructures from natural and synthetic sphingolipids. *Chem. Phys. Lipids* **88**, 21–36.

Hinshaw, J. E., and Schmid, S. L. (1995). Dynamin self assembles into rings suggesting a mechanism for coated vesicle budding. *Nature* **374**, 190–192.

Hinshaw, J. E. (2000). Dynamin and its role in membrane fission. *Annu. Rev. Cell Dev. Biol.* **16**, 483–519.

Marks, B., Stowell, M. H. B., Vallis, Y., Mills, I. G., Gibson, A., Hopkins, C. R., and McMahon, H. T. (2001). GTPase activity of dynamin and resulting conformation change are essential for endocytosis. *Nature* **410**, 231–235.

McNiven, M. A., Cao, H., Pitts, K. R., and Yoon, Y. (2000). The dynamin family of mechanoenzymes: Pinching in new places. *Trends Biochem. Sci.* **25**, 115–120.

Narayanan, R., Leonard, M., Song, B. D., Schmid, S. L., and Ramaswami, M. (2005). An internal GAP domain negatively regulates presynaptic dynamin *in vivo*: A two-step model of dynamin function. *J. Cell Biol.* **169**, 117–126.

Praefcke, G. J., and McMahon, H. T. (2004). The dynamin superfamily: Universal membrane tubulation and fission molecules? *Nat. Rev. Mol. Cell. Biol.* **5**, 133–147.

Sever, S., Muhlberg, A. B., and Schmid, S. L. (1999). Impairment of dynamin's GAP domain stimulates receptor-mediated endocytosis. *Nature* **398**, 481–486.

Shpetner, H. S., and Vallee, R. B. (1992). Dynamin is a GTPase stimulated to high levels of activity by microtubules. *Nature* **355**, 733–735.

Song, B. D., and Schmid, S. L. (2003). A molecular motor or a regulator? Dynamin's in a class of its own. *Biochemistry* **42**, 1369–1376.

Song, B. D., Yarar, D., and Schmid, S. L. (2004a). An assembly-incompetent mutant establishes a requirement for dynamin self-assembly in clathrin-mediated endocytosis *in vivo*. *Mol. Biol. Cell* **15**, 2243–2252.

Song, B. D., Leonard, M., and Schmid, S. L. (2004b). Dynamin GTPase domain mutants that differentially affect GTP binding, GTP hydrolysis, and clathrin-mediated endocytosis. *J. Biol. Chem.* **279**, 40431–40436.

Stowell, M. H. B., Marks, B., Wigge, P., and McMahon, H. T. (1999). Nucleotide-dependent conformational changes in dynamin: Evidence for a mechanochemical molecular spring. *Nat. Cell Biol.* **1**, 27–32.

Sweitzer, S., and Hinshaw, J. (1998). Dynamin undergoes a GTP-dependent conformational change causing vesiculation. *Cell* **93,** 1021–1029.

Takei, K., McPherson, P. S., Schmid, S. L., and De Camilli, P. (1995). Tubular membrane invaginations coated by dynamin rings are induced by GTPγS in nerve terminals. *Nature* **374,** 186–190.

Takei, K., Slepnev, V. I., and De Camilli, P. (2001). Interactions of dynamin and amphiphysin with liposomes. *Meth. Enzymol.* **369,** 478–486.

Tuma, P. L., and Collins, C. A. (1994). Activation of dynamin GTPase is a result of positive cooperativity. *J. Biol. Chem.* **269,** 30842–30847.

Warnock, D. E., Hinshaw, J. E., and Schmid, S. L. (1996). Dynamin self assembly stimulates its GTPase activity. *J. Biol. Chem.* **271,** 22310–22314.

[44] Clathrin-Coated Vesicle Formation from Isolated Plasma Membranes

By ISHIDO MIWAKO and SANDRA L. SCHMID

Abstract

Endocytic clathrin-coated vesicle (CCV) formation is a complex process involving a large number of proteins and lipids. The minimum machinery and the hierarchy of the events involved in CCV formation have yet to be defined. Here we describe an *in vitro* assay for CCV formation from highly purified rat liver plasma membranes. This rapid and easy assay can be used to quantitatively evaluate the different protein requirements for different endocytic receptors.

Introduction

Clathrin-coated vesicle formation is a highly regulated process in which cargo proteins are selectively sequestered. The events involved in CCV formation and the molecules required have been extensively studied and reviewed elsewhere (Conner and Schmid, 2003; Owen *et al.*, 2004; Slepnev and De Camilli, 2000). Core components known to be important for the CCV formation are clathrin, AP2, and dynamin. In addition to these three proteins, dozens of proteins have been reported to be involved in CCV formation. Recently, it has been reported that cargo-specific adaptors such as AP180/CALM, Dab2, β-arrestins, autosomal recessive hypercholesterolemia (ARH) protein, epsin, and HIP1/Hip1R are also important for specific CCV formation (Robinson, 2004; Traub, 2003). In addition, there are several accessory proteins such as endophilin and Hsc70 known to be involved in CCV formation. Some of these accessory proteins including

METHODS IN ENZYMOLOGY, VOL. 404
0076-6879/05 $35.00
DOI: 10.1016/S0076-6879(05)04044-9

amphiphysin, intersectin, SNX9, and synaptojanin interact with dynamin through its C-terminal proline-rich domain. Furthermore, the involvement of actin dynamics in CCV formation through interaction with dynamin and other endocytic accessory proteins has been suggested (Schafer, 2004). However, the hierarchy of events involved in CCV formation still remains to be resolved. To untangle the complexity of the molecular mechanisms involved in CCV formation, it is important to clarify the role of each protein for endocytosis of each endocytic receptor in detail.

Here we describe a simple *in vitro* CCV formation assay to quantitatively measure endocytosis of several endocytic receptors at the same time. The assay provides a potential means to analyze the complicated interconnected networks of adaptor and accessory proteins.

Materials and Reagents

Buffers, Reagents, and Antibodies

XTR transport buffer: 0.25 M sorbitol, 20 mM Hepes, pH 7.4, 150 mM potassium acetate, 1 mM magnesium acetate. XTR transport buffer is filtered through a 0.45 μM filter and stored at 4°.

20-fold ATP-regenerating system: 16 mM ATP, pH 7.0, 100 mM creatine phosphate and 100 units/ml creatine phosphokinase stored in aliquots at −80°.

GTP stock: 10–100 mM GTP in 20 mM Hepes, pH 7.4 stored in aliquots at −80°.

10% BSA.

1000-fold protease inhibitor cocktail (Sigma-Aldrich Co., St. Louis, MO) in DMSO stored at −20°.

Antibodies: Rabbit anti-LRP light chain was obtained from Dr. J. Herz (University of Texas Southwestern Medical Center, Dallas, TX), rabbit anti-SR-BI antibody from Dr. M. Krieger (Massachusetts Institute of Technology, Cambridge, MA), mouse monoclonal anti–human transferrin receptor (TfnR) antibody (H68) from Dr. I. S. Trowbridge (Salk Institute, La Jolla, CA), and mouse monoclonal anti-asialoglycoprotein receptor (ASGPR) antibody from Daiichi Pure Chemicals Co., Ltd (Ibaraki, Japan).

Preparation and Storage of Cytosol, Clathrin, AP2, and Dynamin

Rat and bovine brain cytosol are prepared as previously described (Miwako *et al.*, 2003). To prepare a Cl/AP2 and dynamin-depleted ammonium sulfate supernatant of bovine brain cytosol, cytosol is brought to 30% $(NH_2)_4SO_4$ by adding saturated $(NH_2)_4SO_4$ while gently stirring at 4°.

The suspension is centrifuged for 30 min at 100,000g rpm in a Ti60 rotor and the supernatant is collected and gel-filtered over a G-25 column equilibrated with XTR transport buffer to remove $(NH_2)_4SO_4$. The resulting 30% $(NH_2)_4SO_4$ is >80% depleted of clathrin, AP2, and dynamin, but retains >90% of total cytosolic protein.

Clathrin and AP2 are purified from microsome extracts as previously described (Manfredi and Bazari, 1987; Smythe et al., 1992). Clathrin is stored in 30 mM Tris, pH 7.0 in aliquots at $-80°$. AP2 is stored in XTR transport buffer in aliquots at $-80°$. Recombinant dynamin-1 is prepared as described (Marks et al., 2001).

Isolation of Rat Liver PMs

Rat liver PM sheets are isolated according to a protocol developed by Ann Hubbard and colleagues (Bartles and Hubbard, 1990; Hubbard et al., 1983; Scott et al., 1993) with slight modifications (Miwako et al., 2003). Six male Sprague-Dawley rats (120–170g) are fasted for 24 h. One at a time the rats are placed in an anesthetizing chamber infused with isoflurane gas. When fully unconscious, the rat is opened surgically to locate the heart. An 18-gauge needle is inserted into the left ventricle and ~100 ml ice-cold 0.154 M NaCl is injected to perfuse the liver. The liver is excised, weighed (~6 g/liver), and treated at 0–4° throughout the subsequent procedures. The liver is cut with a razor blade into slivers and transferred into a 40-ml glass dounce. Four volumes of 0.25 M STM containing 0.25 M sucrose, 5 mM Tris HCl, pH 7.4, 0.5 mM MgCl$_2$, protease inhibitor cocktail (Sigma-Aldrich Co.) is added and the liver is homogenized using 10 strokes with a loose-fitting pestle. The homogenate of all the liver tissue is collected in a 250-ml cylinder and adjusted to 20% (liver wet weight to total volume) with 0.25 M STM. The homogenate is filtered through four layers of cheesecloth moistened with 0.25 M STM. The filtrate is transferred to a 50-ml tube (25–30 ml per tube) and centrifuged at 280g (1100 rpm in Beckman Allegra-6R) for 5 min. The supernatant is saved and the pellet is resuspended with 3 strokes of a loose dounce in ½ original homogenate volume of 0.25 M STM. The suspension is again centrifuged as above. The first and second supernatants are combined and centrifuged at 1500g for 10 min (25–30 ml per 50-ml tube). The resulting pellets are pooled and resuspended with 3 strokes of a loose dounce in 1–2 ml of 0.25 M STM per gram initial wet weight liver. 2.0 M STM is added to obtain a density of 1.42 M, determined using a refractometer. Sufficient 1.42 M STM is added to bring the volume to approximately twice that of the original homogenate. Samples are added to cellulose nitrate tubes (~30 ml/tube) and overlaid with 2–4 ml of 0.25 M sucrose. The samples are centrifuged at

82,000g (25,000 rpm, Beckman L7-65, SW-28 rotor, no brake) for 60 min. The pellicle at the interface is collected and resuspended with 3 strokes of a loose dounce. The density of the suspension is adjusted to 1.42 M by adding 2.0 M STM. The sample is transferred to ultracentrifuge tubes ($\sim$30 ml/tube). 1.42 M STM is added to adjust volume if necessary. 2–4 ml of 0.25 M sucrose is overlaid and the floating procedure is repeated again. The pellicle at the interface is collected and resuspended with 3 strokes of a loose dounce. Distilled H_2O is slowly added to obtain a sucrose density of 0.25 M. This suspension is centrifuged at 1500g for 10 min and the final pellet is resuspended with 3 strokes of a loose dounce in 0.25 M sucrose to obtain a final concentration of 2 mg/ml. At each step, the enrichment of the PMs and the protein concentration is measured to calculate the yield and the fold-purification of the PMs.

The enzymatic activity of alkaline phosphodiesterase is used as a PM marker, and measured using a colorimetric assay (Hubbard *et al.*, 1983). 50 μl of 2 mg/ml thymidine 5'-monophosphate p-nitrophenyl ester (Sigma-Aldrich Co.) in 0.1 M Tris, 40 mM $CaCl_2$, pH 9.0 is added to 75 μl of the 100$\sim$1000-fold diluted sample. The sample is incubated for 1 h at 37°. Then the reaction is stopped by adding 1 ml of 200 mM Na_2CO_3, 100 mM glycine, and absorbance at 400 nm is determined. Protein concentration is measured using Coomassie Protein Assay Kit (23200, Pierce Biotechnology, Inc., Rockford, IL). After the first flotation, PMs are obtained in 12–25% yield and 10–15-fold enrichment After the second flotation, the yield is about 10–15% and the enrichment is 15–20-fold. The final PMs have the yield of 10–15% and 20–40-fold enrichment.

100 or 200 μl aliquots of 2 mg/ml PMs in 0.25 M sucrose are frozen in liquid nitrogen and stored at $-80°$. PMs stored in 0.5 M or 0.75 M sucrose in the presence or absence of 10% DMSO at $-80°$ have similar selectivity and activity of LDL receptor-related protein (LRP) uptake in our CCV formation assay; however, PMs stored in XTR buffer have low efficiency of LRP uptake.

Methods

In Vitro CCV Formation Assay from Isolated PMs

Our *in vitro* CCV formation assay is diagrammed in Fig. 1A. 50 μg of 2 mg/ml rat liver PMs are used for each assay. PMs stored at $-80°$ are thawed quickly and collected by centrifugation in a refrigerated Eppendorf centrifuge at 20,000g for 1 min. The pellet is resuspended with XTR transport buffer to a final concentration of 4 mg/ml. Then the sample is centrifuged again to obtain the pellet containing PMs. This procedure

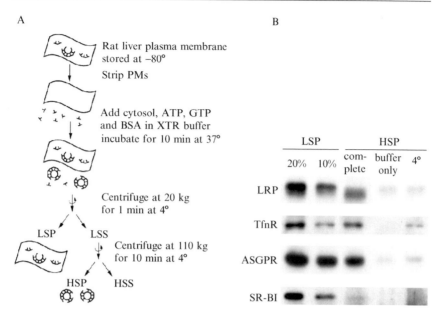

FIG. 1. (A) Diagram of an *in vitro* assay for CCV formation from PMs. LSP: low-speed pellet. LSS: low-speed supernatant. HSP: high-speed pellet. HSS: high-speed supernatant. (B) Temperature-dependent recruitment of endocytic receptors from LSPs to HSPs in the presence of ATP, GTP, and cytosol. Scavenger receptor class B, type I (SR-BI) is a PM marker that is independent of clathrin-mediated endocytosis. (Figures reproduced from Miwako *et al.*, 2003.)

reduces nucleotide-independent LRP recruitment to the high-speed pellet (HSP). The pellet is resuspended with XTR buffer, and the PMs are transferred to another eppendorf tube containing an ATP-regenerating system, 100 μM GTP, 0.2 % BSA, and protease inhibitor cocktail in the presence or absence of cytosol or purified proteins in XTR transport buffer to make a final volume of 40 μl. The samples are incubated at 37° for typically 10–20 min and then the tubes are returned to ice. The eppendorf tubes are tapped mildly to resuspend membranes and reduce trapping of newly formed vesicles. Then they are centrifuged at 20,000g for 1 min to obtain the low-speed pellets (LSP) containing PMs. The supernatants are transferred to 1.5 ml microfuge tubes (357448, Beckman Coulter, Inc., Fullerton, CA) and further centrifuged at 200,000g for 20 min in a TLA-100.3 rotor to obtain the HSPs containing newly formed vesicles. LSPs and HSPs are solubilized in sample buffer at 95° for 5 min and analyzed by SDS-PAGE and immunoblotting with antibodies against endocytic receptors such as LRP, TfnR, and ASGPR. Endocytosis is quantitated by the amount of endocytic receptors recruited in the HSP relative to that remaining in the LSP. Figure 1B is a typical Western blot showing the selective internalization of endocytic receptors depending on cytosol, nucleotides, and temperature.

Inhibition of Membrane Fusion Increases Efficiency

The maximum efficiency of LRP uptake in our *in vitro* assay was about 15%, obtained in the presence of an ATP-regenerating system, 2 mM GTP, and 200 μg cytosol (Fig. 1B). But the efficiency can be enhanced to 20% by inhibiting membrane fusion using the calcium-specific chelator, 1,2-bis (2-aminophenoxy) ethane-N,N,N',N'-tetraacetic acid (BAPTA) (Pryor *et al.*, 2000). In Fig. 2A, 50 μg PMs are incubated in the presence of an ATP regenerating system, 100 μM GTP, and 200 μg bovine brain cytosol with or without 10 mM BAPTA (tetrapotassium salt, Molecular Probes, Inc., Eugene, OR) for the indicated times at 37°. Then LSPs and HSPs are analyzed by SDS-PAGE and Western blotting with the anti-LRP antibody. The amount of LRP recruited into the HSP is quantitated using a densitometer (Personal Densitometer SI; Amersham Biosciences, Piscataway, NJ). In the presence of BAPTA, LRP uptake is increased to 20%. Membrane recycling is also reduced by dilution because increased volume results in the decreased opportunity of vesicles encountering large membrane compartments. 50 μg PMs are incubated in the presence of an ATP regenerating system, 100 μM GTP, and 4 mg/ml bovine brain cytosol in 40–200 μl for 30 min at 37° (Fig. 2B). The efficiency of LRP uptake is enhanced to 20% when the volume is increased to 200 μl. These data suggest that

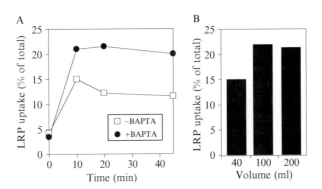

FIG. 2. Efficiency of LRP recruitment to the HSP is increased by inhibiting membrane fusion. (A) Time course of the LRP recruitment to the HSP in the presence or the absence of 10 mM BAPTA. (B) Effect of dilution on the efficiency of LRP recruitment to the HSP.

25% of newly formed vesicles either recycle back and fuse to the PM or fuse to each other forming larger membrane compartments (endosomes) that sediment in the LSP.

Endocytosis of Different Endocytic Receptors

Using this *in vitro* assay, internalization of several endocytic receptors can be measured at the same time to clarify different protein requirements for endocytosis. When two endocytic receptors, LRP and ASGPR, are analyzed, the same blot can be used because of the different molecular weights in an SDS-PAGE gel (LRP light chain is about 95 kDa, ASGPR is 42 kDa). Using purified clathrin, AP2, dynamin, and 30% ammonium sulfate supernatant (AS supt) of bovine brain cytosol, cytosolic protein requirements for LRP or ASGPR uptake are examined. PMs are incubated with an ATP regenerating system, 100 μM GTP, in the presence of physiological quantities of clathrin (0.2 μg), AP2 (0.8 μg), dynamin (2 μg) and 30 μg of AS supt, or in the presence ofexcess clathrin (8 μg), AP2 (8 μg), dynamin (20 μg) alone for 20 min at 37° C (Fig. 3). In the presence of an excess amount of clathrin, AP2, and dynamin, both LRP and ASGPR are

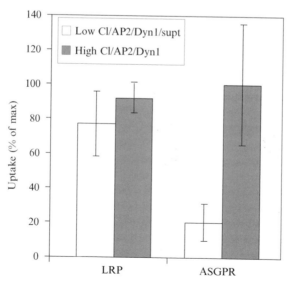

FIG. 3. Different cytosolic protein requirements for the uptake of LRP or ASGPR. The proportion of LRP and ASGPR internalization in the presence of physiological amounts of clathrin, AP2, dynamin, and AS supt (low Cl/AP2/Dyn1/supt), or high concentration of clathrin, AP2, and dynamin (high Cl/AP2/Dyn1) compared to the maximum efficiency of the uptake obtained in the presence of 200 μg bovine brain cytosol.

internalized; however, in the presence of physiological amount of these proteins with AS supt, only LRP uptake is observed. These data show the different protein requirements for different endocytic receptors; the uptake of ASGPR requires some protein(s) excluded from AS supt, whereas clathrin, AP2, dynamin, and AS supt are sufficient for LRP uptake.

Inhibition of LRP Uptake by a Kinase Inhibitor A3

Data in Fig. 4A show the effects of broad-spectrum kinase inhibitors on LRP uptake. PMs are incubated in the presence of an ATP regenerating system, 100 μM GTP, and 120 μg bovine brain cytosol with or without 1 mM H-7 (dihydrochloride, EMD Biosciences, San Diego, CA), 1 mM A3 (hydrochloride, EMD Biosciences), or 1 μM staurosporine (EMD Biosciences) for 30 min on ice. Then samples are incubated for 20 min at 37°. The kinase inhibitor A3 inhibits LRP uptake from PMs, but H-7 and staurosporine do not. These data suggest that a phosphorylation event inhibited by A3 is required for LRP uptake. It has been reported that A3 inhibits the activity of PIP5 kinase (PIP5K) (Arneson *et al.*, 1999). Therefore, to determine whether PIP5K activity could be detected in the isolated rat liver membrane fraction, PMs are incubated in the presence of 500 μM ATP containing [γ-^{32}P]ATP (25 μCi/assay, MP Biomedicals, Inc., Irvine,

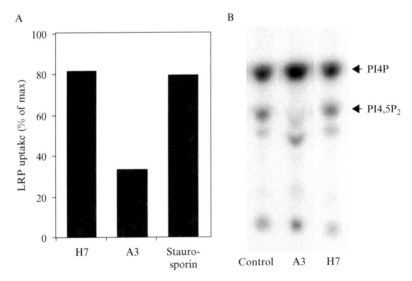

FIG. 4. Effect of membrane-bound phosphatidylinositol-4-phosphate 5-kinase on the LRP uptake. (A) Effect of kinase inhibitors on the internalization of LRP. Assays were performed in the presence of 1 mM H7, 1 mM A3, or 1 μM staurosporine, as indicated. (B) Effect of kinase inhibitors on the production of PI4,5P$_2$ by isolated rat liver PMs.

CA), 100 μM GTP, 120 μg bovine brain cytosol with or without 1 mM H-7, 1 mM A3 for 10 min at 37°. Then the lipids are extracted and analyzed by TLC and the Phosphorimager SI (GE Healthcare) (Arneson *et al.*, 1999). Indeed, PIP5K activity can be detected in isolated rat liver PMs (Fig. 4B) and it is totally inhibited by A3, but not by H-7 (Fig. 4B). Inhibition of LRP uptake by A3 but not H-7 or staurosporine suggests that A3 might reduce LRP uptake by inhibiting the PIPK activity.

The ability to prepare and store large amounts of endocytically active PM substrate and to simultaneously assay endocytosis of several distinct receptors renders this assay useful for dissecting the complex events involved in clathrin-mediated endocytosis.

References

Arneson, L. S., Kunz, J., Anderson, R. A., and Traub, L. M. (1999). Coupled inositide phosphorylation and phospholipase D activation initiates clathrin-coat assembly on lysosomes. *J. Biol. Chem.* **274**, 17794–17805.

Bartles, J. R., and Hubbard, A. L. (1990). Biogenesis of the rat hepatocyte plasma membrane. *Methods Enzymol.* **191**, 825–841.

Conner, S. D., and Schmid, S. L. (2003). Regulated portals of entry into the cell. *Nature* **422**, 37–44.

Hubbard, A. L., Wall, D. A., and Ma, A. (1983). Isolation of rat hepatocyte plasma membranes. I. Presence of the three major domains. *J. Cell Biol.* **96**, 217–229.

Manfredi, J. J., and Bazari, W. L. (1987). Purification and characterization of two distinct complexes of assembly polypeptides from calf brain coated vesicles that differ in their polypeptide composition and kinase activities. *J. Biol. Chem.* **262**, 12182–12188.

Marks, B., Stowell, M. H., Vallis, Y., Mills, I. G., Gibson, A., Hopkins, C. R., and McMahon, H. T. (2001). GTPase activity of dynamin and resulting conformation change are essential for endocytosis. *Nature* **410**, 231–235.

Miwako, I., Schroter, T., and Schmid, S. L. (2003). Clathrin- and dynamin-dependent coated vesicle formation from isolated plasma membranes. *Traffic* **4**, 376–389.

Owen, D. J., Collins, B. M., and Evans, P. R. (2004). Adaptors for clathrin coats: Structure and function. *Annu. Rev. Cell Dev. Biol.* **20**, 153–191.

Pryor, P. R., Mullock, B. M., Bright, N. A., Gray, S. R., and Luzio, J. P. (2000). The role of intraorganellar Ca(2+) in late endosome-lysosome heterotypic fusion and in the reformation of lysosomes from hybrid organelles. *J. Cell Biol.* **149**, 1053–1062.

Robinson, M. S. (2004). Adaptable adaptors for coated vesicles. *Trends Cell Biol.* **14**, 167–174.

Schafer, D. A. (2004). Regulating actin dynamics at membranes: A focus on dynamin. *Traffic* **5**, 463–469.

Scott, L., Schell, M. J., and Hubbard, A. L. (1993). Isolation of plasma membrane sheets and plasma membrane domains from rat liver. *Methods Mol. Biol.* **19**, 59–69.

Slepnev, V. I., and De Camilli, P. (2000). Accessory factors in clathrin-dependent synaptic vesicle endocytosis. *Nat. Rev. Neurosci.* **1**, 161–172.

Smythe, E., Carter, L. L., and Schmid, S. L. (1992). Cytosol- and clathrin-dependent stimulation of endocytosis *in vitro* by purified adaptors. *J. Cell Biol.* **119**, 1163–1171.

Traub, L. M. (2003). Sorting it out: AP-2 and alternate clathrin adaptors in endocytic cargo selection. *J. Cell Biol.* **163**, 203–208.

[45] Nucleotide Binding and Self-Stimulated GTPase Activity of Human Guanylate-Binding Protein 1 (hGBP1)

By Simone Kunzelmann, Gerrit J. K. Praefcke, and Christian Herrmann

Abstract

The synthesis of human guanylate-binding protein 1 (hGBP1) is induced by interferon-γ and its biological function is related to antiviral activity and regulation of proliferation. It interacts with guanine nucleotides, and its catalytic activity on GTP hydrolysis leads to the formation of phosphate ions and both GDP and GMP. Similar to other large GTPases like dynamin, hGBP1 shows higher specific GTP hydrolysis activity with increasing concentration of the protein. This is based on nucleotide-dependent self-association of hGBP1, which leads to self-stimulation of its GTPase activity. In this chapter we describe the characterization of the basic biochemical properties of hGBP1. Essentially, the biological activity of a GTPase is controlled by the type of nucleotide bound. Therefore, nucleotide binding is quantified in terms of affinity and dynamics since both of these aspects are important for the occurrence of hGBP1 bound to the one or the other nucleotide. In addition, we analyze the self-stimulated GTPase activity and show how to extract the hGBP1 homodimer dissociation constant from these data. Finally, with the help of size exclusion chromatography, nucleotide-dependent formation of hGBP1 dimers and tetramers is demonstrated. These biochemical characteristics may help to further understand the biological function of hGBP1.

Introduction

The protein described in this chapter was originally identified by running a lysate from fibroblast cells over GTP-, GDP-, and GMP-agarose, which bound the protein tightly, and it was therefore named human guanylate-binding protein 1 (hGBP1) (Cheng *et al.*, 1983, 1985). Homologues were found in man as well as mouse, chicken, and other vertebrates. The size of 67 kDa does not really qualify this class of proteins as small GTPases, but rather, large GTPases and also due to its biochemical properties (see following) hGBP1 is grouped in the family of dynamin and Mx (Praefcke and McMahon, 2004). The C-terminal CaaX motif suggests farnesylation and membrane localization but only a minor fraction of

METHODS IN ENZYMOLOGY, VOL. 404
0076-6879/05 $35.00
DOI: 10.1016/S0076-6879(05)04045-0

hGBP1 was found at the plasma membrane and the majority in the cytosol (Nantais *et al.*, 1996). Whereas hGBP1 is expressed at very low levels in resting cells, it is one of the most abundant proteins after stimulation of the cell by interferon-γ and other cytokines (Lew *et al.*, 1991; Naschberger *et al.*, 2004). The biological activity of hGBP1 seems to be related to antiviral response (Anderson *et al.*, 1999) while in endothelial cells it has also been shown to be involved in regulation of proliferation and angiogenesis (Guenzi *et al.*, 2001, 2003). Nevertheless, up to now cellular partners are not known, which makes it difficult to further explore the biological role of hGBP1. In this chapter we address the quantitative characterization of the key biochemical properties of this GTPase, namely guanine nucleotide binding, nucleotide-dependent self-association, and GTP hydrolysis activity, which are coupled to each other.

Nucleotide Binding Affinity

The biological function of GTP-binding proteins is controlled by their affinity for guanine nucleotides and by the rates of nucleotide exchange, that is, switching between the GDP and GTP bound state. For many of these proteins it is well established that specific interaction with partner proteins and thereby their biological activity is dependent on the nucleotide bound—and for many one can say they are resting when bound to GDP and active when GTP bound (Vetter and Wittinghofer, 2001). Therefore, it may be misleading to term these proteins GTPases as their GTP hydrolysis activity does not serve substrate turnover but rather switching to the GDP bound form. However, in order to judge the population of hGBP1 in each of the three possible nucleotide bound states, the affinities must be known in the first place. In this section, determination of equilibrium dissociation constants K_d with the help of fluorescent nucleotides is described. As GTP is hydrolyzed rapidly by hGBP1 we used the nonhydrolyzable analogue GppNHp instead.

Experimental Procedure

hGBP1 is synthesized in *E. coli* strain BL21 (DE3) transformed by pQE9 plasmid (Qiagen, Germany) details of which are described previously (Praefcke *et al.*, 1999). From the cell lysate hGBP1 hexa-His tagged at the N-terminus is extracted with the help of Ni-NTA-agarose. Subsequent size exclusion chromatography yields hGBP1 at more than 95% purity in buffer A, which contains 50 mM Tris at pH8, 5 mM MgCl$_2$, and 2 mM dithioerythritol. By centrifugal concentrators (Vivascience, Germany) the protein concentration can be increased up to 1 mM. A stock solution at this

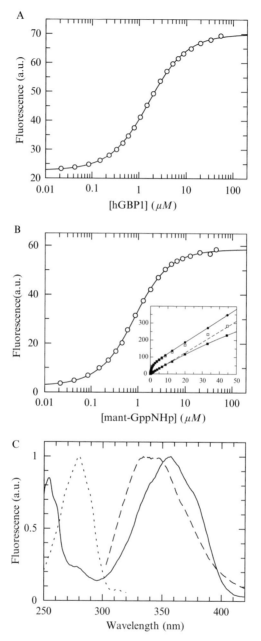

FIG. 1. Fluorescence titration. Measurement of hGBP1/mant-nucleotide equilibrium dissociation constants K_d. (A) hGBP1 is added stepwise to 0.5 μM mant-GppNHp in the cuvette. Fluorescence is excited at 366 nm and detected at 435 nm. The curve represents the fit

high concentration is advantageous for the experiments with hGBP1 where concentration series over a wide range are used. It can be shock frozen in liquid nitrogen and stored at −80°. For calculation of the concentration from the optical density measured at 276 nm an absorption coefficient of 45,400 M^{-1} cm^{-1} is used. All following experiments are carried out in buffer A at 20° if not indicated otherwise.

As there is no change in tryptophan fluorescence detectable when nucleotides bind to hGBP1, nucleotides carrying a fluorescence label at their ribose moiety are used in our experiments. These N-methylanthraniloyl (mant)-nucleotides are synthesized as described by Lenzen *et al.* (1995). The mant-nucleotide leads to good signal to noise ratio of the fluorescence at concentrations in the micromolar range. Mant-GMP, -GDP, or -GppNHp is placed in a standard fluorescence cuvette at 0.5 μM. Using a fluorospectrometer (LS 55, Perkin Elmer, Rodgau, Germany) fluorescence is excited at 366 nm and strong fluorescence emission is observed at 450 nm. From a stock solution of hGBP1, which contains also 0.5 μM of the respective mant-nucleotide aliquots of a few μl are added stepwise and the fluorescence is averaged over one min each. Maximal fluorescence emission is shifted to 435 nm. Alternatively, hGBP1 at 0.5 μM in the cuvette is titrated with the mant-nucleotide.

Results

The titration is started at 0.02 μM of the titrant, which is increased with growing step size up to 50 μM. Figure 1A shows the increase in fluorescence at 435 nm after adding hGBP1 to 0.5 μM mant-GppNHp excited at 366 nm while panel B shows the data for 0.5 μM hGBP1 in the cuvette after addition of increasing amounts of mant-GppNHp as indicated on the abscissa. The fluorescence in panel B is excited at 290 nm and detected at 435 nm and the spectra in panel C show the basis for that. Tryptophan fluorescence of hGBP1 is excited at 290 nm whereas the excitation spectrum

---

according to equation 2. (B) Titration of 0.5 μM hGBP1 in the cuvette with mant-GppNHp. Here, fluorescence energy transfer is employed by excitation at 290 nm and detection at 435 nm. The logarithmic plot shows the titration curve after correction for inner filtering and background substraction (○). The inset shows the raw data (□) and the control experiment (■) with only mant-GppNHp where inner filtering is obvious from the deviation of a linear increase (dotted line). The raw data were corrected for inner filtering according to equation 1 using the coefficient κ determined by curve fitting of the control data. The solid line through the corrected data (●) shows the fitted curve according to equation 2 where a linear slope is included, accounting for the increasing fluorophor concentration. (C) Fluorescence excitation spectra of hGBP1 (dotted line) and mant-GppNHp (continuous line), and emission spectrum of hGBP1 (dashed line) showing large overlap of the two latter.

of the mant-nucleotide has a minimum at this wavelength. As fluorescence emission of the protein and excitation of the mant-nucleotide have large spectral overlap (Fig. 1C) fluorescence energy is transferred from hGBP1 to the bound nucleotide and can be detected at 435 nm. This setup is employed in order to allow the sensitive detection of protein nucleotide interaction at a high background of unbound nucleotide. It is indeed advantageous when a protein/nucleotide reaction system has to be investigated under various conditions, that is, either the enzyme or the substrate in large molar excess (see also kinetics following). Figure 1B shows also the data after correction for inner filtering, which means that the decrease of fluorescence due to optical absorption by the nucleotide is taken into account by the following equation:

$$F = F_0 \cdot 10^{\kappa c} \tag{1}$$

where F_0 represents the measured fluorescence, F the corrected value used in equation 2 below, and c the concentration of mant-nucleotide. The coefficient κ is determined from a control measurement, where only mant-nucleotide is titrated into buffer. This kind of correction is not necessary in Fig. 1A since optical absorption at 366 nm by added hGBP1 can be neglected.

The K_d values as well as the concentrations are in the micromolar range so that a quadratic equation (Eq. 2) is fitted to the data. For the fit of the corrected data in Fig. 1B (filled circles) a linear slope is added in order to account for the increasing fluorescence background of unbound mant-nucleotide.

$$F = F_{\min} + (F_{\max} - F_{\min}) \frac{A_0 + B_0 + K_d - \sqrt{(A_0 + B_0 + K_d)^2 - 4A_0B_0}}{2B_0} \tag{2}$$

A_0 denotes the increasing total concentration of the titrant and B_0 the constant concentration of the other compound in the cuvette that equals 0.5 μM in either type of our titration experiments. The resulting K_d values are summarized in Table I. There is no significant difference in the results comparing titration experiments with the protein (first row each) or the nucleotide in excess (second row). It is worthwhile to note that hGBP1 binds to all three guanine nucleotides with almost the same affinity, which is different from small GTPases in two ways. For instance, Ras binds GDP and GTP tightly, but GMP affinity falls by five orders of magnitude (John *et al.*, 1990). It is speculation only that this may have to do with the possible role of a GMP bound state of hGBP1 or with the fact that GMP is produced by hGBP1 catalyzed GTP hydrolysis. The second feature is the striking difference in affinity between small GTPases and hGBP1, for

TABLE I

EQUILIBRIUM AND KINETIC CONSTANTS OF HGBP1/NUCLEOTIDE INTERACTION

| | k_{on} (μM^{-1} s^{-1}) | k_{off} (s^{-1}) | | K_d (μM) | |
		intercept	displacement	titration	k_{off}/k_{on}
mant-GppNHp	0.46	0.95	0.66	1.1	1.4
	0.53	0.70	0.57	0.7	1.1
mant-GTP	2.4	1.1	n.a.	n.a.	0.46
	2.8	1.0			0.36
mant-GDP	2.3	7.0	8.2	2.4	3.6
	2.2	6.8	5.4	1.8	2.5
mant-GMP	5.3	3.8	1.3	0.53	0.25
	—	—	1.2	0.15	—
GppNHp	0.8^b	n.a.	1.2	1.5^a	n.a.

The value in the upper row of each field is obtained by experiments with the protein in excess, while the values of the lower row are obtained with the respective nucleotide in excess.

[a] This data is from reference (Praefcke *et al.*, 1999) obtained by isothermal titration calorimetry.

[b] This value is calculated from the other two in this row.

n.a., not applicable.

example, K_d values of Ras/GTP and Ras/GDP lie in the picomolar region. In regard to the physiological conditions in a human cell, one can state here that the observed dissociation constants allow binding of hGBP1 to all three guanine nucleotides as their cellular concentrations are above the respective K_d values. Nevertheless, due to its dominant cellular concentration GTP is favored to bind to hGBP1.

Nucleotide Association and Dissociation Kinetics

Of course cellular processes are highly dynamic and nucleotide binding proteins undergo a permanent change of the kind of nucleotide bound (triphosphate versus diphosphate) and consequently a rapid change of partners like various effectors and regulatory proteins. All over, the biological function of GTPases is maintained by a complex balance of nucleotide binding and hydrolysis rate constants. Shifting of these parameters by exchange factors and GAPs allows for signal input by other proteins and leads to interconnection between different cellular pathways. When we investigated the basic underlying nucleotide binding kinetics of hGBP1, we found rates much faster than those observed for small GTPases and therefore employed a rapid mixing technique.

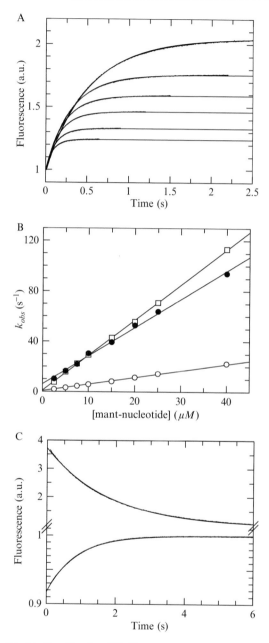

FIG. 2. Nucleotide binding kinetics. (A) Time courses of the fluorescence change in a stopped flow experiment after mixing of 0.25 μM hGBP1 and increasing concentrations of mant-GppNHp from top to bottom, 2.5 μM, 5 μM, 7.5 μM, 10 μM, 15 μM, and 20 μM. Single

Experimental Procedure

For kinetic experiments a stopped flow apparatus (Bio-logic, Grenoble, France) is used where mant-nucleotide and protein are mixed within less than 1 ms. In order to ensure pseudo-first order conditions for the kinetics, either the protein or the nucleotide is in 10-fold or more molar excess. Fluorescence is excited at 366 nm or 295 nm, respectively, and detected by a photomultiplier after passing through a 400 nm cutoff filter. A single exponential equation is fitted to the time traces yielding the observed rate constant k_{obs}.

Results

The nucleotide binding kinetics of hGBP1 are studied for mant-GMP, GDP, -GTP, and -GppNHp. Typically, the mant-nucleotide is placed at 0.5 μM in one of the mixing syringes while increasing concentrations of hGBP1 from 5 μM onwards are placed in the other. In addition, experiments are carried out the other way round having the nucleotide in excess. The k_{obs} value obtained from the time course of the fluorescence change increases in a linear manner up to concentrations of 40 μM (Fig. 2A). Thus, the association rate constant k_{on} for the hGBP1 nucleotide complex is obtained from the linear slope of the k_{obs} data plotted versus concentration (Fig. 2B) while the intercept corresponds to its dissociation rate constant k_{off} according to Eq. 3.

$$k_{obs} = k_{on} \cdot [excess\ compound] + k_{off} \tag{3}$$

When the intercept value is close to zero, k_{off} carries a large relative error. In this case it is preferable to determine k_{off} from a displacement experiment. To this end the preformed complex of hGBP1 and the mant-nucleotide in one syringe is mixed with a 1000-fold molar excess of non-labeled nucleotide from the other syringe (Fig. 2C). As the association of the latter with hGBP1 is fast, and saturating the single exponential time course of the fluorescence change yields the dissociation rate constant k_{off} of the respective mant-nucleotide/hGBP1 complex. Results are summarized in Table I. They show that association as well as dissociation occur on a rapid time scale; particularly, dissociation is orders of magnitude faster than observed for small GTPases. This finding raises the question

exponential curves yielding the k_{obs} value match the experimental traces. (B) plot of k_{obs} vs. concentration of nucleotides; ○ mant-GppNHp, □ mant-GTP, ● mant-GDP. Linear regression yields data for k_{on} and k_{off} as listed in Table I. (C) displacement of GppNHp (5 μM) from hGBP1 (2.5 μM) by large excess of mant-GDP (200 μM) increasing curve, and displacement of mant-GppNHp from hGBP1 (1 μM each) by 1000-fold excess of GMP decaying curve. Single exponential fits are included yielding the k_{off} values in Table I.

if regulatory proteins of hGBP1 accelerating nucleotide exchange are necessary and likely to exist. According to the k_{off} values for all nucleotides in Table I, one might conclude for hGBP1 that dissociation is not a limiting factor for the interchange of different nucleotides, whereas this is a key feature for small GTPases underlying strong regulation by exchange factors, that is, Ras, Rho, Ran, etc. The dynamics of hGBP1 nucleotide association and dissociation are on a time scale of a few seconds, at least in the absence of other proteins. In contrast to small GTPases, regulation of biological function by nucleotide exchange factors seems to be unlikely for hGBP1.

In order to guarantee this important finding about the dissociation rate and to make sure it is not an artefact introduced by the fluorescence-label, we addressed the dissociation rates of the nucleotide without the mant-group. To this end we placed the preformed complex of hGBP1 and nonlabeled GppNHp into one syringe and large molar excess of mant-GDP into the other. Hence the observed kinetics are controlled by the dissociation of GppNHp because the large excess of mant-GDP leads to much faster association of the latter and it prevents rebinding of the former at the same time. Again we take advantage of fluorescence energy transfer (excitation at 295 nm) in order to cope with the large background of unbound mant-GDP (Fig. 2C). The obtained k_{off} value for GppNHp is included in Table I. Clearly, there is only a small difference between the nucleotide with and without the mant-group. Finally, we wanted to demonstrate that nucleotide association is not perturbed either by the mant-label. Calculation of k_{on} according to equation 4 and with help of the K_d value obtained from isothermal titration calorimetry (Praefcke et al., 1999) yields a value of 0.8 $\mu M^{-1}s^{-1}$ similar to the corresponding value for mant-GppNHp. The results collated in Table I show that the mant-group has only marginal influence on the interaction between hGBP1 and GppNHp. In addition, good agreement is found for the K_d values obtained by titration and calculated from the rate constants according to Eq. 4.

$$K_d = \frac{k_{off}}{k_{on}} \tag{4}$$

Nucleotide Hydrolysis Activity

So far it is not known how far nucleotide binding and GTP hydrolysis activity of hGBP1 are related to its biological function. Nevertheless, as pointed out above, the type of nucleotide bound to the GTPase is not only controlled by binding affinities and kinetics but also by the enzymatic activity. For the investigation of catalyzed GTP hydrolysis the enzyme is incubated with the nucleotide at defined concentrations and the reaction is stopped after an appropriate period of time. Selecting well-timed intervals for taking aliquots from the reaction mixture and stopping will yield a reasonable time course and the kinetics of GTP hydrolysis. Basically, the concentration of remaining GTP and the composition and amount of products may be assayed by various methods like measurement of fluorescence or counting radioactivity of the nucleotides, which must be labeled correspondingly. Prior to that, separation of the nucleotide mixture by some chromatographic means is necessary. Here we describe the use of unmodified GTP, which is separated from other nucleotides and quantitatively analyzed by HPLC. Stopping of the reaction is achieved by injection of an aliquot directly onto the HPLC column, which is sufficiently precise as long as time points are separated by minutes rather than seconds.

Experimental Procedure

Commercially available GTP is further purified to more than 99% by the use of Q-sepharose column chromatography and a triethylammonium carbonate buffer gradient as previously described (Lenzen *et al.*, 1995). A solution of 100 μl of 1.0 mM pure GTP in buffer A is equilibrated 37°. 50 μM bovine serum albumin is included in order to stabilize hGBP1 at low concentrations. The reaction is started by adding the protein at the desired concentration from a 0.5 mM stock solution. Aliquots of 20 μl are taken after appropriate time points as indicated in Fig. 3A. They are each injected immediately onto the HPLC system connected to a C18 column. In order to separate the nucleotides with such a hydrophobic HPLC column, the elution buffer contains 7.5% acetonitril, 100 mM potassium phosphate (pH 6.5), and 10 mM tetrabutylammonium (TBA) bromide. The hydrophobic cations of the latter compound form ion pairs with the phosphate groups of the nucleotides. Consequently, GTP carrying the most negative charges elutes last from the column whereas GMP shows the shortest retention time. From the integration of the peak areas for GTP, GDP, and GMP their concentrations can be calculated. As seen in Fig. 3A, GTP decreases from 100% down to 50% with constant rates, without any sign of product inhibition or inactivation of the enzyme. By linear regression the rate constants are calculated and division by the protein concentration yields the specific GTP hydrolysis activity. One of the advantages of using HPLC is the simultaneous detection of all different nucleotides, which is particularly important for the study of hGBP1 as it leads to the production of GDP and GMP (Neun *et al.*, 1996; Praefcke *et al.*, 1999; Schwemmle and Staeheli, 1994; Schwemmle *et al.*, 1996). Furthermore, the

A

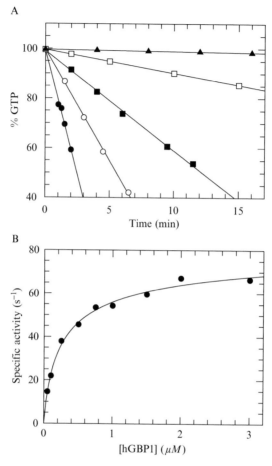

FIG. 3. Concentration-dependent GTPase activity. (A) 1 mM GTP is incubated at 37° with different concentrations of hGBP1, ▲ 0.05 μM, □ 0.25 μM, ■ 0.75 μM, ○ 1.5 μM, and ● 3.0 μM hGBP1. Linear regression of the data and division by the protein concentration yields the specific GTPase activity. (B) Results from A are plotted versus protein concentration and a theoretical curve according to equation 5 is fitted to the data yielding $k_1 = 0.02$ s^{-1}, $k_2 = 1.5$ s^{-1} and $K_{\text{dimer}} = 0.4$ μM.

hydrolytic stability of "nonhydrolyzable" GTP-analogues like GTPγS or GppNHp can be tested that have characteristic retention times. A disadvantage of the system is the detection limit of about 1 μM GTP but it can be increased using fluorescent nucleotides. It must also be mentioned that nucleotide/TBA ion pair formation on the column disallows the use of high salt conditions, which may come into play after chemical quenching of reaction mixtures prior to analysis when shorter time scales of the kinetics are addressed.

Results

Figure 3A shows typical results of GTP hydrolysis catalyzed by hGBP1 at different concentrations. In comparison to small GTPases like Ras or Rho, the turnover rate is much higher. More intriguingly, an enzyme concentration-dependent catalytic activity is observed. The specific GTPase activity of hGBP1 increases with the concentration of the protein to a saturating value of 1.5 s^{-1} at 37° as shown in Fig. 3B. This behavior can be explained by self-association and thereby self-activation of hGBP1, which is favored at higher concentrations. Assuming in a minimum model the formation of a hGBP1 dimer, the following equation can be fitted to the data in Fig. 3B:

$$specific\ activity = k_1 + (k_2 - k_1) \frac{E_0 + K_d/4 - \sqrt{(E_0 + K_d/4)^2 - E_0^2}}{E_0} \quad (5)$$

k_1 is the catalytic activity of monomer hGBP1 whereas k_2 and K_{dimer} represent the GTP hydrolysis rate constant and the dissociation equilibrium constant of the dimer, respectively. A value of 0.4 μM is obtained for K_{dimer} and 1.5 s^{-1} for k_2. While there is a high confidence on these two values the result for k_1 is not so precise as measurement of the GTPase activity becomes inaccurate at hGBP1 concentrations lower than 0.1 μM. Nevertheless, an approximately 10-fold increase of GTPase activity can be estimated for the dimer compared to the monomer. As dynamin shows a similar behavior, the protein concentration-dependent GTPase activity is a hallmark of the dynamin superfamily of large GTPases and related proteins. This assay can be used to explore the role of potential amino acid residues and protein domains in catalytic activity and self-assembly of the protein (Praefcke *et al.*, 2004). Similar to the discussion above about fast nucleotide dissociation kinetics and exchange factors, one must question here the necessity of a GAP. It seems an inappropriate regulatory protein for hGBP1 in the light of high intrinsic GTP hydrolysis activity and self-activation.

Nucleotide-Dependent Self-Association

The concentration dependence of the specific GTPase activity suggests the formation of dimers or higher protein assemblies. In order to understand the catalytic mechanism and other functional properties of hGBP1, it is important to determine the size of the oligomers. Size exclusion chromatography, also termed gel filtration, is widely used for purification and separation of macromolecules according to their molecular size. This results in a linear relationship between the elution volume and the logarithm of the molecular weight of the macromolecule. We were interested in the size of hGBP1 oligomers and in how far it may depend on the type of the bound nucleotide. As described in the first sections, the affinities between hGBP1 and guanine nucleotides are not very high and dissociation kinetics are fast. Therefore, running hGBP1 bound to such a nucleotide over a gel filtration column will lead to dissociation and separation of this complex during chromatography. In order to ensure permanent interaction between hGBP1 and the nucleotide, the buffer on the column must contain saturating concentrations of this nucleotide. Such experimental conditions will allow us to study the self-association of hGBP1, which is coupled to the nucleotide interaction. The transition state of GTP hydrolysis is mimicked by 200 μM GDP with 300 μM $AlCl_3$ and 10 mM NaF, which leads to the formation of $hGBP1.GDP.AlF_x$ complex.

Experimental Procedure

For gel filtration experiments buffer A is used, which contains additionally 200 μM of the respective nucleotide. Prior to injection of 200 μl of 20 μM hGBP1, the column is equilibrated with two column volumes of this buffer. The eluting protein is detected by UV absorption or, more sensitively, tryptophan fluorescence. A wavelength of 295 nm should be selected for detection or excitation, respectively, because the nucleotide in the buffer leads to strong light absorption at lower wavelengths.

Results

Figure 4 shows the calibration of the size-exclusion chromatography column (Supelco GFC-1300, SigmaChrom, $V_0 = 13.25$ ml) with different proteins, the molecular weights of which range from 43 to 669 kDa. The linear regression of the logarithm of the molecular weight (MW) and the elution volume V_e (relative to the total volume V_0 of the column) is shown by the straight line. The elution volume of hGBP1 is dependent on the type of nucleotide present in the buffer. Basically, three different groups can be recognized, represented by the three arrows pointing from the respective abscissa values to the calibration line and from there another arrow each indicates the corresponding log (MW) value on the ordinate. Nucleotide free, GMP- and GDP-bound hGBP1 belong to one of these groups and a MW value of about 62 kDa is identified. The molecular weight of the His6-tagged hGBP1 is 69 kDa as verified by mass spectroscopy. The second group contains the GTP- and GppNHp-bound forms of hGBP1 the average

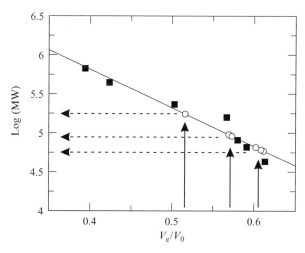

FIG. 4. Nucleotide-dependent self-association. Size exclusion chromatography serves to estimate the molecular size of the proteins. The logarithm of the molecular weight is plotted versus the elution volume of seven different marker proteins (■) and a linear relationship is fitted. Then, in six seperate experiments the elution volume of hGBP1 is determined in the presence of different nucleotides (○), from left to right, GDP.AlFx, GTP, GppNHp, GMP, GDP, and without any nucleotide. On the basis of the calibration line the elution volume allows to estimate the size of the assemblies. The three arrows each indicate the groups of tetramer, dimmer, and monomer hGBP1.

elution volume of which corresponds to a MW of about 93 kDa. Finally, the elution volume of the GDP.AlF$_x$-bound state of hGBP1 yields a MW of 175 kDa. However, for this method not only the MW is important but to some extent also the shape of the macromolecules. Monomer hGBP1 is an elongated protein as known from the X-ray structure, which leads to a relatively large elution volume (Prakash *et al.*, 2000a). In contrast, the dimer and larger assemblies of hGBP1 likely show a more compact shape and therefore our experimental finding of three groups of MWs roughly with the ratios 1.3:2:4 may represent monomer, dimer, and tetramer forms of hGBP1, respectively. The elution peaks of the GTP bound and, even more pronounced, the GDP.AlFx-bound hGBP1 are wider and show some tailing, which is probably caused by dynamic equilibrium between a small fraction of GDP-bound monomers and GTP-bound dimers and GDP.AlF$_x$-bound tetramers, respectively. Thus the molecular weight calculated from the elution volume of the peak maximum represents a lower limit. Furthermore, during the GTPase cycle of hGBP1, the transient assembly of two GTP-bound dimers into one tetramer seems more reasonable than the formation of a trimer. In summary, size exclusion chromatography clearly shows nucleotide-dependent self-association of hGBP1 and, most likely, hGBP1 forms a dimer after binding of GTP. Further transient association to a tetramer is suggested by the observation of GDP.AlF$_x$-bound hGBP1, which is believed to mimic the transition state of GTP hydrolyzing enzymes. Together with the results from the previous section, biochemical properties similar to dynamin are obvious. hGBP1 does not form self-assemblies as large as dynamin or Mx but self-association is clearly dependent on the interaction with GTP. The impact of self-association on the enzymatic activity is even more pronounced for hGBP1 compared to dynamin, which shows only 4-fold increase of GTPase activity in the assembled form in the absence of other factors such as microtubules or lipid vesicles (Binns *et al.*, 1999; Stowell *et al.*, 1999; Warnock *et al.*, 1996).

Conclusions

A paradigm for small GTPases is that specific (inter)action is coupled to the GTP bound state, for example, interaction with effectors leading to their activation. We have shown that GTP binding to hGBP1 leads to self-association and self-stimulation of GTPase activity, while interaction with other proteins could not be demonstrated up to now. Partner proteins may have an impact on catalytic activity, product ratio, and on the nucleotide dissociation constants that can modulate the biological function—or, vice versa, particular nucleotide states of hGBP1 may control the function of the interaction partners. Once partner proteins for hGBP1 are identified, the methods shown in this chapter will facilitate investigation of their functional role.

References

Anderson, S. L., Carton, J. M., Lou, J., Xing, L., and Rubin, B. Y. (1999). Interferon-induced guanylate binding protein-1 (GBP-1) mediates an antiviral effect against vesicular stomatitis virus and encephalomyocarditis virus. *Virology* **256**, 8–14.

Binns, D. D., Barylko, B., Grichine, N., Atkinson, M. A., Helms, M. K., Jameson, D. M., Eccleston, J. F., and Albanesi, J. P. (1999). Correlation between self-association modes and GTPase activation of dynamin. *J. Protein Chem.* **18**, 277–290.

Cheng, Y. S., Becker-Manley, M. F., Chow, T. P., and Horan, D. C. (1985). Affinity purification of an interferon-induced human guanylate-binding protein and its characterization. *J. Biol. Chem.* **260**, 15834–15839.

Cheng, Y. S., Colonno, R. J., and Yin, F. H. (1983). Interferon induction of fibroblast proteins with guanylate binding activity. *J. Biol. Chem.* **258**, 7746–7750.

Guenzi, E., Topolt, K., Cornali, E., Lubeseder-Martellato, C., Jorg, A., Matzen, K., Zietz, C., Kremmer, E., Nappi, F., Schwemmle, M., Hohenadl, C., Barillari, G., Tschachler, E.,

Monini, P., Ensoli, B., and Sturzl, M. (2001). The helical domain of GBP-1 mediates the inhibition of endothelial cell proliferation by inflammatory cytokines. *EMBO J.* **20,** 5568–5577.

Guenzi, E., Topolt, K., Lubeseder-Martellato, C., Jorg, A., Naschberger, E., Benelli, R., Albini, A., and Sturzl, M. (2003). The guanylate binding protein-1 GTPase controls the invasive and angiogenic capability of endothelial cells through inhibition of MMP-1 expression. *EMBO J.* **22,** 3772–3782.

John, J., Sohmen, R., Feuerstein, J., Linke, R., Wittinghofer, A., and Goody, R. S. (1990). Kinetics of interaction of nucleotides with nucleotide-free H-ras p21. *Biochemistry* **29,** 6058–6065.

Lenzen, C., Cool, R. H., and Wittinghofer, A. (1995). Analysis of intrinsic and CDC25-stimulated guanine nucleotide exchange of p21ras-nucleotide complexes by fluorescence measurements. *Methods Enzymol.* **255,** 95–109.

Lew, D. J., Decker, T., Strehlow, I., and Darnell, J. E. (1991). Overlapping elements in the guanylate-binding protein gene promoter mediate transcriptional induction by alpha and gamma interferons. *Mol. Cell. Biol.* **11,** 182–191.

Nantais, D. E., Schwemmle, M., Stickney, J. T., Vestal, D. J., and Buss, J. E. (1996). Prenylation of an interferon-gamma-induced GTP-binding protein: the human guanylate binding protein, huGBP1. *J. Leukoc. Biol.* **60,** 423–431.

Naschberger, E., Werner, T., Vicente, A. B., Guenzi, E., Topolt, K., Leubert, R., Lubeseder-Martellato, C., Nelson, P. J., and Sturzl, M. (2004). Nuclear factor-kappaB motif and interferon-alpha-stimulated response element co-operate in the activation of guanylate-binding protein-1 expression by inflammatory cytokines in endothelial cells. *Biochem. J.* **379,** 409–420.

Neun, R., Richter, M. F., Staeheli, P., and Schwemmle, M. (1996). GTPase properties of the interferon-induced human guanylate-binding protein 2. *FEBS Lett.* **390,** 69–72.

Praefcke, G. J. K., Geyer, M., Schwemmle, M., Kalbitzer, H. R., and Herrmann, C. (1999). Nucleotide-binding characteristics of human guanylate-binding protein 1 (hGBP1) and identification of the third GTP-binding motif. *J. Mol. Biol.* **292,** 321–332.

Praefcke, G. J. K., Kloep, S., Benscheid, U., Lilie, H., Prakash, B., and Herrmann, C. (2004). Identification of residues in the human guanylate-binding protein 1 critical for nucleotide binding and cooperative GTP hydrolysis. *J. Mol. Biol.* **344,** 257–269.

Praefcke, G. J. K., and McMahon, H. T. (2004). The dynamin superfamily: Universal membrane tubulation and fission molecules? *Nat. Rev. Mol. Cell. Biol.* **5,** 133–147.

Prakash, B., Praefcke, G. J. K., Renault, L., Wittinghofer, A., and Herrmann, C. (2000a). Structure of human guanylate-binding protein 1 representing a unique class of GTP-binding proteins. *Nature* **403,** 567–571.

Schwemmle, M., Kaspers, B., Irion, A., Staeheli, P., and Schultz, U. (1996). Chicken guanylate-binding protein. Conservation of GTPase activity and induction by cytokines. *J. Biol. Chem.* **271,** 10304–10308.

Schwemmle, M., and Staeheli, P. (1994). The interferon-induced 67-kDa guanylate-binding protein (hGBP1) is a GTPase that converts GTP to GMP. *J. Biol. Chem.* **269,** 11299–11305.

Stowell, M. H., Marks, B., Wigge, P., and McMahon, H. T. (1999). Nucleotide-dependent conformational changes in dynamin: Evidence for a mechanochemical molecular spring. *Nat. Cell Biol.* **1,** 27–32.

Vetter, I. R., and Wittinghofer, A. (2001). The guanine nucleotide-binding switch in three dimensions. *Science* **294,** 1299–1304.

Warnock, D. E., Hinshaw, J. E., and Schmid, S. L. (1996). Dynamin self-assembly stimulates its GTPase activity. *J. Biol. Chem.* **271,** 22310–22314.

[46] Stimulation of Dynamin GTPase Activity by Amphiphysin

By YUMI YOSHIDA and KOHJI TAKEI

Abstract

Dynamin functions in the fission of endocytic pits in the process of clathrin-mediated endocytosis. Dynamin GTPase activity is essential for its fission activity, and it is stimulated by self-assembly as well as by interacting with its binding partners, such as microtubules, SH3 domain containing proteins, or inositol phospholipids. Amphiphysin 1, SH3 domain-containing binding partner of dynamin 1, is proposed to cooperatively function in endocytosis. Amphiphysin 1 is essential for dynamin-dependent synaptic vesicle recycling in the synapse, and it enhances dynamin-dependent vesicle formation *in vitro*. In order to elucidate the molecular mechanism underlying the amphiphysin's effect, we measured dynamin GTPase activity in the presence of both amphiphysin 1 and lipid membranes. We describe here in detail the procedure of the dynamin GTPase assay and the results demonstrating stimulatory effect of amphiphysin on dynamin GTPase activity, which is highly dependent on the liposome size.

Introduction

Dynamin GTPase, a key molecule implicated in clathrin-mediated endocytosis, functions in the fission process of clathrin coated pits (Slepnev and De Camilli, 2000; Takei and Haucke, 2001). Dynamin polymerizes forming rings and spirals at the neck of the coated pits (Takei *et al.*, 1995) or around lipid tubules *in vitro* (Sweitzer and Hinshaw, 1998; Takei *et al.*, 1998). The fission reaction is GTP-dependent, and it occurs upon GTP hydrolysis. Dynamin is proposed to function as a mechano-chemical enzyme, that is, nucleotide-dependent conformational change of dynamin generates mechanical force that drives fission reaction (Sweitzer and Hinshaw, 1998; Takei *et al.*, 1995, 1998). Consistently, lipid tubules coated with dynamin are constricted by addition of GTP (Danino *et al.*, 2004). Alternatively, a molecular switch model, in which GTP-bound dynamin activates its downstream effector molecule, is also proposed. However, such a molecule remains to be identified (Sever *et al.*, 1999).

Dynamin comprises several functional domains. The GTPase domain of dynamin is located at its N-terminus and is followed by the pleckstrin

METHODS IN ENZYMOLOGY, VOL. 404   0076-6879/05 $35.00
DOI: 10.1016/S0076-6879(05)04046-2

homology (PH) domain, the GTPase effector domain (GED), and C-terminal Proline/Arginine rich domain (PRD). PH domain mediates interaction with phosphatidyl inositol 4,5-bisphosphate (PtdIns(4,5)P$_2$), and the interaction leads to the stimulation of dynamin GTPase activity (Barylko et al., 1998). GED interacts with the GTPase module of the adjacent dynamin molecules within dynamin polymers, and the interaction also enhances dynamin GTPase activity (Muhlberg et al., 1997). PRD binds a variety of Src-homology 3 (SH3) domain containing proteins (Gout et al., 1993).

One of the major dynamin-binding partners that contain SH3 domain is amphiphysin (Fig. 1A), which present primarily in brain as a homo- or hetero-dimmer of two similar isoforms, amphiphysin 1 and 2 (Ramjaun et al., 1999; Slepnev et al., 1998; Wigge et al., 1997). The SH3 domain-mediated interaction is likely physiologically important in nerve terminals, because disruption of the interaction by microinjection of the SH3 domain, or its interacting peptide of dynamin, results in a block in synaptic vesicle recycling (Shupliakov et al., 1997). The central domain contains binding sites for clathrin heavy chain and α-subunit of AP-2, providing a function as a linker between clathrin coat molecules and dynamin (Slepnev et al., 2000). BAR (BIN/Amphiphysin/Rvs) domain at the N-terminus mediates dimerization (Ramjaun et al., 1999; Slepnev et al., 1998; Wigge et al., 1997) as well as binding to acidic phospholipids (Peter et al., 2004; Takei et al., 1999). Crystallographic analysis of the amphiphysin BAR domain revealed a crescent-like shape comprising a dimer of triple-helix arranged in anti-parallel, which represents a configuration conserved in a variety of proteins (Habermann, 2004). A concave portion of BAR domain preferentially binds to relatively small sizes of liposomes, in cooperation with an amphipathic N-terminal helix, suggesting a role of the BAR domain as a curvature-sensor (Peter et al., 2004). This property might be important to sense plasma membrane curvature during coated pit formation and the following fission, and to alter sequential recruitment and dissociation of cytosolic factors during vesicle formation.

Amphiphysin 1 can coassemble with dynamin 1 into rings in solution, or on the lipid tubules. Amphiphysin is thought to function cooperatively with dynamin 1, and it enhances dynamin-dependent vesicle formation in vitro (Takei et al., 1999, 2005; Yoshida et al., 2004). In order to elucidate the molecular mechanism underlying the amphiphysin's effect, we recently analyzed how dynamin GTPase activity is modulated by amphiphysin using methods described in the following sections. Interestingly, dynamin GTPase activity was greatly stimulated by amphiphysin in the presence of lipid membrane, and the effect was highly dependent on the size of liposomes (Yoshida et al., 2004).

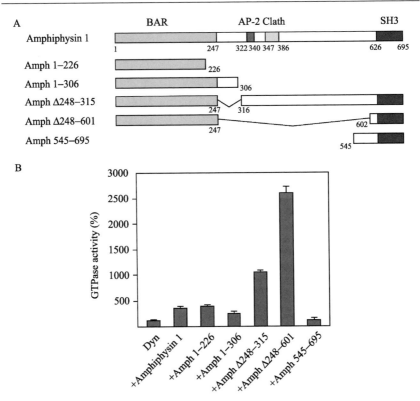

FIG. 1. Dynamin GTPase assay was performed in the presence of truncated amphiphysin. The truncated amphiphysins were designed as shown in (A). Effect of these amphiphysins on dynamin GTPase activity was assayed (B). The data was normalized by the GTPase activity of dynamin. Deletion of the middle domain of amphiphysin (Amph Δ248–315 and Amph Δ248–601) resulted in strong stimulation of the dynamin GTPase activity. Binding sites for AP-2 and clathrin are depicted in (A). (Reproduced with permission from Y. Yoshida *et al.* [2004]. *EMBO J.* **23**, 3483–3491.)

Methods

Purification of Dynamin from Bovine Brain

Dynamin 1 was purified from bovine brain essentially by the method of Liu (1994). The following procedures were performed at 4°. Two fresh bovine brains (600 g) obtained at a slaughterhouse were homogenized with a polytron homogenizer in five volumes of homogenizing buffer (20 mM Tris/HCl, pH7.4, 1 mM dithiothreitol, 1 mM phenylmethylsulfonyl fluoride, and 15 μg/ml leupeptin) containing 1 mM CaCl$_2$. The homogenate was centrifuged at 30,000g for 30 min, and the pellet was washed by the same buffer containing 0.1 mM CaCl$_2$. After the centrifugation, the pellet was washed again by the homogenizing buffer containing 5 mM EGTA and 2 mM EDTA. Dynamin was extracted by homogenization in 150 mM NaCl containing homogenizing buffer and stirring for 30 min. The extract was adjusted to 30% saturation with solid ammonium sulfate, stirred for 1 h, then dynamin was collected by centrifugation at 30,000g for 15 min. The pellet was resuspended in the minimal volume of (<15 ml) buffer A (20 mM NaPO$_4$, pH 7.0, 1 mM EDTA, 0.05% Tween 80) and dialyzed against the same buffer. The dialyzed protein was loaded on a SP-Sepharose column (Amersham Pharmacia Biotech, Piscataway, NJ) pre-equilibrated in buffer A, and eluted with a 0–0.4 M NaCl linear gradient. Fractions containing dynamin are combined and dialyzed against buffer B (20 mM Tris/HCl, pH 7.7), loaded on a MonoQ column (Amersham Pharmacia Biotech), then eluted with a 0–0.3 M NaCl linear gradient. Dynamin containing fractions were concentrated by centrifugation using centriplus YM-50 (millipore, Billerica, MA) and stored at −80°.

Expression and Purification of Amphiphysin and Its Truncation Constructs

Human amphiphysin1 is expressed in *Escherichia coli* BL21 strain as a glutathione S-transferase (GST) fusion protein using pGEX vector (Amersham Pharmacia Biotech) following standard procedures and as described by Slepnev (2000). The cDNA encoding amphiphysin1 and its BAR domains (Amph 1–226, 1–306 in Fig. 1A) were prepared by polymerase chain reaction amplification using specific primers and subcloned into pGEX-6P vector as *Bam*H1-*Eco*R1 fragment. To prepare Amph Δ248–601, a BAR domain (1–247aa) was amplified using specific primers introduced *Bam*H1 site and Nco1 site, and subcloned into *Bam*H1-Nco1 site of full-length amphiphysin1 plasmid described above. Amph Δ248–315, which truncated the intrinsic proline-rich stretch, was subcloned into pGEX-6P vector as follows. The BAR domain (1–247aa) of amphiphysin was amplified and subcloned into pGEX-6P vector as *Bam*H1-*Eco*R1 fragment. Then 316–695aa of amphiphysin was ligated as *Eco*R1-*Eco*R1 fragment into the vector. The SH3 domain (Amph 545–695 in Fig. 1A) was subcloned into pGEX-2T vector.

After verifying the nucleotide sequence of the constructs by DNA sequencing, the constructs were transformed into *E. coli* strain BL21. Transformed cells are grown at 37° in LB medium (10 g/liter tryptone, 5 g/liter yeast extract, 10 g/liter NaCl) supplemented with 100 mg/liter ampicillin (typically 2 L culture). The expression of GST-fusion amphiphysin1 was induced at OD$_{600}$ = 0.8 by the addition of 0.1 mM isopropyl-1-thio-β-D-galactopyranoside at 37° for 6 h.

Bacterial cells were harvested by centrifugation at 5000g and sonicated in the 100 ml of sonication buffer (50 mM Tris/HCl, pH 8.0, 300 mM NaCl, 5 mM EDTA) containing 0.5 mM phenylmethylsulfonyl fluoride, 0.5 μg/ml pepstatine. After centrifugation at 100,000g for 1 h, the supernatant was incubated with 3 ml of Glutathione-Sepharose 4B (Amersham Pharmacia Biotech) for 1 h at 4° on a rotator. Beads were washed 3 times with 10 volumes of sonication buffer in a batch, then transferred into 3 ml column, and further washed with 5 column volumes of sonication buffer. GST-fusion amphiphysin1 was eluted with 20 mM glutathione, and dialyzed against MonoQ column buffer (20 mM Tris/HCl, pH 7.7, 0.2 M NaCl). Dialyzed amphiphysin was concentrated by centrifugation using centriplus YM-50 (millipore). GST was cleaved by Prescission Protease (Amersham Pharmacia Biotech) at 4° for 16 h. Amphiphysin1 was then separated from GST by chromatography on MonoQ column. Amphiphysin1 was eluted with a 0.2–0.5 M NaCl linear gradient. Fractions containing amphiphysin1 were corrected and dialyzed against Tris buffer (20 mM Tris/HCl, pH 7.7) and stored at −80°.

Preparation of Large Unilamellar Liposomes

Bovine brain lipid extract (Folch Fraction I) was purchased from Sigma (St. Louis, MO). Cholesterol was purchased from Avanti Polar Lipids (Alabaster, AL). PtdIns(4,5)P_2 was from Calbiochem (La Jolla, CA). All lipids were solubilized in chloroform at the concentration of 25 mg/ml, 20 mg/ml, and 1 mg/ml, respectively, and stored at −20°.

Large unilamellar liposomes were prepared as described (Takei et al., 1999) with some modifications. Lipid stock solutions containing total 1 mg of lipids ware combined in a 10-ml glass tube at the ratio of cholesterol/brain lipid extract/PtdIns(4,5)P_2 = 20:74:6 (w/w%). A stream of nitrogen gas was passed through the tube to evaporate the solvent. During this process, the test tube was rotated by hand so that the lipid spread evenly forming opaque thin lipid films on the wall. The lipid film was further dried in lyophilizer for 2 h. Then, a stream of water-saturated nitrogen was passed approximately 10 min to hydrate the lipid until opacity of the dried lipid film was slightly lost. After hydration, 2 ml of degassed 0.3 M sucrose in distilled water was gently poured. Nitrogen gas was flushed into the tube without agitating the solution, then the tube was sealed, and left undisturbed for 2 h at 37° to allow spontaneous formation of liposomes. The tube was gently swirled to resuspend the liposomes (concentration, 1 mg/ml). The liposomes were mostly large in size, some exceeding 1 μm in diameter as checked by dynamic light scattering. Prepared liposomes can be stored at 4° for several days. Small liposomes were prepared by sonication of the large unilamellar liposomes in a microcentrifuge tube. Large unilamellar liposome was sonicated on ice for 1 min. Liposomes became clear after the sonication.

Dynamin GTPase Assay

Dynamin GTPase activity was measured essentially by the method of Barylko (2001). GTPase assay was performed in cytosolic buffer (25 mM Hepes-KOH, pH 7.2, 25 mM KCl, 2.5 mM magnesium acetate, 100 mM potassium glutamate) in 2.0 ml microcentrifuge tubes. One hundred μl of reaction mixture contained protein and lipid at the following concentrations.

> 0.2 μM dynamin (2 μg/100 μl)
> 2 μg liposomes (2 μg/100 μl)
> Various concentrations of amphiphysin1 or 0.4 μM truncated amphiphysin

The reaction mixture was pre-incubated in the cytosolic buffer on a heat block at 37° for 5 min. Reactions were initiated by the addition of 20 μl of radioactive GTP solution (5 mM GTP containing [γ-^{32}P]GTP, 148 Bq/μl diluted in cytosolic buffer). After 15 min of incubation, reaction was terminated by the addition of 1 ml of termination buffer (benzene/isobutanol, 1:1 (vol/vol)) into the reaction tubes. Reaction mixture was vigorously vortexed for 30 sec, and spun down. 250 μl of Silicotungstic acid (4% silicotungstic acid [tungstosilicic acid, Sigma] in 3 N sulfuric acid) was added to the reaction tubes and vortexed for 1 min. After spinning down the reaction mixtures, 100 μl of ammonium molybdate (10% ammonium molybdate in ddw) was added to the reaction tubes for the extraction of released ^{32}P$_i$ into the upper, organic phase. After 1 min of vortex and centrifugation for 3 min, 0.5 ml of the organic phase was poured into 5 ml test tubes for liquid scintillation. Then 4.5 ml of Clearsol II (Nacalai Tesque, Tokyo, Japan) were added to the test tube, which was vortexed, and left still at 4° for 30 min. The radioactivity was measured by liquid scintillation counter using Packard 2260XL (Hewlett Packard, Palo Alto, CA).

Stimulation of Dynamin GTPase Activity by Amphiphysin

Addition of increasing amount of amphiphysin1 to a reaction mixture containing large unilamellar liposomes (1779.5 ± 461.7 nm in diameter) produced a prominent increase of the GTPase activity. The activity was maximal when the molar ratio of dynamin to amphiphysin 1 was in the 1:1–2 range (Fig. 2A). In contrast, amphiphysin drastically decreased the GTPase activity of dynamin in the presence of the small liposomes.

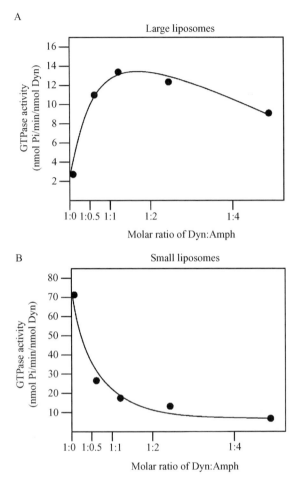

Fig. 2. GTPase activity of dynamin with amphiphysin. Dynamin GTPase activity was assayed in the presence of large unilamellar liposome (A) or small liposome (B). Note that the different effect of amphiphysin on dynamin GTPase activity was observed by the size of liposomes. (Reproduced with permission from Y. Yoshida *et al.* [2004]. *EMBO J.* **23,** 3483–3491.)

Thus, the effect of amphiphysin 1 on dynamin GTPase activity is strongly affected by the size of the liposomes. Absolute GTPase activity of dynamin was much higher with small liposomes than with large liposomes (Fig. 2B), probably because the surface area of liposomes has been increased by sonication.

To identify the domains of the protein responsible for its stimulatory action on the catalytic activity of dynamin, the effect of truncated amphiphysin 1 molecules (Fig. 1A) was investigated in the presence of large liposomes. A roughly similar stimulation to full-length amphiphysin was produced by the BAR domains (Amph 1–226, 1–306). Strikingly, Amph Δ248–601, which contains both BAR domain and SH3 domain, but lacks the central region, produced a 26-fold stimulation of the GTPase activity. The proline-rich stretch following BAR domain is proposed to bind intramolecularly to the SH3 domain, and negatively regulates the interaction of amphiphysin with dynamin (Farsad et al., 2003). Consistently, amphiphysin mutant lacking the proline-rich stretch (Amph Δ248–315) stimulated the GTPase activity about 3-fold over the value obtained with full-length amphiphysin1. Amph 545–695 alone, corresponding to the C-terminal SH3 domain, had no effect (Fig. 1B).

Conclusion

While accumulating evidence supports dynamin's function in GTP-dependent membrane fission, dynamin GTPase activity assays in most of the reported studies have been carried out with little attention to lipid curvature. Here we have used different size of liposomes to assay dynamin GTPase activity, and demonstrated that the GTPase activity was stimulated by amphiphysin only in the presence of large liposomes. Besides amphiphysin, there are several other SH3 domain-containing proteins that interact with dynamin (Peter et al., 2004). Moreover, a subset of endocytic proteins including dynamin and amphiphysin are subject to phosphorylation and dephosphorylation (Tomizawa et al., 2003). All of these factors might potentially regulate dynamin GTPase activity, and the present method would be a strong tool to analyze the regulatory mechanisms.

References

Barylko, B., Binns, D., Lin, K. M., Atkinson, M. A., Jameson, D. M., Yin, H. L., and Albanesi, J. P. (1998). Synergistic activation of dynamin GTPase by Grb2 and phosphoinositides. J. Biol. Chem. 273, 3791–3797.

Barylko, B., Binns, D. D., and Albanesi, J. P. (2001). Activation of dynamin GTPase activity by phosphoinositides and SH3 domain-containing proteins. In "Methods in Enzymology" (W. E. Balch, ed.), Vol. 329, pp. 486–496. Elsevier Inc., San Diego.

Danino, D., Moon, K. H., and Hinshaw, J. E. (2004). Rapid constriction of lipid bilayers by the mechanochemical enzyme dynamin. J. Struct. Biol. 147, 259–267.

Farsad, K., Slepnev, V., Ochoa, G., Daniell, L., Haucke, V., and De Camilli, P. (2003). A putative role for intramolecular regulatory mechanisms in the adaptor function of amphiphysin in endocytosis. Neuropharmacology 45, 787–796.

Gout, I., Dhand, R., Hiles, I. D., Fry, M. J., Panayotou, G., Das, P., Truong, O., Totty, N. F., Hsuan, J., Booker, G. W., Campbell, I. D., and Waterfield, M. D. (1993). The GTPase dynamin binds to and is activated by a subset of SH3 domains. *Cell* **75,** 25–36.

Habermann, B. (2004). The BAR-domain family of proteins: A case of bending and binding? *EMBO Rep.* **5,** 250–255.

Liu, J. P., Powell, K. A., Sudhof, T. C., and Robinson, P. J. (1994). Dynamin I is a Ca(2+)-sensitive phospholipid-binding protein with very high affinity for protein kinase C. *J. Biol. Chem.* **269,** 21043–21050.

Muhlberg, A. B., Warnock, D. E., and Schmid, S. L. (1997). Domain structure and intramolecular regulation of dynamin GTPase. *EMBO J.* **16,** 6676–6683.

Peter, B. J., Kent, H. M., Mills, I. G., Vallis, Y., Butler, P. J., Evans, P. R., and McMahon, H. T. (2004). BAR domains as sensors of membrane curvature: The amphiphysin BAR structure. *Science* **303,** 495–499.

Ramjaun, A. R., Philie, J., de Heuvel, E., and McPherson, P. S. (1999). The N terminus of amphiphysin II mediates dimerization and plasma membrane targeting. *J. Biol. Chem.* **274,** 19785–19791.

Sever, S., Muhlberg, A. B., and Schmid, S. L. (1999). Impairment of dynamin's GAP domain stimulates receptor-mediated endocytosis. *Nature* **398,** 481–486.

Shupliakov, O., Low, P., Grabs, D., Gad, H., Chen, H., David, C., Takei, K., De Camilli, P., and Brodin, L. (1997). Synaptic vesicle endocytosis impaired by disruption of dynaminSH3 domain interactions. *Science* **276,** 259–263.

Slepnev, V. I., Ochoa, G. C., Butler, M. H., Grabs, D., and De Camilli, P. (1998). Role of phosphorylation in regulation of the assembly of endocytic coat complexes. *Science* **281,** 821–824.

Slepnev, V. I., and De Camilli, P. (2000). Accessory factors in clathrin-dependent synaptic vesicle endocytosis. *Nat. Rev. Neurosci.* **1,** 161–172.

Slepnev, V. I., Ochoa, G. C., Butler, M. H., and De Camilli, P. (2000). Tandem arrangement of the clathrin and AP-2 binding domains in amphiphysin 1 and disruption of clathrin coat function by amphiphysin fragments comprising these sites. *J. Biol. Chem.* **275,** 17583–17589.

Sweitzer, S. M., and Hinshaw, J. E. (1998). Dynamin undergoes a GTP-dependent conformational change causing vesiculation. *Cell* **93,** 1021–1029.

Takei, K., McPherson, P. S., Schmid, S. L., and De Camilli, P. (1995). Tubular membrane invaginations coated by dynamin rings are induced by GTP-gamma S in nerve terminals. *Nature* **374,** 186–190.

Takei, K., Haucke, V., Slepnev, V., Farsad, K., Salazar, M., Chen, H., and De Camilli, P. (1998). Generation of coated intermediates of clathrin-mediated endocytosis on protein-free liposomes. *Cell* **94,** 131–141.

Takei, K., Slepnev, V. I., Haucke, V., and De Camilli, P. (1999). Functional partnership between amphiphysin and dynamin in clathrin-mediated endocytosis. *Nat. Cell Biol.* **1,** 33–39.

Takei, K, and Haucke, V. (2001). Clathrin-mediated endocytosis: Membrane factors pull the trigger. *Trends. Cell Biol.* **11,** 385–391.

Takei, K., Yoshida, Y., and Yamada, H. (2005). Regulatory mechanisms of dynamin-dependent endocytosis. *J. Biochem. (Tokyo)* **137,** 243–247.

Tomizawa, K., Sunada, S., Lu, Y. F., Oda, Y., Kinuta, M., Ohshima, T., Saito, T., Wei, F. Y., Matsushita, M., Li, S. T., Tsutsui, K., Hisanaga, S., Mikoshiba, K., Takei, K., and Matsui, H. (2003). Cophosphorylation of amphiphysin I and dynamin I by Cdk5 regulates clathrin-mediated endocytosis of synaptic vesicles. *J. Cell Biol.* **163,** 813–824.

Wigge, P., Kohler, K., Vallis, Y., Doyle, C. A., Owen, D., Hunt, S. P., and McMahon, H. T. (1997). Amphiphysin heterodimers: Potential role in clathrin-mediated endocytosis. *Mol. Biol. Cell* **8,** 2003–2015.

Yoshida, Y., Kinuta, M., Abe, T., Liang, S., Araki, K., Cremona, O., Di Paolo, G., Moriyama, Y., Yasuda, T., De Camilli, P., and Takei, K. (2004). The stimulatory action of amphiphysin on dynamin function is dependent on lipid bilayer curvature. *EMBO J.* **23,** 3483–3491.

[47] Tuba, A GEF for CDC42, Links Dynamin to Actin Regulatory Proteins

By Gianluca Cestra, Adam Kwiatkowski, Marco Salazar, Frank Gertler, and Pietro De Camilli

Abstract

Tuba is a 178kD protein containing four NH_2-terminal SH3 domains, a central Dbl homology (DH) domain followed by a BAR domain, and two COOH-terminal SH3 domains. The four NH_2-terminal SH3 domains bind the GTPase dynamin, a protein critical for the fission of endocytic vesicles. The DH domain functions as a CDC42-specific guanine nucleotide exchange factor and is unique among DH domains because it is followed by a BAR domain rather than a PH domain. The COOH-terminal SH3 domain binds directly to N-WASP and Ena/VASP proteins, key regulatory proteins of the actin cytoskeleton, and recruits a larger protein complex comprising additional actin regulatory factors. The properties of Tuba provide new evidence for a functional link between dynamin, endocytosis, and actin. The presence of a BAR domain, rather than a PH domain, may reflect its action at high curvature regions of the plasma membrane. Its multiple binding sites for dynamin generate an exceptionally high avidity for this GTPase and make the NH2-terminal region of Tuba a very useful tool for the one-step purification of dynamin.

Introduction

Fission of a variety of endocytic buds, including clathrin coated pits, from the plasma membrane is achieved by the concerted action of several membrane associated proteins (Engqvist-Goldstein and Drubin, 2003; Owen *et al.*, 2004; Slepnev and De Camilli, 2000). A key role in this process

METHODS IN ENZYMOLOGY, VOL. 404
0076-6879/05 $35.00
DOI: 10.1016/S0076-6879(05)04047-4

is played by the GTPase dynamin, as first established when dynamin was identified as the protein encoded by the *shibire* gene of *Drosophila*. *Shibire* mutant flies move normally at the permissive temperature, but become rapidly paralyzed at the restrictive temperature due to a block in the fission of the endocytic vesicles that recycle synaptic vesicle membranes in nerve terminals (Koenig and Ikeda, 1989). The precise mechanism of action of dynamin remains to be elucidated (Conner and Schmid, 2003), but strong evidence has linked dynamin to actin, suggesting that its function in endocytosis may involve the actin cytoskeleton (Ochoa *et al.*, 2000; Orth *et al.*, 2002; Schafer, 2004; Schafer *et al.*, 2002).

Dynamin is often found at sites of actin nucleation and its disruption affects actin dynamics (Ezratty *et al.*, 2005; Krueger *et al.*, 2003; Lee and De Camilli, 2002; Orth *et al.*, 2002). Conversely, actin and actin regulatory proteins are present at sites of endocytosis, and perturbation of actin impairs several forms of endocytosis (Engqvist-Goldstein and Drubin, 2003; Merrifield, 2004; Merrifield *et al.*, 2005; Yarar *et al.*, 2005). More importantly, several dynamin and clathrin interacting proteins are either physically or functionally linked to the actin cytoskeleton. These include cortactin, Abp1, intersectin, syndapin/pacsin, HIP, HIPR, Eps15, and epsin (Engqvist-Goldstein and Drubin, 2003). Intersectin represents a particularly striking example of the tight relationship between endocytosis and the actin cytoskeleton, as it brings dynamin and the PI(4,5)P2 phosphatase synaptojanin together with N-WASP and CDC42. Furthermore, its DH domain activates CDC42, that cooperates with PI(4,5)P2 in the activation of N-WASP, a regulator of actin nucleation (Broadie, 2004; Hussain *et al.*, 2001).

Several dynamin interacting proteins contain a BAR (BIN/Amphiphysin/Rvs) domain, a lipid bilayer binding domain that acts both as a curvature-generating and a curvature-sensing module. As shown by X-ray crystallography (Peter *et al.*, 2004), BAR domain dimers form a "bent rod" or elongated "banana"-shaped structure, which was proposed to bind negatively charged phospholipid membranes via positively charged residues found on the concave surface and at the two tips of the "banana." BAR domain containing proteins that bind dynamin are thought to function as adaptors that coordinate the recruitment of dynamin with the membrane deformations underlying vesicle budding and fission. Founding members of this protein family are amphiphysin (which also binds clathrin coat components) and endophilin (David *et al.*, 1996; Farsad *et al.*, 2001; Sakamuro *et al.*, 1996; Takei *et al.*, 1999; Wigge *et al.*, 1997).

Recently, BAR domains have been identified in a large number of proteins in higher eukaryotes (Habermann, 2004; Peter *et al.*, 2004). One of them is Tuba, a 178kD protein with four NH2-terminal SH3 domains, a central DH domain, a BAR domain, and two COOH-terminal SH3 domains (Salazar *et al.*, 2003) (Fig. 1A). Tuba was first characterized because of the unusual and interesting position of its BAR domain immediately downstream of a DH domain. Typically, DH domains, which function as guanyl nucleotide exchange factors for Rho family GTPases, are followed by a PH domain, a phospholipid binding module that helps to recruit and activate the DH-containing protein at the appropriate target membrane. The BAR domain of Tuba may functionally replace the PH domain and favor the action of Tuba at the surface of highly curved bilayers. Tuba was independently identified as a ligand for EVL, an Ena/VASP family protein, in a yeast two-hybrid screen (Salazar *et al.*, 2003).

Three of the four NH2-terminal domains of Tuba bind dynamin with variable affinity (Fig. 1B). As a result, the NH2-terminal region comprising all four SH3 domains has a very strong avidity for dynamin. The DH domain acts specifically on CDC42, thus activating it (Salazar *et al.*, 2003). The BAR domain binds lipids, as expected, although its poor solubility has prevented a detailed characterization of its binding properties (Fig. 1C).

The COOH-terminal SH3 domain provides a strong link to the cytoskeleton, primarily the actin cytoskeleton. It binds directly to N-WASP and Ena/VASP proteins (Salazar *et al.*, 2003). In addition, affinity-chromatography of brain extracts using this SH3 domain results in the isolation of a variety of other actin regulatory proteins possibly reflecting indirect interactions. Targeting of the COOH-terminal SH3 domain of Tuba to the surface of mitochondria via a mitochondrial targeting sequence induced F-actin recruitment to these organelles, indicating that this domain alone can promote F-actin nucleation and/or recruitment. Collectively, the COOH-terminal half of Tuba comprises several domains that cooperate in the regulation of actin nucleation: an SH3 domain that binds directly to proteins that control actin nucleation, a DH domain that activates CDC42 and therefore cooperates with N-WASP in actin nucleation, and a BAR domain expected to mediate the action, and possibly the activation, of Tuba at the surface of curved bilayer domains.

The dynamin-binding properties of the NH2-terminal domain of Tuba provide strong evidence for the functional link between dynamin, endocytosis, and the actin cytoskeleton (Salazar *et al.*, 2003). It is noteworthy, however, that alternative splicing generates a truncated form of Tuba lacking the NH2-terminal dynamin binding region. In addition, two other proteins, similar to this truncated form of Tuba, and referred to as Tuba 2 and Tuba 3 (Fig. 1A), are present in the database (Salazar *et al.*, 2003). Thus the COOH terminal portion of Tuba may have an independent function in actin regulation.

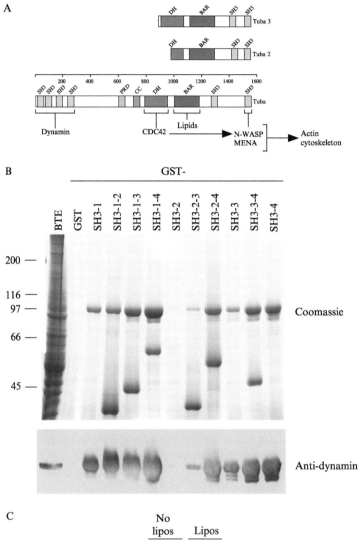

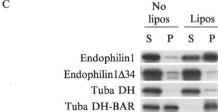

We describe here methods for analyzing the interaction of fragments of Tuba with lipids and with dynamin. The property of the DH-BAR module of Tuba to bind liposomes in a cosedimentation essay (Fig. 1C) supports the hypothesis that the BAR domain of Tuba localizes and/or activates its DH domain at the membrane. The juxtaposition of four SH3 domains that bind dynamin generates a module with a striking avidity for this GTPase. Thus, the NH2-terminus of Tuba is a very convenient tool to deplete dynamin from tissue extracts as well as to affinity-purify this protein from tissue and cell extracts to near purity in one step.

Materials

Antibodies

Anti-dynamin polyclonal antibody DG1 was generated in the lab against a dynamin I GST-fusion protein lacking the proline-rich domain (Butler *et al.*, 1997).

Constructs

To assemble the full-length sequence of human Tuba, the partial clone KIAA1010 (a gift of the Kazusa DNA Res. Institute, Kisarazu, Japan) was used as the starting template. 5'-RACE was performed on human skeletal muscle (Marathon-Ready cDNAs-Clontech, Mountain View, CA) using

---

FIG. 1. (A) Tuba and the closely related proteins Tuba 1 and Tuba 2. Cartoon illustration of Tuba protein domains with corresponding major binding partners (SH3, Src Homology domain 3; PRD, Proline Rich Domain; CC, coiled coil; Dbl Homology domain; BIN/Amphiphysin/Rvs domain) (Modified from [Salazar *et al.*, 2003]). (B) The NH2-terminal SH3 domains of Tuba bind dynamin in a multivalent fashion. GST or GST-SH3 domains of Tuba were used as baits in affinity purification experiments from rat brain Triton X-100 extracts (BTE). The material retained by the columns was subjected to SDS-PAGE and either stained by Coomassie blue (upper panel) or blotted and stained for dynamin (lower panel). Each SH3 domain was tested alone or in combination with other SH3 domains. SH3-1, SH3-3, and SH3-4 bound to dynamin, whereas SH3-2 did not. The strongest binding was achieved with SH3-4. The presence of multiple SH3 domains in the GST-fusion construct increases the recovery of dynamin. (C) The DH-BAR domain of Tuba binds to liposomes in a cosedimentation assay. Full length endophilin1, a fragment of endophilin1 missing the first 34 aa that contain a critical lipid bilayer binding site (endophilin1 Δ34), the DH domain of Tuba, and the DH-BAR module of Tuba were *in vitro* translated in the presence of ^{35}S-methionine. 10% of each *in vitro* translation reaction was incubated with or without sucrose-loaded liposomes and then subjected to centrifugation. The partial sedimentation of the DH-BAR domain in the absence of liposomes reflects the poor solubility of this fragment. Note, however, that no material is left in the supernatant in the presence of liposomes, demonstrating liposome binding [S, supernatant (unbound material); P, pellet (bound material)].

KIAA1010-specific primers and the Advantage 2 PCR Enzyme System (Clontech) to amplify the missing 5′ sequence. The full-length clone was amplified from brain cDNA library (Clontech) using specific oligos and the final construct sequenced completely. Individual Tuba SH3 domain GST fusion constructs or constructs containing combinations of different SH3 domains were generated by subcloning fragments of Tuba into in pGEX-4T1. SH3-1 (aa 1–54), SH3-1-2 (aa 1–127), SH3-1-3 (aa 1–210), SH3-1-4 (aa 1–308), SH3-2 (aa 65–127), SH3-2-3 (aa 65–210), SH3-2-4 (aa 65–308), SH3-3 (aa 144–210), SH3-3-4 (aa 144–308), and SH3-4 (aa 241–308) were all generated by PCR amplification of the corresponding DNA sequence that introduced SalI and NotI sites at the 5′ and 3′ end of each fragment, respectively. SH3-6 (aa 1512–1578) was subcloned using EcoRI and XhoI sites. The DH domain (aa 775–985) was subcloned into the (HA tagged) pcDNA vector introducing BamHI and SalI sites at the 5′ and 3′, respectively. The DH-BAR domain (aa 775–1272) was subcloned into the pcDNA vector using BamHI and SalI. Endophilin constructs were previously generated by Khashayar Farsad in the lab (Farsad *et al.*, 2001).

Methods

Binding of In Vitro *Translated DH-Bar Domain to Lipids in Liposome Sedimentation Assay*

Liposome Preparation. A stock solution of bovine brain lipids (Folch Fraction 1, Sigma, B1502, St. Louis, MO) that contained approximately 10% phosphatidylinositol lipids was prepared by resuspending lipids in 2:1 chloroform/methanol to a concentration of 50 mg/ml. Four mg of the stock solution were resuspended in 4 ml of 2:1 chloroform/methanol and dried on the wall of a 5 ml borosilicate disposable tube under a gentle flow of nitrogen gas (N_2) to generate a lipid film. Lipid films were incubated 2 h at 37° following addition of 4 ml of 0.3 M sucrose to the tubes and then resuspended by gentle vortexing and stored at 4°.

In Vitro Translation. TnT Coupled Reticulocyte Lysate System (Promega, Madison, WI) was used according to the manufacturer's instructions. 25 μl rabbit reticulocyte lysate was added to 2 μl of TNT Reaction Buffer, 1 μl of RNasin ribonuclease inhibitor (40 u/μl), 1 μl of amino acid mixture minus methionine (1 mM), 2 μl [^{35}S] methionine (>1000 Ci/mmol at 10 mCi/ml), 2 μl of DNA template(s) (0.5 μg/μl), 1 μl of T7 polymerase, and 16 μl of nuclease-free water to a final volume of 50 μl. The reaction was incubated at 30° for 1 h.

Binding Assay. 10 μg sucrose-loaded liposomes were added to 10% of ^{35}S-methionine labeled *in vitro* translated reaction product (5 μl) in 50 μl of binding buffer (25 mM Hepes-KOH pH7.4, 25 mM KCl, 2.5 mM Mg Acetate, 150 mM K-glutamate) and incubated at 37° for 20 min. Liposomes were sedimented at 100,000g in a Beckman Coulter TLA 100.3 rotor for 20 min. The supernatant was removed and resuspended in 2% SDS while the pellet was gently washed in binding buffer before being resuspended in 2% SDS. Samples were then subjected to SDS-PAGE and analyzed by autoradiography at −20° for 18 h.

Affinity Purification of Dynamin Using Tuba NH$_2$-Terminal SH3 Domains

Preparation of GST-SH3 Fusion Proteins. BL21CodonPlus-RP *E. coli* strain (Stratagene, La Jolla, CA) was transformed with GST-SH3 fusion constructs and grown overnight in LB medium (100 μg/ml ampicillin, 35 μg/ml chloramphenicol). The following day, bacteria were diluted 1 to 50 in fresh media and cultured to an OD600 of 0.5, at which time protein expression was induced by addition of IPTG (final concentration of 0.5 mM). Cells were then shifted to 30°, grown for 3 h, pelleted, and resuspended in ice cold PBS (NaCl 8g/l, KCl 0.2g/l, KH2PO4 (anhyd) 0.24g/l, Na2HPO4 (anhyd) 1.44g/l, pH 7.4), supplemented with 1% (final) Triton-X-100. Cells were sonicated for 5 min and protein lysates centrifuged (20,000g) for 30 min at 4°. Glutathione-Sepharose 4B (Amersham Pharmacia, Piscataway, NJ) (1 ml of a 50% suspension in 20% ethanol storage solution) was settled in plugged columns. Columns were washed with 50 bed volumes of H$_2$O and with 50 bed volumes of PBS +1% Triton-X-100. Clarified bacterial lysate was added to the column, and the column was incubated at 4° for 45 min under continuous swirling. The column was then drained and washed with 200 bed volumes of PBS +1% Triton.

Preparation of Brain Triton-X-100 Extract. Five frozen rat brains were thawed in 30 ml of lysis buffer [25 mM Tris, pH 7.5, 150 mM NaCl, 1 mM EDTA and a protease inhibitor mixture (Roche Molecular Biochemicals, Indianapolis, IN)] and homogenized. The homogenate was clarified from tissue debris by centrifugation at 1000g for 15 min. The supernatant was supplemented with Triton-X-100 (1% final concentration), incubated at 4° for 1 h under continuous swirling and then pelleted at 100,000g for 45 min. The supernatant of this centrifugation was collected and 50 mg of this Triton-X-100 extract was flowed 10 times over 1 mg of each SH3-domain fusion protein loaded on a disposable gravity column. Columns were washed with 50-bed volumes of ice cold lysis buffer (25 mM Tris, pH 7.5, 150 mM NaCl, 1 mM EDTA, 1% Triton-X-100) and the bound material was recovered by elution in 2% SDS and subjected to SDS-PAGE.

References

Broadie, K. (2004). Synapse scaffolding: Intersection of endocytosis and growth. *Curr. Biol.* **14,** R853–R855.

Butler, M. H., David, C., Ochoa, G. C., Freyberg, Z., Daniell, L., Grabs, D., Cremona, O., and De Camilli, P. (1997). Amphiphysin II (SH3P9; BIN1), a member of the amphiphysin/ Rvs family, is concentrated in the cortical cytomatrix of axon initial segments and nodes of ranvier in brain and around T tubules in skeletal muscle. *J. Cell Biol.* **137,** 1355–1367.

Conner, S. D., and Schmid, S. L. (2003). Regulated portals of entry into the cell. *Nature* **422,** 37–44.

David, C., McPherson, P. S., Mundigl, O., and de Camilli, P. (1996). A role of amphiphysin in synaptic vesicle endocytosis suggested by its binding to dynamin in nerve terminals. *Proc. Natl. Acad. Sci. USA* **93,** 331–335.

Engqvist-Goldstein, A. E., and Drubin, D. G. (2003). Actin assembly and endocytosis: From yeast to mammals. *Annu. Rev. Cell Dev. Biol.* **19,** 287–332.

Ezratty, E. J., Partridge, M. A., and Gundersen, G. G. (2005). Microtubule-induced focal adhesion disassembly is mediated by dynamin and focal adhesion kinase. *Nat. Cell Biol.* **7,** 581–590.

Farsad, K., Ringstad, N., Takei, K., Floyd, S. R., Rose, K., and De Camilli, P. (2001). Generation of high curvature membranes mediated by direct endophilin bilayer interactions. *J. Cell Biol.* **155,** 193–200.

Habermann, B. (2004). The BAR-domain family of proteins: A case of bending and binding? *EMBO Rep.* **5,** 250–255.

Hussain, N. K., Jenna, S., Glogauer, M., Quinn, C. C., Wasiak, S., Guipponi, M., Antonarakis, S. E., Kay, B. K., Stossel, T. P., Lamarche-Vane, N., and McPherson, P. S. (2001). Endocytic protein intersectin-l regulates actin assembly via Cdc42 and N-WASP. *Nat. Cell Biol.* **3,** 927–932.

Koenig, J. H., and Ikeda, K. (1989). Disappearance and reformation of synaptic vesicle membrane upon transmitter release observed under reversible blockage of membrane retrieval. *J. Neurosci.* **9,** 3844–3860.

Krueger, E. W., Orth, J. D., Cao, H., and McNiven, M. A. (2003). A dynamin-cortactin-Arp2/ 3 complex mediates actin reorganization in growth factor-stimulated cells. *Mol. Biol. Cell* **14,** 1085–1096.

Lee, E., and De Camilli, P. (2002). Dynamin at actin tails. *Proc. Natl. Acad. Sci. USA* **99,** 161–166.

Merrifield, C. J. (2004). Seeing is believing: Imaging actin dynamics at single sites of endocytosis. *Trends Cell Biol.* **14,** 352–358.

Merrifield, C. J., Perrais, D., and Zenisek, D. (2005). Coupling between clathrin-coated-pit invagination, cortactin recruitment, and membrane scission observed in live cells. *Cell* **121,** 593–606.

Ochoa, G. C., Slepnev, V. I., Neff, L., Ringstad, N., Takei, K., Daniell, L., Kim, W., Cao, H., McNiven, M., Baron, R., and De Camilli, P. (2000). A functional link between dynamin and the actin cytoskeleton at podosomes. *J. Cell Biol.* **150,** 377–389.

Orth, J. D., Krueger, E. W., Cao, H., and McNiven, M. A. (2002). The large GTPase dynamin regulates actin comet formation and movement in living cells. *Proc. Natl. Acad. Sci. USA* **99,** 167–172.

Owen, D. J., Collins, B. M., and Evans, P. R. (2004). Adaptors for clathrin coats: Structure and function. *Annu. Rev. Cell Dev. Biol.* **20,** 153–191.

Peter, B. J., Kent, H. M., Mills, I. G., Vallis, Y., Butler, P. J., Evans, P. R., and McMahon, H. T. (2004). BAR domains as sensors of membrane curvature: The amphiphysin BAR structure. *Science* **303**, 495–499.

Sakamuro, D., Elliott, K. J., Wechsler-Reya, R., and Prendergast, G. C. (1996). BIN1 is a novel MYC-interacting protein with features of a tumour suppressor. *Nat. Genet.* **14**, 69–77.

Salazar, M. A., Kwiatkowski, A. V., Pellegrini, L., Cestra, G., Butler, M. H., Rossman, K. L., Serna, D. M., Sondek, J., Gertler, F. B., and De Camilli, P. (2003). Tuba, a novel protein containing bin/amphiphysin/Rvs and Dbl homology domains, links dynamin to regulation of the actin cytoskeleton. *J. Biol. Chem.* **278**, 49031–49043.

Schafer, D. A. (2004). Regulating actin dynamics at membranes: A focus on dynamin. *Traffic* **5**, 463–469.

Schafer, D. A., Weed, S. A., Binns, D., Karginov, A. V., Parsons, J. T., and Cooper, J. A. (2002). Dynamin2 and cortactin regulate actin assembly and filament organization. *Curr. Biol.* **12**, 1852–1857.

Slepnev, V. I., and De Camilli, P. (2000). Accessory factors in clathrin-dependent synaptic vesicle endocytosis. *Nat. Rev. Neurosci.* **1**, 161–172.

Takei, K., Slepnev, V. I., Haucke, V., and De Camilli, P. (1999). Functional partnership between amphiphysin and dynamin in clathrin-mediated endocytosis. *Nat. Cell Biol.* **1**, 33–39.

Wigge, P., Kohler, K., Vallis, Y., Doyle, C. A., Owen, D., Hunt, S. P., and McMahon, H. T. (1997). Amphiphysin heterodimers: Potential role in clathrin-mediated endocytosis. *Mol. Biol. Cell* **8**, 2003–2015.

Yarar, D., Waterman-Storer, C. M., and Schmid, S. L. (2005). A dynamic actin cytoskeleton functions at multiple stages of clathrin-mediated endocytosis. *Mol. Biol. Cell* **16**, 964–975.

[48] Expression and Properties of Sorting Nexin 9 in Dynamin-Mediated Endocytosis

By RICHARD LUNDMARK and SVEN R. CARLSSON

Abstract

Sorting nexin 9 (SNX9) is identified as an important regulator of dynamin function in clathrin-mediated endocytosis. SNX9 recruits dynamin to the plasma membrane and promotes its GTPase activity, resulting in membrane constriction and ultimate transport vesicle scission. This chapter describes procedures to express recombinant SNX9, to biochemically characterize the cytosolic complex between SNX9 and dynamin, and to identify additional interacting partners of SNX9. Assays are presented to investigate the requirements for SNX9-dependent membrane recruitment of dynamin *in vitro* and *in vivo*.

METHODS IN ENZYMOLOGY, VOL. 404
0076-6879/05 $35.00
DOI: 10.1016/S0076-6879(05)04048-6

Introduction

Members of the dynamin family of large GTPases are involved in membrane fission events that take place during intracellular trafficking of proteins and lipids (Praefcke and McMahon, 2004). It has been known for some time that dynamin is required for the release of budded vesicles from the plasma membrane during clathrin-mediated endocytosis, and more recently dynamin has also been linked to caveolar and other non-clathrin-dependent uptake processes. Dynamin oligomerization at the base of the bud stimulates the hydrolysis of GTP, which leads to constriction through a conformational change and deformation of the membrane. This activity needs to be restricted to membrane areas that possess high curvature like the vesicular neck, and dynamin itself is therefore not an effective membrane binder. Additional proteins are anticipated to be required to target dynamin to the membrane and also to regulate the GTPase-activating oligomerization.

Studies of the biochemical properties of recombinant and endogenous sorting nexin 9 (SNX9) have brought forward this protein as a recruiter of dynamin to the neck of clathrin-coated vesicles. SNX9 belongs to a sub-family of sorting nexins that in addition to a phosphoinositide-specific PX (phox) domain has a curvature-sensing BAR (Bin/amphiphysin/Rvs) domain (Lundmark and Carlsson, 2004; Carlton et al., 2005). The domain structure of SNX9 is well adapted to fit into the complex network of interacting proteins that come together to generate a clathrin-coated vesicle at the plasma membrane (see Fig. 1). An amino terminal SH3 (src homology 3) domain binds to the proline-rich region of neuronal dynamin-1 and the ubiquitously expressed dynamin-2 (Dyn2) (Lundmark and Carlsson, 2003; Soulet et al., 2005). This interaction is strong enough to form an endogenous complex between SNX9 and Dyn2 already in the cytosol (Lundmark and Carlsson, 2003). To link into the vesicle budding network, SNX9 has an unstructured region (LC region) with motifs for binding to the core components AP-2 and clathrin. The PXBAR unit of SNX9 targets the SNX9-Dyn2 complex to the plasma membrane. In combination with the interactions with AP-2 and clathrin, the membrane-binding properties of SNX9 are anticipated to precisely target SNX9 to the highly curved neck of the clathrin coated pit (Lundmark and Carlsson, 2004; Soulet et al., 2005).

SNX9 is found in the cytosol in an inactive ternary complex together with Dyn2 and the metabolic enzyme aldolase. Upon phosphorylation of SNX9, the SNX9-Dyn2 assembly is released from aldolase and can now bind to the membrane, as shown by a biochemical assay (Lundmark and Carlsson, 2004). The phosphorylation-dependent activation for membrane

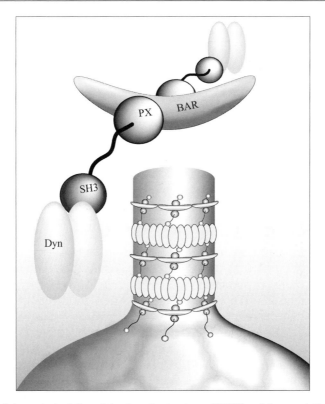

FIG. 1. Schematic depiction of the domain structure of SNX9 and the association between SNX9 and dynamin. Dimeric SNX9 is shown with its SH3-, PX-, and BAR-domains. Dynamin dimers (Dyn) are bound to the SH3-domains. The initial membrane binding occurs via the PX- and BAR-domains of SNX9. Through its unstructured region (curved black line), SNX9 has the ability to attach to the clathrin/AP-2 lattice. The lower part of the picture illustrates that SNX9 and dynamin are thought to co-assemble at the neck of a forming clathrin-coated vesicle.

targeting allows for a temporal and spatial regulation of dynamin assembly. Indeed, time-lapsed TIRF experiments showed a synchronized recruitment of SNX9 and dynamin at the very late stage of vesicle generation, just before vesicle detachment (Soulet *et al.*, 2005). Overexpression of dominant negative constructs of SNX9 affects the uptake of transferrin (Lundmark and Carlsson, 2003), and the use of siRNA to knock down the expression of SNX9 significantly inhibits the membrane recruitment of dynamin (Lundmark and Carlsson, 2004) and decreases the uptake of transferrin (Soulet *et al.*, 2005). These results show that SNX9 is needed for efficient clathrin-mediated endocytosis and the biochemical characterization supports a model for the mechanistic function of SNX9. This chapter describes in detail some of the relevant biochemical methods that are used by us for studies on the structure and function of SNX9.

Isolation of SNX9

Recombinant SNX9

To get material for biochemical characterization, production of antibodies, and as reagents for identifying interaction partners, SNX9 is produced recombinantly as a fusion protein with glutathione S-transferase (GST). The method described in the following section results in good yields of intact full-length protein (10–15 mg/l of culture) and is also used in our lab for the production of individual domains of SNX9.

The SNX9 insert of IMAGE clone 3832234 (Geneservice, www.geneservice.co.uk) is amplified by PCR and cloned into the pGEX-6P-2 vector (Amersham, Uppsala, Sweden). After transfection into BL21 (DE3) pLysS cells, expression is induced with 0.05 mM IPTG at an OD_{600} of 0.5. After 4 h at 37°, the cells are harvested by centrifugation, washed once with 100 mM NaCl, 20 mM Hepes-KOH pH 7.4 (NH-buffer) and pelleted. The pellet (without excess liquid) is frozen in liquid nitrogen and thawed by adding 5 ml of extraction medium per g of cells. The extraction medium consists of CelLytic™ B-II Bacterial Cell Lysis/Extraction Reagent (Sigma, St. Louis, MO), DNAase I (100U/ml, Invitrogen), and 1 mM PMSF. After vortexing and incubation by slow agitation for 15–30 min at 4°, the material yields a homogenous solution. At this step, as additional protease inhibition, EDTA is added to 1 mM, and the extract is centrifuged at 75,000g in a Beckman 25.50 rotor. The clear supernatant is directly incubated with Glutathione-Sepharose beads (Amersham, 3–5 ml of beads per liter of original culture), previously washed with 0.1% Nonidet P-40 (NP40), 1 mM EDTA in NH-buffer (wash buffer). Incubation is performed batch-wise in a tube, and the beads are subsequently transferred to a 10-ml PolyPrep Column (Bio-Rad, Hercules, CA). For large proteins, the yields are significantly improved if the incubation time is increased. We routinely incubate the extract with beads on a slow rotator in the cold room for several hours or overnight. The beads are washed with 5 bead-volumes of wash buffer by gravity flow. Elution is done by the addition 10 mM glutathione in 50 mM Tris-HCl pH 8.8 (made fresh). One-half bead-volume of elution buffer is soaked in, after which the outlet of the column is closed and an additional full bead-volume of elution buffer is added. The beads are resuspended and then incubated on ice for at least 1 h. The eluate is allowed to drop out and collected in tubes, and the column is further washed by step-wise addition of elution buffer. By this procedure, >80% of the recombinant protein is collected in one bead-volume.

GST-SNX9 is highly purified after this step but may contain small amounts of bacterial contaminants, and for certain constructs free GST may also appear in the eluate. We usually add a gel filtration step to get rid of the last contaminants and as a convenient way to exchange the buffer. For GST-SNX9, a 1.5 × 50 cm column of Sephacryl S-300 HR (Amersham) is used with a flow rate of 6 ml/h. The composition of the buffer is crucial for the yield and behavior of constructs containing SNX9 and domains derived from it. We have found that equilibration of the gel filtration column with 100 mM potassium acetate, 20 mM Hepes-KOH pH 7.4 minimizes the formation of oligomers and results in the elution mainly of dimers of GST-SNX9 in good yields. The obtained protein is finally frozen in liquid nitrogen and stored at −80°.

The fusion of SNX9 with GST is used in pull-down experiments to detect binding partners, but it also acts as a good antigen for the production of antibodies. Our experience with several different GST-fusion proteins implies that the GST moiety is an effective trigger of the immune response to yield high titers of antibodies both to the protein of interest and to GST. The latter antibodies are removed either by absorption to immobilized GST or by affinity purification by use of immobilized GST-free recombinant protein. The removal of GST from the recombinant protein is achieved by on-column cleavage by PreScission (Amersham, used as suggested by the manufacturer).

SNX9-Complex in Cytosol

Studies of endogenous SNX9 from cultured cells by immunoprecipitation and Western blotting have revealed important properties of this protein. SNX9 is an intracellular protein that recycles between a cytosolic pool and a pool peripherally associated with membranes. We have mainly utilized human hematopoietic K562 cells that expresses relatively high amounts of SNX9 (0.1–0.2% of the total proteins in the cytosol), and the steady-state distribution of SNX9 between cytosol and membranes is approximately 50:50. When cytosol is prepared from cells in a concentrated form at physiological ionic strength (see protocol below), SNX9 is found in a complex together with Dyn2 and aldolase. The complex sediments by ultracentrifugation at 14.5 S, which corresponds to a molecular weight of around 400 kDa for a spherical complex. The stoichiometry of the components is uncertain, but the complex may consist of a monomer (or a dimer) of SNX9 (Mw 67 kDa), a dimer of Dyn2 (2 × 98 kDa), and a tetramer of aldolase (4 × 41 kDa).

The complex of SNX9-Dyn2-aldolase is very labile after disruption of the cells and can easily escape detection if the cytosol is treated improperly. For example, dilution of cytosol below 5 mg/ml in preparation buffer (KSHM, 100 mM potassium acetate, 85 mM sucrose, 1 mM magnesium acetate, 20 mM Hepes-KOH pH 7.4) shifts the equilibrium towards dissociated SNX9 and Dyn2 (the concentration of proteins in the cytosol of intact K652 cells is close to 50 mg/ml). In washings during immunoprecipitations, we lower the ionic strength to keep the complex together (see following). Also in sedimentation analyses by ultracentrifugation, it is important to omit the salt in the gradient to obtain the full complex. In addition, we have found that the presence of phosphate interferes with the SNX9-aldolase interaction (the concentration of free phosphate is very low intracellularly). This effect is likely related to the nature of the association between the two proteins. Aldolase and SNX9 interacts via positively charged residues close to the active site in aldolase, which are involved in the binding of phosphate groups in substrate and products, and a stretch of negatively charged residues in the LC region in SNX9 (Lundmark and Carlsson, 2004).

A procedure for the preparation of cytosol containing intact SNX9-Dyn2-aldolase complex is described. This preparation is used for immunoprecipitations, sedimentation analyses, and recruitment studies. We have used the method mainly for K562 cells, but it can also be used for other types of cells, such as HeLa cells. Cells grown to high density are collected by centrifugation and washed twice with PBS and twice with KSHM. We usually get 1 ml of packed K652 cells from 700 ml of culture. We centrifuge for 5 min at 350g, except in the last run which is at 850g. One volume of degassed KSHM, containing 1 mM PMSF, 1 mM sodium orthovanadate, 10 μg/ml each of pepstatin, chymostatin, and leupeptin, and 2 μg/ml of aprotinin, is added to one volume of cell pellet. The cells are resuspended and flushed with nitrogen gas to remove oxygen. The cells are quickly frozen in liquid nitrogen, thawed in a 37°-water bath, and immediately put on ice. During this procedure, the cells are opened and cytosol is released together with vesicles and membrane fragments. The cells are centrifuged at 350g for 5 min, and the supernatant is collected (the pellet, which consists of opened cells, may be used for recruitment experiments; see below). After centrifugation at 70,000g for 30 min, the cytosol is collected and immediately frozen in aliquots in liquid nitrogen and stored at −80°. The protein concentration is 15–20 mg/ml.

About half of the content of SNX9 in K562 cells is found in the cytosol prepared by this method. The other half is membrane-bound and found in the pellet obtained after 70,000g centrifugation. Very little SNX9 (<5%) is left in the opened cells. This is also true for Dyn2, AP-2, and clathrin. This probably means that regions of the plasma membrane containing clathrin coats and attached vesicles are disrupted and released during freezing and thawing. If the preparation is done at lower ionic strength, also the membrane-associated material is dissolved. SNX9 in such preparations has a tendency to form large aggregates (Lundmark and Carlsson, 2002).

Due to the fragility of the SNX9-Dyn2-aldolase complex, it cannot easily be purified from the cytosol. In our hands, using conventional chromatography, the complex dissociates and the yields are low. However, a fraction enriched about 10-fold for the complex can be achieved from cytosol by gradient centrifugation. Cytosol (1 ml) prepared as above is layered onto a 12-ml 10–30% continuous sucrose gradient made in 20 mM Hepes-KOH pH 7.4. The gradient is centrifuged for 18 h at 200,000g in a Beckman SW41 rotor and 1-ml fractions are collected from the top. The complex migrates ahead of the majority of cytosolic proteins and is found in fractions 7–9. This preparation can be stored in the refrigerator for a few days but becomes inactive, for example, for Dyn2 membrane-recruitment after freezing.

Identification of SNX9-Interacting Partners

Pull-Downs

Many molecular contacts in the cell are characterized by weak temporal interactions allowing for a dynamic system that is constantly shifting. We have used GST-based constructs of SNX9 to detect binding partners and to localize their binding sites with an efficient method to capture low-affinity partners from cytosol.

Purified GST-fusion protein (1–10 μg) is immobilized on 10 μl of glutathione beads by incubation for 30–60 min in 500-μl Eppendorf tubes. The beads are washed with 0.1% NP40 in 20 mM Hepes pH 7.4 (centrifugation at 1000g for 1 min), and excess liquid is removed with a Hamilton syringe. Cytosol (15–20 mg/ml) is thawed, adjusted with NP40 to 1% and cleared by centrifugation for 10 min in a cold Microfuge, and added to the beads. Depending on the abundance of target protein and the scale of the experiment, cytosol amounts may vary from 0.5 mg to 10 mg of protein. The samples are incubated in the cold by slow agitation on a rotator for 30–60 min and centrifuged. Most of the supernatant is removed, leaving only a few μl above the bead surface. The lids of the tubes are cut off, and the tubes are pressed down, upside down, into 1.5-ml Eppendorf tubes containing a three-phase gradient. The gradient, which is made in advance, consists of (from bottom to top) 0.4 ml of 17% glycerol, 27% sucrose, 0.2 ml of 17% glycerol, and 0.6 ml of 0.5% NP40 (all solutions made in 20 mM Hepes-KOH pH 7.4). The tubes are spun at maximum speed in a Microfuge for 30 sec, and most of the supernatant is removed (the beads are hard to see at this step). The last remaining liquid is removed with a Hamilton syringe, and the beads are heated in sample buffer for SDS-PAGE.

With the method described, the separation of bound and free ligand only takes a few seconds allowing for detection of interactions with high off-rates. It is also a convenient way to change tubes, which is necessary when working with cytosol due to adherence of many proteins to the plastic. The one-step wash procedure may result in some unspecific proteins still bound to the beads. It is therefore important to include negative controls and, if specific bands are detected, to repeat the experiment with additional washings if cleaner results are required, for example, for mass spectrometry identification.

Immunodetection

In order to isolate the SNX9-Dyn2-aldolase complex from cytosol by immunoprecipitation, and to obtain SNX9-depleted cytosol, the following protocol is used. Affinity-purified anti-SNX9 antibodies (1–10 μg of IgG) are immobilized on 10 μl of protein A-agarose beads (Sigma) by incubation for 30–60 min. The beads are washed twice with 0.5% NP40 in 20 mM Hepes-KOH pH 7.4 (wash buffer) and all liquid is removed. Cytosol (15–20 mg/ml), adjusted to 0.5% with NP40 and cleared by centrifugation as described above, is added (at least 0.5 ml of cytosol is required for subsequent detection of the complex components by Coomassie staining). Incubation is performed for 1–3 h by slow agitation in the cold, after which the beads are sedimented by a brief centrifugation. (Note: If the supernatant is to be used in recruitment experiments that require SNX9-depleted cytosol, the NP40 addition is omitted.) The beads are washed three times by incubation in the cold for 5 min in 1 ml of wash buffer followed by a brief centrifugation. The beads are transferred to new tubes before the last wash. Finally, the last trace of liquid is removed with a Hamilton syringe and the sample is prepared for SDS-PAGE. The resulting Coomassie-stained gel should reveal the SNX9-band migrating at approximately 78 kDa, Dyn2 at 110 kDa, and aldolase at 41 kDa, together with IgG heavy (50 kDa) and light (25 kDa) chains.

For *in vivo* visualization of the biochemically identified interactions of SNX9, we have used affinity-purified SNX9 antibodies, together with antibodies to candidate partners, to detect colocalization of endogenous proteins by immunofluorescence. The K562 cells used in biochemical studies are not suitable for microscopy analysis due to their round shape. Instead HeLa cells are used, which express lower level of SNX9. The distribution, however, is shifted more to the membrane-associated pool, which facilitates proper detection. Cells grown on coverslips are washed in PBS, fixed for 20 min in room temperature by addition of ice-cold 2–4% paraformaldehyde in PBS, and opened and blocked with 0.05–0.1% saponin, 5% goat serum in PBS. Primary and secondary antibodies are diluted in wash buffer (2.5% goat serum in PBS) and incubations are performed at room temperature for one h followed by three washes in wash buffer. Mounted coverslips are visualized using confocal microscopy.

Recruitment of SNX9 and Dyn2 to Membranes *In Vitro*

We have utilized two types of *in vitro* assays for analysis of the membrane binding properties of SNX9 and to reveal the role for SNX9 to recruit Dyn2 to the membrane. The first is based on binding to cell membranes using opened and washed cultured cells. In the second method artificial liposomes with defined composition are used as targets for SNX9 and Dyn2 binding.

Recruitment to Cellular Membranes

K562 cells are opened by a freeze/thaw cycle and the cells are collected by centrifugation (see method above). Cells are carefully resuspended in 15 ml of KSHM-buffer, avoiding the material at the bottom of the tube which may contain unbroken cells, transferred to a new tube, and incubated on ice for at least 1 h. After centrifugation at $350 \times g$ for 5 min, the cells are resuspended to 50×10^6 cells/ml in KSHM-buffer containing 0.8% BSA. Trypan Blue staining should reveal >99% opened cells. This material contains negligible amounts of both SNX9 and Dyn2.

The assay is performed in a total volume of 40 μl in Eppendorf tubes. The reaction components (protein, cytosol, nucleotides, inhibitors, etc.) are mixed together on ice and adjusted to 30 μl with KSHM-buffer. The tubes are spun briefly to bring down the reagents, after which 10 μl of the permeabilized cells are added by pipetting into the reagent mixture. The samples are incubated for 20 min on ice or in a 37°-water bath and the cells are collected by centrifugation at $10,000 \times g$ for 5 min in a Microfuge in the cold, rinsed with 200 μl of KSHM-buffer without resuspension, and again centrifuged. The pellets are finally solubilized in 25 μl of 1% NP40 in PBS with protease inhibitors, and after centrifugation the supernatants are transferred to new tubes and processed for SDS-PAGE and western blotting.

Recombinant full-length SNX9 (final concentration 0.1–0.2 μM), as well as truncated constructs including the carboxyl terminal part containing the PXBAR unit, bind effectively to the cellular membranes on ice without the presence of additional reagents. When desalted cytosol is tested (final protein concentration 5 mg/ml) no binding of either SNX9 or Dyn2 is detected when incubated on ice or at 37 °. Addition of ATP and an ATP-regenerating mixture (final concentrations 1 mM ATP, 8 mM creatine phosphate, and 50 μg/ml creatine kinase, added from a 20-fold concentrated stock) to cytosol results in a significant binding of SNX9 and Dyn2 when the mixture is incubated at 37 °, but not on ice. This shows that the cytosolic SNX9-Dyn2-aldolase complex must be activated by an ATP-dependent process to expose the membrane-binding domains in SNX9. This activation was found correlate with the phosphorylation of the LC region of SNX9, most likely releasing a blockage caused by aldolase. If GTPγS (100 μM) and the phosphotyrosine phosphatase inhibitor sodium orthovanadate (1 mM) is added to the cytosol and ATP mixture, the binding of SNX9 to membranes increases to almost 100%. The reason for the positive effect of the latter reagents is not clear.

We routinely use an incubation time of 20 min, although the binding reaction is considerably faster. Initial experiments showed that the rate limiting step was the activation of the SNX9-complex in the cytosol. Using optimal conditions (in the presence of ATP, GTPγS, and vanadate), a 10-min incubation of cytosol results in >80% of SNX9 bound. The binding is slightly increased when the incubation time is extended to 20 min.

Desalting of cytosol is necessary to remove free nucleotides, which otherwise results in a background binding when unsupplemented cytosol is incubated with cellular membranes at 37 °. We perform desalting of cytosol immediately prior to the assay by use of MicroBio-Spin P6 columns (Bio-Rad), equilibrated with KSHM-buffer and handled as recommended by the manufacturer. The procedure will not significantly decrease the protein concentration. The cytosol is finally centrifuged for 10 min in a Microfuge before being added to the assay.

The binding of Dyn2 to the cellular membranes is dependent on the binding of SNX9, resulting maximally to 25% of the amount found in the cytosol. When cytosol is depleted of SNX9 by antibodies, the binding of Dyn2 is completely abolished, although cytosol still contains large amounts of Dyn2 dimers. Addition of recombinant SNX9 to the depleted cytosol restores the binding of Dyn2.

Binding to Liposomes

The results obtained using cellular membranes (see above) can be reproduced in similar assays with artificial liposomes, confirming that the binding reaction is to membrane lipids and not via membrane proteins. In addition, liposomes can be used to investigate the requirements in lipid composition and membrane shape for the binding of proteins under study. We have reported results employing a straightforward assay that makes use of large multilamellar liposomes that sediment in the Microfuge, to investigate the phosphoinositide specificity of recombinant SNX9. This assay does not measure the curvature preference in binding, which requires that size-defined liposomes are produced (Peter *et al.*, 2004). It is not possible to determine the lipid specificity of cytosolic SNX9 since the incubation of liposomes with cytosol, under conditions to activate the SNX9-Dyn2-aldolase complex, alters the lipid composition of the liposomes caused by lipid kinases and phosphatases present in the cytosol. For the preparation and use of liposomes, the reader is referred to a recent series on liposomes in Methods in Enzymology.

Recruitment of SNX9 and Dyn2 to Membranes *In Vivo*

The results from *in vitro* recruitment assays point to a crucial role of SNX9 for the localization of Dyn2 to the membrane. In order to investigate this in living cells, we have made use of the RNAi technique to knock-down the expression of SNX9 in HeLa cells. Two different targets (AAGAGA-GUCAGCAAUCAUGUCU and AACCUACUAACACUAAUCGAU) were used. The sequences were obtained using the Dharmacon siRNA design Website (www.dharmacon.com/sidesign/) and chosen due to the higher T_m in the 5' end. Both targets gave approximautely 85% depletion of SNX9 after one transfection and >95% depletion after an additional transfection. The latter target was slightly more effective and used in subsequent experiments. HeLa cells were transfected twice with siRNA, performed at day 0 and day 2, and the cells were assayed at day 3. The transfection mixture contained siRNA, OptiMEM medium (Gibco), and Oligofectamin (Invitrogen, Grand Island, NY) and was done according to the manufacturer's instructions.

Investigations by immunofluorescence showed that while the membrane localization of Dyn2 was not completely abolished in the absence of SNX9, there was a marked decrease in membrane staining. This result was confirmed biochemically after separation of cytosol and membranes (70,000g supernatant and pellet, described previously), showing that the steady-state distribution of Dyn2 was shifted to the cytosol after depletion of SNX9, without affecting the total amount of Dyn2. The conclusion from our experiments with RNAi is that Dyn2 is dependent on SNX9 for efficient recruitment to membranes in living cells. This result fits nicely with the recent results from Soulet *et al.* (2005), showing that depletion of SNX9 in HeLa cells significantly affects the uptake of transferrin. Together, the investigations point to an essential role for SNX9 in dynamin-dependent scission of clathrin vesicles at the plasma membrane.

References

Carlton, J., Bujny, M., Rutherford, A., and Cullen, P. (2005). Sorting nexins—unifying trends and new perspectives. *Traffic* **6**, 75–82.

Lundmark, R., and Carlsson, S. R. (2002). The beta-appendages of the four adaptor-protein (AP) complexes: Structure and binding properties, and identification of sorting nexin 9 as an accessory protein to AP-2. *Biochem. J.* **362**, 597–607.

Lundmark, R., and Carlsson, S. R. (2003). Sorting nexin 9 participates in clathrin-mediated endocytosis through interactions with the core components. *J. Biol. Chem.* **278**, 46772–46781.

Lundmark, R., and Carlsson, S. R. (2004). Regulated membrane recruitment of dynamin-2 mediated by sorting nexin 9. *J. Biol. Chem.* **279**, 42694–42702.

Peter, B. J., Kent, H. M., Mills, I. G., Vallis, Y., Butler, P. J., Evans, P. R., and McMahon, H. T. (2004). BAR domains as sensors of membrane curvature: The amphiphysin BAR structure. *Science* **303**, 495–499.

Praefcke, G. J., and McMahon, H. T. (2004). The dynamin superfamily: Universal membrane tubulation and fission molecules? *Nat. Rev. Mol. Cell. Biol.* **5**, 133–147.

Soulet, F., Yarar, D., Leonard, M., and Schmid, S. L. (2005). SNX9 regulates dynamin assembly and is required for efficient clathrin-mediated endocytosis. *Mol. Biol. Cell* **16**, 2058–2067.

[49] Rapid Purification of Native Dynamin I and Colorimetric GTPase Assay

By ANNIE QUAN and PHILLIP J. ROBINSON

Abstract

Dynamin I is a large GTPase enzyme required in membrane constriction and fission during multiple forms of endocytosis. The first method described here is for the rapid purification of native dynamin from peripheral membrane extracts of sheep brain using ammonium sulfate precipitation and affinity purification on recombinant SH3 domains. The method greatly enriches for dynamin I at high purity and allows for large-scale

METHODS IN ENZYMOLOGY, VOL. 404
0076-6879/05 $35.00
DOI: 10.1016/S0076-6879(05)04049-8

biochemical and functional studies. The second method is a nonradioactive, high-throughput colorimetric GTPase assay for dynamin activity. The approach is based on terminating incubations with EDTA and the use of malachite green for high-sensitivity detection of inorganic phosphate release. The two methods will facilitate high-throughput screens for potential dynamin inhibitors or activators.

Introduction

Dynamin is a 96-kDa GTPase enzyme involved in membrane constriction and fission during receptor-mediated endocytosis (RME) and synaptic vesicle endocytosis (SVE). At a late stage of the process, dynamin assembles into rings to form a collar or helix around the neck of the invaginated vesicles. Upon GTP hydrolysis it pinches them from the plasma membrane. Dynamin is also needed for most, but not all, forms of non-clathrin-dependent endocytosis, such as phagocytosis, caveolae internalization, and endocytosis of cytokine receptors. Dynamin contains four functional domains: an N-terminal GTPase domain, a pleckstrin homology (PH) domain, a proline-arginine rich domain (PRD), and an assembly domain (Cousin and Robinson, 2001). In a *Drosophila* strain called *shibire*, mutations in dynamin's GTPase domain allow assembly of dynamin helices at the nascent vesicle neck and block a late stage of synaptic vesicle fission (Koenig and Ikeda, 1989). The mutations do not block GTP binding, but block GTP hydrolysis. Overexpression of GTPase-defective dynamin mutants inhibits both RME and SVE in a variety of cells (Marks *et al.*, 2001). Thus the GTPase activity of dynamin is a potentially attractive candidate for development of a specific endocytosis inhibitor. We describe a method for the large-scale rapid purification of dynamin I and a sensitive colorimetric high-throughput GTPase assay that together will facilitate the screening of large chemical libraries for dynamin inhibitors.

Purification of Native Dynamin I

Purification of native dynamin I from sheep brain using affinity purification provides a rapid and reproducible technique for isolation of high levels of biologically active native dynamin I without the use of large-scale ion exchange columns, detergents, or guanosine nucleotides. The affinity ligand is bacterially expressed glutathione S-transferase (GST) fused to the amphiphysin II-SH3 domain, which is easy to produce in large quantities. The McMahon lab demonstrated the utility of using GST-SH3 domains from various proteins to isolate dynamin from brain or cultured cells (Wigge *et al.*, 1997). However, it is a challenge to elute dynamin from most

SH3 domains. The McMahon lab suggested the combination of the amphiphysin II—SH3 domain and 1.2 M NaCl to elute dynamin (Vallis *et al.*, 1999). We use an extract from 200 g of sheep brain to purify dynamin in 3 days with a recovery of 8–15 mg of protein at high purity (greater than 98%). The purification of 15 mg of dynamin I provides enough sample for about 50,000 GTPase assays (at 0.3 μg dynamin per sample).

Expression and Purification of Recombinant GST-Amphiphysin II-SH3 Domain Protein

Not all GST fusion proteins are suitable for the purification of their target protein partner. Most bind so well they cannot be extracted from the glutathione sepharose column without also eluting the GST-based affinity ligand, thus leaching is a significant problem. We found that GST fusion proteins attached to Glutathione-Sepharose 4B (GSH) (Amersham Biosciences, Uppsala, Sweden) beads remain fully bound in the presence of 100 mM glycine pH 3.0, 2% Triton X-100, or 1.5 M NaCl, without leaching of the GST fusion protein, making ideal washing or eluting solutions. We tested 12 SH3 domains for quantitative dynamin I binding in brain extracts and most bind dynamin I with high efficiency. The SH3 domains of amphiphysin I and II and of endophilin I bind dynamin I quantitatively (Fig. 1A). The endophilin I-SH3 still binds strongly in the presence of 0.5–1 M NaCl, and that of amphiphysin I binds less well (Fig. 1A). However, amphiphysin II-SH3 domain does not bind dynamin in the presence of 1 M NaCl (Fig. 1A), making it most suitable for the dynamin I purification.

For our affinity purification of native dynamin I, recombinant GST-amphiphysin II-SH3 is immobilized on GSH beads. The amphiphysin II-SH3 domain construct (in a pGEX2T plasmid) was provided by Pietro de Camilli (Howard Hughes Medical Institute, New Haven, CT). The plasmid is transformed into competent JM109 *E. coli* cells (Promega, Madison, WI) for protein expression using standard transformation procedures. We screened the transformed cells by sodium dodecyl sulfate (SDS) polyacrylamide gel electrophoresis (PAGE) for the greatest expression. For each dynamin purification, three 300-ml bacterial cultures are required to obtain 3 ml of a 50% v/v slurry of GST-amphiphysin II-SH3 GSH beads at 60 mg/ml protein (1 ml of 50% v/v slurry GSH beads per 300 ml culture). The bacterial cells are cultured in three separate 1 l baffled flasks (Nalgene Nunc, Rochester, NY) each with 30 ml of Lennox LB medium (Bio101 Systems, MP Biochemicals, Irvine, CA) and 100 μg/ml ampicillin (Sigma Aldrich, San Diego, CA) overnight at 37° with rotator shaking at 200 rpm. The baffled flasks increase culture aeration, improving the amount of

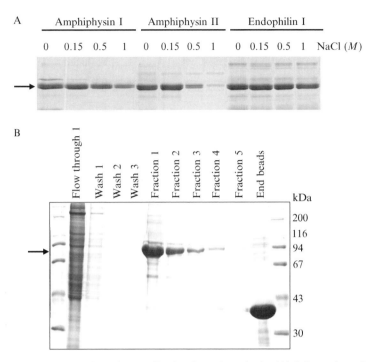

FIG. 1. Summary of dynamin I purification from sheep brain. (A) Salt regulates dynamin binding to SH3 domains. The SH3 domains of the indicated proteins were used as a GST fusion protein on GSH beads to pull-down dynamin I from sheep brain homogenates in the presence of increasing amounts of NaCl. The proteins were separated by SDS-PAGE and stained with Coomassie blue. The arrow indicates dynamin I. Binding to the SH3 domain of amphiphysin II was abolished by high salt concentrations. (B) Summary of the dynamin purification. The Coomassie blue-stained mini-gel shows 15 μl aliquots of fractions obtained during affinity-purification of native dynamin I using peripheral membrane extracts from sheep brain, 35% ammonium sulfate precipitation, and affinity purification on GST-amphiphysin II-SH3 GSH beads. The main protein in the End beads lane is GST-amphiphysin II-SH3.

protein expression. After 16 h of growth add 270 ml of fresh LB medium and 50 μg/ml ampicillin to each of the 30 ml cultures and incubate for a further 2 h at 37° with rotator shaking. Determine the optical density (OD_{600}) of the undiluted cultures hourly from 2 h. An OD in the range of 0.6–0.9 is optimal for inducing protein expression with 1 mM IPTG for 4 h at 37° rotating at 200 rpm. After 4 h induction time harvest the cells by centrifuging at 18,500g for 15 min at 4° in a JLA 10.500 rotor in an Avanti J25I centrifuge (Beckman Coulter, Fullerton, CA). The supernatant is discarded and the three bacterial cell pellets are ready for cell lysis or storage at −80°.

For bacterial cell lysis, the three pellets are each resuspended in 40 ml of ice cold STE buffer (300 mM NaCl, 10 mM Tris, 1 mM ethylenediamine-tetraacetic acid (EDTA), 0.1 mM phenylmethylsulfonyl fluoride (PMSF), pH 8.0), then add 100 μg/ml lysozyme and one complete protease inhibitor EDTA-free tablet (Roche Diagnostics, Mannheim, Germany) and transfer to three 50-ml Falcon tubes. The suspensions are incubated on a rotating wheel for 30 min at 4°. Then add 5 mM 1,4-dithiothreitol (DTT) and 1% Triton X-100 to the suspensions and incubate on rotating wheel for a further 10 min at 4°. When the cells have fully lysed, the suspension will become viscous and turbid. To assist with efficient lysis the suspensions are then twice rapidly frozen in liquid nitrogen and thawed. To shear bacterial DNA, the suspensions (3 × 45–50 ml) are then sonicated twice for 1 min on ice, with a 1 min cooling down period between sonications. The lysates are centrifuged at 36,300g for 10 min at 4° in a Beckman Coulter JA 25.50 rotor. The three supernatants are incubated each with 1 ml of a 50% v/v slurry of GSH beads (the total volume used is 3 ml of GSH beads (50% v/v slurry) for 900 ml of original bacterial culture) in phosphate buffered saline (PBS) for 1 h at 4° rotating gently. In advance, the GSH beads are washed once in PBS to remove manufacturer's storage buffer, once in PBS with 0.1% Triton X-100 to block nonspecific binding, and twice in PBS before resuspending in PBS to make the 50% v/v slurry. After incubation to bind the recombinant protein, each of the 1 ml of 50% v/v slurry of GSH beads is pelleted by centrifuging at 500g for 5 min at 4°. The low speed prevents beads crushing. The protein-bound beads are then washed three times with ice cold PBS with centrifugation at 500g for 5 min at 4° between washes, and finally resuspended in PBS to make the 50% v/v slurry of GST-amphiphysin II-SH3 GSH beads. The concentration of GST-amphiphysin II-SH3 bound GSH beads is estimated by SDS-PAGE. The beads can be stored at 4° for 5 days. Add 0.02% sodium azide as a preservative for storage for 2 weeks, or add glycerol to 20% v/v for storage at −20° for up to 6 months. To maximize dynamin recovery from the affinity purification avoid long storage time.

Purification of Dynamin I from Brain

Buffers and Reagents. All buffers can be made a day earlier and stored at 4° until ready for use, except that protease inhibitors should be added immediately prior to their use.

Buffer A:1 (1000 ml): 20 mM Tris/HCl pH 7.7.

Buffer A:2 (750 ml): 20 mM Tris/HCl, 1 mM CaCl$_2$, 2 mM DTT, 7 mg leupeptin (Merck Biosciences, Darmstadt, Germany), 1 mM PMSF*, pH 7.7.

Buffer A:3 (2000 ml): 20 mM Tris/HCl, 0.1 mM CaCl$_2$, 2 mM DTT, 20 mg leupeptin, 1 mM PMSF*, pH 7.7.

Buffer A:4 (750 ml): 20 mM Tris/HCl, 2 mM ethylene glycol tetraacetic acid (EGTA), 2 mM EDTA, 250 mM NaCl, 2 mM DTT, 3.75 mg leupeptin, 1 mM PMSF*, pH 7.7.

Column Buffer A/Dialysis buffer 1 (4000 ml): 200 mM NaCl, 20 mM HEPES, 1 mM DTT, 1 mM EDTA, 0.1% tween 80, pH 7.3.

Column Buffer B (100 ml): 200 mM NaCl, 20 mM HEPES, 1 mM DTT, pH 7.3.

Column Buffer C (100 ml): 1.2 M NaCl, 20 mM PIPES, 1 mM DTT, pH 6.5.

Dialysis Buffer 2 (2000 ml): 200 mM NaCl, 20 mM HEPES, 1 mM DTT, 50% glycerol, pH 7.3.

* PMSF is added to buffers immediately before use, as it has a half-life of about 40 min in aqueous solutions.

Sheep brain is collected from a commercial abattoir. Brains must be obtained on site within a few minutes of sacrifice, rinsed in cold PBS, diced into 2–4 g cubes, and frozen immediately in liquid nitrogen. Back in the laboratory, the frozen brains should be stored at −80° and used within 12 weeks to obtain dynamin with the greatest specific activity.

Procedure

DAY 1. Add 200 g of frozen sheep brain slowly to 1000 ml of buffer A:1, while keeping the buffer on ice. The brain cubes are defrosted when they sink. Strain the brains with a colander or mesh and place in 750 ml ice cold buffer A:2 for homogenization. Homogenize with a T25 basic Ultra Turrax (IKA, Wilmington, NC) at 11,000 rpm speed setting for 30–60 sec. Avoid excessive frothing of the homogenate, which causes protein oxidation. The calcium is important here to drive the smaller pool of cytosolic dynamin to the larger pool of peripheral membrane dynamin (Liu *et al.*, 1994). The DTT prevents artefactual formation of non-native disulphide bonds in the dynamin tetramer (unpublished observations). Centrifuge the homogenate at 18,500g for 30 min at 4° in a JLA 10.500 rotor. Discard the supernatant and rehomogenize the pellets for 30–60 sec in 750 ml ice cold buffer A:3 and centrifuge at 18,500g for 30 min at 4°. The supernatant is discarded and the resultant pellet is rehomogenized in buffer A:3. Centrifuge at 18,500g for 30 min at 4° again. Now dynamin is extracted from the pellet using a buffer containing 250 mM salt and metal chelators (Buffer A:4). Higher salt concentrations will extract more dynamin, but the additional proteins recovered are more difficult to separate later. The pellet is rehomogenized in 750 ml buffer A:4. The homogenate is stirred slowly for

30 min at 4° and centrifuged at 18,500g for 30 min at 4°. The resultant supernatant is the dynamin extract. Recentrifuge the dynamin extract at 18,500g for 20 min at 4° to remove trace particulate material.

A 35% ammonium sulfate ($(NH_4)_2SO_4$) precipitation of the dynamin extract concentrates and enriches for dynamin I. In a beaker, $(NH_4)_2SO_4$ is slowly added to 35% saturation (20.8 g for every 100 ml extract) with gentle stirring. When all the $(NH_4)_2SO_4$ is dissolved, stir gently for further 30 min at 4°. Centrifuge the solution at 18,500g for 30 min at 4°. The pellets are resuspended gently in no more than 50 ml total volume of column buffer A. After resuspension, dialyze the fraction overnight at 4° using Spectra/Por 4 regenerated cellulose membrane MWCO 12,000–14,000 (Spectrum, Los Angeles, CA) against 2000 ml column buffer A.

DAY 2. Change the dialysis buffer once and dialyze for at least a further 2 h. Centrifuge the dialysate in a JA 25.50 rotor at 75,600g for 20 min at 4°, filter the supernatant through 1 μm syringe filters, followed by 0.45 μm syringe filters. All filters should be pre-equilibrated with column buffer A to reduce nonspecific binding.

Prewash the GST-amphiphysin II-SH3 GSH beads three times with column buffer A, centrifuge at 500g for 5 min at 4° between washes. Incubate the dialysate with the 3 ml of washed 50% v/v slurry of GST-amphiphysin II-SH3 GSH beads for 1 h at 4° on rotator. The dialysate containing the beads is then poured onto an Econo-Pac disposable chromatography column (25 ml capacity, Bio-Rad) at room temperature (22°), collecting the flow through (called flow through 1). Wash the beads in the column with 25 ml of ice cold column buffer B. Occasional stirring of the beads with a spatula will speed up the flow. Repeat the wash step three times, collecting each of the washes (called wash 1, wash 2, wash 3). Dynamin I is batch-eluted from the GST-amphiphysin II-SH3 GSH beads with a minimum of five fractions of column buffer C (Fractions 1–5). For each of the first three fractions, 3 ml column buffer C is added to the beads; the column is plugged and incubated for 10 min before opening the column flow to collect the eluate. This is repeated for two more fractions, but reduce the incubation time to 5 min. The beads are stirred with a small spatula during each incubation time to increase the elution efficiency. The beads are resuspended in PBS and kept for SDS-PAGE to ensure no dynamin remains (called end beads). The flow through 1, washes 1–3, fractions 1–5, and the end beads are analyzed with SDS-PAGE on 12% acrylamide mini-gels to determine which fractions contain eluted dynamin I (Fig. 1B). Most of the dynamin I is eluted in fractions 1–3. Efficient elution should remove greater than 98% dynamin I from the beads. Pool the fractions containing dynamin and dialyze overnight against 1000 ml dialysis buffer 2.

DAY 3. Change the dialysis buffer once and dialyze for at least a further 2 h. The dialysate volume will shrink by half due to the glycerol, which prevents dynamin I precipitating in solution during freezing and storage of the protein. Aliquot the dynamin I into desired microliter volumes, then freeze and store at $-80°$. To test dynamin I purity and state of aggregation, analyze the purified dynamin I samples on SDS-PAGE on 12% acrylamide mini-gels under reducing and nonreducing conditions. If dynamin has oxidized, it will migrate as tetramers at 400 kDa instead of 96 kDa in gels lacking β-mercaptoethanol. Oxidized dynamin is constitutively active and poorly sensitive to normal dynamin activators. Test the biological activity of the purified dynamin I using the colorimetric GTPase assay described below. We routinely recover 8–15 mg of dynamin I that retains high specific activity for at least 1 year on storage. Dynamins II and III are also present at much less than 1%.

Colorimetric GTPase Assay

We and others have developed a sensitive, nonradioactive malachite green based colorimetric assay to measure inorganic phosphate released by dynamin I's GTPase activity (Hill *et al.*, 2004; Song *et al.*, 2004). The most sensitive colorimetric assays for inorganic phosphate are based on the formation of a phosphomolybdate complex at low pH (Hohenwallner and Wimmer, 1973). The colorless phosphomolybdate complex is converted to a colored complex in the presence of an enhancer, which is a basic pH indicator dye like malachite green. The complex with phosphate is quantified spectrophotometrically. Among other dyes used for the enhancer (e.g., crystal violet and quinaldine red) malachite green dye has the highest sensitivity (Cogan *et al.*, 1999). The dye is soluble, yellow, and stable in the presence of 6N acid (Van Veldhoven and Mannaerts, 1987). It has a high molar absorption coefficient when complexed with phosphomolybdate and changes from yellow to blue-green. Other studies have modified the malachite green method to measure inorganic phosphate released from protein (Buss and Stull, 1983), protein phosphatases (Geladopoulos *et al.*, 1991), and lipid phosphatases (Lee *et al.*, 2004).

Dynamin hydrolyzes GTP into GDP and inorganic phosphate (Pi). Its GTPase activity can be stimulated by a number of protein or lipid effectors, or by its own self assembly. This method uses sonicated L-α-phosphatidyl-L-serine (PS, 100%) liposomes. PS liposomes facilitate dynamin self assembly (Lin *et al.*, 1997; Tuma *et al.*, 1993) and an optimal concentration of PS liposomes will stimulate maximal dynamin GTPase activity *in vitro* (Barylko *et al.*, 2001) at very low cost. The protocol offers advantages in speed and safety over traditional assay methods based on ^{32}P release, hence it is suitable for high throughput screening for potential dynamin GTPase inhibitors. In our recent studies we used the method to screen long-chain amines or ammonium salts for dynamin I GTPase inhibitors for use in the study of endocytosis (Hill *et al.*, 2004). The main disadvantage of this method over radiation-based GTPase assays is its sensitivity to contaminating phosphate in buffers or drug solutions. Acceptable assay sensitivity requires appropriate assay controls and blanks to account for this. Alternatively, endogenous phosphate can sometimes be removed by sample pretreatment with Dowex anion exchange resin (Mahuren *et al.*, 2001).

Buffers and Reagents

Malachite Green Reagent: Dissolve 50 mg malachite green (final concentration 1 mg/ml) (Ajax Chemicals, NSW, Australia) and 500 mg ammonium molybdate tetrahydrate (final concentration 100 mg/ml) (Ajax Chemicals) in 50 ml of 1 M hydrochloric acid (HCl) followed by filtration through a 0.45 μm filter. Store the solution at room temperature (22°) in the dark for a maximum of 8 weeks.

5× Phosphate Standards: Sodium dihydrogen orthophosphate monohydrate (NaH$_2$PO$_4$.H$_2$O) (BDH AnalaR Merck Chemicals, Darmstadt, Germany) is dried in an oven at 110° for 3 h or overnight to remove water prior to making up standard solutions. Prepare a 1 mM stock solution and dilute it in water to obtain seven 5× standard concentrations (500, 250, 150, 100, 50, 25, and 5 μM 5× stocks). Store them in multiple aliquots (e.g., 100 μl each) and indefinitely freeze in well-sealed tubes.

10× GTPase Assay Buffer: 100 mM Tris/HCl, 100 mM NaCl, 20 mM MgCl$_2$.6H$_2$O, 0.5% Tween 80, pH 7.4.

5× Dynamin Diluting Buffer: 30 mM Tris/HCl, 100 mM NaCl, 0.1% Tween 80, pH 7.4.

Guanosine 5′ Triphosphate (GTP) (Roche Diagnostics, Mannheim, Germany): Prepare a 100×GTP stock solution (30 mM, i.e., 17 mg in 1 ml of 100 mM Tris/HCl pH 9.0) and store aliquots at −20°. The pH of the GTP solution is generally acidic, but varies with each batch. Adjust the final pH to 7.0 by spotting microliter amounts onto pH indicator strips and adjust the solution with microliter amounts of 100 mM HCl. This is critical because dynamin I GTPase activity is pH dependent. GTP stock solutions should be made fresh every two months. A 10× working solution is prepared in 20 mM Tris/HCl pH 7.4 just before use. The final GTP concentration in the complete assay buffer will be 0.3 mM.

Leupeptin: Prepare a 10× stock solution at 10 μg/ml in water, use at final concentration of 1 μg/ml. Store frozen for up to 24 months.

PMSF: Prepare a 100 mM PMSF stock solution in ethanol and make a 10× PMSF working solution (1 mM) by diluting with water just before use.

L-α-phosphatidyl-L-serine (PS) (Sigma Aldrich) liposomes: PS from the manufacturer comes as 10 mg/ml chloroform:methanol 95:5 solution. Take 40 μl and evaporate dry under nitrogen flow, leaving a volume of about 5 μl. Resuspend in 1 ml of 30 mM Tris/HCl pH 7.4 and sonicate for 1–2 min on ice for a working solution of 400 μg/ml. Freeze/thaw the stock solution twice with vigorous vortexing on each thaw, as this improves liposome formation. It is stored at $-20°$ and can be frozen/thawed many times. The optimal PS concentration should be determined separately for every new purified batch of dynamin I or of PS stock, with a concentration curve from 1 μg/ml to 80 μg/ml (Fig. 2A, for example).

Stop Buffer: 0.5 M EDTA pH 8.0.

Inhibitor Compounds: Make 10× working solution of each potential dynamin inhibitor at the required concentrations. Ideally, prepare primary stock solutions at high concentration in dimethyl sulfoxide (DMSO), then make secondary 10× working stocks in water (if it is soluble) or in up to a maximum of 50% DMSO (preferably use 10% DMSO in 20 mM Tris pH 7.4). This will give a maximum of 5% DMSO in the final assay. Dynamin I can tolerate this level without adverse effects (but, surprisingly, dynamin II cannot tolerate more than 1% DMSO). Only use high-grade DMSO (e.g., Sigma Aldrich Cat no. D2650 - in 5 × 5 ml sealed ampules).

Procedure

The assay is performed in triplicate in round bottom, 96-well polystyrene plates, which are kept on ice while preparing the reactions. The final reaction volume in each well is 40 μl. Each reaction mix consists of a "complete assay buffer" containing: 4 μl of 10× GTPase assay buffer; 4 μl of 10× leupeptin; 4 μl 10× GTP; 4 μl of 10× PMSF; and 16 μl of water (this volume depends on presence of PS and/or an inhibitor, see following). Those samples to be stimulated with PS for dynamin activation have 4 μl of 10× optimal PS concentration (the amount is determined separately, see following) replacing 4 μl of the water. When screening inhibitor compounds dilute them to their appropriate 10× working stock concentrations

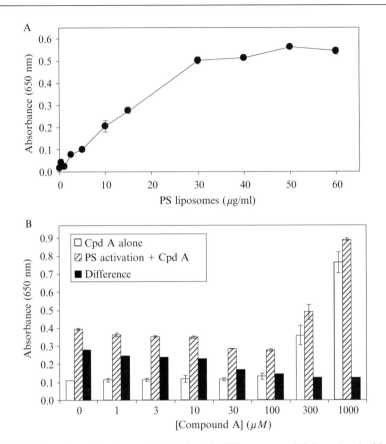

FIG. 2. Examples of the dynamin I colorimetric GTPase assay. (A) L-α-phosphatidyl-L-serine (PS) liposome concentration curve. Stimulation of dynamin I GTPase activity was measured as a function of PS concentration using the colorimetric assay. The curve reaches a plateau at about 30 μg/ml PS in this experiment, but varies between PS preparations. The curve is representative of the level of optimal dynamin activity in the absorbance range 0.4–0.6 OD units. (B) Concentration-dependent effect of compound A on dynamin I GTPase activity stimulated by 40 μg/ml PS. The effect of the drug alone at increasing concentrations is shown in the open bars. The stimulation by PS + Compound A is shown in the hatched bars. Compound A reduces PS-stimulated dynamin I GTPase activity (hatched bars), until at high concentrations there appears to be stimulation. The solid bars show a subtraction of the other values, which, after eliminating the effect of background phosphate, reveals the full inhibitory curve of compound A.

(described previously) then add 4 μl of stocks to the appropriate wells, also replacing 4 μl of the water. The total volume for each reaction at this point (without enzyme) should be 32 μl. The phosphate standards are also prepared in triplicate from the seven 5× stocks (described above) by addition of 8 μl of each 5× phosphate standard to wells containing the complete assay buffer (including PS, but excluding dynamin), bringing their final volume to 40 μl.

Preincubate the plate(s) for 10 min at 30°. Ideally, heating and shaking 96-well plates is performed on an Eppendorf Thermomixer with a microplate attachment (Eppendorf, Hamburg, Germany) at a mixing rate of 300 rpm. During the preincubation time, dilute the purified dynamin I in ice cold 5× dynamin diluting buffer, to 5× its optimal working dilution (determined separately, see following). Only dilute dynamin I when ready for use, as it loses activity when left on ice for extended time in its diluted state. Initiate the reaction with 8 μl of diluted dynamin I at 5 sec intervals to all appropriate wells containing the 32 μl of complete assay buffer. Use of an 8-channel multi-pipettor facilitates the process. Add an equivalent volume of dynamin diluting buffer to the control wells (see following for essential controls required in every experiment). Incubate the plate for 12 min at 30°. Terminate the reaction with 10 μl of stop buffer at 5 sec intervals to all reactions, including the phosphate standards. Move the samples to room temperature (22°) until all plates are completed. The samples are stable in the stop buffer for at least 2 h. Add 150 μl of malachite green reagent to all wells, mix, and allow color to develop for 5 min. The absorbance of the samples is read on a microplate spectrophotometer (Versamax Molecular Devices, Sunnyvale, CA) at a wavelength of 650 nm. The color is stable for up to 2 h, but if left overnight the samples precipitate. The malachite green assay detects phosphate from 1 to 200 μM, and higher concentrations are subject to the limits of the spectrophotometer (which is nonlinear above OD values of 2.5) (Fig. 3). The standard curve is linear in the working range of 1–100 μM (inset to Fig. 3).

The final amount of dynamin in each well is in the range 0.2 to 0.8 μg/well (normally 0.3–0.4 μg/well, or 70–100 nM dynamin I) and is determined empirically. Dynamin is diluted to cover this range and a concentration producing a maximum activity of 0.6 OD units is selected as a compromise between assay sensitivity and linearity, and the need to conserve dynamin stocks. The PS concentration curve is performed (Fig. 2A) and a concentration producing 80–90% of maximum activity is selected. This produces an absorbance reading of 0.4–0.6 at 650 nm. The amounts of dynamin and of PS are determined separately.

Essential Controls

There are essential control conditions for the assay to eliminate contaminating background phosphate. A sample with the complete assay buffer components (but no dynamin I) is required to account for spontaneous

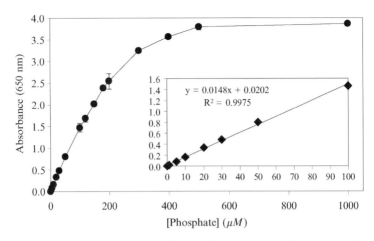

FIG. 3. Phosphate standard curve for the malachite green assay. The assay is linear over the range 1–200 μM phosphate. Maximum sensitivity is limited by the optical density units on the spectrophotometer, which declines above 2.5 units. The inset shows the standard curve for a typical GTPase assay and is linear in the assay sensitivity range. All values are n = 3 and standard deviations are shown, but are normally smaller than the symbols.

GTP hydrolysis. This is used as the blank on the spectrophotometer. The most significant controls are those that test for the presence of background phosphate in samples or test compounds. In a concentration curve, such contamination can produce major assay interference (e.g., Fig. 2B). There are two sets of controls: single samples and concentration curves. For single sample controls, include samples with or without dynamin, PS, and drug vehicle. For concentration curve controls, the test drug is diluted into complete assay buffer (without dynamin) at every concentration used in the stimulated samples (with PS and dynamin present). Each of these values should be subtracted from the final stimulated GTPase activity to determine the net activity. For example, Fig. 2B shows that compound A alone has a significant amount of contaminating phosphate detected by the assay, which can be removed by subtraction.

Acknowledgments

We thank Timothy Hill and Adam McCluskey (The University of Newcastle, Australia) for advice and assistance in developing the malachite green colorimetric assay. We thank Victor Anggono (Children's Medical Research Institute) for Fig. 1A and Ngoc Chau (Children's Medical Research Institute) for her technical assistance. We also thank Byeong Doo Song and Sandra Schmid (The Scripps Research Institute, La Jolla, CA) for initial discussions on the colorimetric assay.

References

Barylko, B., Binns, D. D., and Albanesi, J. P. (2001). Activation of dynamin GTPase activity by phosphoinositides and SH3 domain-containing proteins. *Methods Enzymol.* **329**, 486–496.

Buss, J. E., and Stull, J. T. (1983). Measurement of chemical phosphate in proteins. *Methods Enzymol.* **99**, 7–14.

Cogan, E. B., Birrell, G. B., and Griffith, O. H. (1999). A robotics-based automated assay for inorganic and organic phosphates. *Anal. Biochem.* **271**, 29–35.

Cousin, M. A., and Robinson, P. J. (2001). The dephosphins: Dephosphorylation by calcineurin triggers synaptic vesicle endocytosis. *Trends Neurosci.* **24**, 659–665.

Geladopoulos, T. P., Sotiroudis, T. G., and Evangelopoulos, A. E. (1991). A malachite green colorimetric assay for protein phosphatase activity. *Anal. Biochem.* **192**, 112–116.

Hill, T. A., Odell, L. R., Quan, A., Abagyan, R., Ferguson, G., Robinson, P. J., and McCluskey, A. (2004). Long chain amines and long chain ammonium salts as novel inhibitors of dynamin GTPase activity. *Bioorg. Med. Chem. Lett.* **14**, 3275–3278.

Hohenwallner, W., and Wimmer, E. (1973). The Malachite green micromethod for the determination of inorganic phosphate. *Clin. Chim. Acta* **45**, 169–175.

Koenig, J. H., and Ikeda, K. (1989). Disappearance and reformation of synaptic vesicle membrane upon transmitter release observed under reversible blockage of membrane retrieval. *J. Neurosci.* **9**, 3844–3860.

Lee, S. Y., Wenk, M. R., Kim, Y., Nairn, A. C., and De Camilli, P. (2004). Regulation of synaptojanin 1 by cyclin-dependent kinase 5 at synapses. *Proc. Natl. Acad. Sci. USA* **101**, 546–551.

Lin, H. C., Barylko, B., Achiriloaie, M., and Albanesi, J. P. (1997). Phosphatidylinositol (4,5)-bisphosphate-dependent activation of dynamins I and II lacking the proline/arginine-rich domains. *J. Biol. Chem.* **272**, 25999–26004.

Liu, J. P., Powell, K. A., Südhof, T. C., and Robinson, P. J. (1994). Dynamin I is a Ca^{2+}-sensitive phospholipid-binding protein with very high affinity for protein kinase C. *J. Biol. Chem.* **269**, 21043–21050.

Mahuren, J. D., Coburn, S. P., Slominski, A., and Wortsman, J. (2001). Microassay of phosphate provides a general method for measuring the activity of phosphatases using physiological, nonchromogenic substrates such as lysophosphatidic acid. *Anal. Biochem.* **298**, 241–245.

Marks, B., Stowell, M. H., Vallis, Y., Mills, I. G., Gibson, A., Hopkins, C. R., and McMahon, H. T. (2001). GTPase activity of dynamin and resulting conformation change are essential for endocytosis. *Nature* **410**, 231–235.

Song, B. D., Yarar, D., and Schmid, S. L. (2004). An assembly-incompetent mutant establishes a requirement for dynamin self-assembly in clathrin-mediated endocytosis *in vivo. Mol. Biol. Cell* **15**, 2243–2252.

Tuma, P. L., Stachniak, M. C., and Collins, C. A. (1993). Activation of dynamin GTPase by acidic phospholipids and endogenous rat brain vesicles. *J. Biol. Chem.* **268**, 17240–17246.

Vallis, Y., Wigge, P., Marks, B., Evans, P. R., and McMahon, H. T. (1999). Importance of the pleckstrin homology domain of dynamin in clathrin-mediated endocytosis. *Curr. Biol.* **9**, 257–260.

Van Veldhoven, P. P., and Mannaerts, G. P. (1987). Inorganic and organic phosphate measurements in the nanomolar range. *Anal. Biochem.* **161**, 45–48.

Wigge, P., Vallis, Y., and McMahon, H. T. (1997). Inhibition of receptor-mediated endocytosis by the amphiphysin SH3 domain. *Curr. Biol.* **7**, 554–560.

[50] Assays and Functional Properties of Auxilin-Dynamin Interactions

By Sanja Sever, Jesse Skoch,
Brian J. Bacskai, and Sherri L. Newmyer

Abstract

The large GTPase dynamin is required for budding of clathrin-coated vesicles from the plasma membrane, but its mechanism of action is still not understood. Growing evidence indicates that the GTP-bound form of dynamin recruits downstream partners that execute the fission reaction. Recently, we reported nucleotide-dependent interactions between dynamin and auxilin, which suggested that auxilin cooperates with dynamin during vesicle formation. Here we describe three different *in vitro* assays that monitor auxilin-dynamin interactions, as well as fluorescence lifetime imaging microscopy that identify direct interactions between dynamin and auxilin in cells.

Introduction

During clathrin-mediated endocytosis, the GTPase dynamin is required for the formation of clathrin coated vesicles (Conner and Schmid, 2003; Schmid *et al.*, 1998; Sever, 2002; Sever *et al.*, 2000a) and the cochaperone auxilin cooperates with hsc70 to remove the clathrin coat after vesicle budding (Greener *et al.*, 2000; Schlossman *et al.*, 1984; Umeda *et al.*, 2000; Ungewickell *et al.*, 1995). There are currently two models for dynamin's role in endocytosis. Based on a "classical model," dynamin functions as a mechanochemical enzyme (Praefcke and McMahon, 2004). It uses conformational change upon GTP hydrolysis of self-assembled protein to sever clathrin-coated vesicles. Alternatively, dynamin, like other GTPases, acts as a "molecular switch" that activates downstream effectors (Sever *et al.*, 1999, 2000a,b). However, until recently no downstream effectors for dynamin have been identified. We have reported interactions between dynamin and chaperone machinery hsc70 and auxilin (Newmyer *et al.*, 2003). The interactions are direct and nucleotide-dependent, identifying hsc70 and auxilin as proteins that specifically interact with dynamin in its GTP-bound form. The only other protein that specifically binds dynamin in its GTP-bound form is dynamin itself (Carr and Hinshaw, 1997; Newmyer *et al.*, 2003; Sever *et al.*, 1999). Dynamin-dynamin interactions

METHODS IN ENZYMOLOGY, VOL. 404
Copyright 2005, Elsevier Inc. All rights reserved.
0076-6879/05 $35.00
DOI: 10.1016/S0076-6879(05)04050-4

are promoted by a C-terminally located GTPase effector domain (GED) that also functions as dynamin's intermolecular GTPase activating protein (GAP). Thus, dynamin self-assembly involves GTP-dependent GAP-GAP interactions (Newmyer *et al.*, 2003; Sever *et al.*, 1999), which result in rapid assembly-stimulated GTPase activity. In contrast to dynamin-dynamin interactions, dynamin-auxilin interactions lead to potent inhibition of assembly-stimulated GTP-hydrolysis. Since this inhibition exhibits the same degree of cooperativity as determined for dynamin-dynamin interactions (Tuma, 1994; Warnock *et al.*, 1996), it has been suggested that auxilin may bind GAP and compete with GAP-GAP interactions that promote self-assembly (Newmyer *et al.*, 2003). In order to expand our original findings using biochemical assays, we next examined whether dynamin and auxilin also interact in live cells by looking at fluorescence resonance energy transfer (FRET) using fluorescence lifetime imaging microscopy (FLIM) (Sever *et al.*, 2005).

Here we describe four assays that monitor auxilin-dynamin interactions. Included are a nucleotide-dependent dynamin affinity column, a coprecipitation assay that measures auxilin-dynamin nucleotide-dependent direct binding, a GTPase assay that examines auxilin's ability to inhibit the stimulated rate of GTP hydrolysis, and FRET with FLIM analysis of dynamin-auxilin interaction in cells.

Materials and Reagents

Expression and Purification of His_6-tagged Auxilin-1

Auxilin harboring an N-terminal His_6 tag is expressed in High Five insect cells. The His_6 tag is incorporated via PCR amplification of a 5' auxilin fragment from GST-auxilin (Ungewickell *et al.*, 1995) using the following pair of primers: forward His_6 primer: GAATTCCACCATGCACCA CCACCACCACCACAGCGGCGAAGCCATGGACAGCT CAGGTG CC reverse primer: CTGGAATCCAAAAAGCTTCG.

The PCR product is subcloned into pADtet7-HA-auxilin (described in the following sections) via *Hind*III and *Eco*RI restriction sites. Subsequently, the entire coding sequence is subcloned into pVL1393 utilizing the flanking *Eco*RI and *Not*I restriction sites. The resulting pVL-His_6- auxilin construct is recombined with wild-type Baculovirus and transfected into Sf9 insect cells, maintained in Grace's insect medium (GIBCO-BRL) and 10% fetal bovine serum (FBS), with the Baculogold Transfection Kit (BD Biosciences, San Jose, CA). Sf9 cells are subsequently infected with the recombined viral particles and plaque purification is used to isolate a viral population derived from a single virus. The viral stock is amplified by stepwise infection of Sf9 cells to a high titer virus stock of $\geq 1 \times 10^8$ plaque-forming units (pfu)/ml. For large-scale production of His_6-auxilin, the high titer virus stock is then used to infect 1 L of High Five cells grown to 2×10^6 cells/ml in suspension at 28°, with 5 pfu of virus per cell (usually 50 ml). The cells are harvested by centrifugation (20 min at 1000g) 48–56 h postinfection. The cell pellet from 1 l of cell suspension is resuspended in 100 ml of ice-cooled 1XPBS, and equally divided between two 50-ml Falcon tubes to ensure rapid freezing. Cells are frozen by submerging the tubes into liquid nitrogen, and stored at −80°. Cell pellets are quickly thawed by swirling in a 37° water bath in 10 ml lysis buffer (50 mM HEPES, 300 mM NaCl, 1 mM PMSF, 20 mM Imidazole, pH 7.5), containing one tablet of protease inhibitors (Roche Diagnostics, Manneheim, Germany). The cells are lysed by sonication for 1.5 min (6× 15 sec burst, 20 sec cooling). The cells are supplemented with Triton X-100 to 0.5% and the solution is kept on ice for 15 min with occasional mixing. The extract is clarified by centrifugation at 33,000g for 20 min, and mixed with 1 ml Ni-NTA resin (50% slurry in 1XPBS, Qiagen GmbH, Hilden, Germany) for 2 h at 4° with rotation. The affinity resin is washed twice in batch with the lysis buffer containing 0.5% Triton X-100, transferred to a 10-ml column, and further washed twice with 2 ml of the same buffer and 3 times with 2 ml of the same buffer in which concentration of Imidazole is increased to 50 mM. Auxilin is eluted with three consecutive incubations of 0.3 ml of elution buffer (50 mM HEPES, 200 mM Imidazole, pH 7.5). Elution samples are pooled together and the protein is dialyzed overnight into HCB150 (20 mM HEPES, 1 mM $MgCl_2$, 1 mM EGTA, pH 7.0, 150 mM NaCl) containing 0.1 mM DTT. This procedure results in 4–8 mg of auxilin bound to 1 ml Ni-NTA resin. Auxilin purity is shown in Fig. 2A. Protein is divided into 40-μl aliquots and stored at −80° after freezing in liquid nitrogen. For GST-coprecipitation experiments, auxilin concentration is adjusted to 1 mg/ml with dialysis buffer.

Expression and Purification of GST-dynamin

Human dynamin-1 is expressed in insect cells as a glutathione S-transferase (GST) fusion protein using a Bac-To-Bac baculovirus expression system (GIBCO-BRL, Invitrogen, Mannheim, Germany) and purified by affinity chromatography on glutathione-Sepharose (Amersham Biotech) resin.

The GST domain from a pGEX-3X vector (Pharmacia, Piscataway, NJ) was digested with RsrII and BamHI to release GST and the fragment was cloned in frame within the original pFASTBac-dynamin^wt (Damke et al., 2001) digested with the same restriction enzymes. The transfer vector containing the GST-dynamin-1 cDNA is transformed into DH10Bac strain of *Escherichia coli* (Invitrogen, Grand Island, NY). The transposition reaction and extraction of the viral DNA (bacmid) are performed based on the manufacturer's instruction (GIBCO-BRL). Bacmid DNA is transfected into Sf9 cells using CellFECTIN reagent (Life Technologies) according to manufacturer's protocol. Seventy hours after transfection, the virus-containing medium is harvested and cells lysed in SDS-PAGE sample buffer are analyzed for expression of dynamin by Western blotting using polyclonal anti-dynamin antibodies raised against GAP domain (Sever *et al.*, 1999). The viral stock is amplified by stepwise infection of Sf9 cells maintained in Grace's insect medium and 10% FBS. The viral stock is amplified by exponentially growing Sf9 cells to a high titer virus stock of $\geq 1 \times 10^8$ plaque-forming units (pfu)/ml. For large-scale production of GST-dynamin, the high titer virus stock is then used to infect 1 l of High Five cells grown to 2 x 10^6 cells/ml in suspension at 28°, with 5 pfu of virus per cell (usually 50 ml). To generate affinity columns, 1 l of insect High Five cells are infected with viruses expressing GST-dynamin and dynamin at 2:1 ratio (typically 60 ml GST-dynamin virus and 30 ml of dynamin virus). The cells are harvested by centrifugation (10 min at 1000g) 48 h post infection while allowing no more then 20% of the cells to lyse (as determined by Trypan-blue exclusion) to minimize proteolysis and release of recombinant protein in the media. The cell pellet from 1 l of cell suspension is resuspended in 100 ml of ice cold 1xPBS, and equally divided and pelleted within two 50-ml Falcon tubes to ensure rapid freezing. Cell pellets stored at −80° are quickly thawed by swirling in a 37° water bath in 20 ml HCB100 buffer (20 mM HEPES, 1 mM MgCl$_2$, 1 mM EGTA, pH 7.0, 100 mM NaCl), containing one tablet of protease inhibitors (Roche). The cells are lysed by sonication, 1.5 min (6 × 15 sec burst, 20 sec cooling). The cells are supplemented with Triton X-100 to 0.5% and the solution is kept on ice for 15 min with occasional mixing. The extract is clarified by centrifugation at 33,000g for 20 min. For dynamin affinity columns, all of the extract (20 ml) is incubated with 1 ml Glutathione-Sepharose 4Fast Flow beads (2 ml 50% slurry, Amersham Biotech) at 4° for 1 h. The affinity resin is washed 6 times in batch with 10 ml of HCB100 buffer. This procedure results in 5 mg of GST-dynamin/dynamin bound to 0.5 ml of Glutathione-Sepharose 4B beads.

For pull-down experiments, the extract is stored at this point by dividing it into smaller aliquots (typically 1 ml aliquots), freezing them quickly by submerging the tubes into liquid nitrogen, and storing at −80°. One ml of the extract is thawed quickly by swirling in a 37° water bath, placing it on ice, and mixing it with 0.1 ml Glutathione-Sepharose 4Fast Flow beads (Amersham Biotech) at 4° for 1 h. The affinity resin is washed 6 times in batch with 1 ml of HCB100 buffer. This procedure results in 0.1 mg of GST-dynamin/dynamin bound to 0.1 ml of Glutathione-Sepharose 4B beads.

Expression of Auxilin-1 Using Adenoviral Expression Vector

For adenovirus expression of bovine auxilin-1, the gene was subcloned as two fragments into pADtet7vector (Altschuler *et al.*, 1998; Hardy *et al.*, 1997). Digestion of GST-auxilin (Holstein *et al.*, 1996) and pADtet7 with *Hind*III/*Not*I restriction enzymes enables the insertion of the 3' auxilin fragment into the multiple cloning site. Subsequently a HA tag is added to the 5-coding region of auxilin through PCR amplification with the following primers: forward HA tag primer: GAATTCCACCATGGA GTATGATGTTCCTGATTATGCTCATATGGACAGCTCAGGTGCC reverse primer: CTGGAATCCAAAAAGCTTCG.

The region encoding the HA-tagged N terminus is subsequently subcloned into the pADtet7-auxilin-C terminal construct via *Eco*RI/*Hind*III restriction sites, generating pADtet7-HA-auxilin. The generation of recombinant adenovirus carrying the auxilin gene is as described previously (Altschuler *et al.*, 1998; Hardy *et al.*, 1997). Within this expression system, the gene is introduced under the control of a tetracycline-repressible promoter. In order to express auxilin using this system, cells also need to express a chimeric tetracycline-regulatable transcription activator (tTA) that can be introduced into the cell either by co-infection with adenovirus constructed with tTA (see following) or by employing stable cell lines that express tTA (Altschuler *et al.*, 1998; Hardy *et al.*, 1997).

Methods

Nucleotide-Specific Dynamin Affinity Columns

If dynamin acts as a classical regulatory GTPase (Sever *et al.*, 1999, 2000a), it should be possible to identify effectors that interact with the GTP-bound form of dynamin. Using nucleotide-specific dynamin affinity columns, we have identified two proteins from rat brain cytosol, hsc70 and auxilin, that specifically interact with dynamin:GTP (Newmyer *et al.*, 2003). We would like to point out that we were concerned that tagging dynamin on its N-terminus with the relatively large GST-moiety may impair its interactions with putative effector proteins such as auxilin. To overcome this problem, we have generated dynamin affinity columns by co-infecting cells with baculoviruses carrying the nontagged and GST-tagged dynamin. Since dynamin is a homo-tetramer (Muhlberg *et al.*, 1997), or a dimer (Zhang and Hinshaw, 2001), such procedure resulted in generation of hetero-tetramers/dimers consisting of tagged and nontagged protein, in which GST-tagged dynamin monomers are facilitating binding to the Glutathione-Sepharose 4B beads (Figs. 1 and 2). Expression of dynamin-1 in baculovirus-infected insect cells has been described in detail (Damke

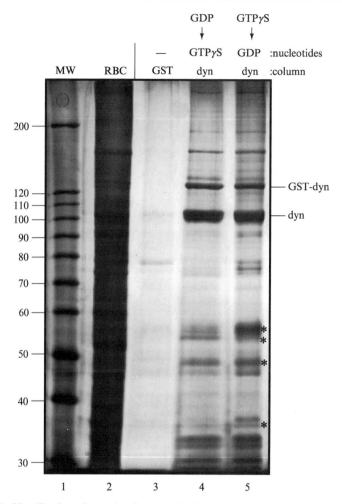

FIG. 1. Identification of proteins from rat brain cytosol that interact specifically with dynamin. Silver staining of the input rat brain cytosol (RBC, lane 2), the material eluted from immobilized GST (lane 3), material eluted from immobilized GDP-dynamin with GTPγS (lane 4), and material eluted from immobilized GTPγS -dynamin with GDP (lane 5). Proteins (lanes 3–5) were eluted in buffer containing 250 mM NaCl as described in the text. Eluted proteins specific to GST-dynamin:GTPγS (lane 5) are labeled with an asterisk (*). Reprinted from Newmyer *et al.*, 2003, with permission from Elsevier.

et al., 2001), and we closely followed the same procedure in regard to expression of GST-dynamin, as described previously.

GST-dynamin/dynamin bound to the glutathione S-transferase beads (approximately 5 mg of dynamin bound to 1 ml 50% slurry) is incubated in

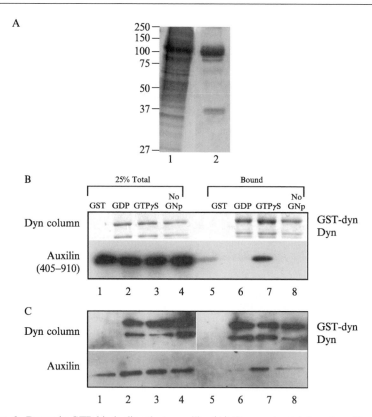

FIG. 2. Dynamin-GTP binds directly to auxilin. (A) Coomassie staining of auxilin from insect cell lysate (lane 1) and after purification as described in text (lane 2). (B and C) GST-dynamin attached to Glutathione-Sepharose was used as bait and incubated with 2 μM of recombinant aux[405–910] (B) or recombinant auxilin (C). All incubations are performed in HCB100 buffer with the indicated nucleotide for 1 h at 4°. 25% of incubation mixture is loaded as a total protein standard (lanes 1–4), and remaining beads are extensively washed and bound protein eluted with boiling (lanes 5–6). 0.5 mM guanine nucleotide, as indicated in the figure, is present at all times. The samples are analyzed by SDS-PAGE and immunoblotting, with the exception of dynamin in panel B, which is stained using Coomassie dye. Note that dynamin bound to Glutathione-Sepharose is a heterotetramer consisting of dynamin and GST-dynamin.

modified HCB100 supplemented with either 0.5 mM GDP or GTPγS for 10 min at room temperature to allow binding of the nucleotides to dynamin. GST bound to the glutathione S-transferase beads is used as a control (5 mg of GST bound to 0.5 ml 50% slurry). Subsequently, beads are incubated for 90 min at 4° with 0.5 ml of RBC (10 mg/ml) obtained as follows: 12 rat brains were homogenized in 25 ml of buffer containing

20 mM HEPES pH 7.5, 1 mM MgCl$_2$, 1 mM DTT, 10 μM pepstatin, 5 μM aprotinin, 1 μg/ml TLCK, 10 μM leupeptin, 1 mM phenylmethylsulfonyl fluoride (PMSF), and calpain inhibitor at 0.5 μg/ml (Calbiochem, La Jolla, CA), using a dounce homogenizer and 10 passes of each of the A and B pestles. The homogenate is centrifuged at 7000g for 15 min. After addition of 1% Triton X100, the supernatant is rocked at 4° for 1 h, followed by a second centrifugation in a Ti45 rotor (Beckman Instruments, Fullerton, CA) at 100,000g for 1 h. Endogenous nucleotides are removed by passing cytosol through PD-10 columns (Amersham Biosciences). GDP or GTPγS are added (0.5 mM), and the extract batch-bound to the affinity resin in HCB150 buffer. After 60 min incubation at 4°, the mixture is poured into a column and washed with 10 volumes of modified HCB100 buffer that contains 0.1% Triton X, 0.5 mM DTT, and 5 mM MgCl$_2$, followed by 10 volumes of modified HCB250 (with 250 mM NaCl), with all washes containing the original nucleotide. Bound proteins are eluted with 1.5 column volume (375 μl) of modified HCB250 buffer incubated with beads for 10 min at 4°, with the nucleotides present in the buffer being swapped. In case of GST affinity chromatography, there is no nucleotide in the buffers. Eluted proteins are precipitated using 1/4 volume of 100% trichloroacetic acid (TCA) (94 μl). Samples are mixed well and left on ice for 30 min. Subsequently, precipitated proteins are collected by centrifugation at 12,000g for 15 min at 4°. With a finely drawn out pasture pipette all of the supernatant is removed, and the pellet is washed with ice-cold acetone (750 μl). Do not vortex, but mix by gently flicking the tube. Spin samples for 10 min at 4°, and once again remove the supernatant using a finely drawn out pasture pipette. Remaining acetone is allowed to evaporate by leaving tubes open at room temperature overnight. Samples are resuspended in 20 μl 1× sample buffer and analyzed by SDS-PAGE. Bound proteins are detected by silver staining using standard protocol. 5 μl of sample is used to detect auxilin using Western blotting and polyclonal anti-auxilin antibodies.

Nucleotide-Specific Pull-Down Assays

Glutathione S-transferase (GST) coprecipitation experiments using recombinant proteins, as described in detail below, demonstrated that dynamin binds both hsc70 and auxilin directly. Furthermore, interactions between dynamin-GTP and auxilin do not require the tensin homology domain, since auxilin fragment, aux[405–910], that lacks this domain but still contains clathrin binding and DNA J-domains exhibits the same binding properties as a full-length protein (Fig. 2B and C). The cDNA construct coding for His$_6$ fusion proteins of aux[405–910], pQE30-Aux54, and purification of the expressed fragment have been described previously (Greene and Eisenberg, 1990).

GST-dynamin/dynamin bound to the glutathione S-transferase beads (5–10 μg of dynamin bound to 10 μl 50% slurry in HCB100 buffer, 0.5 mM DTT) is incubated with recombinant auxilin-1 or aux$^{405–910}$ (3–5 μg of protein in 10 μl HCB150 buffer) in 0.5 ml Eppendorf tubes. This is followed by addition of 1 μl of 20 mM GNP. GTPγS and GDP are acquired from Sigma Chemical Co (St. Louis, MO) and stored for up to 4 months at -80° as a 20 mM stock solution in 10 mM Tris, pH 7.4, 50 mM MgCl$_2$. Finally, 20 μl of modified HCB100 containing 5 mM MgCl$_2$ and 0.1% Triton X is added to the samples. Samples are mixed by gentle flicking and incubated on ice for 1 h. 10 μl of the sample is removed, 5 μl of 5X sample buffer is added and analyzed by SDS-PAGE to control for protein concentrations in the assay. Beads are collected by centrifugation, washed 4 times with 100 μl of the same buffer with respective GNP, then resuspended in 20 μl 1 $\times$ sample buffer and analyzed by SDS-PAGE. Bound proteins are detected by Western blotting using polyclonal anti-auxilin antibodies (Fig. 2B and C). To control for the amount of dynamin bound to Glutathione-Sepharose 4B beads, the membrane is stripped and reprobed with anti dynamin hudy 1 antibodies (Fig. 2C). When binding of auxilin fragment, aux$^{405–910}$, is assayed, it is possible to first cut the SDS-PAGE so that the proteins with molecular weight $\geq$90 kD are stained using Coomassie dye (Fig. 2B, dynamin column), while proteins $\leq$90 kD are transferred and detected using Western blotting (Fig. 2B, auxilin panel).

Auxilin-Inhibited GAP-stimulated GTPase Activity

Dynamin self-assembly stimulates its basal GTPase activity 10–100-fold *in vitro* (Barylko *et al.*, 2001; Stowell *et al.*, 1999; Tuma and Collins, 1994; Warnock *et al.*, 1996). This is due to activation of dynamin's intramolecular GTPase activating protein (GAP, amino acids 618–752 of dynamin), which becomes activated upon dynamin self-assembly (Muhlberg *et al.*, 1997; Sever *et al.*, 1999). Most assays that measure assembly-stimulated GTP hydrolysis by dynamin use templates such as lipids to promote dynamin self-assembly (Barylko *et al.*, 2001; Song *et al.*, 2004; Stowell *et al.*, 1999). In contrast, we have developed a soluble assay that reconstitutes assembly-stimulated GTP hydrolysis by simply adding recombinant isolated GAP domain to unassembled dynamin (Sever *et al.*, 1999). Thus, recombinant GAP domain, on its own, stimulates GTP hydrolysis by dynamin approximately 100-fold (Sever *et al.*, 1999). Pre-incubation of dynamin with auxilin prior to addition of the recombinant GAP domain abolishes GAP's ability to stimulate GTP hydrolysis. The GAP-stimulated assay that we routinely use is described.

Dynamin's intrinsic GTPase activity is directly related to its degree of self-assembly and the assembly of dynamin is dependent on dynamin concentration and ionic strength. Thus, assays for GAP-stimulated GTPase activity is performed at 50 mM NaCl, which results in maximal stimulation of dynamin, in our hands. Assays are performed in 1.5 ml microcentrifuge tubes. [α-^{32}P] (specific activity 400 Ci/mmol; 10 mCi/ml) is purchased from Amersham Pharmacia Biotech (Arlington Heights, IL). GTP is acquired from Sigma Chemical Co (St. Louis, MO) and stored for up to 4 months at $-80°$ as a 50 mM stock solution in 10 mM Tris, pH 7.4, 50 mM MgCl$_2$. For assays 3.2 μl of this stock is diluted to 40 μl in water and 2 μl of [α-^{32}P]GTP is added to generate 4 mM assay stock of [α-^{32}P]GTP. The total volume of a typical assay is 20 μl, which is sufficient to provide samples for TLC. Initially, HCB150 is added to each microcentrifuge tube on ice. The volume of HCB150 in a control sample that does not contain dynamin or effectors is 7 μl. In the assays, this volume is adjusted so that the final volume of dynamin, auxilin, and buffer is 7 μl. In successive order dynamin (0.2 μg in 1 μl, 0.1 μM final concentration), GAP (1 μl in 1 μg, 3 μM final concentration), and auxilin (1–20 μg in up to 5 μl, 0.5–10 μM final concentration) are added. Note that all proteins are diluted in HCB150. Finally, 7 μl of H$_2$O is added so that the concentration of the NaCl at this stage of the assay is 75 mM. Control assays contain only dynamin (basal rate), or dynamin with GAP (maximal stimulated rate). The reaction is gently mixed and incubated on ice for at least 30 min to allow sufficient time for protein-protein binding. Finally, DTT (1 μl of 1 mM DTT), BSA (1.5 μl of 10% BSA) and GTP (2.5 μl of the 4 mM [α-^{32}P]GTP assay stock) is added to each tube. An aliquot (typically 2 μl) is spotted onto a TLC strip as the "zero-minute" time point. The tubes are then incubated in a 37° water bath. Aliquots from the reaction mixtures are removed and spotted onto the TLC place at appropriate time intervals (typically 3 min). Thin-layer chromatography (TLC) is performed on flexible cellulose polyethylene-imine TLC plates (5 × 20 cm, Selecto Scientific, Norcross, GA). The TLC plates are predeveloped within distilled water for 20 min, air-dried overnight, and stored at 4° until needed. Pencil marks are placed at 0.5 cm intervals along a line that is 1 cm from the bottom of the plate to mark the location that will be spotted at the end of each time point. Typically 2 μl is spotted per time interval. When the incubation is complete, the TLC plates are dried with a hair dryer and then placed in developing solution (a 1:1 mixture of 1 M LiCl and 2 M formic acid) where the two solutions are mixed immediately before use. The plates are allowed to develop, then removed from the liquid and dried with a hair dryer. The plates are wrapped in plastic wrap and placed in a PhosphorImager cassette for approximately 1 h. The results are then imaged and quantified using

PhosphorImager image analysis software (Molecular Dynamics, Amersham Biosciences Corp., Piscataway, NJ). Rates of GTP hydrolysis are calculated from a minimum of five time points and expressed as the percentage of GDP/GTP+GDP after subtracting the "zero-minute" background (% hydrolyzed GTP in Fig. 3A). Rate for each auxilin concentration is plotted against the different protein concentrations (Fig. 3B) to obtain the Michaelis-Menten constant Km for auxilin. We prefer to put all the rates from different experiments on the same graph instead of calculating and plotting an average rate for the given auxilin concentration.

Fluorescence Lifetime Imaging Microscopy of Auxilin-Dynamin Interactions

Our previous approaches to detect endogenous complexes of dynamin and auxilin using co-immunoprecipitation approaches were unsuccessful, so we turned to fluorescence lifetime imaging microscopy (FLIM). While fluorescence microscopy provides two- or three-dimensional information about fluorophore concentration, FLIM can reveal spatial differences in fluorophore population lifetimes that are independent of concentration. Besides being useful in fluorophore identification, which transcends issues of spectral overlap, FLIM inherently observes lifetime truncations on a pixel by pixel basis that are induced by fluorescence resonance energy

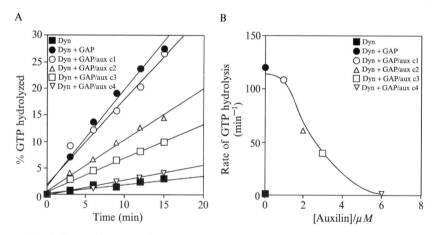

Fig. 3. Recombinant auxilin inhibits GAP-stimulated GTP hydrolysis by dynamin. (A) Time course of GTP hydrolysis by 0.1 μM dynamin (■) in the presence of 3 μM GAP (●) and increasing concentrations of auxilin; 1 μM (○), 2 μM (△), 3 μM (□) and 6 μM (∇) of auxilin. (B) The GTPase activity determined under (A) is plotted against above stated concentrations of auxilin.

transfer (FRET). The fluorescence lifetime of a high energy donor fluorophore may be influenced by its surrounding micro-environment, and is shortened in the immediate vicinity of a lower-energy FRET acceptor fluorophore. Nonradiative energy transfer occurs between a donor and acceptor if these are within the Förster distance (R_0) of roughly 100 Å (10 nm) and results in a shortened fluorescence lifetime of the donor. The degree of lifetime shortening is inversely proportional to the sixth power of the distance between the fluorophores, and it can be displayed with subcellular spatial resolution in a pseudo-colored lifetime image. Thus, in a controlled molecular imaging experiment, detection of shortened lifetimes demonstrates FRET and indicates spatial proximity of the two labeled molecules.

Auxilin overexpression has been found to aberrantly aggregate clathrin and inhibit transferrin internalization (Zhao et al., 2001). Thus, in order to perform FLIM experiments we first sought a cell type where adenoviral expression of wild type auxilin-1 (aux^{wt}) has no negative effect on endocytosis. As shown in Fig. 4, overexpression of aux^{WT} does not inhibit endocytosis of rhodamine-transferrin in mouse epithelial IMCD (inner medullary collecting duct) cells stably expressing the human vasopressin receptor type II, IMCD-V$_2$R (Sever et al., 2005). In addition, aux^{wt} causes no clathrin accumulation into cytosolic granules (Fig. 4A). A detailed protocol follows.

To adenovirally express auxilin, IMCD-V$_2$R cells are grown on coverslips in 6 well dishes (three 12-mm coverslips are placed in one well) to 70% confluency in DMEM media (Cellgro) supplemented with 4 mM glutamine (Gibco), 10% FBS, and 1 mg/ml Geneticin (G418, Gibco). Cells are washed twice with 1 × PBS and infected with 1.2 ml of serum-free DMEM media containing 100 μl transactivator virus, 50 μl virus expressing dynamin-1, and 50 μl virus expressing auxilin-1. After two hours of infection at 37°, media-containing virus is replaced with full DMEM. 18 h postinfection, V2R internalization is induced within the adenovirus-treated IMCD-FlagV$_2$R cells to be able to follow auxilin-dynamin interactions during receptor-mediated endocytosis. The cells are washed twice with 1 × PBS and serum starved in DMEM media containing glutamine but lacking FBS and G418 for 1 h. Cells are washed twice with modified 1 × PBS, and vasopressin is added in the same modified PBS at final concentration of 1 μM (2 ml of media per one well of 6-well dish) for 20 min. Modified PBS is prepared by mixing 5 ml of buffer A (18 mM CaCl$_2$, 70 mM KCl, 18 mM MgCl$_2$, 2.74 mM NaCl) with 80 ml of H$_2$O and adding 5 ml of buffer B (70 mM NaH$_2$PO$_4$, 330 mM Na$_2$HPO$_4$). We chose 20 min for stimulation because the kinetics of V$_2$R internalization in LLC-PK cells established that during 20 min approximately 40% of receptor is

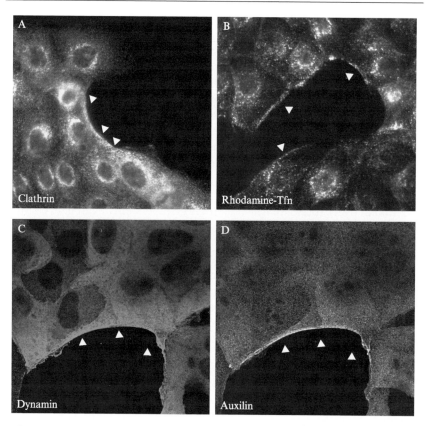

FIG. 4. Overexpression of auxilin does not inhibit Tfn endocytosis in IMCD cells
(A) Steady state distribution of clathrin and (B) rhodamine-Tfn internalization is not affected
in IMCD cells co-expressing auxilin-1 and dynamin-1. IMCD cells expressing Ha-tagged
aux^{WT} and dynamin-1 are grown on coverslips 18 h postinfection. Cells are incubated with
rhodamine-Tfn for 10 min at 37° before fixation, and stained with anti-clathrin light chain
antibody. Cells are analyzed by fluorescence microscopy. (C and D) Auxilin and dynamin
colocalize at the edge of cellular monolayer. IMCD cells expressing HA-tagged aux^{WT} and
dynamin-1 are grown on coverslips 18 h postinfection. After inducing Flag V2R endocytosis,
the cells are fixed, incubated with anti-dynamin (C) and anti-auxilin (D) antibodies, and
analyzed by fluorescence microscopy. Arrows point to the edge of the cellular monolayer that
supports active endocytosis marked by increased presence of clathrin, rhodamine-Tfn,
dynamin, and auxilin.

internalized (Bouley *et al.*, 2003), and thus endocytosis should be fully
active in internalizing the remaining 60% of the receptors still on the cell
surface. All above manipulations are done in 6-well dishes. After addition
of vasopressin, cells are washed briefly in 1 × PBS and fixed with 3%
formaldehyde in 1 × PBS containing 1 mM CaCl$_2$ and 1mM MgCl$_2$ for

30 min. After this point, coverslips are removed from the 6-well dish, and placed upside-down on subsequent drops of solutions starting with permeabilization solution (5% SDS in 1 × PBS) for 3 min. The cells are blocked using 5% normal goat serum and 5% normal donkey serum in 1 × PBS for 1 h. The cells are incubated with affinity purified anti-auxilin antibodies for 1 h at dilution of 1:200. All antibodies are diluted in DakoCytomation (Dako Cytomation, Carpinteria, CA). Coverslips are washed 3 times for 5 min with 1 × PBS, and incubated with secondary goat anti-rabbit antibodies conjugated to Alexa-488 (1:2000 dilution) (Jackson ImmunoResearch Laboratories, Inc., West Grove, PA). Coverslips are washed 3 times for 5 min with 1 × PBS and incubated with monoclonal anti-dynamin antibodies, hudy-1 (Stressgene) that have been diluted 1:200 in DakoCytomation. For control experiments, incubation with hudy 1 antibodies is replaced with incubation with donkey anti-goat antibodies conjugated to Alexa-568 (1:2000 dilution) for the positive interactions, or with goat anti-mouse antibodies conjugated to Alexa-568 (1:2000 dilution) for the lack of interactions. After 3 × 5 min wash with 1XPBS, coverslips are incubated with goat anti-mouse secondary antibodies conjugated to Alexa-568 (1:2000 dilution) (Jackson ImmunoResearch Laboratories, Inc.). After 3 × 5 min wash with 1 × PBS, coverslips are briefly washed with water and mounted using GVA mounting solution (Zymed, San Francisco, CA). Auxilin and dynamin staining is shown in Fig. 4C and FLIM experiments are performed using a femtosecond pulsed (80Mhz) Ti-sapphire laser (Tsunami; Spectra Physics, Mountain View, CA) mode-locked at 800 nm to induce 2-photon excitation of Alexa Fluor 488. Images are captured on commercial multi-photon microscope (Radiance 2000; BioRad, Hercules, CA). A high-speed microchannel plate detector (MCP R3809; Hamamatsu, Japan) and hardware/software package from Becker and Hickl (SPC830; Berlin, Germany) are used to measure the fluorescence lifetimes on a pixel-by-pixel basis via time-correlated single photon counting (TCSPC). Lifetimes from the cells imaged are curve fitted to two exponential decay curves with one exponential fixed at the average lifetime for Alexa488. The remaining variable exponential reveals the presence or absence of a shortened lifetime for each pixel within each image as well as the relative amplitude compared to the amplitude of the fixed lifetime component (Bacskai *et al.*, 2003). The lifetime (τ) of a fluorophore (Alexa488) can be represented by a multiexponential decay function where $t = a_1 e^{-t/\tau_1} + a_2 e^{-t/\tau_2} + a_3 e^{-t/\tau_3} + a_n e^{-t/\tau_n}$. The average lifetime or amplitude for a particular region of interest is calculated from the lifetime values from each pixel in that region or cell. As shown in Table I, in the absence of an acceptor fluorophore the lifetime of a488 is ∼2253 ± 64 ps. If an acceptor fluorophore (a568) is present but remains too distant from the donor, the

TABLE I

FLIM Analysis for Proximity Between Auxilin and Dynamin in IMCD Cells

Donor	Acceptor	Mean lifetime ($\tau1$ in picosec) (Mean ± SD)	Mean lifetime (p) compared to Aux (a488)
Aux (a488)	none	2253 ± 64	
Aux (a488)	GAM (a568)	2183 ± 86	p > 0.1
Aux (a488)	DAG (a568)	869 ± 119	p < 0.05
Aux (a488)	Dyn (a568)	1918 ± 231	p < 0.05

lifetime remains almost unchanged at $\sim$2183 ± 86 ps (Table I). In contrast, in a positive control, addition of antibodies against a488 that are conjugated to Alexa568 (DAGa568) resulted in a very short lifetime of 869 ± 119 ps. Importantly, cells stained with both anti-auxilin (a488) and anti-dynamin (a568) antibodies resulted in a statistically significant reduction of a488 fluroescence lifetime to 1918 ± 231 ps.

Goat anti-mice secondary antibodies conjugated to Alexa-568 (GAM (a568)) are used as a negative control and donkey anti-goat secondary antibodies conjugated to Alexa-568 (DAG (a568)) are used as a positive control. For statistical analysis of residuals, one-way ANOVA followed by a Tamhane post hoc test was performed with SPSS software. A p value of less than 0.05 constitutes significance.

References

Altschuler, Y., Barbas, S., Terlecky, L., Mostov, K., and Schmid, S. L. (1998). Common and distinct functions for dynamin-1 and dynamin-2 isoforms. *J. Cell Biol.* **143,** 1871–1881.

Bacskai, B. J., Skoch, J., Hickey, G. A., Allen, R., and Hyman, B. T. (2003). Fluorescence resonance energy transfer determinations using multiphoton fluorescence lifetime imaging microscopy to characterize amyloid-beta plaques. *J. Biomed. Opt.* **8,** 368–375.

Barylko, B., Binns, D. D., and Albanesi, J. P. (2001). Activation of dynamin GTPase activity by phosphoinositides and SH3 domain-containing proteins. *Methods Enzymol.* **329,** 486–496.

Bouley, R., Sun, T. X., Chenard, M., McLaughlin, M., McKee, M., Lin, H. Y., Brown, D., and Ausiello, D. A. (2003). Functional role of the NPxxY motif in internalization of the type 2 vasopressin receptor in LLC-PK1 cells. *Am. J. Physiol. Cell Physiol.* **285,** C750–762.

Carr, J. F., and Hinshaw, J. E. (1997). Dynamin assembles into spirals under physiological salt conditions upon the addition of GDP and gamma-phosphate analogues. *J. Biol. Chem.* **272,** 28030–28035.

Conner, S. D., and Schmid, S. L. (2003). Regulated portals of entry into the cell. *Nature* **422,** 37–44.

Damke, H., Muhlberg, A. B., Sever, S., Sholly, S., Warnock, D. E., and Schmid, S. L. (2001). Expression, purification, and functional assays for self-association of dynamin-1. *Methods Enzymol.* **329,** 447–457.

Greene, L. E., and Eisenberg, E. (1990). Dissociation of clathrin from coated vesicles by the uncoating ATPase. *J. Biol. Chem.* **265,** 6682–6687.

Greener, T., Zhao, X., Nojima, H., Eisenberg, E., and Greene, L. E. (2000). Role of cycline G-associated kinase in uncoating clathrin-coated vesicles from non-neuronal cells. *J. Biol. Chem.* **275,** 1365–1370.

Hardy, S., Kitamura, M., Harris-Stansil, T., Dai, Y., and Phipps, M. L. (1997). Construction of adenovirus vectors through Cre-*lox* recombination. *J. Virol.* **71,** 1842–1849.

Holstein, S. E. H., Ungewickell, H., and Ungewickell, E. (1996). Mechanism of clathrin basket dissociation: Separate functions of protein domains of the DnaJ-homologue auxilin. *J. Cell Biol.* **135,** 925–937.

Muhlberg, A. B., Warnock, D. E., and Schmid, S. L. (1997). Domain structure and intramolecular regulation of dynamin GTPase. *EMBO J.* **16,** 6676–6683.

Newmyer, S. L., Christensen, A., and Sever, S. (2003). Auxilin-dynamin interactions link the uncoating ATPase chaperone machinery with vesicle formation. *Developmental Cell.* **4,** 929–940.

Praefcke, G. J., and McMahon, H. T. (2004). The dynamin superfamily: Universal membrane tubulation and fission molecules? *Nat. Rev. Mol. Cell Biol.* **5,** 133–147.

Schlossman, D. M., Schmid, S. L., Braell, W. A., and Rothman, J. E. (1984). An enzyme that removes clathrin coats: Purification of an uncoating ATPase. *J. Cell Biol.* **99,** 723–733.

Schmid, S. L., McNiven, M. A., and De Camilli, P. (1998). Dynamin and its partners: A progress report. *Curr. Opin. Cell Biol.* **10,** 504–512.

Sever, S. (2002). Dynamin and endocytosis. *Curr. Opin. Cell Biol.* **14,** 463–467.

Sever, S., Damke, H., and Schmid, S. L. (2000a). Dynamin:GTP controls formation of constricted coated pits, the rate limiting step in clathrin-mediated endocytosis. *J. Cell Biol.* **150,** 1137–1147.

Sever, S., Damke, H., and Schmid, S. L. (2000b). Garrotes, springs, ratchets, and whips: Putting dynamin models to the test. *Traffic* **1,** 385–392.

Sever, S., Muhlberg, A. B., and Schmid, S. L. (1999). Impairment of dynamin's GAP domain stimulates receptor-mediated endocytosis. *Nature* **398,** 481–486.

Sever, S., Newmyer, S., Skoch, J., Ko, D., McKee, M., Bouley, R., Ausiello, D. A., Hyman, B. T., and Bacskai, B. J. (2005). Novel role for auxilin in formation of clathrin coated vesicles. In preparation.

Song, B. D., Leonard, M., and Schmid, S. L. (2004). Dynamin GTPase domain mutants that differentially affect GTP binding, GTP hydrolysis, and clathrin-mediated endocytosis. *J. Biol. Chem.* **279,** 40431–40436.

Stowell, M. H. B., Marks, B., Wigge, P., and McMahon, H. T. (1999). Nucleotide-dependent conformational changes in dynamin: Evidence for a mechanochemical molecular spring. *Nat. Cell Biol.* **1,** 27–32.

Tuma, P. L., and Collins, C. A. (1994). Activation of dynamin GTPase is a result of positive cooperativity. *J. Biol. Chem.* **269,** 30842–30847.

Umeda, A., Meyerholz, A., and Ungewickell, E. (2000). Identification of the universal cofactor (auxilin 2) in clathrin coat dissociation. *Eur. J. Cell Biol.* **79,** 336–342.

Ungewickell, E., Ungewickell, H., Holstein, S. E., Lindner, R., Prasad, K., Barouch, W., Martin, B., Greene, L. E., and Eisenberg, E. (1995). Role of auxilin in uncoating clathrin-coated vesicles. *Nature* **378,** 632–635.

Warnock, D. E., Hinshaw, J. E., and Schmid, S. L. (1996). Dynamin self assembly stimulates its GTPase activity. *J. Biol. Chem.* **271,** 22310–22314.

Zhang, P., and Hinshaw, J. E. (2001). Three-dimensional reconstruction of dynamin in the constricted state. *Nat. Cell Biol.* **3,** 922–926.

Zhao, X., Greener, T., Al-Hasani, H., Cushman, S. W., Eisenberg, E., and Greene, L. E. (2001). Expression of auxilin or AP180 inhibits endocytosis by mislocalizing clathrin: Evidence for formation of nascent pits containing AP1 or AP2 but not clathrin. *J. Cell Sci.* **114,** 353–365.

[51] Assay and Functional Analysis of Dynamin-Like Protein 1 in Peroxisome Division

By MICHAEL SCHRADER and STEPHEN J. GOULD

Abstract

Recent studies have demonstrated that peroxisome division requires at least one dynamin-like protein, Vps1p, in the yeast *Saccharomyces cerevisiae* and DLP1 (DRP1) in mammalian cells. Although the requirement for these proteins in peroxisome division is supported by many lines of evidence, their roles in peroxisome division have yet to be identified. Given the independence of peroxisomes from other organelle systems, the peroxisome system appears to have unique attributes for studying the function of dynamin-like proteins in organelle division. Here, we present methods that have been used for studying the role of DLP1 in peroxisome biogenesis and division.

Introduction

Peroxisomes arise by the growth and division of pre-existing peroxisomes (Lazarow and Fujiki, 1985). Although many studies have been devoted to studying peroxisome biogenesis (reviewed in Purdue and Lazarow, 2001), precious few have identified factors that are essential for peroxisome division. The first was by Hoepfner *et al.* (2001), who established that the division of yeast peroxisomes required Vps1p, one of three dynamin-like proteins of *S. cerevisiae*, but did not require Mdm1 or Dnm1, the two other dynamin-like proteins of yeast. Soon afterwards, a homolog of Vps1, DLP1, was found to be essential for peroxisome division in mammalian cells (Koch *et al.*, 2003, 2004; Li and Gould, 2003). The requirement for such a factor also extends to higher plants, as the dynamin-related protein DRP3A has been implicated in peroxisomal division in *Arabidopsis thaliana* (Mano *et al.*, 2004). The peroxisome-associated dynamin-like proteins have also been implicated in the biogenesis of other organelles, with yeast Vps1 playing an important role in vacuole biogenesis and mammalian DLP1 playing a critical role in mitochondrial division (Pitts *et al.*, 1999; Smirnova *et al.*, 1998, 2001).

Although these large GTPases are the only factors known to participate in peroxisome division, many studies have implicated the PEX11 family of

METHODS IN ENZYMOLOGY, VOL. 404
0076-6879/05 $35.00
DOI: 10.1016/S0076-6879(05)04051-6

proteins in the division process (reviewed in Purdue and Lazarow, 2001). Most recent studies, however, indicate that the PEX11 proteins function upstream of the division process (Koch *et al.*, 2003, 2004; Li and Gould, 2003), perhaps in the import of lipids that are essential to growth of the peroxisome membrane.

This chapter is intended to provide detailed information on methods and strategies used successfully in our laboratories for the analysis of DLP1 function in peroxisome division. We present morphological and biochemical methods to analyze the association of DLP1 with peroxisomes, to inhibit DLP1 function by RNA interference or expression of dominant negative DLP1 mutants, as well as approaches to identify interacting partners. Most of the methods presented take into consideration the particular properties of the peroxisomal compartment, but may be used to examine the influence of DLP1 on other subcellular organelles. We also describe the use of PEX11 proteins for inducing peroxisome membrane growth.

Experimental Procedures

Localization of DLP1 on Peroxisomes

Detection of DLP1 on Elongated Peroxisomes in Pex11pβ-Expressing Cells

BACKGROUND. To examine the association of DLP1 with peroxisomes during the division process, we took advantage of the observation that overexpression of Pex11 proteins induces peroxisome proliferation through a multistep process involving peroxisome elongation and subsequent division (Schrader *et al.*, 1998).

METHOD. HepG2 (or COS-7) cells were maintained in Dulbecco's modified Eagle's medium (DMEM) supplemented with 10% fetal calf serum (FCS) and 100 units/ml each of penicillin and streptomycin at 37° in a 5% CO_2-humidified incubator. Cells were either transfected by electroporation (10 μg DNA, 250 (or 230) V, 1500 μF, 129 Ω, 25–30 ms duration (see the following section on *Inhibition of DLP1 Function*, Method B) or by incubation with polyethylenimine (25 kDa PEI, Sigma, St. Louis, MO). For PEI transfection, an easy and low-cost method (Fischer *et al.*, 1999), 2.4×10^4 cells/cm^2 were seeded on glass coverslips in 6-well plates, and 24 h after plating the medium was changed to a volume of 2.5 ml. For three dishes, 10 μg of a Pex11pβ-myc plasmide and 100 μl of PEI stock solution (0.9 mg/ml in H_2O dest.) were diluted into 750 μl each with 150 mM NaCl.

After 10 min at room temperature, the PEI solution was pipetted to the DNA, and after 15 more min, 500 μl of the PEI/DNA solution were added to the cells. The transfection medium was replaced after 5 h, and cells were processed for indirect immunofluorescence 24–48 h post-transfection.

Cells were fixed for 20 min in 4% paraformaldehyde, 75 mM phosphate-buffer, pH 7.4, at room temperature, and washed with phosphate-buffered saline (PBS). Cells were permeabilized with 25 μg/ml digitonin in PBS for 5 min, and incubated for 10 min in PBS containing 1% bovine serum albumine (BSA) to block free aldehyde groups. Cells were then incubated with primary antibodies directed to DLP1 and to the myc-epitope tag of Pex11pβ-myc (Santa Cruz Biotechnology, Santa Cruz, CA), and with fluorescently labeled secondary antibodies each for 1 h at room temperature in a humidified chamber. Several antibodies directed to DLP1 have been described (Li and Gould, 2003; Pitts *et al.*, 1999) or are commercially available (BD Transduction Laboratories, San Jose, CA). We also used primary antibodies to PMP70 and the myc epitope to label peroxisomes. Antibody incubations were followed by washing three times in PBS. Coverslips were mounted on glass microscope slides in Mowiol 4-88 (Calbiochem, La Jolla, CA) containing 0.5% n-propyl gallate (Sigma) to prevent photobleaching. Samples were examined by phase-contrast/epi-fluorescence microscopy using the appropriate filters and a 100× objective, or by confocal microscopy with appropriate spectrometer settings for each fluorophor. Digital images were recorded and optimized for contrast and brightness using Adobe Photoshop software.

Using this protocol, DLP1 was found to align along elongated peroxisomes in spots. Similar observations were made in controls; however, colocalization was less frequent without a proliferative stimulus.

Notes: (1) Digitonin permeabilization for the localization of Pex11pβ is recommended, because it is lost during Triton X-100 permeabilization. (2) Treatment with 25 μg/ml digitonin for 5 min does not permeabilize the peroxisomal membrane (differential permeabilization). Only epitopes on the cytosolic surface of the peroxisomal membrane can be detected. (3) Negative controls should be performed to assess the level of general background (i.e., incubation with secondary antibodies only). (4) The peroxisomal staining pattern for DLP1 can be improved by expression of a GFP-DLP1 fusion protein. (5) The measurement of peroxisome division by live cell imaging of mammalian cells has yet to yield success. The problems in detecting DLP1 associated with division sites on peroxisomes are related to the relative small size of peroxisomes, their movement, and the relatively low rate at which they divide.

Association of DLP1 with Isolated Peroxisomes

Immunoblotting of Highly Purified Peroxisomal Fractions

METHOD. To collect biochemical evidence for the association of DLP1 with peroxisomes, we used gradient centrifugation to isolate peroxisomes. Peroxisomes from rat liver were enriched by subcellular fractionation and further purified by metrizamide or OptiPrep gradient centrifugation according to standard procedures that have been described in detail (e.g., Völkl *et al.*, 1996). The resulting peroxisomal fraction is usually well separated from other subcellular organelles and highly pure. The purity can be controlled by the determination of marker enzyme activities, by immunoblotting with antibodies directed to organelle-specific marker proteins, and by electron microscopy. The peroxisomal (and other) gradient fractions were analyzed by immunoblotting using DLP1-specific antibodies that detected an 80-kDa protein. Rat liver cytosol and rat brain homogenate served as positive controls. The amount of DLP1 in the peroxisomal fractions was increased after treatment of the animals with the potent peroxisome proliferator bezafibrate (Koch *et al.*, 2003).

Immunoprecipitation of Peroxisomal Membranes

METHOD. Human fibroblasts were transfected with a plasmid designed to express mycDLP1-WT, and a plasmid expressing HA-tagged PMP34, a peroxisomal membrane protein. The cells were cultured under standard conditions and transfected by electroporation (10 μg DNA, 230 V, 1500 μF, 129 Ω, 25–30 ms duration) (see following section *Inhibition of DLP1 Function*, Method B). Cells were harvested 48 h posttransfection by scraping in hypotonic lysis buffer (10 m*M* Tris-HCl, pH 7.5, 1 m*M* EDTA, and protease inhibitor mix) and lysed by passing through a 22-gauge syringe needle five times. Lysates were then centrifuged at 25,000*g* for 15 min to pellet large organelles. The resulting supernatant was diluted in TBS buffer (50 m*M* Tris-HCL, pH 7.5, 150 m*M* NaCl, 1 m*M* EDTA, and protease inhibitor mix). Immunoprecipitation was initiated by adding 15 μl of a 50% slurry of anti-HA monoclonal antibodies coupled to agarose beads (Santa Cruz Biotechnology) into 500 μl of the above supernatant together with 1% BSA. The sample was then incubated with mixing at 4° for 2 h. Anti-HA beads were pelleted by brief centrifugation, washed five times with 1 ml of TBS, and resuspended in SDS sample buffer. The supernatant and corresponding immunoprecipitate sample were then separated by SDS-PAGE, transferred to PVDF membranes, and probed with anti-myc polyclonal antibodies (Santa Cruz Biotechnology).

With this method, it was found that mycDLP1-WT was coimmunopurified with peroxisome membranes. Similar results were obtained with mycDLP1T59A, a dominant negative mutant predicted to lack GTPase activity and to display increased membrane association. As a control for nonspecific binding of peroxisomal membranes to anti-HA beads, parallel experiments were carried out by coexpressing a VSVG-tagged PMP34.

Notes: (1) Other peroxisomal membrane proteins can be used for immunopurification of peroxisomal membranes. (2) Peroxisome-associated DLP1 was found to be increased in Pex11β-expressing cells when immuno-precipitate samples from PMP34–3×HA/mycDLP1-WT and Pex11β-3×HA/mycDLP1-WT–expressing cells were compared.

Inhibition of DLP1 Function

RNA Interference of DLP1

Background. RNA interference (RNAi) technologies have been proven to be a valuable tool to identify and investigate proteins that are involved in peroxisome biogenesis. Gene silencing in mammalian cultured cells can be achieved by DNA-directed RNAi, in which a DNA vector is used to express short hairpin RNAs, or by directly introducing small, interfering RNA duplexes (siRNA) into target cells. To knock down the expression of DLP1 by RNAi, we have chosen chemical synthesis of 21-nucleotide siRNA duplexes (available in an annealed form from Dharmacon Research, Lafayette, Co; or Ambion, Austin, TX), which are easier to transfect than DNA-based vectors that express siRNA. Meanwhile, several siRNA sequences efficiently targeting human DLP1 (Acc. No. NM012062) have been identified (Koch *et al.*, 2003; Li and Gould, 2003; Lee *et al.*, 2004). Efficient knockdown of DLP1 in cells of human and rodent origin was achieved with a conserved DLP1 target sequence corresponding to the coding region 783–803 (sense strand: 5′-UCCGU-GAUGAGUAUGCUUUdTdT-3′) (Koch *et al.*, 2003, 2004; Lee *et al.*, 2004). DLP1 siRNA was transfected into the cells using either chemical transfection or electroporation. Appropriate controls were transfected with siRNA sequences targeting either GL2 luciferase (coding region 153–173) or the nonperoxisomal gene *PXR2* (sense strand: 5′-GCAGGGAAAA-GGCUCUAGGdTdT-3′) (Dharmacon). Analysis of peroxisome morphology and siRNA efficiency is facilitated by the use of cells stably expressing a fluorescent fusion protein bearing a peroxisomal targeting signal (e.g., GFP-PTS1, DsRed-PTS1) (Fig. 1).

Method A: Silencing of DLP1 by Chemical transfection with Oligofectamine

The annealed DLP1 siRNA duplexes were introduced into COS-7 cells by oligofectamine (Invitrogen, Grand Island, NY)-mediated transfection. 24 h before transfection, COS-7 cells were harvested by trypsin/EDTA

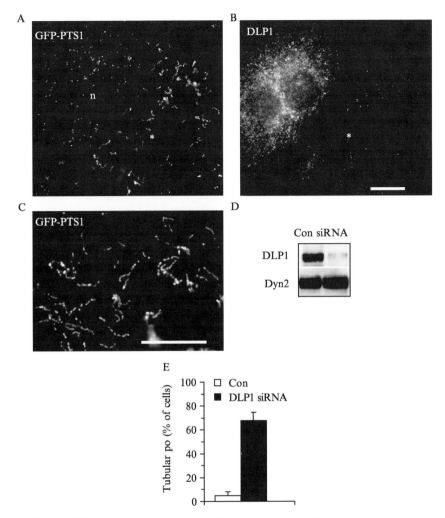

FIG. 1. COS-7 cells stably expressing a GFP construct bearing the C-terminal peroxisomal targeting signal 1 (GFP-PTS1) were transfected with DLP1 siRNA using oligofectamine and processed for immunofluorescence (A–C) and immunoblotting (D) 48 h posttransfection. For visualization of endogenous DLP1, cells were incubated with rabbit anti-DLP1 antibodies, and subsequently with goat anti-rabbit IgG conjugated to TRITC (B). Peroxisomes in silenced cells have a highly elongated, tubular morphology (A, B; asterisks). In contrast, peroxisomes in adjacent, nontransfected cells (A, B; n) or in controls (Con) treated with luciferase siRNA (E) have a spherical morphology. (C) Higher magnification view of tubular and constricted peroxisomes after silencing of DLP1. (D) Cells were transfected with the respective siRNAs and lysed. Equal amounts of protein were separated by SDS-PAGE and transferred to nitrocellulose membranes. Standard immunoblotting was carried out using enhanced chemoluminescence. Expression of DLP1 was determined with an anti-DLP1 antibody. Antibodies against dynamin-2 (Dyn2) control for nonspecific alterations and equal loading. (E) Quantitation of peroxisome morphology after silencing of DLP1. Data are presented as means ± SD (p < 0.01). Bars, 10 μm.

treatment and plated on 6-well plates (2×10^5 cells/well) (plate on glass coverslips for indirect immunofluorescence) in DMEM/10%FCS. The following day, the medium was replaced with 1 ml fresh, serum- and antibiotics-free medium, and 200 μl of transfection mix were added to a well. For preparation of the tranfection mix, 4 μl of oligofectamine were added to 11 μl of serum-free medium (solution 1), and 4 μl of siRNA duplex (20 μM) were mixed with 181 μl of serum-free medium (solution 2). Both solutions were incubated for 5 min at room temperature. Solution 1 was then added dropwise to solution 2 and incubated for 20 min at room temperature. The transfection medium was replaced after 4–5 h by medium containing 10% FCS. Retransfection with DLP1 siRNA 24 h and 48 h after the first transfection is recommended to increase the number of transfected and silenced cells. Cells were usually assayed for DLP1 protein level and peroxisome morphology 3–4 days after plating by indirect immunofluorescence and standard immunoblotting of cell lysates (Fig. 1). Antibodies against control proteins (e.g., anti-tubulin, anti-Pex13, or anti-dynamin 2 antibodies) were applied to control for equal amounts of protein in the samples and for the specificity of the knockdown.

Using the transfection protocol described, we routinely obtain transfection efficiencies between 40 and 80%. After two retransfections the peroxisomes in approximately 70–80% of the cells displayed a highly elongated, tubular morphology (up to 15 μm in length), and DLP1 was barely detectable in those cells by indirect immunofluorescence (Fig. 1). The peculiar morphology is due to a complete block of peroxisomal fission resulting in the accumulation of peroxisomes in an elongated and constricted state (Koch et al., 2004) (Fig. 1). Immunoblots revealed that the DLP1 protein level was reduced to 30–35% of the control level. Since the transfection efficiency is usually about 70–80%, the DLP1 protein level is even lower in those cells that took up the DLP1 siRNA.

Notes: (1) Transfection and silencing conditions are dependent on the transfection reagent, cell type, cell density, and the amount of siRNA used. In order to achieve maximum effectiveness of exogenously introduced siRNA duplexes while maintaining a high level of cell viability, transfection optimization experiments are required. (2) In the beginning, it is advisable to test siRNA efficiency at various time points after transfection. (3) Cell viability can be reduced if the transfection medium is not removed. (4) If cotransfection of DNA-expression vectors together with siRNA duplexes is intended, the use of Lipofectamine 2000 (Invitrogen) is recommended.

Method B: Silencing of DLP1 by Electroporation

For transfection by electroporation COS-7 cells grown to 90% confluency on an area of 75 cm^2 (1×10^7 cells) were harvested by trypsination, washed in 10 ml of HBS solution (21 mM Hepes, 137 mM NaCl, 5 mM KCl,

0.7 mM Na$_2$HPO$_4$, 6 mM dextrose), resuspended in 0.5 ml HBS, and transferred to a 0.4 cm gap, sterile electroporation cuvette containing 20 μl (range: 20–50 μl) of siRNA duplex (20 μM). Electroporation was performed in an electroporator (e.g., Easyject Plus, Peqlab, Erlangen, Germany; or BTX ECM 600, BTX, San Diego, CA) at 230 V (use 250 V for HepG2 cells), 1500 μF, 129 Ω, 25–30 ms duration. After electroporation, cells were immediately resuspended in DMEM/10% FCS and plated on 6-well plates (2×10^5 cells/well). Cells were usually assayed for DLP1 protein level and peroxisome morphology and abundance 3–4 days after plating by indirect immunofluorescence and standard immunoblotting of cell lysates. To examine the activity of other peroxisomal proteins in siRNA-treated cells, the cells were transfected with the desired expression vector (10 μg DNA) 24–48 h after the transfection with siRNA by electroporation or by chemical transfection reagents.

Using the transfection protocol described, we routinely obtain transfection efficiencies between 70 and 80%. Retransfection is not required. Immunoblots showed that the DLP1 protein level was reduced to 20–40% of the control level. Effects on peroxisome morphology were the same as observed with Method A (see Fig. 1).

Notes: (1) Electroporation of siRNA typically requires optimization of various parameters that affect cell uptake and cell viability (e.g., electroporation medium, cell number, voltage, capacitance, resistance). (2) It is advisable to test siRNA efficiency at various time points after transfection.

Expression of Dominant Negative Mutants of DLP1

BACKGROUND. An alternative method to test the involvement of DLP1 in peroxisome division is the expression of wild type (WT) and dominant negative DLP1 mutants. DLP1 mutants have been generated by PCR site-directed mutagenesis of the amino-terminal tripartite GTP binding motif. These mutants exhibit reduced GTP binding affinity (S39N) and reduced GTP hydrolysis activity (T59A, K38A), and are available as GFP- or myc-tagged fusion proteins (Li and Gould, 2003; Pitts *et al.*, 1999). A well-characterized mutant is DLP1-K38A harboring a lysine-to-alanine mutation in GTP binding element 1 (Pitts *et al.*, 1999; Smirnova *et al.*, 1998; Yoon *et al.*, 2001). In addition, several other dominant negative DLP1 mutants have been generated to analyze mitochondrial division (Smirnova *et al.*, 2001).

METHOD. WT and mutant forms of DLP1 (GFP- or myc-tagged) were transfected into human fibroblasts, COS-7 cells, or HepG2 cells using either chemical transfection with PEI (see previous section, *Localization of DLP1 on Peroxisomes*) or electroporation (see section on *Inhibition of DLP1 function,* Method B). Cells were cultured under standard conditions, and processed for indirect immunofluorescence 24–96 h after transfection (see previous section, *Localization of DLP1 on Peroxisomes*). Antibodies specific for the myc epitope tag and PMP70, an integral peroxisomal membrane protein marker, were used for detection. With this method, a reduction of peroxisome abundance and the formation of tubulo-reticular peroxisomes was observed. A massive elongation (hypertubulation) and the formation of tubulo-reticular networks of peroxisomes was observed in HepG2 and COS-7 cells by coexpressing Pex11pβmyc and GFP-DLP1-K38A. In contrast to appropriate controls (expressing Pex11pβmyc/GFP-DLP1-WT, Pex11pβmyc/DLP1-WT, or Pex11pβmyc/GFP-Dyn2-K44A, a mutated form of dynamin II, not involved in peroxisome formation) peroxisomal fission was completely inhibited and elongated peroxisomes accumulated.

Notes: (1) It has to be noted that multiple isoforms of DLP1 are generated by alternative splicing of the initial DLP1 transcript, which may have different activities *in vivo* (Li and Gould, 2003). (2) DLP1-dependent morphological changes of peroxisomes can usually be detected without difficulties by fluorescence microscopy when compared to appropriate controls. However, the effects may vary depending on the cell type used. Peroxisome morphology and abundance was quantitated according to Koch *et al.* (2003) and Li and Gould (2002).

Cross-linking and Immunoprecipitation of DLP1

Background. To determine interacting partners of DLP1 on the peroxisomal membrane, cross-linking and immunoprecipitation studies have been performed. A protocol is presented that has been used to test whether there is a physical association between DLP1 and Pex11 proteins Pex11pα, β, and γ.

Method. Fibroblasts were transfected with plasmids expressing HA-tagged fusion proteins of Pex11α, Pex11β or Pex11γ, and mycDLP1. As a control, cells were transfected with PMP34-3×HA and mycDLP1. Cells were maintained under standard conditions (see previous section, *Localization of DLP1 on Peroxisomes*) and transfected by electroporation (see *Inhibition of DLP1 Function*, Method B). 24 h after transfection, cells were lysed by incubation with mixing in TBS buffer (50 mM Tris-HCl, pH 7.5, 150 mM NaCl, 1mM EDTA) containing 0.2% digitonin at 4° for 1 hour (Li and Gould, 2003). Lysates were subjected to centrifugation at 100,000g for 30 min to pellet cellular membranes. The resulting supernatant was diluted into TBS buffer containing 0.2% digitonin and 1% BSA. Immunoprecipitation was performed by adding 15 μl of a 50% slurry of anti-HA monoclonal antibodies coupled to agarose beads (Santa Cruz

Biotechnology) and incubating with mixing at 4° for 2 h. The beads were pelleted by brief centrifugation, and washed three times in TBS buffer plus 0.2% digitonin, three times in high salt TBS buffer (50 mM Tris-HCl, pH 7.5, 350 mM NaCl, 1 mM EDTA) plus 0.2% digitonin and protease inhibitor mix, and three times in TBS buffer. The beads were boiled in SDS-sample buffer, separated together with the corresponding supernatant on 12.5% acrylamide gels, and probed with anti-myc polyclonal antibodies by standard immunoblotting.

To stabilize potentially weak or transient protein interactions, whole, transfected cells can be subjected to chemical cross-linking prior to lysis. For cross-linking, the cells were incubated for 1 h at room temperature with the membrane-permeant, cleavable, homobifunctional cross-linker dithiobis (succinimidylpropionate) (DSP) (Pierce Chemical Co., Rockford, IL) (1 mM final concentration in PBS). Cross-linking was stopped by the addition of Tris-HCl, pH 7.7 to a 10 mM final concentration. After immunoprecipitation, the beads are boiled in SDS-sample buffer containing 5% β-mercaptoethanol to cleave the cross-linker. This approach has been used to study DLP1 interactions on the mitochondrial membrane (Yoon et al., 2003).

With the method described, no evidence for a physical interaction between DLP1 and one of the Pex11 proteins was obtained. Immunoprecipitation experiments in the presence of a cross-linker or nonhydrolyzable GTP analogs, and interaction studies with a yeast two-hybrid system also failed to provide any evidence for Pex11-DLP1 interaction. However, the homophilic interaction of the Pex11 proteins was retained and could be detected (see following).

Notes: (1) It is advisable to test conditions for solubilizing proteins from the peroxisomal membrane. Pex11 proteins were efficiently solubilized by both 1% Triton X-100 and 0.2% digitonin. (2) As a positive control, it can be addressed whether proteins solubilized under these conditions retain the ability to interact with other, known binding partners (as far as binding partners are already known). For example, homophilic interactions of Pex11 proteins have been described. Although Pex11 proteins were solubilized by either detergent (1), homophilic interactions (e.g., Pex11pα-Pex11pα) were only observed in cells homogenized with 0.2% digitonin.

Analysis of other Biochemical Properties of DLP1

DLP1 self-assembly/polymerization activity (Smirnova et al., 2001; Yoon et al., 2001), membrane tubulating activity (Yoon et al., 2001), and GTPase activity (Yoon et al., 2001) has been analyzed in recent in vitro studies. Detailed protocols are provided in the literature cited. These methods have been used to demonstrate that purified DLP1 can self-assemble into multimeric ring-like structures, and that GTP-bound DLP1 tubulates membranes similar to dynamin. Furthermore, it was demonstrated that DLP1-K38A binds but does not hydrolyze or release GTP.

References

Fischer, D., Bieber, T., Li, Y., Elsasser, H. P., and Kissel, T. (1999). A novel non-viral vector for DNA delivery based on low molecular weight, branched polyethylenimine: Effect of molecular weight on transfection efficiency and cytotoxicity. *Pharm. Res.* **16,** 1273–1279.

Hoepfner, D., van den Berg, M., Philippsen, P., Tabak, H. F., and Hettema, E. H. (2001). A role for Vps1p, actin, and the Myo2p motor in peroxisome abundance and inheritance in *Saccharomyces cerevisiae. J. Cell Biol.* **155,** 979–990.

Koch, A., Schneider, G., Luers, G. H., and Schrader, M. (2004). Peroxisome elongation and constriction but not fission can occur independently of dynamin-like protein 1. *J. Cell Sci.* **117,** 3995–4006.

Koch, A., Thiemann, M., Grabenbauer, M., Yoon, Y., McNiven, M. A., and Schrader, M. (2003). Dynamin-like protein 1 is involved in peroxisomal fission. *J. Biol. Chem.* **278,** 8597–8605.

Lazarow, P. B., and Fujiki, Y. (1985). Biogenesis of peroxisomes. *Annu. Rev. Cell Biol.* **1,** 489–530.

Lee, Y. J., Jeong, S. Y., Karbowski, M., Smith, C. L., and Youle, R. J. (2004). Roles of the mammalian mitochondrial fission and fusion mediators Fis1, Drp1, and Opa1 in apoptosis. *Mol. Biol. Cell* **15,** 5001–5011.

Li, X., and Gould, S. J. (2002). PEX11 promotes peroxisome division independently of peroxisome metabolism. *J. Cell Biol.* **156,** 643–651.

Li, X., and Gould, S. J. (2003). The dynamin-like GTPase DLP1 is essential for peroxisome division and is recruited to peroxisomes in part by PEX11. *J. Biol. Chem.* **278,** 17012–17020.

Mano, S., Nakamori, C., Kondo, M., Hayashi, M., and Nishimura, M. (2004). An Arabidopsis dynamin-related protein, DRP3A, controls both peroxisomal and mitochondrial division. *Plant J.* **38,** 487–498.

Pitts, K. R., Yoon, Y., Krueger, E. W., and McNiven, M. A. (1999). The dynamin-like protein DLP1 is essential for normal distribution and morphology of the endoplasmic reticulum and mitochondria in mammalian cells. *Mol. Biol. Cell* **10,** 4403–4417.

Purdue, P. E., and Lazarow, P. B. (2001). Peroxisome biogenesis. *Annu. Rev. Cell Dev. Biol.* **17,** 701–752.

Schrader, M., Reuber, B. E., Morrell, J. C., Jimenez-Sanchez, G., Obie, C., Stroh, T. A., Valle, D., Schroer, T. A., and Gould, S. J. (1998). Expression of PEX11beta mediates peroxisome proliferation in the absence of extracellular stimuli. *J. Biol. Chem.* **273,** 29607–29614.

Smirnova, E., Griparic, L., Shurland, D. L., and van der Bliek, A. M. (2001). Dynamin-related protein Drp1 is required for mitochondrial division in mammalian cells. *Mol. Biol. Cell* **12,** 2245–2256.

Smirnova, E., Shurland, D. L., Ryazantsev, S. N., and van der Bliek, A. M. (1998). A human dynamin-related protein controls the distribution of mitochondria. *J. Cell Biol.* **143,** 351–358.

Völkl, A., Baumgart, E., and Fahimi, H. D. (1996). Isolation and characterization of peroxisomes. *In* "Subcellular Fractionation: A Practical Approach" (J. Graham and D. Rickwood, eds.), pp. 143–167. Oxford University Press, Oxford, U. K.

Yoon, Y., Krueger, E. W., Oswald, B. J., and McNiven, M. A. (2003). The mitochondrial protein hFis1 regulates mitochondrial fission in mammalian cells through an interaction with the dynamin-like protein DLP1. *Mol. Cell. Biol.* **23,** 5409–5420.

Yoon, Y., Pitts, K. R., and McNiven, M. A. (2001). Mammalian dynamin-like protein DLP1 tubulates membranes. *Mol. Biol. Cell* **12,** 2894–2905.

[52] *In Vitro* Reconstitution of Discrete Stages of Dynamin-dependent Endocytosis

By MATTHEW K. HIGGINS and HARVEY T. MCMAHON

Abstract

Many proteins involved in vesicle budding are able to interact with lipids and to deform membranes. In this review we present three *in vitro* assays that can be used to study the roles of these types of proteins and to reconstitute distinct stages of vesicle budding and tubule formation. In the first assay, the dynamic effect of peripheral membrane proteins on liposome morphology is studied, providing insight into the roles of proteins (e.g., the endocytic proteins; dynamins, epsins, amphiphysins, and endophilins) in deforming membranes and in recognizing membrane curvature. In the second assay, preformed lipid nanotubes are used to mimic the neck of a coated vesicle. These nanotubes form a suitable template to study molecules of the dynamin family, allowing visualization of the different conformations of dynamin that occur during vesicle scission. Finally, lipid monolayers have been used as mimics of the internal leaflet of the plasma membrane to investigate early stages of clathrin-coated pit formation. In principle this can equally be used to study budding mechanisms of other coated vesicle types. A combination of these three assays has given considerable insight into the roles of endocytic proteins and is allowing reconstitution and dissection of the different stages of vesicle formation.

Introduction

The formation of vesicles or trafficking intermediates is a complex process, requiring interplay between numerous soluble and membrane-bound proteins and lipids. Together these components work to generate local deformation of a membrane, capture cargo, and promote the formation of a vesicle or tubule intermediate. In cells, these processes can be examined by deleting individual components or by over-expressing

METHODS IN ENZYMOLOGY, VOL. 404
0076-6879/05 $35.00
DOI: 10.1016/S0076-6879(05)04052-8

proteins or their domains. This provides information about trafficking pathways but seldom yields precise information about the individual function of a protein component. Therefore the *in vitro* approach is valuable, allowing more precise roles to be assigned to proteins and allowing vesicle formation to be studied in molecular detail. However, care is needed in interpreting these data as a limited subset of binding partners is used and the lipid composition of model membranes and the concentration of protein components may not reflect those present *in vivo*. Therefore *in vivo* and *in vitro* approaches should be used in tandem, both resulting in different information and with *in vivo* experiments providing crosschecks to confirm *in vitro* data. Despite the pitfalls of studying isolated protein components, tremendous progress has been made in cell biology by such approaches, and only by understanding the molecular details can we begin to intelligently design therapies.

Method 1: Tubulation of Lipid Vesicles by Endocytic Proteins

Introduction

Many endocytic proteins that associate with membranes can deform the lipid bilayer, often causing tubulation. This can be caused by the shape of the lipid binding face of a protein (Peter *et al.*, 2004) or protein oligomer (Praefcke and McMahon, 2004; Sweitzer and Hinshaw, 1998) or can be due to active insertion into the membrane of a helix that stabilizes membrane curvature (Ford *et al.*, 2002). An example of the former is the tubulation of liposomes by the large GTPase, dynamin (see Fig. 1C). Addition of dynamin, in the presence of nonhydrolyzable GTP analogues, causes liposome tubulation, with formation of tubules driven by assembly of an ordered array of dynamin molecules. These dynamin spirals show sufficient order to allow determination of their structure at 20Å resolution by processing of images of frozen tubules (Zhang and Hinshaw, 2001) and are similar in diameter to the necks of endocytic vesicles. In this case tubulation is driven primarily by oligomerization of dynamin into a helix on the membrane template. Indeed, even in the absence of liposomes, dynamin can assemble into ring and tubules-like arrays, forming the conformation preferred by the protein assembly (Hinshaw and Schmid, 1995; Owen *et al.*, 1998).

Dimers of the amphiphysin BAR domain also have intrinsic curvature and therefore bind most favorably to membranes with a curvature complementary to their own (Peter *et al.*, 2004). BAR domains are found in a wide variety of proteins, including amphiphysins (Peter *et al.*, 2004; Razzaq

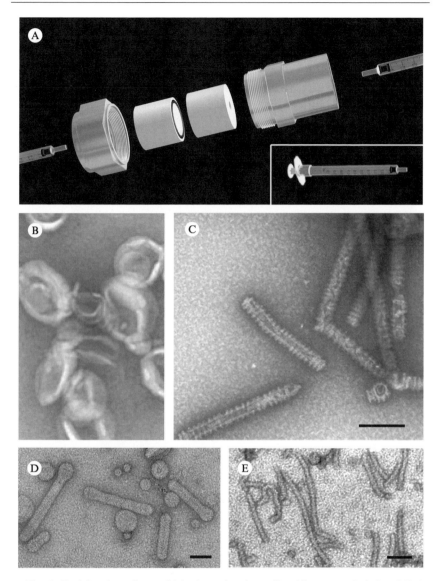

FIG. 1. Studying dynamin, amphiphysin, and epsin mediated liposome tabulation. (A) An "exploded" view of the lipid extrusion devise used for the formation of liposomes. (B) Liposomes formed from Folch brain lipid extract. (C) The effect of dynamin on Folch brain lipid liposomes. (D) Tubules formed by incubating liposomes with *Drosophila* amphiphysin. (E) Tubules formed by incubating the ENTH domain of epsin with liposomes. The scale bars are 100 nm.

et al., 2001; Takei *et al.*, 1999), endophilins (Farsad *et al.*, 2001), and many others (Peter *et al.*, 2004), and the presence of the BAR domain can be tested by observing the ability of a protein to tubulate liposomes (Fig. 1D). A variation of the BAR domain is an N-BAR domain where an additional amphipathic helix inserts into the membrane, taking part in driving or stabilization of membrane curvature. The presence of an amphipathic helix that drives tubulation is also seen in the ENTH domain of epsin (Fig. 1E) (Ford *et al.*, 2002). As many other proteins are predicted to contain N-terminal amphipathic helices, the liposome tubulation assay will prove a valuable test for the folding and insertion of these helices into membranes giving rise to increased membrane curvature.

It should be noted that the ability of an isolated protein to form tubules may not reflect an *in vivo* tubule forming capacity. Instead, the formation of a tubule reflects the ability of a protein to increase the curvature of a liposome. Indeed, a tubule is simply a liposome with high curvature around the circular cross-section, but with a relaxed curvature along its length. Therefore, a clearer view of the role of a protein *in vivo* can be obtained by studying proteins in combination. Whereas epsin can tubulate liposomes, in combination with clathrin it forms invaginated clathrin buds (Ford *et al.*, 2002) and *in vitro* the combination does not form tubules. Therefore the ability of a protein to form tubules may not always represent an *in vivo* tubulation activity, as interactions with other proteins may modify the conformation adopted within the cell. However, the ability of a protein to tubulate liposomes does reveal its capacity for altering and stabilizing a particular membrane curvature.

In the method described below, the processes of bilayer deformation can be studied using synthetic liposomes of any defined lipid composition. Liposomes are incubated with the protein of interest and observed in the electron microscope to reveal changes in liposome conformation accompanied with protein binding. In many of our studies where we did not know the precise lipid composition to use, we have instead used Folch brain lipids (Sigma, St. Louis, MO).

Reagents and Equipment

Lipid Stocks. Phosphatidylinositol and phospatidylinositol-4,5-bisphosphate (PtdIns(4,5)P$_2$) (Avanti Polar Lipids, Alabaster, AL) weredissolved to 1 mg/ml in 3:1 chloroform:methanol. Cholesterol (Avanti), phosphatidylserine (PtdSer), phosphatidylcholine (PtdChol), and phosphatidylethanolamine (PtdEth) (Sigma) were dissolved to 10 mg/ml in chloroform. Folch extract is a total brain lipid extract containing around 10% of a variety of phosphatidylinositol lipids (Sigma, B1502) and is dissolved to

10 mg/ml in a 19:1 mixture of chloroform:methanol. Concentrated lipid stocks were stored at $-80°$ under argon in glass vials with glass or Teflon lids.

Electron Microscopy Grids. First place a droplet of 2% collodian in amyl acetate (TAAB) onto a bath of filtered milli Q water. Allow the solvent to evaporate, leaving a film of plastic on the water surface. Wash a box of 300 mesh carbon electron microscope grids (TAAB) in acetone, air dry, and place onto the carbon surface with the "rough" side in contact with the carbon. Remove the plastic film from the top of the water by laying a piece of parafilm onto the top and pulling it off again. Allow this to dry on the bench. When dry, carbon coat using an E306A carbon evaporator (BOC Edwards, UK), leaving a thin carbon coat on the surface of the grid.

Stain. Prepare 2% uranyl acetate (Biorad, UK) with 0.0025% poly-acrylic acid (Sigma) in water. The presence of polyacrylic acid reduces precipitation of the stain. In the absence of polyacrylic acid, stain should be prepared fresh and filtered before use.

Nucleotides. GDP and GTPγS were purchased from Roche. GTPγS should be fresh to avoid breakdown products, including GDP, which will alter the conformation of bound dynamin.

Extrusion of Liposomes. An extrusion device is used to generate lipo-somes of a defined maximum size (shown in an "exploded" view Fig. 1A). The central element of the device consists of two pieces of Teflon with a 0.6 mm hole passing through each block. These holes widen out at one side to 4 mm to allow the insertion of a 1 ml disposable syringe. At the opposite end is a groove to accommodate a 15 mm o-ring. Two of these pieces are placed together with o-rings in contact to form a seal around a filter. These filters are Whatman Cyclopore filters with pore sizes of 0.1, 0.2, or 0.4 μm. A swagelok male pipe weld connector, bored out to hold the Teflon blocks, is used to hold the arrangement in place. 1 ml dispos-able syringes placed at either end allow the lipid mixture to be extruded through the filter in both directions.

Making Liposomes of a Defined Size

1. Prepare lipid mixtures to a total lipid concentration of 1 mg/ml in a 19:1 mixture of chloroform:methanol in a glass tube. Methanol is necessary to maintain the solubility of some lipids, including PtdIns(4,5)P$_2$. A common lipid mixture consists of 10% cholesterol, 40% PtdEth, 40% PtdSer, and 10% of the lipid specific for the protein of interest (often PtdIns(4,5)P$_2$). An alternative is to use a Folch brain lipid extract or a liver lipid extract.

2. Remove solvent from the lipid mixture by evaporation under a gentle stream of Argon. This generates a thin lipid film on the surface of the glass. When this film becomes white, turn up the flow to spread out the last drops and avoid leaving thick lumps of precipitated lipid. Place the dried lipid in a desiccator under vacuum for 10–15 min to complete drying.

3. Rehydrate the dried lipid mixture by adding 1 ml of filtered buffer (50 mM HEPES pH 7.4, 120 mM NaCl) and leave for 5 min at room temperature. Sonicate in a bath sonicator for 2 min to strip the lipid film off the surface of the glass. Then apply two 2 sec pulses with a small (2 mm) probe sonicator to break lipid particles into liposomes.

4. Filter through 0.1, 0.2, or 0.4 μm polycarbonate filters (Whatman Cyclopore filters, UK) using the extrusion device. The sample should be pushed through the filtration unit between the two 1-ml syringes 11 times. It is important to use an odd number of extrusion steps to ensure that all of the final material has been through the filter and material simply trapped in the extrusion device is not washed back into the sample. A loss of 100–200 μl is usually observed.

Purification of Dynamin and Its Mutants from E. coli

Dynamin can be prepared from a variety of different sources, including rat brain, bacteria, or baculovirus-infected insect cells. The purification protocol relies on the affinity of dynamin for the SH3 domain of amphiphysin 2. The protocol yields approximately 0.5 mg of dynamin from each liter of bacterial culture.

1. To produce dynamin from *E. coli*, the dynamin 1 gene is expressed from pET34b. Grow cells at 37° until log phase, induce with 1 mM IPTG, and incubate overnight at 19°. Harvest cells and resuspend in 150 mM NaCl, 20 mM HEPES pH 7.4, 2 mM EDTA, 2 mM DTT (incubation buffer). Break by French Press and clarify by centrifugation at 200,000g.

2. Lyse 3 l of *E. coli* containing a GST fusion of the Amph2 SH3 domain in incubation buffer. Clarify at 200,000g for 30 min and incubate the supernatant with glutathione beads for 40 min. Wash with incubation buffer.

3. Incubate Amph2 SH3 coated glutathione beads with dynamin containing lysate for 40 min and wash with incubation buffer. Elute dynamin with 1.1 M NaCl, 20 mM PIPES pH6.2, 10 mM $CaCl_2$ and dialyze overnight into incubation buffer. Concentrate using a 50 kDa MWCO Vivaspin concentrator. If the preparation is clean, the sample can be concentrated up to and above 0.5 mg/ml without precipitation.

4. Further purify the dynamin by gel filtration. Use an S200 16/60 gel filtration column (Amersham Biosciences) pre-equilibrated with 150 mM NaCl, 20 mM CHES pH 10, 2 mM EDTA, 2 mM DTT. Use of a pH 10 buffer prevents dynamin oligomerization, allowing greater concentration and purification by gel filtration. Dialyze into 150 mM NaCl, 20 mM HEPES pH 7.4, 0.4 mM DTT for use.

Analyzing the Effect of Proteins on Liposome Conformation

1. Incubate liposomes (0.1 mg/ml lipid) and protein (0.1–0.5 mg/ml) in buffer for 5 min at room temperature.

2. Place 5 μl liposome mixture onto a glow discharged (hydrophilic) electron microscope grid and leave for 1 min to adhere. Blot excess buffer onto Whatman filter paper and stain by pipetting uranyl acetate onto grids, leave for 15 sec, blot, and then repeat. Blot gently, leaving a thin layer of stain on the grids, leave to air dry, and study in a transmission electron microscope.

Method 2: Analysis of Conformational Changes of Dynamin on Preformed Lipid Nanotubes

Introduction

Different lipid molecules associate together to form membrane structures with very different overall curvatures. When liposomes are made in the presence of sufficient concentrations of nonhydroxylated fatty-acid galactoceramides (NFA-GalCer) they form nanotubes (Wilson-Kubalek *et al.*, 1998). We have used these preformed lipid nanotubes to examine the effect of GTP hydrolysis on the conformation of a dynamin helix (Marks *et al.*, 2001; Stowell *et al.*, 1999). Indeed, in the absence of GTP hydrolysis, purified dynamin aggregates to form helices and, as seen above, can tubulate liposomes. Also, dynamin interacts with lipid nanotubes to form ordered arrays with a diameter similar to the constricted necks of clathrin-coats pits observed *in vivo* (Stowell *et al.*, 1999; Takei *et al.*, 1995), suggesting that the nanotube acts as a good mimic of the neck of a vesicle.

The analysis of dynamin helices formed on lipid tubules provides a picture of what happens to a dynamin assembly upon GTP hydrolysis, as dynamin adopts different conformations in the presence of different nucleotides. While the GTP bound state of dynamin (observed in the presence of GTPγS) forms a tight helix with a spacing of 11 nm, in the GDP bound state the spacing has nearly doubled to 20 nm (Stowell *et al.*, 1999).

This suggests a dramatic increase in the pitch of the helix upon GTP hydrolysis that may drive the scission of the vesicle neck: the "poppase" model for dynamin function. Further experiments, with mutants of dynamin, have shown that efficient GTP hydrolysis and the increase in pitch of the dynamin helix are required to allow endocytosis (Marks *et al.*, 2001).

We propose that lipid nanotubes will be important for analysis of the conformational states of other members of the dynamin superfamily. However, it should be noted that these tubules are conformationally stable and may not be as easily "squeezed" as liposomes. Therefore conformational changes perpendicular to the membrane plane may be underestimated. Thus (as with the tubulation assay described in Method 1) conclusions must be verified in live cells, where mutants can be expressed and trafficking monitored (Marks *et al.*, 2001).

Reagents and Equipment

Lipid Stocks and Buffers. Lipid nanotube formation uses the same equipment as liposome formation. The major difference is the use of NFA-GalCer (Sigma) as a replacement for phosphatidylethanolamine and phosphatidylserine. NFA-GalCer stocks are prepared to 10 mg/ml in chloroform and stored at $-80°$ under argon. The buffer used for the study of dynamin on lipid nanotubles is 135 mM NaCl, 5 mM KCl, 20 mM HEPES pH 7.4, 1 mM MgCl$_2$.

Procedure

Making Lipid Nanotubes

1. The key lipid component required to promote nanotube formation is NFA-GalCer (Sigma C 1516). Tubule formation requires in excess of 20% NFA-GalCer with a typical composition of (w/w) 10% cholesterol, 40% NFA-GalCer, 40% PtdChol, 10% PtdIns(4,5)P$_2$. This generates lipid tubes with a diameter of around 28 nm, which are filtered to give a uniform length (Fig. 2A).

2. Make nanotubes using the same protocol as for liposomes. After drying the lipid mixture, resuspend in buffer and sonicate as above. To form a homogenous preparation, extrude through a filter with a pore size of 0.2 μm before use.

Studying the Conformational Stage of Dynamin Spirals

1. Incubate nanotubes (0.1 mM lipid), dynamin (0.5 μM), and nucleotides (1 mM) for 10 min at room temperature. To observe the GTP bound form of dynamin use fresh GTPγS.

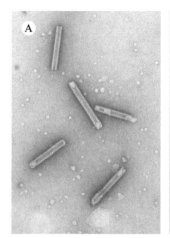
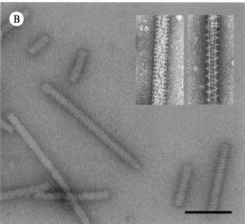

FIG. 2. Studying the conformation of dynamin on a lipid nanotube. (A) Undecorated lipid nanotubes. (B) Lipid nanotubes incubated with dynamin in the presence of GTPγS. The scale bar is 200 nm. The inserts show close-up views of dynamin on lipid tubules in the presence of GTPγS (left hand insert) and GDP (right hand insert) showing that the pitch of the dynamin spiral is greater when GDP bound than when GTPγS bound.

2. Place onto a glow discharged (hydrophilic) electron microscope grid and leave for 1 min to adhere. Blot excess buffer onto Whatman filter paper and stain for two 15 sec incubations with uranyl acetate. Blot gently, leaving a thin layer of stain on the grid surface, air dry, and observe by electron microscopy. Figure 2B shows dynamin spirals observed on lipid nanotubes.

Method 3: Analysis of Early Stages of Endocytosis on a Lipid Monolayer

Introduction

Lipid monolayers have been used for many years as templates for the formation of two-dimensional crystals (Chiu *et al.*, 1997), and for surface tension measurements (Stahelin *et al.*, 2003). We have used them as mimics of the inner leaflet of the plasma membrane to reconstitute the early stages of clathrin-coated vesicle formation (Ford *et al.*, 2001, 2002) and they can equally well be applied to other vesicle and tubule budding mechanisms. A lipid monolayer forms when a droplet of solvent-dissolved lipid is placed onto an aqueous droplet. As the solvent evaporates, lipid molecules become oriented at the air-water interface with head groups in contact with the aqueous mixture and hydrophobic tails extending into the air. Proteins of interest are injected underneath the monolayer and a hydrophobic electron microscope grid is placed onto its hydrophobic surface. The grid can be removed and stained for study in the electron microscope, revealing the structures that form on the aqueous surface of the monolayer. In addition, the grid can be platinum shadowed, with the height of shadows revealing the degree of invagination (or budding) of the membrane.

This assay is well suited for the study of clathrin coat formation as distinctive clathrin arrays can be readily observed by electron microscopy. It was first used to show the role of AP180 in promoting the formation of flat clathrin lattices on a membrane surface containing PtdIns(3,4)P$_2$ and the increased degree of invagination of these lattices upon the addition of the AP2 adaptor complex (Ford et al., 2001). Subsequently, it has been used to investigate the nucleation of clathrin lattices by both epsin (Ford et al., 2002) and disabled-2 (Mishra et al., 2002).

Reagents and Equipment

Buffers and Reagents. Nucleation of clathrin lattices by AP180 or AP2 was studied HKM buffer, consisting of 25 mM Hepes pH 7.4, 125 mM potassium acetate, 5 mM magnesium acetate, and 1 mM dithiothreitol. Lipid stocks and uranyl acetate stocks were prepared as above.

Purified Proteins. When analyzing clathrin lattice formation, clathrin was purified from pig brain as in Smith et al. (1998), dialyzed into HKM buffer, and centrifuged for 20 min at 100,000g immediately prior to use to remove aggregates. AP2 was also purified from pig brain as in Smith et al. (1998) and rat brain AP180 was expressed in baculovirus-infected Sf9 cells and purified as in Ford et al. (2001).

Monolayer Device. This consists of a Teflon block with 60-μl wells (as shown in Fig. 3A and B). Each well has two entrances. The lipid monolayer and electron microscope grid are placed on the large 35 mm diameter top entrance (A) while a smaller side entrance (B) is used for injection of protein samples underneath the monolayer. As traces of lipid contaminants on the Teflon block can result in misleading negative controls, the block should be cleaned thoroughly before and after use by rinsing with hot water, then ethanol, and finally, soaking overnight in a mixture of chloroform/methanol to remove any protein or lipid residue. A humid chamber is used to prevent evaporation during the experiment. A covered container with a wet sponge or paper towel is sufficient.

Carbon Coated Electron Microscopy Grids. These are prepared as described previously; however, if copper grids are used, some copper

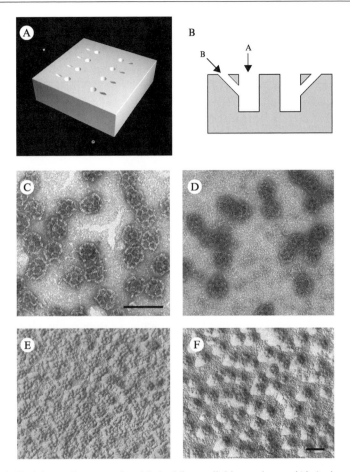

FIG. 3. Studying early stages of vesicle budding on lipid monolayers. (A) A view showing the Teflon block used to investigate the formation of clathrin coats on a lipid monolayer. (B) A cross section through two of the wells. Each well has two entrances. Entrance A is where the lipid monolayer is formed while entrance B is used to inject protein underneath the monolayer. (C) and (D) are images of negatively stained clathrin coats formed in the presence of (C) AP180 or (D) AP180 and AP2. The presence of added AP2 results in formation of more electron dense coats. (E) and (F) are images of platinum shadowed clathrin coats formed in the presence of (E) AP180 or (F) AP180 and AP2. The addition of AP2 leads to invagination of the coats as seen by the increase in the shadow length. Scale bars are 200 nm in length.

leaches from the grid during the incubation with the lipid monolayer. Therefore we use gold electron microscope grids (G204G from Agar Scientific, Stansted, Essex, England) which have been coated using the protocol described above. Forceps are used for handling EM grid, and self-locking spring forceps are especially useful. Whatman filter paper was used for blotting EM grids.

Platinum Shadowing. An E306A carbon evaporator is required. This is used in conjunction with a 10 cm long 0.2 mm diameter piece of platinum wire (TAAB Laboratories, Aldermaston, Berkshire, England) and a 10 cm long, 1 mm thick tungsten wire (also TAAB). Samples are viewed with a transmission electron microscope.

Procedures

Formation of a Lipid Monolayer

1. Prepare lipid mixtures to a total lipid concentration of 0.1 mg/ml in a 19:1 mixture of chloroform:methanol. Methanol is necessary to maintain the solubility of some lipids, including PtdIns(4,5)P$_2$. The lipid mixture can be tailored to the protein of interest, with 10% cholesterol, 40% PtdEth, 40% PtdSer, and 10% PtdIns(4,5)P$_2$ used for the analysis of AP180 and epsin mediated clathrin nucleation. An alternative is to use a brain lipid extract such as Folch extract. Lipid mixtures should be prepared fresh.

2. Arrange the Teflon block in a humid chamber and fill wells with buffer. It is important not to overfill wells, as a concave inner surface will influence monolayer formation.

3. Carefully pipette (or inject with Hamilton syringe) 1 μl of lipid mixture on to the larger entrance of each well (A). As a negative control, pure chloroform or a lipid mixture that will not bind the protein of interest can be injected. These controls are important when studying clathrin coat formation as traces of preformed clathrin cages will otherwise give misleading results. Incubate the block at room temperature for 60 min allowing chloroform to evaporate, leaving a monolayer of lipid on the surface of the buffer.

4. Carefully place one EM grid, carbon side down, onto the top of each buffer droplet. Grids should not be glow discharged before use as a hydrophobic carbon film is required to adhere to the hydrophobic lipid tails of the monolayer.

Analysis of Coat Formation

1. Gently inject proteins into the side injection entrance (B). When analyzing clathrin lattice formation, the final protein concentrations in the well should be 0.5–2 μM for AP180/epsin/adaptor protein, and 30–500 nM for clathrin. Incubate for 60 min at room temperature in the humid chamber.

2. On a piece of parafilm, place 15–20 µl droplets of stain for each grid. Then, to facilitate removal of the grid from the Teflon block, gently inject 30 µl of buffer into the side injection entrance (B in Fig. 3B), thereby raising the grid above the surface of the block. Immediately remove the grid with forceps and lift it vertically off of the droplet. Blot the grid briefly by touching it on a piece of filter paper, then the first stain droplet, and then blot immediately. Touch the grid to the second stain droplet, leave for 30 sec, and blot briefly. This negatively stains the sample by leaving a film of stain on the surface of the grid in which the protein is embedded. However, if the grids is to be platinum shadowed, hold it to the filter paper for several seconds before the stain dries to ensure the removal of excess stain as required for positive staining. Lay the grid on another piece of filter paper to dry.

3. Grids can be examined in a transmission electron microscope immediately, or stored at room temperature. The result of experiments investigating the effect of AP180 and AP2 on clathrin lattice formation in this way is shown in Figure 3C and D.

4. To investigate the degree of invagination of structures formed on the grid surface, a positively stained grid can be platinum shadowed. Set up the vacuum evaporator with a 2 cm long piece of platinum wire coiled tightly around a piece of 1 mm thick tungsten wire. Place the grids to be shadowed on a platform such that the platinum wire maintains an angle of about 10° to the plane of the grid. Place a shield between the grid and the platinum and create a vacuum in the evaporator. Turn on the current and when the platinum wire melts, remove the shield. Often two minutes of shadowing is sufficient, but this depends on the evaporator and should be optimized. The results of single angle platinum shadowing for clathrin lattices formed in the presence of AP180 and AP2 are shown in Fig. 3E and F.

Conclusions and Future Directions

The methods presented here provide snapshots of membrane budding and scission processes. Future developments will make the assays more quantitative rather than qualitative, allowing these studies to address many of the important questions remaining in understanding vesicle formation such as the mechanism of cargo incorporation into a coated vesicle, the energetics of vesicle generation, the use of alternative adaptor molecules, and the regulation of the process. No doubt these techniques will also be applied to many other different membrane budding and scission mechanisms and will help in the assignments of protein functions. The ultimate goal is reconstitution of complete budding pathways *in vitro* as more and more proteins are purified and added into systems like the ones presented.

References

Chiu, W., Avila-Sakar, A. J., and Schmid, M. F. (1997). Electron crystallography of macromolecular periodic arrays on phospholipid monolayers. *Adv. Biophys.* **34,** 161–172.

Farsad, K., Ringstad, N., Takei, K., Floyd, S. R., Rose, K., and De Camilli, P. (2001). Generation of high curvature membranes mediated by direct endophilin bilayer interactions. *J. Cell Biol.* **155,** 193–200.

Ford, M. G., Pearse, B. M., Higgins, M. K., Vallis, Y., Owen, D. J., Gibson, A., Hopkins, C. R., Evans, P. R., and McMahon, H. T. (2001). Simultaneous binding of PtdIns(4,5)P2 and clathrin by AP180 in the nucleation of clathrin lattices on membranes. *Science* **291,** 1051–1055.

Ford, M. G., Mills, I. G., Peter, B. J., Vallis, Y., Praefcke, G. J., Evans, P. R., and McMahon, H. T. (2002). Curvature of clathrin-coated pits driven by epsin. *Nature* **419,** 361–366.

Hinshaw, J. E., and Schmid, S. L. (1995). Dynamin self-assembles into rings suggesting a mechanism for coated vesicle budding. *Nature* **374,** 190–192.

Marks, B., Stowell, M. H.B, Vallis, Y., Mills, I. G., Gibson, A., Hopkins, C. R., and McMahon, H. T. (2001). GTPase activity of dynamin and resulting conformational change are essential for endocytosis. *Nature* **410,** 231–235.

Mishra, S. K., Keyel, P. A., Hawryluk, M. J., Agostinelli, N. R., Watkins, S. C., and Traub, L.M (2002). Disabled-2 exhibits the properties of a cargo-selective endocytic clathrin adaptor. *EMBO J.* **21,** 4915–4926.

Owen, D. J., Wigge, P., Vallis, Y., Moore, J. D., Evans, P. R., and McMahon, H. T. (1998). Crystal structure of the amphiphysin-2 SH3 domain and its role in the prevention of dynamin ring formation. *EMBO J.* **17,** 5273–5285.

Peter, B. J., Kent, H. M., Mills, I. G., Vallis, Y., Butler, P. J., Evans, P. R., and McMahon, H. T. (2004). BAR domains as sensors of membrane curvature: The amphiphysin BAR structure. *Science* **303,** 495–499.

Praefcke, G. J. K., and McMahon, H. T. (2004). The dynamin superfamily: Universal membrane tubulation and fission molecules. *Nat. Rev. Mol. Cell Biol.* **5,** 133–147.

Razzaq, A., Robinson, I. M., McMahon, H. T., Skepper, J. N., Su, Y., Zelhof, A. C., Jackson, A. P., Gay, N. J., and O'Kane, C. J. (2001). Amphiphysin is necessary for organization of the excitation-contraction coupling machinery of muscles, but not for synaptic vesicle endocytosis in *Drosophila*. *Genes Dev.* **15,** 2967–2979.

Smith, C. J., Grigorieff, N., and Pearse, B. M. (1998). Clathrin coats at 21Å resolution: A cellular assembly designed to recycle multiple membrane receptors. *EMBO J.* **17,** 4943–4953.

Stahelin, R. V., Long, F., Peter, B. J., Murray, D., De Camilli, P., McMahon, H. T., and Cho, W. (2003). Contrasting membrane interaction mechanisms of AP180 N-terminal homology (ANTH) and epsin N-terminal homology (ENTH) domains. *J. Biol. Chem.* **278,** 28993–28999.

Stowell, M. H. B., Marks, B., Wigge, P., and McMahon, H. T. (1999). Nucleotide-dependent conformational changes in dynamin: Evidence for a mechanochemical molecular spring. *Nat. Struct. Biol.* **1,** 27–32.

Sweitzer, S. M., and Hinshaw, J. E. (1998). Dynamin undergoes a GTP-dependent conformational change causing vesiculation. *Cell* **93,** 1021–1029.

Takei, K., McPherson, P. S., Schmid, S. L., and De Camilli, P. (1995). Tubular membrane invaginations coated by dynamin rings are induced by GTPγS in nerve terminal. *Nature* **374,** 186–190.

Takei, K., Slepnev, V. I., Haucke, V., and De Camilli, P. (1999). Functional partnership between amphiphysin and dynamin in clathrin-mediated endocytosis. *Nat. Cell Biol.* **1,** 33–39.

Wilson-Kubalek, E. M., Brown, R. E., Celia, H., and Milligan, R. A. (1998). Lipid nanotubes as substrates for helical crystallisation of macromolecules. *Proc. Natl. Acad. Sci. USA* **95,** 8040–8045.

Zhang, P., and Hinshaw, J. E. (2001). Three-dimensional reconstitution of dynamin in the constricted state. *Nat. Cell Biol.* **3,** 922–926.

[53]   A Continuous, Regenerative Coupled GTPase Assay for Dynamin-Related Proteins

By ELENA INGERMAN and JODI NUNNARI

Abstract

Dynamin-related proteins (DRPs) compose a diverse family of proteins that function, through GTPase stimulated self-assembly, to remodel cellular membranes. The molecular mechanism by which DRPs mediate membrane remodeling events and the specific role of their GTPase cycle is still not fully understood. Although DRPs are members of the GTPase superfamily, they possess unique kinetic properties. In particular, they have relatively low affinity for guanine nucleotides and, under conditions that favor self-assembly, they have high rates of GTP turnover. Established fixed time point assays used for the analysis of assembly stimulated GTPase activity are prone to inaccuracies due to substrate depletion and are also limited by lack of time resolution. We describe a simple, continuous, coupled GTP regenerating assay that tackles the limitations of the fixed time point assays and can be used for the kinetic analysis of DRP GTP hydrolysis under unassembled and assembled conditions.

Introduction

Dynamin-related GTPases (DRPs) have evolved in eukaryotic cells to function in such diverse processes as membrane trafficking, organelle division, and resistance to viral infection (Danino and Hinshaw, 2001; Osteryoung and Nunnari, 2003; Praefcke and McMahon, 2004; Song and

METHODS IN ENZYMOLOGY, VOL. 404
0076-6879/05 $35.00
DOI: 10.1016/S0076-6879(05)04053-X

Schmid, 2003). The *S. cerevisiae* genome encodes three DRPs: Dnm1 and Mgm1, which function in the division and fusion of mitochondria, respectively, and Vps1, which functions in the biogenesis of Golgi-derived vesicles and in the division of peroxisomes (Ekena *et al.*, 1993; Gurunathan *et al.*, 2002; Hoepfner *et al.*, 2001; Osteryoung and Nunnari, 2003; Vater *et al.*, 1992). Mammalian cells possess homologs of Dnm1 (Drp1) and Mgm1 (Opa1) that also function in the regulation of apoptosis, in addition to other DRPs, such as dynamin-1, which mediates the scission of clathrin-coated pits from the plasma membrane during endocytosis and is the prototype of the DRP super family (Hinshaw, 2000; Osteryoung and Nunnari, 2003; Schmid, 1997; Youle, 2005).

Recently determined X-ray crystal structures indicate that DRPs possess a core fold similar to that of all regulatory GTPases, thus establishing them as members of the GTPase super family (Klockow *et al.*, 2002; Niemann *et al.*, 2001; Prakash *et al.*, 2000). However, in addition to their relatively large mass, the kinetic properties of DRPs are unique among other classes of GTPases, such as the Ras and heterotrimeric $G\alpha$ families (Song and Schmid, 2003). As well as a GTPase domain, all DRPs possess three other hallmark domains: a middle domain of ~150 amino acids, a divergent sequence called insert B, and a C-terminal assembly or GTPase effector domain (GED) (van der Bliek, 1999). These domains cooperate via both intra- and intermolecular interactions to promote the self-assembly of higher-order DRP structures.

Most of the information concerning DRP function has come from the analysis of dynamin-1. *In vitro*, dynamin self-assembles into spiral-like filaments (Hinshaw and Schmid, 1995). Self-assembly stimulates dynamin's GTPase activity and is required for dynamin's ability to remodel membranes during endocytosis (Damke *et al.*, 2001a; Song *et al.*, 2004; Warnock *et al.*, 1996). In DRPs, GED/GED and GED/GTPase domain interactions are required for self-assembly and assembly- driven GTPase activity (Sever *et al.*, 1999; Smirnova *et al.*, 1999). However, the molecular mechanism of GED-stimulated GTPase activity is currently unknown. The relative affinities of dynamin for GTP and GDP are lower than those of Ras and $G\alpha$ and the rate of GTP hydrolysis of dynamin is significantly higher in the assembled versus unassembled state (Damke *et al.*, 2001a; Eccleston *et al.*, 2002; Warnock *et al.*, 1996). In addition, the dissociation rate of GDP from dynamin is fast in comparison to that of Ras, indicating that GDP release is not likely to be a rate-limiting step in the dynamin GTPase cycle and obviating the need for a nucleotide exchange factor *in vivo* (Damke *et al.*, 2001a; Eccleston *et al.*, 2002).

Based upon its activities and kinetic properties, dynamin has been postulated to play a mechanochemical role in severing endocytic vesicles from the plasma membrane (Hinshaw and Schmid, 1995). However, recent findings suggest that dynamin and other DRPs function as classical signaling GTPases, which, in their GTP-bound form, recruit downstream effectors, such as endophilin, which are responsible for membrane severing (Fukushima *et al.*, 2001; Newmyer *et al.*, 2003; Sever, 2002; Sever *et al.*, 2000a,b, 1999). The *in vitro* kinetic properties of dynamin predict that it would exist *in vivo* predominantly in the GTP-bound form (Song and Schmid, 2003). Thus, in order to function as a classical switch GTPase, other factors, such as interacting proteins, would be predicted to significantly modulate the kinetic properties of DRPs, such that production of the GTP-bound form would be rate-limiting (Song and Schmid, 2003). Thus, while our knowledge of the properties of DRPs is considerable, the exact mechanistic roles played by DRPs in membrane remodeling events are unresolved at present.

Methods for the detailed and accurate kinetic analysis of GTPase activity are essential tools for unraveling the exact molecular role of assembling DRPs in membrane remodeling events (Damke *et al.*, 2001b). Two fundamentally different types of GTPase assays are available for measuring activity: fixed time point and continuous assays. To date, only fixed time point assays have been used to characterize DRP GTPase activity. Specifically, these include a radioactive [^{32}P]GTP assay and a nonradioactive malachite green colorimetric assay, which measures inorganic phosphate release (Lanzetta *et al.*, 1979). In fixed time point assays, we have found that the relatively high rates of GTP hydrolysis by DRPs under self-assembling conditions cause substrate depletion, resulting in inaccurate estimates of K_m, V_{max}, and other kinetic parameters. In addition, fixed time point assays do not have the time resolution required to detect rapid changes in the rates of GTP hydrolysis over time, as might occur upon DRP self-assembly. For example, a lag phase would be overlooked in a fixed time point assay, resulting in inaccurately low estimates of GTP hydrolysis rates.

To overcome these problems, we have employed a continuous GTPase assay, which includes a GTP regenerating system, for the kinetic analysis of Dnm1 GTPase activity. Interestingly, this assay has revealed two unique kinetic parameters that were not detected using standard fixed time point analyses of GTPase activity (Ingerman *et al.*, 2005). First, we found that both Dnm1 and dynamin-1 exhibit significant lags (5 min), before reaching a steady state GTPase activity. In a fixed time point assay, this lag was not detected and, thus, the measured rate of Dnm1's GTP hydrolysis was underestimated. In other self-assembling systems, such as F-actin and mictrotublues, a kinetic lag is indicative of the rate limiting formation of a structure that nucleates self-assembly, which we now postulate to be the case for DRPs. Second, we found that Dnm1 exhibits significant cooperativity, with respect to GTP concentration, during steady state GTPase activity (Hill coefficient = 2.8–1.4), suggesting that an intrinsic GTP dependent switch is functioning to stimulate self-assembly. Here, we describe this assay, which also will be of general use in the characterization of other GTPases possessing relatively high turnover rates for GTP hydrolysis.

Continuous GTPase Assay

Our continuous, coupled substrate regenerating assay has been modified from an assay used previously for measuring the enzymatic activities of ATPases (Renosto *et al.*, 1984). We have adapted this assay for measuring the activity of the dynamin-related GTPase Dnm1, in 96-well plate format (Kiianitsa *et al.*, 2003). Recently, this assay also has been used to characterize the activity of the self-assembling bacterial tubulin-like GTPase, FtsZ (Margalit *et al.*, 2004).

In the continuous, substrate-regenerating assay, Dnm1 hydrolyzes GTP into GDP and inorganic phosphate. As shown in Fig. 1, GTP is regenerated from GDP and phospho(enol)pyruvate (PEP) by pyruvate kinase. The pyruvate produced by this regenerative reaction is reduced to lactate by the enzyme lactate dehydrogenase, using NADH as a cosubstrate. NADH depletion is measured by continuously monitoring a decrease in absorbance at 340 nm over time and is directly proportional to GTP hydrolysis.

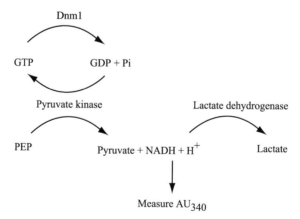

Fig. 1. Schematic representation of the coupled GTP regenerating assay for GTP hydrolysis. See text for a detailed description of each reaction in the assay.

Protein Purification and Storage

Dnm1 is purified from baculovirus-infected Hi5 cells in 25 mM HEPES, 25 mM PIPES, pH 7.0, 500 mM NaCl, 500 mM imidazole, pH 7.4, as described (Ingerman *et al.*, 2005). For maximal protein stability, dimethyl sulfoxide (DMSO) is added to a final concentration (v/v) of 20% to purified protein (~0.8 mg/mL Dnm1), rendering the final freezing buffer concentrations 20 mM HEPES, 20 mM PIPES, 400 mM NaCl, 400 mM imidazole, 20% DMSO. Aliquots of protein are flash frozen in liquid nitrogen and stored at −80°.

Reagents

20× reaction buffer (20× RB) containing 25 mM HEPES, 25 mM PIPES, pH 7.0, 100 mM MgCl$_2$, 20 mM phospho(enol)pyruvate (PEP), and 150 mM KCl, is prepared in advance and stored at −80° in aliquots. Similarly, stock solutions of 50 mM GTP (in H$_2$O, pH adjusted to 7 with 0.5 M NaOH) and 20 mM NADH (in 25 mM HEPES, 25 mM PIPES, pH 7.0) are also made in advance and stored at −80° in aliquots. We routinely verify the concentration of prepared GTP stock solutions spectrophotometrically, by measuring the absorbance of the prepared stock solution at 253 nm (molar absorbtivity of GTP at 253 nm = 13700 M^{-1}cm^{-1}).

Each GTPase assay reaction contains approximately 25 mM HEPES, 25 mM PIPES, pH 7.0, 150 mM NaCl, 7.5 mM KCl, approximately 20 U/mL of pyruvate kinase/lactate dehydrogenase mixture (Sigma P0294, St. Louis, MO), 5 mM MgCl$_2$, 1 mM PEP (Sigma P7002), 600 μM NADH (Sigma N8129), variable amounts of GTP (Sigma G8877), and 0.06 mg/mL of Dnm1. Each reaction also contains 30 mM imidazole and 1.5% DMSO, two additional components of the freezing buffer.

Procedure

For a complete kinetic analysis of Dnm1, we measure the GTPase activity of Dnm1 at the following GTP concentrations: 10, 25, 50, 100, 150, 200, 300, 400, 500, 600, 750, and 1000 μM.

We prepare 10× GTP stock solutions by diluting the 50 mM GTP stock in 25 mM HEPES, 25 mM PIPES, pH 7.0 to 0.1, 0.25, 0.5, 1, 1.5, 2, 3, 4, 5, 6, 7.5, and 10 mM. The use of 10× GTP stock solutions allows us to place a uniform volume of GTP into each reaction tube, resulting in better reproducibility. Each 200 μL GTPase assay reaction contains:

10 μL 20× RB
6 μL 20 mM NADH

12 μL 2M NaCl (this amount accounts for the 15 μL of 400 mM NaCl contained in the Dnm1 freezing buffer and yields a final NaCl concentration of 150 mM)

132 μL 25 mM HEPES, 25 mM PIPES, pH 7.0

5 μL pyruvate kinase/lactate dehydrogenase mixture

20 μL 10× GTP stock solution

15 μL Dnm1 (~0.8 mg/mL Dnm1 in freezing buffer).

For greater accuracy and reproducibility, a master mix containing 20× RB, NADH, NaCl, HEPES/PIPES, and pyruvate kinase/lactate dehydrogenase is prepared. Thus, for every reaction we add 20 μL of 10× GTP stock and 165 μL of master mix. To start the reaction, 15 μL of Dnm1 are added and the contents of the tube are quickly and gently using a pipettor. For better reproducibility and precision, 150 μL of the 200 μL reaction are immediately aliquoted into a 96-well plate.

Absorbance measurements are recorded in a SpectraMAX Pro (Molecular Devices, Sunnyvale, CA) 96-well plate reader that has been preheated to 30° and we collect absorbance readings at 340 nm (λ_{max} for NADH) at 20-sec intervals for 40 min. In order to assess the sensitivity of the assay and to account for auto-hydrolysis/nonspecific GTPase activity, one of the reactions in the 96-well plate should be a "blank," in which 15 μL of freezing buffer is used, in place of Dnm1, to start the reaction.

The values for absorbance versus time are dependent on the concentration of protein being assayed and its GTP turnover rate. At 10 μM GTP, using Dnm1 (MW 87000) at a final assay concentration of 0.057 mg/mL, Dnm1 produces a slope of 2.6×10^{-5} AU$_{340}$/second, which translates into an activity of 1 min^{-1}. At 1 mM GTP, a slope of 1.3×10^{-3} AU$_{340}$/second is typically observed, which translates into an activity of 52 min^{-1}.

In the 96-well plate format, the path length of the assay is dependent upon the volume placed into the well. Determining the precise path length is essential for correctly converting data in the form of AU340 versus time to an accurate GTPase activity. To measure the path length, we first accurately measure the true concentration of NADH in a GTP-free reaction solution (containing 25 mM HEPES, 25 mM PIPES, pH 7.0, 150 mM NaCl, 7.5 mM KCl, approximately 20 U/mL of pyruvate kinase/lactate dehydrogenase mixture, 5 mM MgCl$_2$, 1 mM PEP, approximately 100 μM NADH, 30 mM imidazole, 1.5% DMSO, and 0.06 mg/mL of Dnm1) using a quartz cuvette with a known path length of 1 cm and a standard spectrophotometer. A corresponding NADH-free solution is used as a reference blank. An aliquot of 150 μL of the NADH-free and the NADH-containing solution is placed into separate wells of a 96-well plate. The average AU$_{340}$ measurement of six NADH-free and six NADH-containing wells is determined and the average absorbance of the NADH-free wells is subtracted from the NADH-containing wells. (Net absorbance values are typically 0.235). Using the experimentally determined concentration of NADH and the Beer-Lambert law ($A = \varepsilon \cdot c \cdot l$), we calculate the path length of a 150 μL well volume, which, under our assay conditions, is 0.38 cm.

Notes on the Assay

Excessive amounts of imidazole (>40 mM) inhibit the GTPase activity of Dnm1, as well as that of other DRPs. DMSO, in amounts up to 10%, has a slight stimulatory effect on GTPase activity, likely via an effect on self-assembly. 150 mM NaCl was chosen as a physiological salt concentration. In practice, DRPs should be assayed using a range of NaCl concentrations (20,500 mM), in order to assess their kinetic properties under both assembly and nonassembly conditions.

NADH is not regenerated in this assay because its depletion is required for monitoring protein activity by the coupled assay. Thus, the initial concentration of NADH in the assay, while not directly affecting the activity of a protein, may need to be increased for DRPs with especially high turnovers (>50 min^{-1}), in order to allow continuous measurement of protein activity over longer periods of time (>30 min).

Data Analysis

Using Microsoft Excel, we plot absorbance data versus time. Because Dnm1 exhibits a lag in reaching steady state GTP hydrolysis, we do not utilize data from early time points in our calculation of steady state GTPase velocities. Instead, we determine the hydrolysis rate using absorbance data that are linear with respect to time. Using the "add trendline" function of Microsoft Excel, we fit a line to each plot of absorbance versus time that is obtained for each different GTP concentration. The slope of this line is multiplied by 60 (sec/min) and divided by the molar absorbitivity of NADH (6220 M^{-1}cm^{-1}) and by the previously determined path length (0.38 cm). This operation converts the slope of the plot of absorbance versus time (AU340/sec) to the velocity of GTP hydrolysis (M/min). The velocity is then converted to a protein activity by multiplying the velocity by the molecular weight of Dnm1 and dividing by the concentration of Dnm1 protein in the assay. Activity at each GTP concentration is plotted as a function of GTP concentration and the Genfit function of Mathcad (MathSoft, Inc., Cambridge, MA), which is an algorithm for fitting data to a curve using nonlinear least squares regression, is used to determine k_{cat}, K_m or $K_{0.5}$, and the Hill coefficient.

Acknowledgments

We are grateful to Dr. Irwin Segel for his invaluable advice. We thank members of the lab for thoughtful comments on the manuscript. This work was support by NIH grants to J.N. (1R01EY015924 and 5R01GM062942). E.I. is supported by an NIH training grant (5T32GM007377).

References

Damke, H., Binns, D. D., Ueda, H., Schmid, S. L., and Baba, T. (2001a). Dynamin GTPase domain mutants block endocytic vesicle formation at morphologically distinct stages. *Mol. Biol. Cell* **12**, 2578–2589.

Damke, H., Muhlberg, A. B., Sever, S., Sholly, S., Warnock, D. E., and Schmid, S. L. (2001b). Expression, purification, and functional assays for self-association of dynamin-1. *Methods Enzymol.* **329**, 447–457.

Danino, D., and Hinshaw, J. E. (2001). Dynamin family of mechanoenzymes. *Curr. Opin. Cell Biol.* **13**, 454–460.

Eccleston, J. F., Binns, D. D., Davis, C. T., Albanesi, J. P., and Jameson, D. M. (2002). Oligomerization and kinetic mechanism of the dynamin GTPase. *Eur. Biophys. J.* **31**, 275–282.

Ekena, K., Vater, C. A., Raymond, C. K., and Stevens, T. H. (1993). The VPS1 protein is a dynamin-like GTPase required for sorting proteins to the yeast vacuole. *Ciba Found Symp.* **176**, 198–211; discussion 211–214.

Fukushima, N. H., Brisch, E., Keegan, B. R., Bleazard, W., and Shaw, J. M. (2001). The AH/GED sequence of the dnm1p GTPase regulates self-assembly and controls a rate-limiting step in mitochondrial fission. *Mol. Biol. Cell* **12**, 2756–2766.

Gurunathan, S., David, D., and Gerst, J. E. (2002). Dynamin and clathrin are required for the biogenesis of a distinct class of secretory vesicles in yeast. *EMBO J.* **21**, 602–614.

Hinshaw, J. E. (2000). Dynamin and its role in membrane fission. *Annu. Rev. Cell Dev. Biol.* **16**, 483–519.

Hinshaw, J. E., and Schmid, S. L. (1995). Dynamin self-assembles into rings suggesting a mechanism for coated vesicle budding. *Nature (London)* **374**, 190–192.

Hoepfner, D., van den Berg, M., Philippsen, P., Tabak, H. F., and Hettema, E. H. (2001). A role for Vps1p, actin, and the Myo2p motor in peroxisome abundance and inheritance in *Saccharomyces cerevisiae. J. Cell Biol.* **155**, 979–990.

Ingerman, E., Perkins, E. M., Marino, M., Mears, J. A., McCaffery, M., Hinshaw, J., and Nunnari, J. (2005). Dnml forms spirals that are structurally tailored to fit mitochondria. In preparation.

Kiianitsa, K., Solinger, J. A., and Heyer, W. D. (2003). NADH-coupled microplate photometric assay for kinetic studies of ATP-hydrolyzing enzymes with low and high specific activities. *Anal. Biochem.* **321**, 266–271.

Klockow, B., Tichelaar, W., Madden, D. R., Niemann, H. H., Akiba, T., Hirose, K., and Manstein, D. J. (2002). The dynamin A ring complex: molecular organization and nucleotide-dependent conformational changes. *EMBO J.* **21**, 240–250.

Lanzetta, P., Alvarez, L., Reinach, P., and Candia, O. (1979). An improved assay for nanomole amounts of inorganic phosphate. *Anal. Biochem.* **100**, 95–97.

Margalit, D. N., Romberg, L., Mets, R. B., Hebert, A. M., Mitchison, T. J., Kirschner, M. W., and Ray Chaudhuri, D. (2004). Targeting cell division: Small-molecule inhibitors of FtsZ

GTPase perturb cytokinetic ring assembly and induce bacterial lethality. *Proc. Natl. Acad. Sci. USA* **101,** 11821–11826.

Newmyer, S. L., Christensen, A., and Sever, S. (2003). Auxilin-dynamin interactions link the uncoating ATPase chaperone machinery with vesicle formation. *Dev. Cell* **4,** 929–940.

Niemann, H. H., Knetsch, M. L., Scherer, A., Manstein, D. J., and Kull, F. J. (2001). Crystal structure of a dynamin GTPase domain in both nucleotide-free and GDP-bound forms. *EMBO J.* **20,** 5813–5821.

Osteryoung, K. W., and Nunnari, J. (2003). The division of endosymbiotic organelles. *Science* **302,** 1698–1704.

Praefcke, G. J., and McMahon, H. T. (2004). The dynamin superfamily: Universal membrane tubulation and fission molecules? *Nat. Rev. Mol. Cell Biol.* **5,** 133–147.

Prakash, B., Praefcke, G. J., Renault, L., Wittinghofer, A., and Herrmann, C. (2000). Structure of human guanylate-binding protein 1 representing a unique class of GTP-binding proteins. *Nature* **403,** 567–571.

Renosto, F., Seubert, P. A., and Segel, I. H. (1984). Adenosine 5′-phosphosulfate kinase from Penicillium chrysogenum. Purification and kinetic characterization. *J. Biol. Chem.* **259,** 2113–2123.

Schmid, S. L. (1997). Clathrin-coated vesicle formation and protein sorting: An integrated process. *Annu. Rev. Biochem.* **66,** 511–548.

Sever, S. (2002). Dynamin and endocytosis. *Curr. Opin. Cell Biol.* **14,** 463–467.

Sever, S., Damke, H., and Schmid, S. L. (2000a). Dynamin: GTP controls the formation of constricted coated pits, the rate limiting step in clathrin-mediated endocytosis. *J. Cell Biol.* **150,** 1137–1147.

Sever, S., Damke, H., and Schmid, S. L. (2000b). Garrotes, springs, ratchets, and whips: Putting dynamin models to the test. *Traffic* **1,** 385–392.

Sever, S., Muhlberg, A. B., and Schmid, S. L. (1999). Impairment of dynamin's GAP domain stimulates receptor-mediated endocytosis. *Nature (London)* **398,** 481–486.

Smirnova, E., Shurland, D.-L., Newman-Smith, E. D., Pishvaee, B., and van der Bliek, A. M. (1999). A model for dynamin self-assembly based on binding between three different protein domains. *J. Biol. Chem.* **274,** 14942–14947.

Song, B. D., and Schmid, S. L. (2003). A molecular motor or a regulator? Dynamin's in a class of its own. *Biochem.* **42,** 1369–1376.

Song, B. D., Yarar, D., and Schmid, S. L. (2004). An assembly-incompetent mutant establishes a requirement for dynamin self-assembly in clathrin-mediated endocytosis *in vivo. Mol. Biol. Cell* **15,** 2243–2252.

van der Bliek, A. M. (1999). Functional diversity in the dynamin family. *Trends Cell Biol.* **9,** 96–102.

Vater, C. A., Raymond, C. K., Ekena, K., Howald-Stevenson, I., and Stevens, T. H. (1992). The VPS1 protein, a homolog of dynamin required for vacuolar protein sorting in *Saccharomyces cerevisiae*, is a GTPase with two functionally separable domains. *J. Cell Biol.* **119,** 773–786.

Warnock, D. E., Hinshaw, J. E., and Schmid, S. L. (1996). Dynamin self-assembly stimulates its GTPase activity. *J. Biol. Chem.* **271,** 22310–22314.

Youle, R. J. (2005). Morphology of mitochondria during apoptosis: Worms-to-beetles in worms. *Dev. Cell* **8,** 298–299.

[54] Assay and Properties of the Mitochondrial Dynamin Related Protein Opa1

By Lorena Griparic and Alexander M. van der Bliek

Abstract

Opa1, also known as Mgm1 in yeast, is a mitochondrial member of the dynamin family. Unlike other dynamin family members, Opa1 has an N-terminal mitochondrial targeting sequence, suggesting that this protein is imported into mitochondria. Here, we describe biochemical techniques, such as mitochondrial isolation, digitonin extraction, a protease protection assay, and carbonate extraction, that were used to determine that mammalian Opa1 resides in the intermembrane space where it is tightly bound to the inner membrane. In addition, we describe bacterial expression of the Opa1 GTPase domain, methods for purification, and an *in vitro* assay for GTP hydrolysis.

Introduction

Opa1, also known as Mgm1 in yeast, is a mitochondrial member of the dynamin family. Like other dynamin family members, Opa1 contains a GTPase domain, a middle domain, and a GED (GTPase Effector Domain), or assembly domain (Fig. 1). In addition, Opa1 has an N-terminal mitochondrial targeting sequence, suggesting that this protein is imported into mitochondria. Using immuno-fluorescence and immuno-electron microscopy combined with various biochemical techniques, such as digitonin extraction, a protease protection assay, and carbonate extraction, we determined that mammalian Opa1 resides in the intermembrane space where it is tightly bound to the inner membrane (Griparic *et al.*, 2004). Similar results were also obtained with yeast and by others in mammalian cells (Olichon *et al.*, 2002; Satoh *et al.*, 2003; Wong *et al.*, 2000). Overexpression of both wild type and mutant Opa1 in transfected mammalian cells caused mitochondria to fragment into vesicular structures. The mutation that we used (K301A) is located in the G1 consensus sequence of the Opa1 GTPase domain, similar to mutations that were previously used to manipulate dynamin and Drp1 (Labrousse *et al.*, 1999; Smirnova *et al.*, 1998; van der Bliek *et al.*, 1993). An *in vitro* assay for GTPase activity was used to show that this mutation inhibits GTP hydrolysis. The assay for hydrolysis and the biochemical methods for determining Opa1 localization are described in this chapter.

METHODS IN ENZYMOLOGY, VOL. 404   0076-6879/05 $35.00
Copyright 2005, Elsevier Inc. All rights reserved.   DOI: 10.1016/S0076-6879(05)04054-1

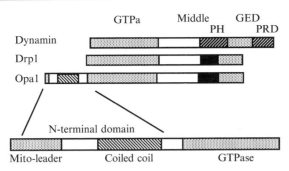

FIG. 1. Selected dynamin family members in mammals. The three dynamin family members that have been studied in our lab are shown here. Dynamin is involved in vesicular traffic. Drp1 contributes to mitochondrial outer membrane division. Opa1 is localized to the mitochondrial intermembrane space where it affects inner membrane morphology and mitochondrial fusion. All family members have GTPase, Middle, and Assembly domains. Genuine dynamins have a pleckstrin homology (PH) domain and a C-terminal proline rich domain (PRD). Other family members have divergent segments at the N-terminus (Opa1) or between the middle domain and GED (black areas in Drp1 and Opa1). The N-terminus of Opa1 has a predicted mitochondrial targeting sequence and coiled coil.

Mitochondrial Isolation

Mitochondria are isolated from bovine brain or liver. The starting material should be fresh in order to obtain intact mitochondria. The tissues are purchased directly from a slaughterhouse, cut into smaller pieces on site, and transported to the lab immersed in ice-cold mitochondrial isolation buffer (see following). This is an adaptation of more generalized protocols for subcellular fractionation by differential centrifugation (Graham and Rickwood, 1997).

Procedure

1. Weigh 50 g of liver or brain tissue and quickly transfer the dish to ice. All subsequent steps are performed on ice.

2. Dice the tissue with a scalpel and transfer to 250 ml isolation buffer (70 mM sucrose, 220 mM mannitol, 2 mM HEPES pH 7.4, and 0.5 mg/ml bovine serum albumin) with protease inhibitor cocktail (Roche Diagnostics, Basel, Switzerland). Homogenize the tissue with three passes of a loose-fitting pestle in a prechilled Potter-Elvejhem homogenizer, followed by five passes of a tight-fitting pestle.

3. Subject the homogenate to centrifugation for 15 min at 1,000g in a prechilled Sorvall SS-34 rotor at +4°. This low-speed pellet (P1), which contains nuclei and unbroken cells, is discarded.

4. The low-speed supernatant (S1) is transferred to a new SS-34 tube and subjected to centrifugation at 10,000g for 15 min, to produce a medium-speed pellet (P2), which contains mitochondria and lysosomes, and a medium-speed supernatant S2, which is discarded. Wash the P2 fraction by resuspending the pellet in isolation buffer and recentrifugation at 10,000g. Repeat the wash step.

5. Optional: If additional cellular fractions are needed, then the S2 fraction can be further processed using sucrose gradients or by centrifugation at a higher speed. For example, centrifugation for 1 h at 100,000g will yield a supernatant (S3) that contains light membranes, such as ER and Golgi, while the supernatant will contain cytosol.

If a more pure mitochondrial fraction is desired, then mitochondria can be separated from lysosomes by Percoll gradient centrifugation. Percoll stock solution can be purchased from GE Healthcare (Piscataway, NJ).

1. Always make sure Percoll is mixed before use, since it will settle over time.
2. Add stock solutions of sucrose, Tris, and EGTA to the Percoll to make 80% Percoll with a final concentration of 250 mM sucrose, 10 mM Tris pH 7.5, 1 mM EGTA (buffer A).
3. Resuspend the P2 fraction in buffer A and add to the 80% Percoll solution with buffer A to bring the final Percoll concentration to 40%.
4. Centrifugate the mixture for 40 min at 66,000g in a Ti70 rotor (Beckman Instruments, Palo Alto, CA) at +4°. Mitochondria collect in a dark brown band.
5. Carefully aspirate the mitochondrial band with a needle and dilute the sample with at least five volumes of buffer A.
6. Pellet the mitochondria by centrifugation for 15 min at 10,000g and wash once again in buffer A. Resuspend the final pellet containing pure mitochondria in a small volume of buffer A.

Digitonin Fractionation of Mitochondria

To determine protein submitochondrial localization of proteins, either hypotonic lysis, which preferentially disrupts the mitochondrial outer membrane, or selective solubilization with digitonin are used. The digitonin method is based on the observation that digitonin preferentially extracts cholesterol, which is more prevalent in the mitochondrial outer membrane (Parsons and Yano, 1967). Low amounts of digitonin make holes in the outer membrane, thus releasing proteins from the intermembrane space.

Increasing digitonin concentrations extract outer membrane proteins as well. Even higher digitonin concentrations make holes in the mitochondrial inner membrane, releasing mitochondrial matrix proteins, while the highest digitonin concentrations also solubilize integral inner membrane proteins. Each preparation of digitonin is different. It is therefore important to titer the amount of detergent over protein (mg digitonin/mg protein) by monitoring the extraction of markers for each of the four different compartments (intermembrane space, outer membrane, matrix, and inner membrane) at different detergent concentrations. Once release conditions are established, the release of the protein of interest can be compared with those of different markers.

We tested cow liver and brain mitochondria for our studies. We noticed that brain mitochondria are resistant to digitonin extraction, so all described experiments were done with the liver P2 fraction, which contains mitochondria. The procedure was based on a protocol described by Greenawalt (1974).

Procedure

1. Prepare a 1.5–2% stock solution of digitonin in isolation buffer (see Introduction) without bovine serum albumin. The detergent is dissolved in isolation buffer brought to the boiling point. Once the digitonin is completely dissolved, place the tube on ice.

2. Prepare a series of dilutions of digitonin in 100 μl aliquots of isolation buffer. The lowest concentration has 0% digitonin, while the highest concentration is that of the undiluted stock solution made in step 1.

3. Using a Bradford assay, determine the protein concentration of the P2 fraction, which was previously resuspended in isolation buffer. To do this, mix 799 μl water with 1 μl P2 and add 200 μl Bio-Rad Protein Assay reagent (Hercules, CA). Incubate for 5 min at room temperature and determine the absorbance at 595 nm. To convert the absorbance value to protein concentration, make a calibration curve with a series of known concentrations of bovine serum albumin. Adjust the concentration of the P2 suspension to obtain 1 mg protein/100 μl solution.

4. Mix the 100 μl aliquots of digitonin from step 2 with 100 μl aliquots of P2 suspension from step 3.

5. Incubate this suspension on ice for 30 min while keeping the mitochondria in suspension by occasionally inverting the tubes.

6. Stop the reaction by diluting the samples with 4 volumes of isolation buffer.

7. Pellet the mitochondria by centrifugation for 10 min at 11,000*g* in a microfuge. Remove the supernatant.

8. Using the same volume as in step 6, resuspend the pellet in isolation buffer with 250 m*M* KCl and incubate for 5 min on ice. KCl is used here to extract peripheral membrane proteins that were exposed by digitonin treatment.

9. Pellet the membrane fraction by centrifugation at 11,000*g* for 10 min. Carefully remove the supernatant.

10. Resuspend the pellet in 160 μl Laemmli sample buffer. This should yield enough material for several SDS-PAGE gels with which to analyze all necessary markers.

11. Load 20 μl samples on a polyacrylamide gel and separate proteins by SDS-PAGE.

12. Transfer the proteins to nitrocellulose by Western blotting and probe with antibodies of choice.

13. Develop the blots with horseradish peroxidase-conjugated secondary antibody and enhanced chemiluminescence reagents (GE Healthcare, Piscataway, NJ).

14. Scan the blots with a densitometer (Molecular Dynamics, Sunnyvale, CA), and quantify the data with Imagequant software (Molecular Dynamics).

Comments

Digitonin concentrations in the range 0–2 mg digitonin/mg protein are generally sufficient to extract proteins from cow liver mitochondria, but it is recommended to start with a wider concentration range and subsequently narrow this range based on the behavior of the mitochondrial markers. As marker for the intermembrane space, we used cytochrome c (antibody from BD Biosciences). As markers for the mitochondrial matrix, we used Hsp60 (antibody from Santa Cruz Biotechnology, Santa Cruz, CA), glutamate dehydrogenase (antibody from Rockland, Gilbertsville, PA), and α-subunit of ATP synthase (antibody from Molecular Probes, Eugene, OR). As marker for the inner membrane, we used TIM23 (antibodies from BD Biosciences) and as marker for the outer membrane, we used porin (antibody from Calbiochem). Polyclonal Opa1 antibodies were prepared in our laboratory as previously described (Griparic *et al.*, 2004). Opa1 was released at a higher digitonin concentration than cytochrome c, porin or any of the soluble matrix proteins, suggesting that Opa1 is tightly bound to the mitochondrial inner membrane (Fig. 2).

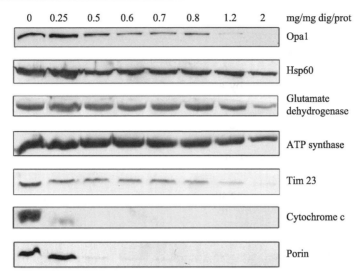

Fig. 2. Extraction of Opa1 from bovine liver mitochondria treated with increasing concentrations of digitonin. Protein that remains in the mitochondrial pellets was analyzed by SDS-PAGE and Western blotting. Hsp60, glutamate dehydrogenase, and ATP synthase were used as mitochondrial matrix markers. Tim23 is an integral membrane protein of the mitochondrial inner membrane. Cytochrome c is a peripheral membrane protein of the intermembrane space and porin is an integral membrane protein of the mitochondrial outer membrane. This experiment is representative of two independent experiments.

Protease Treatment of Digitonin Extracted Mitochondria

To determine whether the bulk of Opa1 is localized to the mitochondrial matrix or to the intermembrane space, we used a protease protection assay in combination with the digitonin extraction procedure described above. Intermembrane space proteins are cleaved by exogenously added proteases when mitochondria are treated with a relatively low concentration of digitonin, while proteolysis of matrix proteins requires a higher concentration of digitonin. This method is adapted from Glick et al. (1992).

Procedure

1. Resuspend 0.5 mg of mitochondrial proteins (P2) in each of 16 Eppendorf tubes with 0.4 ml isolation buffer.
2. Prepare a stock solution of trypsin at 20 mg/ml in water.
3. Add 0, 1, 4, and 8 μl of trypsin to sets of 4 Eppendorf tubes from step 1.

4. Prepare the digitonin stock solution as described in the section *Digitonin Fractionation of Mitochondria*.

5. Make dilutions of the digitonin stock solution with isolation buffer to obtain final concentrations of 0, 0.25, 0.5, and 1.2 mg digitonin/mg protein.

6. Add the digitonin dilutions and isolation buffer to the tubes from step 3 to reach a volume of 0.8 ml and final concentrations of 0, 25, 100, and 200 μg trypsin/ml.

7. Incubate on ice for 30 min.

8. Stop the reaction by adding 266 μl cold 20% trichloroacetic acid. Vortex immediately.

9. Incubate at 65° for 5 min.

10. Incubate on ice for 1 h.

11. Pellet the protein precipitate by centrifugation for 10 min at maximum speed in a microcentrifuge.

12. Carefully suction the supernatant, and add 1 ml of ice-cold acetone.

13. Mix the sample by inverting the tube 6–8 times and repellet the protein precipitate by centrifugation for 5 min at maximum speed in a microcentrifuge.

14. Carefully remove the supernatant and leave the tubes open and on ice to dry for 10 min.

15. Dissolve the pellet in 150 μl 1× Laemmli sample buffer. Add unbuffered 1 M Tris until the solution turns dark blue (usually ~1–2 μl).

16. Load 10 μl of the sample on a mini SDS-polacrylamide gel and subject to SDS-PAGE.

17. Transfer the proteins to nitrocellulose by Western blotting, and probe with appropriate primary antibodies.

18. Develop the blots with horseradish peroxidase-conjugated secondary antibody and enhanced chemiluminescence reagents (GE Healthcare, Piscataway, NJ).

Comments

The mitochondrial pellet P2 used in this assay should be washed in isolation buffer without protease inhibitors after the 10,000g centrifugation step in the mitochondrial isolation procedure. The addition of isolation buffer in step 6 of this procedure may cause some resealing of the outer membrane by diluting digitonin. To prevent artifacts due to resealing of the outer membrane, trypsin should be added to the samples before adding digitonin (Glick *et al.*, 1992). Trypsin will then have access to intermembrane space proteins as soon as outer membrane permeabilization starts to occur. We used the same markers as described for digitonin extractions.

OPA1 is digested at lower digitonin concentrations, suggesting that this protein is exposed to the mitochondrial intermembrane space (Fig. 3).

Alkaline Extraction of the Mitochondrial Membrane

To determine whether OPA1 is an integral or peripheral membrane protein, we extracted the mitochondrial membrane with alkaline sodium carbonate. This method was developed by Lazarow and colleagues to release peripheral membrane proteins from within membrane-bound compartments by converting the limiting membranes into flat sheets (Fujiki *et al.*, 1982). After centrifugation, integral membrane proteins are pelleted, while peripheral membrane proteins are released into the supernatant.

Procedure

1. Dispense 100 μl samples of the P2 fraction, each containing approximately 1 mg of protein, into Eppendorf tubes.
2. Mix with 900 μl ice cold 0.2 M Na$_2$CO$_3$, adjusted to pH 13 with 1 N NaOH.
3. Incubate on ice for 30 min, occasionally mixing the sample by inverting the tube.
4. Pellet the membranes by centrifugation for 1 h at 100,000g in a TLA 100.2 rotor using a Beckman TL-100 ultracentrifuge (Palo Alto, CA).
5. Precipitate the proteins in the supernatant with trichloroacetic acid as described above.
6. Dissolve the pellets in 140 μl 1$\times$ Laemmli sample buffer, and load 20 μl per lane on a mini polyacrylamide gel.
7. Separate the proteins with SDS-PAGE.
8. Transfer the proteins to nitrocellulose by Western blotting, and probe with the appropriate primary antibody.
9. Develop the blots with horseradish peroxidase-conjugated secondary antibody and enhanced chemiluminescence reagents (GE Healthcare, Piscataway, NJ).

Comments

The pH of the carbonate solution was adjusted with 1N NaOH. The original protocol calls for Na$_2$CO$_3$ pH 11.5, but it was observed that a higher pH gives better extraction of some peripheral membrane proteins (Miller *et al.*, 1995). Carbonate at pH 13 releases OPA1 into the supernatant. The α-subunit of the ATP synthase complex, which was used as a peripheral membrane protein marker, is also released. TIM 23, which was used as a bona fide integral membrane protein, was retained in the pellet after carbonate extraction.

GTPase Activity of OPA1

OPA1 belongs to the dynamin family of large GTPases. To test the GTPase activity *in vitro*, we expressed the GTPase domains of wild type and mutant OPA1 in bacteria. The mutant GTPase domain is K301A, which changes a critical lysine in the G1 motif into an alanine. This mutation is predicted to inhibit GTP binding. Bacterially expressed GTPase domains were tested with an *in vitro* GTP hydrolysis assay adapted from Damke *et al.* (1994).

Cloning and Expression of Wild Type and K301A Mutant GTPase Domains of Human OPA1

The pCI-Mgm1 construct was made by recloning the human cDNA KIAA0567 into the mammalian expression vector pCI (Promega, Madison, WI). The wild type GTPase domain was then recloned into a bacterial expression vector by PCR amplification of a 1035 bp product from the pCI-Mgm1 construct using a 5' primer with an NheI restriction site (AGTGC-TAGCATGTATTCTGAAGTTCTTGATG) and a 3' primer with an XhoI restriction site (GCACTCGAGAACCTGGCTCAGACTGA-TAACT) to assist recloning into the pet21d vector (Novagen Inc., Madison, WI). These same primers were used to amplify the GTPase domain from the K301A mutant plasmid pCI-Mgm1(K301A). The resulting PCR product was also recloned into pET21d using NheI and XhoI sites.

Bacterial Expression and Purification

1. Transform the bacterial expression constructs into BL21(DE3) pLysS bacteria (Novagen Inc.). Grow a 500 ml liquid culture starting from a fresh bacterial colony.
2. Induce foreign protein expression with 40 μM isopropyl-1-thio-β-D-galactopyranoside when the bacteria reach an OD of 0.4–0.6.
3. Grow the cells for 2 h at 27° and collect them by centrifugation (10 min at 10,000g in a Sorvall GSA rotor).
4. Resuspend the bacterial cell pellet in 100 ml 50 mM NaH$_2$PO$_4$ pH 8.0, 300 mM NaCl.
5. Add Triton X-100 to 0.1% and phenylmethylsulfonyl fluoride to 1mM final concentrations.
6. Lyse the bacteria using French press set at 18,000 psi.

7. Pellet unlysed bacteria and cell debris by centrifugation for 15 min at 12,000g in an SS-34 rotor (Kendro Laboratory Products, Asheville, NC) prechilled to +4°. The supernatant is stored on ice.

8. Prepare a nickel-nitrilotriacetic acid agarose column (Qiagen, Hilden, Germany) by packing the column with 1 ml nickel resin.

9. Wash with 10 ml wash buffer. The buffer consists of 50 mM NaH$_2$PO$_4$, 300 mM NaCl, 60 mM imidazole, 10% glycerol, and 10 mM β-mercaptoethanol.

10. Load the His-tagged protein onto the column.

11. Wash with 10 ml wash buffer.

12. Elute with 1 M imidazole into Eppendorf tubes. The elution buffer contains 50 mM NaH$_2$PO$_4$ pH 8.0, 300 mM NaCl, 1M imidazole, 10% glycerol, and 10 mM β-mercaptoethanol.

Comments

Inducing protein expression for a relatively short period of time, at a lower temperature, and with a relatively low concentration of IPTG kept the expressed GTPase domains in the soluble fraction and therefore increased the likelihood of purifying native protein. The purity of the expressed protein after nickel-nitrilotriacetic acid agarose column chromatography is >90% as judged by SDS-PAGE.

GTP Hydrolysis

1. Prepare the polyethyleneimine-cellulose-F TLC plate (EM Science, Gibbstown, NJ) by marking positions where the samples will be spotted,

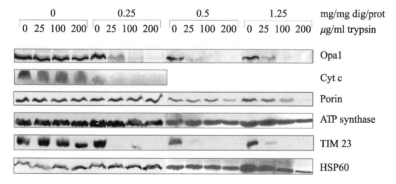

0	0.25	0.5	1.25	mg/mg dig/prot
0 25 100 200	0 25 100 200	0 25 100 200	0 25 100 200	µg/ml trypsin

Opa1

Cyt c

Porin

ATP synthase

TIM 23

HSP60

FIG. 3. Accessibility to trypsin digestion when mitochondria are treated with increasing concentrations of digitonin. After 30 min, the reaction was stopped with trichloroacetic acid and the proteins were analyzed by SDS-PAGE and Western blotting. See the legend of Fig. 2 for assignment of markers to submitochondrial compartments.

approximately 1–2 cm from the bottom of the plate. Prerun the plate by placing it in distilled water.

2. Prepare a reaction mixture containing 2 μl 10× reaction buffer, 12 μl 1mM GTP, 1 μl 10 mCi/ml [α-^{32}P]GTP (GE Healthcare, Piscataway, NJ), and 5 μl water. 10× reaction buffer contains 500 mM Tris/Cl pH 7.0, 20 mg/ml bovine serum albumin, 1 mM dithiothreitol, 5 mM MgCl$_2$.

3. Prepare a sample with all components except for protein, to serve as control for nonenzymatic GTP hydrolysis.

4. Add 1 μg of wild type or mutant GTPase domain to 5 μl reaction mixture, 2.5 μl 10× reaction buffer, adding water to a total volume of 30 μl and incubate at 37°.

5. Immediately after starting the reaction, pipet 2.5 μl of the 0 time point reaction onto the plate and dry with a hair dryer. Repeat with other samples at designated time points.

6. Place the plate in a TLC developing chamber, making sure that the developer level stays below the line along which samples are spotted. The developer is 1 M LiCl$_2$ in 1 M acetic acid.

7. Once the developer solution comes up to 1–2 cm from the top of the plate, the run is stopped by taking the TLC plate out of the developer chamber.

8. Dry the plate with a hair dryer, cover it with Saran wrap, and expose to X-ray film. Quantify the radiolabeled spots with a densitometer and ImageQuant™ software (Molecular Dynamics).

It is necessary to perform the TLC on large plate (20 cm × 20 cm), since shorter plates lack the necessary resolving power. Figure 4 shows a typical

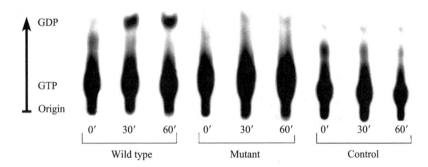

FIG. 4. GTP hydrolysis experiment. Bacterially expressed proteins GTPase domains from wild type Opa1 and Opa1(K301A) were incubated with radiolabeled GTP. The control reactions contained no protein. Samples were taken at the indicated time points and the reaction products were analyzed by TLC. The GTPase domain of wildtype Opa1 is shown to hydrolyze GTP, while Opa1(K301A) does not.

result. GDP has a higher mobility than GTP and is therefore at the top of the plate. The wild type GTPase domain has high GTP hydrolysis activity, whereas the mutant protein is inactive.

References

Damke, H., Baba, T., Warnock, D. E., and Schmid, S. L. (1994). Induction of mutant dynamin specifically blocks endocytic coated vesicle formation. *J. Cell Biol.* **127,** 915–934.

Fujiki, Y., Hubbard, A. L., Fowler, S., and Lazarow, P. B. (1982). Isolation of intracellular membranes by means of sodium carbonate treatment: Application to endoplasmic reticulum. *J. Cell Biol.* **93,** 97–102.

Glick, B. S., Brandt, A., Cunningham, K., Muller, S., Hallberg, R. L., and Schatz, G. (1992). Cytochromes c1 and b2 are sorted to the intermembrane space of yeast mitochondria by a stop-transfer mechanism. *Cell* **69,** 809–822.

Graham, J. M., and Rickwood, D. (1997). Subcellular fractionation. *In* "The Practical Approach Series" (D. Rickwood and B. D. Hames, eds.), p. 339. IRL Press, Oxford.

Greenawalt, J. W. (1974). The isolation of outer and inner mitochondrial membranes. *Meth. Enzymol.* **31,** 310–323.

Griparic, L., van der Wel, N. N., Orozco, I. J., Peters, P. J., and van der Bliek, A. M. (2004). Loss of the intermembrane space protein Mgm1/OPA1 induces swelling and localized constrictions along the lengths of mitochondria. *J. Biol. Chem.* **279,** 18792–18798.

Labrousse, A. M., Zapaterra, M., Rube, D. A., and van der Bliek, A. M. (1999). *C. elegans* dynamin-related protein *drp-1* controls severing of the mitochondrial outer membrane. *Mol. Cell* **4,** 815–826.

Miller, J. D., Tajima, S., Lauffer, L., and Walter, P. (1995). The beta subunit of the signal recognition particle receptor is a transmembrane GTPase that anchors the alpha subunit, a peripheral membrane GTPase, to the endoplasmic reticulum membrane. *J. Cell Biol.* **128,** 273–282.

Olichon, A., Emorine, L. J., Descoins, E., Pelloquin, L., Brichese, L., Gas, N., Guillou, E., Delettre, C., Valette, A., Hamel, C. P., Ducommun, B., Lenaers, G., and Belenguer, P. (2002). The human dynamin-related protein OPA1 is anchored to the mitochondrial inner membrane facing the inter-membrane space. *FEBS Lett.* **523,** 171–176.

Parsons, D. F., and Yano, Y. (1967). The cholesterol content of the outer and inner membranes of guinea-pig liver mitochondria. *Biochim. Biophys. Acta* **135,** 362–364.

Satoh, M., Hamamoto, T., Seo, N., Kagawa, Y., and Endo, H. (2003). Differential sublocalization of the dynamin-related protein OPA1 isoforms in mitochondria. *Biochem. Biophys. Res. Commun.* **300,** 482–493.

Smirnova, E., Shurland, D. L., Ryazantsev, S. N., and van der Bliek, A. M. (1998). A human dynamin-related protein controls the distribution of mitochondria. *J. Cell Biol.* **143,** 351–358.

van der Bliek, A. M., Redelmeier, T. E., Damke, H., Tisdale, E. J., Meyerowitz, E. M., and Schmid, S. L. (1993). Mutations in human dynamin block an intermediate stage in coated vesicle formation. *J. Cell Biol.* **122,** 553–563.

Wong, E. D., Wagner, J. A., Gorsich, S. W., McCaffery, J. M., Shaw, J. M., and Nunnari, J. (2000). The dynamin-related GTPase, Mgm1p, is an intermembrane space protein required for maintenance of fusion competent mitochondria. *J. Cell Biol.* **151,** 341–352.

[55]   Assay and Functional Analysis of Dynamin-Like Mx Proteins

By GEORG KOCHS, MIKE REICHELT, DGANIT DANINO, JENNY E. HINSHAW, and OTTO HALLER

Abstract

Mx proteins are interferon-induced large guanosin triphosphatases (GTPases) that share structural and functional properties with dynamin and dynamin-like proteins, such as self-assembly and association with intracellular membranes. A unique property of some Mx proteins is their antiviral activity against a range of RNA viruses, including influenza viruses and members of the bunyavirus family. These viruses are inhibited at an early stage in their life cycle, soon after host cell entry and before genome amplification. The association of the human MxA GTPase with membranes of the endoplasmic reticulum seems to support its antiviral function by providing an interaction platform that facilitates viral target recognition, MxA oligomerization, and missorting of the resulting multiprotein complex into large intracellular aggregates.

Introduction

Mx proteins are members of the superfamily of high molecular weight GTPases (Haller and Kochs, 2002). Many of these dynamin-like GTPases localize to intracellular membranes and are involved in intracellular trafficking, membrane remodeling, and fission processes (Danino and Hinshaw, 2001; McNiven et al., 2000). Some family members appear to have a different role, in being key components of the early innate immune response against a wide variety of invading pathogens (Haller and Kochs, 2002; Praefcke and McMahon, 2004; Taylor et al., 2004). Accordingly, these GTPases are upregulated by type I (α/β) or type II (γ) interferons (IFN), and they function as intracellular resistance factors capable of restricting the growth of distinct pathogens. The Mx GTPases are expressed exclusively in IFN-α/β–treated cells (Haller and Kochs, 2002).

Human MxA, a 78 kDa protein, accumulates in the cytoplasm of IFN-treated cells and inhibits the replication of a wide range of viruses, including influenza viruses, measles virus, and bunyaviruses (Haller and Kochs, 2002). Biochemical and cell culture studies suggest that MxA interacts

METHODS IN ENZYMOLOGY, VOL. 404
0076-6879/05 $35.00
DOI: 10.1016/S0076-6879(05)04055-3

directly with viral target structures. MxA was shown to bind to the nucleo-capsids of Thogoto virus (THOV), an influenza-like orthomyxovirus. As a consequence, the THOV nucleocapsids were prevented from entering the nucleus where transcription and replication of the viral genome normally occurs (Kochs and Haller, 1999b; Weber *et al.*, 2000). MxA also inhibits the replication of La Crosse virus (LACV), a bunyavirus with a cytoplasmic replication phase. In this case, MxA binds to the viral nucleocapsid (N) protein and forms large copolymers that accumulate in the perinuclear area (Kochs *et al.*, 2002b). Overall, MxA seems to act by interfering with the proper transport of critical viral components to their ultimate target compartments in infected cells.

Membrane association and homo-oligomerization are essential for the biological function of dynamin-like proteins (Danino and Hinshaw, 2001; Praefcke and McMahon, 2004). MxA forms homo-oligomers both *in vitro* and *in vivo* (Accola *et al.*, 2002; Kochs *et al.*, 2002a). Three distinct domains are involved in the self-assembly process: (i) a "self-assembly sequence" (SAS) that is located within the N-terminal G-domain and is conserved in all members of the dynamin-like GTPases (Nakayama *et al.*, 1993); (ii) a central interactive domain (CID) that mediates the association with the C-terminal part of the molecule (Ponten *et al.*, 1997); and (iii) a leucine zipper motif (LZ) at the extreme C-terminus that interacts with the CID (Schumacher and Staeheli, 1998). The interaction between LZ and CID results in increased GTPase activity, indicating that the LZ region acts as a "GTPase effector domain" (GED) (Schwemmle *et al.*, 1995), similar to the GED of dynamin. Furthermore, intra- and intermolecular inter-actions are critical for protein stability and recognition of viral target structures (Flohr *et al.*, 1999; Janzen *et al.*, 2000; Schwemmle *et al.*, 1995). MxA protein localizes to a subcompartment of the smooth endoplasmic reticulum, suggesting that membrane binding and compartmentaliza-tion are important for its antiviral function (Accola *et al.*, 2002; Reichelt *et al.*, 2004).

Our current model proposes that, in IFN-treated cells, MxA forms large membrane-associated self-assemblies that serve as a stable storage pool from which MxA monomers are transiently released. The equilibri-um between assembled and monomeric forms is presumably regulated by the GTPase activity. Upon infection, MxA monomers sense viral target structures and, by binding to them, form new assemblies involving specific viral components. Depending on the local concentration of the binding partners, more and more MxA molecules are recruited into these mem-brane-associated copolymers, leading to mislocalization of the viral components and to viral inhibition (Haller and Kochs, 2002).

Methods

Expression and Purification of MxA

Recombinant human MxA is expressed in *E. coli* as an N-terminal histidine-tagged protein using the pQE-9 vector system (Qiagen, Hilden, Germany), as described (Kochs *et al.*, 2002a; Richter *et al.*, 1995).

E. coli strain M15 carrying the pQE-9-MxA expression plasmid is incubated at 28° in LB (Luria-Bertani) medium containing 100 μg/ml ampicillin and 25 μg/ml kanamycin. At an optical density of 0.3 at 600 nm, MxA expression is induced by adding 0.03 mM isopropyl-β-D-thiogalactopyranoside followed by further incubation for 2 h. The cells are harvested at an optical density of 0.6 to 0.8 by centrifugation. The bacterial pellet can be stored at $-70°$. For lysis, the pellet from a 1 l culture is resuspended in 10 ml of buffer A (50 mM Tris, pH 8.0, 500 mM NaCl, 5 mM MgCl$_2$, 0.1% NP40, 10% glycerol, 2 mM imidazole, and 7 mM β-mercaptoethanol). Lysis is performed on ice by seven cycles of sonication for 30 s each with a Branson B15 sonifier. The homogenate is clarified by centrifugation for 25 min at 20,000g. The resulting supernatant (Fig. 1A and B, lanes 1) is incubated with 800 μl Ni-nitrilotriacetic acid-NTA agarose (Qiagen) by end-over-end rotation for 2 h at 4°. Then the agarose beads are poured into a column of 5 mm in diameter and washed with 30 ml of buffer A containing 30 mM imidazole followed by a 10 ml washing step with buffer B (like buffer A but with 100 mM NaCl and 2 mM imidazole). His-MxA is eluted in 0.5 ml steps with buffer B containing 250 mM imidazole (Fig. 1A and B, lanes 2). For further purification and concentration, the Ni-eluate is directly loaded on a Mono Q anion exchange column (1 ml, 5 × 50 mm; Amersham-Pharmacia, Freiburg, Germany) equilibrated in buffer C (50 mM Tris, pH 8.0, 2 mM MgCl$_2$, 100 mM NaCl, 10% glycerol, 0.05% NP40, 1.4 mM β-mercaptoethanol). After washing with 5 ml of buffer C, His-MxA is eluted with a linear gradient of 5 ml from 0.1 to 1 M NaCl in buffer C. His-MxA typically elutes between 300 to 500 mM NaCl (Fig. 1A and B, lanes 3) with a protein concentration of about 1 mg/ml. The purified protein can be stored at $-70°$.

The GTPase activity of MxA (the hydrolysis of GTP into GDP and inorganic phosphate) is measured as described (Richter *et al.*, 1995). A GTPase inactive mutant with an amino acid exchange in the N-terminal G-domain, MxA(T103A) (Ponten *et al.*, 1997), is used as a negative control. About 1 μg of each purified protein is incubated in 50 μl of buffer D (50 mM Tris, pH 8.0, 5 mM MgCl$_2$, 100 mM KCl, 10% glycerol, 0.1 mM dithiothreitol (DTT)) with 1 mM GTP and [^{32}P]-α-GTP (200 nCi) in the

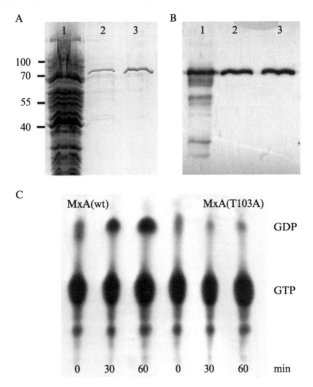

FIG. 1. Purification and GTPase activity of MxA. (A) Coomassie blue-stained SDS-polyacrylamide gel loaded with fractions obtained during the purification of His-MxA expressed in *E. coli*. Lane 1, supernatant after centrifugation; lane 2, eluate from the Ni-NTA affinity column; lane 3, eluate from the Mono Q ion exchange column. (B) Western blot analysis. One-fifth of the fractions shown in (A) were blotted on a PVDF membrane and analyzed using the monoclonal anti-MxA antibody M143 (dilution 1:500) (Flohr *et al.*, 1999). (C) GTPase activity of recombinant wild-type MxA and MxA(T103A), a GTPase-inactive mutant (Ponten *et al.*, 1997). Thin-layer chromatogram of GTP and GDP after incubation with Mono Q-purified wild-type MxA or MxA(T103A) (1 μg) for 0, 30, and 60 min at 37°.

presence of 100 n*M* AMP-PNP. At various time points the reaction is terminated by mixing 10 μl of the reaction mixture with the same volume of 2 m*M* ethylenediaminetetraacetic acid (EDTA) in 0.5% sodium dodecyl sulfate (SDS). To separate the radio-labeled GDP product from the GTP substrate, 1.5 μl of the stopped reaction is spotted onto a polyethylene-imine-cellulose thin-layer chromatography plate (PEI-5725, Merck, Darmstadt, Germany) and developed in 1 *M* LiCl and 1 *M* acetic acid. Figure 1C shows an autoradiography of a dried plate. The result demonstrates that Mx proteins, in contrast to small Ras-like GTPases, exhibit a high intrinsic GTPase activity.

Oligomerization of MxA

MxA Oligomerization in Living Cells

Mx-proteins form high molecular weight oligomers *in vivo* and *in vitro* as shown by chemical cross-linking, gel filtration, and ultrastructural analysis of recombinant proteins (Kochs *et al.*, 2002a; Melen *et al.*, 1992; Richter *et al.*, 1995). To monitor their self-assembly in living cells, we established a nuclear translocation assay (Ponten *et al.*, 1997). The idea of this assay is that in case of homooligomerization, an artificial nuclear form of MxA should be able to drag the normally cytoplasmic MxA protein into the nucleus.

For this approach, MxA is modified by a foreign nuclear translocation signal (NLS). The NLS of the large T-antigen of SV40 is fused to the N terminus of MxA resulting in TMxA that translocates into the nucleus when expressed in eukaryotic cells (Zürcher *et al.*, 1992). To distinguish nuclear TMxA from the normal cytoplasmic MxA, wild-type MxA is N-terminally tagged with a FLAG peptide (Hopp, 1988). Mouse 3T3 cells are seeded onto glass coverslips. After 4 h, the cells are transfected with pHMG-expression plasmids using 2 μl of Lipofectamine transfection reagent (Invitrogen, Karlsruhe, Germany). At 20 h post-transfection the cells are fixed with 3% paraformaldehyde (PFA) in phosphate-buffered saline (pH 7.4) (PBS), permeabilized with 0.5% Triton X-100, and stained using the monoclonal anti-Mx antibody M143 (Flohr *et al.*, 1999) or a monoclonal anti-FLAG antibody (Sigma) in PBS with 5% fish gelatin. After three washing steps with PBS, the antigen-bound primary antibodies are detected with fluorophore (Cy2)-conjugated donkey antibodies (Dianova, Hamburg, Germany) and analyzed with a fluorescence microscope. Figure 2A shows that FLAG-tagged wild-type MxA accumulates in cytoplasmic dots (left panel) and TMxA in the nucleus (right panel). Upon coexpression of both cDNAs, the wild-type MxA, detected by the FLAG antibody, predominantly accumulates in the nucleus, indicating a tight association with TMxA (Fig. 2A, middle panel). Interestingly, a fraction of FLAG-MxA remains in the cytoplasm, presumably bound to intracellular membranes (see following). This approach might also be suitable to study protein-protein interactions of other candidate proteins in living cells.

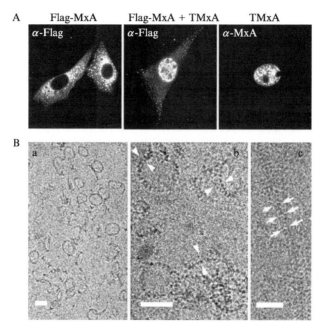

FIG. 2. MxA oligomerization *in vivo* and *in vitro*. (A) TMxA and FLAG-tagged wild-type MxA were transfected either separately (left and right) or together (cotransfection, middle panel) in murine 3T3 cells. The subcellular localization of the recombinant proteins was analyzed by indirect immunofluorescence using either the anti-FLAG or the anti-MxA antibody M143 (dilution 1:500). (B) Cryo-TEM images of MxA self-assemblies. Purified His-MxA dialyzed overnight in low salt buffer with 1 m*M* GMP-PCP self-assembled into rings and open arcs (a). (b) Higher magnification of the rings revealed a structure of two parallel sets of electron dense globular domains (double arrowheads). (c) When dialyzed in the presence of GDP/BeF, MxA self-organized into long, straight, and ordered complexes (arrows). (Bars = 50 nm).

Structural Analysis of MxA Oligomers by Cryo-transmission Electron Microscopy

In vitro, MxA forms ring-like and helical oligomers, similar to the structures formed by dynamin (Accola *et al.*, 2002; Kochs *et al.*, 2002a). These structures are studied at higher resolution using transmission electron microscopy at cryogenic temperatures (cryo-TEM). Cryo-TEM is based on ultra-rapid thermal fixation of the sample that preserves the structures at their native state (Danino *et al.*, 1997, 2004). Therefore, it is a powerful method, better suited to study Mx oligomer formation than conventional chemical fixation methods like negative staining.

Specimens for cryo-TEM are prepared at a controlled temperature of 25° and saturated water atmosphere in a controlled environment vitrification system. An 8 μl drop of purified His-MxA (1 μg/μl) in buffer E (20 mM Hepes, pH 7.2, 2 mM EGTA, 1 mM MgCl$_2$, 1 mM DTT, 50 mM NaCl) and 1 mM nucleotide is placed on a TEM grid covered with perforated carbon film (Ted Pella, USA) held by a tweezers. The drop is blotted with a filter paper forming a film of 100–250 nm in thickness, and dropped into liquid ethane at its freezing point of −183°. The specimen is then transferred to liquid nitrogen for storage. Specimens are examined in a Tecnai G^2 T12 transmission electron microscope operating at 120 kV using a Gatan 626 cryo holder maintained at below −175°. Imaging is done at low dose exposures to minimize radiation damage of the specimens, and images are recorded digitally on a Gatan 791 wide-angle cooled CCD camera at effective magnifications of up to 50,000 using specific imaging procedures as described (Danino *et al.*, 2001).

Ring-like MxA structures form upon dialysis of MxA in the presence of 1 mM GMP-PCP, a nonhydrolyzing analogue of GTP (Fig. 2Ba). Higher magnification images reveal that the rings are composed of two parallel sets of electron dense globular domains that are formed most likely by intra- and interdomain associations of the MxA molecules (Fig. 2Bb). Likewise, long, ordered MxA assemblies displaying a ladder pattern are induced by dialysis of MxA against buffer E in the presence of GDP/BeF (1 mM GDP, 5 mM NaF, 500 μM BeCl2, in 5% ethylene glycol) for 20 h at 4° (Fig. 2Bc). Binding of GDP/BeF mimics a transition state during GTP hydrolysis (Ahmadian *et al.*, 1997). This indicates that cryo-TEM is possibly the method of choice to study the structure and the kinetics of assembly of high molecular weight Mx oligomers.

Membrane Association of MxA

Interestingly, human MxA is a membrane-associated large GTPase, although it lacks obvious membrane interaction domains comparable to the PH domain of dynamin (Accola *et al.*, 2002; Reichelt *et al.*, 2004).

To study the association of MxA with membranes, Vero cells constitutively expressing human MxA (Frese *et al.*, 1995) are analyzed by differential centrifugation. For this, VA3 cells from two 150 cm^2 petri dishes are lysed in 1 ml of buffer F (25 mM Hepes, pH 7.5, 2.5 mM MgCl$_2$, 150 mM NaCl, 1 mM DTT) using a dounce-homogenizer. The homogenate is centrifuged at 1000g (P1) and the resulting supernatant is subjected first to 10,000g (P10) and then to 100,000g centrifugation (P100 and S100). The pellets are resuspended in buffer F and equal amounts of protein are analyzed by Western blotting using the monoclonal anti-Mx antibody M143

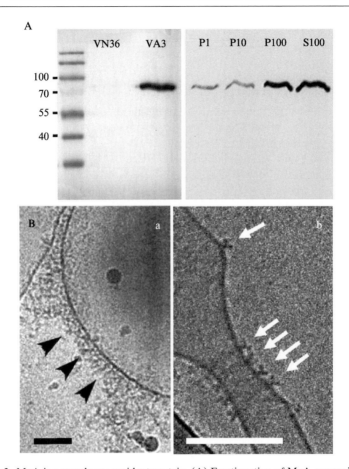

FIG. 3. MxA is a membrane-resident protein. (A) Fractionation of MxA-expressing cells. Vero cells constitutively expressing high levels of recombinant human MxA (VA3) were compared to control cells lacking MxA (VN36) (Frese *et al.*, 1995). Lysates of VA3 cells were centrifuged first at 1000*g* (P1), then at 10,000*g* (P10), and the resulting supernatant at 100,000*g* (P100 and S100). Equivalent amounts of protein were analyzed by Western blotting using the monoclonal anti-MxA antibody M143. (B) *In vitro* interaction of purified His-MxA with liposomes. MxA was incubated with PS-liposomes for 1 h at 37° in the presence of GTP and then analyzed by cryo-TEM. MxA assembles on the surface of a lipid vesicle (a). In the presence of lipid tubes MxA forms ring-like assemblies (b) with a characteristic Y-like appearance (arrows). Scale bar = 100 nm.

(Fig. 3A). Approximately 30% of the total MxA protein content is sedimented with the microsomal pellet, P100, suggesting that part of MxA is associated with membranes.

Structural Analysis of MxA-membrane Association In Vitro

To study *in vitro* association of MxA with membranes, a 10 μg/μl solution of phosphatidylserine (PS, Avanti Polar Lipids, Alabaster, AL) in chloroform is dried under a stream of nitrogen, kept under vacuum overnight, and resuspended to a final concentration of 2 μg/μl in buffer E. Liposomes are generated by extruding the lipid solution 15 times through a 1 μm polycarbonate membrane (Avanti Polar Lipids). Purified MxA is incubated with the liposomes (final concentrations 0.3 and 0.25 μg/μl for MxA and PS, respectively) for 1 h at 37° in the presence of 1 mM GTP, and is subsequently prepared for cryo-TEM as described above. Figure 3Ba shows that MxA associates with the surface of liposomes, indicating that MxA has the potential to directly interact with lipid membranes. In the presence of lipid tubes, sometimes the formation of ring-like MxA structures can be detected. At high magnification, these assemblies show an Y-like shape (Fig. 3Bb), reminiscent of the Y-shaped assemblies of dynamin around lipid tubes (Zhang and Hinshaw, 2001).

MxA Interaction with Viral Target Structures

According to our hypothesis, the antiviral action of Mx proteins is based on their direct interaction with viral components essential for gene expression and genome replication (Haller and Kochs, 2002; Kochs and Haller, 1999a; Kochs *et al.*, 2002b). To show direct interaction of human MxA with viral structures, we analyze infected cells by immunofluorescence and coimmunoprecipitation.

VA3 cells are seeded on glass coverslips and infected with 10 plaque-forming units per cell of La Crosse virus (LACV), a MxA-sensitive bunyavirus (Frese *et al.*, 1996). After 16 h the cells are stained for triple fluorescence analysis as described above, using the monoclonal anti-MxA antibody M143, a goat polyclonal anti-Syntaxin17 antiserum, and a polyclonal rabbit antiserum directed against the nucleoprotein (N) of LACV. Syntaxin17 is a marker for a subcompartment of the smooth endoplasmic reticulum (ER) (Steegmaier *et al.*, 2000). The primary antibodies are detected with fluorophore (Cy2, Cy3, and Cy5)-conjugated donkey secondary antibodies, respectively, and analyzed with a Leica TCS^{SP2} confocal laser scanning microscope. Colocalization of MxA with Syntaxin17 clearly shows that MxA is a membrane-associated protein with a distinct subcellular distribution (Fig. 4Aa and b). The colocalization of MxA with the viral N protein indicates a tight association of membrane-resident MxA with viral nucleocapsids (Fig. 4Aa and c).

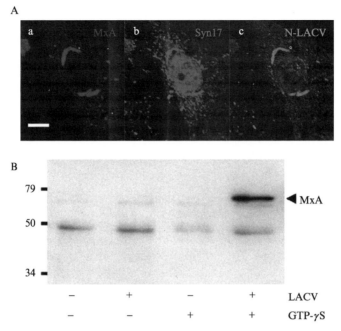

FIG. 4. MxA forms complexes with viral target structures and cellular membranes. (A) Colocalization of MxA/N protein complexes with smooth ER membranes. VA3 cells were infected with LACV for 16 h, fixed as described above, and stained with monoclonal anti-MxA antibody (a, green), goat anti-Syntaxin17 antibody (b, red), and a rabbit antiserum specific for the viral N protein (c, blue). The primary antibodies were detected with fluorophore (Cy2, Cy3, and Cy5)-conjugated secondary antibodies, respectively. The pictures were recorded using a Leica TCS^{SP2} confocal laser scanning microscope. Bar = 8 μm. (B) VA3 cells were infected with LACV or left uninfected. Cells were lysed in the presence or absence of GTP-γS and subjected to immunoprecipitation using the N-specific antiserum. Bound MxA was detected by Western blotting using the monoclonal anti-MxA antibody M143 (from Kochs *et al.*, 2002, © 2002). (See color insert.)

Coimmunoprecipitation of MxA with Viral N Protein

To verify the MxA/N interaction, virus-infected VA3 cells from one 150 cm^2 petri dish are lysed as described above in buffer G (50 mM Tris, pH 7.5, 0.1% Nonidet P-40, 5 mM MgCl$_2$, 0.5 mM DTT) in the presence or absence of 200 μM guanosine 5'-O-[γ-thio]-triphosphate (GTP-γS), a nonhydrolyzable GTP analogue. The lysates are incubated with 2 μl of the polyclonal rabbit anti-N antiserum coupled to 30 μl of protein A-sepharose beads (Amersham-Pharmacia) for 2 h at 4°. After washing the beads several times in buffer G, bound proteins are eluted by incubation in SDS-containing Laemmli sample buffer for 5 min at 95° and analyzed by Western blotting using the monoclonal anti-MxA antibody M143. The Western blot (Fig. 4B) demonstrates that MxA interacts with the viral N protein only in the presence of GTP-γS, indicating that MxA is competent for interaction in its GTP-bound conformation (Kochs *et al.*, 2002b).

References

Accola, M. A., Huang, B., Masri, A. A., and McNiven, M. A. (2002). The antiviral dynamin family member, MxA, tubulates lipids and localizes to the smooth endoplasmic reticulum. *J. Biol. Chem.* **277,** 21829–21835.

Ahmadian, M. R., Mittal, R., Hall, A., and Wittinghofer, A. (1997). Aluminium fluoride associates with the small guanine nucleotide binding proteins. *FEBS Lett.* **408,** 315–318.

Danino, D., Bernheim-Groswasser, A., and Talmon, Y. (2001). Digital cryogenic transmission electron microscopy: An advanced tool for direct imaging of complex fluids. *Colloids and Surfaces A: Physicochem. Engineer. Aspects* **183,** 113–122.

Danino, D., and Hinshaw, J. E. (2001). Dynamin family of mechanoenzymes. *Curr. Opinion Cell Biol.* **13,** 454–460.

Danino, D., Kaplun, A., Lindblom, G., Rilfors, L., Oradd, G., Hauksson, J. B., and Talmon, Y. (1997). Cryo-TEM and NMR studies of a micelle-forming phosphoglucolipid from membranes of Acholeplasma laidlawii A and B. *Chem. Phys. Lipids* **85,** 75–89.

Danino, D., Moon, K. H., and Hinshaw, J. E. (2004). Rapid constriction of lipid bilayers by the mechanochemical enzyme dynamin. *J. Struct. Biol.* **147,** 259–267.

Flohr, F., Schneider-Schaulies, S., Haller, O., and Kochs, G. (1999). The central interactive region of human MxA GTPase is involved in GTPase activation and interaction with viral target structures. *FEBS Lett.* **463,** 24–28.

Frese, M., Kochs, G., Feldmann, H., Hertkorn, C., and Haller, O. (1996). Inhibition of Bunyaviruses, Phleboviruses, and Hantaviruses by human MxA protein. *J. Virol.* **70,** 915–923.

Frese, M., Kochs, G., Meier-Dieter, U., Siebler, J., and Haller, O. (1995). Human MxA protein inhibits tick-borne Thogoto virus but not Dhori virus. *J. Virol.* **69,** 3904–3909.

Haller, O., and Kochs, G. (2002). Interferon-induced Mx proteins: Dynamin-like GTPases with antiviral activity. *Traffic* **3,** 710–717.

Hopp, T. P. (1988). A short polypeptide marker sequence useful for recombinant protein identification and purification. *Bio/Technology* **6,** 1204–1210.

Janzen, C., Kochs, G., and Haller, O. (2000). A monomeric GTPase-negative MxA mutant with antiviral activity. *J. Virol.* **74,** 8202–8206.

Kochs, G., Haener, M., Aebi, U., and Haller, O. (2002a). Self-assembly of human MxA GTPase into highly-ordered dynamin-like oligomers. *J. Biol. Chem.* **277,** 14172–14176.

Kochs, G., and Haller, O. (1999a). GTP-bound human MxA protein interacts with the nucleocapsids of Thogoto virus (*Orthomyxoviridae*). *J. Biol. Chem.* **274,** 4370–4376.

Kochs, G., and Haller, O. (1999b). Interferon-induced human MxA GTPase blocks nuclear import of Thogoto virus nucleocapsids. *Proc. Natl. Acad. Sci. USA* **96,** 2082–2086.

Kochs, G., Janzen, C., Hohenberg, H., and Haller, O. (2002b). Antivirally active MxA protein sequesters La Crosse virus nucleocapsid protein into perinuclear complexes. *Proc. Natl. Acad. Sci. USA* **99,** 3153–3158.

McNiven, M. A., Cao, H., Pitts, K. R., and Yoon, Y. (2000). The dynamin family of mechanoenzymes: Pinching in new places. *Trends Biochem. Sci.* **25,** 115–120.

Melen, K., Ronni, T., Broni, B., Krug, R. M., Vonbonsdorff, C. H., and Julkunen, I. (1992). Interferon-induced Mx proteins form oligomers and contain a putative leucine zipper. *J. Biol. Chem.* **267,** 25898–25907.

Nakayama, M., Yazaki, K., Kusano, A., Nagata, K., Hanai, N., and Ishihama, A. (1993). Structure of mouse Mx1 protein: Molecular assembly and GTP-dependent conformational change. *J. Biol. Chem.* **268,** 15033–15038.

Ponten, A., Sick, C., Weeber, M., Haller, O., and Kochs, G. (1997). Dominant-negative mutants of human MxA protein: Domains in the carboxy-terminal moiety are important for oligomerization and antiviral activity. *J. Virol.* **71,** 2591–2599.

Praefcke, G. J., and McMahon, H. T. (2004). The dynamin superfamily: Universal membrane tubulation and fission molecules? *Nat. Rev. Mol. Cell. Biol.* **5,** 133–147.

Reichelt, M., Stertz, S., Krijnse-Locker, J., Haller, O., and Kochs, G. (2004). Missorting of LaCrosse virus nucleocapsid protein by the interferon-induced MxA GTPase involves smooth ER membranes. *Traffic* **5,** 772–784.

Richter, M. F., Schwemmle, M., Herrmann, C., Wittinghofer, A., and Staeheli, P. (1995). Interferon-induced MxA protein: GTP binding and GTP hydrolysis properties. *J. Biol. Chem.* **270,** 13512–13517.

Schumacher, B., and Staeheli, P. (1998). Domains mediating intramolecular folding and oligomerization of MxA GTPase. *J. Biol. Chem.* **273,** 28365–28370.

Schwemmle, M., Richter, M. F., Herrmann, C., Nassar, N., and Staeheli, P. (1995). Unexpected structural requirements for GTPase activity of the interferon-induced MxA protein. *J. Biol. Chem.* **270,** 13518–13523.

Steegmaier, M., Oorschot, V., Klumperman, J., and Scheller, R. H. (2000). Syntaxin 17 is abundant in steroidogenic cells and implicated in smooth endoplasmic reticulum membrane dynamics. *Mol. Biol. Cell* **11,** 2719–2731.

Taylor, G. A., Feng, C. G., and Sher, A. (2004). p47 GTPases: Regulators of immunity to intracellular pathogens. *Nat. Rev. Immunol.* **4,** 100–109.

Weber, F., Haller, O., and Kochs, G. (2000). MxA GTPase blocks reporter gene expression of reconstituted Thogoto virus ribonucleoprotein complexes. *J. Virol.* **74,** 560–563.

Zhang, P., and Hinshaw, J. E. (2001). Three-dimensional reconstruction of dynamin in the constricted state. *Nat. Cell Biol.* **3,** 922–927.

Zürcher, T., Pavlovic, J., and Staeheli, P. (1992). Mechanism of human MxA protein action: Variants with changed antiviral properties. *EMBO J.* **11,** 1657–1661.

Author Index

A

Aaronson, S. A., 218
Abagyan, R., 563, 564
Abe, T., 529, 530, 534
Abraham, C., 268
Acar, C., 468
Accola, M. A., 633, 637, 638
Acharya, J. K., 307
Acharya, U., 307
Achiriloaie, M., 563
Achstetter, T., 184
Acker, J., 176, 186, 187
Adachi, M., 217, 218, 229
Adamik, R., 176, 177, 178, 179, 181, 185, 187, 190, 191, 198, 203, 359, 366
Adamson, P., 470, 476
Aderem, A., 222
Adesnik, N., 481, 483
Adkinson, A. L., 491, 496
Admon, A., 148, 150
Aebi, U., 59, 633, 634, 636, 637, 641
Agarwal, N., 59
Aggarwal, A. K., 297
Agnello, D., 186
Aguilar, R. C., 165, 207, 317, 319
Ahmadi, K., 253
Ahmadian, M. R., 638
Ahn, H. J., 186
Ahnert-Hilger, G., 250
Aikawa, Y., 422, 425, 427, 428, 429
Aivazian, D., 481, 483
Akiba, T., 612
Akiyama, S. K., 136
Akkerman, J. W. N., 328
Al-Awar, O., 243, 267
Albanesi, J. P., 108, 389, 491, 496, 526, 529, 563, 578, 612
Alber, S., 109, 110
Alberti, S., 44, 47, 297, 298, 300, 301, 304, 308, 312
Albertinazzi, C., 247, 268, 269, 273
Albini, A., 513

Albright, E., 118
Alessi, D. R., 108, 415
Al-Hasani, H., 581
Alizadeh, A. A., 20, 22
Allaman, M. M., 317
Allan, D., 110
Allen, R., 583
Allen, W. E., 380
Alpar, A., 250
Altan-Bonnet, N., 44
Altrock, W. D., 272, 276
Altschuler, Y., 308, 389, 424, 574
Al-Tuwaijri, M., 136
Alvarez, L., 613
Amaral, L. P., 1
Amherdt, M., 58, 59, 64, 67, 72, 75, 77, 81, 83, 85, 95, 98, 100, 109, 165, 346, 351, 353, 417
Amor, J. C., 150
Ampe, C., 469
An, Z., 136
Andersen, J., 268, 269
Anderson, G. A., 1
Anderson, J. M., 334
Anderson, K. E., 148
Anderson, M., 118
Anderson, R. A., 398, 427, 510, 511
Anderson, S. L., 513
Ando, K., 175
Andrade, J., 148, 149, 217, 411, 412, 414
Andreev, J., 148, 217
Aneja, R., 149, 150, 412, 415, 416
Anraku, Y., 86
Ansley, S. J., 432, 454
Ansorge, W., 28
Antoine, J. C., 478
Antonarakis, S. E., 185, 192, 218, 222, 538
Antonny, B., 58, 60, 67, 73, 75, 77, 78, 84, 89, 91, 95, 98, 100, 101, 102, 149, 150, 159, 165, 175, 176, 186, 187, 189, 418
Antony, C., 30, 442, 444
Antoshechkin, I., 454
Aoyama, A., 136

645

M

Wheler, P., 468
White, J., 10, 31
Whitford, K. L., 243
Whitney, J. A., 335
Wieland, F. T., 95, 96, 100, 129, 353, 411, 412, 414, 415, 416, 418
Wiemann, S., 2, 6, 9, 10, 12, 18
Wiessman, A. M., 198
Wigge, P., 227, 491, 499, 526, 529, 538, 557, 558, 578, 598, 603
Wildi, G., 192
Williams, R. L., 453
Williger, B. T., 165, 359
Willison, K. R., 468, 469, 470, 471, 474, 475, 476, 478
Wilm, M., 62, 271
Wilson, C. J., 1
Wilson, J. M., 233, 239, 242, 243, 247
Wilson-Kubalek, E. M., 603
Wimmer, C., 59
Wimmer, E., 563
Winand, N. J., 9
Wing, S. S., 198, 199
Winistorfer, S. C., 317
Winklehner, P., 186
Witczak, O., 385
Witke, W., 116, 300
Wittinghofer, A., 100, 326, 455, 513, 515, 516, 521, 525, 612, 634, 636, 638
Wlodarska, I., 192
Wolde, M., 116, 300
Wong, E. D., 620
Wong, E. T., 59
Wong, S. H., 439
Wong, V., 335
Woolfe, T. F., 118
Worm, U., 268
Wortsman, J., 564
Woscholski, R., 253
Wright, A. F., 468
Wu, C. H., 20, 22
Wu, C. Y., 186
Wu, M., 440
Wu, W. J., 347, 355

X

Xia, G., 180
Xiao, B., 359
Xing, G. C., 148

Xing, L., 513
Xin-Hau, Z., 176, 181
Xu, G., 481, 483
Xu, K. F., 179, 180, 181
Xu, L., 400
Xu, W., 345
Xu, X., 360
Xu, Y., 442

Y

Yabe, D., 483
Yager, P., 499
Yagi, R., 217, 218, 220, 221, 222, 223, 224, 225, 226, 227, 228, 229, 424, 425
Yahara, N., 373
Yalcin, A., 1, 9, 225, 226
Yamada, A., 135, 216, 217, 218, 220, 221, 222, 223, 224, 225, 226, 227, 228, 229, 424, 425
Yamada, H., 529
Yamada, Y., 136, 144
Yamaji, R., 176, 177
Yamamoto, A., 108
Yamamoto, H., 217, 218, 220, 221, 222, 223, 224, 225, 226, 227, 228, 229, 424, 425
Yamamoto, T., 233
Yamamura, K., 242
Yamazaki, M., 165, 225, 243, 428
Yan, J. P., 175, 177
Yang, C., 398, 425, 427
Yang, J.-S., 301, 305
Yang, M., 136
Yang, S. F., 400
Yang, W., 29
Yang-Feng, T. L., 59
Yano, H., 216, 217, 222, 224, 225, 227
Yano, Y., 622
Yarar, D., 491, 496, 500, 538, 546, 547, 556, 563, 612
Yarrow, J. C., 385
Yashar, B. M., 468
Yasuda, T., 529, 530, 534
Yazaki, K., 633
Yazaki, Y., 400
Yeh, L. S., 20, 22
Yeh, Y., 135, 136, 137, 144
Yeung, T., 59, 64, 67, 75, 77, 81, 83, 85, 95, 98, 109, 417
Yewdell, J. W., 184
Yin, F. H., 512

Subject Index

A

Actin cytoskeleton, *see* Arf1; *β*cap73;
    Dynamin; Tight junction
γ-Adaptin homology-Golgi associated
    Arf-binding proteins, *see* GGA proteins
ADP-ribosylating factors, *see specific Arfs*
AGAP1
    assays
        Arf1-GTP
            GTP loss assay, 157–159
            substrate preparation, 151–153
        GTP-to-GDP conversion
            Arf1 tryptophan fluorescence assay,
                159, 161
            radioassay, 155, 157
        large unilamellar vesicle
            preparation, 151
    general properties, 150
    purification, 153–154
AMAP1
    ASAP1 similarities, 217
    functional overview, 217–218
AMAP2
    Arf6 binding studies
        ArfGAP domain–glutathione
            S-transferase fusion protein
            preparation, 218–219
        guanine nucleotide status analysis, 221
        prospects for study, 229
        pull-down assays
            domain binding, 219–221
            full-length protein binding, 221–222
        subcellular colocalization, 223, 225
    Arf6-related protein binding, 222
    functional overview, 217–218
    overexpression effects on Tac and
        transferrin uptake, 226–227
    RNA interference studies of receptor
        endocytosis, 225–226
Amphiphysin
    AMAP2 interactions, 222
    BAR domain, 529

dynamin interactions
    assay
        GTPase stimulation, 533–535
        large unilamellar liposome
            preparation, 532–533
        protein preparation, 530–532
    coassembly, 529
    purification of recombinant protein from
        bacteria, 531–532
AP-1, Arf6 interactions during clathrin/AP-2
    coat recruitment to membranes, 396–397
ARD1
    assays
        cholera toxin-catalyzed
            ADP-ribosylagmatine formation,
            204
        GTPase activation, 203
        GTP binding, 202–203
        ubiquitin ligase activity, 204
    discovery, 196
    purification of recombinant protein from
        bacteria, 199–202
    structure
        ARF domain, 197
        GD1 region, 198
        GTPase-activating domain, 198
        RING domain and ubiquitin ligase
            activity, 198–199
    subcellular distribution, 199
    tissue distribution, 199
    Western blot, 202
Arf1
    CopI function, 96, 108, 216, 368
    cytoskeletal regulation at Golgi apparatus
        brefeldin A inhibition studies, 355–356
        cytoskeletal protein identification, 356
        Golgi membrane binding and vesicle
            budding reactions
            bovine brain cytosol preparation, 348
            incubation conditions, 348
            rat liver membrane preparation, 348
            two-stage reactions to characterize
                GTPase involvement, 349

A GFP-based time-lapse confocal microscopy
   followed by chemical fixation.

B Immunoperoxidase labeling and
   embedding in resin.

C Cutting of serial sections.

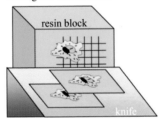

D Identification of the structure
   in sections for EM.

E Digital 3D reconstruction.

MIRONOV ET AL., CHAPTER 5, FIG. 1. The main steps in the CVLEM procedure. A. The structure of interest (circled; in this case a transport carrier) is monitored *in vivo* using an appropriate marker tagged with GFP and time-lapse confocal microscopy. The cell is then fixed at a time chosen by the experimenter (e.g., during budding, translocation, or fusion; in this case during translocation). B. The cell is labeled by the immuno-HRP technique with an antibody against the GFP-tagged protein marker. The patterns of peroxidase labeling and of GFP fluorescence must coincide. C. The cell in A and B is identified in the resin block by using the system of spatial coordinates (also described in Polishchuk *et al.*, 2000), and serial sections are cut. D. The carrier previously monitored *in vivo* (circled) is identified in each section by using specific cellular structures as spatial landmarks (see example in Fig. 2). E. The images in the serial sections are used for the 3-D computer-aided reconstruction.

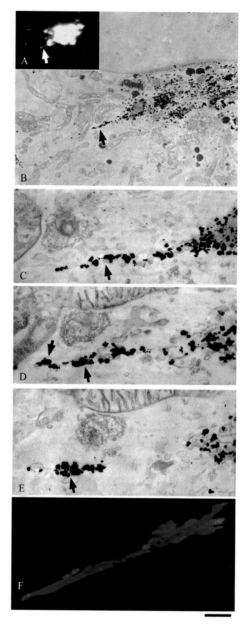

MIRONOV *ET AL.*, CHAPTER 5, FIG. 2. Ultrastructure of a Golgi-to-plasma-membrane carrier (GPC) formation site. A. A VSVG-GFP-transfected Cos7 cell was fixed at the moment of formation of a GPC (arrow), labeled with an antibody against VSVG using the immuno-gold protocol, and processed for CVEM. B–E. The same GPC formation site (arrows and arrowhead) at low EM magnification (B) and under higher magnification in serial sections (C–E). F. 3D reconstruction from the serial sections. Bar: 10.5 μm (A), 2.1 μm (B), 490 nm (C–E).

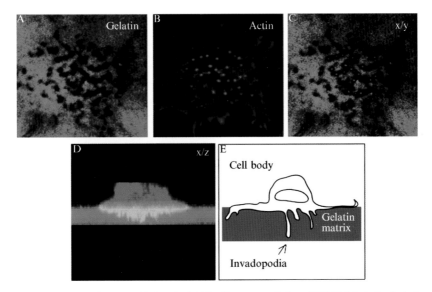

HOOVER *ET AL.*, CHAPTER 14, FIG. 1. LOX cells were seeded on CFDSE-labeled gelatin for 24 h, fixed and stained for actin with rhodamine-phalloidin. Images are representative single confocal sections of an invading cell. (A) Single confocal plane of gelatin with degradation spots and actin (B) along the x/y axis at tips of invadopodia. (C) Merged x/y images of A and B. (D) Stacked side projection of same cell along x/z axis showing invadopodia extending into the gelatin matrix. (E) Schematic representation of D.

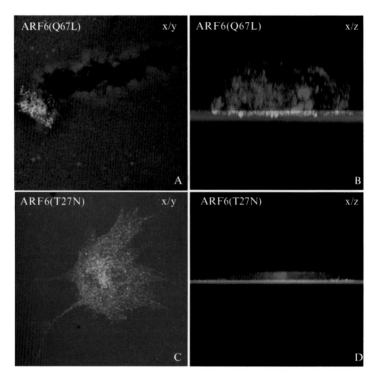

HOOVER *ET AL.*, CHAPTER 14, FIG. 2. LOX cells were transfected with HA-tagged ARF6 mutants, seeded on CFDSE-labeled gelatin and immunofluorescently labeled red. Images are shown along the x/y or x/z axis. (A) Representative confocal image of an ARF6(Q67L)-transfected LOX cell exhibiting an aggressive degradation trail. (B) Image of an ARF6 (Q67L)-transfected LOX cell with a rounded phenotype and invadopodia extending into the gelatin matrix. (C and D) Images of an ARF6(T27N)-expressing cell that shows a spread phenotype and does not form invadopodia or degrade gelatin.

A

$$\text{Arf·GTP} \xrightarrow{\text{GAP}} \text{Arf·GDP+Pi}$$

B

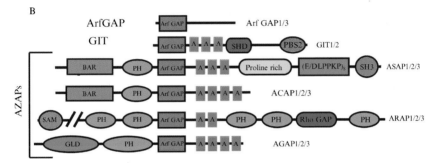

CHE *ET AL.*, CHAPTER 15, FIG. 1. Proteins with Arf GAP activity: (A) Reaction catalyzed by Arf GAPs. (B) Schematic of proteins with Arf GAP activity. Key: A, ankyrin repeat, BAR, Bin Amphiphysin Rsv161,167 domain; GLD, GTP-binding protein-like domain; PBS2, paxillin binding site; PH, pleckstrin homology domain; SAM, sterile α-motif domain; SH3, Src-homology 3 domain; SHD, spa homology domain. Accession #s: rat Arf GAP1 = U35776; human Arf GAP3 = NM_014570; rat Git1 = AF085693; mouse Git2 = NM_019834; mouse ASAP1 = AF075461, AF075462; human ASAP2 = NM_003887; human ASAP3 = NM_017707; human ACAP1 = D30758, NM_014716; human ACAP2 = AJ238248; human ACAP3 = KIAA1716; human ARAP1 = AB018325, AY049732; human ARAP2 = AY049733; human ARAP3 = NM_022481; human AGAP1 = NM_014770; human AGAP2 = NM_014914; human AGAP3 = AF359283.

A

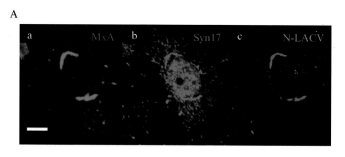

B

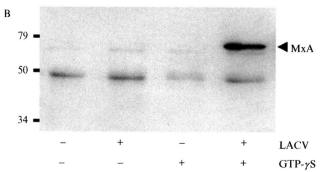

KOCHS *ET AL.*, CHAPTER 55, FIG. 4. MxA forms complexes with viral target structures and cellular membranes. (A) Colocalization of MxA/N protein complexes with smooth ER membranes. VA3 cells were infected with LACV for 16 h, fixed as described above, and stained with monoclonal anti-MxA antibody (a, green), goat anti-Syntaxin17 antibody (b, red), and a rabbit antiserum specific for the viral N protein (c, blue). The primary antibodies were detected with fluorophore (Cy2, Cy3, and Cy5)-conjugated secondary antibodies, respectively. The pictures were recorded using a Leica TCS^{SP2} confocal laser scanning microscope. Bar = 8 μm. (B) VA3 cells were infected with LACV or left uninfected. Cells were lysed in the presence or absence of GTP-γS and subjected to immunoprecipitation using the N-specific antiserum. Bound MxA was detected by Western blotting using the monoclonal anti-MxA antibody M143 (from Kochs *et al.*, 2002, © 2002).